Soft and Biological Matter

"Soft and Biological Matter" is a series of authoritative books covering established and emergent areas in the realm of soft matter science, including biological systems spanning all relevant length scales from the molecular to the mesoscale. It aims to serve a broad interdisciplinary community of students and researchers in physics, chemistry, biophysics and materials science.

Pure research monographs in the series, as well as those of more pedagogical nature, will emphasize topics in fundamental physics, synthesis and design, characterization and new prospective applications of soft and biological matter systems. The series will encompass experimental, theoretical and computational approaches. Topics in the scope of this series include but are not limited to: polymers, biopolymers, polyelectrolytes, liquids, glasses, water, solutions, emulsions, foams, gels, ionic liquids, liquid crystals, colloids, granular matter, complex fluids, microfluidics, nanofluidics, membranes and interfaces, active matter, cell mechanics and biophysics.

Both authored and edited volumes will be considered.

More information about this series at http://www.springer.com/series/10783

Gertz I. Likhtenshtein

Biological Water

Physicochemical Aspects

Springer

Gertz I. Likhtenshtein
Department of Chemistry
Ben-Gurion University of the Negev
Beer-Sheva, Israel

Institute of Problems of Chemical Physics
Russian Academy of Science
Chernogolovka, Russia

ISSN 2213-1736 ISSN 2213-1744 (electronic)
Soft and Biological Matter
ISBN 978-3-030-82502-7 ISBN 978-3-030-82503-4 (eBook)
https://doi.org/10.1007/978-3-030-82503-4

This Springer imprint is published by the registered company Springer Nature Switzerland AG
The registered company address is: Gewerbestrasse 11, 6330 Cham, Switzerland

Preface

The role water, which is a crucial key component in the development of all aspects of life on Earth, in chemistry, biology, industry, and everyday life of human being cannot be overemphasized. Water provides an important chemical contribution to the functioning and degradation of biological systems. Water represents on average 55–60% of the human body weight and is distributed throughout the body and organs. Health, life expectancy, daily comfort, and mood of every human being largely depend on the body water balance. Water plays a central role in regulating cell volume, nutrient transport, waste removal, and thermal regulation and provides a medium for biological reactions Adequate hydration is crucial in maintaining optimal physical and mental functioning, and the need for a fast and reliable hydration status assessment in behavioral medicine research has become increasingly important.

Water is fascinating in all its phases, forms, and states of aggregation and is unique in its number of anomalous properties. Seventy four anomalous properties of liquid are connected to its ability to form up to four well-defined hydrogen bonds allowing for different local structural arrangements. Classical and modern physical chemistry, chemical physics, chemical kinetics, organic, inorganic, and quantum chemistry provide an arsenal of physical methods and establish a basis for the investigation of structure and action mechanism of processes involving water.

Research in this area which combines fundamental importance of human welfare and intellectual fascination with exploring the unknown will promote solving exciting and complicated problems in biochemistry, biophysics, and medicines related to biological water (здесь еще бы подумать, звучит не очсны).

The book is a view on the subject from a physical chemist with the long and broad expertise in chemical biophysics, enzyme catalysis, and physical methods of studying structure and dynamics of biological systems including hydrated biological molecules, in particular. The book is not intended to provide an exhaustive survey of each topic but rather a discussion of their theoretical and experimental background and recent developments. The literature on water is so vast, with so many scientists having made important contributions in the area that it is impossible in the space allowed for this book to give a representative set of references. In fact, for each section in this area, one can write several books. The author apologizes to those he has not been able to include.

Chapter 1 of the monograph is a brief outline of water basic properties and activity. Chapters 2 and 3 present a brief review of physical methods used for the study of biological water. Common and specific features of structure and molecular dynamics of neat supercooled and supercritical water and ice are discussed in Chaps. 4 and 5. Hydration of simple biological molecules and ions is a subject of Chap. 6. Chapter 7 describes protein hydration. Water in enzyme catalysis as a promoter and chemical reagent is in focus of Chap. 8. Hydration of nucleic acids, cells, and tissues biomembranes constitute the contents of Chaps. 9–11. Chapter 12 is devoted to clinical aspects of water.

The book, as a whole, and separate chapters can be used as a subsidiary manual for instructors and graduate and undergraduate students of university chemistry, physics, biophysics, biochemistry, physiology, and medicine.

The department of Chemistry, Ben-Gurion University of the Negev and Institute of Problem of Chemical Physics, Russian Academy of Science, provided excellent conditions for writing this book for which the author is extremely grateful.

Beer-Sheva, Israel
Chernogolovka, Russia

Gertz I. Likhtenshtein

Contents

Chapter 1
Water Basic Properties and Activity

Abstract The role of water in chemistry, biology, industry, and everyday life of human being impossible to overstate (Szent-Györgyi in Science 124:873–875, 1952 [1]; Bagchi in Water in Biological and Chemical Processes: From Structure and Dynamics to Function, Cambridge University Press, 2013 [2]; Purdie Salas in Water Can Be…, Millbrook Press, 2014 [3]; Paul in Water Is Water: A Book About the Water Cycle (Hardcover), Roaring Brook Press, 2015 [4]; Mondal et al. in J. Phys. Chem. Lett. 8:4878–4882, 2017 [5]; Ball in Nature 452:291–292, 2008 [6]; Popkin et al. in Nutr. Rev. 68:439–458, 2010 [7]; Emoto in Secret Life of Water, Simon & Schuster, 2011 [8]; Makulski et al. in Phys. Chem. Chem. Phys. 20:22468–224769, 2018 [9]). Water is a crucial key component of the development of all aspects of life on earth. It is a very good solvent for salts and all major components of cells like proteins, deoxyribonucleic acid (DNA), or polysaccharides. Water provides an important chemical contribution to the functioning and degradation of biological systems. Most of the physical properties of water are quite atypical. Water has unusually large values of viscosity, surface tension, heat of vaporization, entropy of vaporization, and all in all more than 70 abnormalities. These abnormalities can be ascribed to the extensive hydrogen-bonding interactions present in liquid water. The "open" structure of ice that allows for maximum hydrogen bonding can explain why solid water is less dense than liquid water. Water molecules are prone to under coordination in clusters, droplets, nanocavities, nanotubes, and glasses. A huge number of technological, agricultural, and medical research activities are based on the use of water, water solutions, or water embedded in different environments.

1.1 Introduction

Despite the exceptional simplicity of the chemical structure of water due to the branched system of hydrogen bonds, its physicochemical properties are many-sided (Fig. 1.1) and their understanding requires an integrated approach using a whole cascade of methods (Fig. 1.2) [1–9].

G. I. Likhtenshtein, *Biological Water*, Soft and Biological Matter,
https://doi.org/10.1007/978-3-030-82503-4_1

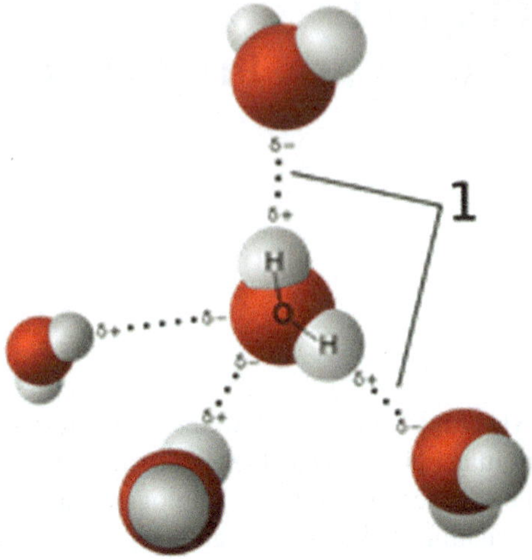

Fig. 1.1 Model of hydrogen bonds between molecules of water. https://en.wikipedia.org/wiki/Water

Figure 1.3 gives an overview of the time windows of several physical techniques. Combining the results by all these and others methods can give a phenomenological picture of the structure and cooperative dynamics in neat water, ice, and state of water in biological objects.

1.2 Water and Life

Water, the main basis for life on Earth continuously circulates through one of Earth's most powerful systems, that is, the water cycle (Fig. 1.4). Water flows endlessly between the ocean, atmosphere, and land. Earth's water is finite, meaning that the amount of water in, on, and above our planet does not increase or decrease.

Since ancient times of the Greek philosophers and during all stages of human civilization, the problems of water have attracted the minds of great scientists and philosophers (Fig. 1.5).

Water represents on average 55–60% of the human body weight and is distributed throughout the body and organs (Fig. 1.6). Health, life expectancy, daily comfort, and mood of an each and every human being largely depend on the water balance of his body.

1.3 Quantum Mechanics of Water

Molecular orbital diagram of water and their corresponding energy is presented in Fig. 1.7.

"and so this men of Indostan
Disputed loud and long,
Each in his own opinion
Exceeding stiff and strong,
Though each was partly in the right
And all were in the wrong!"

Fig. 1.2 Stanza and representation of the poem, The Blind Men and the Elephant by John Godfrey Saxe. The elephant can be thought of as a protein and those men as the various experimental probes that provide information that are true but partial. We, hence, need to combine all possible techniques in order to understand the real scenario [5]. With permission from American Chemical Society

In the diagram, at the bottom, the low-energy orbitals are occupied by two electrons each in the ground state of the molecule while at the top are listed some unoccupied high-energy orbitals. These are mostly antibonding (or non-bonding). The highest occupied orbital ($1b_1$) is non-bonding and highly localized on the oxygen atom. The next lowest orbital ($2a_1$) is a non-bonding orbital. Above the highest unoccupied molecular orbital (HOMO) is the largest unoccupied molecular orbital (LUMO), which does not contain any electrons in the ground state, as is the case for the orbitals at higher energies. Data on the investigation of the valence orbital electron momentum distributions of H_2O were described [12].

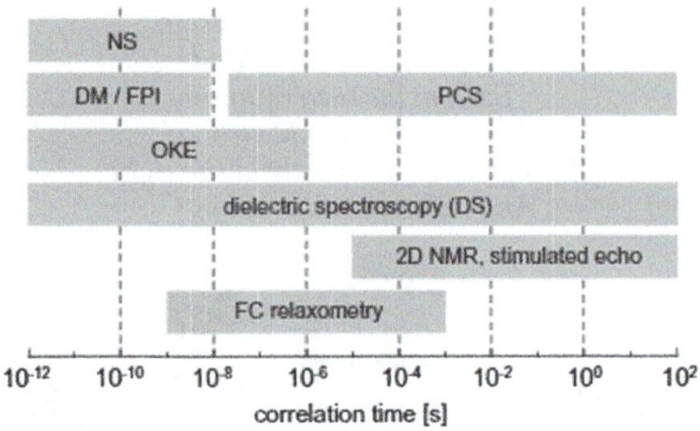

Fig. 1.3 Time windows of the most important methods probing dynamics in dense molecular liquids and comparison with NMR methods: neutron scattering (NS), double monochromator/Fabry–Perot interferometry (DM/FPI), photon correlation spectroscopy (PCS), and optical Kerr effect (OKE) [10]. With permission from Elsevier

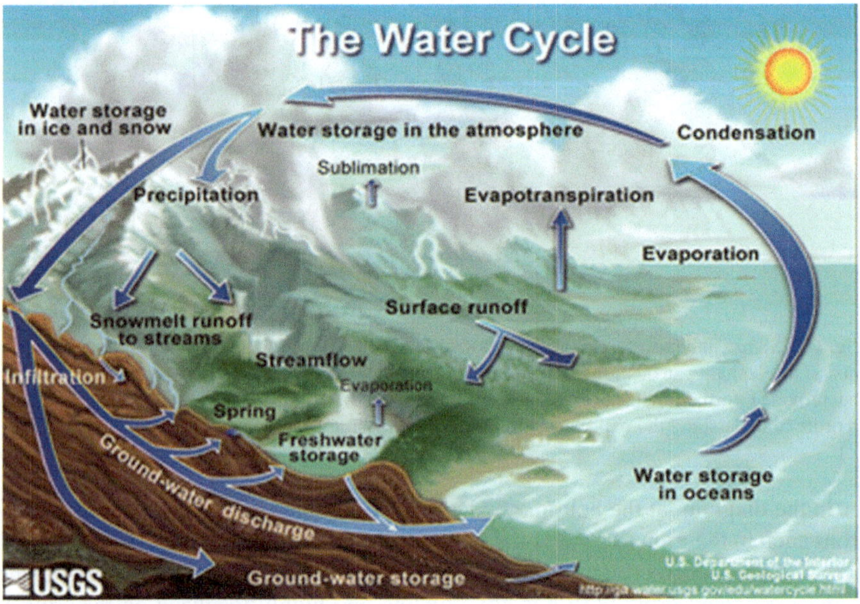

Fig. 1.4 Water circle in the Earth. https://www.usgs.gov/special-topic/water-science-school/science/water-cycle?qt-science_center_objects=0#qt-science_center_objects

1.4 Water Basic Physicochemical Properties

Main [1–9] are molar mass 18.0151 g per mole, melting point 0.00 °C, boiling point, 100.00 °C, maximum density (at 3.98 °C) 1.0000 g per cm^3, heat capacity (C) 75.385 J/(mol K) vapor pressure (25 °C) 23.75 torr, heat of fusion (0 °C) 6.010 kJ per

Fig. 1.5 Great philosophers and scientists who have made a fundamental contribution to understanding the role of water and its physicochemical properties. With permission of Dr. Martin Chaplin

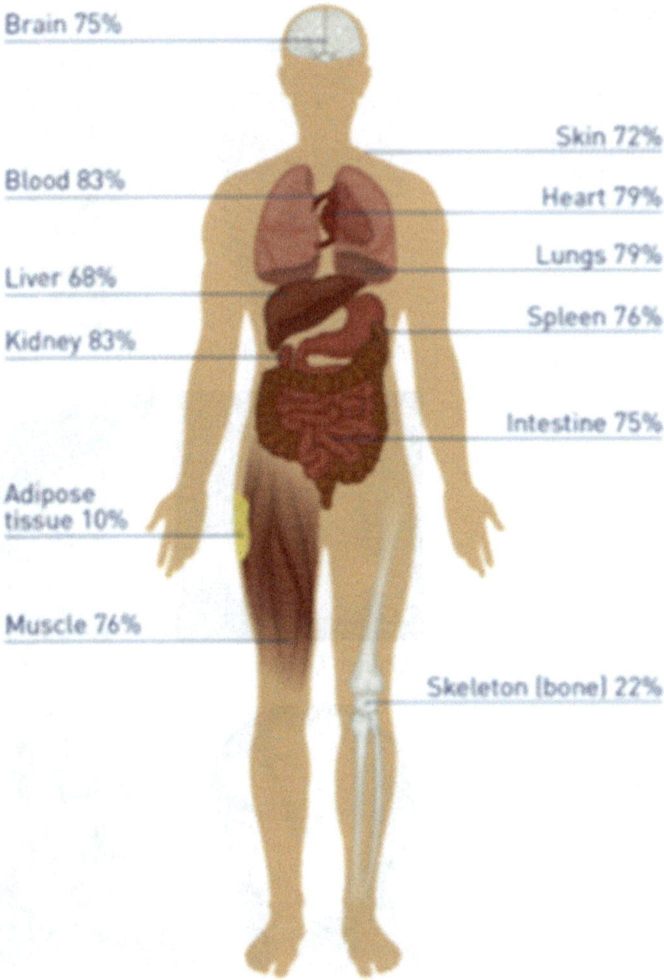

Fig. 1.6 Water composition of tissues and organs by weight [5]. Adapted from Pivarnik and Palmer [11]. Taylor permission request has send

mole, heat of vaporization (100 °C) 40.65 kJ per mole, heat of formation (25 °C), heat of formation (25 °C) 285.85 kJ per mole, entropy of vaporization (25 °C), viscosity, surface tension (25 °C) 71.97 dynes per cm, density 999.9720 kg/m $\approx$ 1 t/m^3 = 1 kg/l = 1 g/cm^3 (liquid, maximum, at ~4 °C) 917 kg/m^3 (solid), magnetic susceptibility (χ) -1.298×10^{-5} cm^3/mol (20 °C, 1 atm), thermal conductivity 0.58 W/m K, refractive index (*nD*) 1.3325, viscosity 1 cP (20 °C), dipole moment 1.85 *D*, dielectric constant 80 (20 °C), specific heat capacity (*C*) 75.375 $\pm$ 0.05 J/mol K, std. molar entropy (69.95 J/mol K) enthalpy of formation (ΔfHo 298) -285.83 kJ/mol, Gibbs free energy ($\Delta fG°$) -237.24 kJ/mol, thermal conductivity 0.6065 W/(m K), *ionic product* of pure water, 10^{-14} at 25 °C, *acidity* (pK_a), and *basicity* (pK_b) 13.995,

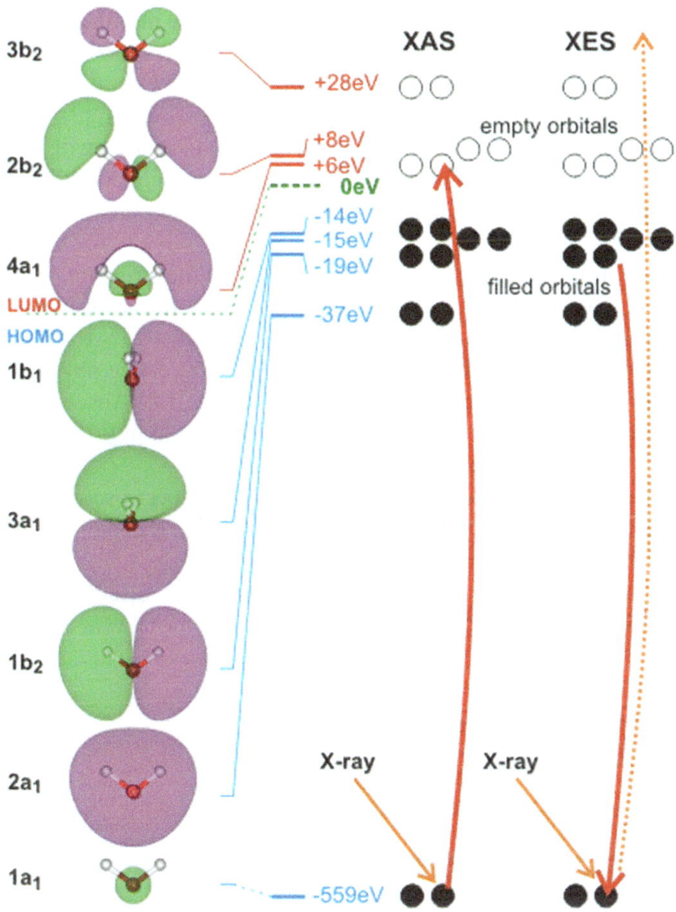

Fig. 1.7 Molecular orbital diagram of isolated water. q=molecular+orbitals+of+water+molecule&oq=molecular+orbitals++of+watervb&gs_l=psy ab.1.1.0i19j0i22i30i19.110325.119281..122686...1.0..0.220.2145.0j12j1..0.1j2..gws-wiz.10..35i39i19j35i362i39j0i203j0i10i203j0i22i30j0i13i30i19.SdJgISkro5k

and ice dielectric constant (ε_r) ~ 3.15 [https://en.wikipedia.org/wiki/Properties_of_water].

[https://www.cambridge.org/core/journals/journal-of-glaciology/article/dielectric-properties-of-ice-and-snowa-review/A4F950358E513838DAE111F6A07EC077/core-reader] "Guideline on the Use of Fundamental Physical Constants and Basic Constants of Water".

Water with one deuterium atom HDO occurs naturally in ordinary water in low concentrations (~0.03%) and D_2O in far lower amounts (0.000003%) Oxygen also has three **stable isotopes**, with ^{16}O present in 99.76%, ^{17}O in 0.04%, and ^{18}O in 0.2% of water molecules [13].

Water is a good solvent, due to its **polarity**. Substances that will mix well and dissolve in water (e.g., salts, ethanol, DMSO) are known as hydrophilic substances, while those that do not mix well with water (e.g., fats and oils), are hydrophobic substances.

Timescales of physical methods used to investigate the structure of liquid water and aqueous and biomolecular solutions are as follows: X-ray absorption ($\approx 10^{-15}$ s), infrared, Raman ($\approx 10^{-14}$–10^{-13} s), pump–probe laser ($\approx 10^{-12}$–10^{-10} s), magnetic resonance (10^{-2}–10^{-11} s), and microwave (10^{-9}–10^{-13} s). Some physical processes and its corresponding timescales are indicated in Fig. 1.8. Absorption, infrared, Raman scattering, and NMR of water spectra are presented in Figs. 1.9, 1.10, 1.11 and 1.12.

The experimental spectra of water obtained by some spectroscopic methods.

The overall hydration of the main biological molecules is depicted schematically in Fig. 1.13.

Various aspects related to biological water are discussed in reviews [1–4, 7, 14, 15].

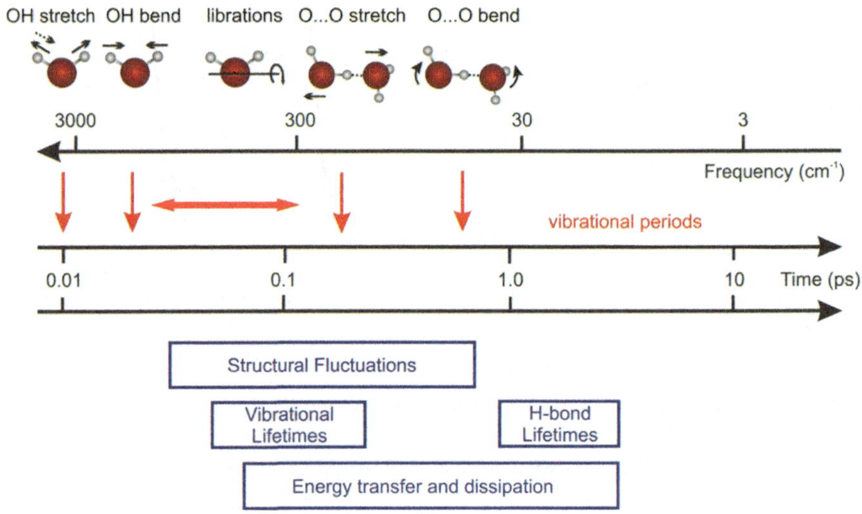

Fig. 1.8 Timescales and processes in bulk H_2O. Red arrows above the logarithmic time axis mark the periods of vibrational and librational degrees of freedom schematically illustrated at the top of the figure. Horizontal boxes below the time axis illustrate the time range covered by particular processes of bulk water dynamics [14]. With permission from American Chemical Society

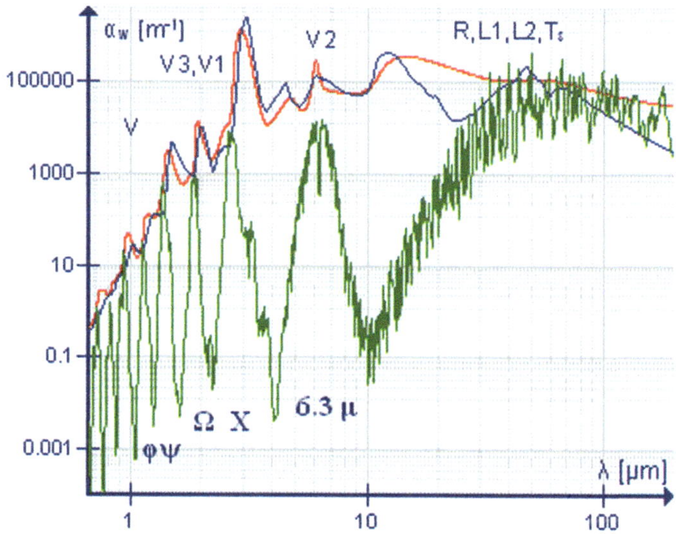

Fig. 1.9 Absorption spectrum (*attenuation coefficient* vs. wavelength) of liquid water (red), atmospheric *water vapor* (green), and ice (blue line) between 667 nm and 200 μm. The plot for vapor is a transformation of data synthetic spectrum for gas mixture "Pure H_2O" (296 K, 1 atm) retrieved from *Hitran* on the Web Information System. https://en.wikipedia.org/wiki/Electromagnetic_absorption_by_water

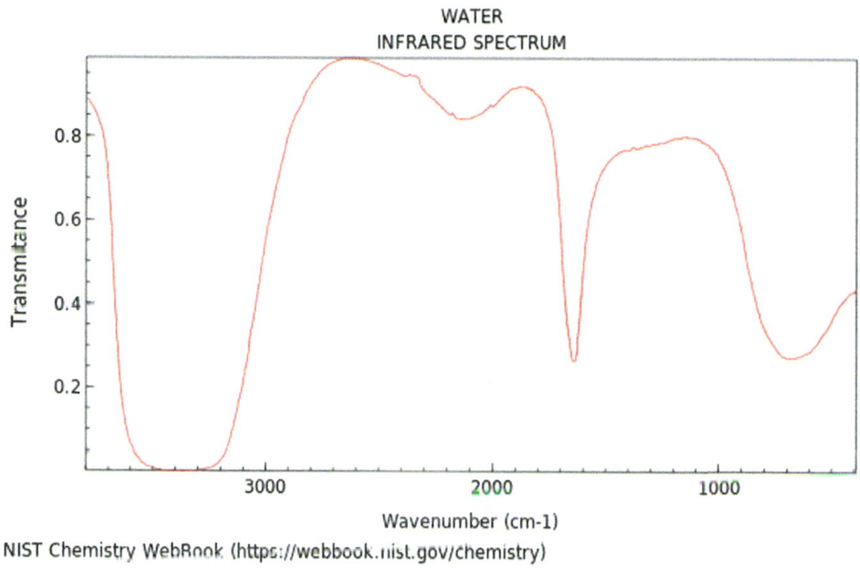

Fig. 1.10 Water infrared spectrum. https://en.wikipedia.org/wiki/Electromagnetic_absorption_by_water

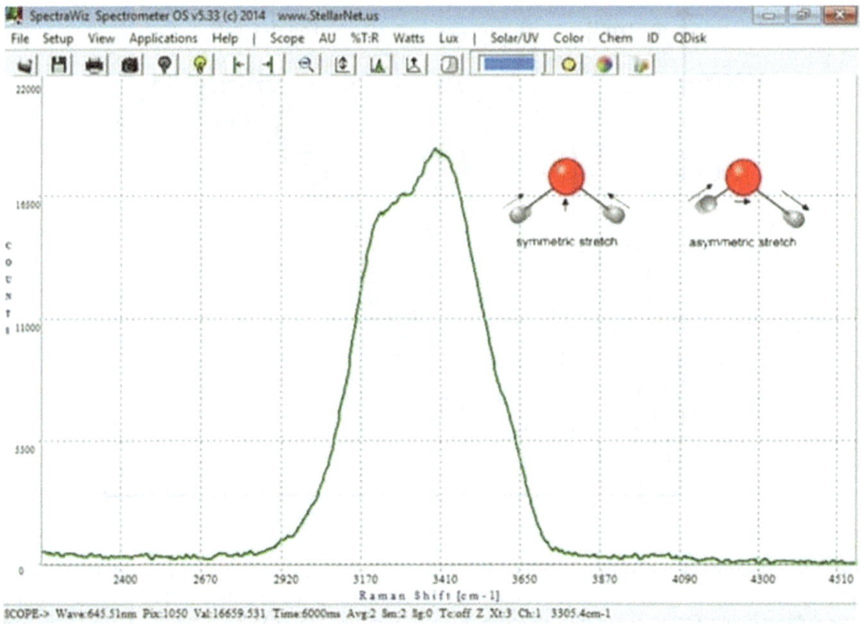

Fig. 1.11 Raman scattering spectrum of pure water. The shoulder at ~3250 cm^{-1} corresponds to the asymmetric –OH stretch, and the most intense feature at ~3410 cm^{-1} corresponds to the symmetric –OH stretch. Even more intriguing, there is a small shoulder at ~3630 cm^{-1} which corresponds to the –OH stretch of a water molecule which is only partially involved in hydrogen bonding—a particularly rare event. https://www.stellarnet.us/application-note-raman-spectrum-water/

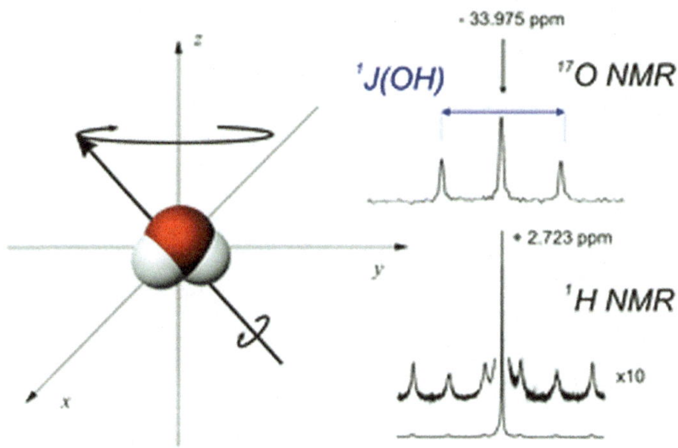

Fig. 1.12 ^{17}O and ^{1}H NMR spectra of isolated water molecules [9]. With permission of Royal Society of Chemistry

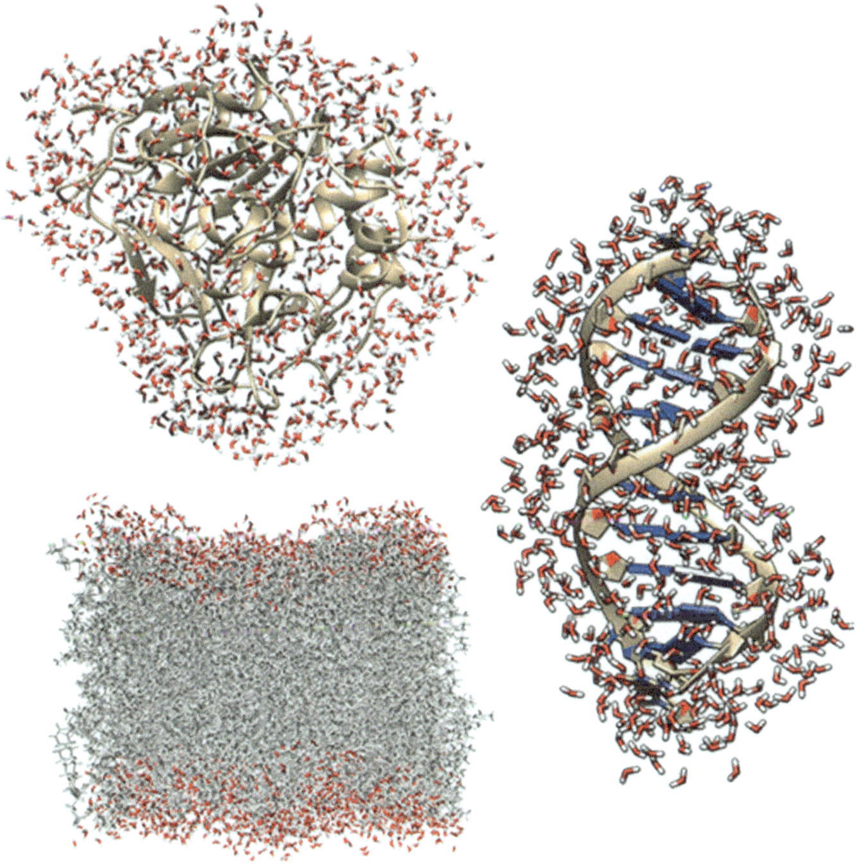

Fig. 1.13 Hydration of a protein, nuclear acid fragment, and biomembrane [13]. With permission from American Chemical Society

References

1. A. Szent-Györgyi, Bioenergetics. Science **124**, 873–875 (1952)
2. B. Bagchi, *Water in Biological and Chemical Processes: From Structure and Dynamics to Function* (Cambridge University Press, 2013)
3. L. Purdie Salas, *Water Can Be*... (Millbrook Press, 2014)
4. M. Paul, *Water Is Water: A Book About the Water Cycle (Hardcover)* (Roaring Brook Press, 2015)
5. S. Mondal, S. Mukherjee, B. Bagchi, Protein hydration dynamics: much ado about nothing? J. Phys. Chem. Lett. **8**, 4878–4882 (2017)
6. P. Ball, Water an enduring mystery. Nature **452**, 291–292 (2008)
7. B.M. Popkin, K.E. D'Anci, I.H. Rosenberg, Water, hydration and health. Nutr. Rev. **68**, 439–458 (2010)
8. M. Emoto, *Secret Life of Water* (Simon & Schuster, 2011)
9. W. Makulski, M. Wilczek, K. Jackowski, ^{17}O and ^{1}H NMR spectral parameters in isolated water molecules. Phys. Chem. Chem. Phys. **20**, 22468–22476 (2018)

10. D. Kruk, A. Herrmann, E.A. Rissler, Field-cycling NMR relaxometry of viscous liquids and polymers. Prog. Nucl. Magn. Reson. Spectrosc. **63**, 33–64 (2012)
11. J.M. Pivarnik, R.A. Palmer, Water and electrolytes during exercise, in *Water and Electrolyte Balance During Rest and Exercise*, ed. by J.F. Hickson, I. Wolinski (Boca Raton, CRC Press, 1994), pp. 245–262
12. C.G. Ning, B. Hajgató, Y.R. Huang, S.F. Zhang, K. Liu, Z.H. Luo, S. Knippenberg, J.K. Deng, M.S. Deleuze, High resolution electron momentum spectroscopy of the valence orbitals of water. Chem. Phys. **343**, 119–130 (2008)
13. E.H. Hardy, A. Zygar, M. Zeidler, M. Holz, F.D, Sacher, Isotope effect on the translational and rotational motion in liquid water and ammonia. J. Chem. Phys. **114**, 3174–3181 (2001)
14. D. Laage, T. Elsaesser, J.T. Hynes, Water dynamics in the hydration shells of biomolecules. Chem. Rev. **117**, 10694–10725 (2017)
15. EFSA Panel on Dietetic Products Nutrition and Allergies (NDA), Scientific opinion on dietary reference values for water. EFSA J. **8**, 1459–1507 (2010)

Chapter 2
Physical Methods for the Study of Biological Water. Magnetic Resonance

Abstract This chapter describes the fundamentals of electron paramagnetic (spin) magnetic resonance (EPR, ESR), and nuclear magnetic resonance (NMR). The focus is on phenomena whose quantitative parameters are directly used to characterize the molecular structure, polarity and proticity and dynamics of water and ice, namely NMR chemical shift, the dipolar coupling between nuclear spins, fast-field-cycling (FFC) NMR effect, the isotropic Fermi contact, the nuclear Overhauser enhancement or effect (NOE), dynamic nuclear spin polarization, electron spin dipole–dipole interaction, free induction decay (FID), nuclear and electron spin–lattice and spin–spin relaxation, nuclear magnetization of chemically exchanging systems. The above physical phenomena form the basis of methods such as electron–electron double resonance (ELDOR) or double electron–electron resonance (DEER) or pulse electron nuclear double resonance (ENDOR) or multifrequency ESR (MF ESR) techniques, two-dimensional ESR (2D ESR), two-dimensional electron–electron double resonance (2D-ELDOR), ENDOR with circularly polarized radio frequency fields (CP-ENDOR), electron nuclear–nuclear resonance (double ENDOR), proton electron double resonance imaging (PEDRI), and electron nuclear–nuclear triple resonance (TRIPLE).

2.1 Introduction

Electron and some nuclear possess corresponding intrinsic angular momentums that are quantized. In an external magnetic field, there is a high-energy state and a low-energy state depending on the relative orientation of the magnet to the external field, and in thermal equilibrium, the low-energy orientation is preferred [1–19]. In a constant magnetic field of strength H_0, the electron spin resonance (ESR) phenomena involve the resonance absorption or dispersion of a microwave frequency (0.3–250 GHz) of electromagnetic field (ν) by a system of particles with the intrinsic spin moment of an unpaired electron [1–13].

The absorption leads to magnetization in the excited state of the system. Accordingly, the electron magnetic resonance condition is:

G. I. Likhtenshtein, *Biological Water*, Soft and Biological Matter,
https://doi.org/10.1007/978-3-030-82503-4_2

$$h\nu = g_e \beta_B \boldsymbol{H}_0, \quad (2.1)$$

where g_e is a g-factor, characterizing the value of the intrinsic electron spin moment (free electron g-value is 2.002319), β_B is the Bohr magneton ($9.27400968 \times 10^{-24}$ J T^{-1}). The values of g-factors and magnetic field strength H_0 dictate position of resonance frequencies in the NMR and ESR spectra.

The nucleus-bearing spin also has different energies in a nonzero magnetic field [14–19]. The energy difference between the two states with the nuclear quantum number $m = 1/2$ and gyromagnetic relation γ is:

$$\Delta E = h\nu = \gamma(h/2\pi)H_0 \quad (2.2)$$

and the NMR phenomena involve the resonance absorption or dispersion of a frequency range from around 20 kHz to around 300 GHz of electromagnetic field (ν).

Spin relaxations times are important parameters of electron spin resonance and nuclear magnetic resonance. Two approaches have been used for describing relaxation phenomena. In the frame of the first approach, the characteristic time of recovery of the induced magnetization to the ground state on account of a transfer of magnetic energy to energy of media (lattice) is noticed as spin–lattice relaxation ($1/T_1$). Another type of relaxation, spin–spin or spin–phase relaxation ($1/T_2$), is related to time of return of the spin system to equilibrium in the exited magnetic state as a result of spin–spin interaction with environment.

The second approach is based on the properties of the electron Larmor precession, which is the precession of the magnetic moments of electrons about external magnetic field (Fig. 2.1). The angular momentum vector $\overrightarrow{J}$ processes about the external field axis with an angular frequency known as the Larmor frequency,

$$\omega = -\gamma B \quad (2.3)$$

where ω is the angular frequency, and γ is the gyromagnetic ratio. Magnetic resonance occurs when the frequency of the electromagnetic radiation w_r coincides with the Larmor frequency. In NMR and ESR, the Bloch equations [20] are used to calculate the dynamic of spin magnetization $M = (M_x, M_y, M_z)$, after the microwave resonance absorption, as a function of relaxation times T_1 and T_2. The relaxation times T_1 and T_2 are called as longitudinal and transverse relaxation times, being synonyms of spin–lattice and spin–spin relaxation time, respectively.

The interaction of an unpaired electron, by way of its magnetic moment, with nearby nuclear spins, results in additional allowed energy states. This phenomenon leads to the superfine (hipper fine) splitting of the ESR signal into N lines in accordance with the quantum mechanical rule $N = 2I + 1$. Two common mechanisms by which electrons and nuclei interact are: the Fermi contact interaction and

Fig. 2.1 Larmor precession

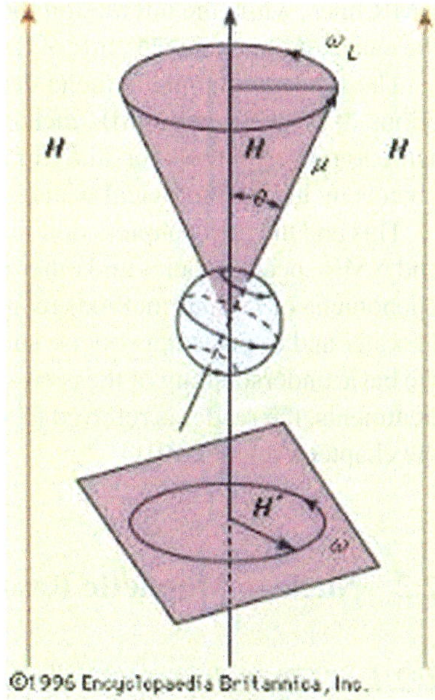

dipolar interaction. The isotropic Fermi contact interaction is the quantum mechanical magnetic interaction between an electron and an atomic nucleus when the electron is inside that nucleus. This interaction leads to the appearance of spin density on s-orbits of atoms bearing nucleus and applies to the case of isotropic interactions which are independent of sample orientation in a magnetic field. The anisotropic dipole–dipole interaction between spin electron and spin of nucleus depends on the sample orientation. Both phenomena serve as a tool for the investigation of structure paramagnetic spices and its rotational or dynamics.

Dynamic nuclear spin polarization is the third mechanism for interactions between an unpaired electron and a nuclear spin. Spin polarization is the degree to which the spin is aligned with a given direction. The alignment of electron spins at a given magnetic field and temperature is described by the Boltzmann distribution under the thermal equilibrium. Dynamic nuclear polarization (DNP) relates to transferring spin polarization from electrons to nuclei, resulted in aligning the nuclear spins to the extent that electron spins are aligned [10–12, 14].

Interaction between two electron spins can occur according to two fundamental mechanisms. The spin dipole–dipole interaction arises because the magnetic dipole of one paramagnetic center induces a local magnetic field at the site of another paramagnetic group. This interaction manifests itself in the dipolar energy splitting (dipolar coupling) and change of spin–lattice (T_1) and spin–spin (T_2) relaxation times. In the frame of the spin exchange mechanism, electron spins are involved in a quantum–mechanical interaction due to the overlapping orbitals bearing spins. The dipolar coupling between nuclear spins through space also resulted in splitting of the

NMR lines, while the interaction through intermediate covalent bonds (shielding) is the cause of the chemical shift of the NMR lines.

The fundamental phenomena described above form the basis of numerous variations of the ESR and NMR methods and their combinations. These methods can serve as powerful tools for studying the molecular structure and dynamics of myriad objects including biological water.

This and the next chapters don't intend to present a rigorous derivation of the ESR and NMR spectroscopies and other physical methods, but rather to describe physical phenomena of various methods related to study of structure and molecular dynamics of water and to present a concise summary of the equations and relations needed for the basic understanding of the corresponding experiment. For more rigorous in-depth treatments, the reader is referred to the excellent books and reviews cited throughout the chapter's text [1–19].

2.2 Nuclear Magnetic Resonance

2.2.1 General

Nuclear magnetic resonance spectra are characterized by a number of parameters available for experimental measurements. Among them, the most important are the position of the spectra, their fine structure, intensity, and relaxation times T_1 and T_2. The combination of these parameters in the object of interest allows to measure the distance between the nuclei, the distance between the electron and the nuclei, and the value of the correlation time τ_c, which is a quantitative characteristic of molecular mobility (Fig. 2.2), chemical exchange and other dynamic processes.

Properties of three isotopes of water nuclei used in NMR and ESR are shown in Table 2.1.

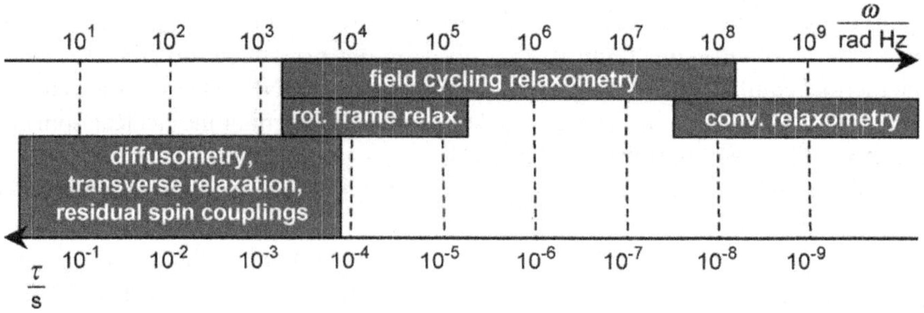

Fig. 2.2 Schematic representation of the correlation time *t* and angular frequency scales covered by diverse NMR techniques. The ranges indicated refer to proton resonance [21]. With permission from Elsevier

Table 2.1 Properties of three isotopes of water nuclei used in NMR and EPR

Isotopes	^1H	^2H	^{17}O
Spin states	+½, −½	1, 0, −1	5/2, 3/2, 1/2, −1/2, −3/2, −5/2
Gyromagnetic ratio, rad × T^{-1} × s^{-1}	26.7522×10^7	4.1066×10^7	-3.6281×10^7
Frequency at 2.3488 T (MHz)	100	15.3506	13.5565

https://en.wikipedia.org/wiki/Isotopes_of_hydrogen, https://en.wikipedia.org/wiki/Isotopes_of_oxygen

2.2.2 NMR Chemical Shift

Shielding is a barrier made of inner-shell electrons, and it decreases the nucleus pull on the outer electrons [13, 14, 16, 22]. Electrons orbiting around the nucleus generate a small magnetic field that opposes B_0, and electrons are shielding the nucleus from B_0. Due to the addition of electron density, magnetic induction, or other effects chemical shift has been decreased. Thus, the higher the electron density around the nucleus, the higher the opposing magnetic field to B_0 from the electrons, the greater the shielding. When an atom is deshielded, it means that nucleus chemical shift has been increased due to the removal of electron density, magnetic induction, or other effects. Important factors influencing chemical shift are electron density, electronegativity of neighboring groups and anisotropic induced magnetic field effects. The nucleus also feels stronger magnetic field.

The detected frequencies (in Hz) for ^1H, ^{13}C, and ^{29}Si nuclei are usually referenced against TMS (tetramethylsilane). Chemical shift δ is expressed in parts per million (ppm) by frequency. For example, the chemical shift positions of the water signal in several common solvents are: acetone (2.8), benzene (0.4), chloroform (1.6), dimethyl sulfoxide (3.3), methanol (4.8), pyridine (4.9), and D_2O (4.8). The most deshielding solvents are the protic H-bonding ones. Next in effectiveness are those with oxygen atoms with lone electron pairs available for H-bonding. An additional effect on water shifts takes place due to the aromatic ring current [22]. The following results of ab initio density functional theory calculations of the NMR chemical shift of liquid water and hexagonal ice were reported, depending on the structural model used: (1) the calculated isotropic shift of ice Ih with respect to the gas phase is –8.0 or 8.1 ppm for the proton, and −48.6 or −48.1 ppm for oxygen; (2) the proton anisotropy is −33.4 or −33.6 ppm; a gas-to liquid shift of −5.8 ppm for hydrogen, and −36.6 ± 0.5 ppm for oxygen; and (3) molecules beyond the first solvation shell influence the proton chemical shift predominantly via the electric field generated by their permanent electric dipole moment.

2.2.3 Proton Spin Relaxation!

Due to the fundamental importance of the study of the structure and molecular dynamics of a huge number of objects in chemistry, physics, and biology, a whole cascade of methods for measuring and calculating the quantitative parameters of relaxation T_1 and T_2 was proposed and developed for 70 years [14, 16, 19–32].

Free induction decay (FID) [20] by the rotating component of the magnetization vector in the x–y plane is a short-lived sinusoidal electromagnetic signal which appears following the 90° pulse in a form $[\sin \omega_0 t]\, \mathrm{e}^{-t/T_2^*}$ (Fig. 2.3). The time dependence of the FID measurement is governed by the relaxation rate $1/T_2^* = 1/T_2 + 1/T_{2\mathrm{IH}}$ where $1/T_2$ is the transverse relaxation rate, while $T_{2\mathrm{IH}}$ describes signal loss due to dephasing in the presence of an inhomogeneous background magnetic field.

The first sequence of two-pulse nuclear spin-echo detection was developed by Hahn [20]. The value of T_2 can be measured by performing a series of spin-echo experiments (Fig. 2.4). In a spin-echo experiment, a $\pi/2$ pulse is applied, followed after a time τ, by a π pulse. A free induction decay is observed immediately after the initial $\pi/2$ pulse. Monitoring the decay of the spin-echo intensity

$$I(2\tau) = I_0 \exp(-2\tau/T_\mathrm{M}) \tag{2.4}$$

allows one to determine T_M, the phase memory time.

In the Hahn three-pulse sequence $\pi/2$–τ–$\pi/2$–T–$\pi/2$–τ–echo, after the first 90° pulse, the magnetization vector spreads out forming a disk in the x–y plane (Fig. 2.4). The spreading continues for a time τ, and then a second 90° pulse is applied such that the "disk" is now in the x–z plane. After a further time T a third pulse is applied and a stimulated echo is observed after waiting a time τ after the last pulse. The decay of the echo intensity is fitted to an exponential function of T as

$$I(2\tau + T) = I_0 \exp(-T/T_1) \tag{2.5}$$

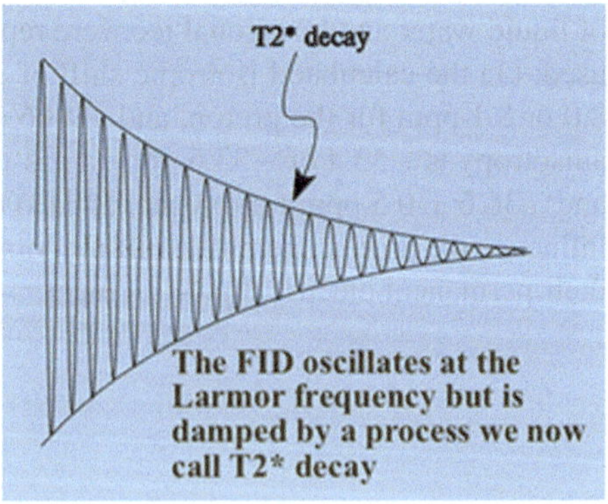

Fig. 2.3 Free induction decay signal. http://mriquestions.com/free-induction-decay.html

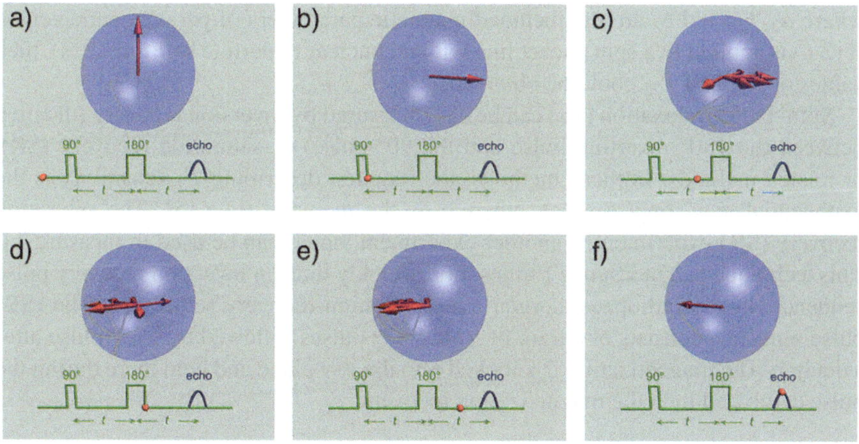

Fig. 2.4 Hahn spin-echo sequence in the rotating frame. **a** Magnetization onto z-axis in the magnetic field; **b** $\pi/2$ pulse along the x-axis in the rotating frame leads to the magnetization M onto the axis; **c** because the magnetic field is inhomogeneous, different spins process at different rates. The fastest processing spins are processing to the right in the rotating frame, and the slowest processing spins are processing to the left; **d** at time τ, the application of a π pulse (along the x-axis) rotates the spins about the x-axis by 180°. The spins continue to process at the same speed (and direction) as before, **d** at time 2τ all the spins converge along the x-axis, resulting in a spin-echo; **e** Progressively, the fast moments catch up with the main moment and the slow moments drift back toward the main moment; **f** Complete refocusing has occurred and at this time, an accurate T2 echo can be measured with all T2* effects removed. Quite separately, return of the red arrow towards the vertical (not shown) would reflect the T1 relaxation. 180 degrees is π radians so 180° pulses are often called π pulses. https://en.wikipedia.org/wiki/Spin_echo

Alternative method to measure T_2 is Carr–Purcell and Carr–Purcell–Meiboom–Gill (CPMG) sequence. The Carr–Purcell sequence [23] allows one to determine the value of T_2 in a single measurement and can minimize the error caused by diffusion. In this experiment, at time $t = 0$ an initial $\pi/2$ pulse is applied, and at times $t = \tau$, 3τ, 5τ, etc., a π pulse is applied. This results in a series of echoes at times 2τ, 4τ, 6τ, etc. The height of each successive echo as a function of the time of that echo allows one to determine the value of T_2. The method was approved using the Carr–Purcell–Meiboom–Gill (CPMG) sequence [24], in which the π pulses are applied along the y-axis in the rotating frame, instead than the x-axis. In this case, errors in a given pulse will be mostly canceled out by the succeeding π pulse.

Spin–lattice and spin–spin relaxation rates can be determined from NMR or ESR spectral saturated curves. The ability of NMR or ESR signals to become saturated in an applied microwave RF or MW field is determined by the value of the MW amplitude H_1 and the spin relaxation parameters ($T_{1e,n}$ and $T_{2e,n}$). For example, the dependence of the ESR absorption signal (A) on H_1 is described by (2.6) for a saturated line shape:

$$A = \frac{a_{2T_{2e}H_1}}{1 + b_2 T_{2e}^2 + c_2 H_1^2 T_{2e} T_{1e}} \tag{2.6}$$

where a_2, b_2, and c_2 are the included magnetic parameters of paramagnetic center. If (2.6) is applied to a spin packet line with no nuclear hyperfine splitting (hps), then values of T_{1e} and T_{2e} could be obtained.

Spin–lattice relaxation (T_1) can be also measured by inversion recovery (the time between the 180° inverting pulse and the 90°-pulse) or saturation recovery (SR), or related methods. In these methods, the function describing T_1 sensitivity is the exponential, e.g., $\exp(-\tau/T_1)$, where τ is the recovery time [5]. The saturation recovery (SR) experiment is another experiment which can be used to measure T_1. This technique can measure T_1 times more quickly than an inversion recovery pulse sequence. https://radiopaedia.org/articles/saturation-recovery-sequences. The (SR) pulse sequence consists of a train of $\pi/2 - \pi/2$ pulses followed by a $p/2$ pulse after a delay τ. The magnetization is knocked into the x–y plane and held there during the pulse comb and then the magnetization recovers.

2.2.4 Spin Relaxations as Tools in Investigation of Structure and Molecular Dynamics of Water

Figure 2.5 schematically illustrates model of a water molecular motion in the magnetic field used for a quantitative analysis of the dipolar coupling effects.

The importance of dipolar coupling to NMR relaxation was first recognized in the seminal work of Bloembergen, Purcell, and Pound (BPP) [25]. The BPP theory made the assumption that the autocorrelation function of the microscopic fluctuations causing the relaxation with the correlation time, τ_c, explains the relaxation constant of a pure substance in correspondence with its state, taking into account the effect of tumbling motion of molecules on the local magnetic field disturbance. From the *t* BPP equation, one can get $1/T_1$ and $1/T_2$ for magnetic dipolar relaxation, depending on the Larmor frequency in correspondence with the strength of the main magnetic field, the time of the molecular tumbling motion, magnetic permeability of free space,

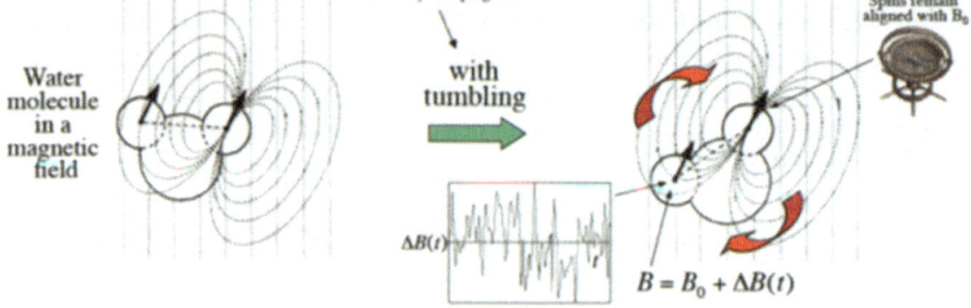

Fig. 2.5 Model of water molecular tumbling in the magnetic field. https://web.stanford.edu/class/rad226b/Lectures/Lecture5-2016-Dipolar-coupling.pd

the reduced Planck constant, γ the gyromagnetic ratio of species of nuclei, and r the distance between the two nuclei carrying magnetic dipole moment.

According to the classic equations provided by Solomon, for identical spins, e.g., water, both spin relaxation rates ($1/T_1$ and $1/T_2$) depend on the spin–spin distance r, the dynamic correlation time τ_c, the proton Larmor frequency ω_0, and magnetic moment μ_0 (Fig. 2.6) [26].

By measuring T_1 and T_2, it is possible to determine the spin–spin distance r and the dynamic correlation time τ_c which is different in media of different viscosity and magnetic field, and for species of various size (Fig. 2.7).

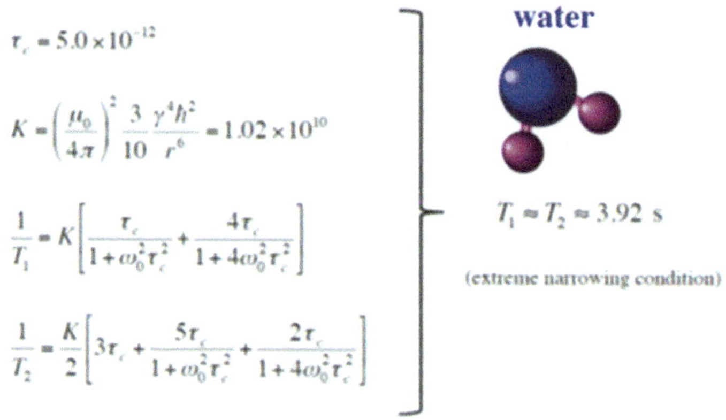

Fig. 2.6 Solomon equations. μ_0 is the magnetic moment, γ gyromagnetic ratio; r is the internuclear (fixed intramolecular) distance, and ω_0 is the Larmor frequency. https://web.stanford.edu/class/rad226b/Lectures/Lecture5-2016-Dipolar-coupling.pd

Fig. 2.7 T_1 and T_2 as function on the correlation time τ_c at different conditions. Creative Commons, PO Box 1866, Mountain View, CA 94042

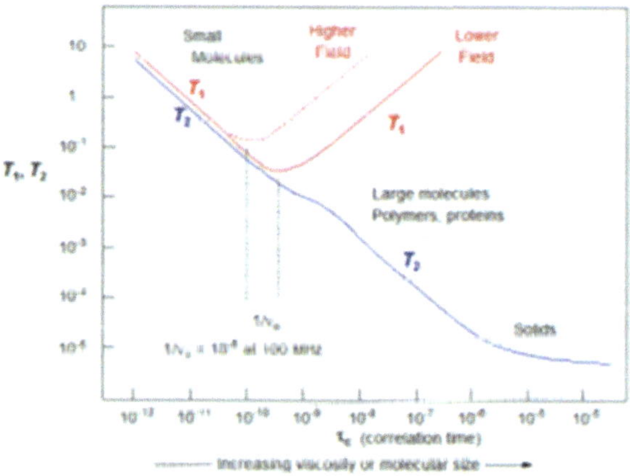

2.2.5 Fast-Field-Cycling Relaxometry

Fast-field-cycling (FFC) NMR relaxometry deals with the variation of the spin–lattice relaxation times (T_1) in a complex system, as the strength of the applied magnetic field is changed [21, 28–30]. The total frequency ranges for proton and deuteron spin–lattice relaxation covered in this way are 10^3 Hz $< \nu_{\text{proton}} < 10^9$ Hz, 10^2 Hz $< \nu_{\text{deuteron}} < 10^8$ Hz. In special cases including conventional NMR spectrometers, the range can be extended from 1 kHz to 500 MHz. Applying the FFC technique (Fig. 2.7) information about molecular dynamics can be achieved.

In the context of field-cycling NMR relaxometry, spin–lattice relaxation in multi-spin 1/2 systems can be treated as a sum of two-spin 1/2 relaxation rates of a reference spin i interacting with all other spins j in pairs [21]. The magnetization after the relaxation interval employing a typical relaxometer was recorded in the form of a free induction decay (FID) after a radio frequency (RF) pulse or a spin–echo pulse sequence in the detection field. The repetition time amounted to several times the spin–lattice relaxation time in the polarization field. The effective spin–lattice relaxation rate of dipolar coupled spins 1/2 thus reads

$$\frac{1}{T_1} = \sum_{j \neq i} \frac{1}{T_1^{(i,j)}}, \tag{2.7}$$

where $T_1^{(i,j)}$ 1 is the two-spin 1/2 spin–lattice relaxation time of the "tagged" spin i interacting with a spin j in an ensemble of multispin 1/2 systems. The spin–lattice relaxation rate of spins directly reflects the spin transition probabilities per time unit for single and double-quantum transitions and hence is proportional to a linear combination of spectral densities in the form

$$\frac{1}{T_1} = C_{\text{coupl}}[I_1(\omega_0) + 4I_2(2\omega_0)], \tag{2.8}$$

where $\varpi_0 = \gamma B_0$ is the resonance angular frequency depending on the gyromagnetic ratio g of the nuclei, I_2 is the spectral density and the external magnetic flux density B_0: The prefactor C_{coupl} is a constant specific for the type of the dominating spin coupling.

The complete expressions for the spin–lattice relaxation rates in the laboratory and rotating frames and, for comparison, the transverse relaxation rate including the full expressions for the dipolar coupling constants of two-spin 1/2 systems are

$$\frac{1}{T_1} = \left(\frac{\mu_0}{4\pi}\right)^2 \frac{3}{10r^6}\gamma^4\hbar^2\left[\frac{\tau_c}{1+\omega_0^2\tau_c^2} + \frac{4\tau_c}{1+4\omega_0^2\tau_c^2}\right] \quad (\rightarrow \text{lab. frame})$$

$$\frac{1}{T_{1\rho}} = \left(\frac{\mu_0}{4\pi}\right)^2 \frac{3}{20r^6}\gamma^4\hbar^2$$

$$\times \left[\frac{3\tau_c}{1 + 4\omega_1^2\tau_c^2} + \frac{5\tau_c}{1 + \omega_0^2\tau_c^2} + \frac{2\tau_c}{1 + 4\omega_0^2\tau_c^2} \right] (\rightarrow \text{rot. frame})$$

$$\frac{1}{T_2} = \left(\frac{\mu_0}{4\pi}\right)^2 \frac{3}{20r^6}\gamma^4\hbar^2 \left[3\tau_c + \frac{5\tau_c}{1 + \omega_0^2\tau_c^2} + \frac{2\tau_c}{1 + 4\omega_0^2\tau_c^2} \right]$$
$$(\rightarrow \text{lab. frame}). \tag{2.9}$$

The symbols are defined as follows: μ_0 is the magnetic field constant, γ gyromagnetic ratio; h is Planck's constant divided by 2π; r is the internuclear distance, $\varpi - = \gamma B_0$ and $\varpi = \gamma B_1$. The magnetic flux densities B_0 and B_1 refer to the external field and the amplitude of the RF field, respectively. As an example of a proton T_1-nuclear magnetic resonance dispersion (NMRD) study, water proton spin–lattice relaxation rate $R1(\omega_I)$-NMRD were measured in snowflakes and ordinary hexagonal ice at proton Larmor frequencies ranging from 1 to 30 MHz and at different temperatures ranging from −2 to −10 °C. The spin–spin relaxation rate $1/T_2(\omega_I)$ was determined at a single Larmor frequency of 16.3 MHz. The high-field wing of the proton $R_1(\omega_I)$-NMRD profile was characterized by two parameters: a correlation time τ_c which described the dipole–dipole spectral density, and the relaxation rate at low fields $R_{\text{max real}}$ which was determined from T_2.

In another example of successive application of ^{1}H fast-field-cycling NMR relaxometry, measurements of intracellular water residence time in cultured cells (mammary adenocarcinoma TS/A) were performed [29]. The extracellular (R_{1ex}) and the intracellular one (R_{1in}) were carried out by measuring T_1 at fixed field (0.2 T) before and after the addition of a paramagnetic Gd complex (Prohance) at different concentrations in the external medium. MR images at 1 T were acquired with an Aspect M2-High-Performance MRI System (Aspect Magnet Technologies Ltd, Netanya, Israel). Without the addition of paramagnetic contrast agents (CA), water exchange between the two compartments occurs much faster than the recovery of magnetization and the time evolution of the magnetization in the IR experiment is described by the following equation:

$$M_z = M_0 \cdot \{1 - 2 \cdot \exp[-t(V_{ex} \cdot R_{1ex} + V_{in} \cdot R_{1in})]\}$$
$$M_z = M_0 \cdot \{1 - 2 \cdot \exp[-t(V_{ex} \cdot R_{1ex} + V_{in} \cdot R_{1in})]\} \tag{2.10}$$

In the presence of a Gd-HPDO3A, the recovery of longitudinal magnetization after inversion is no monoexponential and is expressed as

$$M_z = M_0 \cdot \left\{1 - 2\left[(1 - a_S) \cdot \exp(-t \cdot R_{1L}) + a_S \cdot \exp(-t \cdot R_{1S})\right]\right\}$$
$$M_z = M_0 \cdot \left\{1 - 2\left[(1 - a_S) \cdot \exp(-t \cdot R_{1L}) + a_S \cdot \exp(-t \cdot R_{1S})\right]\right\} \tag{2.11}$$

where M_z is the instantaneous magnetization, M_0 is its Boltzmann equilibrium value, a_L and R_{1L} are the fraction and rate constant for the apparent component with the longer $T_1\left(T_{1L} = R_{1L}^{-1}\right)$, a_S and R_{1S} are the fraction and rate constant for the apparent component with the shorter $T_1\left(T_{1S} = R_{1S}^{-1}\right)$, and t is the running time for recovery by

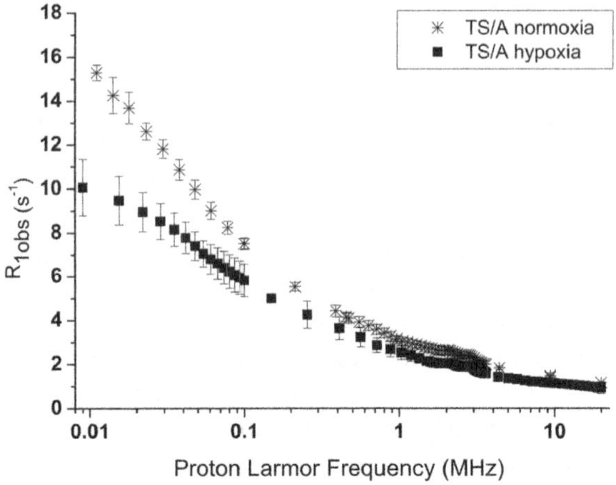

Fig. 2.8 FFC NMR profiles acquired on TS/A cells grown in normoxic (black stars) and hypoxic (black filled square) conditions. The T_1 was measured at 25 °C; the mean ± SD is calculated from at least three independent experiments [29]. Published under the terms of the Creative Commons CC BY license

relaxation. The observed dispersion curve (Fig. 2.8) is characteristic of diamagnetic tissue relaxation. Obtained findings suggested that T_1 measurements at low and variable magnetic field strengths, giving access to the assessment of intracellular water lifetime.

Obtained findings suggested that T_1 measurements at low and variable magnetic field strengths, giving access to the assessment of intracellular water lifetime.

Some academic and industrial applications showing the relevance of exploiting FFC NMR relaxometry of ^{1}H. ^{2}H, ^{13}C, ^{31}P, ^{19}F in complex materials to study the molecular dynamics was reviewed [30]. The basic nuclear magnetic resonance dispersion features were outlined in representative examples of magnetic resonance imaging (MRI) contrast agents, porous media, proteins, and food stuffs. The review focused on the new directions and perspectives for the FFC technique, such as the latest Wide Bore FFC NMR relaxometers, which allow probing the dynamics of confined surface water contained materials including biological structures. The use of the latest field-cycling technology with a new cryogen-free variable-field electromagnet, which enhances the range of available frequencies in the 2D T_1–T_2 correlation spectrum for separating oil and water was also confirmed.

In the frame of FFC NMR relaxometry, by changing the intensity of an applied magnetic field, the fluctuation in longitudinal relaxation time values (T_1) of the observed nuclei (in this case ^{1}H) was observed [33]. The dispersion of the T_1 values occurs when each frequency in the distribution of magnetic fields (DMF) generated by the motional fluctuations matches the Larmor frequencies (ω_L) of the observed nuclei which is related to the frequency of the water motion. The experimental design of FFC NMR experiments was based on both a prepolarized (PP) and a non-polarized (NP) sequence recognizing the polarization, relaxation, and acquisition steps. For example, FFC experiment was carried out in the following sequential consequence and corresponding conditions: (1) in the PP sequence, a polarization field (B_{POL}) set at the ω_{L} of 10 MHz was applied for a period of polarization time (T_{POL}); (2) in the NP sequence, a relaxation each field (B_{RLX}) was applied at ^{1}H Larmor frequency (ω_{L}) values ranging from 0.01 to 10 MHz for a period τ; (3) at the end of each B_{RLX} field, a ^{1}H 90° pulse was applied into an acquisition magnetic field (B_{ACQ}) held for a fixed time at the ω_{L} value of 7.2 MHz make magnetization observable, and the free induction decay (FID) acquirable; (4) a time domain of 100 μs sampled with 1000 points was applied; and (5) the crossover field between NP and PP sequences was approximately retrieved when the relaxation field intensity was half that of the polarization field [33].

For heterogeneous systems, the longitudinal relaxation time (T_1) values of the observed nuclei were obtained for each B_{RLX} by changing the τ values. The relationship between signal intensity and τ was modeled by (2.12):

$$I(\tau) = I_0 \exp\left[-(\tau/T_1)^k\right] \tag{2.12}$$

here, $I(\tau)$ is the ^{1}H signal intensity at each fixed BRLX, I_0 is the ^{1}H signal intensity at the thermal equilibrium, T_1 is the average proton spin–lattice relaxation time, and k is a heterogeneity parameter related to the stretching of the decay process, which describes the superposition of exponential contributions, related to distribution in T_1. The longitudinal nuclear relaxation rate R_1 is given as

$$R_1 = \sum_{1=1}^{n} c_n \frac{\tau_n}{1 + (2\omega_{\mathrm{L}}\tau_n)^2} \tag{2.13}$$

The obtained six fit parameters ($c_1, c_2, c_3, \tau_1, \tau_2, \tau_3$) were used to have an average exchange correlation time

$$\langle\tau\rangle = \frac{\sum_n c_n \tau_n}{\sum_n c_n} \tag{2.14}$$

In work [34], the time evolution of the nuclear magnetization of chemically exchanging systems in liquids was calculated for the prepolarized fast-field-cycling sequence of nuclear magnetic resonance relaxometry (Fig. 2.9). The obtained parameter expressions of the magnetization made it possible to derive the longitudinal relaxation rates and the residence times of the exchanging sites. The prepolarized PP sequence was found to be a typical FFC sequence of successive different field values. The PP sequence allows one to explore the evolution of the nuclear magnetization at low field and to derive the intrinsic relaxation rates of the studied nuclei and their lifetimes in their different sites.

Bi-exponential relaxation domain, in the space of logarithmic coordinates is displaced in Fig. 2.10 (R^+_{r} and R^-_{r} are effective relaxation rates).

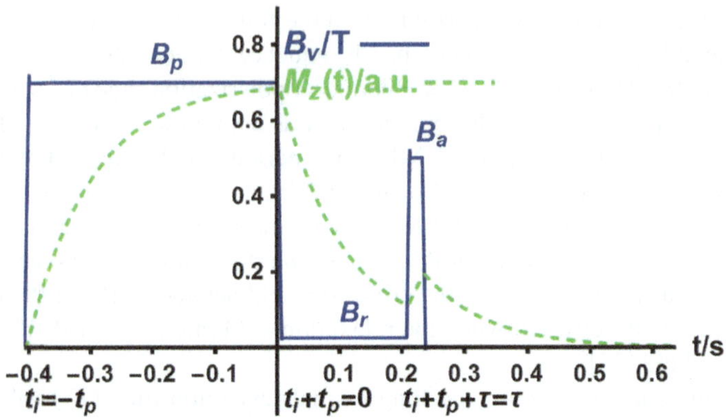

Fig. 2.9 Basic cycle of the magnetic field $Bv(t)$ (continuous line) in a prepolarized (PP) sequence. The field Bv successively takes the polarization value Bp for a polarization time t_{p}, the relaxation value Br for an evolution time, the signal acquisition value Ba for an acquisition time ta, and the zero value for a repetition delay taken to be equal to t_{p}. The nuclear magnetization $Mz(t)$ (dotted line) in arbitrary units tends to its instantaneous equilibrium value $\chi Bv(t)$ so that it evolves as the field $Bv(t)$ with some delay. The time origin is conveniently taken to be at the start of the evolution period τ [34]. Published under the terms of the Creative Commons CC BY license

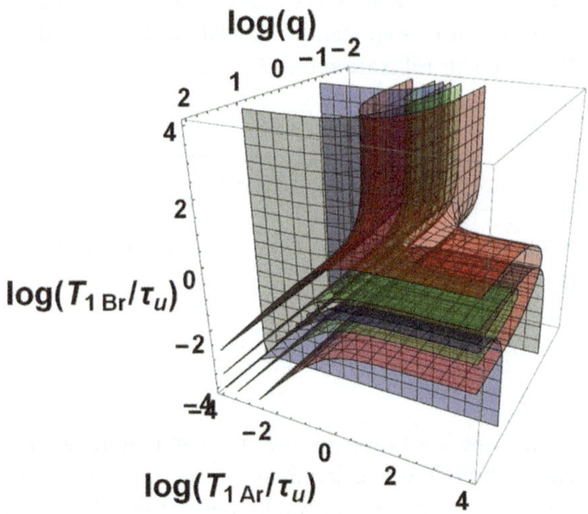

Fig. 2.10 Bi-exponential relaxation domain, $0.1 \leq w+/w- \leq 10$, in the space of logarithmic coordinates log q with $q = p_B/p_A$, $\log(T_{1Ar}/\tau_u)$, $\log(T_{1Br}/\tau_u)$ for immediate signal acquisition $s = 0$. This domain is defined by its boundaries $w+/w- = 0.1$ (gray contour) and $w+/w- = 10$ (blue contour). The surface contours $R^+{}_r/R^-{}_r = 3$ (green surface) and $R^+{}_r/R^-{}_r = 20$ (red surface) are also displayed. See details in [34]. Published under the terms of the Creative Commons CC BY license

Two dynamic models were used in the analysis, an isotropic rotation diffusion model and the six discrete site exchange model [31]. For applied to the isotropic rotation diffusion model, the result was an expression for the water proton T_1 NMRD profile of ice reading:

$$R_{1,\mathrm{diff}}^{\mathrm{theory}}(\omega_0) = \frac{4}{3}\left[\frac{1}{T_2\tau_{\mathrm{c,diff}}^2\omega_0^2}\right]. \tag{2.15}$$

For the six-site exchange model, the corresponding equation reads:

$$R_{1,\mathrm{exch}}^{\mathrm{theory}}(\omega_0) \approx \left[\frac{1}{T_2\tau_{\mathrm{c,exch}}^2\omega_0^2}\right]. \tag{2.16}$$

The proton relaxation time T_1 and the effective correlation time τ_c, assuming $\omega_0\tau_c >> 1$, are equal

$$T_1(T) = C(\omega_0) \times \tau_c(T), \tag{2.17}$$

where the numerical factor $C(\omega_0)$ is frequency-dependent and for the six-site exchange model, it reads

$$C_{\mathrm{exch}}(\omega_0) = \frac{1.79\omega_0^2}{K_{\mathrm{real}}^{\mathrm{DD}}} \tag{2.18}$$

here $K^{\mathrm{DD}} = 1.7/T_2\tau_{\mathrm{c,exch}}$.

2.2.6 Water Diffusion Profiles

The spins in a sample can diffuse to a new position during the experiment [32].

By the application of a gradient in H_0 across the sample, the motion of a given spin will result in a frequency shift, and the diffusive motion of the spins can be detected by the reduction in the echo lifetime as measured for example by a two-pulse or three-pulse Hahn echo sequence [16]. In the presence of a magnetic field gradient, a spin that moves around will have a precession frequency that varies with time. A detailed calculation yielded the following result for the magnetization: $M(t) = M_0 \exp(-\gamma^2 G^2 D t^3)$ where G is the magnetic field gradient and D is the diffusion constant. Thus, by measuring the echo lifetime in the presence of a known magnetic field gradient, one can determine the diffusion constant of the spins (i.e., the water molecules) in the sample.

To measure one-dimensional (1D) spatially resolved diffusion profiles, a standard pulsed gradient stimulated echo (PGSE) sequence [31, 32, 34–36] with the addition of a read imaging gradient (Fig. 2.11) was used. These data supported that fluid diffusion

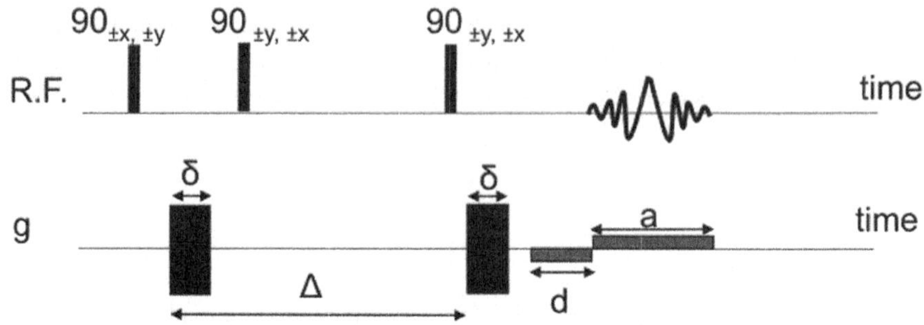

Fig. 2.11 Timing diagram for the PGStE sequence with a read gradient. In this sequence, a gradient g is applied for δ, and displacement observation time of Δ is used to encode for diffusion as a read gradient g_r with a negative lobe of duration d and positive lobe of duration $2a$ to encode for position. A 16 point phase cycle is used [35]. With permission of John Wiley & Sons

is sensitive to the structure of cartilage while biopolymer diffusion is sensitive to the molecular weight of the biopolymers.

Effectivity of high-field nuclear magnetic resonance (NMR) relaxometry and diffusometry along with magnetic resonance imaging was demonstrated by example of monitoring phase transition molecular dynamics during hydrate formation occurring in water droplets dispersed in liquid cyclopentane [37]. 1D T_2 relaxation measurements indicate the extent of hydrate formation as well as a reduction in water droplet size with progression of hydrate. In addition, a novel signal rise observed in two-dimensional T_1–T_2 relaxation correlation experiments showed complex diffusion dynamics due to coupling between regions with varying relaxation and diffusion.

2.2.7 *The Nuclear Overhauser Enhancement or Effect (NOE)*

The general Overhauser effect was first predicted in 1953 by Overhauser [38] who observed that saturation of the conduction electron spin resonance by microwave irradiation causes a polarization of nuclear spins. The nuclear Overhauser enhancement or effect (NOE) is the cross-relaxation of spin polarization from one spin to another induced by dipole–dipole interaction. Because NOE amplitude depends on the separation of the two spins, it can be used to measure the distance between them and dynamic correlation time τ_c.

In a system consisting of two spins of value ½, I and S, their spin components along the z-axis (parallel to the polarizing field) are "up" or "down" are designated by corresponding eigenstates α and β. The four eigenstates of the system are $\alpha\alpha$, $\alpha\beta$, $\beta\alpha$ and ®$\beta\beta$®$\alpha\sigma$ are indicated in Fig. 2.12. The transition probabilities per unit time between the states are W_{1I}, W_{1S}, W_{0IS}, and W_{2IS}.

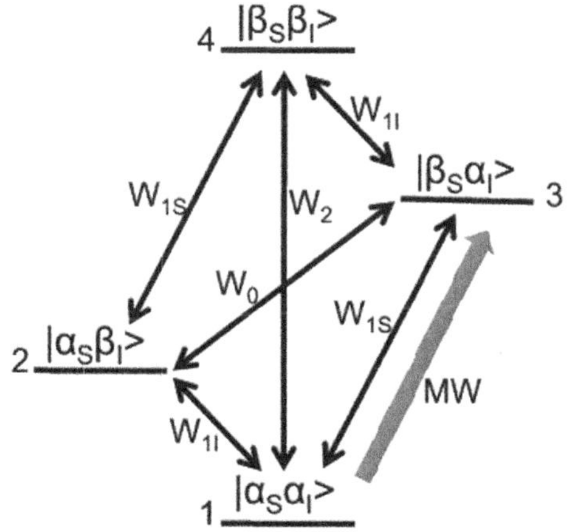

Fig. 2.12 Energy-level diagram of a two-spin system. The transition probabilities between the eigenstates are indicated by red, blue, and purple arrows, for spin I, spin S and double flips, respectively. The populations are represented by the gray balls. beat.voegeli@phys.chem.ethz.ch

In the NOE, the variation of a resonance when the equilibrium of spin populations is interfered with spin transitions, the magnitude of this transition is stated as a relative intensity change according to the following equation [39].

$$\sigma_{IS} = (I - I_0) \times 100\% \quad (2.19)$$

where σ_{IS} shows the NOE of spin I once it has been perturbed by spin S, I is the intensity in the presence of NOE, and I_0 is the equilibrium intensity. The enhancement is referred to the intensity variation that can be potentially either negative or positive depending on the motional characteristics (molecular size and solvent viscosity), sign of the spin's gyromagnetic ratio, and the spectrometer frequency. For more details on the theory, readers are referred to reviews [40, 41]. For a situation in which I is initially at equilibrium and S is not, the relaxation rate depends on the type of nuclei involved, on their distance, and on the correlation time for the IS vector [40].

As an example of the NOE application, the ability of solute–solvent NOE differences measured of the same protein in different chemical environments to reveal hydration changes on the relative timescale was recently demonstrated [41]. The author employed the time–correlation function (TCF) which is relevant to dipole–dipole magnetization transfer NOE captures related to the average dynamics of the vectors r_{IS} joining a protein atom (typically hydrogen) with spin I and the water protons with spin S and takes in consideration angle (θ_{IS}) which swept by the connecting vector during time t. The Fourier transform of $G_{IS}(t)$ yields the spectral density function (SDF) $J(\omega)$ whose linear combinations can be measured experimentally as the self- and cross-relaxation rates ρ and σ. Under certain conditions, assuming exponential correlation decay.

Protein–water σ_{NOE} rate of 0.01 m s^{-1} between a protein UBQ surface proton was measured. The measurement indicated that water can yield a spin–spin vector relaxation time τ close to the tumbling time of the protein itself referring to the extreme case of a water molecule sticking strongly to the surface. Using long-ranged distance dependencies $1/r^n$ with n ranging from 1 to 3, values of the water residence times of tens of picoseconds were obtained. The authors concluded, that the presence of a complexing macromolecule or a confining surface wall modulates the water mobility in the vicinity of the probe protein, hence revealing which residues of the said protein are proximate to the foreign interface and which are chemically unaffected.

2.3 Electron Spin Resonance

2.3.1 Continuous-Wave EPR Measurements

Continuous-wave ESR (CW ESR) is an experimental technique during which a sample is continuously illuminated with microwave radiation at a fixed frequency and the strength of the applied magnetic field is swept, while observing changes in the microwave absorption [1–18]. Usually, for technical reasons, the collected spectrum is plotted in an X–Y format where the X-axis displays the magnetic field strength in Gauss and the Y-axis displays the first derivative of the absorption in arbitrary units (Fig. 5.1).

The fundamental ESR rule (2.1) predicts strong dependence of the ESR spectra position of samples of different g-factors spin precession frequencies (microwave linewidths) and therefore on the strength of magnetic field (Fig. 2.13). This phenomenon forms the base for multifrequency spectroscopy which convenient to precisely measure of values of g-factors and hyperfine splitting (hfs).

The main ESR features related to solving problems of structure and molecular dynamics of biological water are: (1) the position of the spectral features depending on the value of the g-factor according to the resonance condition; (2) spin–lattice (T_1) and spin–spin (T_2) relaxation times; (3) the ESR spectra hyperfine splitting attributed to the contact and dipolar interactions of electron spin with nuclear spin I; and (4) the degree of electron spin polarization. Measurement of these features in combination with other chemical, physicochemical and biochemical methods provides a powerful tool for the detailed investigation of the structure and dynamics of molecular objects including water in biological systems of high complexity.

These features can be measured and characterized by suitable parameters through an arsenal of experimental CW ESR methods and advanced pulse techniques [1, 3, 4, 8, 10]. The CW technique is used for collecting information of interest through the analysis of ESR spectra or saturation curves, obtained from electron–electron double resonance (ELDOR) or double electron–electron resonance (DEER), electron nuclear double resonance (ENDOR) or multifrequency ESR (MF ESR) techniques,

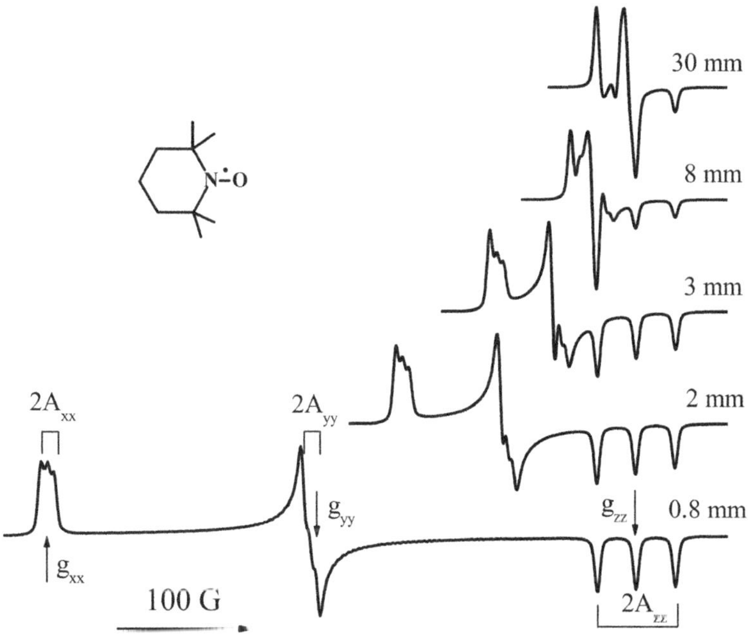

Fig. 2.13 EPR spectra of nitroxide radical calculated for different spin precession frequencies (microwave linewidths). http://hf-epr.awardspace.us/index.htm

two-dimensional ESR (2D ESR), two-dimensional electron–electron double resonance (2D-ELDOR), ^{15}ENDOR with circularly polarized radio frequency fields (CP-ENDOR), electron nuclear–nuclear resonance (double ENDOR), proton electron double resonance imaging (PEDRI)$_{17}$, and electron nuclear–nuclear triple resonance (TRIPLE).

2.3.1.1 Methods Based on Analysis of the Electron–Nuclear Spin Interactions

The electron–nuclear spin interaction is originated from the coupling of the nuclear magnetic moment to the magnetic field generated by the electron magnetic moment or equivalently from the coupling of the electron magnetic moment to the magnetic field generated by the nuclear magnetic moment [42, 43].

The total Hamiltonian of the system of an electron and many nuclear spins is given by

$$\hat{H} = \hat{H}_{\mathrm{e}} + \hat{H}_{\mathrm{N}} + \hat{H}_{\mathrm{eN}} + \hat{H}_{\mathrm{NN}}, \tag{2.20}$$

and composed of the single spin Zeeman energies, Hamiltonian for interaction in the applied magnetic field along the z-axis, the hyperfine interaction and the intrinsic nuclear–nuclear interaction, respectively. Both the spin and orbital angular momentum of the electron contribute to electron magnetic moment. The hyperfine interaction between the electron and the nuclear spins consists of the isotropic Fermi contact interaction and the anisotropic dipole–dipole interaction. For example, the isotropic ^{14}N-hyperfine coupling, a_{0N}, of nitroxyl spin labels depends linearly on the unpaired electron spin density on the nitrogen atom ($\rho_{\pi N}$) and to a lesser extent on that on the oxygen atom ($\rho\pi$O). The spin density distribution, and hence the hyperfine coupling, is perturbed linearly by the reaction field from the polar environment.

The contact interaction is given by [42, 43]

$$\hat{H}_{eN} = \sum_n a_n \hat{\mathbf{S}}_e \cdot \hat{\mathbf{J}}_n, \tag{2.21}$$

where $\hat{S}_e$ and $\hat{J}_n$ are the spin and nucleus operators, respectively.

And

$$a_n = \frac{\mu_0}{4\pi}\gamma_e\gamma_n\frac{8\pi}{3}|\Psi(\mathbf{R}_n)|^2, \tag{2.22}$$

here μ_0 is the vacuum magnetic permeability, $\mathbf{R}_n$ denotes the coordinates of the nth nucleus, γ_n and γ_e are the nuclear gyromagnetic ratio, and the electron gyromagnetic ratio, respectively; $|\psi(R_n)|^2$ is the spin density of s-electron on the nucleus. The electron nuclear dipole–dipole spin interactions for an electron in an s-symmetry orbital and in a non-s-symmetry orbital are expressed as follows:

$$\begin{aligned} H_{HF} &= \frac{16\pi}{3}\gamma_I\mu_B\mu_N\delta(\mathbf{r})[\mathbf{S}\cdot\mathbf{I}], \quad l = 0 \\ H_{HF} &= \frac{2\gamma_I\mu_B\mu_N}{r^3}\left[(\mathbf{L}-\mathbf{S})\cdot\mathbf{I} + 3\frac{(\mathbf{S}\cdot\mathbf{r})(\mathbf{I}\cdot\mathbf{r})}{r^2}\right], \quad l \neq 0. \end{aligned} \tag{2.23}$$

According to theory [44], deviation of g-factor from g_e for free electron (g_e = 2.0023) depends on the spin density ρ_0, the spin density coupling constant (λ_0), and the value of the energy gap between the orbital of the unpaired electron residence and nearest orbital (ΔE(pp*)):

$$\Delta g_x = \lambda_0\rho_0/\Delta E(\text{pp}*) \tag{2.24}$$

The polarity dependence of the principal z-element of the hyperfine tensor (A_{zz}) is determined by both the isotropic and anisotropic terms. Obviously, for a paramagnetic molecule, for example, for a nitroxide radical, the values of r, ρ_0, $|W(\text{Rn})|^2$, and ΔE(pp*) must be influenced by environmental properties such as dielectric constant (ε) and the ability to form hydrogen bonds. The dependence g-factors and A for nitroxides on spin density and ΔE(pp*) allows them to be used as spin probes to provide quantity information of local polarity, proticity and pH. The physical reason for such a dependence is that the contribution of resonance structure II

$$-R(2)N - O \leftrightarrow R(2)N^{+} - O- \tag{2.25}$$

causes an increase in spin density on the nitrogen atom and therefore increase of A_{iso} values [4, 8].

A_{zz} values are large, and g_{xx} values are less for the ionic structure because of the reduction of the electron nucleus distance and increase ΔE(pp*) as result of removing electron from the orbital bearing the spin. A negative correlation between a decrease of g_X and an increase in A_Z has been observed for nitroxides in simple liquids and labeled proteins [45]. The spin density distribution, and hence the hyperfine coupling, is perturbed linearly by the reaction field from the polar environment. The polarity dependence of the principal z-element of the hyperfine tensor is determined by both the isotropic and anisotropic terms: [46]

$$A_{zz} = a_{0N} + 2|T_{\perp}^{d}| \tag{2.26}$$

where $|T_{\perp}^{d}|$ is the perpendicular element of the traceless hyperfine tensor that arises from the electron–nuclear dipolar interaction. A polarity effect (up to $\Delta A_{iso} = 0.13$ mT) was observed for the TEMPOL radical and for the formation of H-bond with N–O group of the nitroxides ($\Delta A_{iso} = 0.15$ mT) [47 and references therein]. The g-factor values were also reported to be sensitive to the medium.

A spin pH probes techniques for the measurement of local acidity in biological and non-biological objects including organs and even living organisms was invented in the work of Khramtsov et al. in 1982 [48] and then widely used [49, 50]. EPR spectra of stable nitroxides of the imidazoline and imidazolidine types

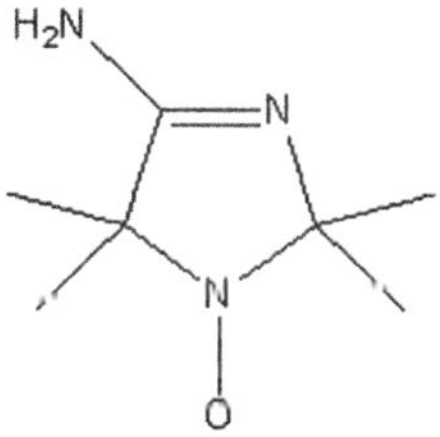

R1

have been shown to be sensitive to pH and serve as spin pH probes. Data on the synthesis of a wide set of pH-sensitive nitroxides are a pH spin probe (in the range from 6 to 8.0 with accuracy −0.05) of different sensitivity, stability to reduction, lipophilicity, covalent binding properties to macromolecules, and applications have been reported [49, 50]. Typical pH-dependence of the observed hyperfine splitting constant, a_N, DNP spectra of the nitroxide NR1 acquired in acid, pH 4.98, and slightly alkaline, pH 7.62, and the effect of protonation of the radical heterocycle resulting in decreasing unpaired electron density at the nitrogen nucleus of the N–O fragment were presented. pH dependencies of the hyperfine splitting, a_N, for R1 probe measured using L-band EPR spectroscopy is shown in Fig. 2.14.

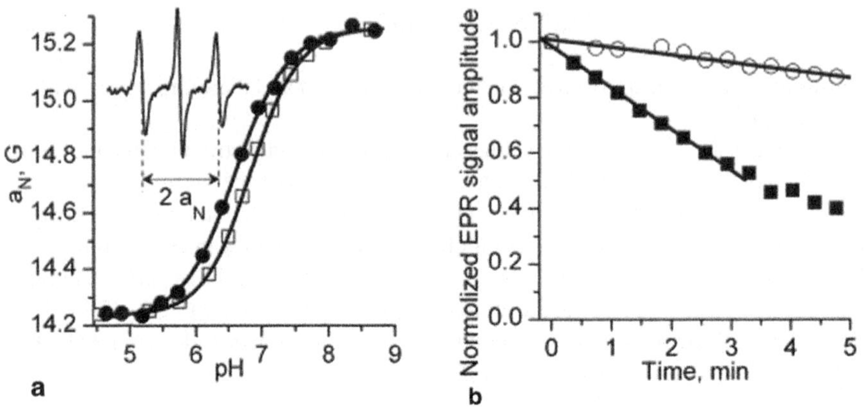

Fig. 2.14 **a** pH dependencies of the hyperfine splitting, a_N, for R1 probe measured using L-band EPR spectroscopy. Temperature-dependent shift of the titration curve was observed with p*K*a values being equal to 6.84 (□, 23 °C) and 6.60 (●, 37 °C). Insert: L-band EPR spectrum of the R1 probe measured in vivo after injection (10 μL, 10 mM) in mammary tumor tissue. The hyperfine splitting, a_N, was found to be equal to 14.72 *G* which corresponds to the value of pH = 6.52 assuming tumor tissue temperature 34 °C and p*K*a = 6.65. **b** EPR signal decay of the probe R1 after injection (10 μL, 10 mM) in mammary glands (○) and mammary tumors (■). The analysis of the initial part of the kinetics yields the rates of the EPR signal reduction, k_{red}, in extracellular media of the tissues being equal to 0.5×10^{-3} s^{-1} and 2.5×10^{-3} s^{-1}, respectively [50]. With permission of John Wiley & Sons

2.3.1.2 Electron Spin Nuclear Spin Interaction and Dynamical Phenomena

Radical motion of a spin label (rotation, tumbling, libration, wobbling) can partially or completely average the dipolar interaction depending on frequency, anisotropy, and mechanism of the motion. Because the spin label motion is dictated by molecular dynamics of media, spin labeling can be a tool for study of dynamical state of objects involving water. Combination of multifrequency radiospectroscopy with other microwave (MW) techniques allows one to characterize the molecular motion of a nitroxide over a wide range of correlation times ($\tau_c = 10^6$–10^{-10} s). The stochastic Liouville equation is commonly used for describing the effect of motion on the line shape of ESR [51].

For X-band (30 mm) spectra, one can estimate rotational correlation times in the region of the fast motion ($\tau_c = 5 \times 10^{-9}$–$10^{-10}$ s) via the following equation [51]

$$\frac{1}{\tau_c} = \frac{3.6 \times 10^9}{\left(\sqrt{\frac{h_0}{h_{-1}}} - 1\right)\Delta H_0} \quad (2.27)$$

where h_0, h_{-1} are the heights of the ESR spectrum hyperfine components, respectively, and ΔH_0 is the line width of the middle hyperfine component. The detailed theory of nitroxide ESR spectra for nitroxide motion in the slow motion region (τ_c

$= 10^{-7}$–10^{-8} s) was developed by Freed [52] using the stochastic Liouville equation. In the slow motion region, the following equation can be used at X-band (30 mm):

$$\tau_c = ax\left(1 - \frac{A_{zz}}{A^0_{zz}}\right)^b \tag{2.28}$$

where A^0_{zz} and A_{zz} are the z-components of A-tensor for immobilized and mobile nitroxide, respectively. The coefficient a was found to be 5.4×10^{-10} and 2.6×10^{-10} s for systems modeling isotropic and anisotropic Brownian diffusion, respectively, and b was found to be -1.36 and -1.39 for the respective models. The effect of saturation transfer between the component of the ESR spectra of nitroxide radicals provides the basis for an approach to the study of very slow motion, with correlation time $\tau_c = 10^{-3}$–10^{-6} s. The advantages of the multifrequency ESR technique for investigations of the mobility of stable radicals were demonstrated in [2, 4–7, 53]. The increased sensitivity to the g tensor which allows high-frequency EPR to distinguish rapid rotations about the g_x, g_y, or g_z tensor directions, so that at high frequencies one can resolve from the spectrum the nature of anisotropic motions, is shown in Fig. 2.15.

2.4 Nuclear Relaxation Rates and Polarization Enhancement by Electron Spin

2.4.1 Relaxation Processes in a System of Two Spins

The electron–nucleus spin dipolar interaction causes R_1 (longitudinal) and R_2 (transverse) nuclear relaxation because of its modulation. The nuclear relaxations are enhanced by the presence of paramagnetic ions and radicals and the major source of relaxation of nuclei I may be the coupling with the unpaired electrons. For the Solomon and Curie contributions, proton relaxation enhancement (PRS) is described by equations [26, 55]:

$$R_2^{\mathrm{PRE}} = \frac{k_{\mathrm{Solomon}} + k_{\mathrm{Curie}}}{r^6} \qquad (3)$$

$$k_{\mathrm{Solomon}} = \left(\frac{\mu_0}{4\pi}\right)^2 \frac{\gamma_I^2 g_e^2 \mu_B^2 S(S+1)}{15}\left[4\tau_c + \frac{\tau_c}{1+(\omega_I-\omega_S)^2\tau_c^2} + \frac{3\tau_c}{1+\omega_I^2\tau_c^2} + \frac{6\tau_c}{1+(\omega_I+\omega_S)^2\tau_c^2} + \frac{6\tau_c}{1+\omega_S^2\tau_c^2}\right]$$

$$k_{\mathrm{Curie}} = \frac{1}{5}\left(\frac{\mu_0}{4\pi}\right)^2 \frac{\omega_I^2 g_e^4 \mu_B^4 S^2(S+1)^2}{(3kT)^2}\left[4\tau_{\mathrm{Curie}} + \frac{3\tau_{\mathrm{Curie}}}{1+\omega_I^2\tau_{\mathrm{Curie}}^2}\right] \tag{2.29}$$

where ω_I is the nuclear Larmor frequency, ω_S is the electron Larmor frequency, r is the electron–nucleus distance, g_e is the electron g-factor, μ_B is the electron Bohr

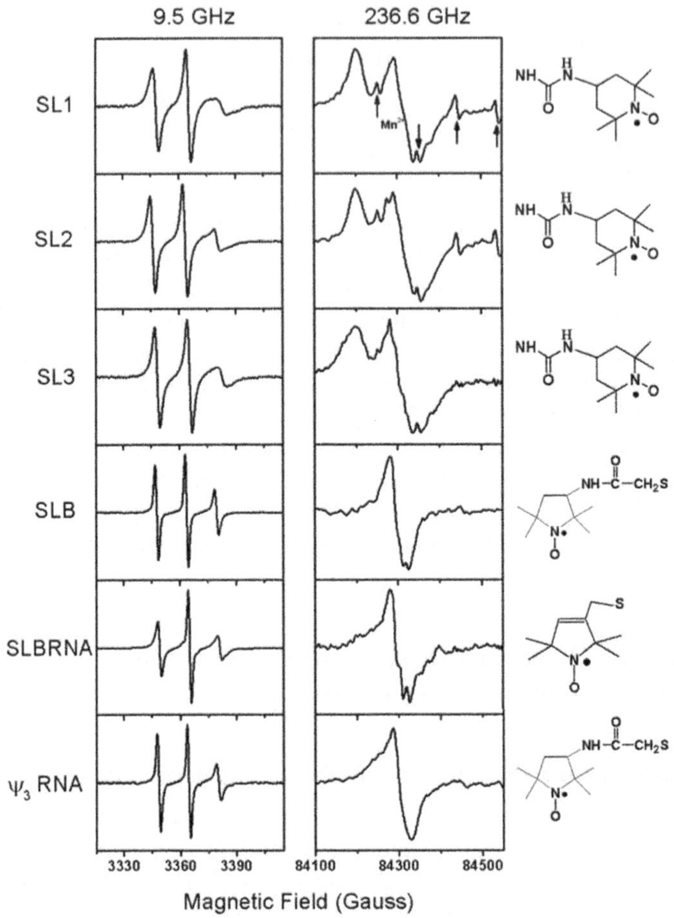

Fig. 2.15 ESR spectra of mini c TAR DNA labeled at positions SL1, SL2, SL3, SLB, mini TAR RNA labeled at its 3′ (SLB) terminal, and ψ_3 RNA labeled at the 5′ terminal. Concentrations of labels were approximately 100 μM (except SL3 which was 250 μM) in pH 7.5 20 mM Hepes, 20 mM NaCl, 0.2 mM Mg^{++}, temperature 20 °C. The spectra are compared at EPR frequencies of 9.5 and 236.6 GHz to show the resolving power of high-frequency EPR for differences in motion. EPR conditions are given in the Methods section. The features labeled Mn^{++} are due to low-level Mn^{++} impurities whose line shapes are vastly sharpened at high frequency and provide effective internal field markers [54]. With permission from American Chemical Society. SEND 18/05

magneton, g_I is the nuclear magnetogyric ratio, k is the Boltzmann constant, T is the temperature, S is the electron spin quantum number. The correlation times related to the Solomon and Curie mechanisms responsible for relaxation are:

$$\tau_c^{-1} = \tau_e^{-1} + \tau_r^{-1} + \tau_M^{-1} \tag{2.30}$$

and

$$\tau_{\text{Curie}}^{-1} = \tau_{\text{r}}^{-1} + \tau_{\text{M}}^{-1} \tag{2.31}$$

If the proton exchange rate τ_{M}^{-1} is negligible, τ_{c} is determined by both the reorientation time of the electron–nucleus vector τ_{r} and the electron relaxation time τ_{e}, whereas τ^{Curie} is determined by τ_{r} only.

2.4.2 Contrast Enhancement of Water Protons in Magnetic Resonance Imaging

Contrast enhancement in Magnetic Resonance Imaging (MRI) [55] is defined as the difference in intensity between the same pixel or the same collection of pixels of region of interest in the precontrast (Ipre) and post-contrast (Ipost) magnetic resonance imaging (MRI). The ability of paramagnetic probes to increase the relaxation rates r of solvent water is a linear function of the concentration of the contrast agent [CA], the slope of which is known as r^{CA} relaxivity.

$$r = r^0 + r^{\text{CA}}\ [\text{CA}] \tag{2.32}$$

where r^0 represents the relaxation rate of the water protons in the absence of contrast agent (gadolinium(III) containing molecules, iron oxide and iron platinum nanoparticles, manganese chelates, barium sulfate, and enzyme-activated MR agents). The electron spin interacts through dipole–dipole coupling with the nuclear spin residing in the paramagnetic complex (the inner-sphere case) as well as outside of it (the outer-sphere case).

The measured ^{1}H relaxation rates in the presence of paramagnetic complexes were given by a sum of two terms:

$$T_{\text{1P}}^{-1} = \frac{P_{\text{M}}q}{T_{1I} + \tau_{\text{M}}} + T_{\text{1OS}}^{-1} \tag{2.33}$$

The first and the second terms of the equation correspond to the inner- and outer-sphere contribution to the PRE, respectively. $P_{\text{M}q}$ is the mole fraction of ligand protons in the bound position and q is the number of bound ligand molecules, T_{1I} is the spin–lattice relaxation time for the nuclear spin I in the complex, τ_{M} is the chemical exchange lifetime.

According to the theory of Hwang and Freed [56], the spin–lattice relaxation rate of proton nuclei, $1/T1\text{p}$, upon encounter with a "radical" free electron, are dominated by dipolar interaction because of low electron spin density at the van der Waals distance. Slopes $k - d(1/T1\text{p})/d[R\cdot]$ of the dependence of proton spin lattice relaxation rate on the concentration of the nitroxide probes $[R\cdot]$ can characterize the dipole–dipole interaction that dominates in an encounter between a proton and paramagnetic species. To explore the effect of electrostatic interaction on the dipole–dipole relaxation rate, in the presence of nitroxide of different charges the ratios of experimental values k^{+}/k^{0} or k^{-}/k^{0} were used to estimate the electrostatic potential, $U(R0)$, in the vicinity of a proton in biologically important molecules which can affect on interactions with a contrast agent [57]. In addition, the magnitude of the potential and its distribution along a biological molecule can significantly influence the state of water in the immediate vicinity.

2.4.3 Electron Nuclear Double Resonance

Electron nuclear double resonance (ENDOR) was introduced by Feher [58]. ENDOR spectra contain information on the type of nuclei in the vicinity of the unpaired electron, on the distances between nuclei, on the spin density distribution, and on the electric field gradient at the nuclei (Q). The nuclear quadrupole interaction is present in nuclei with $I > 1/2$. The four terms in the Feher equation describe the electron Zeeman interaction, the nuclear Zeeman interaction, the hyperfine interaction (hfs), and the nuclear quadrupole interaction (Q), respectively (Fig. 2.16).

In a steady-state ENDOR experiment, an EPR transition is partly saturated by microwave radiation of amplitude B_1. A driving radio frequency (rf) field of amplitude B_2 induces nuclear transitions. While the magnetic field is swept through the EPR spectrum, the rf frequency follows the Zeeman frequency of the nucleus. Transitions occur at frequencies ω_1 and ω_2 and obey the NMR selection rules $\Delta M_I = \pm 1$ and $\Delta M_S = 0$. These NMR transitions are detected by ENDOR via the intensity changes to the simultaneously irradiated EPR transition. The ENDOR spectrum can reveal both the hyperfine coupling constant (a) and the nuclear Larmor frequencies (ν_n).

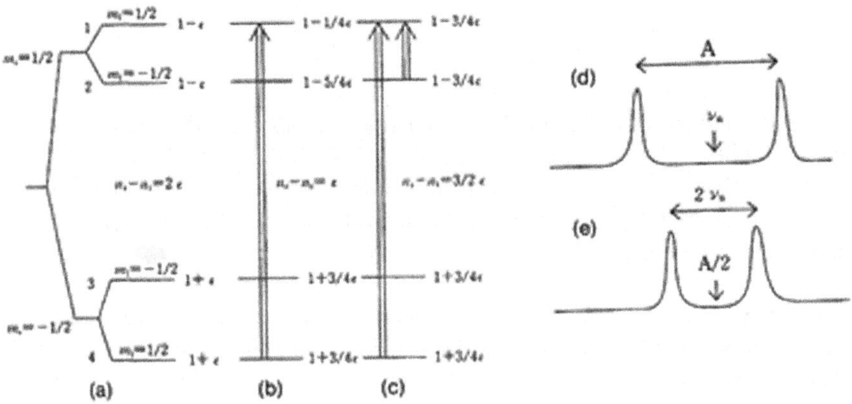

Fig. 2.16 ENDOR sublevels for $S = 1/2$ and $I = 1/2$. **a** At thermal equilibrium, **b** after ESR transition 4 → 1, **c** added by NMR transition. The population difference for $n_4 - n_1 = 2\varepsilon$ (**a**) and $n_4 - n_1 = \varepsilon$ (**b**) [3]. With permission of WILEY–VCH, Weinheim

$$\nu_1 = |\nu_n - a/2|$$
$$\nu_2 = |\nu_n + a/2| \quad (2.34)$$

Due to the fact that ESR is much more sensitive than NMR, ENDOR allows to study objects in concentrations that are at least two orders of magnitude lower than in the case of the "classical" NMR.

Detailed theory and typical applications of ENDOR were described in the comprehensive review [59].

2.5 Pulse EPR Measurements

2.5.1 General

In a pulse experiment, the magnetization is typically perturbed by applying a brief, intense, sinusoidal electromagnetic field. Once the pulse is removed, one can then measure the emitted microwave signal which is created by the sample magnetization as it recovers to its initial value. Over the years, a vast variety of pulse sequences have been developed [2, 3, 8, 10, 14, 60–66]. Using them, it is possible to gain extensive knowledge of the structural and dynamical properties of paramagnetic compounds.

Some of the more important developments in the area include [2–67]: Fourier transform ESR (FT ESR), two-dimensional Fourier transform ESR (2D FT ESR), electron spin echo (ESE), electron spin-echo envelope modulation (ESEEM), hyperfine sublevel correlation spectroscopy (HYSCORE), double nuclear coherence transfer hyperfine sublevel correlation electron spin echo (DONUT-HYSCORE), electron spin-echo detected magnetization transfer (ESE MT), two-dimensional electron spin-echo correlation spectroscopy (2D SECSY), pulse electron nuclear double resonance (pulse ENDOR), pulse electron–electron double resonance (PELDOR), electron spin transient nutation (ESTN), two-dimensional electron spin transient nutation (2D ESTN), phase inverted echo amplitude detected nutation (PEANUT), saturation recovery ESR (SR ESR), pulse multifrequency ESR (PMF ESR), magnetic isotope effect (MEF), quantum beats effect (QBE), double-quantum coherence pulsed ESR (DQC ESR), ESR spectra hole burning, and ESR nutation, dynamic nuclear polarization (DNP).

In this section, pulse ENDOR, Hyperfine-correlated electron nuclear double resonance spectroscopy (HYEND), the Fourier transform ESR (FT ESR), two-dimensional Fourier transform ESR (2D FT ESR), electron spin echo (ESE), electron spin-echo envelope modulation (ESEEM), hyperfine sublevel correlation spectroscopy (HYSCORE), double nuclear coherence transfer hyperfine sublevel correlation electron spin echo (DONUT-HYSCORE), as methods which appeared to be the relevant for applications for study structure and molecular dynamics of water, are briefly discussed.

Many qualitative insights into the behavior of EPR-active spins can be gleaned from precession phenomena. The effects of the magnetic field on the spin magnetization vector can be considered in the laboratory frame as well as in the rotating frame. In the laboratory frame, the static magnetic field B_0 is assumed to be parallel to the z-axis and the microwave field B_1 parallel to the x-axis.

2.5.2 Pulse ENDOR

Pulse ENDOR and ESEEM spectra contain information on the type of nuclei, on the distances between nuclei and on the spin density distribution (hyperfine interaction) and on the electric field gradient at the nuclei (nuclear quadrupole interaction) [11, 59–66]. Pulse ENDOR techniques, make use of a combination of microwave and radio frequency pulses. Pulse ENDOR can investigate the interactions with more distant nuclei than that CW EPR. The microwave pulse sequence generates an electron spin echo which is modulated by manipulation of the nuclear spins due to the radio frequency pulses (Fig. 2.17). In Davies ENDOR of an $S = 1/2$, $I = 1/2$ system, a microwave (mw) inversion recovery pulse sequence (π–T–$\pi/2$–τ–π–τ–echo) is used (Fig. 2.17) [11]. Davies ENDOR is useful for systems with large HFIs. In Mims pulse ENDOR [60, 61], both EPR transitions are excited by the applied stimulated echo mw pulse sequence ($\pi/2$–τ–$\pi/2$–T–$\pi/2$–τ–echo). Applications of this method are limited to relatively small hfi constants ($B_1 > a$). Due to the influence of the rf field on the electron spin, the effective field experienced by the nuclear spins becomes dependent on the electron spin (m_S) and on the HFI strength, which leads to a change of the ENDOR line intensity.

Hofer group [62] introduced a time-domain pulsed method in which polarization transfer ENDOR schemes can be performed with selective as well as nonselective mw pulses. In these experiments, the mixing period consists of two rf $\pi/2$ pulses separated by a variable time interval T. Hyperfine-correlated electron nuclear double resonance spectroscopy (HYEND), where the nuclear transition frequencies are correlated with

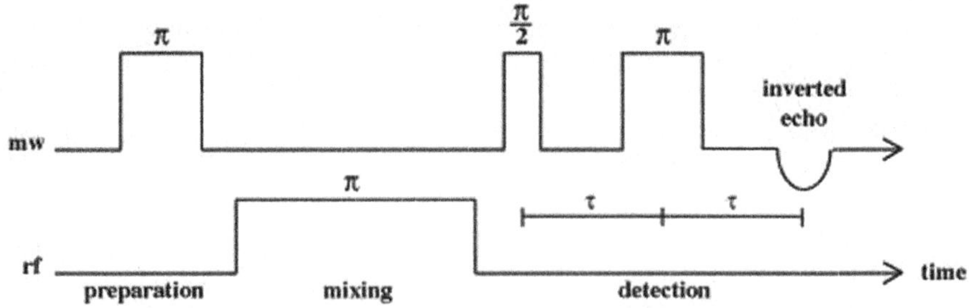

Fig. 2.17 EPR spectra of nitroxide radical calculated for different spin precession frequencies (microwave linewidths). http://hf-epr.awardspace.us/index.htm

the corresponding hyperfine frequencies, leads to a considerable simplification of the interpretation of the spectra.

2.5.3 Electron Spin-Echo Envelope Modulation

Electron spin-echo envelope modulation (ESEEM) pioneered by Mims [61] arises from a coupling between the EPR-active electron being probed and nuclei within the system when there is state mixing of the hyperfine levels [2, 7, 10, 11, 61–66]. In ESEEM experiments, the modulation of the echo intensity is monitored as a function of the delay between the microwave pulses. The resulting ENDOR spectrum is an ESR-enhanced NMR spectrum of the nuclei that are coupled to the electron being probed. In ESEEM experiments, time-resolved data acquisition is achieved on ESR timescales, which is essentially much more sensitive than the NMR timescale. The two most common forms of ESEEM experiments use either a two-pulse or three-pulse echo sequences, which are 1000 times faster than NMR.

Because echo envelope modulation arises when there is state mixing of the hyperfine levels ESEEM can be observed in systems where the nuclear quadrupole coupling is relatively large compared to the hyperfine coupling, i.e., remote ^{14}N couplings. A quadrupolar nucleus is one that has a quantum spin number greater than ½. Such nuclei have a lower symmetry than spin-½ nuclei and the quadrupole moment that varies between nuclei is a measure of this asymmetry. An optimum Zeeman field value exists for each nucleus and hyperfine coupling, as has been demonstrated by multifrequency ESEEM experiments on $S = 1/2$, $I = 1/2$ as well as on $S = 1/2$, $I = 1$ systems [66]. The principles, procedures, and applications of ESEEM have been reviewed in the comprehensive books [11, 63, 65]. In the pulse sequences used for two-, three-, and four-pulse ESEEM and HYperfine Sublevel COrRElation spectroscopy (HYSCORE) experiments the NMR transitions are observed due to mixing of the frequencies of the semi-forbidden and allowed EPR transitions respectively, which have been coherently excited using short, intense microwave pulses, separated by times τ, T, τ_1, and τ_2 (Fig. 2.18) [68].

For the two-pulse technique, the modulation envelope $E2p(t)$ **is written as (Elsevier)**

$$\text{Two-pulse: } E_{2p}(\tau) = -k/4\left[2 - 2\cos(\omega_\alpha \tau) - 2\cos(\omega_\beta \tau) + \cos\left((\omega_\alpha + \omega_\beta)\tau\right) + \cos\left((\omega_\alpha - \omega_\beta)\tau\right)\right] \tag{2.35}$$

$\omega_{\alpha,\beta}$ are the fundamental nuclear frequencies and the modulation depth k is

$$k = 4c^2 s^2 \tag{2.36}$$

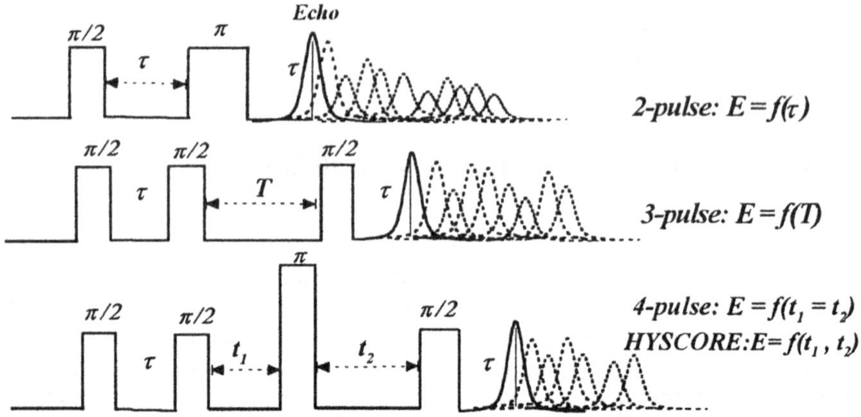

Fig. 2.18 Pulse sequences used for two-, three-, and four-pulse ESEEM and HYSCORE experiments. In the two-pulse experiment at time *t* after the second pulse, a spin echo is generated. The modulation envelope is obtained as *t* is incremented. In the three-pulse sequence, two pulses separated by time *t* are applied, followed by a third pulse after time *T*, and the stimulated echo is observed at time *t* after the third pulse. The echo envelope obtained as *T* is incremented. Four-pulse ESEEM and HYSCORE are based on a three-pulse sequence with an additional *p*-pulse applied between the second and third *p*: two pulse. As in the three-pulse sequence, the first two *p*: two pulses separated by the time interval *t*, generate nuclear coherences. Then, the *p*-pulse transfers populations from one *MS* manifold to the other and mixes nuclear coherences in one *MS* manifold with those of the other manifold. The last *p*: two-pulse leads to the formation of the stimulated echo. In the four-pulse experiment, the nuclear coherences evolve during the two variable time intervals t_1 and t_2, stepped under the constraint t_1_t_2_T. In the HYSCORE experiment, the echo intensity is recorded as a function of the interpulse time t_1, e.g., between the time the second and third pulses, and t_2, e.g., between the time the third and fourth pulses, where *ms*__1:2 and _1:2 correspond to the *a* and *b* spin manifolds, respectively [68]. With permission from Elsevier

with c_2 and s_2 being the transition probabilities of the allowed and semi-forbidden transitions, respectively. In three-pulse technique (Fig. 2.19), the various parameters have the same meaning and definitions as in the two-pulse case [68].

HYSCORE, a two-dimensional ESEEM technique based on a four-pulse sequence (Fig. 2.19), was introduced by Hofer et al. [69].

For $S = 1/2$, $I = 1/2$, the expression of the time-domain HYSCORE related to the cross-peaks and the modulation envelope is presented as

$$\begin{aligned}E_{\text{HYSCORE}}(\tau, t_1, t_2) \sim k/4B\big\{&c^2\cos(\omega_\alpha t_1 + \omega_\beta t_2 + \delta_+)\\ &- s^2\cos(\omega_\alpha t_1 - \omega_\beta t_2 + \delta_-)c^2\cos(\omega_\alpha t_2 + \omega_\beta t_1 + \delta_+)\\ &-s^2\cos(\omega_\alpha t_2 - \omega_\beta t_1 + \delta_-)\big\}\end{aligned} \tag{2.37}$$

Four-pulse ESEEM is based on a three-pulse sequence with an additional *p*-pulse applied between the second and third π: two-pulse, As an example of application of three- and four-sequence echo, in work [70], the normalized echo decay $V(2\tau, T)$

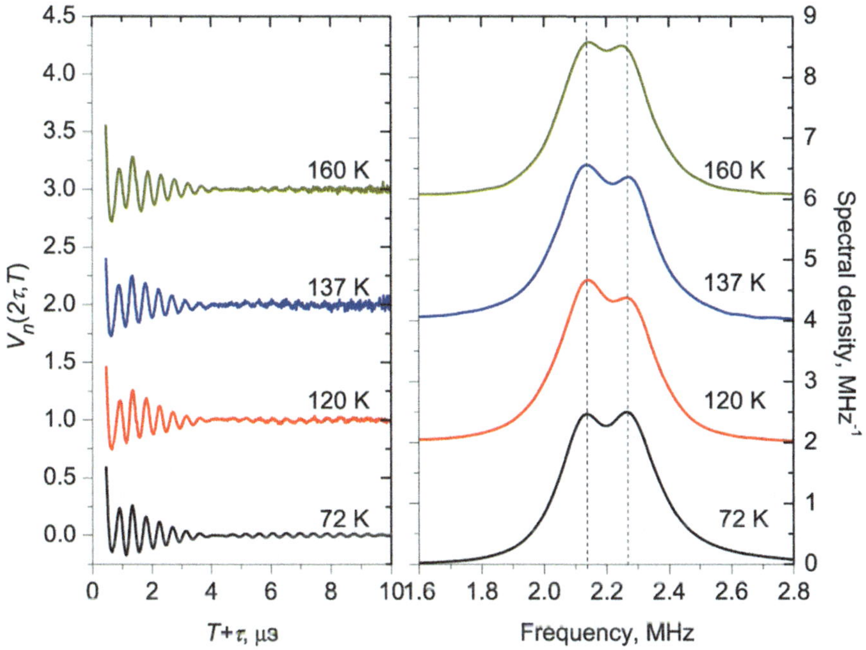

Fig. 2.19 Time-domain (left) and frequency-domain (right) ^2H ESEEM spectra for the glassy water/DMSO mixture. Data are shifted along the vertical axis by a dimensionless value of 1.0 (left) and by 2 MHz^{-1} (right). Spectra are normalized to the unit value of their total intensity in the frequency range of 1.6–2.8 MHz. The dotted lines show the spectral positions of singularities at 72 K. The lipid POPC or DPPC with spin-labeled lipid Tempo-PC at a molar ratio of 100:1 was dissolved in chloroform and evaporated under the nitrogen stream [70]. With permission from American Chemical Society

$$V_n(2\tau, T) = \frac{V(2\tau, T)}{\langle V(2\tau, T)\rangle} - 1 \tag{2.38}$$

was separated due to the spin relaxation and ESEEM effect ^2H ESEEM spectra for the glassy water/DMSO mixture. Time-domain (left) and frequency-domain (right) ^2H ESEEM spectra for the glassy water/DMSO mixture are presented in Fig. 2.19.

For computation in the case of R1-^{14}N, the system of two interacting spins (one electron, $S = 1/2$, and one nitrogen nucleus, $I = 1$) exposed to the static magnetic field of the spectrometer, B_0, was described by the spin Hamiltonian (in frequency units):

$$\hat{K} = (\beta/h)\vec{B}_0\overline{\overline{g}}\vec{\hat{S}} - (\gamma_N/2\pi)\vec{B}_0\vec{\hat{I}} + \vec{\hat{S}}\overline{\overline{A}}\vec{\hat{I}} + \vec{\hat{I}}\overline{\overline{Q}}\vec{\hat{I}} \tag{2.39}$$

where h, Planck constant; β, Bohr magneton; γ_N, gyromagnetic ratio for ^{14}N nucleus; $\hat{S}$, $\hat{I}$, electron and nuclear spin vector operators. The individual terms of (2.44) represent the electron and nuclear Zeeman, hyperfine, and nuclear quadrupole interactions,

g, Aj, Q are the correspondent tensors of ^{14}N nitrogen nucleus quadrupole moment. DFT calculations and W-band ESEEM experiments showed that: (1) Q_{yy} has the strongest dependence on different environmental situations compared with Q_{xx} and Q_{zz}, being essentially different to the behavior of the g- and A-tensors; (2) the Q_{yy} value is especially sensitive to the proticity and polarity of the nitroxide environment in H-bonding and nonbonding situations; and (3) when using Q_{yy} as a testing probe of the environment, its ruggedness toward temperature changes represents an important advantage over the g_{xx} and A_{zz} parameters which are usually employed for probing matrix effects on the spin-labeled molecular site.

As an example of effective use of the Overhauser dynamic nuclear polarization technique, the measurements of water density and diffusion dynamics in the vicinity of a spin label attached to a cysteine in the Tyr71 → Cys GroES mutant of the chaperonin GroEL–GroES complex (Fig. 2.20), obtained using two magnetic resonance techniques: electron spin-echo envelope modulation and ODNP were performed [71]. ATP-dependent binding of the chaperonin GroEL to its cofactor GroES forms a cavity in which encapsulated substrate proteins can fold in isolation from bulk solution (Fig. 2.20). Measurements of water density and diffusion dynamics in the vicinity of the spin label revealed that residue 71 in GroES is fully exposed to bulk water in free GroES and that water confines within the cavity of the GroEL–GroES complex. The data indicated that bulk water-exposed and cavity-confined GroES surface water share similar properties. Interestingly, the diffusion dynamics of water near the GroES surface are found to be unusually fast relative to other protein surfaces studied.

In the determination of water accessibility to spin-labeled amino acid residues in the light-harvesting complex of green plants LHCIIb, three-pulse ESEEM measurements were performed with a $p/2$–t–$p/2$–T–$p/2$–t–echo pulse sequence with a standard phase cycle [71, 72]. The ESEEM water **accessibility** parameter $P(D_2O)$ that quantifies the average number of water molecules in van der Waals contact with the label was characterized by the average density of exchangeable protons at distances in the range of ~6–20 Å.

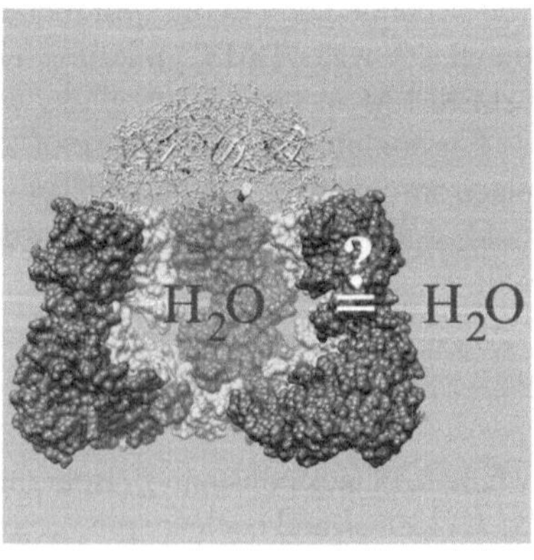

Fig. 2.20 Hydrated chaperonin GroEL [71]. With permission from American Chemical Society

The ESEEM modulation depth, k_D, is related to an orientation average <k> of the modulation depth, k, in the theoretical description of three-pulse ESEEM. According to ESEEM theory, k depends on the number, n, and distances, r_i, of nuclei coupled to the electron spin. The relation is given as

$$k_D = \frac{\langle k \rangle}{2}[1 - \cos(2\pi \nu_D \tau)], \tag{2.40}$$

where time τ in the second factor on the right-hand side (suppression factor) is the delay between the first two pulses. Due to its r^{-6} dependence, the modulation depth is strongly biased toward distances of closest approach, i.e., to water molecules in van der Waals contact with a spin label. The suppression factor

$$[1 - \cos(2\pi \nu_D \tau)] \tag{2.41}$$

can range between 0 and 2. By fitting with fixed frequency ν_D, only deuterium nuclei contribute to k_D. A three-pulse ESEEM-based water accessibility parameter $\Pi(D_2O)$ characterizing the average density of exchangeable protons at distances in the range of ~6–20 Å, was defined as

$$\Pi(D_2O) = \frac{2k_D}{[1 - \cos(2\pi \nu_D \tau)]} \times \left(\frac{\nu_D}{2\,\text{MHz}}\right)^2. \tag{2.42}$$

This parameter in combination other ESR technique has provided complementary information on water accessibility of a spin label in a membrane protein LHCIIb.

DONUT-HYSCORE (double nuclear coherence transfer hyperfine sublevel correlation) is a two-dimensional experiment, which was designed to obtain correlations between nuclear frequencies belonging to the same electron spin manifold [73]. The sequence employed is $\pi/2$–τ_1–$\pi/2$–t_1–π–τ_2–π–t_2–$\pi/2$–τ_1–echo, and the echo was measured as a function of t_1 and t_2, whereas τ_1 and τ_2 are held constant. A major improvement in the resolution of the ESEEM frequencies and in their assignment has been achieved with these methods.

2.5.4 Dynamic Nuclear Polarization

The general Overhauser effect was first predicted by Albert Overhauser who observed that saturation of the conduction electron spin resonance by microwave irradiation causes a polarization of nuclear spins [38]. Dynamic nuclear polarization (DNP) results from transferring spin polarization from electrons to nuclei, thereby aligning the nuclear spins to the extent that electron spins are aligned. Those electrons can be also aligned to a higher degree of order by other preparations of electron spin order such as chemical reactions leading to chemical-induced DNP, CIDNP, optical pumping, and spin injection [38–41, 74–83]. DNP has been used for studying the detection distances between unpaired electron and nuclear spins and for enhancing the sensitivity of NMR signals.

A method to obtain local water dynamics through Overhauser dynamic nuclear polarization (DNP) which allows to measure local water dynamics within 5 Å distances, utilizing the modulation of the nuclear spin relaxation rate of water protons through their time-dependent dipolar coupling to paramagnetic probes, was introduced [80, 81]. The outcome of the approach is the quantitative determination of the key DNP parameter ρ, ξ, the coupling factor, which provides local translational diffusion dynamics of the solvent within 5 Å of a spin label. A highly amplified proton nuclear magnetic resonance signal allows the use of minute microliter sample volumes and 100 μM sample concentrations.

The NMR signal enhancement, *E*, upon irradiation of the electron spin resonance (ESR) transition is given by [80, 81]

$$E = 1 - \rho f s \frac{|\gamma_S|}{\gamma_I} \tag{2.43}$$

where $\boldsymbol{\rho}$ is the coupling factor describing the efficiency of the dipolar coupling between the two spins, *f* is the leakage factor which corrects for nuclear spin relaxation caused through other mechanisms than the time-dependent dipolar field of the unpaired electron and differences in electron spin concentration, *s* is the saturation factor describing the degree to which the electron spin populations have been equilibrated, and γ_S and γ_I are the magnetogyric ratios of the electron and proton, respectively, $f = 1 - T_1/T_{10}$, where T_1 is the ^{1}H relaxation time in the presence of the free radical and T_{10} the ^{1}H relaxation time in the absence of dissolved radical [81].

The coupling factor ξ depends on the transition probabilities between the electron nucleus spin states in a hyperfine coupled system and is a function of κ, the electron relaxation time τ_c, the diffusion correlation time τ_D, and diffusion spectral density functions $J(\omega_{I,S}, \tau_c, \tau_D)$ [81]. The correlation time can be described as the lifetime of the dipolar interaction between the electron spin of the spin probe and the proton spin of the water molecule, and it is related to the diffusivity between the two [82]:

$$\tau_c = d^2/(D_{local} + D_{SL}) \tag{2.44}$$

where D_{local} is the diffusivity of the hydration water within 0.5–1.5 nm of the spin probe, D_{SL} is the diffusivity of the spin probe, and *d* is the distance of approach between the water and the spin probe. ξ is function from $J(x, s)$, the Lorentzian spectral density function

$$J(\omega, \tau) = \frac{\tau}{1 + \omega^2 \tau^2} \tag{2.45}$$

where τ_c is the correlation time for the dipole–dipole interaction, equal to

$$\tau_c^{-1} = \tau_S^{-1} + \tau_R^{-1} + \tau_M^{-1}, \tag{2.46}$$

here τ_S is the electron relaxation time, τ_R is the molecular reorientation time, and τ_M is the nuclear lifetime in bound position. The dipole coupling factor k is proportional to the squared dipole–dipole energy. The coupling factor ξ—determines the local efficiency of the ODNP polarization transfer, depends only on the motion of the water molecules relative to the spin probe and can be given as ratio [83].

$$\xi = \sigma/\rho = k_\sigma/k_\rho, \tag{2.47}$$

where local self-relaxivity, k_ρ, related to the self-relaxation rate, was defined as follows

$$\rho = k_\rho C, \tag{2.48}$$

where C *is* the spin probe concentration, a local cross-relaxivity parameter, k_σ, related to the concentration dependence of the cross-relaxation rate, σ,

$$\sigma = k_\sigma C, \tag{2.49}$$

and

$$\xi = \sigma/\rho = k_\sigma/k_\rho. \tag{2.50}$$

Both k_σ and k_ρ arise from time-dependent dipolar interactions that scale as r^{-6}:

$$k = \frac{2}{15}\left(\frac{\mu_0}{4\pi}\right)^2 \frac{\gamma_I^2 g_e^2 \mu_B^2 S(S+1)}{r^6} \tag{2.51}$$

and depend only on the dynamics and amount of water that very closely (0.5–1.5 nm) approaches the spin probe.

Advances that have helped develop ODNP into a tool for mapping the dynamic landscape of hydration water with subnanometer locality were reviewed [83]. The analytical model and extending the description of the ODNP-derived NMR signal enhancements were suggested. This extended model highlights several aspects of ODNP data analysis, including the unique ability of ODNP to probe the electron-^{1}H cross-relaxation process, which is very sensitive to fast (tens of ps) dynamical processes. Specifically, it was shown that weak protein surface water attraction will be reflected in small τ_c and large D and large k_σ values while strong protein surface water attraction will present the opposite trend of large τ_c and small D and small k_σ values. In addition, there can be contributions from strongly bound water on protein surfaces with lifetimes exceeding ~1 ns, whose presence would be reflected in a large k_{low} value that increases as the rotational tumbling of the protein is slowed, for example, upon immobilization or immersion in a viscous solvent [83]. Measurement

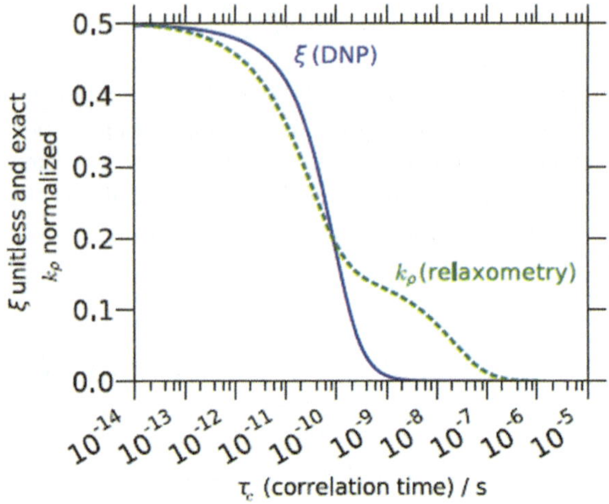

Fig. 2.21 ODNP, which measures the coupling factor, ξ (solid blue line) exhibits a better selectivity for high-frequency (typically translational) motions of the solvent than an NMR relaxometry measurement that measures the self-relaxivity of the water in the presence of the spin probe, k_ρ (dashed green line). The calculations shown here use the force-free hard sphere diffusion model at a field strength of 0.35 T (15 MHz NMR resonance); for both measurements, the range of correlation times probed will scale inversely with field [83]. With permission from Elsevier

of the coupling factor $\xi((2\pi)15\ \text{MHz}/\gamma_H;\ \tau_c)$ as a function of the parameter τ_c presented in Fig. 2.21, allowed to find the value of τ_c.

Outline of the complete procedure for ODNP data processing is indicated in Fig. 2.22. In the figure, R_1 and $R_{1,0}$ refer to the inverses of the NMR spin–lattice relaxation times, i.e., $R_1 = T_{1-1}$ and $R_{1,0} = T_{1,0-1}$. The spin-label-driven proton self-relaxation rate constant, k_ρ, was obtained by subtracting $R_{1,0}$ from $R_{1(p=0)}$ which is the microwave power-dependent relaxation rate $R_{1(p)}$. The spin-label-driven proton self-relaxivity, k_ρ, was then obtained from the self-relaxation rate by normalizing against the spin label concentration (CSL). The cross-relaxivity was determined from the data in Fig. 2.22. The ratio of the relaxivities k_σ and k_ρ yields the coupling factor, ξ ($\xi = k_\sigma/k_\rho$).

In order to better understand the contribution of partially bound waters, the value of ξ the contribution from the fast waters (i.e., k_σ) was subtracted from the

$$k_{\text{low}} = \frac{5}{3}k_\rho - \frac{7}{3}k_\sigma \tag{2.52}$$

where k_{low} describes the slower timescale (~15 MHz) fluctuations of the dipolar interaction. The value of ξ is related to the ratio between k_σ and k_{low}:

$$\xi = \frac{5k_\sigma/k_{\text{low}}}{3 + 7k_\sigma/k_{\text{low}}} \tag{2.53}$$

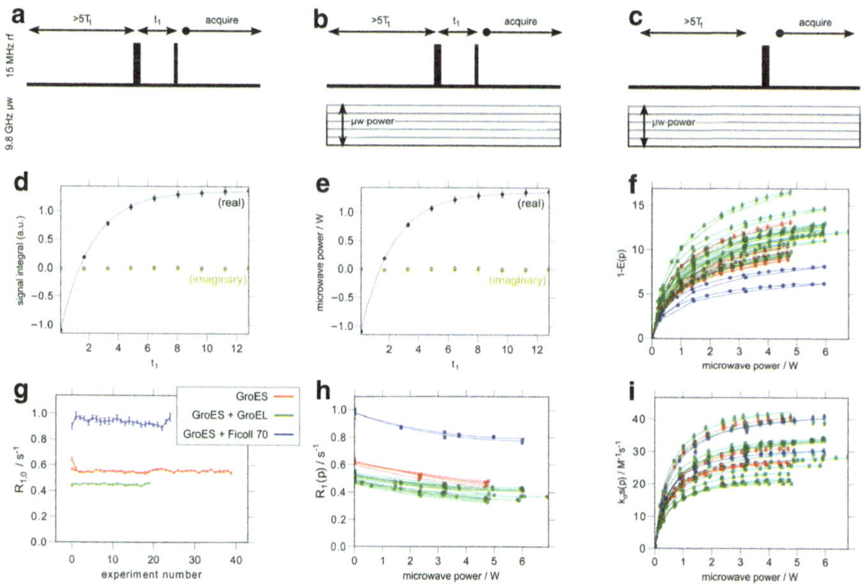

Fig. 2.22 Outline of the complete procedure for ODNP data processing is shown for representative data. First, a variety of NMR measurements is carried out including an inversion recovery sequence acquired on a sample without spin label (**a**), a series of inversion recovery sequences acquired with spin label and different microwave powers (**b**), and a simple NMR spectrum acquired at different powers of ESR-resonant microwaves (**c**). The data corresponding to these pulse sequences are shown in panels **d**–**f**. The inversion–recovery curves (**d**, **e**) are fitted to determine the NMR relaxation rates $R_{1,0}$ (**g**) and $R_{1(p)}$ (**h**). The latter multiplies $1 - E(p)$ (**f**) to yield $k_{\sigma s(p)}$ (**i**), which are fitted to an asymptotic curve (shown as a solid line), allowing us to extrapolate it to full saturation of the ESR transition and determine $k_\sigma \approx k_{\sigma \mathrm{smax}}$. The multiple curves in panels **f**, **h** and **i** are for repeated experiments as indicated by the color code in panel **g** [71]. With permission from American Chemical Society

where $0 \leq k_\sigma / k_{\mathrm{low}} \lesssim 1$.

The value of translational correlation time, τ_c, can be obtained using the plot of the coupling factor measurement, ξ, as a function of the modeled time, τ_c (Fig. 2.23).

The purpose of review [78] was to share the technical know-how of the hardware, theory, measurement, and data analysis method needed to successfully utilize and disseminate the ODNP technique. Figure 2.24 shows the change in the spin population of different levels in the course of a simplified ODNP experiment on an $S = 1/2$, $I = 1/2$ spin system: the thermal equilibrium state (Fig. 2.24a) and transient state (Fig. 2.24b). The $\beta e \alpha n$ and $\beta e \beta n$ states have higher population as predicted by the Boltzmann distribution. The latter is achieved after mw irradiation on resonance with the EPR transitions is applied, so that $T_{1e} < t_{mw} < T_{zq}, T_{dq}, T_{1n}$, where t_{mw} denotes the duration of the mw irradiation [78].

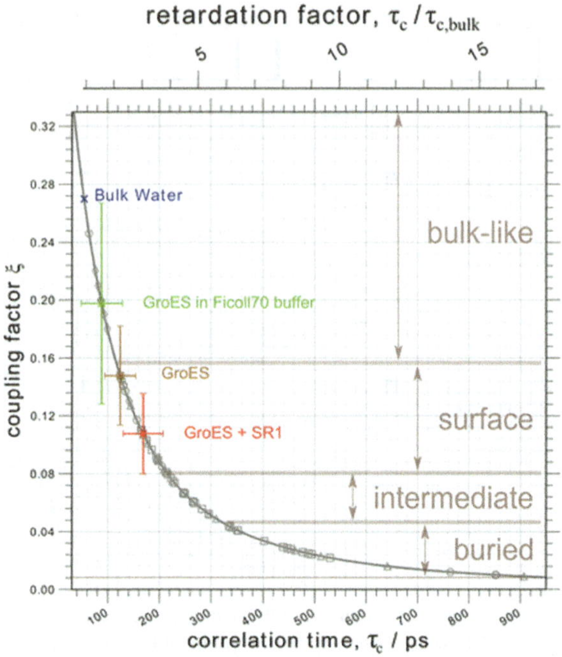

Fig. 2.23 Plot of the coupling factor measurement, ξ, as a function of the modeled translational correlation time, τ_c. The data points for the ODNP measurements for free GroES, GroES in complex with SR1, and GroES in the presence of Ficoll 70 are in brown, red, and green, respectively. The FFHS model gives a fixed relationship, $\xi(\tau_c)$, for measurements at 0.35 T (corresponding to 15 MHz nuclear Larmor frequency) that is illustrated by the solid gray line. The gray symbols indicate previous ODNP measurements for a variety of proteins, small peptides, lipids, and DNA that are grouped (in brown text, to the right) according to the location of the spin label. As explained previously, 22 measurements in the zone designated "buried" were for labels attached within the core of a lipid bilayer, globular protein, or compact polymer system; in the zone designated "surface" for labels attached to the surfaces of proteins or other polymer; in the "intermediate" zone for labels attached near but not at the surface of, for example, a lipid bilayer; and in the "bulk" zone for small molecule nitroxides freely dispersed in water or certain highly charged polymers such as DNA [81]. With permission from American Chemical Society

An example of proceeding raw ODNP dataset is obtained for a spin label attached to a surface-exposed cite of a globular protein The full ODNP measurement consists of three separate series of experiments, two of which performed on the sample with a radical: (1) NMR enhancement as function of mw power, (2) series of inversion recovery experiments to obtain T_1 relaxation as function of applied mw power, and (3) inversion recovery to obtain $T_{1,0}$, is performed on a sample of identical composition, except without the paramagnetic radical.

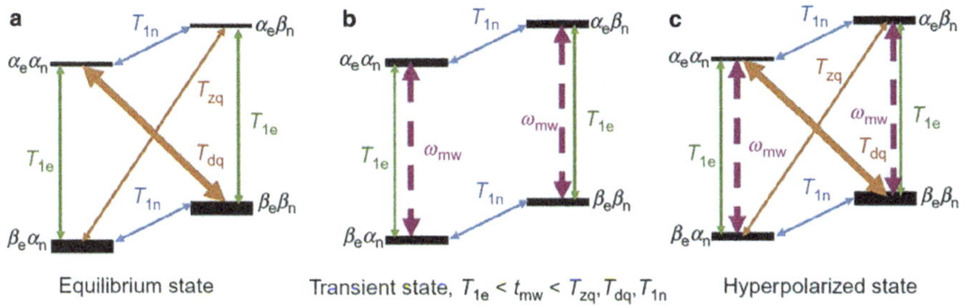

Fig. 2.24 Energy level and populations diagram for the S¼1/2, I¼1/2 spin system. Thickness of the black bars represents the population of the corresponding level at each stage of the experiment. **a** At thermal equilibrium (no mw irradiation). **b** At transient state where irradiation time (t_{mw}) is short such that electron transitions are saturated but redistribution due to cross-relaxation is still negligible. **c** At steady state with mw irradiation [78]. With permission from Elsevier

References

1. W.R. Hagen, *Biomolecular EPR Spectroscopy* (CRC, 2008)
2. G.R. Eaton, S.S. Eaton, D.P. Barr, R.T. Weber, *Quantitative EPR* (Springer, Berlin, 2010); A. Savitsky, K. Mobius, *High-Field EPR Spectroscopy on Proteins and Their Model Systems* (Wiley-VCH Verlag GmbH, 2009)
3. J. Jamauci, Fundamentals of electron spin resonance, in *Nitroxides: Application in Chemistry, Biomedicine, and Materials Science*, ed. by G.I. Likhtenshtein, J. Yamauchi, S. Nakatsuji, A. Smirnov, R. Tamura (WILEY-VCH, Weinheim, 2008), pp 71–120
4. G.I. Likhtenshtein, *Electron Spin in Chemistry and Biology: Fundamentals, Methods, Reactions Mechanisms, Magnetic Phenomena, Structure Investigation* (Springer, Berlin, 2016)
5. H. Wang, M. Zhao, J.L. Ackerman, Y. Song, Saturation-inversion-recovery: a method for T_1 measurement. J. Magn. Reson. **274**, 137–143 (2017); O.Ya. Grinberg, A.A. Dubinskii, V.F. Shuvalov, L.G. Oranskii, V.I. Kurochkin, Ya.S. Lebedev, EPR submillimeter spectroscopy of free radicals. Dokl. Akad. Nauk SSSR **230**, 884–887 (1976)
6. J.H. Freed, New technologies in electron spin resonance. Ann. Rev. Phys. Chem. **51**, 655–689 (2000)
7. K. Moebius, A.N. Savitsky, *High-Field EPR Spectroscopy of Proteins and Their Model Systems: Characterization of Transient Paramagnetic States* (Royal Society of Chemistry, 2009)
8. G.I. Likhtenshtein, *Nitroxides: Brief History and Recent developments* (Springer, 2020)
9. O.Y. Grinberg, A.A. Dubinskii, V.F. Shuvalov, L.G. Oranskii, V.I. Kurochkin, Y.S. Lebedev, EPR submillimeter spectroscopy of free radicals. Dokl. Akad. Nauk SSSR **230**, 884–887 (1976)
10. G.R. Eaton, S.S. Eaton, K.M. Salikhov, *Foundations of Modern EPR* (World Scientific, Singapore, 1998)
11. A. Schweiger, G. Jeschke, *Principles of Pulse Electron Paramagnetic Resonance* (Oxford University Press); J.A. Weil, J.R. Bolton, *Electron Paramagnetic Resonance Spectroscopy: Elementary Theory and Applications*, 2nd edn. (Wiley-Interscience, 2007)
12. J.B. Lambert, E.P. Mazzola, C.D. Ridge, *Nuclear Magnetic Resonance Spectroscopy: Introduction to Principles, Applications, and Experimental Methods*, 2nd edn. (Wiley, 2019)
13. F. Bloch, Nuclear induction. Phys. Rev. **70**, 4604–4673 (1946)
14. P. Hore, *Nuclear Magnetic Resonance*, 2nd edn. (Oxford Chemistry Primers, 2015)
15. M.H. Levitt, *Spin Dynamics: Basics of Nuclear Magnetic Resonance*, 2nd edn. (Wiley, 2008); A. Abragam, *Principles of Nuclear Magnetism* (International Series of Monographs on Physics) (Oxford University Press, USA, 1983)

16. J.B. Lambert, E.P. Mazzola, C.D. Ridge, *Nuclear Magnetic Resonance Spectroscopy: Introduction to Principles, Applications, and Experimental Methods*, 2nd edn. (Wiley, 2019); E.L. Hahn, Spin echoes. Phys. Rev. **80**, 580–594 (1950); Recovery: a method for T_1 measurement. J. Magn. Reson. **274**, 137–143 (2017)
17. D.C. Apperley, R.K. Harris, P. Hodgkinson, *Solid State NMR: Basic Principles & Practice Solid State NMR* (Momentum Press, 2012)
18. K.H. Hausser, H.R. Kalbitzer, *NMR in Medicine and Biology: Structure Determination, Tomography, In Vivo Spectroscopy* (Springer Verlag, 2012)
19. Attaur-Rahman, M. Iqbal Choudhary, *Applications of NMR Spectroscopy*, Vol. 7 (Bentham Books, 2019)
20. E.L. Hahn, Free nuclear induction. Phys. Today 4–9 (1953)
21. R. Kimmicha, E. Anoardo, Field-cycling NMR relaxometry. Prog. Nucl. Magn. Reson. Spectrosc. **44**, 257–320 (2004)
22. B.G. Pfrommer, F. Mauri, S.G. Louie, NMR chemical shifts of ice and liquid water: the effects of condensation. J. Am. Chem. Soc. **122**, 123–129 (2000)
23. H.Y. Carr, E.M. Purcell, Effects of diffusion on free precession in nuclear magnetic resonance experiments. Phys. Rev. **94**, 630–638 (1954)
24. S. Meiboom, D. Gill, Modified spin-echo method for measuring nuclear relaxation times. Rev. Sci. Instrum. **29**, 688–691 (1958)
25. N. Bloembergen, E.M. Purcell, R.V. Pound, Relaxation effects in nuclear magnetic resonance absorption. Phys. Rev. **73**, 679 (1948)
26. I. Solomon, Relaxation processes in a system of two spins. Phys. Rev. **99**, 559 (1955)
27. P. Conte, G. Alonzo, P. Conte, G. Alonzo, Environmental NMR: fast-field-cycling relaxometry. eMagRes **2** (2013) https://doi.org/10.1002/9780470034590.emrstm1330
28. D. Kruk, A. Herrmann, E.A. Rössler, Field-cycling NMR relaxometry of viscous liquids and polymers. Prog. Nucl. Magn. Reson. Spectrosc. **63**, 33–64 (2012)
29. M.R. Ruggiero, S. Baroni, S. Aime, S.G. Crich, Relaxometric investigations addressing the determination of intracellular water lifetime: a novel tumour biomarker of general applicability. Mol. Phys. **117**(7–8), 968–997 (2018)
30. R.M. Steele, J.P. Korb, G. Ferrante, S. Bubici, New applications and perspectives of fast field cycling NMR relaxometry. Magn. Reson. Chem. **54**, 502–509 (2016)
31. P.-O. Westlund, A proton T_1-nuclear magnetic resonance dispersion study of water motion in snowflakes and hexagonal ice. Mol. Phys. **117**, 960–967 (2019)
32. T. Sleator, Pulsed nuclear magnetic resonance and spin echo. Adv. Exp. Phys. V85.0112/G85.2075 (2013)
33. F. Parrino, P. Conte, C. De Pasquale, V.A. Laudicina, V. Loddo, L. Palmisano, Influence of adsorbed water on the activation energy of model photocatalytic reactions. J. Phys. Chem. C **121**, 2258–2267 (2017)
34. P.H. Fries, E. Belorizky, Theory of fast field-cycling NMR relaxometry of liquid systems undergoing chemical exchange. Mol. Phys. **117**, 849–860 (2019)
35. S.E. Mailhiot, S.L. Codd, J.R. Brown, J.D. Seymour, R.K. June, Pulsed gradient stimulated echo (PGStE) NMR shows spatial dependence of fluid diffusion in human stage IV OA cartilage. Magn. Reson. Med. **80**, 1170–1177 (2018)
36. E.O. Stejskal, J.E. Tanner, Spin diffusion measurements: spin echoes in the presence of a time-dependent field gradient. J. Chem. Phys. **42**, 288–292 (1965)
37. L.W. Thrane, J.D. Seymour, S.L. Codd, Probing diffusion dynamics during hydrate formation by high field NMR relaxometry and diffusometry. J. Magn. Reson. **303**, 7–16 (2019)
38. A.W. Overhauser, Polarization of nuclei in metals. Phys. Rev. **92**, 411–441 (1953)
39. R.R. Gil, A. Navarro-Vazquez, Application of the nuclear Overhauser effect to 578 the structural elucidation of natural products, Chap. 1, in *Modern NMR Approaches to the Structure 579 Elucidation of Natural Products: Volume 2: Data Acquisition and Applications to Compound 580 Classes* (The Royal Society of Chemistry, 2017)
40. M.P. Williamson, Applications of the NOE in molecular biology. Annu. Rep. NMR Spectrosc. **65**, 77–109 (2009)

41. P. Honegger, O. Steinhauser, The protein–water nuclear Overhauser effect (NOE) as an indirect microscope for molecular surface mapping of interaction patterns. Phys. Chem. Chem. Phys. **22**, 212 (2020)
42. H.M. McConnell, R.E. Robertson, Isotropic nuclear resonance shifts. J. Chem. Phys. **29**, 1361–1366 (1958)
43. M. Karplus, Contact electron-spin coupling of nuclear magnetic moments. J. Chem. Phys. **30**, 11–16 (1959)
44. O.N. Griffith, P.C. Host, *Spin Labeling, Theory and Application*, Vol. 1, ed. by L. Berliner (Academic Press, New York, 1976), pp. 448–569
45. V.I. Krinichnyi, *2-mm Wave Band EPR Spectroscopy of Condensed Systems* (CRC Press, Boca Raton, FL, Ann Arbor, London, Tokyo, 1995)
46. D. March, Spin-label EPR for determining polarity and proticity in biomolecular assemblies: transmembrane profiles. Appl. Magn. Reson. **37**, 435–454 (2010)
47. P. Fajer, D.D. Thomas, J.B. Feix, J.S. Hyde, Measurement of rotational molecular motion by time-resolved saturation transfer electron paramagnetic resonance. Biophys. J. **50**, 1195–1202 (1986)
48. V.V. Khramtsov, L.M. Weiner, I.A. Grigor'ev, L.B. Volodarsky, Proton exchange in stable nitroxyl radicals. ESR study of the pH of aqueous solutions. Chem. Phys. Lett. **91**, 69–72 (1982)
49. V.V. Khramtsov, In vivo spectroscopy and imaging of nitroxide probes, in *Nitroxides: Theory, Experiment and Applications*, ed. by A.I. Kokorin (2012), pp. 317–346
50. I.A. Grigor'ev, A. Samouilov, V.V. Khramtsov, In vivo monitoring of pH, redox status, and glutathione using L-band EPR for assessment of therapeutic effectiveness in solid tumors. Magn. Reson. Med. 67, 1827–1836 (2012)
51. D. Kivelson, Theory of ESR line with of free radicals. J Chem. Phys. **33**, 1094–1106 (1960)
52. J.H. Freed, Theory of the ESR spectra of nitroxides, in *Spin Labeling. Theory and Applications*, ed. by L. Berliner, Vol. 1 (Academic Press, New York, 1976)
53. S.K. Misra, J.H. Freed. Molecular motions, *Multifrequency Electron Paramagnetic Resonance*, ed. by S.K. Misra, Vol. 11 (Wiley-VCH, New York, 2011), pp. 497–544
54. Y. Sun, Z. Zhang, V.M. Grigoryants, W.K. Myers, F. Liu, K.A. Earle, J.H. Freed, S.P. Charles, The internal dynamics of mini c TAR DNA probed by electron paramagnetic resonance of nitroxide spin-labels at the lower stem, the loop, and the bulge. Biochemistry **51**, 8530–8541 (2012)
55. D.W. McRobbie, E.A. Moore, M.J. Graves, M.R. Prince, *MRI from Picture to Proton* (Cambridge University Press, 2007)
56. L.P. Hwang, J.H. Freed, Dynamics effect of pair correlation function on spin-relaxation by translational diffusion in liquids. J. Chem. Phys. **63**, 4017–4025 (1975)
57. G.I. Likhtenshtein, I. Adin, A. Krasnoselsky, I. Vaisbuch, A. Shames, Glaser, NMR and ESR studies of electric field distribution around biologically important molecules. Biophys. J. **77**, 443–454 (1999)
58. G. Feher, Observation of nuclear magnetic resonances via the electron spin resonance line. Phys. Rev. **103**, 834–835 (1956)
59. L. Kulik, W. Lubitz, Electron-nuclear double resonance. Photosynth. Res. **102**, 391–401 (2009)
60. W.B. Mims, Pulsed ENDOR experiments. Proc. R. Soc. Lond. Ser. A Math. Phys. Sci. **283**, 452–457 (1965)
61. M. Mehring, H. Hofer, A. Grupp, Pulsed electron nuclear double and triple resonance schemes. Ber. Bunsenges. Phys. Chem. **91**, 1132–1137 (1987)
62. W.B. Mims, Amplitudes of superhyperfine frequencies in electron-spin-echo-envelope. Phys. Rev. B **6**, 3543–3545 (1973)
63. S.A. Dikanov, Y.D. Tsvetkov, *Electron Spin Echo Envelope Modulation (ESEEM) Spectroscopy* (CRC Press, Inc., Boca Raton, FL, 1992)
64. L.G. Rowan, E.L. Hahn, W.B. Mims, Electron-spin-echo envelope modulation. Phys. Rev. **137**, A61 (1965)

65. L. Kevan, M.K. Bowman, *Modern Pulsed and Continuous Wave Electron Spin Resonance* (Wiley, New York, 1990)
66. A.A. Savitsky, A.A. Dubinskii, M. Plato, Y.A. Grishin, H. Zimmermann, A.K. Möbius, High-field EPR and ESEEM investigation of the nitrogen quadrupole interaction of nitroxide spin labels in disordered solids: toward differentiation between polarity and proticity matrix effects on protein function. J. Phys. Chem. B **112**, 9079–9090 (2008)
67. P.T. Callaghan, PGSE-MASSEY, a sequence for overcoming phase instability in very-high-gradient spin-echo NMR. J. Magn. Reson. **88**, 493–500 (1990)
68. Y. Deligiannakis, M. Louloudi, N. Hadjiliadis, Electron spin echo envelope modulation (ESEEM) spectroscopy as a tool to investigate the coordination environment of metal centers. Coord. Chem. Rev. **204**, 1–112 (2000)
69. P. Hofer, A. Grupp, H. Nedenführ, M. Mehring, Hyperfine sublevel correlation (hyscore) spectroscopy: a 2D ESR investigation of the squaric acid radical. Chem. Phys. Lett. **132**, 279 (1986)
70. V.N. Syryamina, S.A. Dzuba, Dynamical transitions at low temperatures in the nearest hydration shell of phospholipid bilayers. J. Phys. Chem. B **121**, 1026–1032 (2017)
71. J.M. Franck, M. Sokolovski, N. Kessler, E. Matalon, M. Gordon-Grossman, Song-i Han, D. Goldfarb, A. Horovitz, Probing water density and dynamics in the chaperonin GroEL cavity. J. Am. Chem. Soc. **136**, 9396−9403 (2014)
72. A.B. Volkov, C. Dockter, T. Bund, H. Paulsen, G. Jeschke, Pulsed EPR determination of water accessibility to spin-labeled amino acid residues in LHCIIb. Biophys. J. **96**, 1124–1141 (2009)
73. D. Goldfarb, V. Kofman, J. Libman, A. Shanzer, R.S. Rahmatouline, A. Schweiger, Double nuclear noherence transfer (DONUT)-HYSCORE: a new tool for the assignment of nuclear frequencies in pulsed EPR experiments. J. Am. Chem. Soc. **120**, 7020–7029 (1998)
74. C.-Y. Cheng, S. Han, Dynamic nuclear polarization methods in solids and solutions to explore membrane proteins and membrane systems. Annu. Rev. Phys. Chem. **64**, 507–532 (2013)
75. C. Griesinger, M. Bennati, H.M. Vieth, C. Luchinat, G. Parigi, P. Hofer, F. Engelke, S.J. Glaser, V. Denysenkov, T.F. Prisner, Dynamic nuclear polarization at high magnetic fields in liquids. Prog. Nucl. Magn. Reson. Spectrosc. **64**, 4–28 (2012)
76. Q.Z. Ni, E. Daviso, T.V. Can, E. Markhasin, S.K. Jawla, T.M. Swager, R.J. Temkin, J. Herzfeld, R.G. Griffin, High frequency dynamic nuclear polarization. Acc. Chem. Res. **46**, 1933–1941 (2013)
77. P. Miéville, S. Jannin, L. Helm, G. Bodenhausen, NMR of insensitive nuclei enhanced by dynamic nuclear polarization. CHIMIA Int. J. Chem. **65**, 260–263 (2011)
78. I. Kaminker, R. Barnes, S. Han, Overhauser dynamic nuclear polarization studies on local water dynamics, in *Methods in Enzymology*, Vol. 564 (Elsevier Inc., 2015). ISSN 0076-687. https://doi.org/10.1016/bs.mie.2015.06.040.82
79. L.J. Berliner, E. Bagryanskaya, Chemically induced electron and nuclear polarization, in *Multifrequency Electron Paramagnetic Resonance*, ed. by S. Misra (Willey-VCH, 2011)
80. K.H. Hausser, D. Stehlik, Dynamic nuclear polarization in liquids. Adv. Magn. Reson. **3**, 79 (1968)
81. B.D. Armstrong, S. Han, Overhauser dynamic nuclear polarization to study local water dynamics. J. Am. Chem. Soc. **131**, 4641–4647 (2009)
82. J.H. Freed, Dynamic effects of pair correlation functions on spin relaxation by translational diffusion in liquids. II. Finite jumps and independent T_1 processes. J. Chem. Phys. **68**, 4034–4037 (1978)
83. J.M. Franck, A. Pavlova, J.A. Scott, S. Han, Quantitative cw Overhauser dynamic nuclear polarization for the analysis of local water dynamics. Prog. Nucl. Magn. Reson. Spectrosc. **74**, 33–56 (2013)

Chapter 3
Physical Methods for the Study of Biological Water—Miscellaneous

Abstract Understanding the various properties of water in different states of aggregation, pressure, and temperature requires the use of a wide range of modern physical methods taken in combination. These methods are: infrared and Raman spectroscopy, femtosecond pump–probe, photon-echo, two-dimensional infrared spectroscopic studies, sum-frequency generation (SFG) and second harmonic generation (SHG), Compton scattering and backscattering, THz time-domain spectroscopy (THz-TDS), time-resolved THz spectroscopy (TRTS), THz emission spectroscopy (TES), and THz attenuated total reflection (THz ATR), time-dependent fluorescence Stokes shift (TDFSS), three-pulse photon-echo peak shift (3PEPS), neutron scattering and diffraction, neutron spin echo (NSE), small-angle scattering using either X-rays (SAXS) neutrons (SANS) scattering techniques, Compton scattering and backscattering, X-ray absorption and X-ray Raman scattering, Kerr electro-optic effect, or DC Kerr effect, and the optical Kerr effect (OKE), or AC Kerr, dielectric relaxation (DR), extended X-ray absorption fine structure (EXAFS), near-edge X-ray absorption fine structure (NEXAFS), Auger spectroscopy, and ultrasound. A brief description of the physical principles of these methods and the corresponding examples of their application is provided.

3.1 Introduction

Despite the simplicity of a water molecule as a basic "brick," the formation of four hydrogen bonds creates a unique opportunity for constructing a variety of structures including clusters. These possibilities are expanding significantly due to the provision of donor–acceptor bonds with environment molecules. The molecular dynamics of these structures is carried out by a wide range of correlation times and amplitudes which is an additional challenge for the study of water in biological objects, in particular [1–10].

For this reason, along with the magnetic resonance, a wide arsenal of independent methods should be used to investigate the structure of liquid water and hydrated biomolecule, tissues, organs, animals, and human body (Figs. 3.1 and 3.2). Although each method is often promoted as producing accurate results, the different methods

G. I. Likhtenshtein, *Biological Water*, Soft and Biological Matter,
https://doi.org/10.1007/978-3-030-82503-4_3

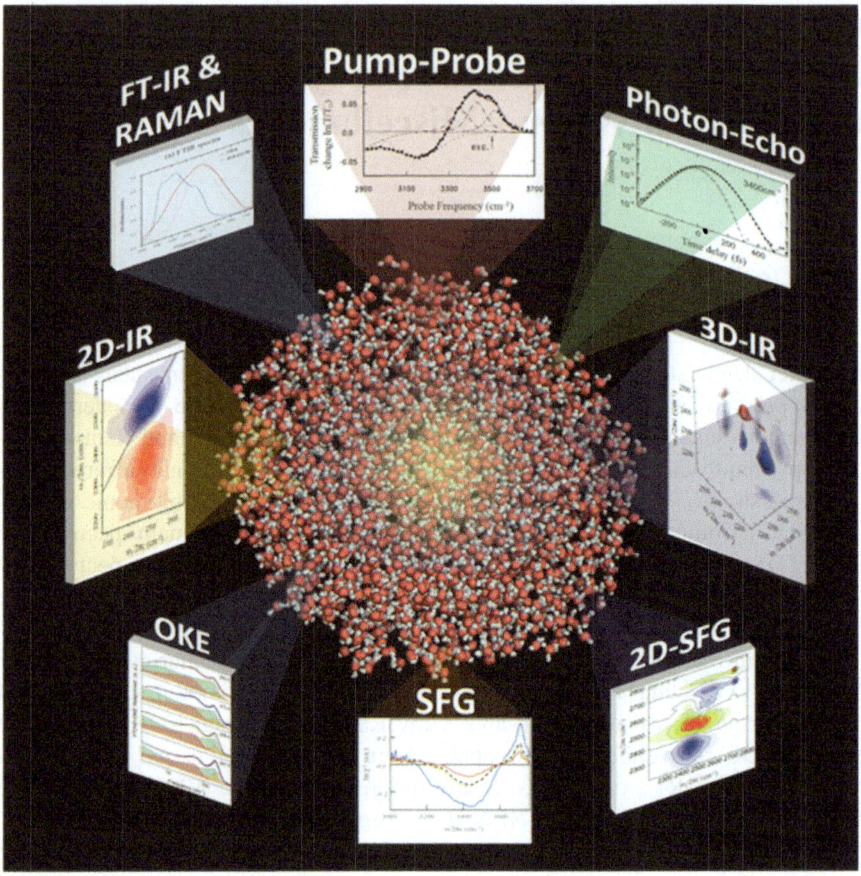

Fig. 3.1 A broad range of optical methods of water investigations [1]. With permission from American Chemical Society

can produce apparently conflicting conclusions, and consequently, the results of their application at a specific object should be analyzed and discussed within the framework of a systematic approach. Timescales of optical methods used to investigate the structure of liquid water and aqueous and biomolecular solutions are from 10^{-15} to 10^{-2} s (Figs. 3.2 and 3.3). A second group of techniques includes dielectric relaxation (DR), depolarized light scattering (DLS), optical Kerr effect (OKE), ultrasound, and terahertz (THz) spectroscopies, which probe different collective dynamics of water molecules and the biomolecule itself.

In review [1], a broad range of methods of water investigations from conventional infrared and Raman spectroscopy to femtosecond pump–probe, photon-echo, optical Kerr effect, sum-frequency generation, and two-dimensional infrared spectroscopic studies was addressed. Novel approaches, such as two-dimensional sum-frequency

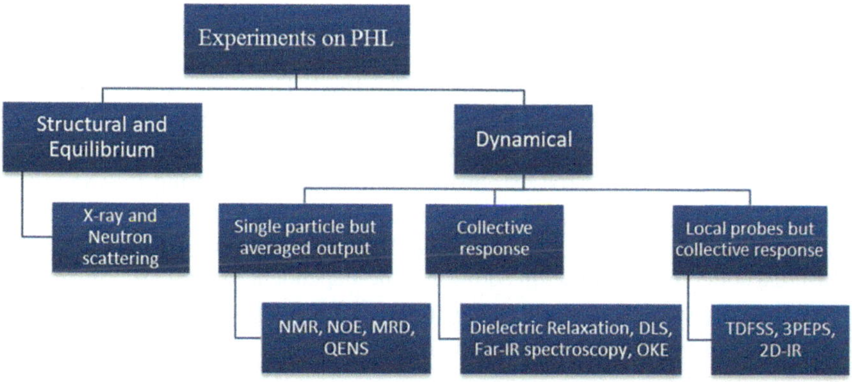

Fig. 3.2 Summary of the different experimental techniques that are employed to study the structure and dynamics of the protein hydration layer (PHL). Different techniques may give rise to different information regarding timescales of relaxation because of the nature of the response that they measure as outputs [2]. With permission from American Chemical Society

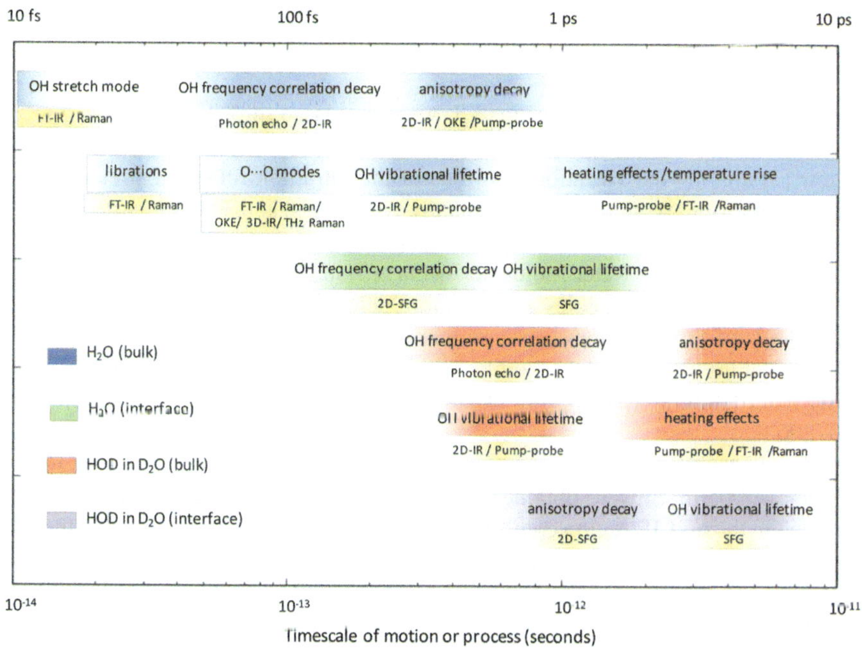

Fig. 3.3 Schematic overview of timescales of molecular motions, vibrational mode period, energy redistribution mechanisms, and spectroscopic observables obtained by multidimensional vibrational spectroscopy. The processes are distinguished between those taking place in bulk water and those at interface, as well as between the cases of neat H_2O and isotope-diluted HOD in D_2O. Under each motion or process, the corresponding technique is indicated, as discussed in detail throughout the review [1]. With permission from American Chemical Society

generation, three-dimensional infrared, and two-dimensional Raman terahertz spectroscopy were considered, and the complementary aspects of various linear and nonlinear spectroscopic techniques were compared. Different experimental techniques that are used to obtain information about the structure and dynamics of the protein hydration layer, such as NMR, magnetic resonance dispersion (MRD), nuclear Overhauser effect (NOE), quasi-elastic neutron scattering (QENS), were summarized in Fig. 3.2 [2]. From other hand, dielectric relaxation, depolarized light scattering (DLS), the optical Kerr effect (OKE), and far-infrared (FIR) spectroscopy measure a collective response. In addition, time-dependent fluorescence stokes shift (TDFSS), three-pulse photon-echo peak shift (3PEPS), and two-dimensional infrared (2D-IR) 27 can act as local probe techniques.

3.2 Vibrational Optical Spectroscopy

3.2.1 Conventional Absorption and Raman Spectroscopies [1–8]

The spectroscopic observables (Fig. 3.3) were discussed throughout review [1].

Different regimes in time for the vibrational structural dynamics, considered in review [3] are: (1) the vibrational periods of the intramolecular OH stretching and bending vibrations, which are approximately 10 (0.01 ps) and 20 fs; (2) on a timescale between several tens and several hundreds of femtoseconds, librational (hindered rotational) motions of water molecules occur; (3) at short times these are localized on a single or a few molecules and are more delocalized over groups of molecules at hundreds of femtoseconds, representing truly intermolecular modes; and (4) H-bond stretching vibrations, i.e., motions of the two outer oxygen atoms changing the H-bond length, have a period of 200 fs, while H-bond bending has periods of 600–800 fs.

Absorption spectra of liquid water at various temperatures in the whole IR region ($0 < \tilde{\nu} < 4000$ cm^{-1}) were obtained from attenuated total reflection (ATR) spectra recorded in the mid-IR region combined with absorption spectra measured in the FIR region [8]. The peak at about 3400 cm^{-1} was attributed symmetric and asymmetric stretching. The spectra of the H_2O and D_2O molecules were in accordance with following thermodynamic values: (1) H_2O (D_2O) molecules of lower enthalpies H^g, greater than the enthalpy of H_2O or D_2O molecules in ices by 1.3 kJ mol^{-1}; (2) display vibrations that resemble those of H_2O (D_2O) molecules in ice; and (3) molecules with enthalpies H^e reorient without breaking their H-bonds as H^e appears to be greater than the potential barrier for rotations around their symmetry axes.

The process of Raman scattering is the transition of a molecule from its ground state to an excited vibrational state, accompanied by the simultaneous absorption of an incident photon and emission of a Raman scattered photon (Fig. 3.4) [7]. The Raman scattered light can be collected by a spectrometer and displayed as intensity

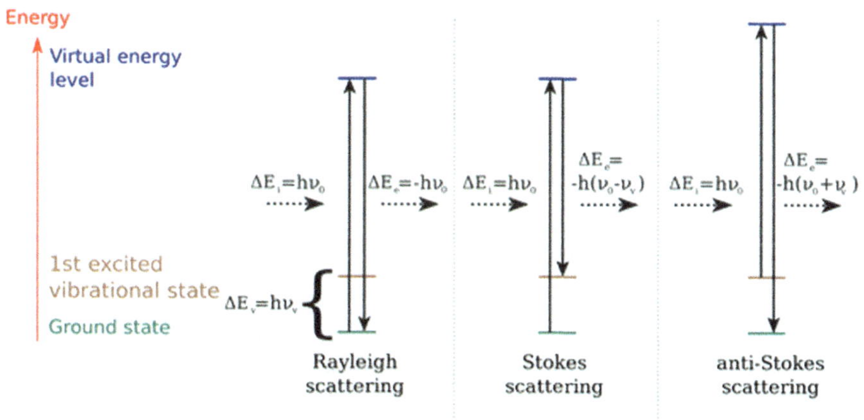

Fig. 3.4 Process of Raman scattering can be considered as the *transition of a molecule from its ground state to an excited vibrational state*. This transition is accompanied by the simultaneous absorption of an incident photon and emission of a Raman scattered photon. The horizontal lines indicate vibrational energy levels. Ground electronic state and excited electronic states are shown. The diagrams show how a molecule in the ground state can make a transition from the lowest vibrational level to the first excited vibrational level by means of Raman scattering. Black up-arrows indicate the frequency of the laser excitation light; black down-arrows indicate frequency of the Raman scattered light. The difference in length between black up- and down-arrows indicates molecular vibration frequency. It should be noted here that Raman scattering at different excitation wavelengths, UV, visible, and near-IR, produces the same change in vibrational energy. Therefore, the excitation wavelength can be chosen to avoid spectral interference by fluorescence. For visible excitation, the fluorescence light frequency and Raman scattered light frequency are similar. This leads to intense fluorescence background in visible excitation Raman spectra. Near-IR light has too low a frequency to excite fluorescence, while for UV excitation, the fluorescence light frequency is much lower than the Raman scattered light frequency. Hence, fluorescence background in the Raman spectrum can be reduced by using UV or near-IR excitation [7]. With permission from Elsevier

versus frequency change. The Raman spectrum of a particular species consists of a series of peaks each shifted by one of the vibrational frequencies characteristic of that molecule. The different possibilities of light scattering are: Rayleigh scattering (no exchange of energy: incident and scattered photons have the same energy), Stokes Raman scattering (atom or molecule absorbs energy: scattered photon has less energy than the incident photon) and anti-Stokes Raman scattering (atom or molecule loses energy and scattered photon has more energy than the incident photon).

Fourier transform infrared (FT-IR) and Raman spectroscopy both reflect the distributions of vibrational frequencies of water oscillators and affected by intra- and intermolecular couplings (Fig. 3.5) [1]. Vibrational modes are coupled in a complex manner, which strongly affects the OH stretching mode spectra of liquid water, amorphous ice, and crystal ice. FTIR and Raman spectra of OH stretch vibrations of liquid neat H_2O and comparison between FTIR and Raman spectra at 298 K illustrate sensitivity of the vibration spectroscopy to state of water and temperature (Fig. 3.5).

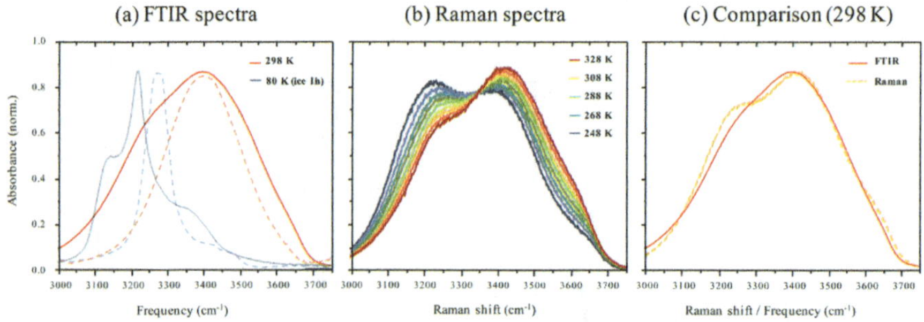

Fig. 3.5 **a** FTIR spectra of OH stretch vibrations of liquid neat H_2O (red solid), liquid isotope-diluted 5% HOD in D_2O (red broken), ice Ih of neat H_2O (blue solid), and ice Ih of 5% HOD in D_2O (blue broken). **b** Raman spectra of OH stretch temperatures [1]. With permission from Elsevier

In another example, a glancing-angle Raman spectroscopy probe was proved to be sensitive to changes within the upper 50 monolayers of water and is thus capable of detecting changes occurring at the air–ice interface [11]. Glancing-angle is the angle between the beam and the surface. In Raman spectra of the OH stretching region of room-temperature bulk water (Fig. 3.6), the large shoulder on the low-wavenumber side (3200 cm^{-1}) of the solid trace was assigned to vibrations involving more highly hydrogen-bonded ("4-coordinate") molecules, by analogy with the spectrum of bulk ice.

3.2.2 Pulse Vibrational and Rotational Spectroscopies

In one-dimensional (1D) time-resolved nonlinear spectroscopy, a pair of pulses is used to excite the sample and to probe the transient modifications of the optical response induced by the excitation [1]. Then, by changing the time delay between the two pulses, one can extract dynamical information. In the pump–probe experiment (Figs. 3.7 and 3.8) two ultrashort pulses (~100 fs) of IR light, tuned to the frequency of the vibrational mode of interest, produce a very broad spectrum related to all of the vibrational oscillators. The experiment is performed in principal stages: (1) the first pump pulse excites some of the vibrations; (2) a second probe pulse, at variable delay times after the pump the probe, is brought in so that it overlaps spatially in the sample with the volume excited by the pump; (3) when the pump pulse precedes the probe pulse in time, some of the vibrations have been excited into the first vibrationally excited state; and (4) the excitation of the vibrations leaves fewer molecules in the ground state to absorb the probe, so the transmission of the probe is increased. A plot of the signal versus time delay of the probe will give the vibrational lifetime that is sensitive to the environment of the molecule of interest. By selecting the appropriate polarizations of the pump and probe beams, it is possible to measure the vibrational lifetime even if the molecules are rotating.

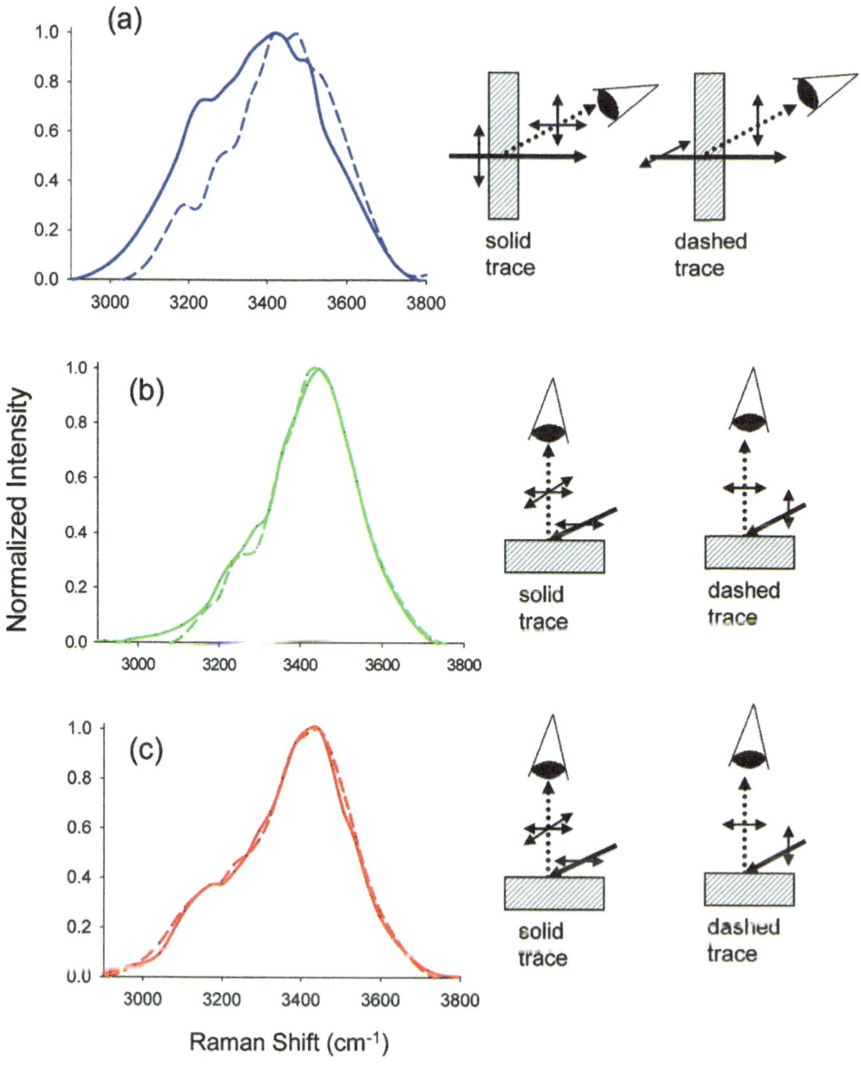

Fig. 3.6 Raman spectra of the OH stretching region of **a** room-temperature bulk water, **b** the air–water interface, and **c** the air–ice interface at −15 °C. Solid traces correspond to vertical polarization in the bulk and s-polarized radiation at the surface, and dashed traces correspond to horizontally polarized light in the bulk and p-polarized light at the surface. The insets illustrate the experimental geometry used in each case. Here the solid line shows the input laser and the dotted line shows the scattered light [11]. With permission from American Chemical Society

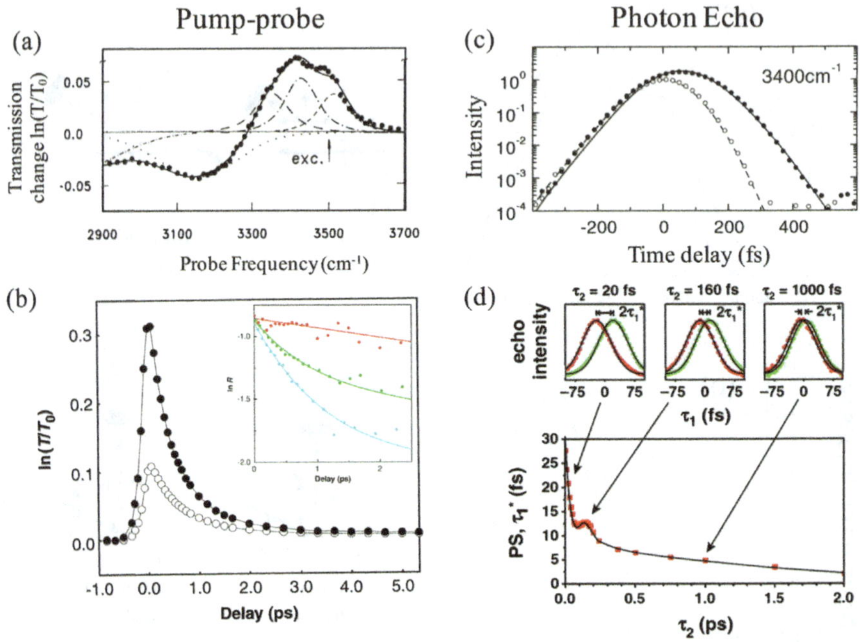

Fig. 3.7 Measurements on liquid water by **a**, **b** pump–probe and **c**, **d** photon-echo spectroscopy. **a** One of the earliest pump–probe spectra of the OH stretch of HOD in D_2O. **b** Corresponding vibrational population relaxation, measured with higher temporal resolution (● and ○ indicate different polarization geometries), and anisotropy decay (inset; red, green, and blue lines denote pump frequencies of 3320, 3400, and 3500 cm^{-1}, respectively). **c** Photon-echo intensity of liquid HOD in D_2O as a function of delay time (○, CaF_2; ●, water response). **d** Corresponding vibrational dephasing times [1]. With permission from American Chemical Society

Figure 3.7 illustrates principle and typical results of pump–probe and photon-echo spectroscopy [1]. In the pump–probe spectrum of liquid water (HOD in D_2O), as it is a difference spectrum between pumped and unpumped absorbance spectra, the positive feature was associated with ground-state contributions (stimulated emission and bleach), and the negative peak arises from the excited-state absorption. The variation of signal as a function of the pump–probe delay time (Fig. 3.7b) depends on the polarization geometry and was suggested to reflect either the vibrational population relaxation or the anisotropy decay. Photon-echo spectroscopy (Fig. 3.7c, d) allows the separation of homogeneous and inhomogeneous broadenings and therefore can measure directly the vibrational dephasing time, which originates from intermolecular dynamics and contains multiple relaxation.

Femtosecond pump–probe infrared spectroscopy has been used to measure the reorientation dynamics of water molecules in aqueous protein α-Lactalbumin solutions [12]. Following an initial vibrational excitation by a polarized pulse, the anisotropy decay was measured with a femtosecond time resolution. Since this decay is the sum of the rotational relaxations of all water molecules within the sample, it

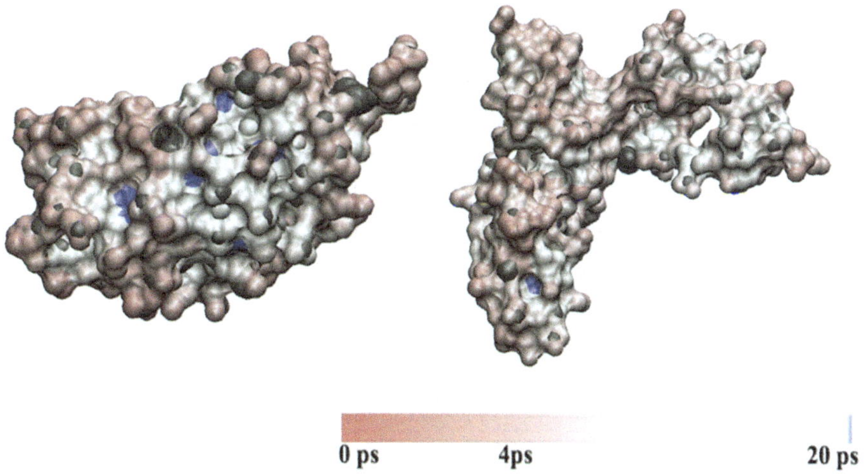

Fig. 3.8 Simulation: water reorientation decay times and the nature of the heavy atoms projected on a protein surface for the two systems: **a** 5% SPC/E (left) and **b** misfolded #1 (right). Red indicates fast water reorientation, white moderate, and blue slow. Gray indicates no water population close to that site. Note that the misfolded configuration has fewer blue sites, indicating relatively faster water reorientation [12]. With permission from American Chemical Society

was possible to determine simultaneously the number of perturbed water molecules and their respective reorientation times [12]. The rate of orientational relaxation can be also measured by making two time-dependent measurements, one with the probe parallel to the pump polarization and one with the probe perpendicular to the probe polarization. Figure 3.8 visualizes the water reorientation time as a function of protein site. It can be seen that some buried residues are hydrated by very slow internal waters, whereas at the surface, the water reorientation dynamics is relatively faster.

Pump–probe time-dependent **IR** spectroscopy probed the decay time related to intramolecular (H–O or D–O) vibrational phonon lifetime [13]. This lifetime features the rate of population decay or vibration energy dissipation depending on the solution viscosity and Stokes–Einstein molecular drift diffusivity, the solute concentration, and resolved Jones–Dole viscosity. The Jones–Dole equation is an empirical expression that describes the relationship between the viscosity of a solution and the concentration of solute within the solution. To understand the meaning and scope of solvation dynamics, physical picture of the dynamical process involved for a solute molecule in a polar solvent was visualized (Fig. 3.9) [14]. A change in the solute is made at time $t = 0$, by femtosecond excitation, which leads to the creation of a dipole giving rise to an instantaneous electric field on the solvent molecules. Because of the interaction of the solvent permanent dipoles with the electric field, the free energy minimum of the solvent shifts to a nonzero value of the polarization. Since the solute is excited instantaneously (Franck–Condon principle), the solvent molecules at $t = 0$

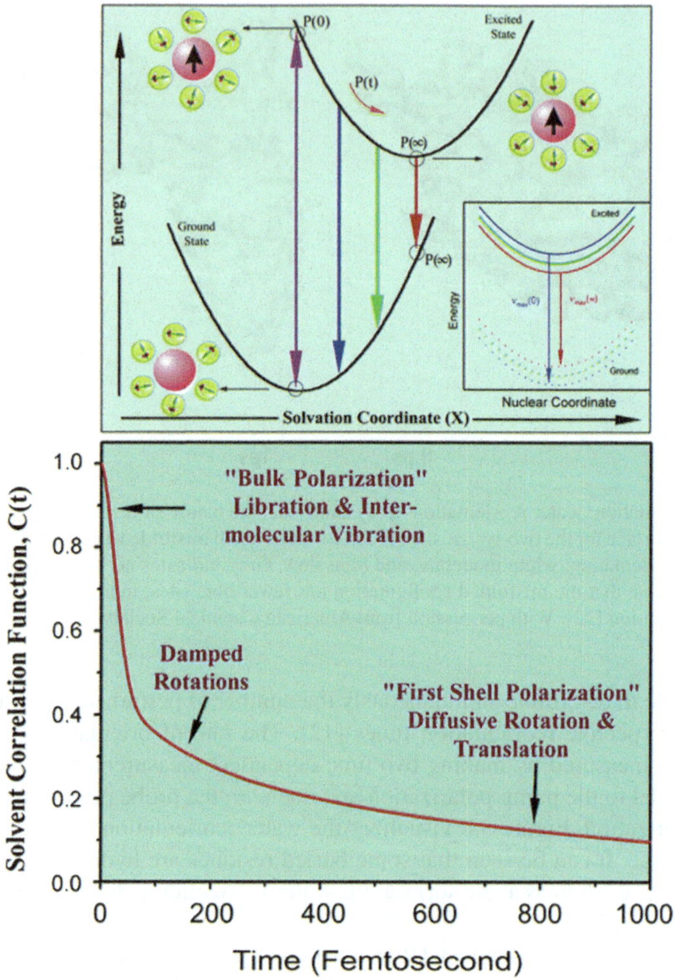

Fig. 3.9 (Upper) Schematic illustration of the potential energy surfaces involved in solvation dynamics, showing the water orientational motions along the solvation coordinate together with instantaneous polarization *P*. Inset: The change in the potential energy along the intramolecular nuclear coordinate. As solvation proceeds, the energy of the solute decreases, giving rise to a red shift in the fluorescence spectrum. Note the instantaneous *P*, e.g., $P(\infty)$, on the two connected potentials. (Lower) A typical solvation time correlation function for water is shown. The time correlation function exhibits three distinct regions: the initial ultrafast decay, an intermediate decay of about 200 fs, and the last slow decay with a time constant of 1 ps. The physical origin of each region is indicated on the plot itself [14]. With permission from American Chemical Society

find themselves in a relatively high-energy configuration. Subsequently, the solvent molecules begin to move and rearrange themselves to reach their new equilibrium positions (Fig. 3.9).

3.2.3 2D Infrared Spectroscopy

Two-dimensional infrared spectroscopy (2D IR) is a nonlinear infrared spectroscopy technique that has the ability to correlate vibrational modes in condensed-phase systems by spreading the vibrational information along multiple axes, yielding a frequency correlation spectrum [9, 15–19]. The 2D IR spectroscopy with femtosecond time resolution has been proved to be a powerful tool for investigation of a vast number of chemical and biological systems such as the hydrogen-bond network of water, for example. A transient 2D IR (t-2D IR) experiment involves the use of a short, actinic pulse prior to collecting the 2D IR spectrum, in order to track the response of the system (Figs. 3.10 and 3.11) [16]. The t-2D IR spectrum is a difference 2D IR spectrum between the initial and transient species. By varying the time delay between the initiating pulse and the infrared probing sequence, t-2D IR experiments can be performed from initiation to a point where the system approaches its equilibrium state. Negative peaks are due to depleted initial population, and positive signals result from the newly formed transient photoproduct.

A schematic of the experiment of 2D-IR experiments and the pulse sequence are pictured in Fig. 3.11 [18]. Three mid-IR pulses (wave vectors $\mathbf{k}_1$, $\mathbf{k}_2$, and $\mathbf{k}_3$) that can be time-delayed are focused into the sample in a non-collinear geometry. The signal is emitted in a unique phase-matched direction ($\mathbf{k}_s$) $\mathbf{k}_2 + \mathbf{k}_3 - \mathbf{k}_1$ and is then overlapped with a local oscillator (LO) pulse for heterodyne detection.

As an example, experiments using ultrafast 2D-IR vibrational echo spectroscopy and polarization-selective IR pump–probe experiments were performed on the OD hydroxyl stretch of dilute HOD in H_2O. The vibrational echo experiments measured spectral diffusion, and the pump–probe experiments measured orientational relaxation. Both experimental observables are directly related to the structural dynamics of water's hydrogen-bond network.

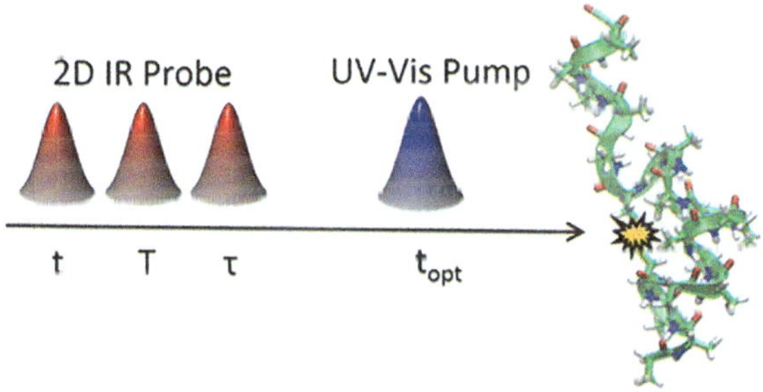

Fig. 3.10 Sequence of pulses transient 2D IR (t-2D IR) [17]. With permission from American Chemical Society

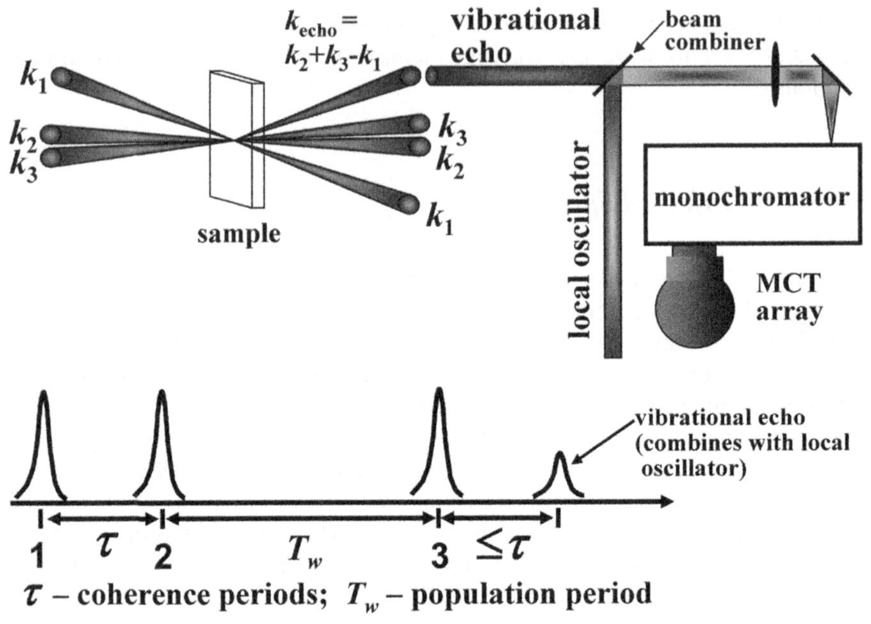

Fig. 3.11 Schematic illustration of the 2D-IR vibrational echo experiment and pulse sequence. Three pulses enter the sample with the timing given by the pulse sequence shown at the bottom of the figure. The nonlinear interactions give rise to a fourth pulse, the vibrational echo, which is then combined with the local oscillator pulse to obtain phase and amplitude information. The combined vibrational echo and local oscillator are frequency resolved and detected with an IR MCT array. For fixed Tw, ô is scanned. The Fourier transformations produce a 2D spectrum. Spectra are recorded as a function of Tw [18]. With permission from American Chemical Society

Figure 3.12 depicts a sample t-2D IR setup with pulses providing 4–7 μm mid-IR pulses and a temporally stretched and tunable excitation/triggering pulse probe IR beams for the 2D IR measurement in the pump–probe geometry [17]. The 2D IR measurement involves (1) the first arm generates an IR probe beam which often has a time delay range of 120 ps; (2) the second arm is diverted into an acoustic optic modulating pulse shaper to generate two pump pulses separated by specific time delays; and (3) the resulting echo signals of the 2D IR measurement allow the structural determination of an object of interest. By determination of coupling through these transient 2D IR measurements, changes of angles and bond distances within a structure during the time evolution can be measured.

By inserting various groups that can be triggered via an ultrafast light pulse into a larger biomolecule, phototriggers, and photoswitches, it was possible to observe quasi- or non-equilibrium conformational changes on the order of 1–10 ps and beyond [17]. These various timescales and photoinitiation methods are summarized in Fig. 3.13.

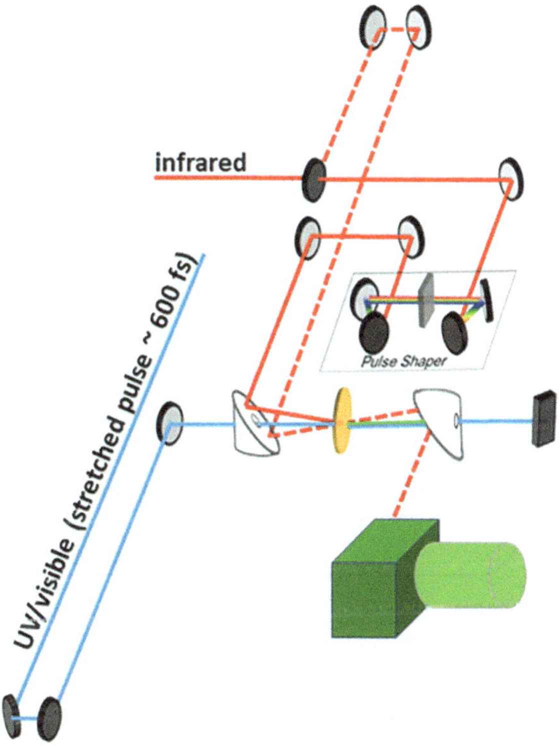

Fig. 3.12 Typical transient 2D IR optical design with a pulse shaping [17]. With permission from American Chemical Society

Equilibrium 2D IR spectroscopy setup involves the production of three phase-locked IR femtosecond pulses in a boxcar or pump–probe configuration sent at specified time intervals to acquire a 2D IR spectrum [16]. The 2D IR echo signal is due to the third-order macroscopic polarization generated by the excitation of vibrational energy states allowing the full three-dimensional response of the system to be measured at the picosecond or faster timescale. Cartoons in Fig. 3.14a, b demonstrate a 2D IR spectrum of a single vibrational mode in a molecule and a 2D IR spectrum of a single vibrational mode in a molecule [16]. The positive peak corresponds to the $v = 0 \rightarrow 1$ transition, where v is the vibrational quantum number for the mode and the negative peak corresponds to the $v = 1 \rightarrow 2$ transition. The combination of high resolution and ultrafast timescale was proved to be of critical importance in observing chemical and physical processes such as solvent-molecule dynamics, structural changes, and chemical exchange. Thus, 2D IR spectroscopy can resolve equilibrium processes on picosecond or faster timescales through either dynamics of conformational distributions or mapping vibrational coupling between molecular groups to obtain relative positions and orientations [16].

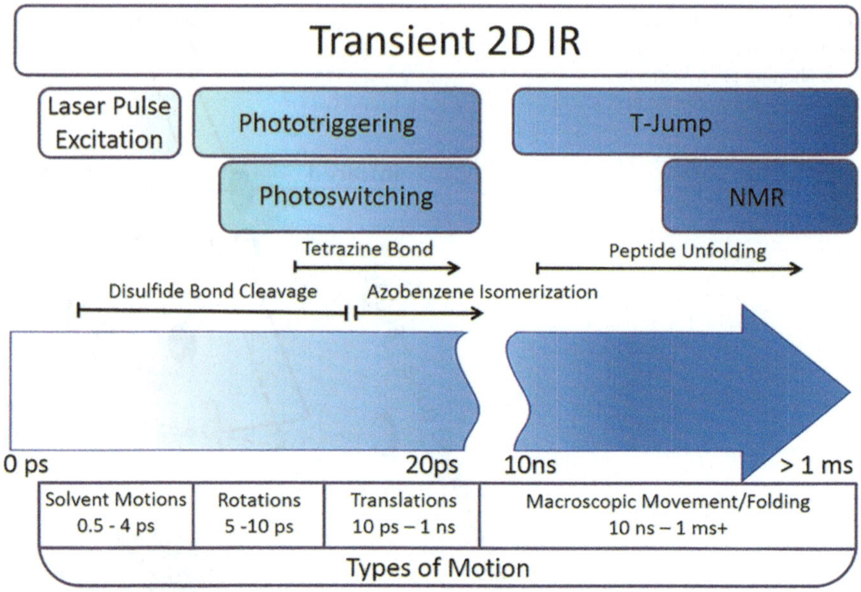

Fig. 3.13 Various effective timescales of techniques for observing conformational changes in biomolecules [17]. With permission from American Chemical Society

3.2.4 Sum-Frequency Generation

Sum-frequency generation (SFG) and second harmonic generation (SHG) surface spectra are spectroscopic methods for investigating the surface of water without interference from the bulk liquid [1, 14, 20–25]. Sum-frequency generation vibrational spectroscopy (SFS) is a nonlinear optical technique which provides vibrational spectra of molecules solely at interfaces. The spectra may be analyzed to provide the polar orientation, molecular conformation, and average tilt angle of the adsorbate to the surface normal.

Schematics describing the beam geometry for sum-frequency generation from an interfacial system between media I and II is presented in Fig. 3.15 [14]. In the typical SFS experiment [20–22]: (1) light is emitted at the sum of the two incident frequencies, i.e., $\omega_{SF} = \omega_{VIS} + \omega_{IR}$ (Fig. 3.16); (2) the intensity of the light is resonantly enhanced when the frequency of the tuneable infrared beam coincides with a vibrational mode of the molecules at the interface; and (3) by detecting the sum-frequency (SF) light as a function of infrared frequency, a vibrational spectrum is obtained, which is up-shifted into the visible region of the electromagnetic spectrum. Importantly, the selection rules for SF activity differ from those for linear vibrational spectroscopy because for a molecular vibrational mode (resonance) to be SF active, it must be in an asymmetric environment. As the bulk aqueous solution and the gas phase are isotropic, they are not active in this spectroscopy, leaving just the anisotropic interfacial layer to provide the spectra, due to a net orientation of surface

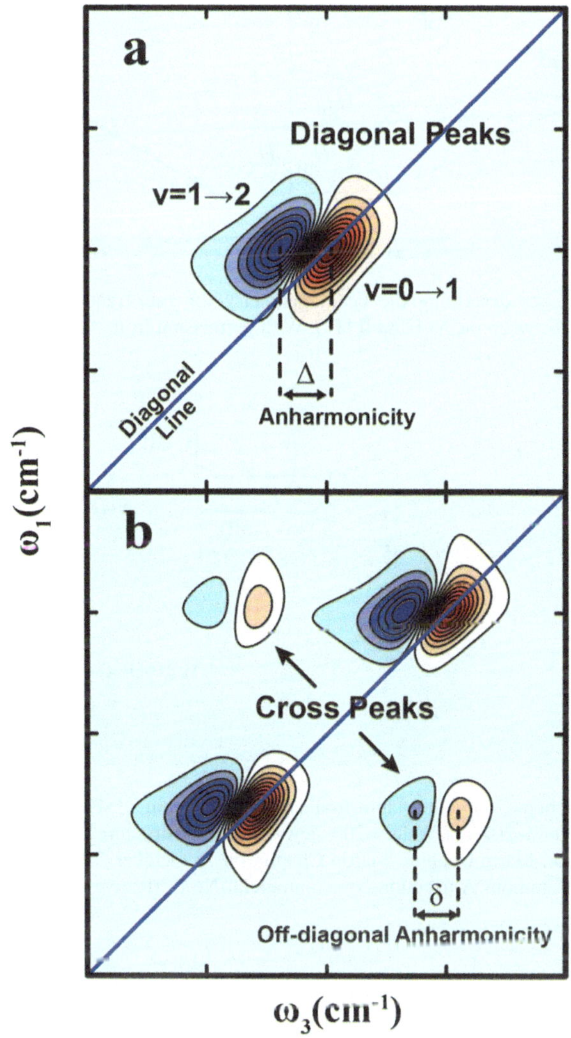

Fig. 3.14 Typical features of a 2D IR spectrum. **a** Diagonal peaks and **b** cross-peaks. The spectra shown were simulated from the third-order response functions [16]. With permission from American Chemical Society

molecular dipoles. Introduction of an interface into an isotropic bulk phase gives rise to a plane of asymmetry and the SF activity of exclusively the interfacial molecules, and the electric dipole-allowed bulk contribution to SFG can be suppressed and the corresponding surface contribution relatively enhanced.

The positive sign refers to co-propagating SHG beams (visible and infrared beams arriving from the same *x* direction) and the negative sign to counter-propagating beams (visible and infrared beams arrive from opposite *x* directions, as in Fig. 3.15).

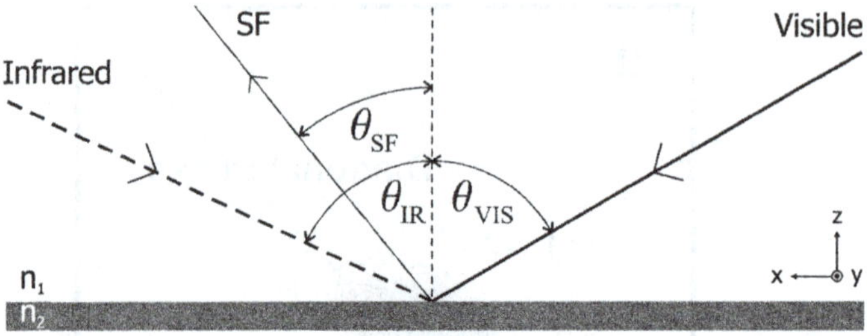

Fig. 3.15 Schematics describing the beam geometry for sum-frequency generation from an interfacial system between media I and II [14]. With permission from American Chemical Society

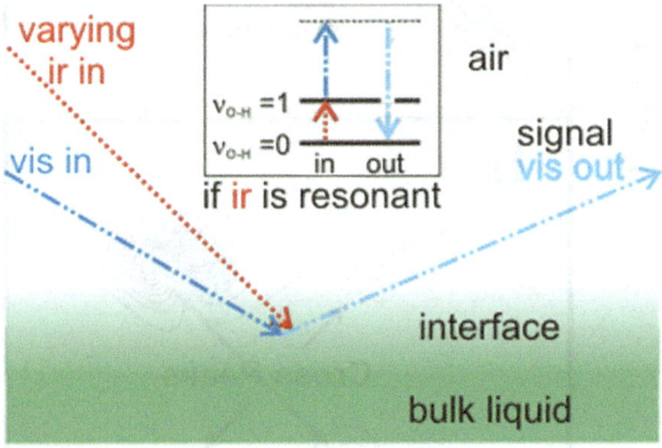

Fig. 3.16 A scheme of the sum-frequency generation (SFG). file:///C:/Users/gertz/Desktop/Book%20on%20water/Water%20reviews/Water%20structure%20methods.html. With permission from Dr. Marin Chaplin. Martin Chaplin on 25 October, 2019. This work is licensed under a Creative Commons Attribution-Noncommercial-No Derivative Works 2.0 UK: England & Wales License

SF light is both reflected from and transmitted into the surface. The more accessible of these two beams is usually detected.

According to the basic theory of optical sum-frequency generation from an interfacial system [23]. Both the interface and the bulk generally contribute to the sum-frequency output. Two different approaches to specify bulk nonlinearity that includes electric-quadrupole and magnetic-dipole contributions were shown to yield the same sum-frequency output if surface nonlinearity were taken into account. Truly bulk and truly surface nonlinear susceptibilities defined and separately deduced from measurements of transmitted and reflected sum-frequency generation.

The following SFG applications for investigation of the O–H stretch mode are exemplified in Fig. 3.17 [24]: (1) the SFG intensity spectrum ($|\chi(2)|^2$) of The O–H stretch contains two feature, namely, a broad 3000–3500 cm^{-1} band and a sharp 3700 cm^{-1} peak (Fig. 3.17a); (2) experimental Im($\chi(2)$) data at the water–air interface showed positive 3700 cm^{-1}, negative 3400 cm^{-1}, and positive 3200 cm^{-1} bands

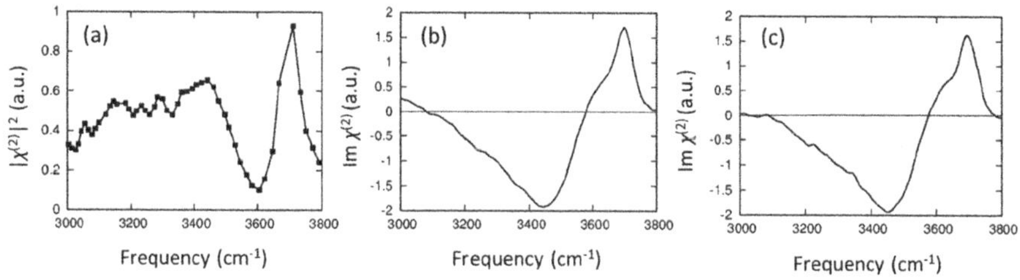

Fig. 3.17 **a** Homodyne-detected SFG intensity spectra ($|\chi(2)|^2$). Reproduced with permission from [18]. Copyright 1993 American Physical Society. **b, c** Heterodyne-detected spectra (Im($\chi(2)$)) for the O–H stretch mode of neat water at the water–air interface. The spectrum in **b** was obtained using a quartz reference signal and **c** with D_2O as reference. Reproduced with permission from [26]. Copyright 2015 AIP. The data are reproduced from [18] for **a** and [26] for **b** and **c**, respectively. Note that the spectrum in **a** is corrected for Fresnel factors, while the spectra in **b** and **c** are not [24]. With permission from American Chemical Society

for the O–H stretch mode (Fig. 3.17b); and (3) positive–negative–positive feature of the O–H stretch mode to a negative–positive signal by recognizing the contamination in the reference signal (Fig. 3.17c).

3.2.5 *Phonons*

A phonon is a collective excitation in a periodic, elastic arrangement of atoms or molecules in condensed matter, in which a lattice of atoms or molecules uniformly oscillates at a single frequency [27–32] https://ru.wikipedia.org/wiki/Phonon. In the optical mode, two adjacent atoms move against each other, while in the acoustic mode they move together. Phonons have particle-like properties, in a way related to the wave–particle duality of quantum mechanics. The developed strategies have enabled clarification of origins of the phonon frequency over the full temperature range [27]. In the phonon relaxation spectrometrics, Raman spectral peak features the Fourier transformation of all bonds vibrating in the same frequency from the real space [28].

A parameter characterizing a specific phonon band is the peak frequency that shifts upon perturbation, upward or downward, along the ω axis. The $\Delta\omega_x$ shift is proportional to the square root of the curvature of the dimer potential at equilibrium in the harmonic approximation.

Raman phonon frequency shift $\Delta\omega_x$ probes (in unit of wavenumber cm^{-1}) is

$$\Delta\omega_x \propto \sqrt{E_x/\mu_x}/d_x \propto \sqrt{k_x + k_c/\mu_x} \tag{3.1}$$

The subscript $x = L$ denotes the O:H non-bond characterized by the stretching vibration frequency at ~200 cm^{-1} and $x = H$ denotes the H–O bond with characteristic phonon frequency of ~3200 cm^{-1} in the bulk water; the k_x and k_C are the force constants or the second differentials of the intra/inter molecular interaction and O–O Coulomb coupling potentials. The phonon spectrometrics probe the bond relaxation in length d and energy E in the form of $\Delta\omega^2 \propto E/(\mu d^2)$ of an oscillator on average with the reduced mass of $\mu = m_1 m_2/(m_1 + m_2)$. Figure 3.18c, d exemplify the differential phonon spectrometrics (DPS) measured under mechanical compression of liquid water and skin molecular under coordination of water and ice [28].

Comparison the segment lengths of the O:H–O bond as a function of $(H_2O)_N$ cluster size (N) (Fig. 3.19), optimized using the PW and the OBS algorithms of results, derived from these two DFT methods has been performed [29]. The molecular CN reduction shortens the H–O bond and lengthens the O:H and the shortening (lengthening) of the H–O bond is coupled to the lengthening (shortening) of the O:H was confirmed. Figure 3.20 shows size dependence of the phonon spectra of $(H_2O)_N$ clusters.

Figure 3.21 illustrates the basic structural unit of the segmented "$O^{\delta-}$:$H^{\delta+}$–$O^{\delta-}$" hydrogen bond in hydrogen-bonded networks. The following conclusions were formulated: (1) the hydrogen bond comprises both the O:H van der Waals (vdW) bond and the H–O polar-covalent bond, as opposed to either of them alone; (2) the

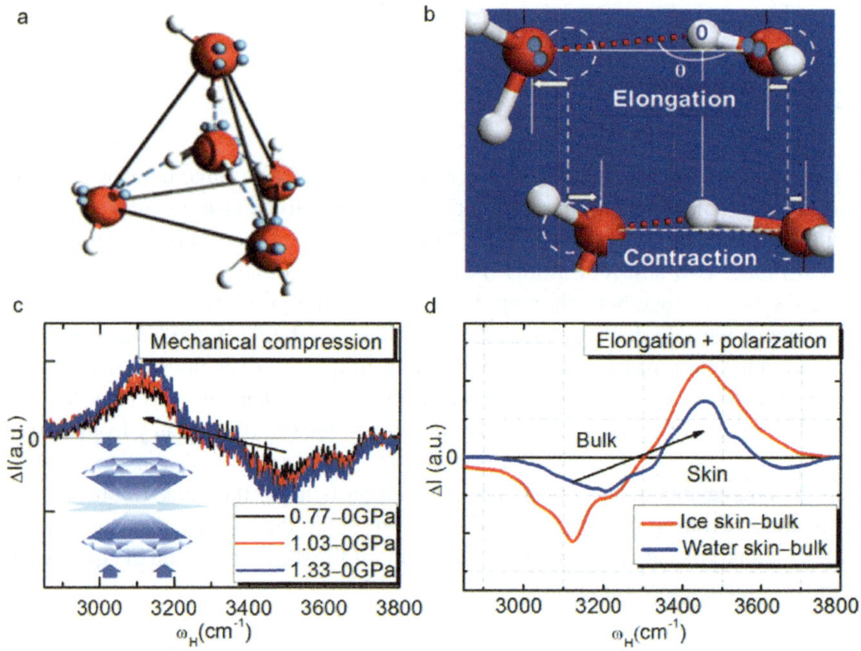

Fig. 3.18 **a** Water structure and the **b** cooperative relaxation of the segmented O:H O bond. The pairing dots represent the non-bonding lone pairs on oxygen. **c** H–O phonon frequency shift of 25 °C water under different pressures. Inset shows the schematic diagram of mechanical compression. **d** The differential Raman spectra between skin and bulk of 25 °C water and of −20 °C ice. Compression lengthens and softens the H O bond, but the skin bond-order deficiency shortens and stiffens the H O bond [28]

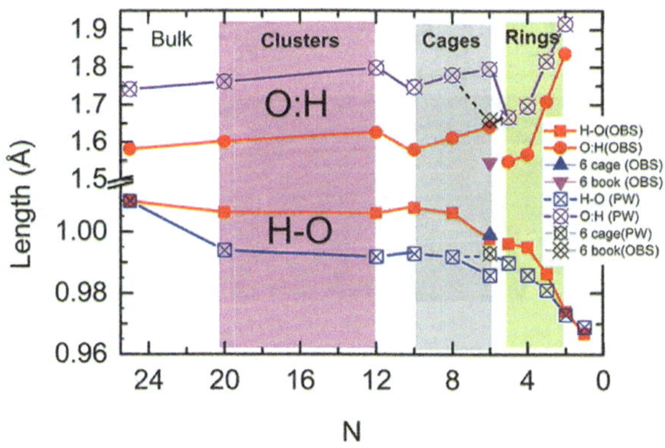

Fig. 3.19 Cluster size dependence of the segmental lengths in the $(H_2O)_N$ clusters. The bond lengths were optimized using the PW and the OBS methods. $N = 6$ gives the three "cage," "book," and "prism" hexamer structures, all with nearly identical binding energy [29]. With permission from American Chemical Society

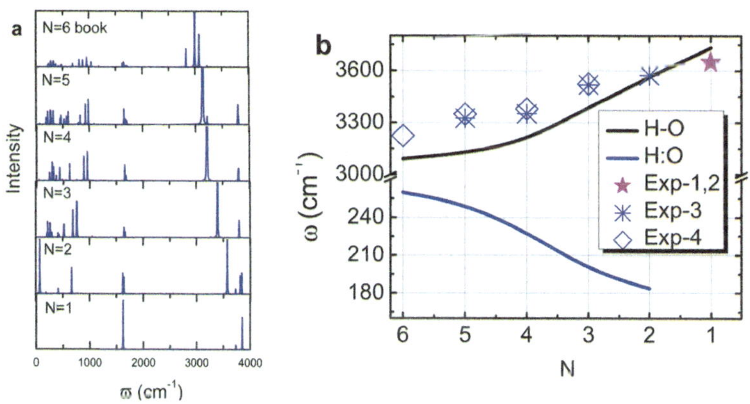

Fig. 3.20 **a** Size dependence of the phonon spectra of $(H_2O)_N$ clusters ($N \leq 6$). **b** Calculated (solid line) ω_H blue shift has a similar trend as the measured frequencies (scattered data) of the H–O phonons of $(H_2O)_N$, shown as Exp-1, Exp-2, Exp-3, and Exp-4. Measurements of the ω_L red shift are not presently available due to experimental limitations [2]. With permission from American Chemical Society

H–O bond is much shorter, stronger, and stiffer (0.1 nm, 4.0 eV, 3000 cm^{-1}) than the O:H bond (0.17 nm, 0.1 eV, 200 cm^{-1}); (3) the bond energy characterizes the bond strength, while the vibration frequency characterizes the bond stiffness; and (4) in addition to the short-range interactions within the O:H and the H–O segments, Coulomb repulsion between the bonding electron pair and the non-bonding electron lone pair.

Typical phonon spectra for ambient water probed using spectroscopies of Fourier transform infrared (FTIR) absorption, Raman reflection, and neutron diffraction are

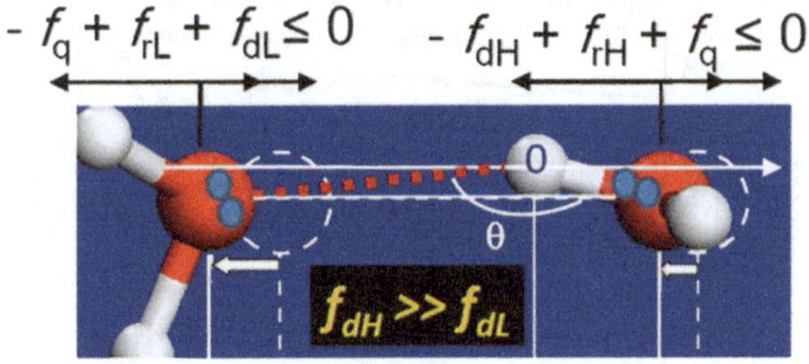

Fig. 3.21 Forces of interelectron-pair (pairing dots) Coulomb repulsion f_q, resistance of deformation f_{rx}, and under-coordination-induced bond contraction f_{dx} as well as the direction and degree of displacement for each O atom (in red) with respect to the H atom as the coordination origin. Subscripts L and H represent the O:H and the H–O segments of the O:H–O hydrogen bond, respectively [29]. With permission from Elsevier

presented in Fig. 3.22 [30]. The characteristic frequency (or energy) features correspond to stretching and bending vibration modes are: (1) at $\omega_H = 3450$ and $3200\ cm^{-1}$ correspond to the H–O stretching phonons in the skin and in the bulk, respectively; (2) the peak at ω_{B1} at 500–700 cm^{-1} is the O:H–O bending mode; (3) the peak of ω_{B2} at 1600–1750 cm^{-1} is the librational mode of H–O–H bending; peaks at $\omega_L <$ 300 cm^{-1} arise from O:H stretching; and (4) water molecules in the gaseous phase exhibit a peak in the vicinity of $\omega_{are} = 3650\ cm^{-1}$. Thus, monitoring the cooperative relaxation of high-frequency ω_H and low-frequency ω_L phonons allows to examine the cooperativity of the O:H–O bond under excitation. The authors concluded that the advantage of phonon spectroscopy is the Fourier transformation between the real and the energy spaces.

Using molecular dynamics simulations, dispersive optical phonon-like modes in the librational and OH stretching bands were computed [5]. On subpicosecond timescales, these modes propagate through water's hydrogen-bond network over distances of up to 2 nm. It was found that the lowest frequency librational Raman peak (435 cm^{-1}) is a transverse optical phonon-like mode while the highest frequency peak (770 cm^{-1}) is a longitudinal optical phonon-like dispersive modes which are due to optical phonons that travel along the H-bond network of water. Comparison of peak positions in longitudinal and transverse dielectric susceptibilities often revealed longitudinal–transverse *(LO–TO)* splitting which indicates the presence of long-range dipole–dipole interactions in the system. In a technique to measure the decoherence time of optical phonons in a solid [31] phonons are excited with a pair of time-delayed 80 fs near-infrared pulses via spontaneous transient Raman scattering. The spectral fringe visibility of the resulting Raman pulse pair, as a function of time delay, was used to measure the phonon dephasing time. The dephasing time of phonons created in bulk diamond is measured to be $\tau = 6.8$ ps ($\Delta\nu =$ 1.56 cm^{-1}). Raman and FTIR measurements probed phonon spectra over the full frequency range of water during cooling, compressing, heating, and salting were carried out [32]. A combination of these two database sets was used to verify predictions of the O:H–O bond cooperative relaxation and associated phonon attributes and

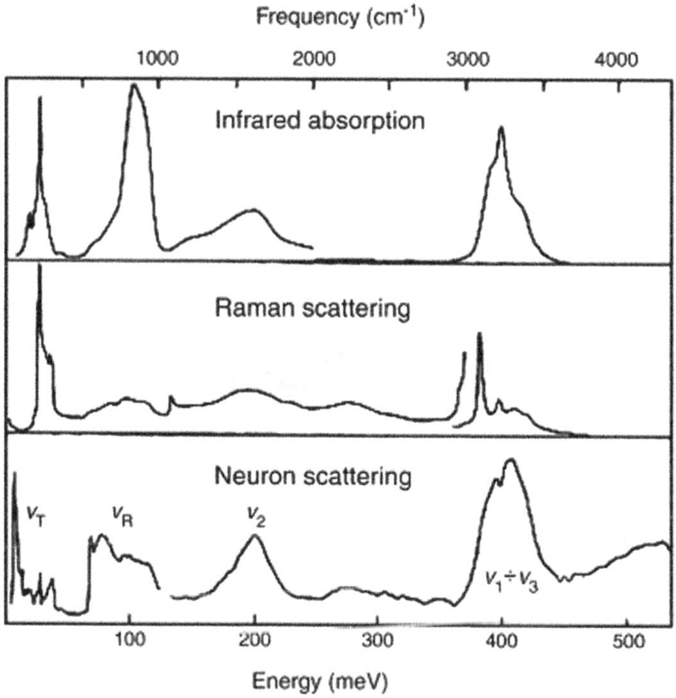

Fig. 3.22 Typical vibrational spectra measured from bulk water at ambient temperature. Peaks at $\omega_L = 300$ cm^{-1} correspond to O:H stretching; the peak at, $\omega_{B1} = 500$–700 cm^{-1} is from O:H–O bending; features $\omega_{B2} = 1600$–1750 cm^{-1} are the H–O–H bending libration mode; features centered at $\omega_H = 3200$ and 3450 cm^{-1} are the H–O stretching mode in the bulk and in the skin of water [30]. Reprinted with permission from Elsevier

electron entrapment and polarization under various stimuli. The developed spectral differential strategy enabled the distillation of the phonon spectral features by differencing the spectra collected before and after conditioning. This strategy was also applied to calculations to discriminate the characteristic bond length d_x or phonon frequency ω_x peaks obtained under different conditions.

3.3 Other Methods

3.3.1 Terahertz Spectroscopy

Nowadays, terahertz spectroscopy is a well-established technique and recent progresses in technology demonstrated that this new technique is useful for both fundamental research and industrial applications [26, 33–42]. The variety of

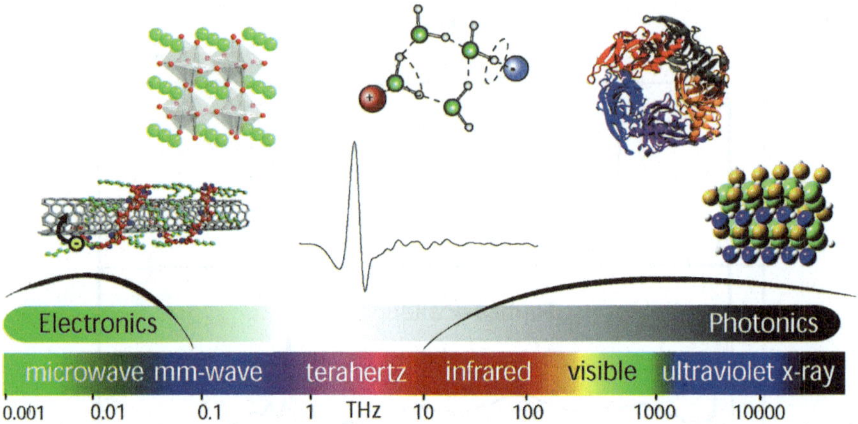

Fig. 3.23 Electromagnetic spectrum on the THz scale, with images highlighting the wide variety of molecules, materials, and phenomena that can be explored using THz spectroscopy. A few examples include charge carrier dynamics in semiconductors, nanomaterials, and correlated electron materials; hydration of solutes and librational modes in water and other liquids; and collective modes in proteins and amino acid crystals [37]. With permission from American Chemical Society

phenomena that can be investigated using terahertz radiation is displayed schematically in Fig. 3.23 [37]. Monitoring reorientation of dipoles in liquids on a picosecond (ps) timescale, by the THz technique, is essential for structure/dynamics and function of biological molecules. Water strongly absorbs THz radiation and must be contained within a thin (~100 μm) sample cell if measured in transmission. The dielectric responses of liquids have been extensively studied in the THz range and modeled using two Debye relaxation processes—a slow one (few ps) and a fast one less than femtoseconds (fs)—depending on the temperature [40].

Terahertz (THz) radiation (1 THz = 10^{12} Hz) are electromagnetic waves in the range 0.3–10 THz (Fig. 3.24) [33]. In other units, 1 THz is equivalent to 33.3 cm^{-1} (wavenumbers), or 0.004 eV photon energy, or 300 μm wavelength. Low-energy interactions can be studied by these waves in a large number of materials, including water, ice and objects containing water. THz spectroscopy is a powerful tool for characterizing vibrational modes (e.g., rotational, torsional, phonon, intramolecular, and intermolecular). These data may be complemented to the knowledge of material behavior obtained with far-infrared (FIR) and Raman spectroscopy. Importantly, interaction with THz waves is non-destructive to allow probing biological materials hydration including reorientation of dipoles in liquids on a picosecond (ps) timescale.

Typical laser-based experiments of THz spectroscopy are: THz time-domain spectroscopy (THz-TDS), time-resolved THz spectroscopy (TRTS), THz emission spectroscopy (TES), and THz attenuated total reflection (THz ATR) [26, 34, 39, 41, 42]. In conventional benchtop experiments, THz pulses were created and detected using short-pulsed Ti:sapphire lasers with pulse widths ranging from ~10 to 100 fs and center wavelengths around 800 nm [37]. THz pulses have a duration of ~1 ps, and coherent detection in the time domain that allows measurement of the transient

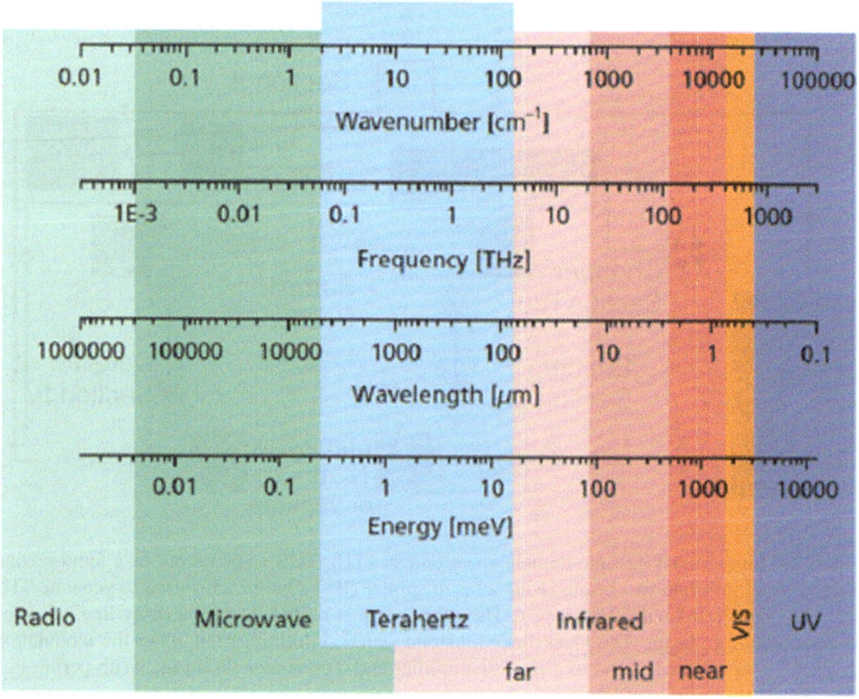

Fig. 3.24 Electromagnetic spectrum: in blue, the terahertz region [33]. With permission from Elsevier

electric field intensity. Fourier transform of the transient electric field enables direct determination of both the amplitude and the phase of each of the spectral components that make possible to obtain directly the complex-valued permittivity of the sample.

Employing the THz time-domain spectrometer (THz-TDS) (Fig. 3.25) is based on sampling the unknown THz field with a known femtosecond laser pulse, which is referred to as the read-out pulse [42]. THz-TDS uses the convolution of a short read-out pulse with the longer THz pulse. The signal $S(t)$ was described as

$$S(t) = I_{opt}(t)E_{THz}(t), \tag{3.2}$$

with $I_{opt}(t)$ being the intensity profile of the laser pulse and $E_{THz}(t)$, the electrical field of the THz pulse at time t. The measured instantaneous signal detected with subpicosecond resolution corresponds to the THz field amplitude at a single point in time (t_1). The next step is to measure the signal at all time points. One of the beams is used to generate THz radiation, and the other one is the read-out beam that detects it.

Figure 3.26 displays schematically a measured THz time trace and the resulting complex fast Fourier transformation (FFT) in the frequency domain.

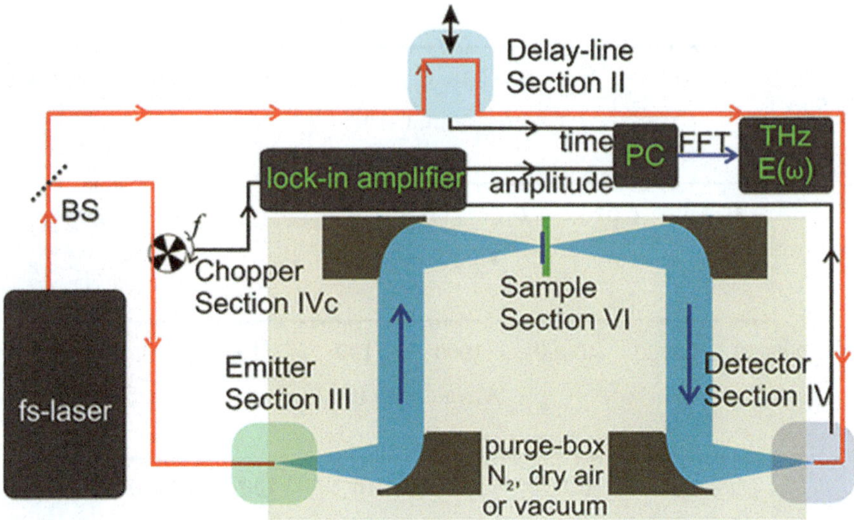

Fig. 3.25 Simplified THz time-domain spectrometer (THz-TDS). The output of a femtosecond laser pulse is split into two beams using a beam splitter (BS). One beam is used to generate THz radiation, modulated with frequency f. The other beam is routed through a delay line and used to detect the THz beam. The black lines illustrate signal connections, inputting the modulation frequency and detected signal to a lock-in amplifier and a computer (PC) [42]. With permission from American Institute of Physics

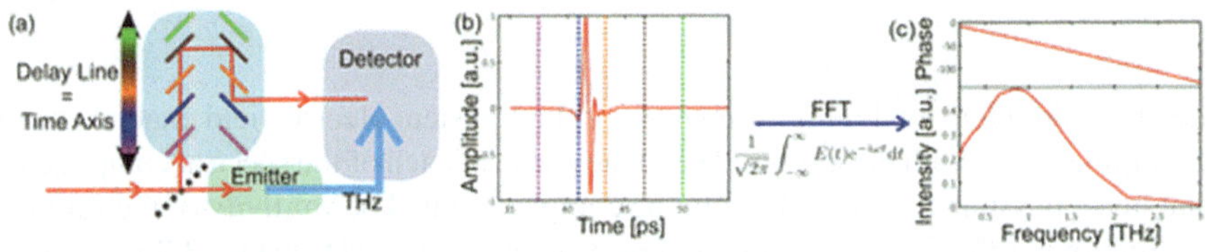

Fig. 3.26 **a** Simplified illustration of TDS. The different colors correspond to different delay-line positions. These positions correspond to different time points in **b**. **b** Measured THz time-domain signal. **c** Fast Fourier transformation (FFT) of the signal in the frequency domain. The upper panel shows the unwrapped phase of the complex Fourier transform, and the lower shows the amplitude [42]. With permission from American Institute of Physics

3.3.2 Dielectrics Measurements

The dielectric properties of water are important for understanding its properties as a solvent and absorber of electromagnetic radiation. Application of dielectric spectroscopy of applications in materials in which electrical charges play in the molecular interactions are broad and diverse and cover presently the electromagnetic frequency range, ranging from about 10^{-6} Hz to almost 10^{15} Hz [43–49]. A dielectric is an electrical insulator that can be polarized by an applied electric field. The relative permittivity, or dielectric constant ε, the factor by which the electric field between the charges is decreased relative to vacuum, is expressed as a ratio relative to the vacuum permittivity $\varepsilon_0 = 8.8541878128(13) \times 10^{-12}\,\mathrm{F\,m^{-1}}$. Polarization P is the vector field that expresses the density of permanent or induced electric dipole moments. The electric susceptibility χ_e is a dimensionless proportionality constant that indicates the degree of polarization P of a dielectric material in response to an applied electric field E: A capacitor is a device which stores electrical energy in an electrical field. The capacitance is the capacity of the capacitor to store charge in it. In a capacitor, two conductors are separated by an insinuator (dielectric) and when an electric field is applied, electrical energy is stored in it as charge.

The dielectric spectroscopy technique parameters are:

Parameter	Time domain $S(t)$	Frequency domain $S(\omega)$
Large ω_0	Fast oscillations	High frequency
Small ω_0	Slow oscillations	Low frequency
Large	Fast decay	Broad linewidth
Small	Slow decay	Narrow linewidth

https://en.wikipedia.org/wiki/Dielectric_spectroscopy.

In review [46], frequency-domain and time-domain techniques, low-frequency, high-frequency and microwave measurement systems, broadband and spot-frequency modes of operation, reflection, transmission and resonant methods were described. The basic principles of dielectric spectroscopy and modern methods of measurement the dielectric properties of liquid samples over the frequency range from about 10^{-6} to 10^{12} Hz were summarized in the comprehensive review [47].

Debye relaxation is the dielectric relaxation response of a non-interacting population of dipoles to an alternating external electric field. Relative permittivity (ε) characterizes the capacitance ratio of a condenser filled with an insulating material and with vacuum. In the microwave region, the complex spectrum of water can be represented by the Debye relaxation spectral function [44] with the extrapolated high-frequency permittivity $\varepsilon(\infty)$ and the extrapolated low-frequency ("static") permittivity $\varepsilon(0)$:

$$\begin{aligned} \varepsilon' &= \varepsilon_\infty + \frac{\varepsilon_S - \varepsilon_\infty}{1 + \omega^2\tau^2} \\ \varepsilon'' &= \frac{(\varepsilon_S - \varepsilon_\infty)\omega\tau}{1 + \omega^2\tau^2} \end{aligned} \tag{3.3}$$

where τ is the relaxation time, ε_∞ represents the value of the real part of the dielectric permittivity of polar molecules at very high frequencies, at very-low-frequency relative permittivity is given by ε_S, ω is the angular frequency ($\omega = 2\pi f$), and τ is the molecule relaxation time. A critical review on the Debye absorption peak of liquid water was reported, and the excess response found on the high-frequency side of the Debye peak was discussed [41]. The electric displacement D under effect of

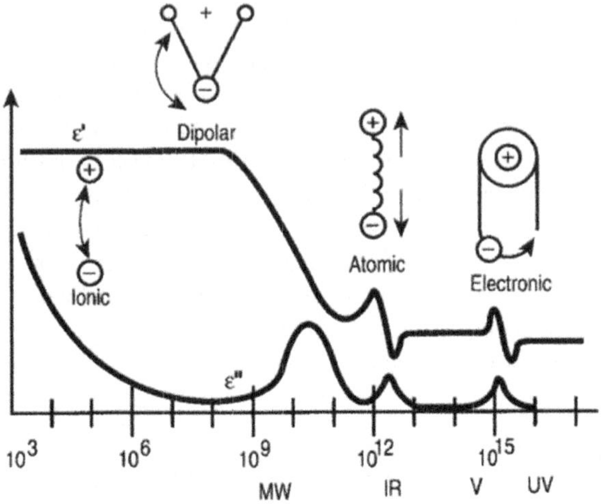

Fig. 3.27 A dielectric permittivity spectrum over a wide range of frequencies. The real and imaginary parts of permittivity are shown, and various processes are depicted: ionic and dipolar relaxation, and atomic and electronic resonances at higher energies. https://web.archive.org/web/20010307184808/http://www.psrc.usm.edu/mauritz/dilect.html

a time-dependent field $E = E_0 \cos(\omega t)$ will have time dependence with a phase lag. Dielectric spectroscopy measures the dielectric properties of a medium as a function of frequency (Fig. 3.27) https://en.wikipedia.org/wiki/Dielectric_spectroscopy. The dielectric responses of liquids have been extensively studied in the THz range and modeled using two Debye relaxation processes—a slow one (few ps) and a fast one less than femtoseconds (fs) both depending on the temperature [40].

As an example of THz application, real and imaginary parts of the complex dielectric function of water and 10 different ethanol–water mixtures were detected (Fig. 3.28) [43]. These data allowed to establish a link between the dielectric function of water–ethanol mixtures and some of their thermodynamic properties in these systems.

To better understand the molecular origin of Debye peak large, scale molecular dynamics simulations was run and several different distance-dependent decompositions of the low-frequency dielectric spectra were performed. It was found that processes take place on scales of 1.5–2.0 nm. The results were consistent with a proposed model [38] which suggested that Debye relaxation is due to the migration of Bjerrum-like defects in the hydrogen-bond network. A Bjerrum defect is a crystallographic defect in ice, when a hydrogen bond non-normally has either two protons (D defect) or no proton (L defect). According the theory, the mechanism of Debye relaxation in liquid water is similar to the mechanism in ice, but the heterogeneity and power-law dynamics of the H-bond network in water results in excess response on the high-frequency side of the peak. Specifically, the dielectric relaxation in water can be driven by the migration of defects through the H-bond network, leading to a Debye-like peak in the lower frequencies. The deviation from the Debye law in the higher sub-THz frequencies is traced to a global fluctuation of the same H-bond network, clearly evident in the Raman Spectra.

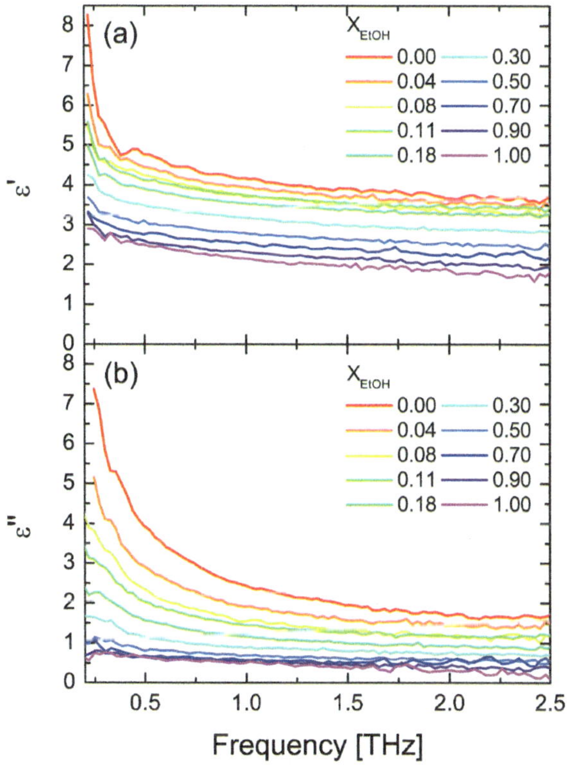

Fig. 3.28 **a** Real and **b** imaginary parts of the complex dielectric function of ten different ethanol–water mixtures at 298 K [43]. With permission of Optical Society of America

In time-domain spectrometry (TDS) technique, the complex permittivity spectrum of the sample was obtained from the scattering coefficient, which is the ratio of Fourier transforms of the responding and the incident signal wave form [47–49]. The methods in use may be subdivided into three categories: (1) slow frequency with limit of about 1 MHz, (2) fast frequency with limit below 25 GHz, and (3) femtosecond THz-pulse spectroscopy. Fast response time-domain spectrometry (TDS) consists in the observation of the change in shape of step voltage pulses after reflection by or transmission through the sample cell. In frequency-domain methods, the sample liquid is exposed to a monochromatic electric field $E(\upsilon)$ and the responding polarization $P(\upsilon)$ is determined as a function of frequency υ [46]. The sample properties are expressed in terms of the frequency-dependent complex (electric) permittivity

$$\epsilon(\upsilon) = \epsilon'(\upsilon) - i\epsilon''(\upsilon) = \frac{1}{\epsilon_0}\frac{P(\upsilon)}{E(\upsilon)} + 1, \tag{3.4}$$

where e_0 denotes the electric field constant and $i^2 = -1$, $e(\upsilon)$ is the real component of the polarization that is in phase with $E(t)$, $e(\upsilon)$ is the (negative) imaginary part representing the contribution with a $\pi/2$ phase shift. As an example, in the frequency range of measurements from 100 MHz to some THz, $e(\upsilon)$ and $e(\upsilon)$ of the dielectric spectrum of water at 20 °C, the water spectrum displays two dispersion/dielectric loss regions with relaxation characteristics, that is, one is correspondent with the nearly exponential slow part in the dielectric decay function and an additional relaxation with relaxation frequency around 1000 GHz.

Local capacitance measurements for water confined between two atomically flat walls separated by various distances h down to 1 nm were reported [50]. Experiments revealed the presence of an interfacial layer with vanishingly small polarization such that its out-of-plane dielectric constant is approximately 2. Experimental setup for dielectric imaging consists of the top layer and side walls, graphite serving as the ground electrode. The three-layer assembly covers an opening in a silicon nitride membrane. The channels are filled with water. Experiments for more than 40 devices with h ranging from ~1 to 300 nm were performed. $\varepsilon_{\perp}$ as a function of h was observed. The bulk behavior ($\varepsilon_{\perp} \approx 80$) was recovered only for water as thick as ~100 nm, which indicating that confinement could affect the dielectric properties of even relatively thick water layers. The results showed a way to investigate the dielectric properties of other fluids and solids under extreme confinement.

3.3.3 Neutron Scattering and Diffraction

Neutron scattering is the dispersal of free neutrons by matter, which understanding and manipulating is fundamental to the applications used in crystallography, physics, physical chemistry, biophysics, and materials research [51–60]. Thermal neutrons have energies of several tens of meV (e.g., 0.025 eV), comparable to the thermal energies $k_B T$ at room temperature. These neutrons are proofed to be probes of low-energy excitations such as lattice vibrations and spin waves with energies in the meV range. Cold neutrons have energies 0.0–0.025 eV. Neutrons are uncharged and can penetrate deeply into most materials and neutron scattering is a bulk probe, traveling long distances in material without being scattered or absorbed and penetrating to depths of 0.1–0.01 m. When a neutron beam interacts with a sample, three possible phenomena can take place: Either the neutrons just go through the sample transmission—or they are captured by the sample—absorption—or their propagation is deviated with respected to the original direction—scattering.

In incoherent scattering there is no interference between the scattered neutron waves from different centers, and only single atoms are followed. The coherent scattering depends on the correlation between the positions of different nuclei at different times which causes interference between scattered neutron waves from different scattering atoms. Thus, coherent scattering gives the information on the relative atomic positions. The scattering process can be either elastic with no exchange of energy between neutrons and sample (neutron diffraction) or inelastic (INS) with exchange of energy with the sample (neutron spectroscopy). Neutrons can lose or gain energy to or from the atom when they bounce against during scattering. Therefore, inelastic scattering can reveal the atomic motion. Quasi-elastic neutron scattering (QENS) is characterized by energy transfers being small compared to the incident energy of the scattered particles where dynamics in the sample (such as diffusive dynamics) lead to a broadening of the incident neutron spectrum. INS and QENS are alternatively and complementary vibrational spectroscopy which provides direct information about the vibrational density of states (VDOS), spanning the entire infrared region for typical molecular vibrations. In a typical neutron scattering experiment, scattered neutrons are measured as a function of change in their direction (related to momentum transfer, Q) and energy E.

Neutron diffraction is a form of elastic scattering where the neutrons possess almost the same energy as the incident neutrons. The technique is similar to X-ray diffraction but the different type of radiation gives complementary information. Encountering matter, X-rays more directly intermingle with the electron clouds surrounding atoms whereas neutrons, not affected by the charged electrons, interact with the atomic nuclei. The conceptual experiment and theory is the same for X-rays and neutrons and follows to Bragg's law, which describes the condition for constructive interference from successive crystallographic planes of the crystalline:

$$2d \cdot \sin(\theta) = n\lambda \tag{3.5}$$

where n is an integer determined by the order given, θ is the scattering angle, and λ is the wavelength (Fig. 3.29).

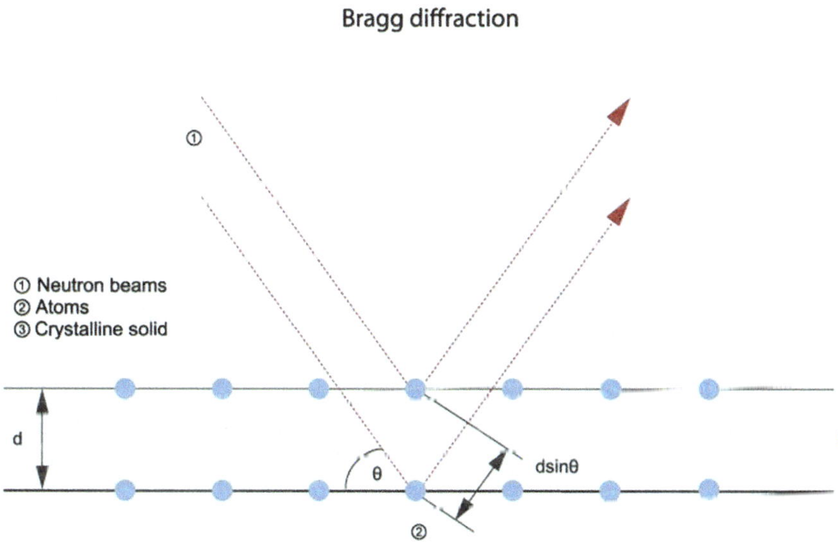

Fig. 3.29 Bragg's low to neutron diffraction. https://nmi3.eu/neutron-research/techniques-for-/structural-research.html#:~:text=Neutron%20diffraction%20or%20elastic%20neutron.provides%20information%20of%

A simplified scheme of inelastic neutron scattering experiment outline is displayed in Fig. 3.30.

Inelastic and quasi-elastic neutron scattering experimental techniques can reveal the displacements of atoms and molecules over distances spanning from angstroms to a few tens of nanometers, on timescales ranging from a fraction of picoseconds to microseconds [56]. The different INS and QENS machines (time-of-flight (ToF), backscattering (BS) and neutron spin echo (NSE)) can be utilized for measurement of the scattering vector (Q) dependence of the systems characteristic time. In radiation scattering experiments, a neutron scattering experiment consists of illuminating a sample with a neutron beam of wave vector k_0, wavelength λ_0, and energy E_0 (Fig. 3.31). Under the effect of the interaction $V(r, t)$ between the neutron and the

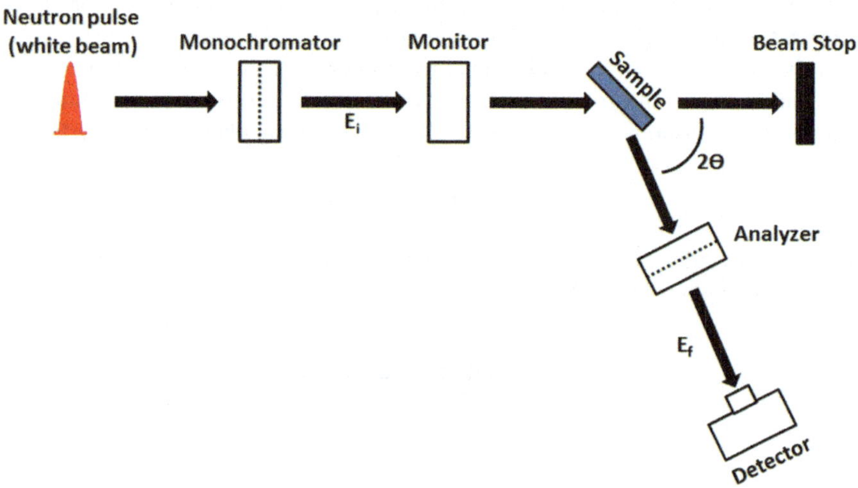

Fig. 3.30 A very basic schematic of an inelastic neutron scattering experiment. http://large.stanford.edu/courses/2013/ph241/jiang2/

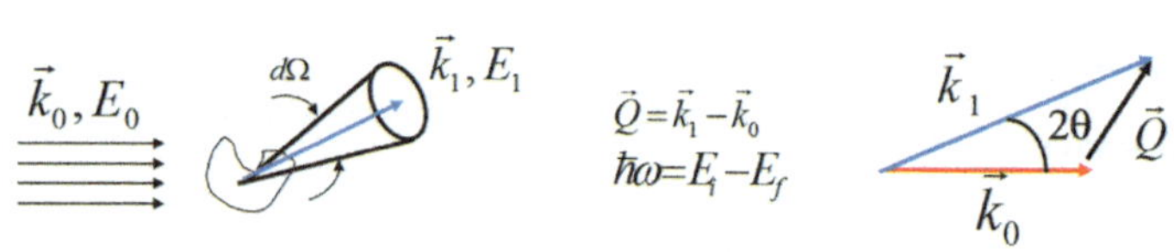

Fig. 3.31 Principle of a neutron scattering experiment. As a result of interaction with the sample, incident neutrons (k_0 and ω_0) can undergo a wave vector and energy changes: respectively $Q = k_1 - k_0$ and $h\omega = h(\omega_0 - \omega_1)$. In this general case of an inelastic interaction (ω_0), Q is a function of θ and ω (3.3). For an elastic scattering process $k_1 = k_0$ (i.e., $\omega_0 = \omega_1$), the elastic scattering vector is $Q_{el} = 4\pi/\lambda \sin(\theta)\cdot u$, where u denotes a unit vector (not shown). In the reciprocal domain, i.e., the domain where the neutron detector directly operates, the two working quantities are Q and ω. The two dual quantities in the real space, i.e., the domain of the sample, are the time, t, and space, r. One can switch from one domain to the other by Fourier transform over r and t [56]. Published under the terms of the Creative Commons Attribution License (http://creativecommons.org/licenses/by/4.0)

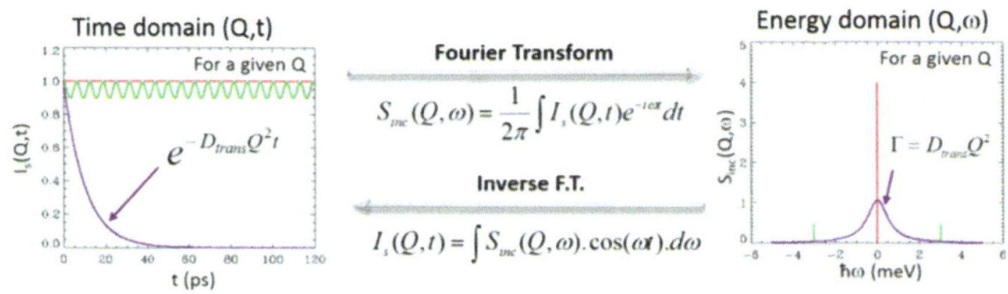

Fig. 3.32 Schematic representation of the inelastic spectra of different dynamical modes. We consider the autocorrelation function $I_S(Q, t)$ of a particle in the (Q, t) domain. The switch to the (Q, ω) domain is achieved by time Fourier transform of $I_S(Q, t)$. By definition of an autocorrelation function, whatever the dynamics of the particle, its position at $t = 0$ is known i.e. $I_S(Q, t = 0) = 1$. If the particle is immobile, whatever t, its position is also known, i.e., $I_S(Q, t) = 1$ (see red curve). The Fourier transform in time of this horizontal straight line is a Dirac function in energy positioned at $h\omega = 0$. This physically corresponds to a situation where neutrons do not exchange any energy with the particle. One talks of "elastic scattering." If now a particle experiences a vibrational mode at frequency $h\omega_0$ the autocorrelation function $I_S(Q, t)$ is a sinusoidal function of time. Its Fourier transform materializes as an elastic Dirac along with two Diracs centered at $h\omega_0$ and $h\omega_0$. These two Diracs of the inelastic zone of the spectrum $(h\omega \neq 0)$ represents the simpler expression of $g(\omega)$, the vibrational density of states (see Sect. 2.5.3). In the case of a particle experiencing a long-range translational diffusion (c.f. Sect. 2.5.1), $I_S(Q, t)$ takes an exponential form. The corresponding dynamical structure factor is a Lorentzian line centered around $h\omega = 0$ with HWHM, Γ [61].

system, downstream of the sample, the radiation is characterized by a wave vector k_1 and an energy E. The diffusion angle, 2θ, is defined as the angle between the vectors k_1 and k_0.

Inelastic and quasi-elastic neutron scattering in both the time and energy domains of key typical dynamical modes are summarized on Fig. 3.32.

The time-of-flight (ToF) technique (Fig. 3.33) allows the study of dynamical phenomena of nuclear or magnetic origin [56]. The quasi-elastic region, which extends up to a few meV around the elastic peak (zero energy transfer point: $\omega =$ 0 meV), is characteristic of phenomena and/or relaxations processes in the system. Over the whole energy range, from 0 to several hundreds meV, the signal can correspond to the exchange of quanta of energy between the neutron and the sample.

In review [53], the differences between INS spectroscopy and conventional infrared and Raman spectroscopies were briefly summarized. A simplified form of the expression for the scattered intensity is:

$$S(Q, n\omega_i) \propto \frac{(QU_i)^{2n}}{n!} \exp\left(-(QU_{\text{Tot}})^2\right)\sigma_{\omega_i} \tag{3.6}$$

where ω_i is the ith mode at transition energy ω, $n = 1$ for a fundamental, 2 for a first overtone or binary combination, 3 for a second overtone or ternary combination, etc., Q is the momentum transfer, U_i is the root mean square displacement of the atoms in the mode, and σ is the inelastic scattering cross section of the atom. U_{Tot} is the

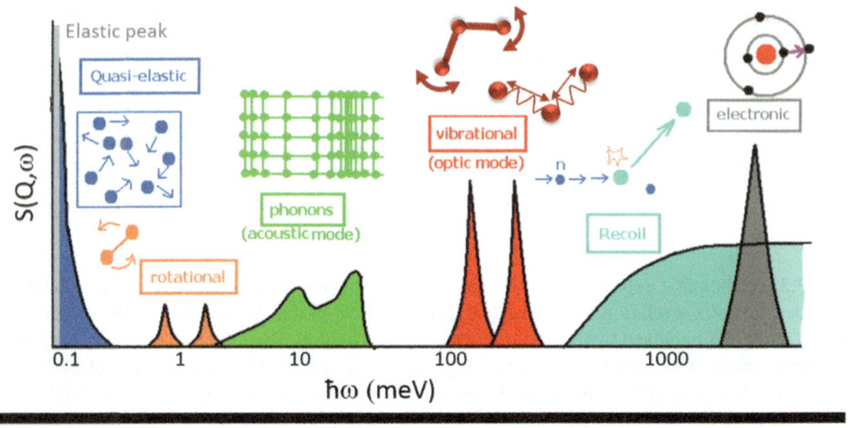

Fig. 3.33 Schematics of the dynamical structure factors, $S\ (Q,\ \omega)$, of a selection of phenomena accessible by quasi-elastic and inelastic neutron scattering [56]. Published under the terms of the Creative Commons Attribution License CC-BY-NC

total root mean square displacement of all the atoms in all the modes (both internal and external). The transition energy and the momentum transfer ($\boldsymbol{Q}$, $|\boldsymbol{Q}| = Q$, Å^{-1}) are

$$\boldsymbol{Q} = \boldsymbol{k} - \boldsymbol{k} \quad \text{where } k = 2\pi/\lambda \tag{3.7}$$

k (Å^{-1}) is the wave vector; λ (Å) is the wavelength of the neutron, i and f refer to the incident and final neutron, respectively.

One of the basic functions used to characterize atomic dynamics is the vibrational density of states (VDOS) can be extracted from an analysis of the isotropic incoherent dynamical structure factor $S_S(Q, E)$, which in hydrogen-rich samples is proportional to the double differential cross section, measured by neutron scattering experiments [62]. E is the exchanged energy and $\hbar Q$ is the modulus of the exchanged momentum. The typical energies of thermal neutrons are similar to the vibrational mode energies (1–500 meV) and the neutron wavelength is comparable to the interatomic distances in condensed materials. The inelastic incoherent neutron scattering (IINS) spectrum is directly proportional to the VDOS, written as

$$G(E) = \Sigma j|\boldsymbol{e}|\delta(E - \hbar\omega)Mjj \tag{3.8}$$

where e_j is the polarization vector of normal modes at the hydrogen sites and M is the hydrogen atomic mass. In the one-phonon approximation the VDOS is related to the dynamical structure factor which depends on $W(Q)$, the Debye–Waller factor which is used in condensed matter physics to describe the attenuation of X-ray scattering or coherent neutron scattering caused by thermal motion, and $n(E)$ (the Bose–Einstein factor, a binomial coefficient of Bose–Einstein statistics). In the energy range up to 150 meV, the collective dynamics in water corresponds to high-frequency modes with wavelengths close to interatomic separations and, therefore, can manifest themselves as inelastic peaks in the measured coherent signal. The dispersion curves can be measured by inelastic X-ray scattering (IXS) and inelastic neutron scattering (INS) for bulk water by reporting the excitation energies of these modes as a function of Q.

Below the application of INS to water is illustrated by two examples. An ability of the inelastic neutron scattering to study the vibrational dynamics of H_2O and OH sorbed on SnO_2 nanoparticles was demonstrated in [63]. The experiment and ab initio calculated molecular dynamic reveal very strong surface H-bonds, with a formation enthalpy twice that of liquid water. It was evidenced that: (1) the first layer of adsorbed water interacts so strongly with the surface that this L1 water is largely (60–90%) dissociated to OH groups; (2) in the second layer (L2) water molecules are also very strongly H-bonded to L1 species; and (3) in the third layer (L3) and subsequent layers water are weakly bound or disconnected from the near-surface layers producing an apparent effect of weak interaction in the vibrational density of states.

The INS vibrational spectra of free and partially constrained water clusters (up to $n = 17$) were calculated with DFT methods using published energy-minimized structures [64]. Specifically, the following INS spectra were calculated: (1) spectra in the librational region for free planar water clusters; (2) vibrational spectra of three-coordinate water molecules for the tetramer; (3) spectra for the three water hexamers: cage, trigonal pyramid, and cyclic; (4) spectra for water clusters 12d (Fig. 3.34); (5) spectra of the librational region for a free water trimer with one water molecule or two molecules fixed; (6) spectra of three water heptamers, prism, cage, and chair spectra for a variety of model "bound" water clusters; (7) spectra of water 17-mer clusters.

Quasi-elastic neutron scattering (QENS) designates a case of inelastic neutron scattering, characterized by energy transfers being small compared to the incident energy of the scattered particles [65–71]. QENS denotes scattering processes where dynamics in the sample (stochastic, diffusive processes, such as overdamped vibrations, localized diffusion) lead to a broadening of the incident neutron spectrum. QENS is investigated on high-resolution spectrometers (neutron backscattering, neutron time-of-flight scattering, neutron spin echo). In the QENS technique, the energy spectrum of the autocorrelation function $S_{\mathrm{self}}(Q, E)$

$$S_{\mathrm{s}}(Q, E) = \mathrm{e}^{-\langle u^2\rangle Q^2/3}\left[A(Q)\frac{\Gamma_{\mathrm{T}}(Q)}{E^2+\Gamma_{\mathrm{T}}^2(Q)} + B(Q)\frac{2D_{\mathrm{R}}+\Gamma_{\mathrm{T}}(Q)}{E^2+[2D_{\mathrm{R}}+\Gamma_{\mathrm{T}}(Q)]^2}\right] \tag{3.9}$$

was measured as a function of the wave vector Q providing access to translational and rotational relaxation phenomena in the system, by analyzing its Q dependence [65–71]. It was found [65] that diffusion-like motions exhibit strong dependence of the decay rate Γ (or relaxation time τ) on Q, while for local relaxation processes the characteristic Γ (or timescale τ) is independent of Q, at least at large Q. Within this assumption of a roto-translational decoupling and of a multiexponential time decay

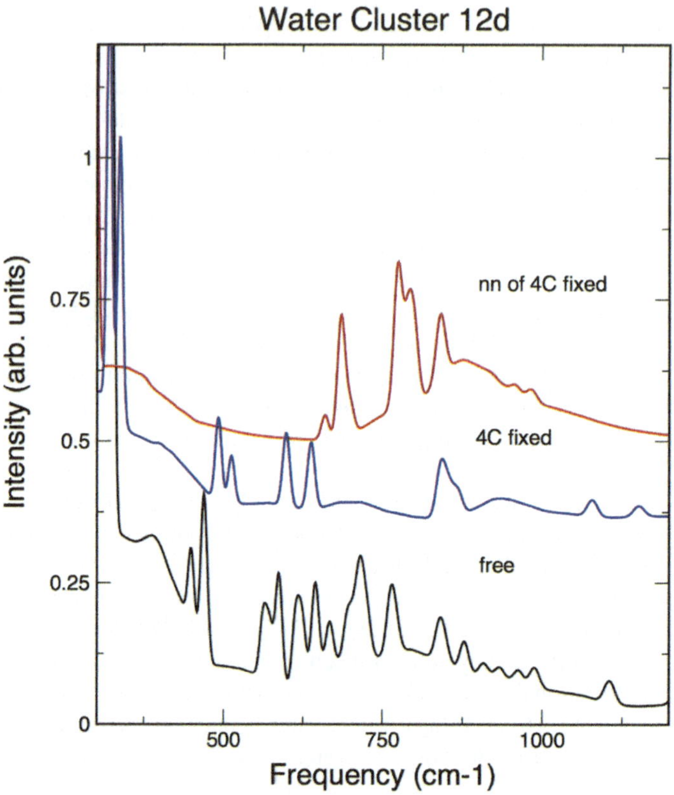

Fig. 3.34 Calculated INS spectra for water clusters 12d which contain three 4C molecules: nearest neighbors of 4C molecules fixed (red), 4C molecules only fixed (blue), free cluster (black) [64]. With permission from Elsevier

of the density–density correlation function, the (self) incoherent dynamic structure factor can be cast into the following form in which the exponential term is the analogue of the Debye–Waller factor of solids, $\langle u^2 \rangle$ represents the single-proton vibrational amplitude, $j_l(x)$ is the spherical Bessel function of l-order, a is the radius of rotation (for water fixed to the O–H distance), Γ_T is the translational half-width at half-maximum decay rate (or relaxation time τ), and D_R is the rotational diffusion coefficient, E is the exchanged energy [66].

Pressure dependence of the translational DT and of rotational DR diffusion coefficients along the isotherm 400 K as measured by QENS (red dots) and by MD simulations performed by using TIP4P2005 potential for water (blue squares) is presented in Fig. 3.35.

In another example of the QENS application, water diffusion was investigated, employing the BASIS spectrometer at ORNL [67]. The spectrometer resolves two distinct quasi-elastic components corresponding to localized water diffusion (on length scales lower than the water domain thickness) and to extended water diffusion

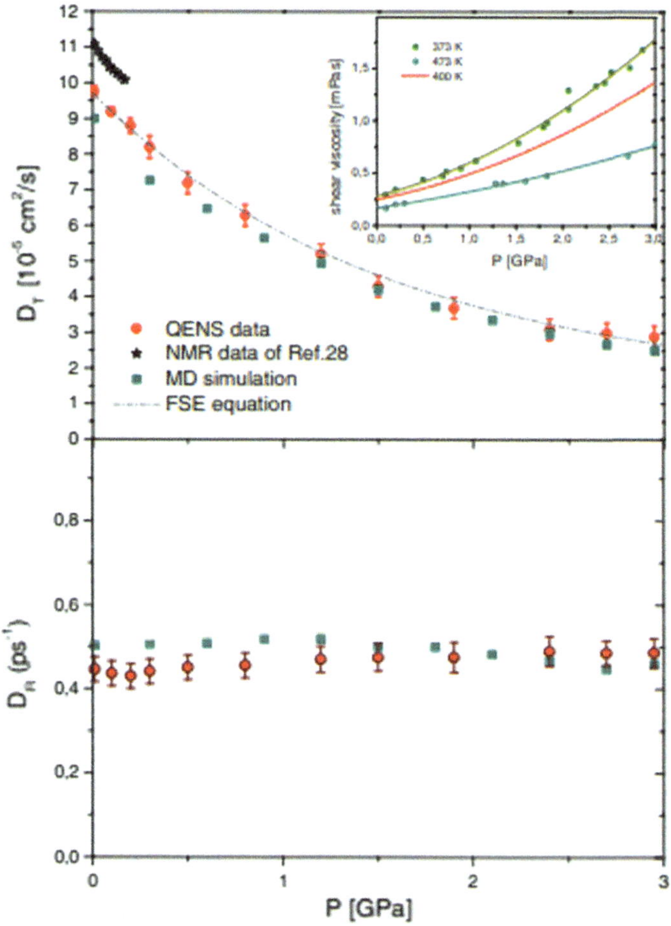

Fig. 3.35 Pressure dependence of the translational DT and of rotational DR diffusion coefficients along the isotherm 400 K as measured by QENS (red dots) and by MD simulations performed by using TIP4P2005 potential for water (blue squares). Black stars represent the available low-pressure data [on DT as measured by NMR at 403 K. The gray curve represents the FSE equation] prediction with a fractional coefficient $\xi = 0.76$. In the inset, we show the shear viscosity dependence on pressure along the 400 K isotherm (red curve), interpolated from data of [9] at 373 K (green dots), and at 437 K [65]. With permission from American Physical Society

(on length scales approaching the water domain thickness). The experiment indicated that water diffusion coefficients in membranes are time and length scale dependent. Water exchanges between diffusion on a localized and extended volume increment were also quantitatively investigated.

Employing QENS, the dynamic behavior of hydration water in phospholipid membranes has been studied to understand the relationship between water and biological molecules [68]. The measurable ranges from the 10^{-12} to 10^{-9} s timescale, and the 10^{-11} to 10^{-8} m length scale was shown to be suitable to investigate the slowing down of water molecules due to their interaction with lipid membranes.

QENS experiments with the LET spectrometer at ISIS were performed on oriented d54-DMPC layers on Si wafers hydrated with H_2O vapor and revealed that water exhibits an anisotropic and subdiffusive behavior in nanometer-length scales [69]. The QENS experiments on two samples, perdeuterated DMPC (d67-DMPC) with H_2O and protonated DMPC with D_2O were carried out at an energy resolution of 3.6 μeV [70]. Data analysis showed that the hydration water can be categorized into three types: (1) free water with dynamic behavior is slightly different from that of bulk water; (2) loosely bound water with dynamic behavior is one-order of magnitude slower than that of free water; and (3) tightly bound water with dynamic behavior is comparable to that of DMPC molecules.

Small-angle scattering can be carried out using either X-rays (SAXS) or neutrons (SANS) scattering techniques [72–79]. In combination with deuterium labeling and solvent-contrast variation (H_2O/D_2O exchange), SANS provides information on individual components in large heterogeneous complexes [77]. The major advantages of SANS are connected with the possibilities of using the contrast variation in the studied systems by hydrogen/deuterium substitution, as well as with neutron magnetic scattering. Two types of experiments with the neutron spin opposite (scattering intensity $I+$) and parallel (scattering intensity $I-$) to the direction of beam were performed. The progress and developments of biological applications of SANS in the past decade were reviewed [77].

Figure 3.36 displays a simplified schematic overview of a SANS experiment in solution [74]. In solution, SANS experiment can proceed through several steps: (1) after production at a neutron reactor or spallation source, neutrons are moderated and guided toward an instrument; (2) optical elements select the wavelength and direction of the incoming neutrons prior to their interaction with the biological sample; (3) after the interaction process with the sample, scattered neutrons are registered by a detector as a function of the scattering vector Q, which is defined as the difference between the incoming wave vector k and the outgoing wave vector k_0; and (4) the scattered signal can be integrated by a one-dimensional intensity, $I(Q)$, which can be used for structural analysis of the solubilized biomolecules.

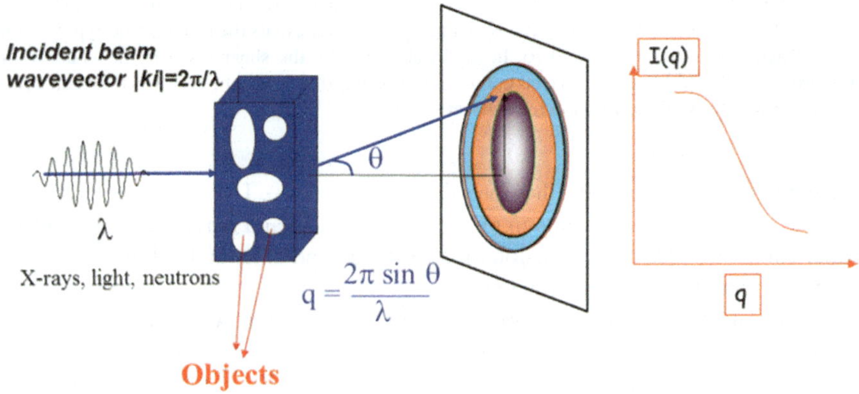

Fig. 3.36 Principle of the small-angle neutron scattering (SANS) experiment [74]. Published under the terms of the Creative Commons Attribution License CC-BY-NC

For non-interacting particles, the measured intensity can be described by a three-dimensional Fourier transform of the particle contrast $\Delta\rho$ with respect to the solvent, integrated over the solvent-excluded particle volume V and averaged over all possible orientations,

$$I(Q) \propto \left\langle \left| \int \Delta\rho \cdot \exp(i\mathbf{Q} \cdot \mathbf{r})\mathrm{d}V \right|^2 \right\rangle. \tag{3.10}$$

$\Delta\rho$ is defined as the difference between the scattering length density (SLD) of the solubilized particle, r_{part}, and the surrounding solvent, ρ_{solv} which calculated from the chemical composition of a molecule, $\rho = \left(\sum b_i\right)/V$, where b_i are the scattering lengths of the atoms composing the molecule.

The nature of the micropores with amorphous solid water and their collapse using neutron scattering of gram-quantities of D_2O–ASW of internal surface areas up to 230 ± 10 $m^2\ g^{-1}$ prepared at 77 K were characterized employing small-angle neutron scattering [78]. The model-free interpretation of the small-angle scattering data suggests micropores, which remain stable up to 120–140 K and then experience a sudden collapse. Analysis of the small-angle neutron scattering signal using the Guinier–Porod model suggested that a sudden transition from three-dimensional cylindrical pores with 15 Å radius of gyration to two-dimensional lamellae at about 140 K is the mechanism underlying the pore collapse. The experimental evidence for the low-temperature density minimum of water confined within cylindrical pores of ordered nanoporous materials of different pore size, which is based on combined data from in situ small-angle scattering of X-rays (SAXS) neutrons (SANS) and additional wide-angle X-ray scattering (WAXS), was given in [79]. A suggested two-step density model reveals that: (1) a wall layer covers approximately two layers of water molecules with higher density than the residual core water in the central part of the pores; (2) the temperature-induced changes of the scattering signal from both X-rays and neutrons are consistent with a minimum of the average water density; (3) the temperature at which this minimum occurs depends monotonically on the pore size; (4) for water confined in the smallest pores of only 2 nm in diameter; and (5) the density minimum can be explained in terms of a structural transition of the surface water layer closest to the hydrophilic pore walls. Various aspects of the use of small-angle scattering for studying the properties of water in complex objects systems are considered in publications [75–78].

3.3.4 *Compton Scattering and Backscattering*

Compton scattering is the scattering of a photon by a charged particle, usually an electron, resulting in a decrease in energy (increase in wavelength) of the an X-ray photon. Part of the energy of the photon is transferred to the recoiling electron (Fig. 3.37) [80–85].

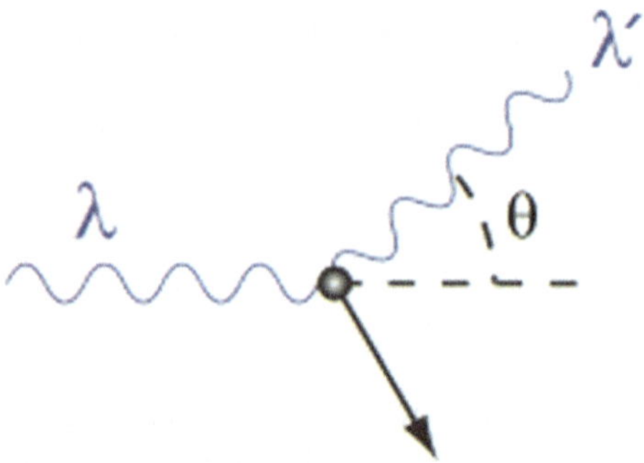

Fig. 3.37 Incident ray of wavelength λ scattered by free electron (black sphere) that receives momentum (black arrow); $\Delta\lambda = \lambda' - \lambda > 0$. https://www.tau.ac.il/~tsirel/dump/Static/knowino.org/wiki/Compton_effect.html

A method for removing the inelastic scattering from neutron data was introduced, and the effect of Compton scattering on X-ray data was discussed [83, 84]. Unlike in a crystal, the atoms and molecules in a liquid are in a state of rapid and diffusive motion. A local density of atoms and molecules which fluctuates rapidly with time was given as,

$$n(\mathbf{r}, t) = \sum_i \delta(\mathbf{r} - \mathbf{R}_i(t)), \tag{3.11}$$

where $\mathbf{R}i(t)$ is the position of atom i at time t [83]. Figure 3.38 illustrates the relative strengths of the single atom and Compton scattering for the X-ray experiment on ambient water.

The electron momentum density of water can be measurable in an inelastic X-ray scattering experiment in the form of the Compton profile which is sensitive to the precise intra- and intermolecular geometries [81]. A detailed quantitative computational study of the sensitivity of Compton scattering to the different terms arising from the intramolecular structure, nearest neighbor interactions (hydrogen bonds), and many-body properties (cooperativity and anticooperativity effects) was performed. The computation indicated that the Compton profile is predominantly sensitive to the internal O–H bond length and the hydrogen-bond geometry, being less affected by the internal H–O–H angle and the many-body properties. Figure 3.39 illustrates typical Compton profile in a water pentamer.

In another example, data on electron momentum density in ice were used as a tool to quantify order–disorder transitions by comparing Compton profiles differences of ice VI, VII, VIII, and XII with respect to ice Ih (Fig. 3.41) [82]. The determination of the fraction of electron directly involved in phase transitions as well as the use of position space signatures for quantifying oxygen site disorder were reported. These measurements allowed to calculate the fraction of electrons directly involved in the different ice phase transitions. The Compton profile within the impulse approximation **10** is given by

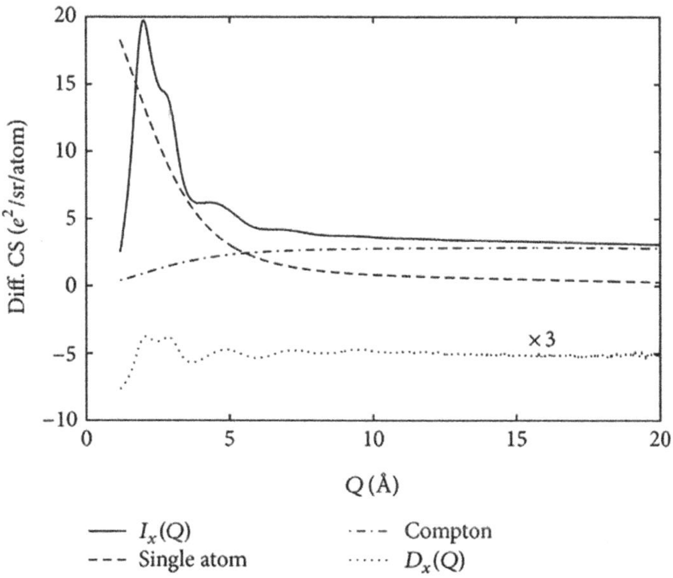

Fig. 3.38 Illustration of the relative contributions of single atom scattering and Compton scattering. The X ray intensity data (solid line) in the units of electrons per steradian per atom are from the experiment described in [85]. The single atom scattering and the Compton scattering in the same units are shown as long dashed and dot-dashed lines, respectively. The interference scattering, $D_x(Q)$, using normalization scheme II is shown as the dotted line, multiplied by a factor of 3 and shifted downward for clarity. The intramolecular interference scattering has not been subtracted from these data [83]. With permission from Hindawi

$$\begin{aligned} J(q, \mathbf{e}) &= \int n(\mathbf{p})(\mathbf{p} \cdot \mathbf{e} - q)\mathrm{d}\mathbf{p} \\ &= \sum_{\substack{\text{occupied} \\ \text{states}}} \int \chi^*(\mathbf{p})\chi(\mathbf{p})\delta(\mathbf{p} \cdot \mathbf{e} - q)\mathrm{d}\mathbf{p}, \end{aligned} \tag{3.12}$$

where **e** is the unit vector along the scattering vector K, $n(p)$ is the electron momentum density, q a scalar momentum variable, $\chi(\mathbf{p})$ is the electron wave function in momentum space, i.e., the Fourier transform of the wave function in real space (Fig. 3.40).

Backscattering technique is based on the scattering diffuse reflection of waves, particles, or signals back to the direction from which they came (Fig. 3.41) [86–89]. Coherent backscattering is observed when coherent radiation (such as a laser beam) propagates through a medium which has a large number of scattering centers (such as milk) of size comparable to the wavelength of the radiation.

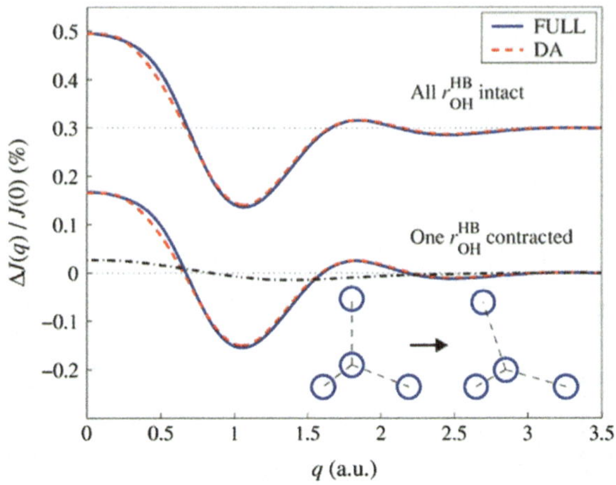

Fig. 3.39 Difference Compton profile between a water pentamer with distorted hydrogen bonds (see text) and the symmetric reference pentamer ($r_{OH} = 0.97$ Å, $R_{OO} = 2.75$ Å, $b = 0°$ for all hydrogen bonds). Top: all $r_{OH} = 0.97$ Å in the distorted pentamer. Bottom: same as top, except one O–H bond of the central molecule is contracted ($r_{OH} = 0.9572$ Å) in the distorted pentamer. Solid line: full calculation, dashed line: DA, dash-dotted line: difference between the top and the bottom cases for the full calculation. A schematic pentamer geometry is also shown (view from the side). The curves at the top are offset vertically for clarity [81]. With permission from American Physical Society

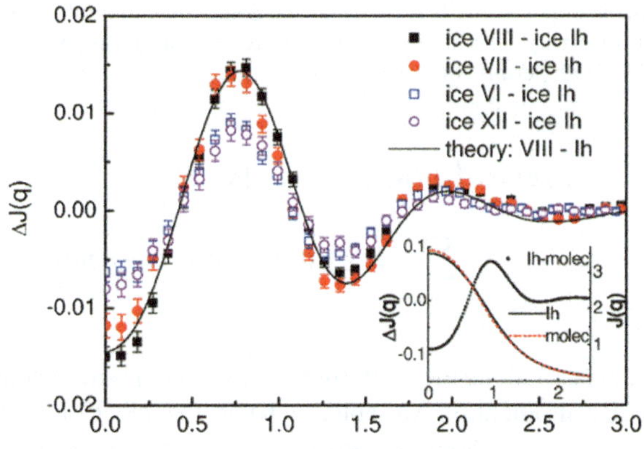

Fig. 3.40 Experimental Compton profile (CP) differences between ice VIII, ice VII, ice VI, ice XII, and ice Ih taken as reference. The theoretical CP difference between ice VIII and ice Ih is also shown. In the inset, calculated isolated water molecule and ice Ih CPs are shown, together with their difference. The water molecule CP is narrower than the ice Ih CP showing a higher localization of electronic density in momentum space, i.e., higher delocalization in position space due to the lack of the repulsive antibonding interaction [82]. With permission from American Physical Society

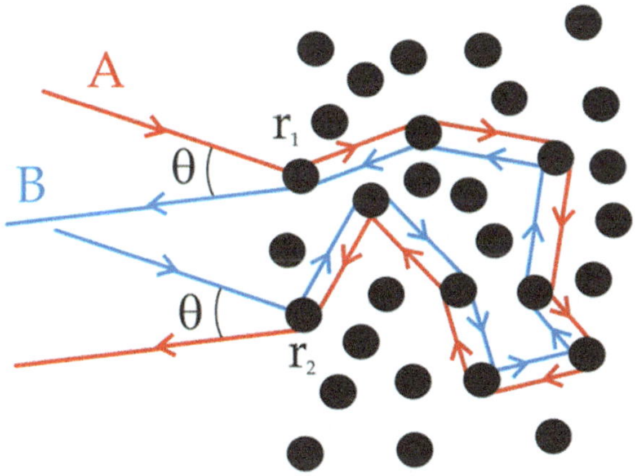

Fig. 3.41 Propagation of two rays in a random medium. Since one can be obtained from the other by time inversion, they interfere coherently when the angle θ goes to zero. https://en.wikipedia.org/wiki/Coherent_backscattering

3.3.5 X-Ray Absorption and X-Ray Raman Scattering

The four different measurement methods of X-ray absorption (XAS) [90–98] and X-ray Raman scattering (XRS) are schematically depicted in Fig. 3.42.

X-ray absorption spectroscopy (XAS) is an electronic absorption technique for which the initial state is a deeply buried core level (Figs. 3.43 and 3.44). XAS data are obtained by tuning the photon energy, to a range where core electrons can be excited (0.1–100 keV). XA spectra can provide detailed information on the local chemical and geometric environment of the target atom [90–97]. Water X-ray absorption spectrum of water has been established to high accuracy providing the perspective to understand the spectra of water including changes with temperature, mass of the water molecule, and presence of monovalent ions [90].

Studies probing the region from the absorption onset to several tens of eV above the edge are referred to as near-edge X-ray absorption fine structure (NEXAFS) [90, 91, 94]. NEXAFS experiments probe excitations from core levels into the unoccupied energy levels accessible to the target atom, producing both an excited electron and a core hole. The processes involved in excitation and de-excitation are illustrated in Fig. 3.44.

Experimentally, the water oxygen K-edge NEXAFS spectra exhibits three distinct spectral regions highlighted in Fig. 3.45: a small pre-edge feature around 535 eV and larger main edge (centered between 537 and 538 eV) and post-edge (centered between 541 and 542 eV) features [94]. It was proposed that the pre-edge feature intensity was related to the number of broken hydrogen bonds. DFT calculations indicated that broken hydrogen-bond donor sites1 1relate to asymmetry that enhances the p-character of the 4a1 state and increase the oscillator strength of transitions from the oxygen and, therefore, the pre-edge feature intensity was related to the number of broken hydrogen bonds.

Near the oxygen K-edge (~530 eV), a fine-structure pattern of liquid water micro-jets was observed when the surface-selective total ion yield (TIY) was measured.

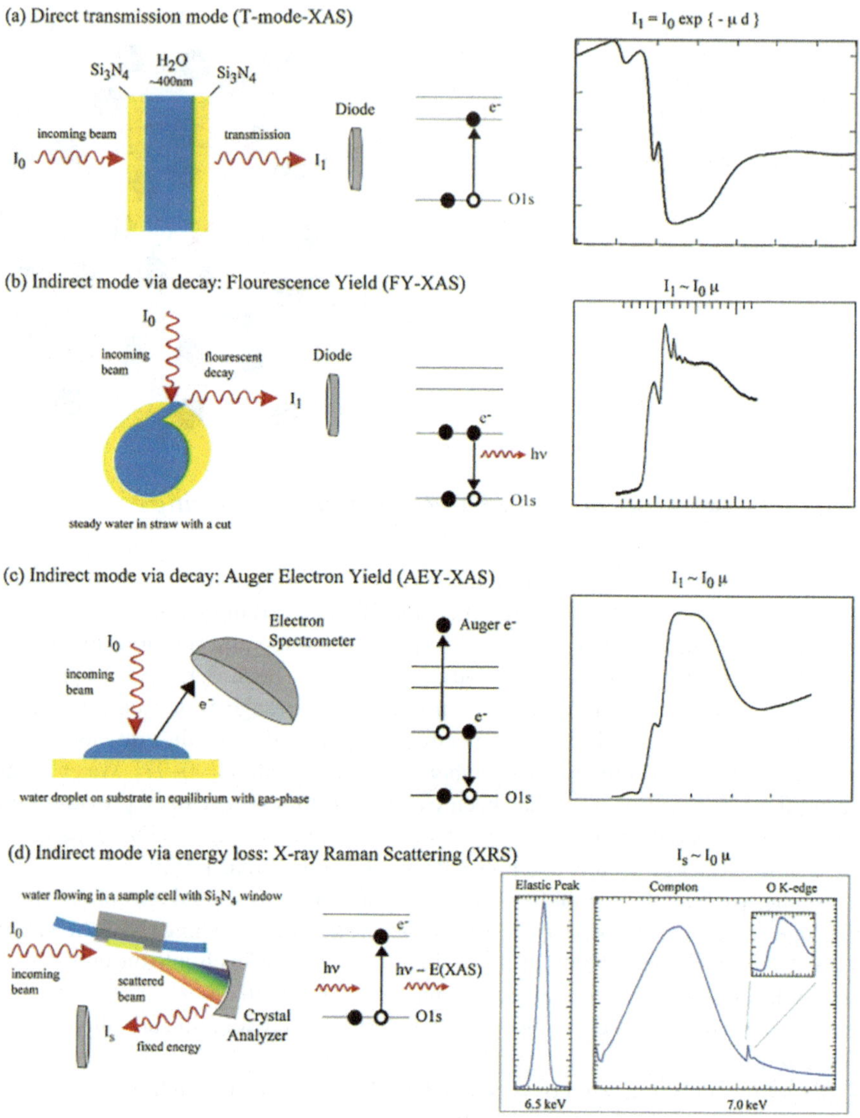

Fig. 3.42 Four different measurement methods of XAS and XRS of water shown in three columns: experimental setup, physical process, and detected spectrum of liquid water for **a** transmission, **b** fluorescence yield, **c** Auger yield, and **d** X-ray Raman as an inelastic process [90]. With permission from Elsevier

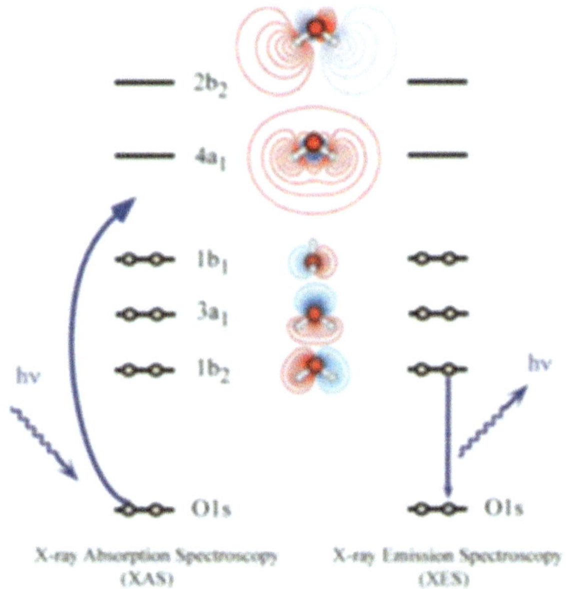

Fig. 3.43 Schematic orbital diagram of a free water molecule illustrating the principles for X-ray absorption and emission spectroscopy [90]

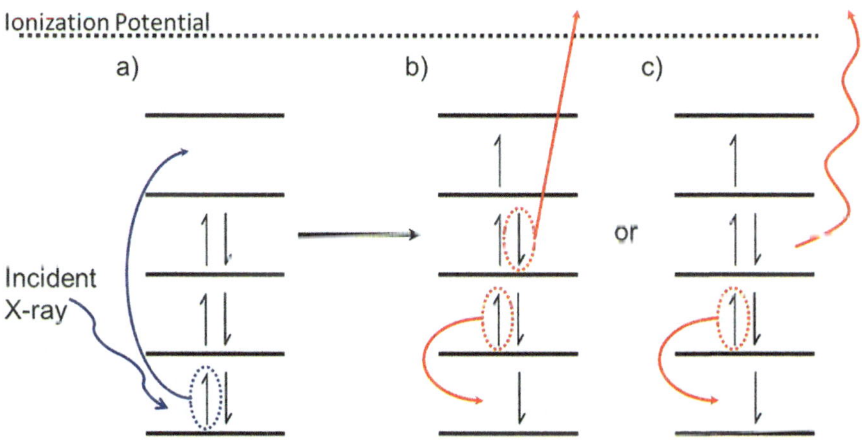

Fig. 3.44 NEXAFS-relevant electronic transitions: **a** absorption of an X-ray photon results in excitation of a core electron into an unoccupied state. **b** Non-radiative relaxation occurs via the Auger process. **c** Radiative relaxation occurs via emission of a fluorescent X-ray photon. As depicted in these energy-level diagrams, relaxation of the excited electron is slow relative to relaxation of the core hole [91]. With permission from Elsevier

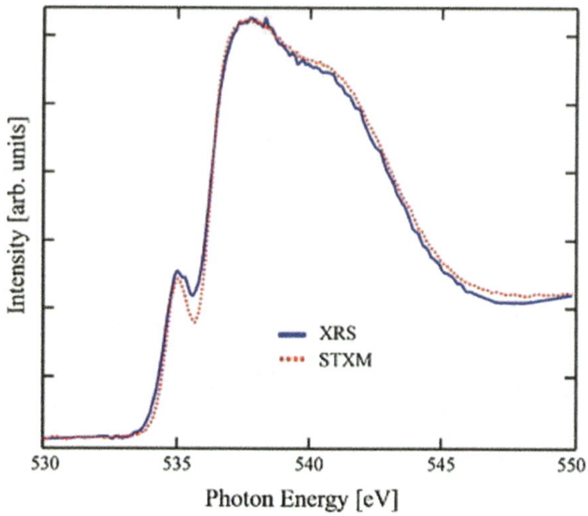

Fig. 3.45 Comparison of the oxygen K-edge NEXAFS spectra of liquid water to gas-phase water vapor and water ice indicating the spectral regions comprising the pre-edge, main edge, and post-edge of the liquid water spectrum. The dotted line in the ice spectrum represents the spectral contribution of ice Ih, with contributions above this line arising from amorphous ice and coinciding with features of the liquid spectrum. With permission from Elsevier

Similar finding was found to be similar to that for gaseous water monomers (Fig. 3.46) [95]. The results evidence that: (1) a broadened and blue-shifted spectrum emerges when detecting the bulk-sensitive total electron yield (TEY); (2) TIY EXAFS measurements produce a nearest neighbor O–O distance for surface molecules (3.00 Å) slightly longer than that of the isolated water dimer (2.98 Å); (3) the O–O distance extracted from TEY EXAFS corresponds to that accepted for bulk water (2.85 Å); and (4) an equilibrium liquid water surface dominated by water molecules interacting weakly at longer distances than in the bulk.

Another version of XAS, extended X-ray absorption fine structure (EXAFS) produces excitations into the continuum, creating photoelectrons and probing the local structural environment. EXAFS spectra appear as low-amplitude oscillations in the signal intensity in the decaying baseline observed at energies above an absorption edge. These oscillations are originated from interference between the wave functions of photoelectrons and backscatter from nearby atoms. The oscillation intensity $\chi(k)$ is dependent on the number of nearest neighbors (Nj), decreasing amplitude function ($f\ j(k)$), the interatomic separation (dRj), (φj) a phase shift function arising from the atomic potential (φj), and the Debye–Waller factor (σj), which is used to describe the attenuation of X-ray scattering or coherent neutron scattering caused by thermal motion. Utilizing a simple single-scattering formalism, the authors found a nearest neighbor O–O radius of 2.85 ± 0.05 Å.

In work [93], O K-edge X-ray absorption spectra data were obtained by tuning the photon energy to a range where core electrons was excited (0.1–100 keV). The effects of isotope substitution in liquid water were probed by X-ray absorption spectroscopy at the O K-edge as measured in transmission mode (Fig. 3.47). It was found that in this condition the D_2O spectrum blue shifted with respect to H_2O, and the D_2O spectrum is less broadened. It was suggested that the shift is related to the difference in ground-state zero-point energies between D_2O and H_2O, while the difference in broadening is related to the difference in ground-state vibrational zero-point distributions. The experimentally extracted values of blue shift and broadening were served as a test for calculations of ground-state zero-point energies and vibrational zero-point distributions in liquid H_2O and D_2O.

O K-edge X-ray absorption spectra of liquid were performed employing the scanning transmission X-ray microscopy (STXM) (Fig. 3.48).

Review [91] summarized the experimental techniques employed in XAS studies of liquid samples and computational methods used for interpretation of the resulting spectra and summarizes experiments and results obtained in the XAS investigations of liquids. X-ray Raman scattering (XRS), non-resonant inelastic scattering of X-rays from core electrons, is analogous to vibrational Raman scattering, with the difference being that the wavelengths of the exciting photons fall in the X-ray regime and the corresponding excitations are from deep core electrons [90, 98, 99]. A high-energy X-ray photon gives energy to a core electron, exciting it to an unoccupied state. In non-resonant IXS experiments, the incident X-ray photon energy is far away from any electron binding energy of the system.

As an example, the isotope effect on the local structure of liquid water at room temperature was studied by X-ray Raman spectroscopy [99]. The difference between the room-temperature spectra of liquid H_2O and D_2O was compared to the difference spectrum between liquid H_2O at 22 and 2 °C (Fig. 3.49). The experimental findings hold that: (1) the isotope effect, in addition to a temperature dependence, change of populations as a result of H-bond breakage, and/or weakening, was attributed to an increase of asymmetry mainly of species in H_2O; (2) the local asymmetry in the hydrogen-bonded network, hydrogen-bonding configurations, was found to be more asymmetric on the donor side for H_2O than for D_2O; and (3) the spectral changes arised due to the increased asymmetry were illustrate in the frame of a cluster model. Thus, quantum effects contribute to the formation of asymmetrical species in the liquid.

A compilation of procedures for planning and performing X-ray Raman scattering (XRS) experiments and analyzing data obtained from them was presented [100]. Prediction the overall shape of the spectra, estimating detection limits for dilute samples, and how to normalization the recorded spectra to absolute units, processing data from multiple-crystal XRS spectrometers with imaging capability were described in details.

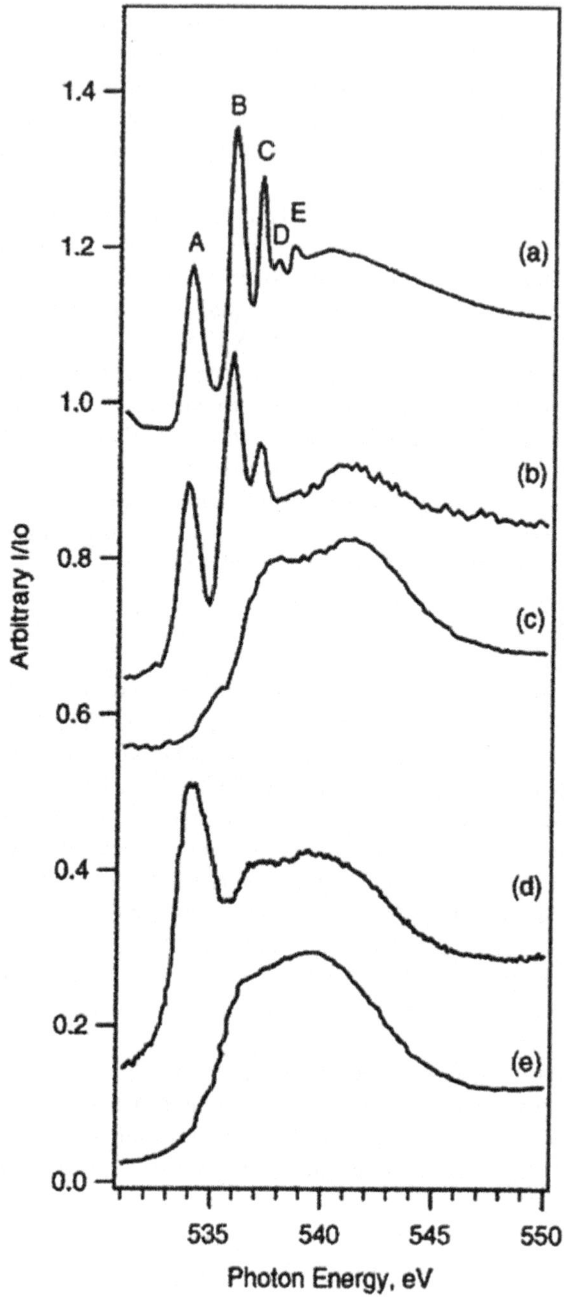

1.4
1.2
1.0
0.8
0.6
0.4
0.2
0.0
Arbitrary I/Io
A
B
C
D
E
(a)
(b)
(c)
(d)
(e)
535
540
545
550
Photon Energy, eV

◀**Fig. 3.46** NEXAFS spectra of **a** TIY gas-phase H_2O, **b** TIY, and **c** TEY from a 20 μm liquid microjet. Peaks A and B have been previously assigned to transitions from the $1a_1$ to the mixed $4a_1/3sa_1$ and $2b_2$ states, respectively. C is assigned as a transition to $3pb_1$ and $3pa_1$ symmetry Rydberg orbitals. D is assigned to the $4pb_2$ state with E consisting of transitions to many different orbitals. The peaks labeled A, B, and C in the gas-phase spectrum are clearly present in TIY of the liquid microjet and are significantly broadened and blue shifted in the TEY spectra. Conversely, peaks D and E are absent in the TIY of the microjet. Oxygen K-edge NEXAFS spectra of ice monolayers measured by **d** H+ and **e** Auger electron yield [95]. With permission from American Chemical Society

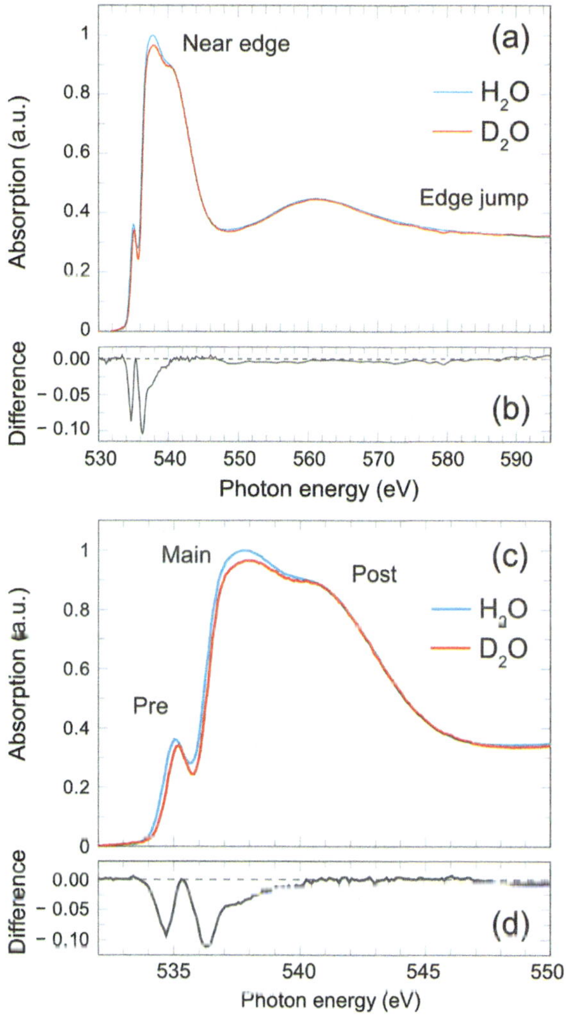

Fig. 3.47 **a** Spectra of liquid H_2O and D_2O measured in transmission mode at room temperature. The spectra are normalized to the same intensity at 580–600 eV. **b** D_2O–H_2O difference of the spectra in **a**. **c** and **d** Zooms into the near-edge regions of the spectra in **a** and **b** [93]. With permission from American Institute of Physics

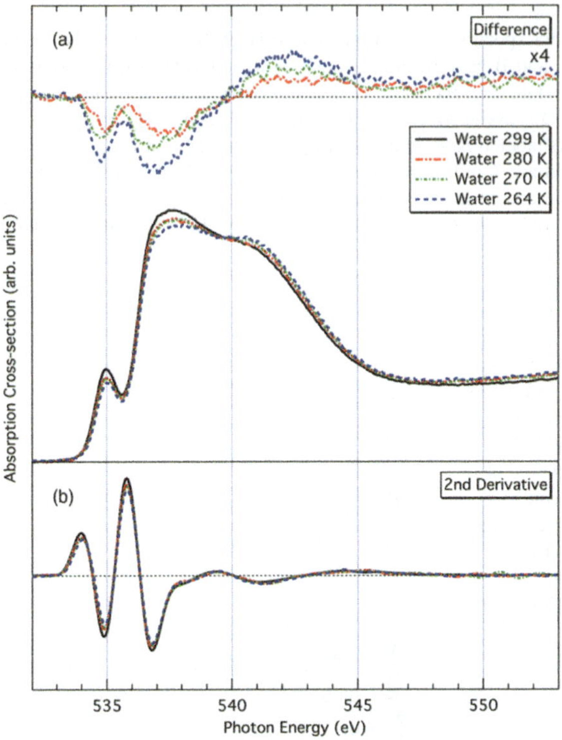

Fig. 3.48 **a** O K-edge X-ray absorption spectra of liquid water as a function of temperature (raw data). STXM transmission spectra of bulk water measured at 299 K (solid black curve), 280 K (double-dotted red curve), 270 K (single-dotted green curve), and 264 K (dashed blue curve) of 500–700 nm thick H_2O samples contained between Si_3N_4 windows. All spectra were background subtracted and area normalized between 532.5 and 550 eV. The top section shows the difference in absorption cross section compared to the STXM transmission spectrum of bulk H_2O (*l*) measured at 299 K. The difference in absorption cross section was magnified by a factor of 4 to highlight the spectral changes; the dotted black line denotes no difference. **b** Second derivatives of O K-edge X-ray absorption spectra of liquid water with identical markers and colors as in **a**. Data were smoothed as described in Sec. II C. Positive peaks in the original intensity spectra are visible as negative minima in the second derivatives of the intensity spectra. The black dashed line denotes zero in the second derivative, corresponding to inflection points in the absorption cross section. Effect of hydrophobic surfaces, such as BaF_2, was manifested through a reduction of the absorption cross section at 535 eV, which is associated with distorted hydrogen bonds. A connection, made between the observed increase in spectral intensity between 544 and 548 eV and high-symmetry points in the electronic band structure, was attributed to a hydrogen-bond network. A double-peak feature in the absorption cross section between 540 and 543 eV is identified as a characteristic of the crystalline phase. It was suggested that the increased tetrahedral ordering in the H-bonded network in liquid water upon cooling is reflected through a decrease in pre-edge absorption cross section [97]. With permission from American Institute of Physics

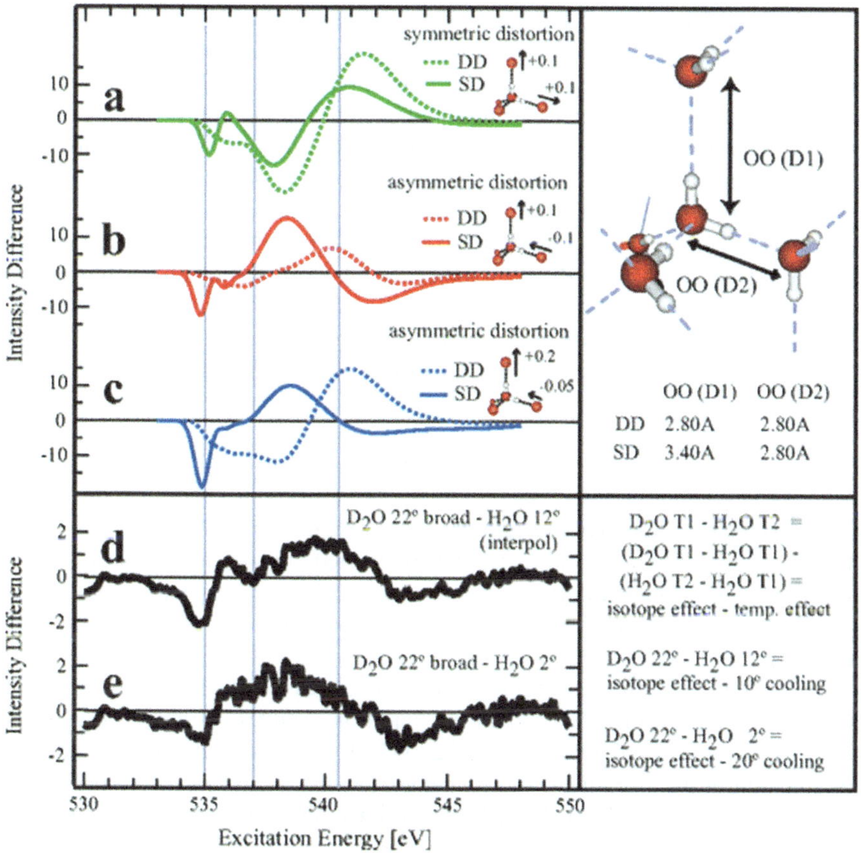

Fig. 3.49 Comparison of model calculations with measured spectra. (Top) Calculated spectra illustrate the effect of changes to the H-bond lengths starting from two different geometries; an icelike DD cluster dotted lines with O O distances of 2.80 Å and an asymmetric SD cluster solid lines generated by increasing the H-bond distance of one of the donors by 0.6 Å O–O D1. Difference spectra of the distorted minus original cluster are shown for **a** elongation of both donating H bonds by 0.1 Å and **b** elongation of the weak donor H bond O–O D1 by 0.1 Å and shortening of the strong donor H bond O–O D2 by 0.1 Å for DD species, the original D1 and D2 bonds are equivalent. **c** Similar to **b** but with a 0.2 Å elongation and 0.05 Å shortening. (Bottom) Experimental difference spectra of **d** D_2O at 22 °C minus H_2O at 12 °C linearly interpolated from the spectra at 2 and 22 °C and **e** D_2O at 22 °C minus H_2O at 2 °C, representing the isotope effect with a temperature effect subtracted [99]. With permission from American Institute of Physics

3.3.6 *Kerr Effect*

The Kerr effect is a change in the refractive index of a material in response to an applied electric field [101–113]. The time-resolved optical Kerr effect (TR-OKE) experiment is commonly used for spectroscopy of very fast molecular motions in fluids. Two special cases of the Kerr effect are considered, these being the

Kerr electro-optic effect, or DC Kerr effect, and the optical Kerr effect (OKE), or AC Kerr effect https://www.britannica.com/science/Kerr-electro-optic-effect. Kerr electro-optic effect is the inducement of double refraction of light in a transparent substance when a strong electric field is applied in a direction transverse to the beam of light. In double refraction, the index of refraction (a measure of the amount the ray is bent on entering the material, and hence the wave velocity of light vibrating in the direction of the electric field) is slightly different from the index of refraction of the vibration perpendicular to it. In the classic Kerr DC effect, an applied varying external electric field causes different indices of refraction for light polarized parallel to or perpendicular to the applied field. The Kerr effect is related to birefringence which is the optical property of a material having a refractive index that depends on the polarization and propagation direction of light. The difference in index of refraction, Δn, is given as [102]

$$\Delta n = \lambda K E^2 \tag{3.13}$$

where λ is the wavelength of the light, K is the Kerr constant, and E is the strength of the electric field. The values of K depend on the medium and are about 9.4×10^{-14} m V^{-2} for water. [This difference in index of refraction causes the material to act like as an optical device that alters the polarization state of a light wave travelling through.]

The optical Kerr effect, or AC Kerr effect, is the case in which the electric field is due to the light itself. This provides a variation in index of refraction which is proportional to the local irradiance of the light. A time-resolved optical Kerr effect investigation of the vibrational and relaxation processes in supercooled bulk water was described in papers [104–113]. The experiment measures the water intermolecular vibrations and the structural relaxation process in an extended temperature range. A mode-coupling analysis of the experimental data allows the characterization of the intermolecular vibration.

Time-resolved optical Kerr effect (OKE) spectroscopy is a nonlinear technique exploiting the transient birefringence induced by the electric field of radiation in an optically transparent and isotropic medium. In experiments performed in [108], the high-frequency (>10 cm^{-1}) part of the light-scattering spectrum (corresponding to the short time OKE signal) and the very-low-frequency part, corresponding to the long-time relaxation tail of the time-domain data, were measured. In a time-resolved optical Kerr effect measurement: (1) the oscillating electric field of a polarized short pulse of monochromatic light (the pump) induces optical birefringence in the sample and induces birefringence which reflects the relaxation and vibrational response of the molecules; (2) a second pulse (the probe), spatially superimposed within the sample with the pump pulse and characterized by a different polarization; and (3) the intensity of probe light leaking through a polarizer, placed after the sample, measures the material response as a function of the time delay between pump and probe which extended from tens of femtoseconds to hundreds of nanoseconds.

As an example, the results of an extensive investigation by means of heterodyne-detected OKE experiments (HD-OKE) of the relaxation in liquid and supercooled water were presented [104]. The authors concluded that the OKE and depolarized-light scattering techniques are mostly sensitive to the intermolecular dynamics of water. In the case of optical heterodyne detection, the delay time-dependent signal is directly proportional to the material response (relaxation) function, $R(t)$, convoluted with instrumental function, $G(t)$ (Figs. 3.50 and 3.51) [104]. The response $R(t)$ is directly connected to the time derivative of the time-dependent correlation function of the dielectric susceptibility

$$R(t) \propto \partial/\partial t\langle\chi(t)\chi(0)\rangle \tag{3.14}$$

where χ is the dielectric susceptibility. The Fourier transform of $R(t)$ corresponds to the frequency-dependent response, the observable measured in depolarized light scattering (DLS) experiments.

The imaginary part of the Fourier transform of the HD-OKE data and the calculated response were shown [104]. The experimental spectra indicate the characteristic vibrational features of liquid water: the two broad bands around 50 and 200 cm^{-1}, and the low-frequency wing. Thus, using an improved experimental setup, it was able

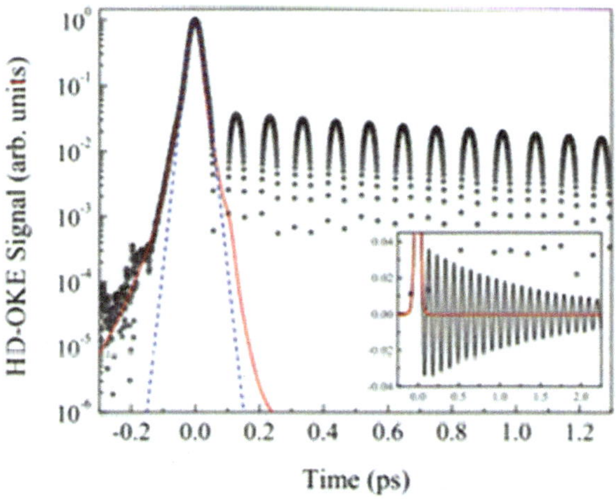

Fig. 3.50 Fourier transform of the time-dependent HD-OKE response, $\tilde{R}(\omega)$, can be directly compared to the frequency-dependent spectrum measured in the DLS experiments, SDLS(ω). The imaginary part of the Fourier transform of the HD-OKE data was detected, after that the contribution of the instrumental function has been removed by deconvolution and of SMC fits at 247–258–273–313 K. The SMC model allows to disentangle the dynamical features that generate the response function used in the fit. The contribution of the very high-frequency components ($\nu > 400$ cm^{-1}) of the response is included in the fit by means of a simple damped harmonic oscillator. This does not affect the response function in the lower frequency range, where the dynamics relevant to our investigation takes place. Thus, using an improved experimental setup, it was able to measure fast vibrational dynamics in wide range and the slow structural relaxation in a single dataset characterized by low noise and wide dynamic range [104]. https://www.researchgate.net/publication/236155023_Optical_Kerr_effect_of_liquid_water_a_new_insight_into_the_vibrational_and_structural_dynamics

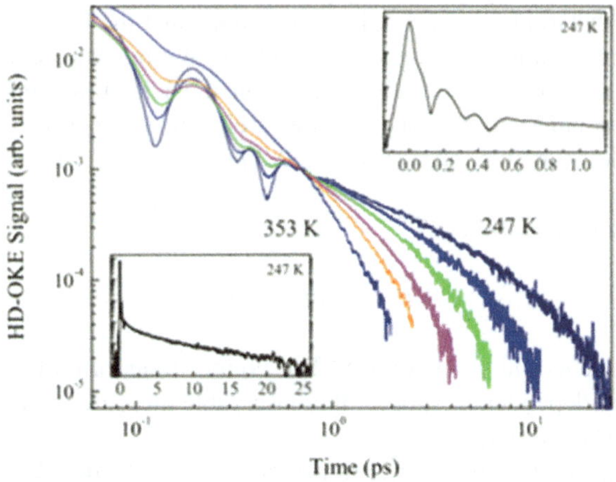

Fig. 3.51 Heterodyne-detected OKE HD-OKE signal as function of temperature in log–log plot. At short delay times (<1 ps), the HD-OKE signal shows oscillations due to vibrational dynamics that become more defined as the temperature is lowered below the melting point. At longer delay times, the signal is characterized by a monotonous decay that has a strong temperature dependence and a clear non-exponential nature. For a readable display, only 6 temperatures (247–258–273–293–313–353 K) out of the 11 measured (the previous ones plus 253–263–268–278–333 K) are reported; the intensities are normalized at 0.7 ps. The upper and lower insets show, in a log-linear plot, the 247 K HD-OKE data at short and long delay times, respectively, system is a self-mode-locked Ti:sapphire laser producing pulse of 20 fs duration with an energy of 3 nJ [104]

to measure fast vibrational dynamics in wide range and the slow structural relaxation in a single dataset characterized by low noise and wide dynamic range.

Molecular dynamics simulations of the optical Kerr effect in liquid and supercooled water were compared with time-resolved Kerr spectroscopy measurements [111]. The short time features of the Kerr response, characterized by peaks near 15, 60, and 160 fs, are weakly temperature dependent. The long-time decay was described by a stretched exponential with a nearly constant stretch parameter and relaxation times that follow a power law ~$(T - \mathrm{TS})^{-\gamma}$, with TS = 198.3 K and $\gamma = 2.35$. A nonlinear optical Kerr effect (OKE) microscope was developed and used to elucidate the ultrafast diffusive motions of intracellular water molecules on a picosecond timescale. In the OKE microscope [112], a pump-induced birefringence was sensed by a delayed probe pulse within a spatially confined volume that measures 0.5 μm in the lateral direction and 4.0 μm along the axial coordinate. This microscope allowed the recording of time-resolved Kerr signals, which reflect the ultrafast structural relaxation of the liquid, exclusively from intracellular aqueous domains [113]. Therefore, the OKE and depolarized-light scattering techniques are sensitive to the intermolecular dynamics of water.

In review [114], aspects of the ultrafast optically-heterodyne-detected optical Kerr effect (OHD-OKE) were discussed, the principles and practice of the OHD-OKE method were outlined, and the method application in recording the dynamics of several molecular liquids was described. It was demonstrated that reorientational motion, both librational and diffusive, is responsible for a significant fraction of the dynamics. The application of OHD-OKE measurements in interpreting ultrafast studies of the optical dynamics of solutions and its ability to record the dynamics of more complex, heterogeneous, media were described.

3.3.7 Ultrasound

Ultrasound is sound waves with frequencies higher than the upper audible limit of human hearing which is approximately 20 kHz. Ultrasound devices operate with frequencies from 20 kHz up to several gigahertz (Fig. 3.52) [114–125]. The ultrasound speed in neat water was found to be 1496 ms^{-1}. Ultrasonication generates alternating low-pressure and high-pressure waves in liquids, leading to the formation and violent collapse of small vacuum bubbles. The speed of sound for a material depends on properties such as compressibility, shear modulus and density for solids and pressure, density, temperature, composition and heat capacity for non-ideal gases and liquids. The propagation of sound in liquids has been widely studied in the past decades [114–125].

The ultrasonic wave is spread into the specimen and is reflected to the receiver, transforming the wave into an electric signal [125]. This signal is then sent to the amplifier and from there to the cathode ray (tube, which displays the signal as peaks proportional to the time). Specifically, the injected short-duration pulse travels at D (m) and back in a time t (μs) is

$$T = 2Dv\ (\mu s) \tag{3.15}$$

where v (m/μs) is the speed at which the pulse travels along the cable and is determined almost totally by the permittivity of the cable dielectric ε, according to the formula.

$$\varepsilon_v = 300\varepsilon\ (m/\mu s) \tag{3.16}$$

where 300 m/μs is the velocity of light in vacuo.

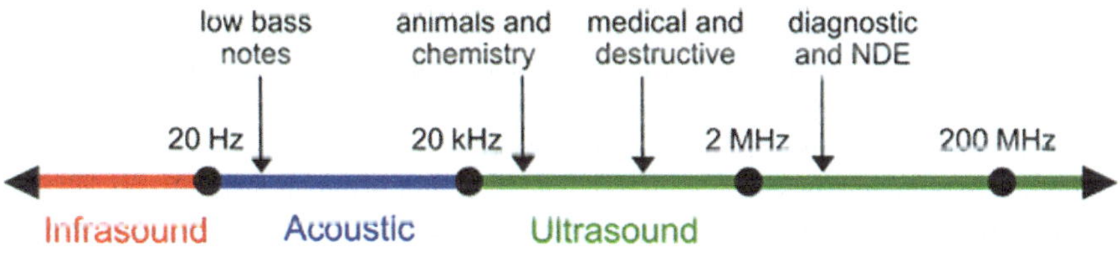

Fig. 3.52 Approximate frequency ranges corresponding to ultrasound, with rough guide of some applications. https://en.wikipedia.org/wiki/Ultrasound

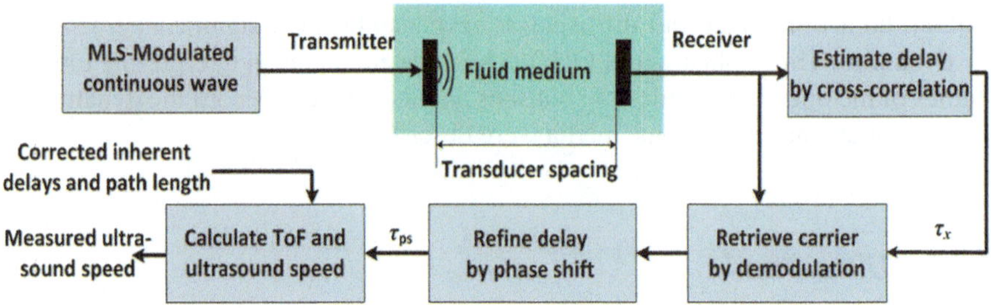

Fig. 3.53 Scheme of ultrasound speed measurement [124]. With permission from Elsevier

Paper [124] presented a method using the maximum length sequence-modulated continuous wave. The time delay between the received signal and the transmitted signal was estimated by combining the cross-correlation, parabolic interpolation, and phase shift technique. Ultrasound speed measurement using an MLS-modulated continuous wave was schematically illustrated in a Fig. 3.53.

Brillouin light scattering is mediated by the refractive index dependence on the material properties of the medium when the index of refraction of a transparent material changes under deformation (compression-distension or shear-skewing) [119, 120]. This method is a useful technique to measure ultrasonic velocity in liquid states in the frequency range from several hundred MHz to GHz. The light scattered by laser-induced acoustic phonons was detected as a heterodyne signal in the lock-in detection. A heterodyne is a signal frequency that is created by combining or mixing two other frequencies using a signal processing technique. In a phase-coherent light scattering method called coherent Brillouin scattering (COBS), measurement, complex resonance spectra of laser-induced acoustic phonons were measured.

In work [121], Brillouin light spectra in water were experimentally investigated by a scanning Fabry–Perot interferometer which is an optical cavity made from two parallel reflecting surfaces (i.e., thin mirrors). Optical waves can pass through the optical cavity only when they are in resonance with it. The optical properties of the scattering media were modulated by a plane ultrasound wave. The dependence of the Brillouin signal amplitude on the ultrasound frequency, amplitude and the position of the ultrasound field based to the light beam was investigated experimentally and showed an enhancement up to 10 times of the Brillouin signal by modulation with the ultrasound wave.

Time-domain Brillouin scattering is an all-optical experimental technique based on ultrafast lasers applied for generation and detection of coherent acoustic pulses on time durations of picoseconds and length scales of nanometers [119]. The transient optical reflectivity of the sample was recorded by the probe beam as the acoustic nanopulse propagates in space contains information on the acoustical, optical, and acousto-optical parameters of the material under study. The experimental method is based on a heterodyning where weak light pulses scattered by the coherent acoustic phonons interfere at the photodetector with probe light pulses of significantly

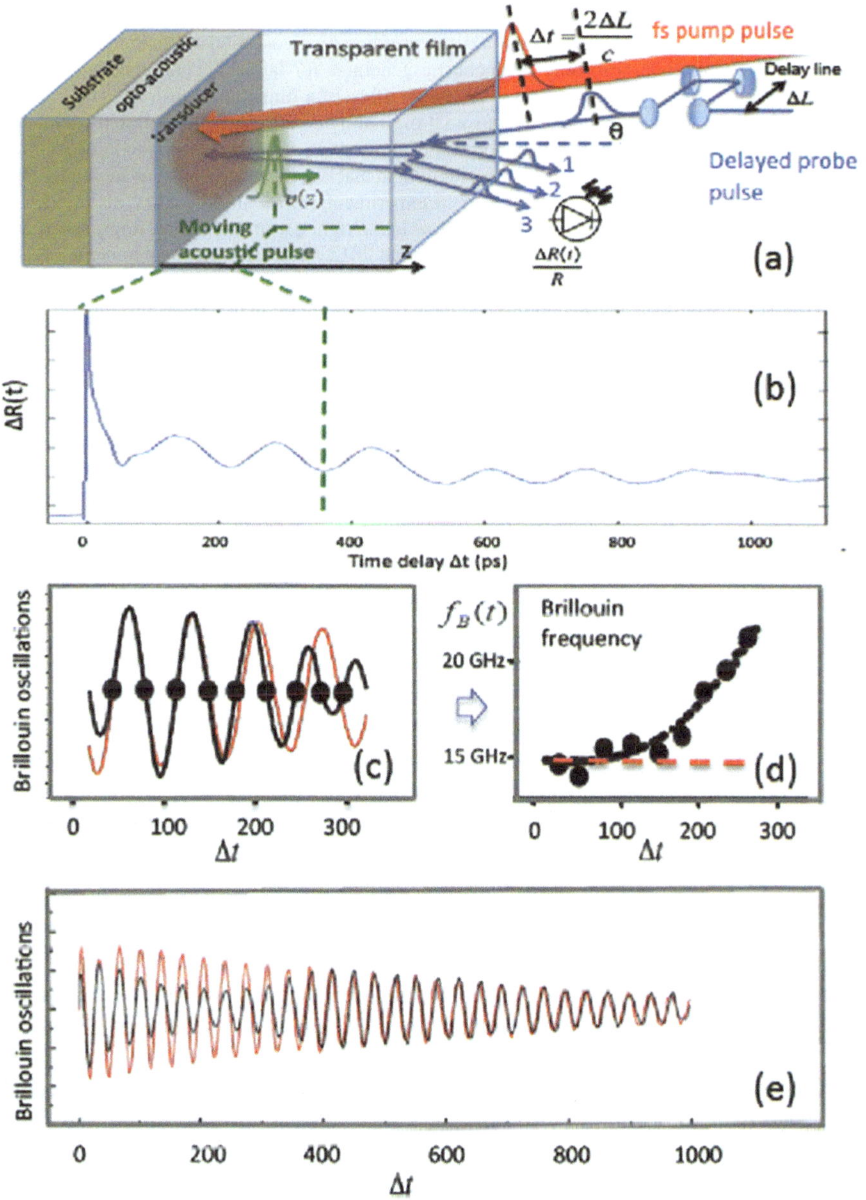

Substrate
opto-acoustic transducer
Transparent film
fs pump pulse
Delay line
Delayed probe pulse
Moving acoustic pulse
(a)
(b)
ΔR(t)
Time delay Δt (ps)
Brillouin oscillations
(c)
Brillouin frequency
20 GHz
15 GHz
(d)
(e)

◂**Fig. 3.54** Principles of opto-acousto-optic depth profiling by picosecond acoustic interferometry. **a** A thin film pumplight-absorbing optoacoustic generator for launching the coherent acoustic pulses into transparent samples is located on the interface of a film and a substrate. The transient optical reflectivity signal **b** contains a sinusoidal oscillating component, the Brillouin oscillation. The Brillouin frequency in an inhomogeneous film can change with time 28. In the left part of **c** a comparison of the experimental Brillouin oscillation, in black, with model constant-frequency Brillouin oscillation, in red, reveals an increase in the experimental Brillouin frequency, also evidenced by the Fourier transform in a moving time window in the right part of the figure. Reproduced with permission from ACS Nano 6, 1410 (2012). Copyright 2012 American Chemical Society. In **d** the Brillouin oscillation amplitude in an inhomogeneous medium, in black, exhibits variations in addition to those caused by the probe photons and the CAPs absorption in homogeneous media, shown in red 42, while the Brillouin frequency is fairly constant [119]. With permission from American Institute of Physics. Reproduced from Appl. Phys. Lett. 94, 111910 (2009), with the permission of AIP Publishing [119]. With permission from American Institute of Physics

higher amplitude reflected from various interfaces of the sample. Principles of opto-acousto-optic depth profiling by picosecond acoustic interferometry were outlined in Fig. 3.54. The applications of time-domain Brillouin scattering for imaging various materials including vegetable and animal cells were reviewed.

Two-dimensional stimulated Brillouin scattering microscopy was demonstrated for the first time using low power continuous-wave lasers tunable around 780 nm [122]. The low power and optical wavelengths in the water transparency window were used in this setup provide a bioimaging technique for probing the mechanical properties of hard and soft tissues. To perform 2D imaging, stimulated Brillouin scattering (SBS) was applied. Brillouin scattering originates from a non-elastic light interaction with acoustic waves in a medium that are generated by thermodynamic fluctuations. The scattered light is shifted in frequency according to the relation

$$\Omega_B = 2n\omega\frac{v}{c}\sin(\theta/2), \tag{3.17}$$

where n is the index of refraction, ω is the frequency of the incident light, v is the speed of sound in the material, c is the speed of light in vacuum, and θ is the angle between the incoming and backscattered light. A conceptual picture of stimulated Brillouin for counterpropagating pump and probe beams is shown in Fig. 3.55.

At the propagation of ultrashort sound pulses in water, studying by the picosecond ultrasonic technique sound pulse was generated when light is absorbed in a metal transducer film deposited onto a substrate, propagates across a thin layer of water, and then reflected back to the surface at which it was generated [118]. In the measurement, the two orthogonally polarized beams pass through the glass slide and were focused onto the same spot of about 20 μm in diameter on the 5 nm Al film. The pump beam is modulated at 1 MHz by an electro-optic modulator. Because of the acoustic attenuation in water, a sinusoidal sound wave decreases in amplitude as it propagates through water and its waveform as a function of time and position is given by

$$\exp[i2\pi f(t - z/c)]\exp(-\alpha z), \tag{3.18}$$

where f is the frequency of the sound wave, c the sound velocity in water, z the propagation distance, and α the attenuation coefficient. Experiments performed in the MHz frequency range and Brillouin frequency range indicated that α varies with the square of the frequency

$$\alpha/f^2 = \times 10^{-15}\,\mathrm{s^2/m} \text{ at } 30\,°\mathrm{C}. \tag{3.19}$$

The propagation distances of the acoustic pulses of 1.4, 2.8, 4, 5.6 and 6.8 μm were estimated from the echo arrival time and the measured low-frequency sound velocity. The width of the echo increases with the propagation distance, indicating that the sound pulse is broadened as it propagates through the water due to the acoustic attenuation.

Measurements of the sound velocity as a function of temperature in water employing the picosecond acoustic interferometry technique were carry out. This method can be used to make velocity measurements on very small sample volumes [117]. The speed of sound in high-purity water were measured in the temperature range 253–473 K at pressures up to 400 MPa utilizing a double-path pulse-echo method with a single 5-MHz ultrasound transducer placed between two unequally spaced reflectors [123]. The density and isobaric specific heat capacity of water were obtained in this condition by thermodynamic integration of the sound-speed data subject to initial values computed from IAPWS-95 on the isobar at $p = 0.1$ MPa.

The aim of work [116] was to provide a complete set of accurate speed of sound results in covering the pressure range. Speed of sound was determined by **the** multiple reflections method between 0.1 and 700 MPa at temperatures from 253.15 to 348.15 K. The ultrasound cell was designed to fit with the characteristics of laboratory high-pressure equipment. The different elements of the cell are shown in Fig. 3.56.

In study [126], structures and behaviors of acoustic cavitation bubbles induced by a high-intensity focused ultrasound (HIFU) transducer, operating at its resonance

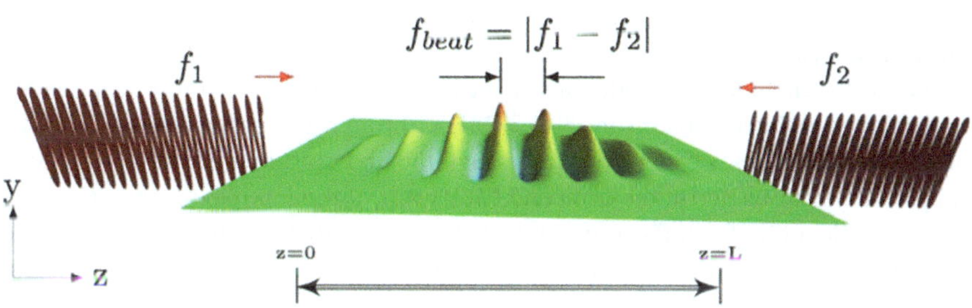

Fig. 3.55 Conceptual picture illustrating stimulated Brillouin scattering. The two fields are incident from the left and right at frequencies f_1 and f_2 respectively. In the center, the waves in the medium represent the density variations (acoustic waves). L is the total interaction length in the sample [122]

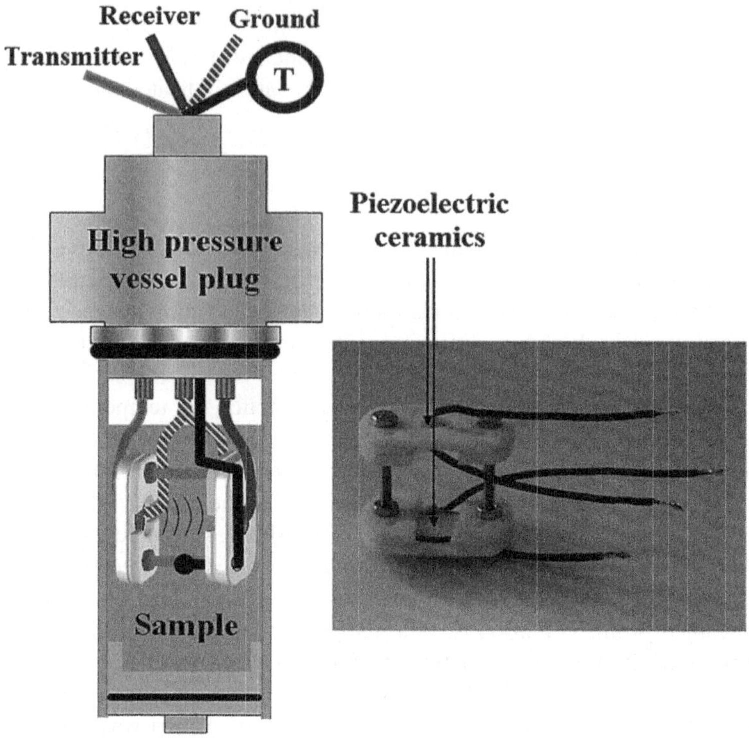

Fig. 3.56 Ultrasound measurement cell [115]. With permission from American Chemical Society

frequency of 250 kHz, were experimentally explored with corresponding observations captured by a high-speed video camera system. Study [127] provided measurements for improving the high-pressure and low-temperature representation of liquid water. Sound speeds of pure liquid water were reported between 0.1 and 700 MPa, from 353 K down to the melting curves of the ice phases. Measurement of travel times of acoustic waves within the sample chamber of known length to give the sound speeds included the following stages: (1) the ultrasonic transducer sends a frequency-modulated pulse from 3 to 7 MHz, through the buffer rod and into the sample chamber; (2) waves traveling through the end closure are partially reflected and partially transmitted at the front and back walls of the sample; (3) the time delay between reflections from these two interfaces corresponds to the round-trip travel time within the sample cavity; (4) repeated reflections in the sample chamber result in a series of echoes of decreasing intensity, all multiples of the sample travel time; (5) the reflections are collected by the transducer and recorded using a digital oscilloscope; (6) the collected waveform can be deconvolved using the source signal as the reference to obtain tightly time-constrained reflections; and (7) the travel time is determined as the average of the series of individual sample reflections. Analyses of measurements are accomplished with a flexible computational thermodynamic framework.

References

1. F. Perakis, L. De Marco, A. Shalit, F. Tang, Z.R. Kann, T.D. Kühne, R. Torre, M. Bonn, Y. Nagata, Vibrational spectroscopy and dynamics of water. Chem. Rev. **116**, 7590–7607 (2016)
2. S. Mondal, S. Mukherjee, B. Bagchi, Protein hydration dynamics: much ado about nothing? J. Phys. Chem. Lett. **8**, 4878–4882 (2017)
3. D. Laage, T. Elsaesser, J.T. Hynes, Water dynamics in the hydration shells of biomolecules. Chem. Rev. **117**, 10694–10725 (2017). X.J. Jordanides, M.J. Lang, X. Song, G.R. Fleming, Solvation dynamics in protein environments studied by spectroscopy. J. Phys. Chem. B **103**, 7995–8005 (1999)
4. H.-S. Tan, I.R. Piletic, M.D. Fayer, Polarization selective spectroscopy experiments: methodology and pitfalls. J. Opt. Soc. Am. B **22**, 2009–2017 (2005). S.K. Pal, J. Peon, A.H. Zewail, Biological water at the protein surface: dynamical solvation probed directly with femtosecond resolution. Proc. Natl. Acad. Sci. U.S.A. **99**, 1763–1768 (2002)
5. D.C. Elton, M. Fernandez-Serra, The hydrogen-bond network of water supports propagating optical phonon-like modes. Nat. Commun. **7**, 10193 (2015). https://doi.org/10.1038/ncomms10193. J.T. King, E.J. Arthur, C.L. Brooks III, K.J. Kubarych, Site specific hydration dynamics of globular proteins and the role of constrained water in solvent exchange with amphiphilic cosolvents. J. Phys. Chem. B **116**, 5604–5611 (2012)
6. S. Mondal, S. Mukherjee, B. Bagchi, On the origin of multiple time scales in the relaxation of protein hydration layer: a molecular dynamics simulation study. arXiv preprint arXiv:1708.01210 (2017)
7. V.I. Mikla, V.V. Mikla, *Raman Spectroscopy in Medicine in Medical Imaging Technology* (Elsevier, 2014), pp. 129–141
8. Y. Maréchal, The molecular structure of liquid water delivered by absorption spectroscopy in the whole IR region completed with thermodynamics data. J. Mol. Struct. **12**, 146–155 (2011)
9. P. Hamm, M.T. Zanni, *Concepts and Methods of 2D Infrared Spectroscopy* (Cambridge University Press, Cambridge, New York, 2011)
10. Q. Sun, Local statistical interpretation for water structure. Chem. Phys. Lett. **568–569**, 90–94 (2013)
11. T.F. Kahan, J.P. Reid, D.J. Donaldson, The first uses glancing-angle Raman spectroscopy to study the extent of hydrogen bonding in the QLL via the shape of the water OH-stretching band. J. Phys. Chem. A **111**, 11006–11012 (2007)
12. Z.F. Brotzakis, C.C.M. Groot, W.H. Brandeburgo, H.J. Bakker, P.G. Bolhuis, Dynamics of hydration water around native and misfolded α-lactalbumin. J. Phys. Chem. B **120**, 4756–4766 (2016)
13. C.Q. Sun, Unprecedented O:⇔:O compression and H↔H fragilization in Lewis solutions. Phys. Chem. Chem. Phys. **21**, 2234–2250 (2019)
14. S.K. Pal, A.H. Zewail, Dynamics of water in biological recognition. Chem. Rev. **104**, 2099–2124 (2004)
15. J. Zheng, K. Kwak, M.D. Fayer, Ultrafast 2D IR vibrational echo spectroscopy. Acc. Chem. Res. **40**, 75–83 (2007)
16. A. Ghosh, J.S. Ostrander, M.T. Zanni, Watching proteins wiggle: mapping structures with two-dimensional infrared spectroscopy. Chem. Rev. **117**, 10726–10759 (2017)
17. D.G. Hogle, A.R. Cunningham, M. Tucker, Equilibrium versus nonequilibrium peptide dynamics: insights into transient 2D IR spectroscopy. J. Phys. Chem. B **122**, 8783–8795 (2018)
18. S. Park, D.E. Moilanen, M.D. Fayer, Water dynamics: the effects of ions and nanoconfinement. J. Phys. Chem. B **112**, 5279–5290 (2008)

19. S. Park, K. Kwak, M.D. Fayer, Femtosecond mid-infrared system. The laser system employed in the experiments has been described in detail elsewhere. Laser Phys. Lett. **4**, 704 (2007)
20. A.G. Lambert, P.B. Davies, J. Neivandt, Implementing the theory of sum frequency generation vibrational spectroscopy: a tutorial review. Appl. Spectrosc. Rev. **40**, 103–145 (2005)
21. C.D. Bain, Sum-frequency vibrational spectroscopy of the solid–liquid interface. J. Chem. Soc. Faraday Trans. **91**, 1281–1296 (1995)
22. Y.R. Shen, Surface properties probed by second-harmonic and sum-frequency generation. Nature **337**, 519–525 (1989)
23. Y.R. Shen, Basic theory of surface sum-frequency generation. J. Phys. Chem. C **116**, 15505–15509 (2012)
24. T. Fujie, T. Ohto, S. Sun, J.R. Rouxel, S. Imoto, E.H.G. Backus, S. Mukamel, M. Bonn, Y. Nagata, Molecular structure and modeling of water–air and ice–air interfaces monitored by sum-frequency generation. Chem. Rev. **120**(8), 3633–3667 (2020)
25. R. Shen, *Fundamentals of Sum-Frequency Spectroscopy* (Cambridge University Press, 2016)
26. D. Grischkowsky, S. Keiding, M. van Exter, C. Fattinger, Far-infrared time-domain spectroscopy with terahertz beams of dielectrics and semiconductors. J. Opt. Soc. Am. B **7**, 2006–2015 (1990)
27. Y.X. Zhang, Z.S. Ma, Y.C. Zhou, W.T. Zheng, J. Zhou, C.Q. Sun, Hydrogen-bond relaxation dynamics: resolving mysteries of water ice. Coord. Chem. Rev. **285**, 109–165 (2015)
28. Y.X. Zhang, Y. Xu, Y. Zhou, Y. Gong, Y. Huang, C.Q. Sun, HCl, KCl and KOH solvation resolved solute-solvent interactions and solution surface stress. Appl. Surf. Sci. **422**, 475–481 (2017)
29. C.Q. Sun, X. Zhang, J. Zhou, Y. Huang, Y. Zhou, W. Zheng, Density, elasticity, and stability anomalies of water molecules with fewer than four neighbors. J. Phys. Chem. Lett. **4**, 2565–2570 (2013)
30. R. Aswani, J.C. Li, A new approach to pairwise potentials for water–water interactions. J. Mol. Liq. **134**, 120–128 (2007)
31. F.C. Waldermann, B.J. Sussman, J. Nunn, V.O. Lorenz, K.C. Lee, K. Surmacz, U.-H. Lee, D. Jaksch, I.A. Walmsley, P.G. Spizzirri, P. Olivero, S. Prwer, Measuring phonon dephasing with ultrafast pulses using Raman spectral interference. Phys. Rev. B **78**, 55201 (2019)
32. Y. Huang, X. Zhang, C.Q. Sun, Hydrogen-bond relaxation dynamics: resolving mysteries of water ice. Coord. Chem. Rev. **285**, 109–165 (2015)
33. J.E. Haddad, B. Bousquet, L. Canioni, P. Mounaix, Review in terahertz spectral analysis. Trends Anal. Chem. **44**, 98–105 (2013). https://doi.org/10.1016/j.trac.2012.11.009
34. S.L. Dexheimer, *Terahertz Spectroscopy: Principles and Application* (Taylor Francis Group, New York, USA, 2008)
35. P.U. Jepsen, D.G. Cooke, M. Koch, Terahertz spectroscopy and imaging—modern techniques and applications. Laser Photon. Rev. **5**, 124 (2011)
36. E.P.J. Parrott, Y. Sun, E. Pickwell-Macpherson, Structural applications of terahertz spectroscopy. J. Mol. Struct. **1006**, 66–76 (2011)
37. J.B. Baxter, G.W. Guglietta, Terahertz spectroscopy. Anal. Chem. **83**, 4342–4368 (2011)
38. P.B. Popov, A. Ishai, A. Khamzin, Y. Feldman, The mechanism of the dielectric relaxation in water. Phys. Chem. Chem. Phys. **18**, 13941 (2016). J.T. Kindt, C.A. Schmuttenmaer, Far-infrared dielectric properties of polar liquids probed by femtosecond terahertz pulse spectroscopy. J. Phys. Chem. **100**, 10373 (1996)
39. N.J.J. van Hoof, S.E.T. ter Huurne, J. Gómez Rivas, A. Halpin, Time-resolved terahertz time-domain near-field microscopy. Opt. Express **26**, 32118–32129 (2018)
40. D. Mittleman, *Sensing with Terahertz Radiation* (Springer, 2013)
41. D.C. Elton, The origin of the Debye relaxation in liquid water and fitting the high frequency excess response. Phys. Chem. Chem. Phys. **19**, 1873 (2017)

42. J. Neu, C.A. Schmuttenmaer, Tutorial: an introduction to terahertz time domain spectroscopy (THz-TDS). J. Appl. Phys. **124**, 231101 (2018)
43. U. Moller, D.G. Cooke, R. Tanaka, P.U. Jepsen, Terahertz reflection spectroscopy of Debye relaxation in polar liquids. J. Opt. Soc. Am. B Opt. Phys. **26**, A113 (2009)
44. P. Debye, *Polar Molecules* (Chemical Catalog, New York, 1929)
45. G.M. Hale, M.R. Querry, Optical constants of water in the 200-nm to 200-micrometer wavelength region. Appl. Opt. **12**, 555–563 (1973)
46. U. Kaatze, Y. Feldman, Broadband dielectric spectrometry of liquids and biosystems. Meas. Sci. Technol. **17**, R17–R35 (2006)
47. F. Kremer, A. Schonhals, W. Luck, *Broadband Dielectric Spectroscopy* (Springer-Verlag, 2002)
48. M.L.T. Asaki, A. Redondo, T.A. Zawodzinski, A.J. Taylor, Dielectric relaxation of electrolyte solutions using terahertz transmission. J. Chem. Phys. **116**, 8469 (2002)
49. C. Rønne, S.R. Keiding, Low frequency spectroscopy of liquid water using THz-time domain spectroscopy. J. Mol. Liq. **101**, 199–218 (2002)
50. L. Fumagalli, A. Esfandiar, R. Fabregas, S. Hu, P. Ares, A. Janardanan, Q. Yang, B. Radha, T. Taniguchi, K. Watanabe, G. Gomila, K.S. Novoselov, A.K. Geim, Anomalously low dielectric constant of confined water. Science **360**, 1339–1342
51. I.L. Smolsky, P. Liu, M. Niebuhr, K. Ito, T.M. Weiss, H. Tsuruta, Biological small angle X-ray scattering facility at Stanford Synchrotron Radiation Laboratory. J. Appl. Cryst. **40**, 453–458 (2007)
52. P.C.H. Mitchell, S.F. Parker, A.J. Ramirez-Cuesta, J. Tomkinson, *Vibrational Spectroscopy with Neutrons With Applications in Chemistry, Biology, Materials Science and Catalysis* (World Scientific Publishing Co. Pte. Ltd., Singapore, 2005)
53. S.F. Parker, D. Lennon, P.W. Albers, Vibrational spectroscopy with neutrons: a review of new directions. Appl. Spectrosc. **65**, 1325 (2011)
54. C.M. Brown, A.J. Ramirez-Cuesta, M.R. Johnson, V. García-Sakai, Chemical spectroscopy using neutrons. Chem. Phys. **427**, 1 (2013)
55. L. Shi, J.L. Skinner, Mixed quantum/classical approach to OH-stretch inelastic incoherent neutron scattering spectroscopy for ambient and supercooled liquid water and ice Ih. J. Chem. Phys. **143**, 014503/1–014503/6 (2015)
56. Q. Berrod, K. Lagrené, J. Ollivier, J.-M. Zanotti, Inelastic and quasi-elastic neutron scattering. Application to soft-matter. EPJ Web Conf. **188**, 05001 (2018)
57. D.I. Svergun, M.H.J. Koch, Small-angle scattering studies of biological macromolecules in solution. Rep. Prog. Phys. **66**, 1735 (2003)
58. A. Furrer, J. Mesot, T. Strässle, *Neutron Scattering in Condensed Matter Physics* (World Scientific, 2009)
59. S.W. Lovesey, *Theory of Neutron Scattering from Condensed Matter; Volume 2: Condensed Matter* (Clarendon Press, Oxford, 1984)
60. G.L. Squires, *Introduction to the Theory of Thermal Neutron Scattering*, 2nd edn. (Dover Publications, New York, 1996)
61. J.M. Zanotti, Inelastic and quasi-elastic neutron scattering. Application to soft-matter. J. Phys. IV (Proc.) **130**, 87 (2005)
62. D.L. Price, K. Sköld, *Methods in Experimental Physics. Neutron Scattering* (Academic Press, Orlando, FL, 1986)
63. H.-W. Wang, M.J. Dello Stritto, N. Kumar, A.I. Kolesnikov, P.R.C. Kent, J.D. Kubicki, D.J. Wesolowski, J.O. Sofo, Vibrational density of states of strongly H-bonded interfacial water: insights from inelastic neutron scattering and theory. J. Phys. Chem. C **118**, 10805–10813 (2014)
64. J. Eckert, Computational study of inelastic neutron scattering vibrational spectra of water clusters and their relevance to hydration water in proteins. Biochem. Biophys. Acta **1861**, 3564–3572 (2017)

65. L.E. Bove, L.S. Klotz, T. Strassle, M. Koza, J. Teixeira, A.M. Saitta, Translational and rotational diffusion in water in the gigapascal range. Phys. Rev. Lett. **111**, 185901 (2013)
66. D. Vural, X. Hu, B. Lindner, N. Jain, Y. Miao, X. Cheng, Z. Liu, L. Hong, J.C. Smith, Quasielastic neutron scattering in biology: theory and applications. Biochim. Biophys. Acta (BBA) Gen. Subj. **1861**, 3638–3650 (2017)
67. J.P. Melchior, N.H. Jalarvo, A quasielastic neutron scattering study of water diffusion in model anion exchange membranes over localized and extended volume increments. Phys. Chem. C **123**, 14195–14206 (2019)
68. T. Yamada, H. Seto, Quasi-elastic neutron scattering studies on hydration water in phospholipid membranes. Front. Chem. **24** (2020). https://doi.org/10.3389/fchem.2020.00008
69. L. Toppozini, F. Roosen-Runge, R.I. Bewley, R.M. Dalgliesh, T. Perring, T. Seydel, H.R. Glyde, V.G. Sakai, M.C. Rheinstädter, Anomalous and anisotropic nanoscale diffusion of hydration water molecules in fluid lipid membranes. Soft Matter **11**, 8534 (2015)
70. T. Yamada, N. Takahashi, T. Tominaga, S. Takata, Y. Seto, Dynamical behavior of hydration water molecules between phospholipid membranes. J. Phys. Chem. B **121**, 8322–8329 (2017)
71. J.P. Embs, F. Juranyi, R. Hempelmann, Introduction to quasielastic neutron scattering. Z. Phys. Chem. **224**, 5–32 (2010)
72. A.E. Whitten, S. Cai, J. Trewhella, mMULCh: modules for the analysis of small-angle neutron contrast variation data from biomolecular assemblies. J. Appl. Crystallogr. **41**(Part 1), 219–221 (2008)
73. E. Mahieu, F. Gabel, Biological small-angle neutron scattering: recent results and development. Acta Crystallogr. Sect. D **74**(Part 8), 715–726 (2018)
74. J.-L. Béchade, M.H. Mathon, Y. de Carlan, Neutron analyses for nuclear materials: texture, residual stresses and small angle scattering. Eur. Phys. J. Web Conf. **104** (2015). https://doi.org/10.1051/epjconf/201510401008
75. A.E. Whitten, S. Cai, J. Trewhella, MULCh: modules for the analysis of small-angle neutron contrast variation data from biomolecular assemblies. J. Appl. Crystallogr. **41**, 219–221 (2008)
76. M.V. Avdeev, E. Dubois, G. Mériguet, E. Wandersman, V.M. Garamus, A.V. Feoktystov, R. Perzynski, Small-angle neutron scattering analysis of a water-based magnetic fluid with charge stabilization: contrast variation and scattering of polarized neutrons. J. Appl. Cryst. **42**, 1009–1019 (2009)
77. E. Mahieu, F. Gabel, Biological small-angle neutron scattering: recent results and development. Acta Crystallogr. Sect. D **74**, 715–726 (2018)
78. C. Mitterdorfer, M. Bauer, T.G.A. Youngs, D.T. Bowron, C.R. Hill, H.J. Fraser, J.L. Finney, T. Loerting, Small-angle neutron scattering study of micropore collapse in amorphous solid water. Phys. Chem. Chem. Phys. **16**, 16013–16020 (2014)
79. M. Erko, D. Wallacher, A. Hoell, T. Hauß, I. Zizak, O. Paris, Density minimum of confined water at low temperatures: a combined study by small-angle scattering of X-rays and neutrons. Phys. Chem. Chem. Phys. **14**, 3852–3858 (2012)
80. A.H. Compton, A quantum theory of the scattering of X-rays by light elements. Phys. Rev. **21**, 483–502 (1923)
81. M. Hakal, K. Nygard, S. Manninen, L.G.M. Pettersson, K. Hamalainen, Intra- and intermolecular effects in the Compton profile of water. Phys. Rev. B **73**, 035432 (2006)
82. Ch. Bellin, B. Barbiellini, S. Klotz, T. Buslaps, G. Rousse, Th. Straessle, A. Shukla, Oxygen disorder in ice probed by X-ray Compton scattering. Phys. Rev. B **83**, 094117 (2011)
83. A.K. Soper, The radial distribution functions of water as derived from radiation total scattering experiments: is there anything we can say for sure? ISRN Phys. Chem. **2013**, Article ID 279463, 67 pages (2013)
84. A.K. Soper, Water: two liquids divided by a common hydrogen bond. J. Phys. Chem. B **115**, 14014–14022 (2011)
85. G. Kaplan, B. Barbiellini, A. Bansil, Compton scattering beyond the impulse approximation. Phys. Rev. B **68**, 235104 (2003)

86. G. Campbell, S.R. Phinn, Measuring the particulate backscattering of inland waters: a comparison of techniques. ISPRS Ann. Photogramm. Remote Sens. Spat. Inf. Sci. **I-7** (2012), in *2012 XXII ISPRS Congress*, 25 Aug–01 Sept 2012, Melbourne, Australia
87. J.M. Sullivan, M. Twardowski, J.R.V. Zaneveld, C.C. Moore, Measuring optical backscattering in water (2013). https://doi.org/10.1007/978-3-642-21907-86
88. M.S. Twardowski, H. Claustre, S.A. Freeman, D. Stramski, Y. Huot, Optical backscattering properties of the "clearest" natural waters. Biogeosciences **4**, 1041–1058 (2007)
89. J. Lin, Z. Lee, M. Ondrusek, X. Liu, Hyperspectral absorption and backscattering coefficients of bulk water retrieved from a combination of remote-sensing reflectance and attenuation coefficient. Opt. Express **26**, A157 (2018)
90. A. Nilsson, A.D. Nordlund, D. Waluyo, N. Huang, Y. Ogasawara, S. Kaya, U. Bergmann, L.-Å. Näslund, Y. Öström, P. Wernet, K.J. Andersson, H. Topsøe, T. Schiros, L.G.M. Pettersson, X-ray absorption spectroscopy and X-ray Raman scattering of water and ice; an experimental view. J. Electron. Spectrosc. Relat. Phenom. **177**, 99–129 (2010)
91. J.W. Smith, R.J. Saykally, Soft X-ray absorption spectroscopy of liquids and solutions. Chem. Rev. **117**, 13909–13934 (2017)
92. J. Yano, V.K. Yachandra X,-ray absorption spectroscopy. Photosynth. Res. **102**, 241–254 (2009). https://doi.org/10.1007/s11120-009-9473-8. PMC 2777224. PMID 19653117.
93. S. Schreck, P. Wernet, Isotope effects in liquid water probed by transmission mode X-ray absorption spectroscopy at the oxygen K-edge. J. Chem. Phys. **145**, 104502 (2016). https://doi.org/10.1063/1.4962237
94. G.N.I. Clark, C.D. Cappa, J.D. Smith, R.J. Saykally, T. Head-Gordon, The structure of ambient water. Mol. Phys. **108**, 1415 (2010)
95. K.R. Wilson, B.S. Rude, T. Catalano, R.D. Schaller, J.K. Tobin, D.T. Co, D.R.J. Saykally, X-ray spectroscopy of liquid water microjets. J. Phys. Chem. B **105**, 3346–3349 (2001)
96. Y. Pertot, C. Schmidt, M. Matthews, A. Chauvet, M. Huppert, V. Svoboda, A. von Conta, A. Tehlar, D. Baykusheva, J.-P. Wolf, H.J. Wörner, Time-resolved X-ray absorption spectroscopy with a water window high-harmonic source. Science **355**(6322), 264–267 (2017)
97. J.A. Sellberg, S. Kaya, V.H. Segtnan, C. Chen, T. Tyliszczak, H. Ogasawara, D. Nordlund, L.G.M. Pettersson, A. Nilsson, Comparison of X-ray absorption spectra between water and ice: new ice data with low pre-edge absorption cross-section. J. Chem. Phys. **141**, 034507 (2014)
98. C.J. Sahle, A. Mirone, J. Niskanen, J. Inkinen, M. Krisch, S. Huotari, Planning, performing and analyzing X-ray Raman scattering experiments. J. Synchrotron. Radiat. **22**, 400–409 (2015)
99. U. Bergmann, Nordlund, P. Wernet, M. Odelius, M.L.G.M. Pettersson, A. Nilsson, Isotope effects in liquid water probed by X-ray Raman spectroscopy. Phys. Rev. B Condens. Matter Mater. Phys. **76**, 024202 (2007)
100. C.J. Sahle, A. Mirone, J. Niskanen, J. Inkinen, M. Krisch, S. Huotari, Planning, performing and analyzing X-ray Raman scattering experiments. J. Synchrotron. Radiat. **22**(Pt 2), 400–409 (2015)
101. J. Kerr, A new relation between electricity and light: dielectrified media birefringent. Philos. Mag. **50**(332), 337–348 (1875)
102. J. Xu, G.S.D. Gordon, T.D. Wilkinson, C. Peucheret, Experimental observation of non-linear mode conversion in few-mode fiber, in *Conference on Lasers and Electro Optics (CLEO)*, San Jose, California (2015), pp. 1–3. https://doi.org/10.1364/CLEO_SI.2015.SM2L.3
103. R. Vaziri, Comment on "Nonlinear refraction measurements of materials using the moiré deflectometry." Opt. Commun. **357**, 200–201 (2015)
104. A. Taschin, P. Bartolini, R. Eramo, R. Righini, R. Torre, Optical Kerr effect of liquid water: a new insight into the vibrational and structural dynamics (2013). https://www.researchgate.net/publication/236155023_Optical_Kerr_effect_of_liquid_water_a_new_insight_into_the_vibrational_and_structural_dynamics

105. K. Winkler, J. Lindner, P. Vohringer, Low-frequency depolarized raman-spectral density of liquid water from femtosecond optical kerr-effect measurements: lineshape analysis of restricted translational modes. Phys. Chem. Chem. Phys. **4**, 2144–2155 (2002)
106. A.B. Ratajska-Gadomska, B. Biakowski, W. Gadomski, Cz. Radzewicz, Ultrashort memory of the quasicrystalline order in water by optical Kerr effect spectroscopy. Chem. Phys. Lett. **429**, 575–580 (2006)
107. G.C. Vebber, V.C. Ferreira, R.R.B. Correia, Surface time-resolved optical Kerr effect on liquid-solid interfaces. Front. Opt. (2019). Washington, DC, 15–19 Sept 2019. ISBN: 978-1-943580-67-5
108. R. Torre, P. Bartolini, R. Righini, Structural relaxation in supercooled water by time-resolved spectroscopy. Nature **428**, 296–299 (2004)
109. P. Bartolini, A. Taschin, R. Eramo, R. Righini, R. Torre, Optical Kerr effect measurement on supercooled water: the experimental perspective. J. Phys. Conf. Ser. **177**, 012009 (2009)
110. R. Torre, *Time-Resolved Spectroscopy of Complex Liquids: An Experimental Perspective* (Springer, New York, 2008)
111. M.S. Skaf, M.T. Sonoda, Optical Kerr effect in supercooled water. Phys. Rev. Lett. **94**, 137802 (2005)
112. E.O. Potma, W.P. de Boeij, D.A. Wiersma, Femtosecond dynamics of intracellular water probed with nonlinear optical Kerr effect microspectroscopy. Biophys. J. **80**, 3019–3024 (2001)
113. S. Palese, L. Schilling, R.J. Dwayne Miller, P. Randall Staver, W.T. Lotshaw, Femtosecond optical Kerr effect studies of water. J. Phys. Chem. **98**, 6308–6316 (1994)
114. N.A. Smith, S.R. Meech, Optically-heterodyne-detected optical Kerr effect (OHD-OKE): applications in condensed phase dynamics. Int. Rev. Phys. Chem. **21**, 75–100 (2010)
115. D. Ensminger, *Ultrasonics: Data, Equations, and Their Practical Uses*, Vol. 10 (CRC Press Taylor & Francis Group, Boca Raton, 2009)
116. H. Baltasar, E.M. Taravillo, M.V.G. Baonza, P.D. Sanz, B. Guignon, Speed of sound in liquid water from (253.15 to 348.15) K and pressures from (0.1 to 700) MPa. J. Chem. Eng. Data **56**, 4800–4807 (2011)
117. L.J. Shelton, F. Yang, W.K. Ford, H.J. Maris, Picosecond ultrasonic measurement of the velocity of phonons in water. Phys. Stat. Sol. **242**, 1379 (2005)
118. F. Yang, T. Atay, C.H. Dang, T.J. Grimsley, S. Che, J. Ma, Q. Zhang, A.V. Nurmikko, H.J. Maris, Study of phonon propagation in water using picosecond ultrasonics. J. Phys. Conf. Ser. **92**, 012024 (2007)
119. V.E. Gusev, P. Ruello, Advances in applications of time-domain Brillouin scattering for nanoscale imaging. Appl. Phys. Rev. **5**, 031101 (2018)
120. T. Sakurai, T. Matsuoka, S. Koda, H. Nomura, Coherent Brillouin scattering for ultrasonic measurements in liquids. J. Acoust. Soc. Jpn. (E) **19**, 4 (1998)
121. C. Esen, G. Schweiger, Ultrasonic enhanced Brillouin light scattering in water. Opt. Commun. **265**, 441–445 (2006)
122. C. Ballmann, J. Thompson, Traverso, Z. Meng, M.O. Scully, V.V. Yakovlev, Stimulated Brillouin scattering microscopic imaging. Sci. Rep. **5**, 18139 (2016)
123. C.-W. Lin, J.P.M. Trusler, The speed of sound and derived thermodynamic properties of pure water at temperatures between 253 and 473 K and at pressures up to 400 MPa. J. Chem. Phys. **136**, 094511 (2012)
124. W. Dai, Y. Wang, C. Qiao, Z. Huang, C. Zhou, Accurate ultrasound speed measurement using an MLS-modulated continuous wave. Measurement **92**, 150–156 (2016)
125. J. Krautkrämer, H. Krautkrämer, *Ultrasonic Testing of Materials*, 4th fully revised edition (Springer-Verlag, Berlin, 1990), pp. 167–221

126. N. Liu, N.B. Khoo, A. Zhang, Study on the structure and behaviour of cavitation bubbles generated in a high-intensity focused ultrasound (HIFU) field. Eur. Phys. J. E **42**, 70 (2019)
127. O. Bollengier, J.M. Brown, G.H. Shaw, Thermodynamics of pure liquid water: sound speed measurements to 700 MPa down to the freezing point, and an equation of state to 2300 MPa from 240 to 500 K. J. Chem. Phys. **151**, 054501 (2019)

Chapter 4
Neat Water

Abstract Despite the exceptional simplicity of the chemical structure of water due to the branched system of hydrogen bonds, its physicochemical properties are many-sided and their understanding requires an integrated approach using a whole cascade of methods. The ability to form up to four well-defined hydrogen bonds allows for different local structural arrangements, thus creating conditions for about 80 unique anomalies that are related to temperature and pressure. The wide variety of possible hydrogen bonds and dipole–dipole and van der Waals interactions has given rise to a variety of water models. The microscopic origin of water's anomalies is determined by the hydrogen-bond network (HBN) and the spatial organization of low-density liquid (LDL) like and high-density liquid (HDL) fractions. The set of intermolecular interactions gives rise to many of the thermodynamic and kinetic behaviors observed in water, particularly in the liquid phase, which significantly differs from many other liquids. Investigation of corresponding processes allows understanding the fundamental molecular transport properties, such as water mass diffusion, hydrogen network relaxation, proton spin relaxation, and proton tunneling. Examples of the successful application of a set of physical methods described in the third chapter for a detailed study of the vibrational, orientational, and phonon properties of neat water are given

4.1 Introduction

Water is fascinating in all its phases, forms, and states of aggregation and is unique in its number of anomalous properties. At high temperature, it behaves like most simple liquids; however, close to ambient temperatures, properties begin to deviate and do so increasingly on further cooling [1–13]. A total of 74 anomalous properties of liquid water are connected to its ability to form up to four well-defined hydrogen bonds allowing for different local structural arrangements (Figs. 4.1, 4.2, 4.3 and 4.4). At ambient temperatures, most molecules favor a closer packing than tetrahedral, with strongly distorted hydrogen bonds. Upon cooling of water, the amount of molecules participating in tetrahedral structures and the size of the tetrahedral patches increase. The two local structures are connected to the liquid–liquid critical point hypothesis

G. I. Likhtenshtein, *Biological Water*, Soft and Biological Matter,
https://doi.org/10.1007/978-3-030-82503-4_4

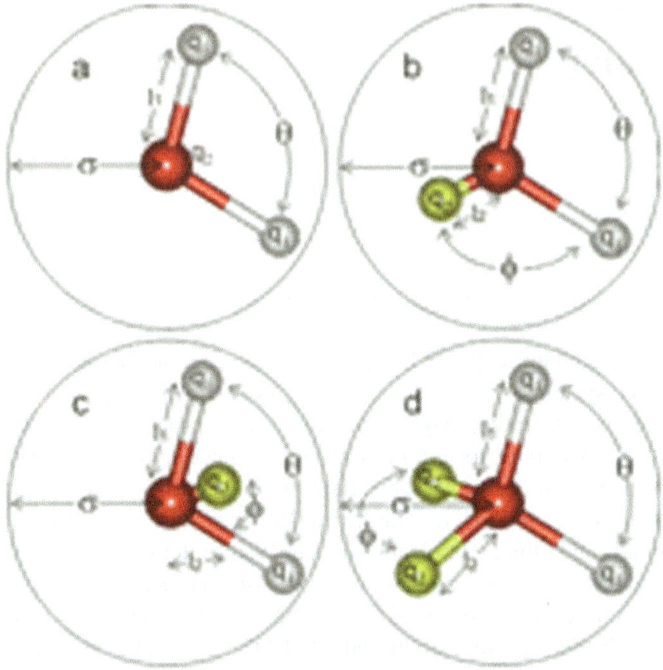

Fig. 4.1 Model types **a**, **b**, and **c** are all planar, whereas type, **d** is almost tetrahedral. The midpoint site ($_M$) in (**c**), and the lone-pair sites ($_L$) in **d** are labeled q_2. file:///C:/Users/gertz/Desktop /GOOD%20reviews/Water%20models.html Martin Chaplin on 22 May, 2020. This work is licensed under a Creative Commons Attribution-Noncommercial-No Derivative Works 2.0 UK: England and Wales License. With permission from Dr. Martin Chaplin

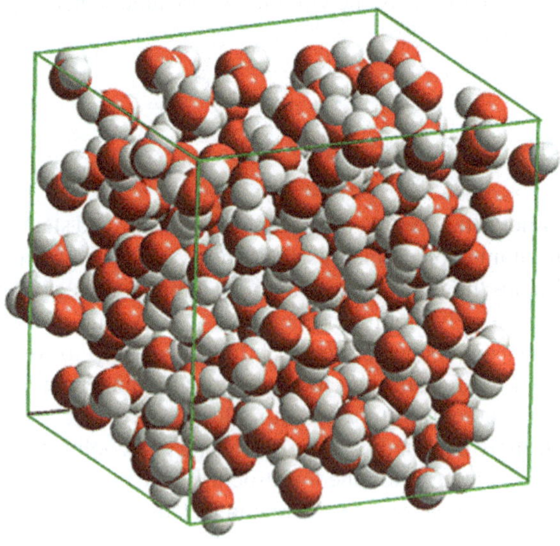

Fig. 4.2 A computational model of a small periodic box containing 216 water molecules Martin Chaplin on May 22, 2020. This work is licensed under a Creative Commons Attribution-Noncommercial-No Derivative Works 2.0 UK: England and Wales License. With permission from Dr. Martin Chaplin

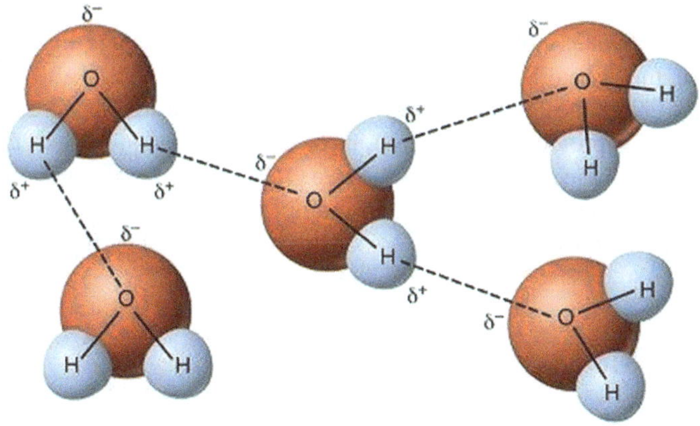

Fig. 4.3 The strong polar bond between water molecules creates water cohesion.https://www.google.com/search?q=water+hydrogen+bonding&oq=water+hydrogen+bond&aqs=chrome.1.69i57j0i22i30l9.30750j0j7&sourceid=chrome&ie=UTF-8

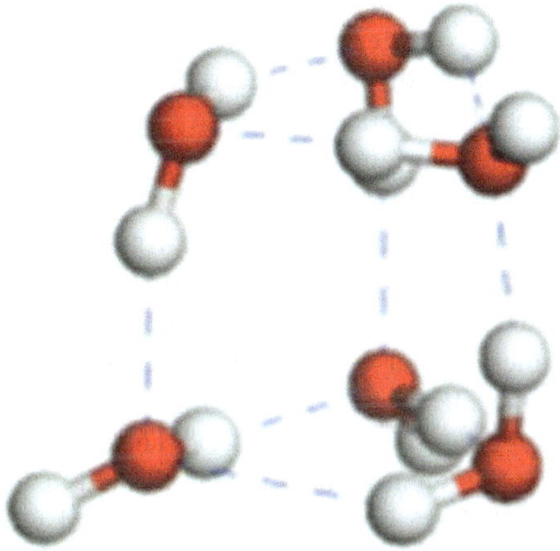

Fig. 4.4 DFT optimized $(H_2O)N$ crystal structures showing the ring-like ($N <$ 6), cage-like ($N = 6$–10), and solid clusters ($N = 12$–20) [2]. With permission from American Chemical Society

in supercooled water corresponding to high-density liquid (HDL) and low-density liquid (LDL). In addition to the short-range interactions within the O:H and the H–O segments, Coulomb repulsion between the bonding electron pair "–" and the non-bonding electron lone pair ":" is of vital importance to the relaxation of the O:H–O bond angle–length–stiffness under external stimulus. When water is in contact with a hydrophobic medium, the hydrogen-bond (H-bond) network of water is interrupted. As a result, the configuration of water at the interface differs significantly from that of the bulk water.

Figure 4.3 displays segmented hydrogen bond, non-bonding lone pair, and electron pair repulsion of water molecule [1]. The optimized ring-like, cage-like, and clusters $(H_2O)_N$ structures are presented in Fig. 4.4 [2].

In review [2], the following aspects of the neat water structure were considered: (1) the simulation protocols and definitions of the geometry on the water dimer; (2) several geometry-based definitions of the free O–H groups; (3) the orientation of the free O–H groups; (4) the free O–H groups number fluctuation dynamics; and (4) the fraction of interfacial water molecules with different H-bond partners.

4.2 Water Models

The wide variety of possible hydrogen bonds and dipole–dipole and van der Waals interactions has given rise to a variety of water models that are the subject of intense debate [4–10]. In accordance with Bernard-Fowler rule [4] which allows the rotation of water molecules within the Ice Ih lattice by hydrogen atoms jumping sites, the presence of hydrogen bond between water molecule leads to localized tetrahedral ordering that restricts molecular reorientation (Fig. 4.5a) [3]. This movement requires that all hydrogen atoms move simultaneously. Thermal fluctuation of some molecules can distort this localized structure, breaking some of the H-bonds and creating bifurcated (Fig. 4.5b, c), trifurcated H-bonds and therefore to lead to tetrahedral ordering defined as orientation defects. These defects in water were considered as the smeared L–D Bjerrum defects in ice. A hydrogen bond with a Bjerrum defect can have either two protons (D defect) or no proton (L defect).

Classical Monte Carlo simulations have been carried out for liquid water at 25 °C and 1 atm using six of the simpler intermolecular potential functions for the water dimer: Bernal–Fowler (BF), SPC, ST2, TIPS2, TIP3P, and TIP4P [5]. The key feature of the TIP4P model is that the site carrying the negative charge is not located at the oxygen atom but on the H–O–H bisector at a distance of 0.15 A° [5]. TIP5P suggested that the negative charge was located at the position of the lone-pair electrons [6], TIP4P/2005 is a rigid four-site model which consists of three fixed point charges and one Lennard-Jones center [7]. Calculated properties, covering a temperature range from 123 to 573 K, and pressures up to 40,000 bar include a variety of thermodynamic properties of the liquid and solid phases, the phase diagram involving condensed phases, properties at melting and vaporization, dielectric constant, pair distribution function, and self-diffusion coefficient. Some examples of the calculation are presented in Figs. 4.6 and 4.7.

Development of an improved four-site water model TIP4P-Ew for biomolecular simulations was reported [10]. The calculated values of densities thermal expansion coefficient were plotted together with the experimental data. The new TIP4P-Ew potential has a density maximum at 1 degrees C and reproduces experimental bulk-densities and the enthalpy of vaporization, ΔH(vap), from − 37.5 to 127 °C at

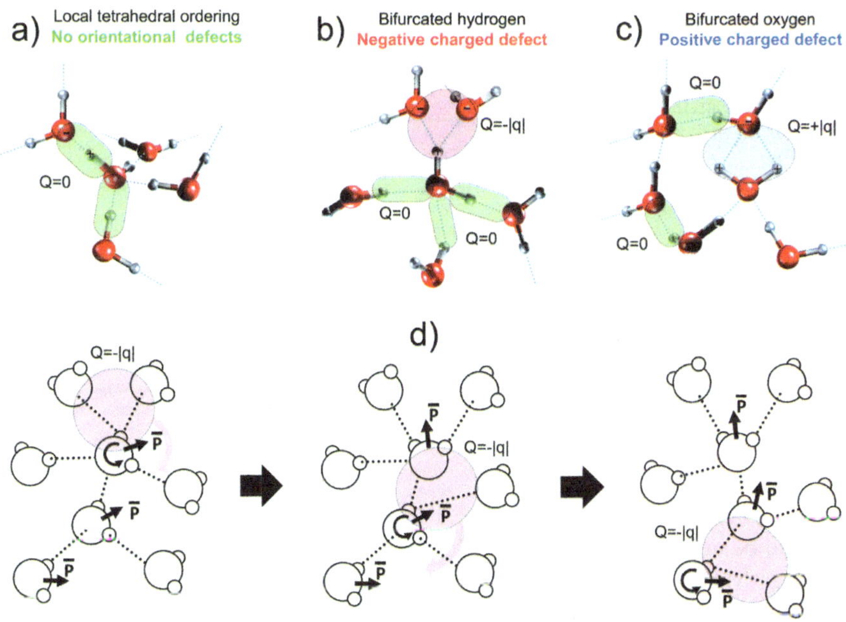

Fig. 4.5 Orientation defects **a** local tetrahedral ordering of the water molecules without defects. **b** The defect of the H-bond network due to bifurcated hydrogen. The negative charge, $Q = -|q|$, can be associated with this defect type (pink shaded area) regarding normal H-bond (green shaded area). **c** The defect of the H-bond network due to bifurcated oxygen. The positive charge, $Q = +|q|$, can be associated with this defect type (blue shaded area) regarding normal H-bond (green shaded area). **d** Schematic representation of the migration of orientation defects (pink-shaded area) with consecutive changing of the dipole moment's direction of water molecules [3]. With permission of Royal Society of Chemistry

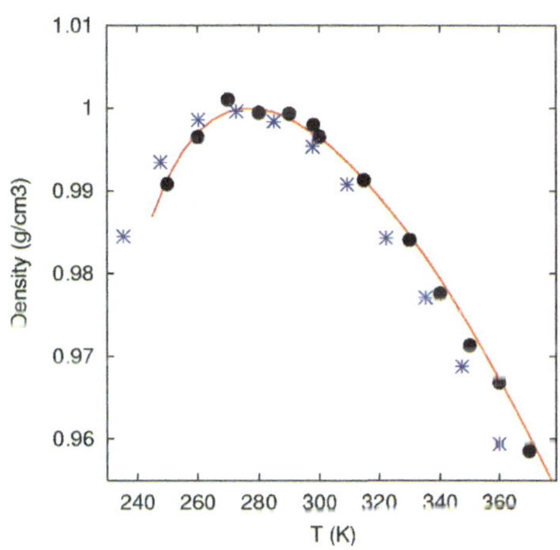

Fig. 4.6 Densities of the TIP4P/2005 model circles_ at $p = 1$ bar compared to the reported values for TIP4P/Ew_(stars)_ and experimental data _full line. The values for TIP4P/Ew are also included for comparison [7]. With permission from American Institute of Physics

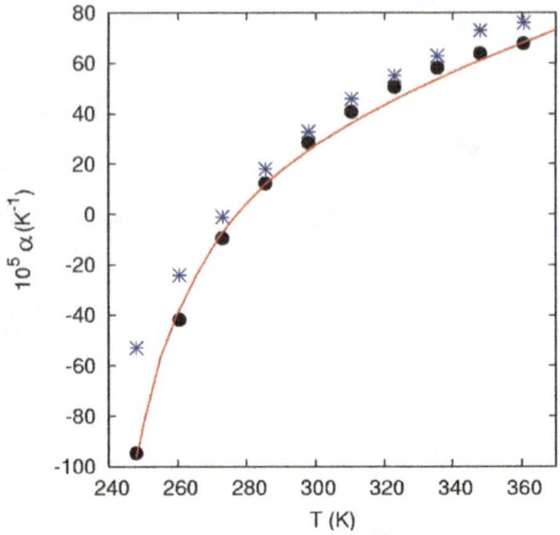

Fig. 4.7 Variation with temperature of the thermal expansion coefficient at $p =$ 1 bar. Symbols are the simulation results _stars: TIP4P/Ew; circles:TIP4P/2005_. The full line represents the experimental data [7]. With permission from American Institute of Physics

1 atm. Obtained structural properties at temperatures between 0 and 77 °C were shown to be in good agreement with X-ray scattering intensities and self-diffusion coefficient value.

In review [8], the performance of several water models (TIP3P, TIP4P, TIP5P, and TIP4P/2005) was analyzed. Ten different properties of water were predicted namely: (1) vapor–liquid equilibria (VLE) and critical temperature; (2) surface tension; (3) densities of the different solid structures of water (ices); (4) phase diagram; (5) melting-point properties; (6) maximum in the density of water at room pressure and thermal coefficients alpha; (7) structure of liquid water and ice; (8) equation of state at high pressures; (9) self-diffusion coefficient; and (10) dielectric constant. In all rigid, non-polarizable models of water, the positive charge located on the hydrogen atoms and a Lennard-Jones interaction site on the position of the oxygen are considered. Nevertheless, they differ in the location of the negative charge. Figure 4.8 shows an example of the prediction of the vapor–liquid equilibria for TIP3P, TIP4P,

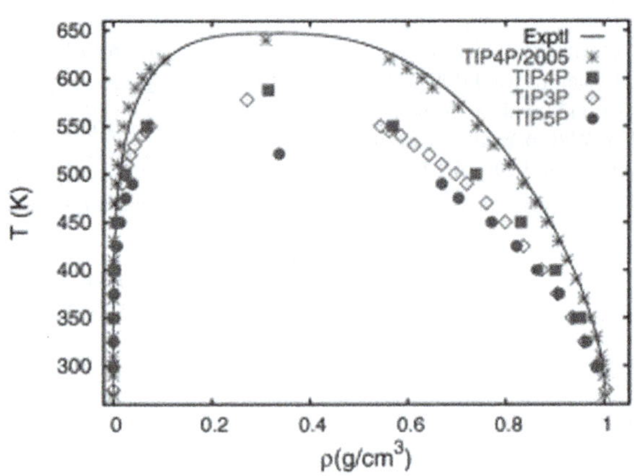

Fig. 4.8 Vapor–liquid equilibria for TIP3P, TIP4P, TIP5P, and TIP4P/2005 models [8]. With permission from American Institute of Physics

TIP4P/2005, and TIP5P [8]. The lowest critical temperature corresponds to that of the TIP5P model followed by that of the TIP3P and TIP4P. TIP4P/2005 provides a critical point in agreement with experiment. As it can be seen in the figure, TIP4P/2005 also provides an excellent description of the coexistence densities along the orthobaric curve.

TIP4P/2005 was found to be a rigid four-site model which consists of three fixed point charges and one Lennard-Jones center. The calculated properties covering a temperature range from 123 to 573 K and pressures up to 40,000 bar include a variety of thermodynamic properties of the liquid and solid phases, the phase diagram involving condensed phases, properties at melting and vaporization, dielectric constant, pair distribution function, and self-diffusion coefficient. The model gives predictions for the densities at 1 bar with a maximum density at 278 K and an averaged difference with experiment of $7 \times 10(-4)$ g/cm^3. The phase diagrams for TIP3P and TIP5P were determined using Hamiltonian Gibbs–Duhem integration. The results are presented in Fig. 4.9.

The aim of paper [11] was to establish whether the proposed is capable of reproducing a number of anomalies. Using molecular dynamics simulations, both the maximum in density and the minimum in the isothermal compressibility along a number of isobars were determined and the decrease in the temperature of the density maximum with increasing pressure was described. The model exhibits an additional minimum in density at a temperature of about 200 K at atmospheric pressure and also presents a minimum in the isothermal compressibility close to 310 K.

By performing a study combining mean-field calculations and Monte Carlo simulations, four scenarios have been proposed for the low-temperature phase behavior of liquid water. The proposed scenarios for the pressure–temperature (P–T) phase diagram are: (1) the stability limit (SL) scenario hypothesizes that the superheated liquid–gas spinodal at negative pressure reenters the positive P region below TH(P); (2) The liquid–liquid critical point (LLCP) scenario hypothesizes a first-order phase transition line between two liquids, a low-density liquid (LDL), and a high-density liquid (HDL) which terminates at LLCP; (3) The singularity-free (SF) scenario hypothesizes that the low-anticorrelation between volume and entropy is sufficient to cause the response functions to increase up on cooling and display maxima at nonzero T; and (4) The critical point-free (CPF) scenario hypothesizes an order–disorder transition, with possibly a weak first-order transition character, separating two liquid phases and extending to $P < 0$ down to the superheated limit of stability of liquid water. Thus, common physical mechanism underlies each of the scenarios, while all of them predict fundamentally different behavior.

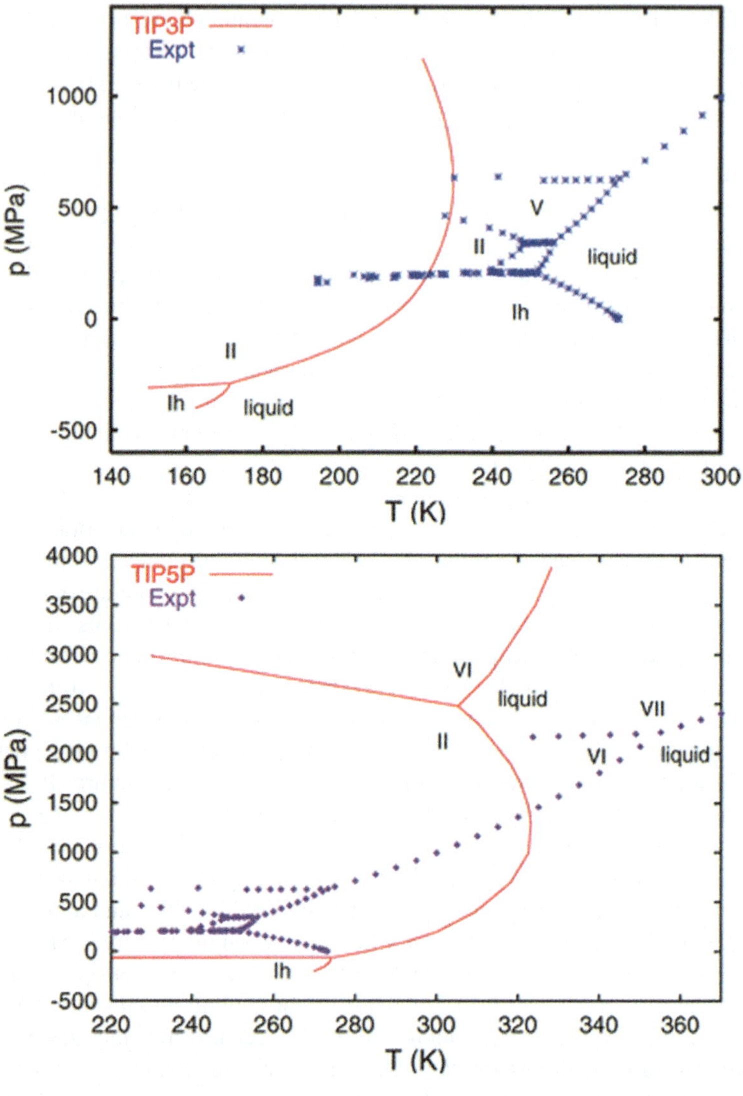

Fig. 4.9 Phase diagram as obtained in this work for TIP3P and TIP5P. Symbols: experimental phase diagram, lines: computed phase diagram. Top: results for the TIP3P model, bottom: results for the TIP5P model [8]. With permission from American Institute of Physics

4.3 Anomalous Properties of Water

Many of the properties of water are quite different from those expected from those of other liquids. In an excellent review [file:///C:/Users/gertz/Desktop/Anomalous%20properties%20of%20water.html], the following anomalies were described: (1) low-temperature anomalies of liquid water; (2) range of anomalous properties of water related to temperature and pressure, (3) water phase anomalies, (4) supercool water and ice anomalies, (5) water density anomalies, (6) water material anomalies, (7) water thermodynamic anomalies, and (8) water physical anomalies.

Figure 4.10 indicates some of the anomalous properties of liquid water that are related to temperature. The water's different behaviors at high and low temperatures are in contrast to the behavior of simple liquids, whose thermodynamic properties vary in a monotonic manner with temperature.

A review [12] analyzed some available experimental data on such properties of water as heat capacity, speed of sound, dynamic viscosity, and thermal conductivity. The results of the analysis were presented in the corresponding figures: (1) the regions of phase diagram where critical fluctuations can induce anomalous behavior; (2) examples of anomalous behavior of several properties of water such as (a) isobaric heat capacity; (b) isochoric heat capacity; (c) adiabatic speed of sound; (d) thermal conductivity under pressure of 4000 bar; (3) the points of different anomalies placed on the phase diagram of water; (4) comparison of equations of state from NIST database and obtained in simulations with TIP4P; (5) comparison of heat capacities cP and cV from NIST database and obtained in simulations with TIP4P at $P = 6000$ bar isobar; and (6) phase diagram and anomalous properties of water. These results allowed to authors to establish an existence of several novel anomalous properties of water which take place in the region of high pressure and relatively low temperature. A molecular dynamics simulations in this region with two common water models—SPC/E and TIP4P—were also performed. The simulation showed that these models fail to describe the novel anomalous region.

Work [13] reviewed several anomalous properties of liquid water, including the dielectric constant and the proton spin–lattice relaxation, and draw attention to a bilinear behavior defining a crossover in the temperature range 50 ± 10 °C between

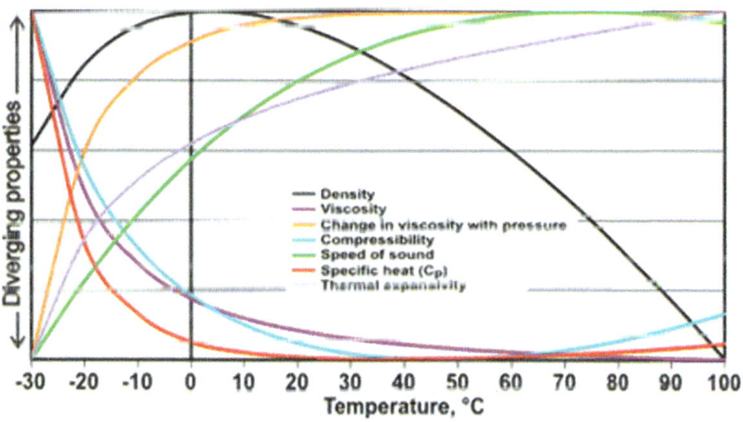

Fig. 4.10 Anomalous properties of liquid water that are related to temperature. file:///C:/Users/gertz/Desktop/Anomalous%20properties%20of%20water.html

two possible states in liquid water. In particular, the temperature dependence of the dielectric constant ε_d at 0.1 MPa displays a bilinear behavior in the 0–100 °C temperature range, with a crossover temperature T^* of approximately 60 °C. This behavior of dielectric constant ε_d as extracted from four different experimental data in the 0–100 °C temperature range is demonstrated in Figs. 4.11 and 4.12. Assuming that a Curie–Weiss law which describes the magnetic susceptibility χ holds for water in this temperature range, the dipole moment of liquid water μ was determined for both temperature intervals. It was found that in the low-temperature range 0–60 °C a

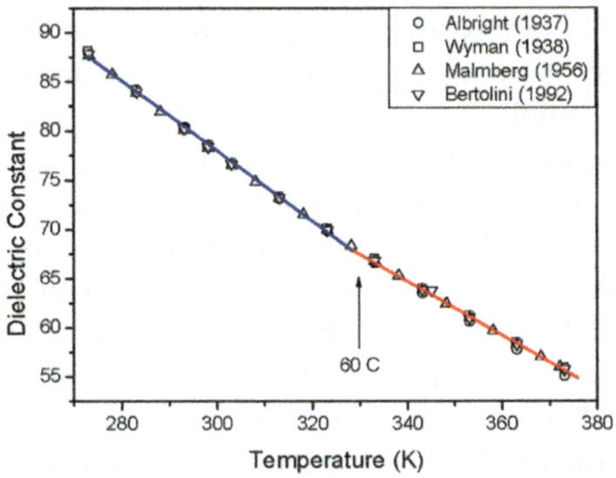

Fig. 4.11 Temperature dependence of the dielectric constant at 0.1 MPa in the temperature range 25–90° [13]. With permission from Inderscience

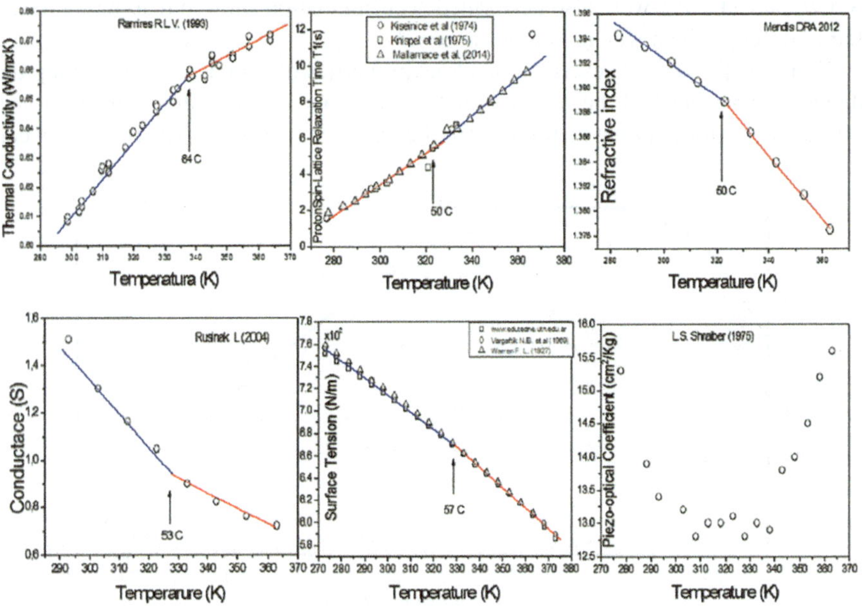

Fig. 4.12 Temperature dependence of thermal conductivity, proton spin–lattice relaxation time, refractive index, conductivity, surface tension, and piezo-optical coefficient at 0.1 MPa in the temperature range 0–100 °C [13]. With permission from Inderscience

value of $\mu_1 \approx 2.3$ Ds, which is close to the value for ice. In the temperature interval 60–100 °C, a value of μ_2 was calculated to be close to 1.8 Ds, which is comparable to that reported for the vapor phase. This suggests that there are of two states in bulk liquid water that differ in the amount of their dipole moment [13].

Figure 4.12 indicates the temperature dependence of thermal conductivity (TC), proton spin–lattice relaxation time (T1), refractive index (RI), conductivity (C), surface tension (ST), and the piezo-optical coefficient (POC) in the 0–100 °C interval. With the exception of POCs, a bilinear dependence defining a crossover temperature T^* (64 ± 5 °C for TC, 50 ± 5 °C for $T1$, 50 ± 5 °C for RI, 53 ± 5 °C for C, and 57 ± 5 °C for ST) was observed. These results confirm that in the 0–100 °C range, liquid water presents a crossover temperature in many of its properties close to 50 °C.

Article [10] presented a structure-based two-state model that successfully describes the experimentally measured structural, thermodynamic, and dynamic properties of liquid water in a unified manner. This model predicts: (1) a special line of dynamical fluctuation maximum, the dynamic Schottky line; (2) the crucial point is that this line is located at a substantially lower temperature than the Widom line at ambient pressure; and (3) the location of the LLCP around 184 K and 173 MPa for real water.. Practical guidance for the future experimental detection of the criticality in liquid water was also provided. The fraction of locally favored tetrahedral structure (LFTS) in water models is depicted in Fig. 4.13.

Other properties of liquid water also showed variations in this temperature range: (1) the thermal expansion coefficient (TEC) presents a crossover temperature at 42 ± 5 °C; (2) at the same temperature, the isothermal compressibility shows a minimum; and (3) the speed of sound in liquid water, which is correlated with the TEC, also presents a maximum at approximately 60 °C.

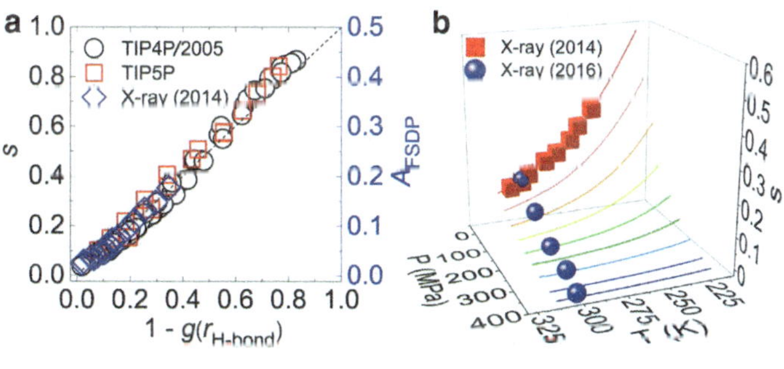

Fig. 4.13 **a** Fraction of LFTS, s, in water models obtained by the structural descriptor ζ versus 1−g(rH–bond)1−g(rH–bond) for TIP4P/2005 (circles) and TIP5P (squares) water. The integrated intensity (AFSDPAFSDP) of the FSDP in SOO(*k*)SOO(*k*) versus 1−g(rH–bond)1−g(rH–bond) obtained from X-ray scattering data of liquid water (is shown by diamonds (see the right vertical axis). **b** The fraction of LFTS, *s*, of liquid water directly estimated from the experimental X ray scattering data as a function of temperature *T* and pressure *P*. The solid lines are the two-state fits [14]. With permission from the PNAS article containing the requested material published under a CC BY-NC-ND license

4.4 Neat Water Structure

Early neutron diffraction measurements on H_2O–D_2O mixtures provided separated partial pair distribution functions $g_{OO}(r)$, $g_{OH}(r)$, and $g_{HH}(r)$ assuming H–D equivalence [15]. Experiment showed that the dominating O–O contribution exhibits peaks at 2.8 and 4.5 Å. These $g_{\alpha\beta}(r)$ patterns indicated and later improved measurements provided which detailed local structure information and a reference for models of liquid water [16].

The technic improvement is illustrated with several examples Figs. 4.14, 4.15, and 4.16 [17–20] which demonstrates total differential scattering cross sections of pure H_2O and pure D_2O water measured using different techniques; oxygen–oxygen pair distribution function patterns; and O–O radial distribution functions of liquid water at ambient conditions.

For water, set r_{ij} as the intermolecular distance of the oxygen atoms of water is schematically depicted in the left panel of Fig. 4.17 [2]. When the sum-frequency generation spectroscopy (SFG) signal was computed from the truncated correlation function formalism (is converged with r_t = 3.3–5.6 Å), the **SFG** signal can be attributed to the contribution of the water molecules within the first and second hydration shell of water, respectively (right panel of Fig. 4.17).

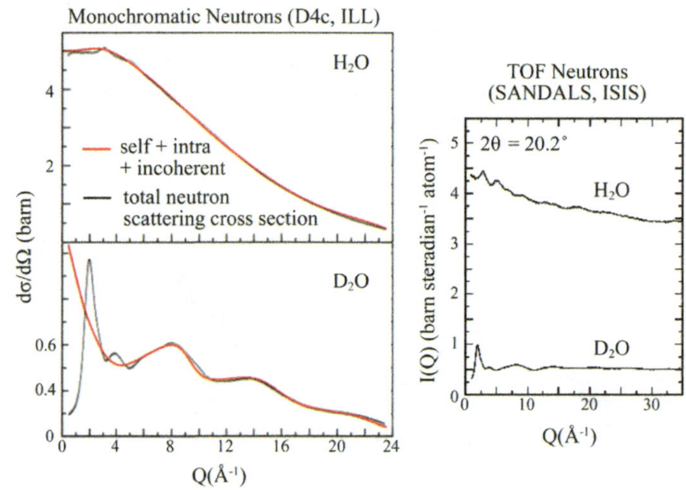

Fig. 4.14 Total differential scattering cross sections of pure H_2O and pure D_2O water as measured by two different neutron scattering techniques. (left) Monochromatic neutrons (λ = 0.5 Å) using the reactor source ILL (gray lines). ($Q = 4\pi \sin(\theta)/\lambda$) The red line corresponds to the sum of self plus intramolecular plus incoherent contributions (right) Time-of-flight neutron diffraction (TOF) at the pulsed neutron source ISIS, which uses a wide band of incident neutron energies ($\lambda \sim 0.1 -$ 3 Å) (black lines) [17].

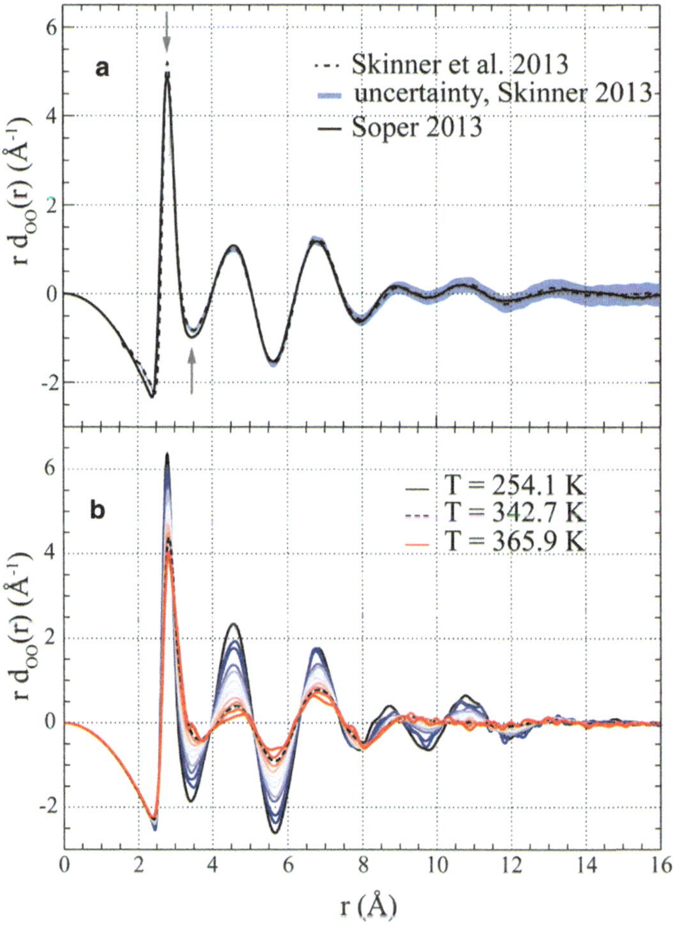

Fig. 4.15 O–O pair distribution function patterns ($rd_{OO}(r) = 4\pi\rho r^2[g_{OO}(r) - 1]$), which emphasize the high-r structure compared to the $g_{OO}(r)$ function. **a** Room-temperature patterns from Soper27 (solid black) and Skinner et al. [11] (black dashed). The light blue area is the error range from [11], and the arrows indicate the two r-regions where the two measurements slightly disagree. **b** $rd_{OO}(r)$ patterns as a function of temperature from [21]. The black-dashed line is at 342.7 K (the hottest containerless measurement), and the solid black line is at 254.2 K (the minimum temperature measured). The other colored lines correspond to a successive change of temperature, color coded from blue to red, at 263.2, 268.2, 277.1, 284.5, 295.2, 307.0, 312.0, 323.7, 334.2, 354.9, and 365.9 K [19]. With permission from American Chemical Society

The geometrical relation of the water dimer, using an intermolecular O···O distance R or intermolecular O···H distance r, and an angle of H–O···O or O–H·· is depicted in Fig. 4.18. The figure indicates simulated SFG spectrum of O–H stretch mode of HOD in D_2O at the isotopically diluted water–air interface ($\mathrm{Im}(\chi_{ssp}^{(2)})$ and the simulated $\mathrm{Im}(\chi_{ssp,fOH}^{(0)})$ with the optimal criteria and snapshot of

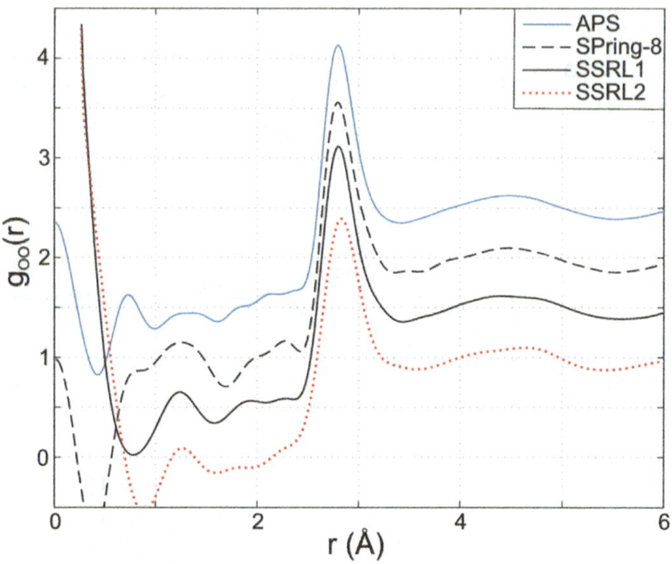

Fig. 4.16 **a** Comparison between oxygen–oxygen radial distribution functions of liquid water at ambient conditions derived from X-ray diffraction measurements (black) 107 and calculated from molecular dynamics simulations performed with the coarse-grained mW model (blue) 99 and the empirical E3B models (green and orange) 0.38106, **b** comparison between the experimental and calculated temperature dependence of the density of liquid water at ambient pressure. The experimental values (black) are from 304 [20]. With permission from American Institute of Physics

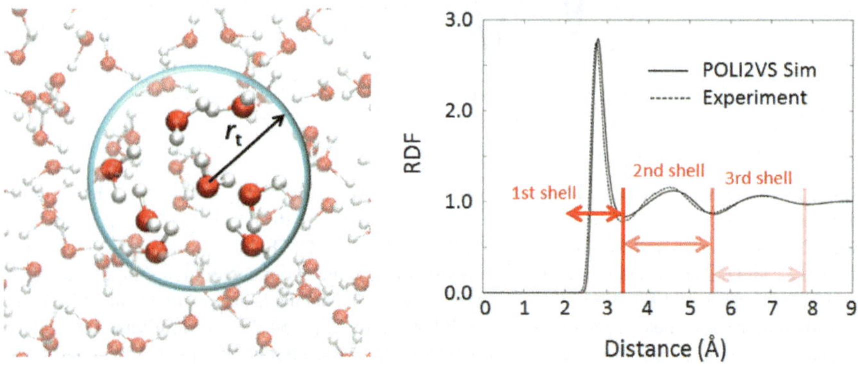

Fig. 4.17 (left) Schematic picture for calculating the cross-correlation term of the chromophores within the cutoff sphere with its radius of r_t in the truncated correlation function formalism. The central water molecule is considered here, and the highlighted water molecules are within the cutoff sphere and thus included in the spectral calculation, while the half transparent water molecules are excluded. (right) Radial distribution function (RDF) of water RDF of oxygen–oxygen atoms in bulk water [2]. With permission from American Chemical Society

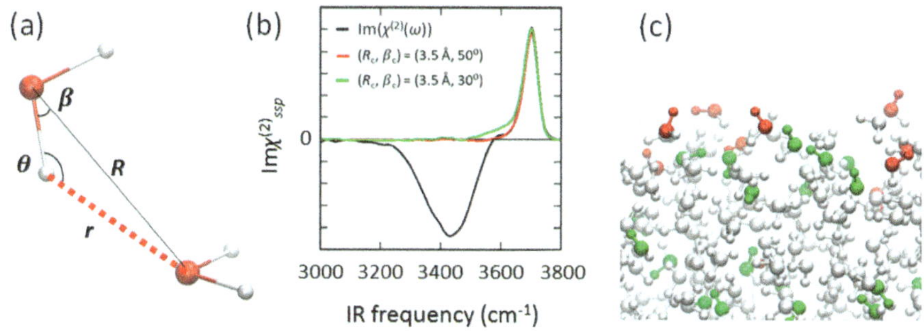

Fig. 4.18 **a** Schematic of the water dimer conformation where the water dimer are connected by H-bond. **b** Simulated SFG spectrum of O–H stretch mode of HOD in D_2O at the isotopically diluted water–air interface ($\mathrm{Im}(\chi_{\mathrm{ssp}}^{(2)})$), as well as the simulated $\mathrm{Im}(\chi_{\mathrm{ssp,fOH}}^{(2)})$ with the optimal criteria of (R_c, β_c) = (3.5 Å, 50°) and (R_c, β_c) = (3.5 Å, 30°).(184), **c** snapshot of the water molecules at the water–air interface with non-H-bonded free O–H groups defined with (3.5 Å, 50°) (red) and (R_c, β_c) = (3.5 Å, 30°) (red + green) [2]. With permission from American Chemical Society

the water molecules at the water–air interface with non-H-bonded free O–H groups, respectively (Fig. 4.19).

The simulation of SFG spectrum of neat water at 300 K with the POLI2VS model consisted of 500 HOD water molecules in the 26.6 Å × 26.6 Å × 160 Å cell HOD in D_2O was performed in the canonical ensemble [2]. From the MD trajectory, the values of molecular dipole moment, the polarizability, the time correlation function of the dipole moment, and polarizability were derived. The simulated dipole moment and polarizability of water molecules are located in the region of $|z| > 4$ Å (or $|z| > |z_{ds}| - 6.7$ Å). The autocorrelation function of the molecular dipole moment and

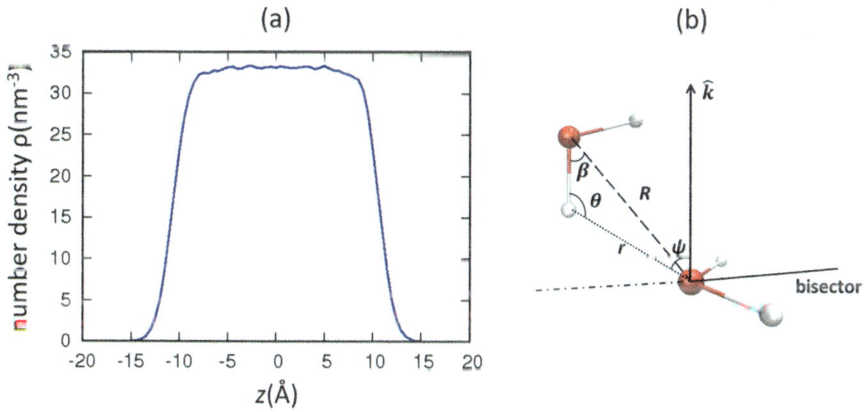

Fig. 4.19 **a** Axial profile of the number density of water. **b** Schematic of the water dimer conformation, the distances (R and r), and the angles (β, θ, ψ) discussed in the text. The $\hat{k}$ axis is perpendicular to the plane of the type H-bond-accepting water molecule. A red (white) sphere represents an oxygen (a hydrogen) atom) [2]. With permission from American Chemical Society

polarizability at the neat HOD–air interface, which is correspondent to the SFG spectra of the O–H stretch mode of HOD in D_2O, the contributions to the SFG response from the HOD molecules with the free O–H groups, and the axial profile of the number density of water (4.19) were calculated.

An application of first principles approaches to study of water and ice was reviewed [9].

The first principles idea is to break down complicated problems into basic elements and then reassemble them from the ground up. The following aspects of the approach were considered: empirical force fields, Ab initio methods, Ab initio force fields and many-body expansion, rigid versus flexible molecules, cluster interaction energies at characteristic points, dimer potentials, fitting the potentials, trimer potentials, dimer spectra and virial coefficients, spectra of the trimer, and larger clusters in liquid water and ice. As an example, the results for the oxygen–oxygen radial distribution function (RDF g_{OO}) calculated by different methods were presented.

The following consequence of physical effects in water condensed state was suggested: [1]. (1) Goldschmidt-Pauling contraction of the H–O polar-covalent bond elongates and polarizes the O:H van der Waals bond through the Coulomb repulsion between the electron pairs of adjacent oxygen (O–O); (2) this process enlarges and stiffens those molecules having fewer than four neighbors; (3) the shortening of the H–O bond raises the local density of bonding electrons, which in turn polarizes the lone pairs of electrons on oxygen; and (4) the stiffening of the shortened H–O bond increases the magnitude of the O1s binding energy shift and causes the blue shift of the H–O phonon frequencies. The combination of these effects in molecular clusters, hydration shells, and the surface skins of water and ice gives rise to their elastic, hydrophobic, highly polarized, ice-like, and low-density behavior at variety temperatures, pressure, and other conditions. Figure 3.22 illustrates the play of different factors affecting on the properties of the basic structural unit of the segmented "$O^{\delta-}$:$H^{\delta+}$–$O^{\delta-}$" hydrogen bond in hydrogen-bonded networks. The ":" represents the electron lone pair of the sp3-hybridized oxygen, and δ is a fraction that denotes the polarity of the H–O polar-covalent bond. The hydrogen bond comprises both the O:H van der Waals (vdW) bond and the H–O polar-covalent bond. The bond energy characterizes the bond strength, while the vibration frequency characterizes the bond stiffness. In addition to the short-range interactions within the O:H and the H–O segments, Coulomb repulsion between the bonding electron pair "–" and the non-bonding electron lone pair ":" takes place.

As it is seen in Fig. 4.20a which plots the number of water molecules in cluster (N) dependence of the O–O distance, the O–O distance expands by 8% when N is reduced from 20 to 3. This value is compatible with the value of 5.9% measured in the water surface at 25 °C. Figure 4.20b indicates the predicted N-dependence of the melting point (T_{mN}) elevation and the O1s core-level shift (ΔE_{1sN}). Both T_{mN} and ΔE_{1sN} are proportional to the covalent bond energy in the form of: $T_{mN}/T_{mB} = \Delta E_{1sN}/\Delta E_{1sB} = E_{HN}/E_{HB} = (d_{HN}/d_{HB})^{-4}$. Subscripts B and N denote the bulk and cluster. The H–O bond and lengthens the O:H and) the shortening (lengthening) of the H–O bond is coupled to the lengthening (shortening) of the O:H.

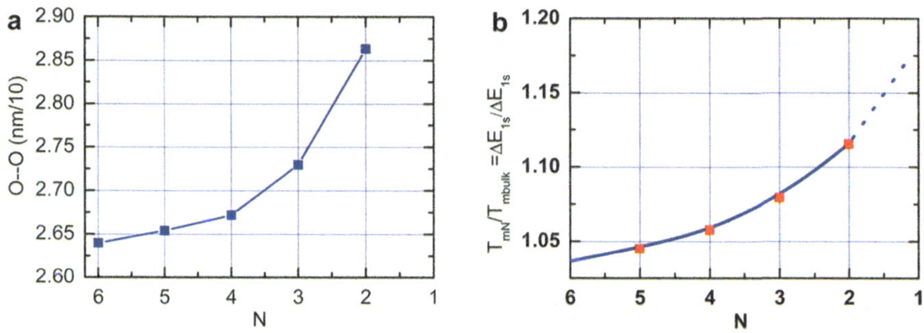

Fig. 4.20 N dependence of **a** the O–O distance, **b** the melting point, T_{mN} (to $N = 2$ for dimers), and the O1s core-level shift (to $N = 1$ for gas monomers) of **$(H_2O)_N$ clusters** based on DFT-derived dHN/dHB values [1]. With permission from American Chemical Society

Results of molecular dynamics simulations of the properties of water clusters in the 70–290 molecule size range were reported [22]. The simulations employed an intermolecular potential model that describes the water–water interactions. The calculated vibrational spectra for clusters containing 90 or 159 molecules show red-shifted shoulders that are not present in the calculated spectrum for the cluster containing 70 molecules.

Structures of cage, prism, and book isomers of water hexamer were derived from broadband rotational spectroscopy (BRS) [23]. Theory predicted the water hexamer to be the smallest water cluster with a three-dimensional hydrogen-bonding network as its minimum energy structure. ***BRS****in* A pulsed supersonic expansion identified the cage, book, and cyclic isomers and their oxygen framework structures by means of oxygen-18–substituted water ($H_2{}^{18}O$) Relative isomer populations at different expansion conditions established that the cage isomer is the minimum energy structure. Rotational spectra consistent with predicted heptamer and nonamer structures were also identified.

The inelastic neutron scattering (INS) vibrational spectra of free and partially constrained water clusters up to $n = 17$ were calculated with DFT methods using published energy-minimized structures [24]. Structures of the water clusters under interest are indicated in Fig. 4.21. Small clusters in the inner shell, being in focus of the investigation, were studied by the neutron diffraction. The calculation showed that INS spectra of the water in librational region, calculated in a combination of model bound clusters, indicated that the in intensity near 500 cm^{-1} is connected to increasing numbers of four-coordinate water molecules. Thus, an analysis of experimental vibrational spectra of free water clusters together with optical spectroscopies computations in the OH stretching region allowed to characterize H bonding in the formation of such clusters and to determine the point at which ice-like structural motifs may form as a function of cluster size.

The calculated INS spectra of the water molecules in structures composed of chains, namely the trimer, tetramer, pentamer, and cyclic hexamer (Fig. 4.22), which have only two contacts (2C) in the region of water librations (i.e., above 300 cm^{-1}),

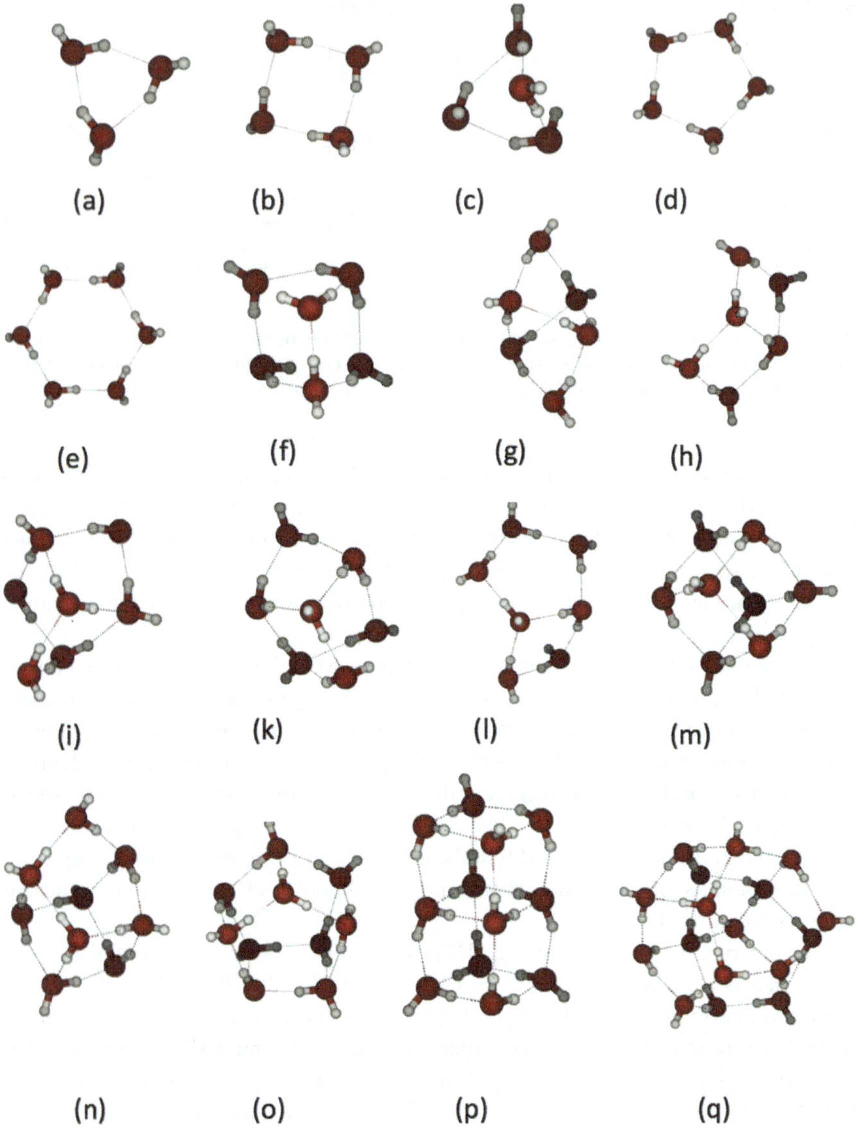

Fig. 4.21 Structures of water clusters used in this work: **a** trimer, **b** tetramer, **c** tetramer pyramid, **d** pentamer; hexamers: cyclic (**e**), prism (**f**), cage (**g**), book (**h**); heptamers: prism (**i**), cage (**k**), chair (**l**); octamer (**m**); nonamer (**n**), decamer (**o**); 12-mer (**p**), and 17-mer sphere (**q**) [24]. With permission from Elsevier

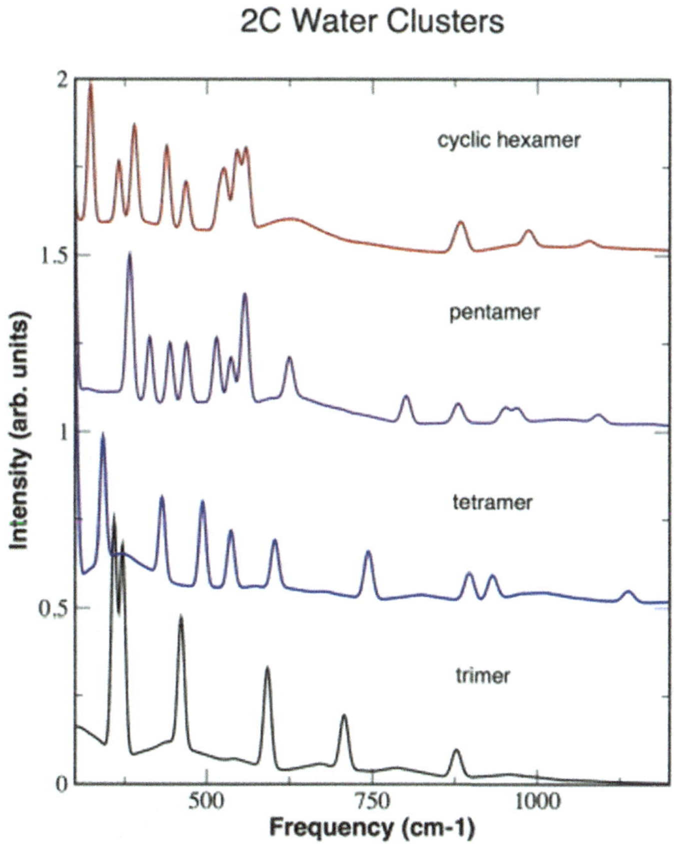

Fig. 4.22 Calculated INS spectra in the librational region for free planar water clusters as indicated in figure formed by two-coordinate water molecules [24]. With permission from Elsevier

indicated the results of coupling between the motions of neighboring molecules, as the spectra of the larger clusters become more complex with shifting and splitting of some of the peaks.

It was concluded that: (1) at least 280 water molecules involved in a free cluster with a size of approximately 3 nm; (2) the spectral region of whole body librations of the water molecules (300–1200 cm^{-1}) is indicative of number of H-bonds formed; (3) librational frequencies are highest for water in the tetrahedral geometry, involved in four H-bonds, two as donor, and two as acceptor; (4) the librational spectrum at around 550 cm^{-1} corresponds to the onset of librations of the tetrahedral water molecules in the highly symmetric ice structures; (5) the smallest minimum energy cluster containing an interstitial tetrahedral (4C) water molecule at the center is the one with $n = 17$; (6) trimers, tetramers, and pentamers all have only two-coordinated (2C) water molecules, as does the cyclic hexamer; (7) three-coordinated (3C) water molecules can be found in low-energy hexamers, such as the trigonal prism and book structures, and make up the entire octamer cube; (8) a four-coordinated water molecule is at the center of a bow tie structure of the heptamer; and (9) 4C water molecules are also present in some $n = 12$ and $n = 15$ structures as linker units.

The reverse Monte Carlo modeling technique was used to fit double H-bond donors (DD) and single donors (SD) models for water to available X-ray and neutron diffraction data in qq space as well as to the electric field distribution as a representation of the OH stretch Raman spectrum of diluted HOD in D_2O [16]. For example, maximizing the number of hydrogen bonds results in the DD model with 18 and 81% SD model which reproduced the experimental data. Within the tetrahedral DD structure model, the assumed connection between spectrum position and H bonding indicates ultrafast dynamics in terms of breaking and reforming H-bonds, while in the strongly distorted model the observed frequency shifts do not necessarily imply H-bond changes. The many-body features of the quantum Drude oscillator (QDO) model, which enables simultaneous determination of the coexistence densities along both branches of the binodal curve over the full range of coexistence, were demonstrated [25]. A short-range pair potential was determined from a single-dimer energy surface completing the QDO picture of water. The suggested strategy captures water's properties from an atomistic perspective by incorporating individual water molecules are depicted in in Fig. 4.23.

The electronic structure rearrangements that occur in water's condensed phase along coexistence are reflected in its dielectric responses such as the permittivity given in from experiment and the QDO model (Fig. 4.24).

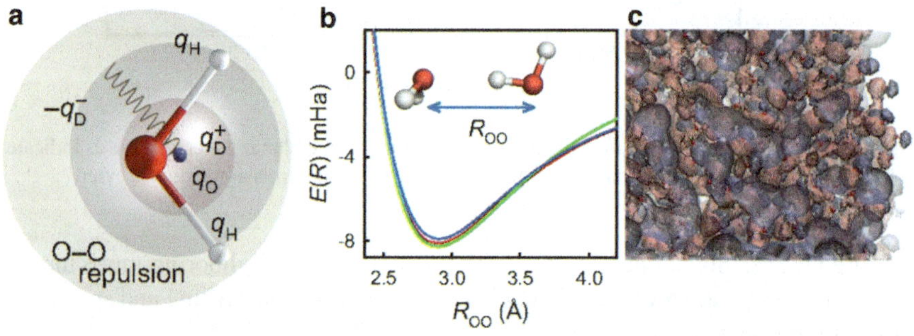

Fig. 4.23 QDO water model and predictions. **a** Schematic of the QDO water model, where a coarse-grained electronic structure, the Drude oscillator, is embedded in a rigid molecular frame decorated with point charges. The point charges capture the low-order electrostatic moments of the isolated molecule. The oxygen charge is placed on the M site down the symmetry axis, which is represented as a blue dot. The QDO is tethered to the M site. **b** The ground-state energy surface of the water dimer as a function of O–O distance with the molecular orientation fixed in that of the minimum energy geometry. QDO model results (red) are compared with high-level ab initio data (green) and polarizable Thole-type model potential TTM3-F, v. 3.0 (51) (blue). **c** Coarse-grained description of the changes in the electron distribution arising at the air–water interface; pink regions denote an increase and blue regions a decrease in electron density [25]. Published with permission of PNAS article containing the requested material published under a CC BY-NC-ND license

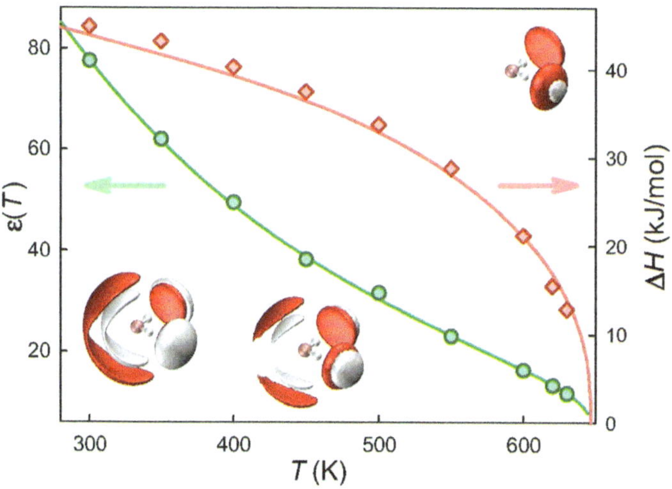

Fig. 4.24 Water properties along the coexistence line. The lines are the reference IAPWS equation of state, and the symbols are the results of our simulations with the errors of the order of symbol size. Circles and left axis (indicated by arrows), relative static dielectric permittivity of water along the binodal; diamonds and right axis, enthalpy of vaporization of water. (*Insets*) Disruption of tetrahedral structure with increase of temperature along the binodal ($T = T = 300$, 450, and 600 K are shown) [25]. Published with permission of PNAS article containing the requested material published under a CC BY-NC-ND license

The following general considerations were highlighted: (1) suggested models have properties defined entirely in terms of isolated molecules—unbiased towards any thermodynamic state point or condensed phase; (2) the resulting interactions contain the complete hierarchy of many-body inductive and dispersive forces as they are solved in strong coupling; (3) the interactions are capable of properly generating the balance between hydrogen-bonding, electrostatic, and van der Waals interactions, (4) the simplification of the electronic problem inherent in the QDO model permits linear-scale sampling methods to be used; and (5) the liquid–vapor density differences are very large near ambient conditions inducing extreme molecular-scale alterations, both in local spatial order (hydrogen bonding) and electronic structure.

Molecular dynamics (MD) simulations were used to compute SFG response of the water–air interface, and the molecular origin of the SFG features was clarified by decomposing the SFG spectra into the individual contributions in terms of the H-bond number of the interfacial water molecule [2] and references therein. The geometry-based definition of the free O–H groups by calculating the SFG spectra of the O–H stretching mode with selected O–H groups at the neat HOD–air interface was optimized. SFG spectra of water at the water–air interface showed a sharp peak at 3700 cm^{-1}, which has been assigned to the free O–H groups of water. It was established that the subensemble of the non-H-bonded O–H groups generated with the H-bond definition contributes to both a 3700 cm^{-1} peak as well as a weakly H-bonded stretch band at $\omega \sim 3600$ cm^{-1}. The computation also identified the fraction of interfacial water molecules with free O–H groups of 28%, a vibrational lifetime of the free

O–H groups of 1.3 ps, and the angle formed by the free O–H groups and the surface normal of 67° at the water–air interface. When water is in contact with a hydrophobic medium, the hydrogen-bond (H-bond) network of water is interrupted [2]. As a result, the configuration of water at the interface differs significantly from that of the bulk water. The structure of interfacial water at the water-hydrophobic medium has been studied using vibrational sum-frequency generation (SFG) spectroscopy. It was found that the rotational dynamics of the O–H group of water near the amphiphilic solute molecule is monitored by decomposing the O–H groups interacting with the hydrophilic and hydrophobic parts of the solute molecule.

In a recent overview [26], the different simulation protocols available for SFG spectra calculations were suggested. The findings through combined SFG experiments and simulations, future challenges for SFG experiments, and simulations at different aqueous interfaces were accounted. It was demonstrated that, combining the SFG spectroscopy with simulation techniques, the experimental data with the simulated SFG spectra allowed to unveil the molecular-level structure of water–air and ice–air interfaces.

Core-level Auger spectroscopy (Fig. 4.25) is a technique to obtain element-specific information of the electronic structure around an absorption site and are thereby suitable tools to study the chemical state, local geometric structure, nature of chemical bonding, and dynamics in electron transfer processes centered around one atomic site. https://www-ssrl.slac.stanford.edu/nilssongroup/corelevel.html

When an atom is probed by a photon or a beam of electrons with energies in the range of several eV to 50 keV, a core-state electron can be removed leaving behind a hole. The core hole can be filled by an outer shell electron, whereby the electron moving to the lower energy level loses an amount of energy equal to the difference in orbital energies. The transition energy can be coupled to a second outer shell electron, which will be emitted from the atom, and an emitted electron will have a kinetic energy.

Electron delocalization and its rates in liquid water and ice determined using core hole decay spectroscopy were examined [27]. The OKLL Auger electron decay, where L corresponds to molecular orbitals with O2p character, was measured. The nature of the wave functions of the two most extreme cases, the pre-edge for molecules with a broken H-bond leading to a fully localized state on the free OH of the excited molecule and the post-edge where the water is fully H-bonded, was identified. The former was shown to lead to a very localized state and the latter to a delocalized state where the residence time of the excited electron on the local molecule is less than **500 attoseconds** (attosecond is 1×10^{-18} of a second). Broken or weak hydrogen bonds—in the liquid or at the surface of ice—provide states where the electron remains localized longer than 20 fs. Plots of the wave functions of these two excited states are seen in Fig. 4.26.

As depicted in Fig. 4.27, electron transfer rates on attosecond timescales are found in liquid water in electronic states extending over the H-bond network, similar to that in bulk ice (steps 1 and 2). The electron hopping rate (< 500 *as*) is then extremely fast compared to the molecular motion in the liquid and the structural rearrangement of water.

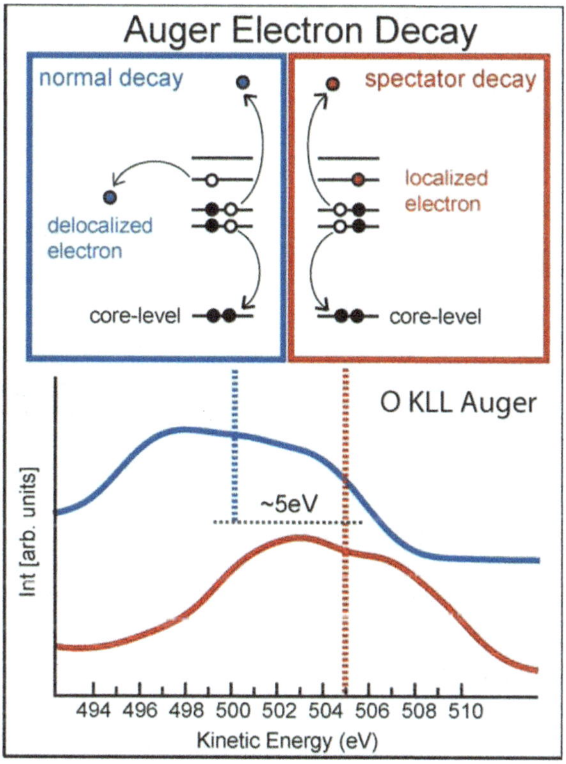

Fig. 4.25 Auger electron decay processes used to derive site-specific timescales in this work. (Top) Normal Auger decay (left), where the excited electron is delocalized from the decaying atom and spectator Auger decay (right), where the excited electron remains localized during the decay. (Bottom) Simulated spectra of a cluster model (see text) indicate a shift of 5 eV between the two decay channels due to screening by the spectator electron [27]

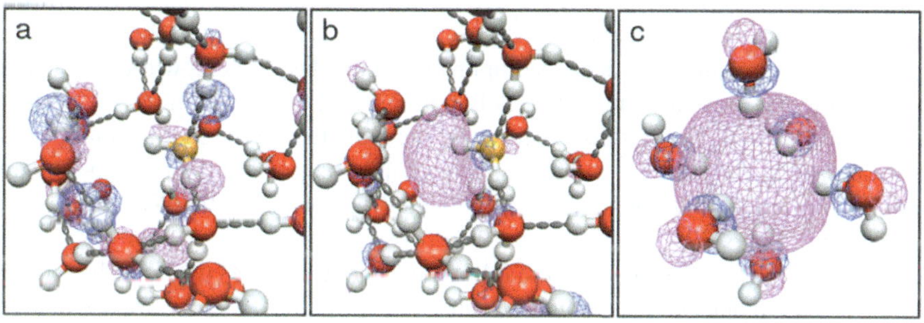

Fig. 4.26 Orbital plots of core-excited states in **a** the post-edge region and **b** the pre-edge region compared to **c** the solvated electron, represented by the relaxed HOMO orbital of a − 1e charged cluster of six water molecules oriented around a central cavity [27]

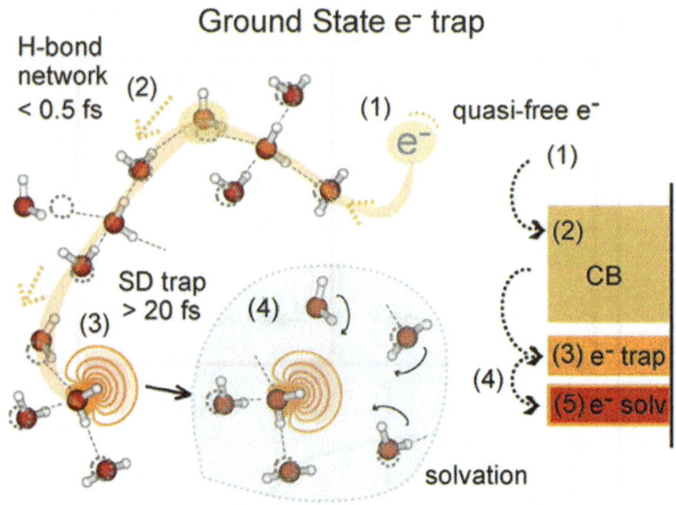

Fig. 4.27 Schematic diagram of electron trapping in liquid water: 1 and 2 an excess electron propagating along the H-bond network, 3 trapping at a weakly H-bonded OH group of a single-donor site followed by 4 electron solvation dynamics to reach the solvated electron [27]

4.5 Two States of Water

The microscopic origin of water's anomalies by inspecting the hydrogen-bond network (HBN) and the spatial organization of low-density liquid (LDL) like and high-density liquid (HDL) like environments was investigated via classical molecular dynamics simulations [28]. Cooling of a sample composed of 512 water molecules from ambient to deeply undercooled conditions at three pressures, namely 1 bar, 400 bars, and 1000 bars, was considered. Some examples of the simulation are illustrated in Figs. 4.28 and 4.29. The score function for the short-range order (SRO) measured using, as a reference, a perfect tetrahedron is indicated in Fig. 4.28. For a given temperature, the S_{SRO} decreases upon increasing the pressure, while it increases upon fixing the pressure and isobarically cooling. The increment in the S_{SRO} upon isobarically cooling the sample is caused by the reduction of the thermal energy and is measured by the increase in the intensity of the first peak in the oxygen–oxygen radial distribution function. Two dimensional slices of the **LDL-like** network are displayed in Fig. 4.29.

All in all, the simulation led to the following conclusions: (1) the HDL-like dominating cluster undergoes fragmentation caused by the percolation of LDL-like aggregates following a spinodal-like kinetics; (2) such fragmentation always occurs at a critical concentration of 20–30% in LDL; (3) the HBN within LDL-like environments is characterized by an equal number of pentagonal and hexagonal rings; (4) the spatial organization of HDL-like environments shows a marked variation; and (5) the inspection of the global symmetry indicates that the intermediate-range order decreases.

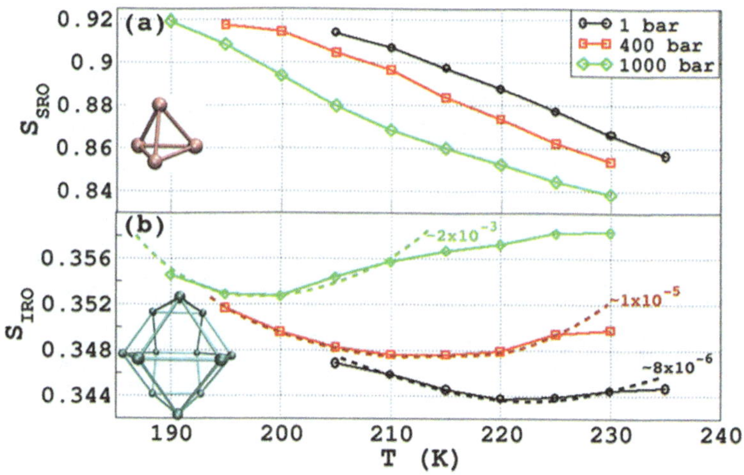

Fig. 4.28 **a** Short-range order using, as a reference, a perfect tetrahedron, for the cooling at 1 bar (black circles), at 400 bars (red squares), and at 1000 bars (green diamonds), respectively. **b** Intermediate-range order using, as a reference, the anticuboctahedron. Dashed lines correspond to the quadratic fittings [28]. With permission from American Institute of Physics

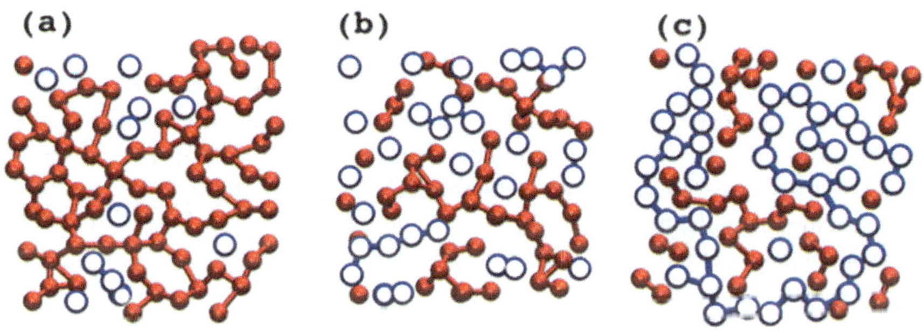

Fig. 4.29 Two-dimensional slices of the LDL-like network (blue, open spheres) and of the HDL-like network (red, filled spheres) at 400 bars and three temperatures: $T = 230$ K (**a**), $T = 215$ K (**b**), and $T = 195$ K (**c**). Becomes more pronounced upon increasing the pressure, hence supporting the hypothesis of a liquid–liquid critical point. Results reveal and rationalize the complex microscopic origin of water's anomalies as the cooperative effect of several factors acting synergistically. Beyond implications for water, our findings may be extended to other materials displaying anomalous behaviors [28]. With permission from American Institute of Physics

Local structure analysis revealed that a majority of the water molecules at ambient conditions are situated in locally disordered high-density-like environments molecules which are characterized by more ordered local environments (Fig. 4.30) [29]. The ratio between high-density and low-density-like molecules was found to be a 3:1, with the latter forming small connected clusters amid the predominant population of high-density-like molecules. Analysis of a local structure index autocorrelation function uncovered a persistence time of $\sim$ 4 ps.

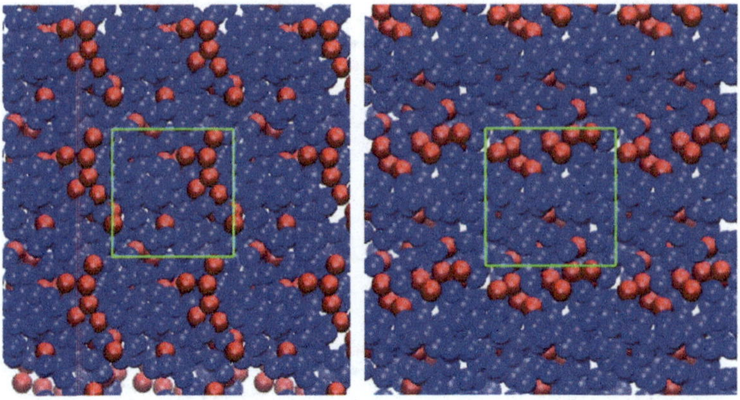

Fig. 4.30 Snapshots of typical liquid water structures (configurations) found in the inherent potential energy surface (IPES) IPES of ambient liquid water computed with the PBE0 + TS-vdW(SC) XC potential at 330 K. The blue spheres represent water molecules characterized by $I < I_{\min}$ (i.e., high-density-like water molecules), whereas the red spheres represent water molecules characterized by $I > I_{\min}$ (i.e., low-density-like water molecules). The green box refers to the simulation cell containing 128 water molecules with only the oxygen atoms shown above for clarity [29]. With permission from Taylor & Francis Group

The diversity of local microscopic environments found in ambient liquid water was characterized by the probability density distribution of the local structure index (LSI) order parameter ($P(I)$) (Fig. 4.31). The $P(I)$ in ambient liquid water obtained from various DFT-based AIMD simulations are all unimodal in shape and rapidly decaying from maxima located between 0.013 Å^2 at 330 K and 0.023 Å^2 at 300 K. Thus, a majority of the molecules are situated in high-density-like locally disordered environments in ambient liquid water.

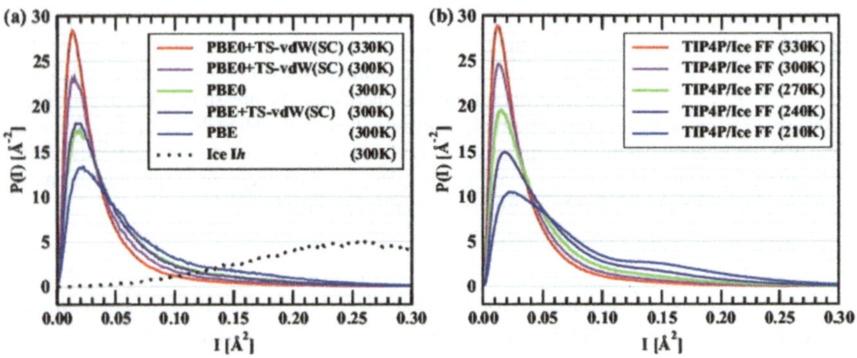

Fig. 4.31 Probability density distributions of the local structure index ($P(I)$) in liquid water obtained from **a** DFT-based AIMD simulations and **b** classical force field (TIP4P/Ice)-based simulations [29]. With permission from Taylor & Francis Group

Coexistence of two types of local structures in the coordination number distribution is justified in Fig. 4.32 [14]. The figure shows that the distribution $P(N_{fs})$ of the coordination number N_{fs} for TIP5P/2005 water, where N_{fs} is the number of water molecules in the spherical first-shell volume V_{fs} of radius 3.5 Å, is proportional to the local number density, N_{fs}/V_{fs},. Both $P(N_{fs})$ and $P(\zeta)$ can be characterized by two

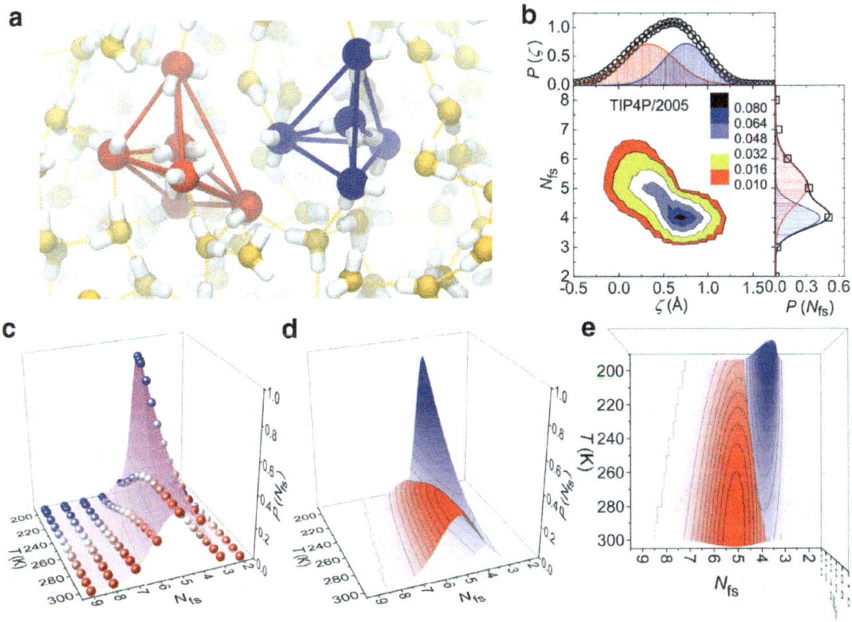

Fig. 4.32 Structural bimodality in the coordination number distribution of liquid water. **a** Snapshot of liquid TIP4P/2005 water at 1 bar and 240 K, where $s \approx 0.5$, i.e., near the Schottky temperature, $T_s = 1/2$. Two types of local structures, LFTS and DNLS, are highlighted by blue and red colors, respectively. LFTS has four H-bonded nearest neighbors with tetrahedral coordination, whereas DNLS typically has five or more nearest neighbors (typically three of them being H-bonded) with broken tetrahedral symmetry. **b** Correlation between the number of water molecules in the spherical first-shell volume of radius of 3.5 Å, Nfs, i.e., the coordination number, and structural descriptor ζ in TIP5P/2005 water at 1 bar and 240 K. The distributions of Nfs and ζ are shown in the right and top panels, respectively. Both distributions show clear bimodal features, which can be properly described by the sum of two Gaussian functions (black lines), with blue and red shades corresponding to LFTS and DNLS, respectively. **c** Distribution of coordination number Nfs as a function of temperature, P(Nfs, T), in TIP5P/2005 water shown by colored balls with red and blue for higher and lower temperatures, respectively. The colored surface is the fit to two Gaussian functions (eqs S1–S5). **d** and **e** T dependence of the two Gaussian components of P(Nfs, T) in the (**d**) side and (**e**) top views. One Gaussian component located at Nfs $\approx$ 4 corresponds to LFTS, whereas the other one at Nfs $\simeq$ 5 corresponds to DNLS, in agreement with the snapshot in (**a**). The fraction of the two Gaussian components is consistent with the fraction determined by the ζ distribution in (**b**) and the theoretical two-state model. A microscopic structural descriptor ζ is the difference in the distance to the central water molecule from the nearest non-H-bonded neighbor and that from the farthest H-bonded one [14]. With permission from American Chemical Society

Gaussian functions following the prediction of the thermodynamic two-state model (see details in the figure capture).

Additional evidences for the coexistence of two types of local structures in the scattering function are depicted in Fig. 4.33: (1) tetrahedral unit produces four peaks in the range of $0.5 \leq k\hat{r}/2\pi \leq 3.0$ with peak wave numbers labeled as k_{Ti} $(i = 1 - 4)$

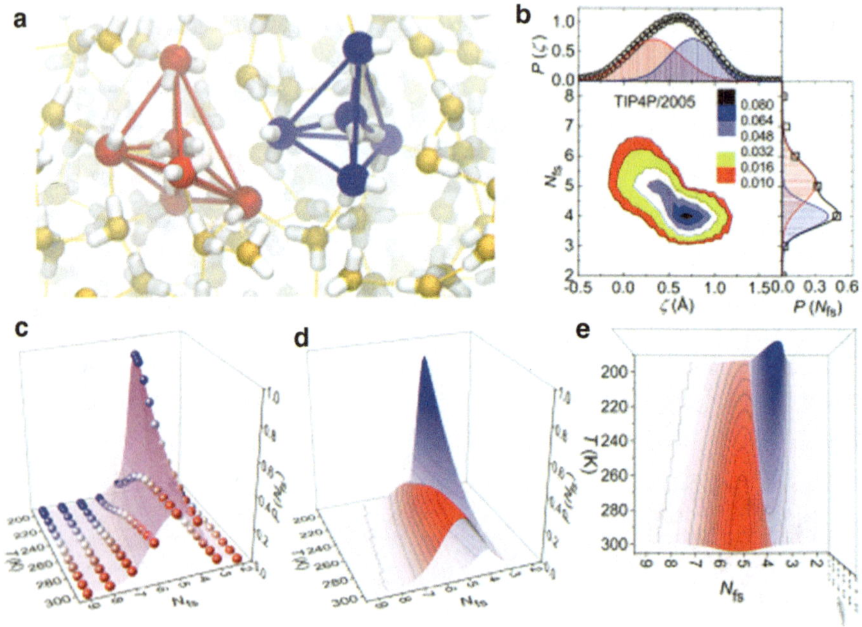

Fig. 4.33 Structural bimodality in the coordination number distribution of liquid water. **a** Snapshot of liquid TIP4P/2005 water at 1 bar and 240 K, where s ≈ 0.5, i.e., near the Schottky temperature, Ts=1/2. Two types of local structures, LFTS and DNLS, are highlighted by blue and red colors, respectively. LFTS has four H-bonded nearest neighbors with tetrahedral coordination, whereas DNLS typically has five or more nearest neighbors (typically three of them being H-bonded) with broken tetrahedral symmetry. **b** Correlation between the number of water molecules in the spherical first-shell volume of radius of 3.5 Å, Nfs, i.e., the coordination number, and tructural descriptor ζ in TIP5P/2005 water at 1 bar and 240 K. The distributions of Nfs and ζ are shown in the right and top panels, respectively. Both distributions show clear bimodal features, which can be properly described by the sum of two Gaussian functions (black lines), with blue and red shades corresponding to LFTS and DNLS, respectively. **c** Distribution of coordination number Nfs as a function of temperature, P(Nfs,T), in TIP5P/2005 water shown by colored balls with red and blue for higher and lower temperatures, respectively. The colored surface is the fit to two Gaussian functions (eqs S1−S5). **d** and **e** T dependence ofthe two Gaussian components of P(Nfs, T) in the (**d**) side and (**e**) top views. One Gaussian component located at Nfs ≈ 4 corresponds to LFTS, whereas the other one at Nfs ≃ 5 corresponds to DNLS, in agreement with the snapshot in (**a**). The fraction of the two Gaussian components is consistent with the fraction determined by the ζ distribution in (**b**) and the theoretical two-state model. A microscopic structural descriptor ζ is the difference in the distance to the central water molecule from the nearest non-H-bonded neighbor and that from the farthest H-bonded one [14]. With permission from American Chemical Society

from low to high k where k is the wave vector; (2) the O–O partial structure factor of TIP4P/2005 water at low T (locally favored tetrahedral structure LFTS-dominant) and high P (disordered normal-liquid structure DNLS-dominant), respectively, were revealed; and (3) the ζ-dependent O–O partial structure factor $S(k, \zeta)$ of TIP4P/2005 water at 1 bar and 240 K, where water has an equal amount of LFTS and DNLS (or s = 1/2). It was revealed that liquid water provides crucial information on the fraction of LFTS and DNLS.

It shows a characteristic peak of normal disordered systems at $k_{\mathrm{D1}} = k_{r\mathrm{OO}}/2\pi \simeq 1$ (see the arrow). (*D* and *E*) ζ-dependent O–O partial structure factor $S(k, \zeta)$ of TIP4P/2005 water at 1 bar and 240 K, calculated by Debye's scattering equation (3), in (*D*) side and (*E*) top views. The characteristic peaks, k_{T1} and k_{D1}, are highlighted by blue and red circles, respectively (*F*). The distribution of ζ shows two Gaussian components corresponding to LFTS (blue shade) and DNLS (red shade), respectively. The wavenumbers in *B*, *C*, *D*, and *E* are scaled by nearest neighbor distance r_{AA} for all cases (A = C, Si, Ge, and O in water).

4.6 Neat Water Dynamics

4.6.1 General

The water O-H stretching vibrations give rise to absorption bands with band origins at 3657 and 3756 cm^{-1} in the gas phase. In the microscopic structure of water in condensed states, complexity in structure and dynamics arises due to a subtle balance between directional hydrogen-bonded interactions and weaker van der Waals (vdW) interactions which are non-directional (17, 29–41). The presence of hydrogen bonds' HB and the related local tetrahedral structure directly affect the structural relaxation of the liquid, the molecule diffusivity, and the reorientational timescale at the molecular level. This set of intermolecular interactions gives rise to many of the thermodynamic and kinetic behaviors observed in water, particularly in the liquid phase, which significantly differs from many other liquids. Investigation of corresponding processes allows to understand the fundamental molecular transport properties, such as water mass diffusion, hydrogen network relaxation, proton spin relaxation, and proton tunneling (17, 29–41). The physical methods described in Chaps. 2 and 3 are involved in solving these problems. In the next section, we focus primarily on the application of time-resolved vibrational spectroscopies to water and compare results obtained for liquid water with those obtained for crystalline ice Ih and amorphous ice.

Water structure is intimately linked to the dynamics of water molecules, which can be measured using Raman and infrared spectroscopies. The assignment of spectral peaks depends on whether they are collective modes or single-molecule motions. Vibrational modes in liquids are usually considered to be associated to the motions of single molecules or small clusters.

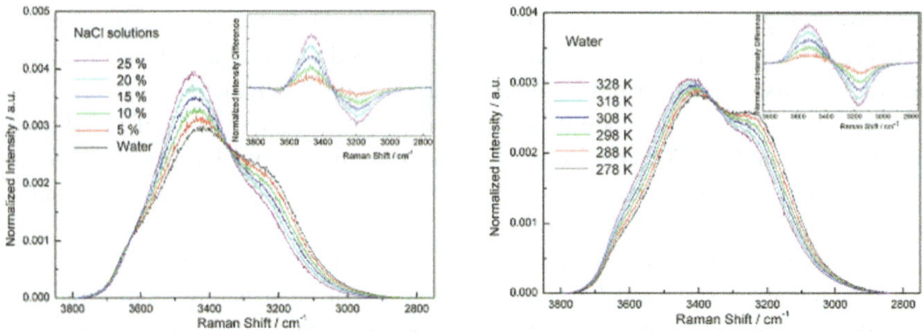

Fig. 4.34 **a** Raman OH stretching band of water at various temperatures from 5 to 55 °C [31]. **b** The Raman OH stretching band of NaCl solutions [31]. The inserts in (**a**) and **b** show the normalized difference spectra [30]. With permission from Elsevier

A comparison between temperature and NaCl induced effects on the Raman OH stretch spectra is shown in Fig. 4.34 [30]. As it is seen in the figure, the 3200 cm^{-1} feature drops in intensity, whereas the 3450 cm^{-1} peak increases in intensity both with increasing temperature and with NaCl concentration. It was proposed that the 3200 cm^{-1} feature is related to tetrahedral structures, while the 3450 cm^{-1} peak is connected to the distorted species.

Results of the investigation of water dynamics as a function of pressure at 273 K and at 297 K employing a nonlinear time-resolved spectroscopic technique were reported [32]. The experiment was based on the heterodyne-detected optical Kerr effect (HD-OKE). The HD-OKE signal, measured at two different pressures and temperatures (Fig. 4.35), indicates the intermolecular vibrational dynamics, extending up to about 1 ps, merging into the structural relaxation that lasts longer. The structural relaxation changes with the applied pressure, and the relaxation gets faster as pressure increases, while it slows in consequence of further pressure increments.

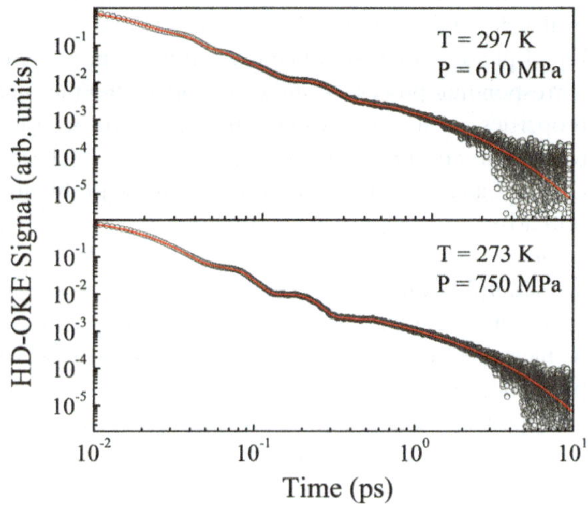

Fig. 4.35 HD-OKE signals measured at two different pressures for the two isotherms. The signal shows, at short times (< 1 ps), fast oscillations due to the intermolecular vibrational dynamics and, at longer times, the monotonic decay due to structural relaxation [32]. With permission from American Chemical Society

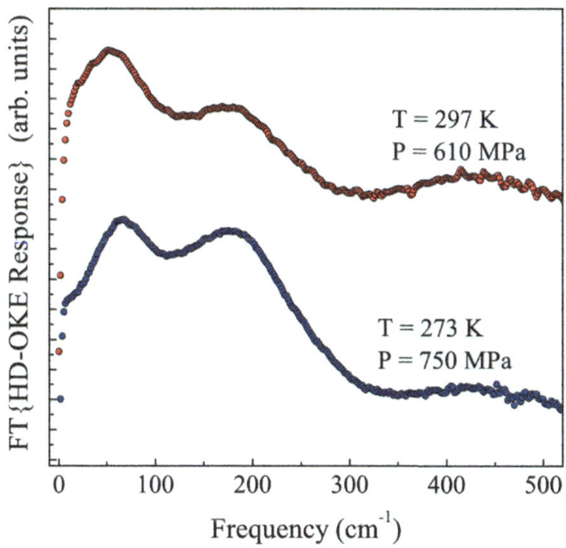

Fig. 4.36 HD-OKE response in the frequency domain, Im[$\tilde{R}(\omega)$], obtained as the Fourier transform of the time-dependent data, after the contribution of the instrumental function has been removed in the deconvolution process [32]. With permission from American Chemical Society

The frequency-domain response function $\tilde{R}(\omega)$ was obtained as the Fourier transform of the HD-OKE signal $S(t)$. Two bands of the frequency response function Im[$\tilde{R}(\omega)$] spectra (Fig. 4.36) were associated to intermolecular vibrations. Further analysis of data on the pressure influence on water dynamics studied by femtosecond time-resolved optical Kerr effect suggested the presence in the water phase diagram of a crossover area that divides two regions characterized by different dynamic regimes, which appear to be related to two liquid forms, one dominated by the high density water and the other by the low density water.

The frequency of a vibrational mode of water monitored by sum-frequency generation (4.37) provides information on the local environment of the interfacial water molecules [26]. The O–H stretch spectrum contains two features; a broad 3000–3500 cm^{-1} band and a sharp 3700 cm^{-1} peak. The broad 3000–3500 cm^{-1} feature evidenced the H bonding of the interfacial water, while the sharp 3700 cm^{-1} peak arises from the dangling O–H groups sticking out of the water interface. The SFG and visible beams were not in resonance with a vibrational transition, while the IR beam was resonant with a vibrational transition of the molecule. The Fresnel factors of visible and SFG are featureless, while the Fresnel factors of IR shows significant frequency dependence (Fig. 4.37). The Fresnel factor is defined as a macroscopic or mesoscopic property, which describes the relationship of the electric field in two different media. The selected simulated spectra together with the experimental spectra measured with and without the Fresnel factor corrections showed the scattered ratio of the amplitudes of the 3700 and 3400 cm^{-1} peaks. Proposed model of dynamic of the basal surface of ice–air is displayed in Fig. 4.38.

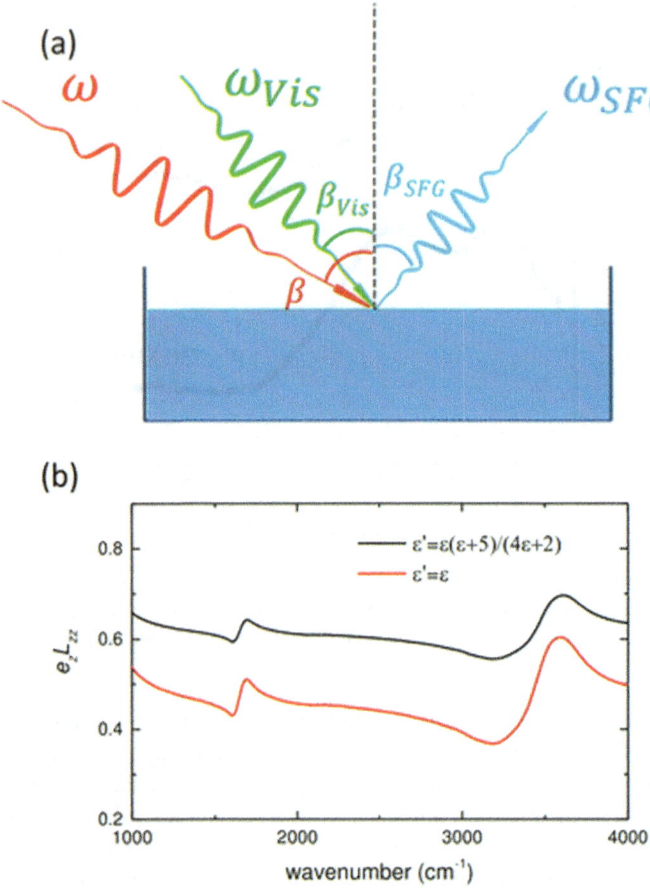

Fig. 4.37 **a** Geometry of input/output beams in a SFG experiment. **b** Frequency-dependent IR Fresnel factor at the water–air interface, $e_z L_{zz}$, calculated with $\varepsilon' = \varepsilon$ and $\varepsilon' = \varepsilon(\varepsilon + 5)/(4\varepsilon + 2)$98 assuming an IR incident angle of 62° [26]. With permission from American Chemical Society

4.6.2 Phonon and Collective Dynamics

- The hydrogen-bond network of water supports propagating optical phonon-like modes [33–38]. Monitoring the cooperative relaxation of high-frequency ω_H and low-frequency ω_L phonons would suffice to examine the cooperativity of the O:H–O bond under excitation. The advantage of phonon spectroscopy is the Fourier transformation between the real and the energy spaces. Each spectral feature represents all bonds with the same vibrational attributes
- A technique to measure the decoherence time of optical phonons in a solid using Raman spectral interference was presented [33]. Phonons are excited with a pair

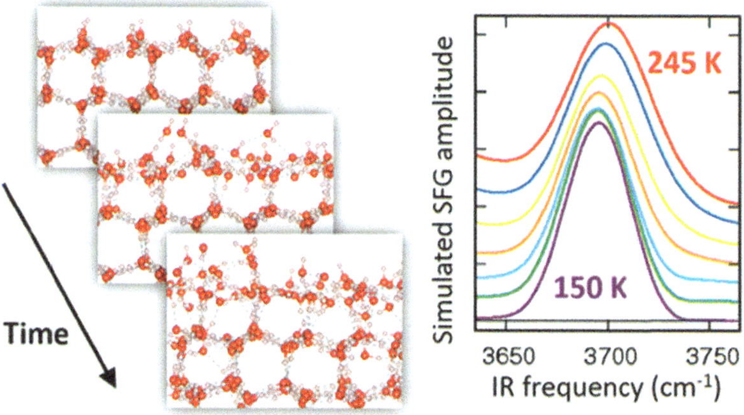

Fig. 4.38 Schematic of the basal surface of ice–air with its perfect form at different time [26]. With permission from American Chemical Society

of time-delayed 80 fs near-infrared pulses via spontaneous transient Raman scattering. The spectral fringe visibility of the resulting Raman pulse pair, as a function of time delay, was used to measure the phonon dephasing time. The dephasing time of phonons created in bulk diamond was measured to be $t = 6.8$ ps ($\Delta\upsilon = 1.56$ cm^{-1}). Raman and FTIR measurements of phonon spectra over the full frequency range of water during cooling, compressing, heating, and salting were probed [34]. A combination of two database sets of Raman and FTIR measurements probed phonon spectra over the full frequency range of water was employed to verify predictions of the O:H–O bond cooperative relaxation. Particularly, a spectral differential strategy enabled the distillation of the phonon spectral features due to the conditioning of skin and defects, heating, salting, etc., by differencing the spectra collected before and after conditioning. This strategy was also applied to calculations to discriminate the characteristic bond length d_x or phonon frequency ω_x peaks obtained under different conditions.

Typical phonon spectra for ambient water probed using spectroscopies of Fourier transform infrared (FTIR) absorption, Raman reflection, and neutron diffraction are exposed in Fig. 4.39 [34]. The characteristic frequency features centered at $\omega_H \approx$ 3450 and 3200 cm^{-1} correspond to the H–O stretching phonons in the skin and in the bulk, respectively. The peak at ω_{B1} at 500–700 cm^{-1} is the O:H–O bending mode. The peak of ω_{B2} at 1600–1750 cm^{-1} is the librational mode of H–O–H bending. Peaks at $\omega_L < 300$ cm^{-1} arise from O:H stretching. Water molecules in the gaseous phase exhibit a peak in the vicinity of $\omega_H \approx 3650$ cm^{-1}. The intensity of the peaks below 500 cm^{-1} is weak, but it is sensitive to a stimulus; any perturbation, even sunlight irradiation, changes the spectra of liquid water. Thus, monitoring the cooperative relaxation of high-frequency ω_H and low-frequency ω_L phonons allowed to examine the cooperativity of the O:H–O bond under excitation.

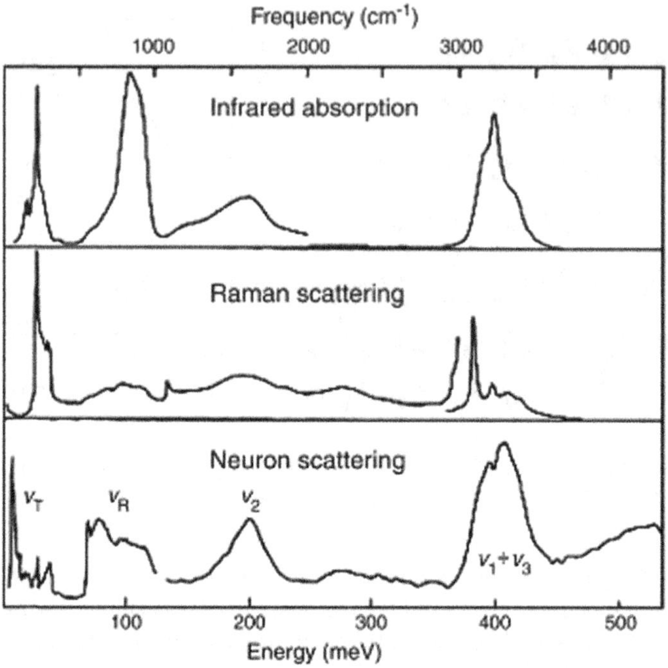

Fig. 4.39 Typical vibrational spectra measured from bulk water at ambient temperature. Peaks at $\omega_L = 300\ cm^{-1}$ correspond to O:H stretching; the peak at, $\omega_{B1} = 500$–$700\ cm^{-1}$ is from O:H–O bending; features $\omega_{B2} = 1600$–$1750\ cm^{-1}$ are the H–O–H bending libration mode; features centered at $\omega_H = 3200$ and $3450\ cm^{-1}$ are the H–O stretching mode in the bulk and in the skin of water. Reprinted with permission from [35]

The advantage of phonon spectroscopy is the Fourier transformation between the real and the energy spaces, and each spectral feature represents all bonds with the same vibrational attributes, regardless of their number or location in real space. Molecular dynamics simulations, using polarizable model (TTM3F) and the non-polarizable model TIP4P/2005f, predicted that on subpicosecond timescales dispersive optical phonon-like modes in the librational and OH stretching bands propagate through water's hydrogen-bond network over distances of up to 2 nm [36]. Figure 4.40 shows the imaginary part of the longitudinal and transverse susceptibility for model TTM3F. In the longitudinal case, the librational peak is seen to shift with $k = 2\pi/\lambda$, while in the transverse case, the lower frequency portion of the band is seen to shift slightly with k (Fig. 4.41).

Results of a comparison of the non-polarizable model TIP4P/2005f, the polarizable model TTM3F, and experiment at $k = 0$ the effect of polarization at low frequencies are presented in Fig. 4.42. As it is seen from the figure, the location of the peaks in TTM3F are in good agreement with the experimental data at 298 K.

High-resolution, inelastic X-ray scattering measurements of the dynamic structure factor $S(Q)$ of liquid water have been performed for wave vectors Q between 4 and

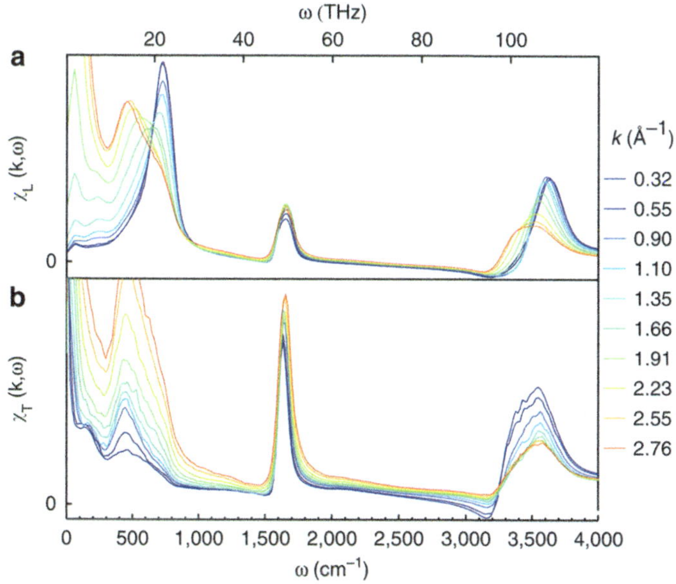

Fig. 4.40 **a** Transverse and **b** longitudinal susceptibility from simulations with TTM3F at 300 K. In the longitudinal spectra, both the librational ($\sim$ 750 cm^{-1}) and OH stretching peak ($\sim$ 3500 cm^{-1}) peaks exhibit dispersion [36]

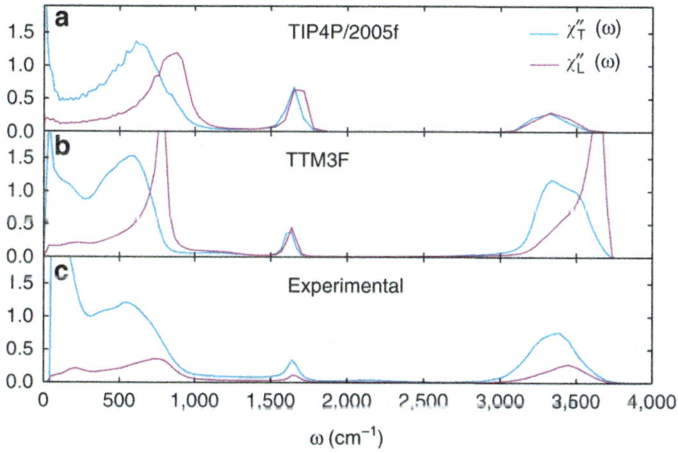

Fig. 4.41 Imaginary parts of dielectric susceptibility **a**, **b**, **c** relate to the non-polarizable model TIP4P/2005f **a**, model (TTM3F) **b** and experimental data **c**, correspindingly [36]

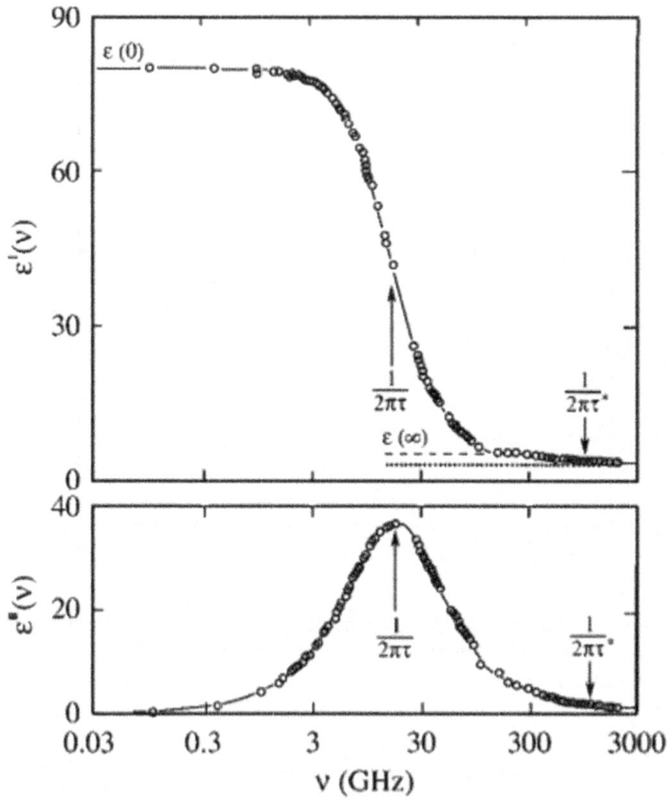

Fig. 4.42 Complex dielectric spectrum of water at 20 °C, some data have been measured at 19 °C. The dashed line indicates the extrapolated high-frequency permittivity $\varepsilon(\infty)$ of the principal relaxation of water, corresponding with the nearly exponential slow part in the dielectric decay function. The dotted line shows the extrapolated high-frequency permittivity of an additional relaxation with relaxation frequency $(2\pi\tau *^{-1})$ around 1000 GHz [39].

30 nm^{-1} at $T = 263$–420 K, ambient pressure at $P = 2$ kbar [37]. The study of their temperature and Q-dependent two inelastic contributions (one dispersing with Q and the other non-dispersive) provided strong support for a dynamics of liquid water controlled by the structural relaxation process. A viscoelastic analysis of the Q-dispersing mode, associated with the longitudinal dynamics, revealed that the sound velocity underwent a complete transition from the adiabatic sound velocity (viscous limit) to the infinite-frequency **sound velocity** (elastic limit). The second excitation was considered as to be a signature of with the transverse dynamics with a longitudinal symmetry component.

A detailed description of the vibrational dynamics of water as measured by inelastic neutron scattering was reported [19]. In particular, when the energy range up to 150 meV, the contribution to the vibrational density of states (VDOS) mainly comes from intermolecular vibration and librational motion [19]. The bulk water

VDOS is characterized by the presence of three main bands: (1) at 6 meV, a broad peak represents the collective modes of hydrogen-bond bending (HBB), perpendicular to the line formed by the hydrogen bond O–H···O; (2) the region between 20 and 35 meV is characterized by a very weak band, which is attributed to hydrogen-bond stretching (HBS) modes parallel to the line formed by the hydrogen bond O–H·· O; and (3) an intense band in the range 50–130 meV (LIB) which is associated with the higher energies of librational motion of water molecules. Thus, HBB and HBS bands correspond to the vibrational, phonon-like, and translational dynamics of water.

Joint inelastic neutron and X-ray scattering (INS and IXS) measurements have been performed on heavy water across the melting point [38]. The instrument used to collect neutron spectra was the new Cold Neutron Chopper Spectrometer (CNCS) operating at the Spallation Neutron Source (SNS) of Oak Ridge National Laboratory, Tennessee, USA. The IXS measurements were carried out at the HERIX spectrometer of the Sector 30 beamline at the Advanced Photon Source (Argonne National Laboratory, Illinois, USA) The spectra bear evidence of low- and high-frequency inelastic shoulders related to transverse and longitudinal modes, respectively. The crossover between the viscoelastic regime and an elastic one occurs whenever both the characteristic frequency and the line width of the low-frequency mode match the inverse of the structural relaxation time. The momentum Q dependence of τ at $T =$ 295 K is determined either by the full iterative fitting procedure of both IXS and INS spectra or from the unconstrained fit of the IXS ones only. It was also observed that the frequency of the transverse mode undergoes a discontinuity across the melting.

4.6.3 Dielectric Properties and Dynamics

Nowadays, the development of broadband dielectric spectroscopy (BDS), which includes the microwave range and THz region, together with the far-infrared (FIR) spectroscopy enables the evaluation of continuous dielectric spectra over an extremely broad frequency band. Complex dielectric spectrum of water at 20 °C is displayed in Fig. 4.42.

Data on dielectric loss of water and dielectric permittivity of water between 0 and 100 °C are collected in Fig. 4.43.

The review [40] described experimental permittivity data for pure water at standard atmospheric pressure, over the frequency range 0–25 THz and the temperature range 0–100 °C ($0 \leq \nu \leq 25$ THz – $0 \leq \nu \leq 25$ THz and $0 \leq t \leq 100°$ C $0 \leq t \leq 100$ °C). For example, at 25 °C, the relaxation frequencies are 18.56, 167.83, 167.83 GHz, 1.944, and 1.944 THz, and the two resonance frequencies are 4.03 and 14.48 THz. Based on this data, an interpolation function that precisely represents $\varepsilon(\nu, t)$ $\varepsilon(\nu, t)$ was derived, relaxations and resonances processes were characterized, and the temperature dependence of the relaxation and resonance parameters were determined. The main results of the work were formulated as follow:

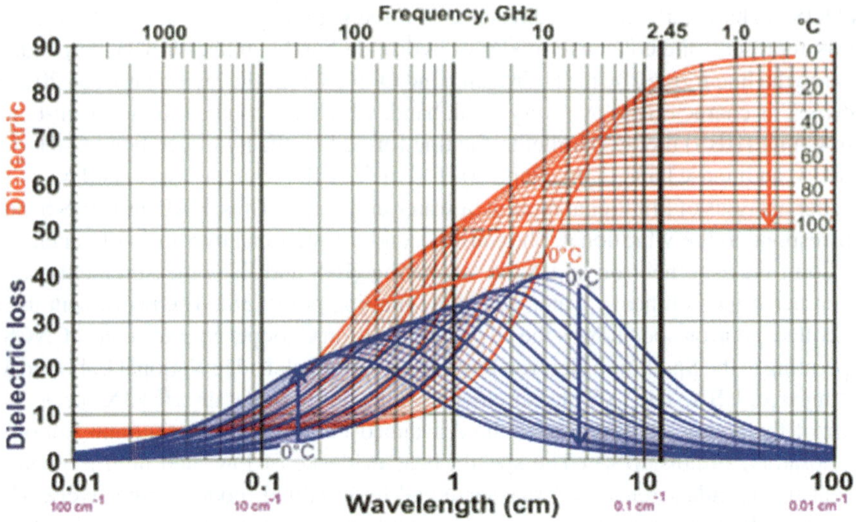

Fig. 4.43 Data on dielectric loss of water and dielectric permittivity of water between 0 and 100 °C, the arrows showing the effect in increasing temperature or increasing water activity. This page was established in 2001 and last updated by Martin Chaplin on March 9, 2020. This work is licensed under a Creative Commons Attribution—Noncommercial-No Derivative Works 2.0 UK: England and Wales License. With permission of Dr. Martin Chaplin

(1) in the long wavelength limit, these optical modes exhibit longitudinal–transverse splitting, indicating the presence of coherent long-range dipole–dipole interactions, as in ice; (2) in the Raman librational band, three peaks were assigned to the three librational motions of the water molecule—twisting ($\approx 435\ cm^{-1}$), rocking ($\approx 600\ cm^{-1}$), and wagging ($\approx 770\ cm^{-1}$); (3) Infrared spectra show two peaks at 380 and 665 cm^{-1}, and similarly dielectric spectra showed peaks at 420 and 620 cm^{-1}; (4) the lowest frequency librational Raman peak ($\approx 435\ cm^{-1}$) is a transverse optical phonon-like mode; and (5) the highest frequency peak ($\approx 770\ cm^{-1}$) is a longitudinal optical phonon-like mode and that the transverse dispersive modes are due to optical phonons that travel along the H-bond network of water.

The results on combined experimental and computational study of sub-THz spectra of liquid water, employing Fourier transform terahertz **(THz)** spectroscopy with spectral resolution of 0.25 cm^{-1} were reported [41]. The molecular dynamical (MD) simulations of water were performed employing Amber 8 and the TIP3P, SPCE (Extended Single Point Charge) and using TIP4P water models and were compared with the theoretical predictions. The dielectric constants of water and heavy water (at 0.2–3.5 THz) from 270 to 362 K were determined to decompose them into four components, namely slow relaxation, fast relaxation, intermolecular stretching vibration, and intermolecular libration which were determined [42].

The absorption spectra of liquid water and various aqueous solutions were analyzed in **a** terahertz frequency domain (from 6 to 200 cm^{-1}) which characterize

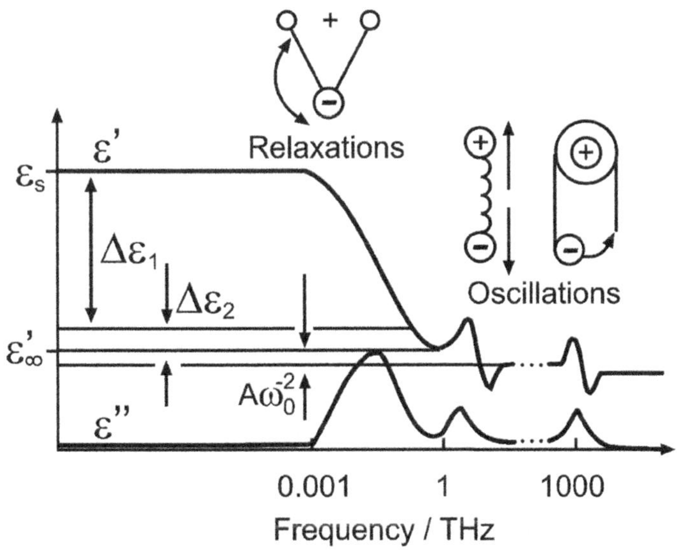

Fig. 4.44 Polarization of water molecules in the electric field [43]. With permission from American Chemical Society

the collective dynamics of water molecules [43]. Particular attention was paid to the relaxation process in the range of ~ 6–80 cm^{-1} which was be interpreted as a monomolecular relaxation of free water molecules. In a typical experiment, after applying the electric field to water, the polarization processes can be divided into four groups: (1) orientational polarization of the molecular skeleton interconnected with hydrogen bonds, (2) orientational polarization of free molecules, (3) number of faster polarization processes including the displacement of nuclei; and (4) polarization of electron shells of molecules (Fig. 4.44). Contribution of the orientation polarization of free water molecules into the permittivity is determined by the $\Delta\varepsilon 2$ amplitude (Fig. 4.44). It was concluded that the amplitude $\Delta\varepsilon 2$ of the high-frequency relaxation band R2 directly correlates with the extent of water structuredness, and in water solutions, the greater the value of $\Delta\varepsilon 2$, the less the structuredness of water and vice versa.

The obtained data allowed to interpret process under investigation as a monomolecular relaxation of free water molecules. Several ranges of the absorption spectrum of water can be related to the specific types of intermolecular dynamics which were considered (Fig. 4.45), namely. (1) Debye relaxation (R1), $1/\lambda =$ 0.6 cm^{-1}; (2) T1 around 50 cm^{-1}—attributed to bending mode between two water molecules connected by the H-bond; and (3) The stretching mode between two water molecules connected by the H-bond (T2) with $1/\lambda$ around 180 cm^{-1}. It was suggested that if the R2 band is caused by the relaxation of free water molecules, the $\Delta\varepsilon 2$ amplitude should be directly related to the number of free molecules in the liquid phase.

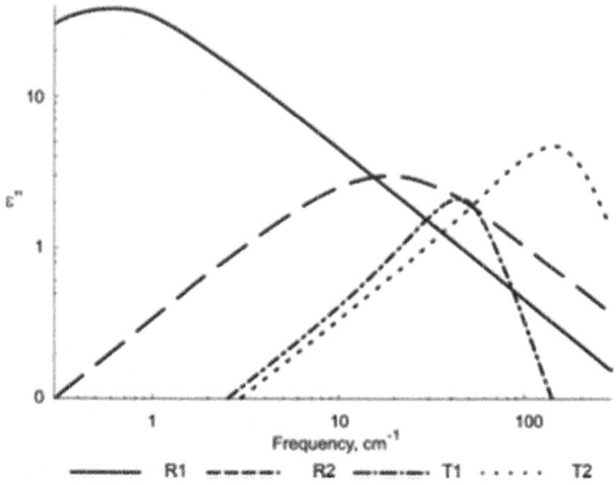

Fig. 4.45 Dielectric losses in aqueous solutions: R1 and R2 are the relaxation bands; T1 and T2 are the bands related to bending and stretching modes, respectively [43]. With permission from American Chemical Society

A high precision, tunable far-infrared terahertz frequency-domain dielectric spectrometer for studying the dynamics of biomolecules in aqueous solutions in the gigahertz-to-terahertz frequency was developed [44]. The system provides a coherent radiation source with power up to 20 mW in the gigahertz-to-terahertz region, and the dynamic range of the instrument reaches 10^{12} s^{-1} and also the measurements of the absorption and refractive index for liquid water in the frequency range from 5 GHz to 1.12 THz (0.17–37.36 cm^{-1}). Absorption coefficient and refractive index of liquid water were found to be strong functions of frequency, monotonically increasing and decreasing, respectively, with rising frequency over this entire spectral range (Fig. 4.46) [44]. The dielectric response from water at 20 °C is shown in Fig. 4.47.

In work [45], it was argued that in the overlap region of the Raman and dielectric spectra (Fig. 4.48) attributed to the dielectric relaxation in water is driven by the migration of defects through the H-bond network. This effect leads to a Debye-like peak in the lower frequencies and the deviation from the Debye law in the higher sub-THz frequencies and is traced to a global fluctuation of the same H-bond network.

The complete dielectric spectra of water in the wide frequency range (up to 10 THz) was described by the model taking in consideration two mechanisms: defect migration and global fluctuations of the H-bond network. In the framework of this model: (1) the main Debye-like peak is attributed to the defect migration; (2) the dielectric loss peak of water in the sub-THz region can be described without involving the secondary fast Debye process; (3) the origin of this deviation from the Debye law probably is an additional energy absorption due to the H-bond network vibrations; and (4) the main Debye-like peak is attributed to the defect migration. Fitting dielectric spectra and analyzing different ways of fitting the high frequency excess led to conclusion to include one or two additional Debye peaks librational region. In contrast, these data are consistent with the theory that Debye relaxation is due to the migration of Bjerrum-like defects in the hydrogen-bond network and the mechanism of Debye relaxation in liquid water is similar to the mechanism in ice.

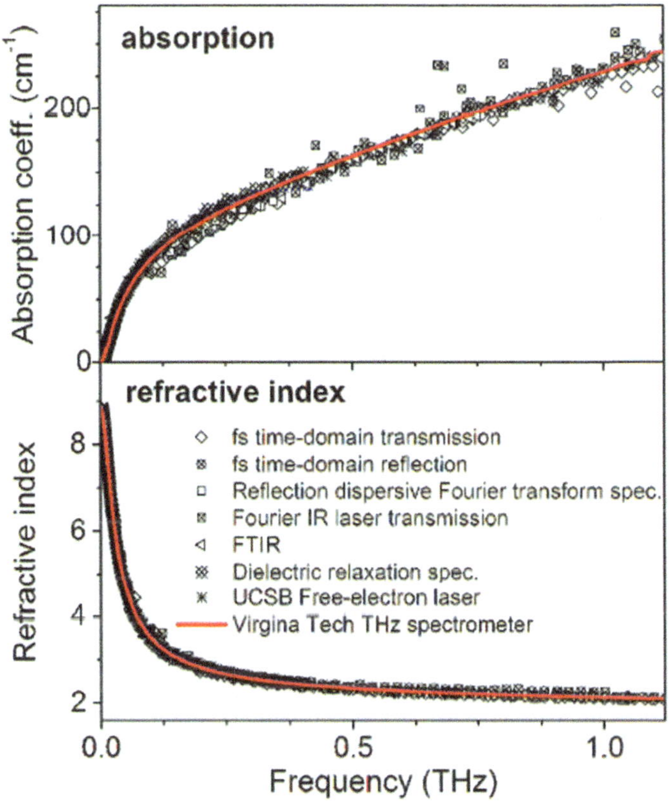

Fig. 4.46 Red, continuous lines in these two plots are spectra of water collected with described instrument. The error bars of absorption and refractive index measurements are within the thickness of the lines. Superimposed on these are data collected from the literature including measurements using FTIR interferometer (∇8 and J9), reflection-dispersive Fourier transform spectroscopy (⦶10 and U11), far-infrared lasers (N12 and I13), free-electron laser (•), 14 THz time-domain transmission (22, 23 and 24) and reflection (⊗25 and Δ26) spectroscopies, and dielectric relaxation [44]. With permission from American Institute of Physics

The literature on the Debye absorption peak of liquid water and the excess response found on the high-frequency side of the Debye peak was critically reviewed and the origin of the Debye relaxation in liquid water was discussed [46]. Under suggested model, the Debye relaxation time τ_D is related to the rotational friction constant ξ_via $\tau_D = \xi/2k_BT$. According to the Stokes equation for a sphere of radius a rotating in a medium with shear viscosity η, the rotational friction is given by $\xi = 8\pi\eta\ R^3$. This equation fits the experimental data for $\xi(T)$ and $\tau_D(T)$, with a value of $R =$ 1.44 A° at 0 °C which is about the right radius for a single water molecule. To better understand the molecular origin of Debye peak, large-scale molecular dynamics simulations and performing several different distance-dependent decompositions of the low-frequency dielectric spectra were ran. The process was found to take place on

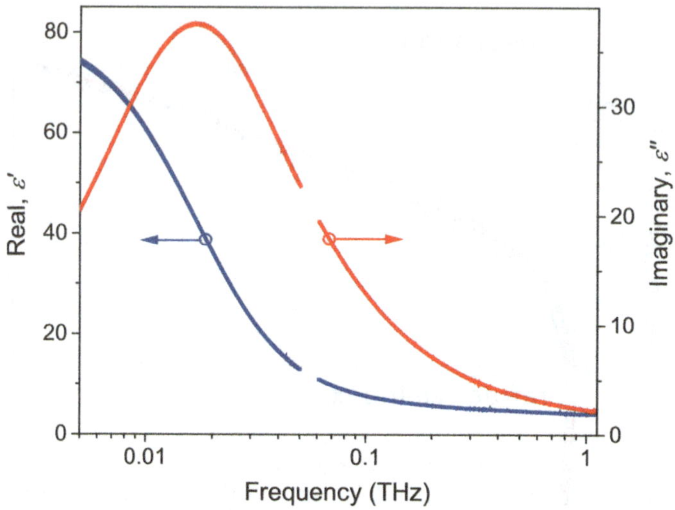

Fig. 4.47 Dielectric response from water at 20 °C is converted from the absorption coefficient and refractive index measurements. The error bars for the calculated dielectric response are within the thickness of the lines [44]. With permission from American Institute of Physics

scales of 1.5–2.0 nm. The *k*-dependence of the Debye relaxation was also calculated, finding it to be highly dispersive. Because obtained findings are inconsistent with models that relate Debye relaxation to local processes such as the rotation/translation of molecules after H-bond breaking, a model which includes one or two additional Debye peaks was analyzed (Fig. 4.49).

Figure 4.50 confirms that data on transverse and longitudinal fits with three Debye relaxations, one DHO peak for H-bond stretching, and three DHO peaks for the librational region.

The work findings are consistent with the theory of Popov et al. [3] that Debye relaxation is due to the migration of Bjerrum-like defects in the hydrogen-bond network. The author stressed that the mechanism of Debye relaxation in liquid water is similar to the mechanism in ice, but the heterogeneity and power-law dynamics of the H-bond network in water results in excess response on the high-frequency side of the peak.

4.6.4 High Amplitude Motion

Ultrafast IR spectroscopy revealed that the molecular orientation relaxation (residing) time in the neat water is 2.6 ps [47]. An analysis of molecular mechanism of water reorientation based on molecular dynamics simulations and the analytic framework of the extended jump model (EJM) was performed [48]. The water reorientation in liquid water was shown to occur through large-amplitude angular jumps due to the exchange

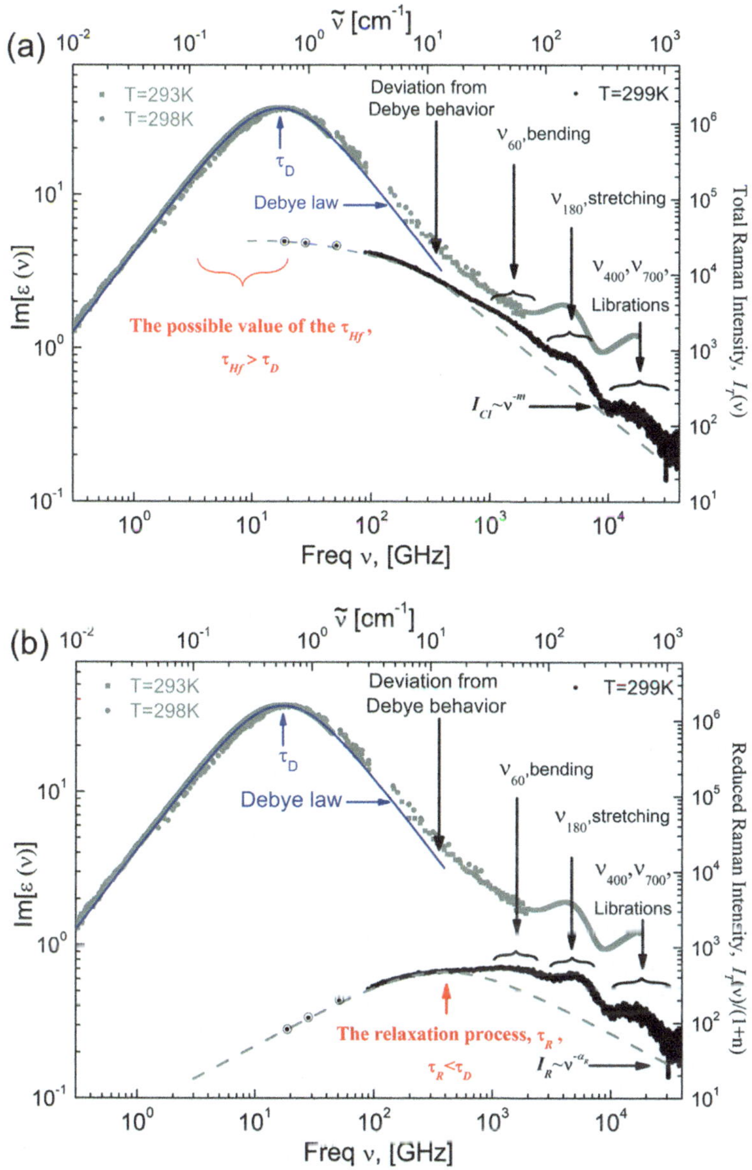

Fig. 4.48 Dielectric and Raman spectra. Both figures represent the dielectric spectra at 293, 298 K (gray squares and circles) and Raman spectra at 299 K (black circles). The additional three points measured in Raman spectra below 3 cm^{-1} are circled. On the left panel (**a**) the total Raman intensity $I_T(\nu)$ is given, whereas on the right panel (**b**) the total intensity corrected by the Bose Einstein correction factor $I_T(\nu)/(1 + n(\nu))$ is plotted. The dashed line in the panel (**a**) defines the baseline ICI which is attributed to collision-induced phenomena. The dashed line in the panel (**b**) defines the relaxation mode, which was fitted by the Cole–Cole law [3]. With permission from American Chemical Society

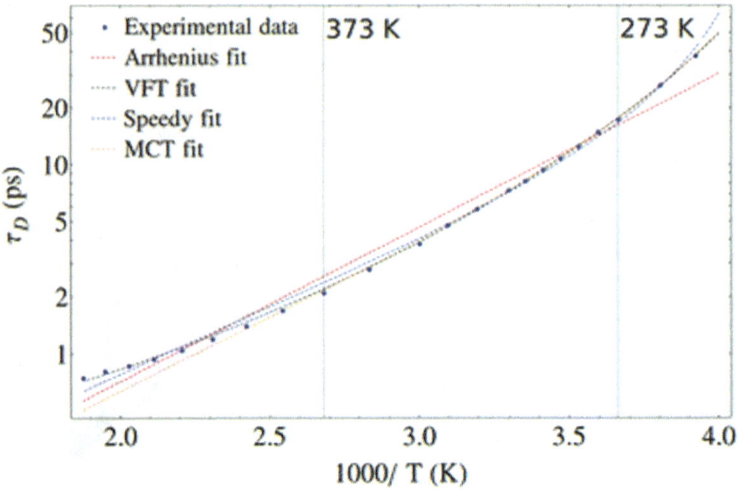

Fig. 4.49 Temperature dependence of Debye [46]. With permission of Royal Society of Chemistry

of hydrogen (H)-bond acceptors, with a minor contribution from the diffusive H-bond frame reorientation between these exchanges. To identify the jump rate-limiting step as the translational motion in which the initial H-bond of the reorientating water is elongated, a detailed analysis of the pathway-free energetics for the water reorientation was carried out. A model presented in Fig. 4.51 allowed to describe the H-bond exchange event in the water orientation. The O*–H* bond orientation proceeded by the θ angular coordinate was considered as a fast coordinate responding nearly adiabatically to the reorganization of the slower environment.

Inspection of trajectory results for the time-dependent **orientation** of individual water O–H bonds indicated the occurrence of sudden large-amplitude angular incompatible with the very small amplitude assumption (Fig. 4.52).

Table 4.1 summarizes values of the water orientation times derived from MD simulations, different models and experiments.

Proposed five-step molecular mechanism for water r**eorientation** includes the following principle stages (Fig. 4.53): (1) the environment first rearranges and reaches the symmetric configuration where O^a and O^b are equidistant from O*; (2) it provides two energetically equivalent H-bond acceptors for H*; (3) then the initial O*H*· · ·O^a H-bond breaks concertedly with the formation of the new O*H*· · ·O^b H-bond, inducing an angular jump in the O*H* bond orientation; and (4) the transition-state (TS) structure for both the slow environment and fast angular coordinates therefore corresponds to a totally symmetric bifurcated H-bond configuration, where H* donates a H-bond to both O^a and O^b. Temperature dependence of the water τ_2 reorientation time determined from the long-time component of the reorientational decay was simulated and compared with the integrated reorientation time and experiments (Fig. 4.53). In a subsequent study using NMR and ultrafast infrared spectroscopies, it was confirmed that that water reorientation is not a diffusive process and occurs through large angular jumps [49] (Fig. 4.54).

Fig. 4.50 Example transverse (top) and longitudinal (bottom) fits. ω_0 and γ are all reported in cm^{-1}. The fit contained three Debye relaxations, one DHO peak for H-bond stretching, and three DHO peaks for the librational region [46]. With permission of Royal Society of Chemistry

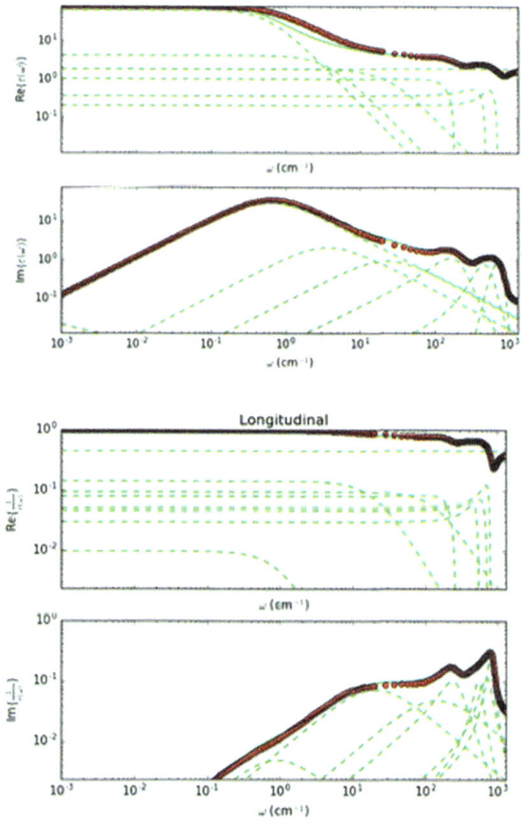

name (Transverse)	f	ω_0	τ (ps)	γ	σ
Debye	65.0	0.60	8.88		
2nd Debye	4.27	3.71	1.43		
3rd Debye	1.86	22.5	0.24		
Brendel Hbond Str	3.43	180	0.186	117	53.2
Brendel L1	1.28	474	0.070	107	120
Brendel L2	0.747	691	0.048	100	150
Brendel L3	0.010	689	0.048	227	46.8
eps inf	1.88				
name (Longitudinal)	f	ω_0	τ (ps)	γ	σ
Debye	0.01	0.9	5.72		
2nd Debye	0.14	20.3	0.26		
3rd Debye	0.10	146	0.04		
Brendel Hbond Str	0.31	259	0.129	221	2.85
Brendel L1	0.18	704	0.047	284	12.1
Brendel L2	0.20	782	0.043	183	7.20
Brendel L3	0.12	538	0.062	163	21.7
eps inf L	0.45				

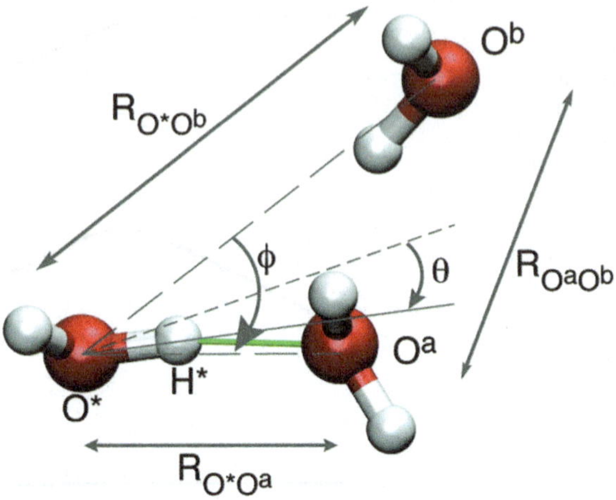

Fig. 4.51 Definition of the different geometric coordinates employed to describe the H-bond exchange event; $R_{\mathrm{O^*Oa}}$, $R_{\mathrm{O^*Ob}}$, and R_{OaOb} are the distances between the different water molecules, θ is the angle between the O*H* bond and the bisector plane of, and φ is the angle [48]. With permission from American Chemical Society

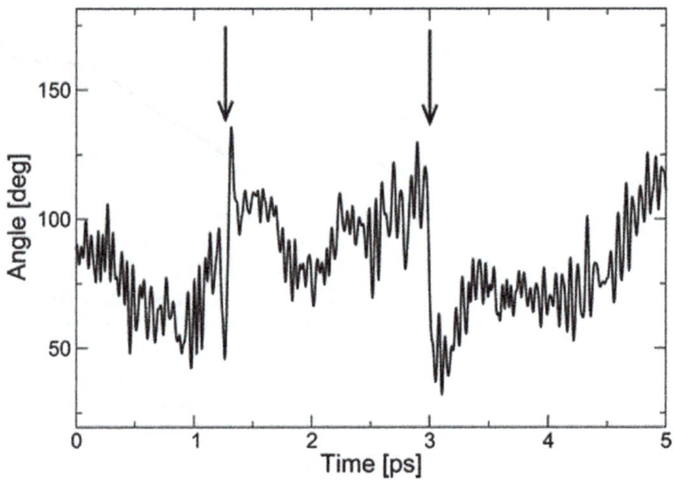

Fig. 4.52 Time evolution of the angles between a selected OH-bond and a laboratory frame axis. The arrows indicates the presence of large-amplitude jumps [48]. With permission from American Chemical Society

Combining diamond anvil cell technology, creating high pressures, ultrafast pump–probe infrared spectroscopy, and classical molecular dynamics simulations indicated that the liquid structure and orientational dynamics are intimately connected, identifying the P–T range of the low-density water (LDW) and high-density water (HDW) regimes [50]. The comparison of the spectroscopic data with

Table 4.1 Reorientation times obtained, respectively, from MD simulations, the diffusive model, the simple jump model, the frame reorientation between the jumps, the EJM, and the experiments [48]

	τ_D (ps)	τ_1 (ps)	τ_2 (ps)
MD (SPCE)	9.7	4.9	2.5
OH diffusion ($A_{DH} = 0.45\ ps^{-1}$)		1.1	0.4
Jump model ($A0 = 68°$, $t_0 = 3.3$ ps)		8.1	3.6
00 frame reorientation		15.5	5.6
Extended jump model (7)		5.3	2.2
Experiment	8.5	2–7.5	1.9–3.0

With permission from American Chemical Society

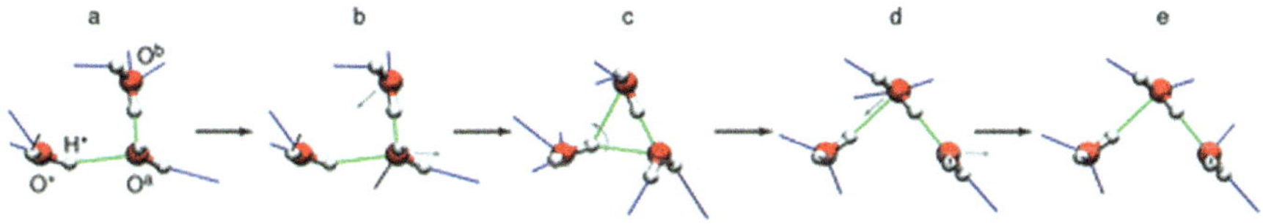

Fig. 4.53 Proposed five-step molecular mechanism for water reorientation. **a–d** are the principle stages of water reorientation [48]. With permission from American Chemical Society

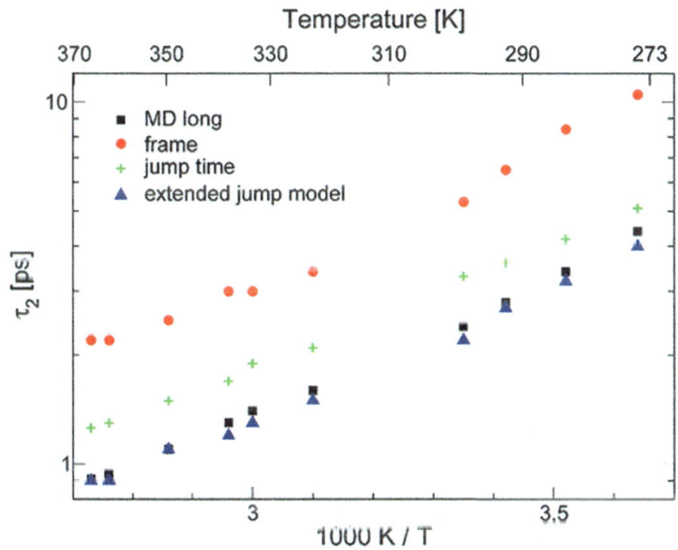

Fig. 4.54 Temperature dependence of the water τ_2 reorientation time determined from the long-time component of the reorientational decay in the simulations, from the integrated reorientation time, and from NMR experiments, together with an Arrhenius fit with a 3.5 kcal/mol activation energy [48]

the computational results highlighted how the structural evolution of liquid water with pressure at room temperature corresponds to a dynamical configuration change involving nearest neighbor molecules without altering the local hydrogen-bonded tetrahedral configuration. The connection between experiments and computations was represented by the determination of the **rotational anisotropy** decay with τ_R computed with the two potentials (Fig. 4.55) which share the behavior seen in experiments.

Thus, ultrafast infrared pump–probe spectroscopy was proved to be a technique able to provide detailed information on the orientational motion of water molecules and on the ultrafast dynamics of the hydrogen-bond network. In particular, it was found that the modification induced in liquid water by changing pressure and temperature is better described as the crossover between two regimes, for which the usual designation HDW and LDW can be maintained.

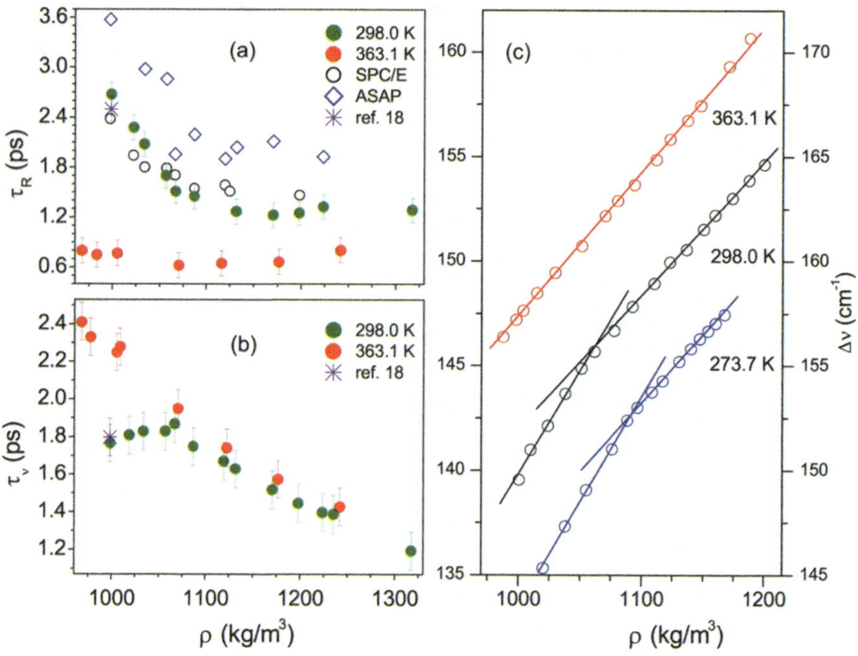

Fig. 4.55 **a** Experimental (full symbols) and ambient temperature computed (open symbols) rotational anisotropy time constant as a function of density at two different temperatures determined by a single exponential fit of $R(t)$. The computed rotational anisotropy decay has been obtained by the time correlation function of the second rank Legendre polynomial of the dipole moment direction unit vector for each HOD molecule.18, **b** OD stretching lifetime obtained by experimental $\Delta\alpha iso(t)$; **c** FWHM of the OD stretching mode as a function of density. For sake of better comparison, the left FWHM scale refers to the 273.7 and 298.0 K traces, and the right scale to the 363.1 K data. Full lines are linear regressions of the experimental data [50]. With permission from American Chemical Society

First measurements of the self-dynamics of liquid water in the GPa range specially conceived for quasi-elastic neutron scattering studies were reported [51]. A direct measurement of both the translational and rotational **diffusion** coefficients of water along the 400 K isotherm up to 3 GPa, corresponding to the melting point, was performed and compared with molecular dynamics simulations. The dynamics of liquid water at 400 K up to 3 GPa, indicating the pressure dependence of the translational D_T and of rotational D_R diffusion coefficients, is presented in Fig. 4.56. D_T was

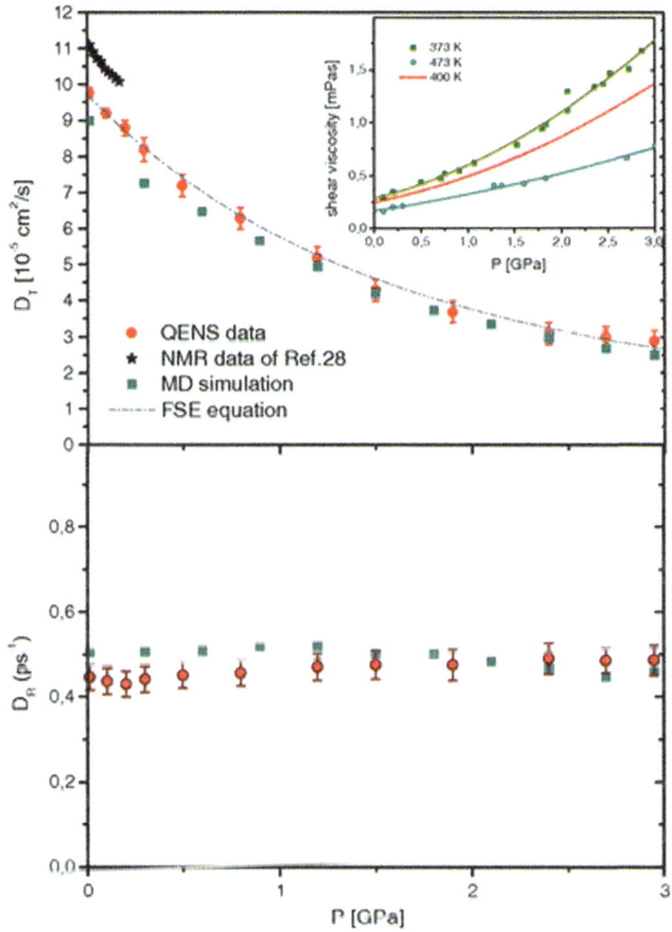

Fig. 4.56 Pressure dependence of the translational D_T and of rotational D_R **diffusion** coefficients along the isotherm 400 K as measured by QENS (red dots) and by MD simulations performed by using TIP4P2005 potential for water (blue squares). Black stars represent the available low-pressure data on D_T as measured by NMR at 403 K. The gray curve represents the FSE equation prediction with a fractional coefficient $\xi = 0{:}76$. In the inset, we show the shear viscosity dependence on pressure along the 400 K isotherm (red curve), at 373 K (green dots), and at 437 K (cyan dots) [51]. With permission from American Physical Society [51]

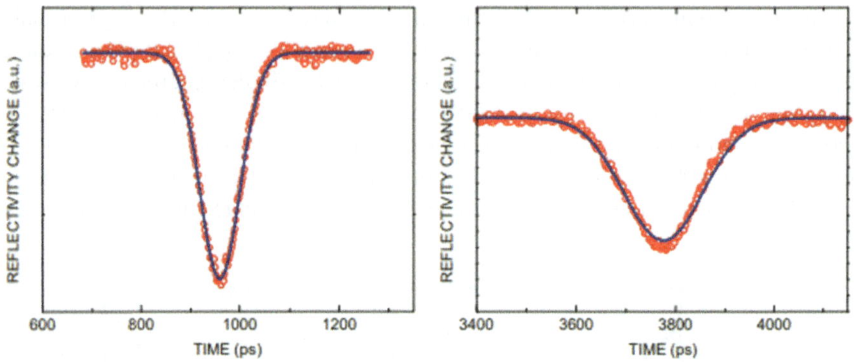

Fig. 4.57 Reflectivity change as a function of time (solid lines) and the corresponding echo obtained experimentally (open circles). The left and right panels show the echoes with transit time 960 ps and 3770 ps, respectively [53]. Published under a CC-BY licence

compared with the macroscopic diffusion coefficient measured by NMR at the same temperature and lower pressures. As it is seen in the figure, a marked D_T dependence on pressure in the explored P range exists, but its variation is almost twice slower than that of the shear viscosity. The observed trend of translational diffusion with pressure (Fig. 4.57) with a fractional coefficient ξ of 0.76 matches the prediction of the entropic barrier hopping theory [52]. Through comparison with structural data and molecular dynamics simulations, it was suggested that this is a consequence of the rigidity of the first neighbors shell and of the invariance of the number of hydrogen bonds of a water molecule under high pressure.

4.6.5 *Ultrasonic Measurements*

The propagation of sound in liquids has been widely studied in the past decades [53–59]. A coherent experiment was performed on liquid D_2O at room temperature. The collective high-frequency sound mode predicted in 1974 [54] by computer molecular dynamics simulations was experimentally observed. Representation of the dispersion curve gave a velocity equal to 3310 m s^{-1} which was interpreted as being a mode propagating within the hydrogen-bonded patches existing in liquid water and distinct from the ordinary sound wave. The velocity (1390 ms^{-1}) was also measured up to 5 GHz by Brillouin light scattering.

Inelastic X-ray scattering measurements in water and ice showed that sound waves with wavelengths between 0.5 and 3 nm propagate at the same velocity in both phases [55]. Specifically, the obtained velocity ~ 3200 ms^{-1} in the Q range 4–14 nm^{-1} was lower than the value of 4000 m s^{-1} observed in polycrystalline ice Ih at ultrasonic frequencies, and more than twice the value of ~ 1500 ms^{-1} found in water in the hydrodynamic limit. Thus, the dynamical response of the two phases is similar at very short wavelengths and in the momentum-transfer range ($\boldsymbol{Q}$) investigated. Collective longitudinal excitations propagate with the same velocity in the liquid and solid phase of H_2O. In the low range ($\mathbf{Q} < 0.003$ Å^{-1}), bulk water at ambient conditions exhibits a single longitudinal acoustic (LA) mode, propagating at a velocity equal to the adiabatic one [56]. At higher Q values (0.1–0.4 Å^{-1}), a large increase of the speed of sound up to more than 3000 m/s was observed by INS. The result suggested that the fast sound represents the propagation of a collective mode in a solid-like network of molecules occurring at frequencies larger than the reciprocal of the liquid relaxation timescale.

The velocity and attenuation of phonons in water in the MHz frequency range has been measured by the pulse-echo ultrasonic technique. Brillouin scattering and picosecond acoustic interferometry measurements have been made for some liquids in the frequency range 1–10 GHz [53]. The light source used for calculated pump and probe pulses was a diode-pumped mode-locked Ti:sapphire laser with pulse width about 200 fs (FWHM), repetition rate 80 MHz, and wavelength 800 nm. The sound pulse generated when light was absorbed in a metal transducer film deposited onto a substrate propagated across a thin layer of water and is then reflected back to the surface and was detected using a resonant optical cavity. The returning sound pulses contain a broadband of frequencies up to 10 GHz. This method extended the frequency band in which quantitative measurements can be made to 20 GHz or higher. Figure 4.57 illustrates application of the method by example of the reflectivity change as a function of calculated time and the corresponding echo obtained experimentally.

Inelastic X-ray scattering showed two branches in water, one with a strong dispersion and one with a weak dispersion visible [57]. The first one has been suggested to describe the longitudinal collective excitations bends up from a slope comparable to the normal velocity of sound to a slope of 3200 ms^{-1} for momentum transfers larger than 4 nm^{-1} and was assigned as fast sound. A quasi-elastic neutron scattering (QENS) study of single-particle dynamics in pure water, measured at temperatures between 256 and 293 K along an isobaric (at a constant pressure) path at 200 MPa, was carried out [58]. Specifically, analysis of the spectral line shapes indicated a departure from simple models of continuous or jump diffusion, with such an effect becoming stronger at lower temperatures. Such a diverging trend of dynamical quantities upon cooling resembles the anomalous compressibility observed in water by small-angle diffraction. These results reveal a complex interplay between single-particle diffusion and local structural rearrangements, both bearing witness of the well-known water anomalies. A fit of dynamical parameter $\tau(T)$ by a Vogel–Tammann–Fulcher law provided a critical temperature of about 220 K, close to the hypothesized position of the second critical point of water and to the Widom line which is the crossover between liquid-like and gas-like behavior in supercritical fluids. The Vogel–Fulcher–Tammann (VFT) equation is given as [59]

$$\tau(T) = \tau_0 \exp[(D_{\mathrm{T}} T_0)/(T - T_0)] \tag{4.1}$$

where D_T denotes the fragility strength coefficient and T_0 is so-called Vogel divergence temperature ($T_0 < T_g$ and $T > T_g$), being T_g the glass transition temperature. The reflectivity change as a function of time and the corresponding echo obtained experimentally is indicated in Fig. 4.57.

4.7 Miscellaneous Water Properties

Review [19] focused on the advances in X-ray and neutron scattering studies of water structure, from ambient temperature to the deeply supercooled and amorphous states, and on problems of water diffusive and collective dynamics, in various thermodynamic conditions and environments presented by 2016. The overview covers the following topics: (1) the derivation of detailed O–O pair correlation functions; (2) the diffusive dynamics of water in a wide range of temperatures (from 400 K down to supercooled water) and pressures (from ambient up to multiple gigapascals); (3) quasi-elastic neutron scattering under high pressure compared with the existing data from nuclear magnetic resonance, dielectric and infrared measurements, and modeling; (4) description of the vibrational dynamics of water as measured by inelastic neutron scattering; (5) the dependence of the water vibrational density of states on temperature and pressure, and in the presence of biological molecules; and (6) the collective dynamics of water and its dispersion curves as measured by coherent inelastic neutron scattering and inelastic X-ray scattering in different thermodynamic conditions. Taking in combination, these approaches made it possible to extend regions of the domain of existence of the liquid state, including metastable conditions, in both supercooled and supercritical water.

As an example, data on translational and rotational diffusion coefficients for liquid water as a function of temperature (left panel) and of pressure (right panel), as derived from different techniques are summarized in Figs. 4.58 and 4.59 [19].

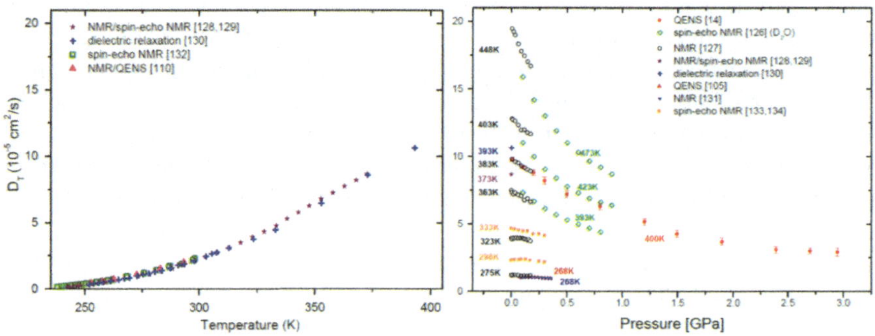

Fig. 4.58 Translational diffusion coefficient for liquid water as a function of temperature (left panel) and of pressure (right panel), as derived from different techniques [19]. With permission from American Chemical Society

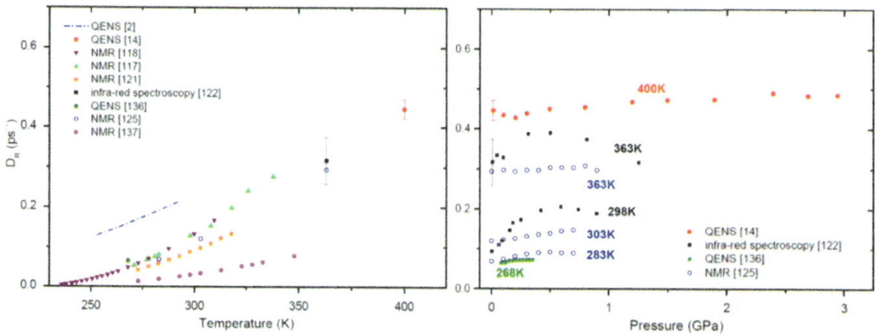

Fig. 4.59 Rotational diffusion coefficient for liquid water as a function of temperature (left panel) and of pressure (right panel), as derived from different techniques [19]

Different aspects that affect the diffusion process, including the molecular models, the system-size effects, the temperature and pressure conditions, and the type of confinement were considered [31]. The self-diffusion coefficient of water in the bulk phase and under confinement, which directly associated with the calculations of tracer or mutual diffusion coefficient of mixtures was discussed. The self diffusivity of water at the thermodynamic limit, D_∞, was calculated from (4.2):

$$D_\infty = D_{\mathrm{MD}} + k_{\mathrm{B}} T x / 6phL \tag{4.2}$$

where D_{MD} is the self-diffusivity obtained from molecular dynamic MD simulations, k_{B} is the Boltzmann constant, T is the absolute temperature, ξ is a dimensionless constant which is approximately equal to 2.837 for cubic simulation boxes, η is the shear viscosity of water, and L is the length of the simulation box. D_{MD} was computed based on the hydrodynamic theory for a spherical particle in a Stokes flow with periodic boundary conditions. The self-diffusivity of water was shown as a function of the system sizes used in the MD simulations, for four of the most widely used force fields. The experimental diffusion coefficient of water was found to be not dependent on the number of molecules, and this finding was confirmed by MD simulations. Water self-diffusion coefficient as a function of temperature at 1 bar is indicated in Fig. 4.60.

4.8 Proton and Electron Mobility in Water

Proton-charge mobility in nanoscopic water droplets in volumes less than 5 nm in diameter with tunable size was examined [60]. The diffusion of confined proton charges causes a dielectric relaxation process with a maximum loss frequency determined by the diffusion constant. For diameters < 1 nm, the diffusion constant was found to be about 100 times smaller than in bulk water. Proton-charge mobility in

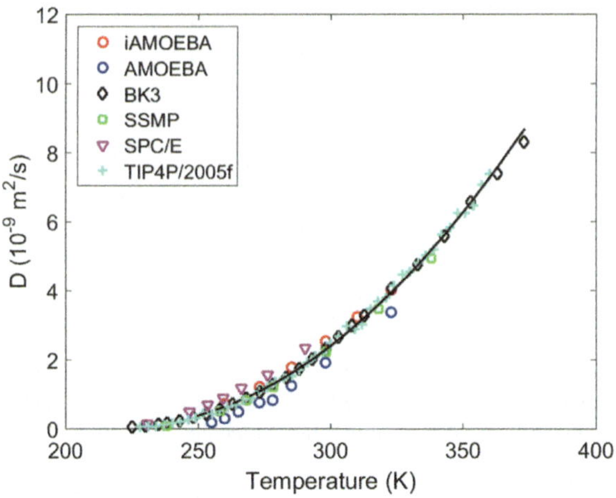

Fig. 4.60 Water self-diffusion coefficient as a function of temperature at 1 bar. Symbols denote MD studies that have included system-size corrections in the calculations by different programs. The black lines denote Speedy–Angel-type correlations of experimental data (solid line: experimental data in the temperature range 273–373 K), dashed line: experimental data in the temperature range 253–293 K [31]. With permission from Taylor & Francis Group

bulk water and nanoscopic water droplets with tunable size was investigated detecting imaginary and real parts of the dielectric difference spectra (Fig. 4.61). The protons confined in the nanodroplets exhibit the signature of a dielectric relaxation process, which occurs for all investigated droplet sizes. As an example, the calculated time is dependent on proton distribution upon switching on a DC electric field at $t = 0$ (Fig. 4.62). Experiments indicated that the proton-charge diffusion constant depends strongly on the size of the nanoscopic water volume (Fig. 4.63): (a) for the smallest investigated water-pool diameter (1.2 nm), the diffusion constant is about 100 times smaller than in bulk water and (b) for large droplet sizes, the proton-charge diffusion constant converges to the bulk value. The low mobility was suggested to be results from the more rigid hydrogen-bond network of nanoconfined water, since proton-charge mobility in water relies on collective hydrogen-bond rearrangements.

The properties of excess hydrated electrons in liquid water at room temperature were studied via coupled quantum-classical simulations in which the system evolves dynamically on the adiabatic potential energy surface with the electron maintained in the ground state throughout the process [61]. The simulations revealed that (1) the diffusion constant of the hydrated electron under field-free conditions is found to be the same as that obtained, via the Nernst–Townsend–Einstein relation, from the electron mobility simulated for a system under an electric field of 3.2×10^6 V/cm; (2) the mode of migration of the excess electron is polaronic in nature; (3) the intramolecular degrees of freedom of the water molecules influence on the hydrated electron transport properties; (4) the electron diffusion constant obtained in simulations under field-free conditions with rigid-water molecules [$D^0{}_e = (3.7 \pm 0.7) \times 10^{-5}$ cm^2/s] is larger than that obtained from simulations where a flexible-water

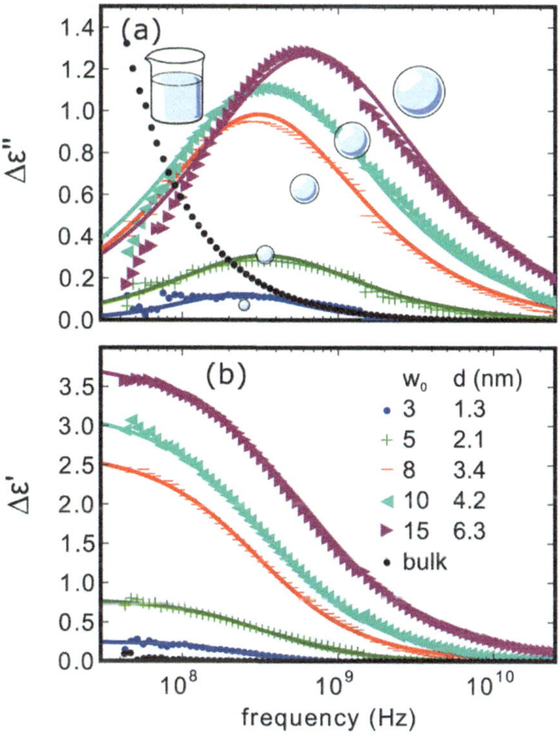

Fig. 4.61 Imaginary (**a**) and real (**b**) parts of the dielectric difference spectra $\Delta\varepsilon_{H+} = \varepsilon_{HCl(aq)} - \varepsilon_{H2O}$ showing the response of protons confined in spherical reverse micelles with different water-pool diameters. Upon increasing the reverse-micelle size, both the intensity and the center frequency increase. The black dots represent the water-background corrected response of 1M HCl solution (vertically scaled for better comparison) [60]. With permission from American Institute of Physics

model potential is employed [$D^0{}_e = (1.9 \pm 0.4) \times 10^{-5}$ cm^2/s] and smaller than the experimental estimated value obtained from conductivity measurements (4.9×10^{-5} cm^2/s); (5) the difference between the diffusion constants calculated for the two models is correlated with a marked enhancement of the probability of reversal of the direction of motion of the migrating electron in flexible water; and (6) the self-diffusion constant of water using the rigid-molecules model [$D_s = (3.6 + 0.4) \times 10^{-5}$ cm^2/s] larger than that found for the flexible-water molecule model [$D_s = (2.3 \pm 0.2) \times 10^{-5}$ cm^2/s], with the latter in agreement with the experimental value ($D_s = 2.3 \times 10^{-5}$ cm^2/s).

A detailed frequency-resolved pump–probe study of hydrated electron dynamics, performed with 5 fs pulses, was examined [62]. The experimental data were described with a model in which the excited state lifetime is ~ 50 fs in regular water and ~ 70 fs in heavy water. The deuteration effect on the lifetime suggests that OH-vibrational modes in the first solvation shell act as accepting modes for energy relaxation.

Mixed quantum–classical molecular dynamics simulations of the relaxation of a ground-state excess electron at interfaces of different phases of water with air that is systems included ambient water/air, supercooled water/air, IhIh ice/air, and amorphous solid water/air interfaces [63]. The possible connections of the examined interfacial systems to finite size cluster anions and the three-dimensional infinite, fully hydrated electron were taken in account. The main results of the simulation are: (1)

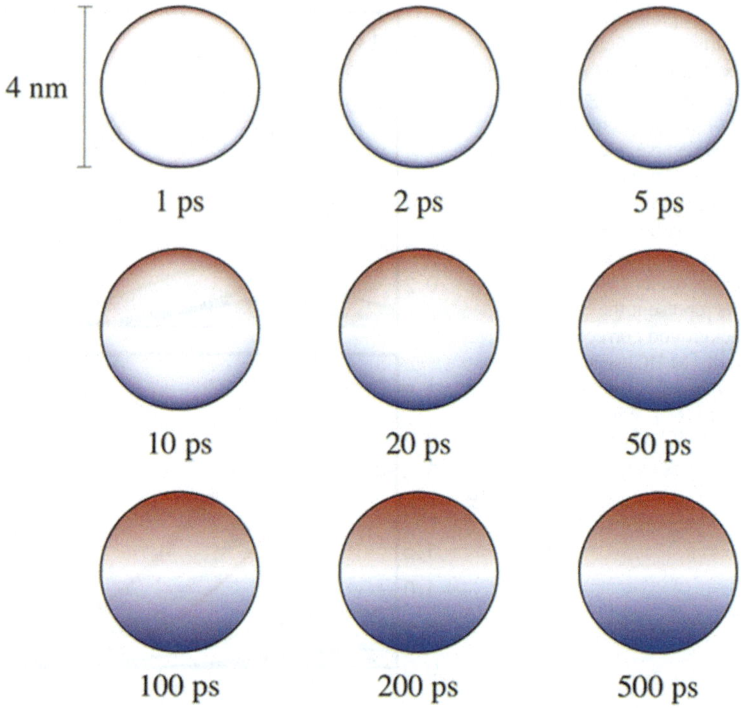

Fig. 4.62 Protons confined in the nanodroplets exhibit the signature of a dielectric relaxation process, which occurs for all investigated droplet sizes As an example, the calculated time is dependent on proton distribution upon switching on a DC electric field at $t = 0$ [60]

Fig. 4.63 Proton-charge diffusion constant in nanodroplets as a function of the water-pool diameter. The spheres above the x-axis indicate the relative sizes. The dashed line indicates the bulk proton-charge diffusion constant, and the solid line is a guide to the eye [60]. With permission of American Institute of Physics [60]

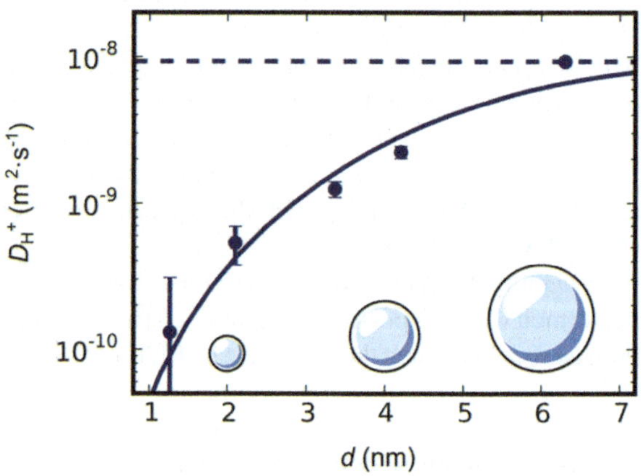

localization site analyses indicate that in the absence of nuclear relaxation, the electron localizes in a shallow potential trap on the interface in a diffuse, surface-bound (SB) state; (2) with relaxation, the weakly bound electron undergoes an ultrafast localization and stabilization on the surface with the concomitant collapse of its radius; (3) in the case of the ambient liquid interface, the electron on the 10 ps timescale diffuses into the bulk to form an interior-bound state; (4) the excess electron persists on the interface in SB states; (5) the relaxation dynamics occur through distinct SB structures which are easily distinguishable by their energetics, geometries, and interactions with the surrounding water bath; (6) the systems exhibiting the most stable SB excess electron states (supercooled water/air and Ihlh ice/air interfaces) are identified by their characteristic hydrogen-bonding motifs which are found to contain double acceptor-type water molecules in the close vicinity of the electron; and (7) these surface states correlate reasonably with those extrapolated to infinite size from simulated water cluster anions.

It was established that the excess electrons in small (i.e., $n \leq 7$) $(H_2O)_n^-$ clusters as well as larger water clusters, which serve as precursors to the hydrated electron, are bound in the dipole field of the neutral cluster and electrostatic interactions with the "free" OH-bonds of the cluster dominate the binding of the excess electron in both clusters [64]. In this way, a quantum Drude model was used to examine selected $(H_2O)_n^-$ clusters in the $n = 12$–24 size range with the goal of elucidating different possible binding motifs. Drude particles are model oscillators used to simulate the effects of electronic polarizability. In addition to the known surface and cavity states, a new binding motif, where the excess electron permeates the hydrogen-bonding network, was identified. Interestingly, electrostatic interactions dominate the binding of the excess electron only for isomers with large dipole moments, whereas in isomers without large dipole moments, polarization and correlation effects dominate, and for the network-permeating states, the excess electron binds even in the absence of electrostatic interactions.

The photoelectron spectroscopy of water cluster anions using density functional theory-based Born Oppenheimer molecular dynamics designed to mimic the different ways that water cluster anions are produced experimentally was studied [65]. This method allowed to calculate electron binding energies for this system. It was found that the simulations are able to accurately reproduce the experimentally observed vertical detachment energies (VDEs) for cluster anions of different sizes, with different VDE distributions. For cluster anion sizes up to 20 water molecules, it was seen that the excess electron always resides on the surface of the cluster. Clusters that are less thermally equilibrated have surface waters that tend to make single H-bonds to the electron and result in lower VDEs, while clusters that are more thermally equilibrated have surface waters that prefer to make two H-bonds to the electron, resulting in higher VDEs.

Electronic absorption spectra of water cluster anions $(H_2O)_n^-$, $n \leq 200$, at $T = 80$ K were obtained by photodissociation spectroscopy [66]. The results were compared with simulations from literature and experimental data for bulk hydrated electrons. For cluster sizes $20 \leq n \leq 40$, almost isoenergetic electron binding motifs were seen and assigned to surface and partially embedded isomers. With increasing cluster size, the surface isomer becomes less populated, and for $n \geq 50$, the partially embedded isomer prevails. In size of $n \approx 100$, the absorption spectrum is similar to that of the bulk hydrated electron but is slightly red-shifted.

Various aspects of the structure and dynamics of water are discussed in the works [67–77].

References

1. C.Q. Sun, X. Zhang, J. Zhou, Y. Huang, Y. Zhou, and W. Zheng, Density, elasticity, and stability anomalies of water. Molecules with fewer than four neighbors. J. Phys. Chem. Lett. **4**, 2565–2570 (2013)
2. F. Tang, O. Hasegawa, W.J. Xie, L. Xu, M. Bonn, Y. Nagata, Definition of free O-H groups of water at the air-water interface. J. Chem. Theory Comput. **14**, 357–364 (2018)
3. I. Popov, P.B. Ishai, A. Khamzin, Y. Feldman, The mechanism of the dielectric relaxation in water. Phys. Chem. Chem. Phys. **18**, 13941 (2016)
4. J.D. Bernal, R.H. Fowler, A theory of water and ionic solution, with particular reference to hydrogen and hydroxyl ions. J. Chem. Phys. **1**, 515 (1933)
5. W.L. Jorgensen, J. Chandrasekhar, J.D. Madura, R.W. Impey, M.L. Klein, Comparison of simple potential functions for simulating liquid water. J. Chem. Phys. **79**, 926 (1983)
6. M.W. Mahoney, W.L. Jorgensen, A five-site model for liquid water and the reproduction of the density anomaly by rigid, nonpolarizable potential functions. J. Chem. Phys. **112**, 8910 (2000)
7. J.L.F. Abascal, C. Vega, A general purpose model for the condense phases of water: TIP4P/2005 J. Chem. Phys. **123**, 234505 (2005)
8. C. Vega, J.L.F.Abascal, M.M. Conde, .J.L. Aragones, What ice can teach us about water interactions: a critical comparison of the performants of different modes. What ice can teach us about water interactions: a critical comparison of the performants of different modes. Faraday Disc. **141**, 251 (2009)
9. K. Szalewicz, C. Leforestier, A. van der Avoird, Comparison of simple potential functions for simulating liquid water. Chem. Phys. Lett. **482**, 1–14 (2009)
10. R. Shi, H. Tanaka, The anomalies and criticality of liquid water. PNAS **117**, 26591–26599 (2020); H.W. Horn, W.C. Swope, J. W. Pitera, J.D. Madura, T.J. Dick, G.L. Hura, T. Head-Gordon, Development of an improved four-site water model for biomolecular simulations: TIP4P-Ew. J. Chem. Phys. **120**, 9665 (2004)
11. H.L. Pi, J.L. Aragones, C. Vega, E.G. Noya, J.L.F. Abascal, M.A. Gonzalez, C. McBride, Anomalies in water as obtained from computer simulations of the TIP4P/2005 model: density maxima, and density, isothermal compressibility and heat capacity minima. Mole. Phys. Int. J. Interf. Chem. Phys. **107**, 365–374 (2009)
12. Y.D. Fomin, E.N. Tsiok, V.N. Ryzhov, V.V. Brazhkin, A novel anomalous region of water. Soft Condensed Matter. arXiv:1706.02923 (1917); S.K. Pal, A.H. Zewail, Dynamics of water in biological recognition. Chem. Rev. **104**, 2099–2124 (2004)
13. L.M. Maestro,M.I. Marqués, E. Camarillo, D. Jaque, J. García Solé, J.A. Gonzalo, F. Jaque, J.C. Del Valle, F. Mallamace, H.E. Stanley, On the existence of two states in liquid water: impact on biological and nanoscopic systems. Int. J. Nanotechnol. **13**, 667–677 (2016)
14. R. Shi, H. Tanaka, Direct evidence in the scattering function for the coexistence of two types of local structures in liquid water. J. Am. Chem. Soc. **142**(6), 2868–2875 (2020)
15. A.H. Narten, W.E. Thiessen, L. Blum, Atom pair distribution functions of liquid water at 25 °C from neutron diffraction. Science **217**, 1033–1034 (1982)
16. M. Leetmaa, K. T. Wikfeldt, M. P. Ljungberg, M. Odelius, J. Swenson, A. Nilsson, LG.M. Pettersson, Diffraction and IR/Raman data do not prove tetrahedral water. J. Chem. Phys. **129**, 084502 (2008)

17. A. Zeidler, P.S. Salmon, H.E. Fischer, J.C. Neuefeind, J.M. Simonson, T.E. Markland, Isotope effects in water as investigated by neutron diffraction and path integral molecular dynamics. J. Phys. Condens. Matter **24**, 284126 (2012)
18. H.E. Fischer, A.C. Barnes, P.S. Salmon, Neutron and X-ray diffraction studies of liquids and glasses. Rep. Prog. Phys. **69**, 233–299 (2006)
19. K. Amann-Winkel, M.-C. Bellissent-Funel, L,E. Bove, T. Loerting, A. Nilsson, A. Paciaroni, D. Schlesinger, L. Skinner, X-ray and neutron scattering of Water. Chem. Rev. **116**, 7570–7589 (2016)
20. L.B. Skinner, C.J. Benmore, J.C. Neuefeind, J.B. Parise, The structure of water around the compressibility minimum. J. Chem. Phys. **141**, 214507 (2014)
21. M.L. Cowan, B.D. Bruner, N. Huse, J.R. Dwyer, B. Chugh, E.T.J. Nibbering, T. Elsaesser, R.J.D. Miller, Ultrafast memory loss and energy redistribution in the hydrogen bond network of liquid H_2O. Nature **434**, 19 (2005)
22. D.R. Moberg, D. Becker, C.W. Dierking, F. Zurheide, B. Bandow, U. Buck, A. Hudait, V. Molinero, F. Paesani, T. Zeuch, The end of ice I. Proc. Natl. Acad. Sci. U.S.A. **116**, 24413–24419 (2019)
23. C. Perez, M.T. Muckle, D.R. Zaleski, N.A. Seifert, B. Temelso, G.C. Shields, Z. Kisiel, B.H. Pate, Structures of cage, prism, and book isomers of water hexamer from broadband rotational spectroscopy. Science **336**, 897 (2012)
24. J. Eckert, Computational study of inelastic neutron scattering vibrational spectra of water clusters and their relevance to hydration water in proteins. Biochem. Biophys. Acta. **1861**, 3564–3572 (2017)
25. V.P. Sokhana, A.P. Jones, F.S. Cipcigan, J. Craina, G.J. Martyna, Signature properties of water: their molecular electronic origins. PNAS **112**, 6341–6346 (2015)
26. F. Tang, T. Ohto, S. Sun, J.R. Rouxel, S. Imoto, E.H.G. Backus, S. Mukamel, M. Bonn, Y. Nagata, Modeling of water-air and ice-air interfaces monitored by sum-frequency generation. Chem. Rev. **120**, 3633–3667 (2020)
27. D. Nordlund, H. Ogasawara, H. Bluhm, O. Takahashi, M. Odelius, M. Nagasono, L.G.M Pettersson. A. Nilsson, Probing the electron delocalization in liquid water and ice at attosecond time scales. Phys. Rev. Lett. **99**, 217406 (2007)
28. F. Martelli, Unravelling the contribution of local structures to the anomalies of water: the synergistic action of several factors. J. Chem. Phys. **150**, 094506 (2019)
29. B. Santra, R.A. Jr DiStasio, F. Martelli, R. Car, Local structure analysis in *ab initio* liquid water. Mol. Phys. **113**, 2829–2841 (2015)
30. Q. Sun, Raman spectroscopic study of the effects of dissolved NaCl on water. Vib. Spectrosc. **62**, 110 (2012)
31. I.N. Tsimpanogiannis, A.O.A. Moultos, L.F.M. Franco, M.B. de Spera, M. Erdős, I. G. Economou, Self-diffusion coefficient of bulk and confined water: a critical review of classical molecular simulation studies. Mole. Simulation **45**, 425–453 (2019)
32. A. Taschin, P. Bartolini, S. Fanetti, A. Lapini, M. Citroni, R. Righini, R. Bini, R. Torre, Pressure effects on water dynamics by time-resolved optical kerr effect. J. Phys. Chem. Lett. **11**, 3063–3068 (2020)
33. F.C. Waldermann, B.J. Sussman, J. Nunn, V.O. Lorenz, K.C. Lee, K. Surmacz, K.H. Lee, D. Jaksch, I.A. Walmsley, P.G. Spizzirri, P. Olivero, S. Prwer, Measuring phonon dephasing with ultrafast pulses using Raman spectral interference. Phys. Rev. B **78** (2008)
34. Y. Huang, X. Zhang, C.Q. Sun, Hydrogen-bond relaxation dynamics: resolving mysteries of water ice. Coord. Chem. Rev. **285**, 109–165 (2015)
35. R. Aswani, J.C. Li, A new approach to pairwise potentials for water–water interactions. J. Mol. Liq. **134**, 120–128 (2007)
36. D.C. Elton, M. Fernández-Serra, The hydrogen-bond network of water supports propagating optical phonon-like modes. Nat. Commun. **7**, 10193 (2016)
37. E. Pontecorvo, M. Krisch, A. Cunsolo, G. Monaco, A. Mermet, R. Verbeni, F. Sette, G. Ruocco, G. High Frequency longitudinal and transverse dynamics in water. Phys. Rev. **71**, 011501 (2005)

38. A. Cunsolo, C.N. Kodituwakku, F. Bencivenga, M. Frontzek, B.M. Leu, A.H. Said, Transverse dynamics of water across the melting point: a parallel neutron and X-ray inelastic scattering study. Phys. Rev. B Condens. Matter Mater. Phys. **85**, 174305 (2012)
39. U. Kaatze, Y. Feldman, Broadband dielectric spectrometry of liquids and biosystems. Meas. Sci. Technol. **17**, R17–R35 (2006)
40. W.J. Ellison, Permittivity of pure water, at standard atmospheric pressure, over the frequency range 0–25 THz and the temperature range 0–100 °C. J. Phys. Chem. Ref. Data **36**, 1–18 (2007)
41. T. Globus, A. Bykhovski, T. Khromova, B. Gelmont, L.K. Tamm, L.C. Salay, Low-terahertz spectroscopy of liquid water, in *Proceedings, Terahertz Physics, Devices, and Systems II*, vol. 6772, pp. 67720S (2007)
42. H. Yada, M. Nagai, K. Tanaka, Origin of the fast relaxation component of water and heavy water revealed by terahertz time domain attenuated total reflection spectroscopy. Chem. Phys. Lett. **464**, 166–170 (2008)
43. N. Penkov, N. Shvirst, V. Yashin, E. Fesenko, E. Fesenko Jr., Terahertz spectroscopy applied for investigation of water structure. J. Phys. Chem. B **119**, 12664–12670 (2015)
44. D.K. George, A. Charkhesht, N.Q. Vinha. New terahertz dielectric spectroscopy for the study aqueous solutions. Rev. Sci. Instrument. **86**, 123105 (2015)
45. Q. Sun, Local statistical interpretation for water structure. Chem. Phys. Lett. **568–569**, 90–94 (2013)
46. D.C. Elton, The origin of the Debye relaxation in liquid water and fitting the high frequency excess response. Phys. Chem. Chem. Phys. **19**, 18739–18749 (2017)
47. S. Park, D.E. Moilanen, M.D. Fayer, Water dynamics: the effects of ions and nanoconfinement. J. Phys. Chem. B **112**, 5279–5290 (2008)
48. D. Laage, J.T. Hynes, On the molecular mechanism of water reorientation. J. Phys. Chem. B **112**, 14230–14242 (2008)
49. D. Laage, Reinterpretation of the liquid water quasi-elastic neutron scattering spectra based on a nondiffusive jump reorientation mechanism. J. Phys. Chem. B **113**, 2684–2687 (2009)
50. S. Fanetti, A. Lapini, M. Pagliai, M. Citroni, M. Di Donato, S. Scandolo, R. Righini, Bini, structure and dynamics of low- density and high-density liquid water at high pressure. J. Phys. Chem. Lett. **5**, 235–240 (2014)
51. L.E. Bove, S. Klotz, T. Strassle, M. Koza, J. Teixeira, A.M. Saitta, Translational and rotational diffusion in water in the gigapascal range. Phys. Rev. Lett. **111**, 185901 (2013)
52. S.R. Becker, P.H. Poole, F.W. Starr, Fractional stokes-einstein and debye-stokes-einstein relations in a network-forming liquid. Phys. Rev. Lett. **97**, 055901 (2006)
53. F. Yang, T. Atay, C.H. Dang, T.J. Grimsley, S. Chel, J. Ma, Q. Zhang, A.V. Nurmikko, H.J. Maris, Study of phonon propagation in water using picosecond ultrasonics. J. Phys. Conf. Ser. **92**, 012024 (2007)
54. A. Rahman, F.H. Stillinger, Propagation of sound in water A molecular-dynamics study. Phys. Rev. A **10**, 368 (1974)
55. G. Ruocco, F. Sette. V. Mazzacurati, G. Signorelli, R. Verbeni, Equivalence of the sound velocity in water and ice at mesoscopic wavelengths. Nature **379**, 521–523 (1996)
56. B.J. Berne, R. Pecora, *Dynamic Light Scattering: With Applications to Chemistry, Biology and Physics* (Dover Publications, Mineola, NY, USA, 2000)
57 E. Burkel, Phonon spectroscopy by inelastic x-ray scattering. Rep. Prog. Phys. **63**, 171 (2000)
58. A. Cunsolo, A. Orecchini, C. Petrillo, F. Sacchetti, Interplay between microscopic diffusion and local structure of liquid water. J. Phys. Chem. B **114**, 16713–16717 (2010)
59. K.L. Ngai, *Relaxation and Diffusion in Complex Systems* (Springer, Berlin, 2011)
60. T.H. van der Loop, N. Ottosson, T. Vad, W.F. Sager, H.J. Bakker, S. Woutersen, Communication: slow proton charge diffusion in nanoconfined water. J. Chem. Phys. **146**, 131101, (2017)
61. R.N. Barnett, U. Landman, A. Nitzan, The properties of excess hydrated electrons in liquid water, at room. J. Chem. Phys. **93**, 8187 (1990)
62. M.S. Pshenichnikov, A. Baltuška, D.A. Wiersma, Hydrated-electron population dynamics. Chem. Phys. Lett. **389**, 171 (2004)

63. Á. Madarász, Excess electron relaxation dynamics at water/air interfaces. J. Chem. Phys. **126**, 234707 (2007)
64. T. Sommerfeld, K.D. Jordan, Electron binding motifs of $(H_2O)_n^-$ clusters. J. Am. Chem. Soc. **128**, 5828–5833 (2006)
65. C.-C. Zho, V. Vlček, D. Neuhauser, B.J. Schwartz, Thermal equilibration controls H-Bonding and the vertical detachment energy of water cluster anions. J. Phys. Chem. Lett. **9**, 5173–5178 (2018)
66. A. Herburger, E. Barwa, M. Ončák, J. Heller, C. van der Linde, D.M. Neumark, M.K. Beyer, Probing the structural evolution of the hydrated electron in water cluster anions $(H_2O)_n^-$, $n \leq 200$, by electronic absorption spectroscopy. J. Am. Chem. Soc. **141**, 18000–18003 (2019)
67. C.Q. Sun, Relaxation of the Chemical Bond, pp. 108–807 (Springer-Verlag, 2014)
68. C.Q. Sun, Y. Sun, The attribute of water: single notion. Multiple Myths. **113**, 494 (2016)
69. C.Q. Sun Unprecedented O:⇔:O compression and H↔H fragilization in Lewis solutions. Phys. Chem. Chem. Phys. **21**, 2234–2250 (2019)
70. M. Havenith, It is water what matters: THz absorption spectroscopy as a new tool to study solvation dynamics, in *2015 40th International Conference on Infrared, Millimeter, and Terahertz waves (IRMMW-THz)*, Hong Kong, 2015, pp. 1–1. https://doi.org/10.1109/IRMMW-THz.2015.7327887
71. A. Cunsolo, A. Orecchini, C. Petrillo, F. Sacchetti, Interplay between microscopic diffusion and local structure of liquid water. J. Phys. Chem. B **14**, 16713–16717 (2010)
72. C.A. Angell, Supercooled water: two phases? Nat. Mater. **13**, 673–675 (2014)
73. D. Laage, Reinterpretation of the liquid water quasi-elasticneutron scattering spectra based on a nondiffusive jump reorientation mechanism. J. Phys. Chem. B **113**, 2684–2687 (2009)
74. H.J. Bakker, J.L. Skinner, Vibrational spectroscopy as a probe of structure and dynamics in liquid water. Chem. Rev. **110**, 1498–1517 (2010)
75. C.J. Fecko, J.D. Eaves, J.J. Loparo, A. Tokmakoff, P.L. Geissler, Ultrafast hydrogen-bond dynamics in the infrared spectroscopy of water. Science **301**, 1698–1702 (2003)
76. F. Perakis, L. De Marco, A. Shalit, F. Tang, Z.R. Kann, T.D. Kühne, R. Torre, M. Bonn, Y. Nagata, Vibrational spectroscopy and dynamics of water, fives, vibrational spectroscopy and dynamics of water. Chem. Rev. **116**, 7590–7607 (2016)
77. B.B. Nandi, Dielectric relaxation of biological water. J. Phys. Chem. B **101**, 10954–10961 (1997). https://doi.org/10.1021/jp971879g

Chapter 5
Supercooled and Supercritical Water and Ice

Abstract The uniqueness of the supramolecular structure of the water ensembles is the reason for the formation and special properties of supercooled and supercritical water and ice. One of the most intriguing anomalies of water is its ability to exist as distinct amorphous ice forms, glass polymorphism, or polyamorphism. Supercooled water contains a greater number of tetrahedral hydrogen-bonding water molecule and pentametric water clusters. Two liquid forms, low-density-liquid (LDL)-like and high-density-liquid (HDL)-like, take place also in the supercooled state. A supercritical water is a substance at a temperature and pressure above its critical point, where distinct liquid and gas phases do not exist. Supercritical liquid–gas boundaries (Widom line) are lines in the pressure–temperature (pT) diagram that delimit more liquid-like and more gas-like states of a supercritical fluid, and 19 recognized forms of ice crystals are composed of water molecules held in a regular tetrahedral arrangement by hydrogen bonding. Virtually all ice in the biosphere is ice I_h. Other states of ices can be formed with changes in temperature and pressure and are characterized by molecular and supermolecular details. Low-density amorphous ice and high-(HDA) to low-density amorphous ice (LDA) can be created by cooling the liquid rapidly below T_g or by vapor deposition onto a cold plate, kept below T_g at near-atmospheric pressure (Loerting et al. in J. Phys. Chem. B 115:14167, 2011 [1]). Alternatively, amorphous ice has been formed by temperature- or pressure-induced transformations. The use of physical methods to study the properties of various forms of water is illustrated by typical examples.

5.1 Introduction

Water that remains liquid below its melting point can be in supercooled and cold metastable glassy states (Fig. 5.1) [2–14]. In this condition, the substance crystallization can be avoided if the cooling occurs fast enough [2, 3]. Below melting temperature T_m, in region, where liquid water becomes thermodynamically less stable than ice, it is possible to perform measurements in the liquid state for times significantly longer than its equilibration time but still shorter than the crystallization time when nucleation of ice prevails [4]

G. I. Likhtenshtein, *Biological Water*, Soft and Biological Matter,
https://doi.org/10.1007/978-3-030-82503-4_5

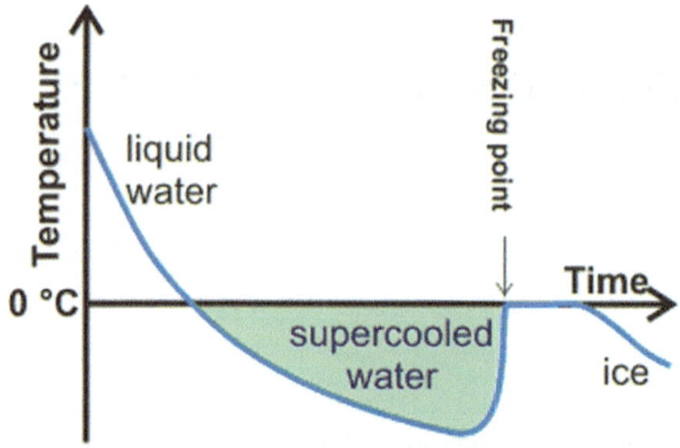

Fig. 5.1 Schematic presentation of a process of water supercooling. It was last updated by Martin Chaplin on January 3, 2020. This work is licensed under a Creative Commons Attribution-Noncommercial-No Derivative Works 2.0 UK: England & Wales License. With permission from Dr. Martin Chaplin

Supercooled water contains a greater number of tetrahedral hydrogen-bonding water molecule and pentametric water clusters (Fig. 5.2), which increase as the temperature is lowered. Supercool water exhibits a number of anomalous physical properties.

A supercritical water (SCW) is a substance at a temperature and pressure above its critical point, where distinct liquid and gas phases do not exist. The water critical point occurs at 647.096 K (373.946 °C; 705.103 °F) and 22.064 MPa (3200.1 psi; 217.75 atm) [8]. In the vicinity of the critical point, the physical properties of the liquid and the vapor change dramatically. Under supercritical conditions the fraction

Fig. 5.2 Pentametric water cluster. It was last updated by Martin Chaplin on January 3, 2020. This work is licensed under a Creative Commons Attribution-Noncommercial-No Derivative Works 2.0 UK: England & Wales License. With permission from Dr. Martin Chaplin

of hydrogen bonds (HB) molecules greatly decreases with respect to ambient P and T, the HB network is substantially distorted. The physical properties of SCW such as density, viscosity, dielectric constant, hydrogen bonding, and ionic product are vastly different from those of normal water and steam [9]. Supercritical water is of importance both for fundamental research and industrial applications as a cheap inorganic and non-toxic solvent, having a wide applicability in chemistry processes due to the singular physical and chemical properties, and as a promising medium for various emerging chemical, biological, and geological processes including chemical synthesis.

Many crystalline structures of water, **17** recognized forms of ice, are known [7]. Ice crystals are composed of water molecules held in a regular tetrahedral arrangement by hydrogen bonding. The angle between bonds in the crystal lattice is close to the tetrahedral angle of 109.5°, which is also quite close to the angle between hydrogen atoms in the gas phase water molecule, which is 105°. The most familiar ice is the hexagonal ice that forms snowflakes. Metastable cubic ice seeds more readily than the hexagonal arrangement. As pressure and temperature vary, the water molecules adapt their arrangement to minimize energy, producing the different phases of ice. The structure and the molecular processes in ice are subject of computer simulations or quantum-chemistry calculations over the whole temperature and pressure range. Amorphous, non-crystalline, ice can be produced either by *rapid cooling* of liquid water or by compressing ordinary ice at low temperatures and has a lack of long-range order in its molecular arrangement.

5.2 Supercooling Water

5.2.1 Water Phase Diagrams

Phase diagrams (Figs. 5.3 and 5.4) summarize data on the effects of temperature and pressure on water states. The T P region in which nucleation prevents the observation of metastable liquid states has been named "no-man's land." In the diagrams, the border to the no-man's land is ill-defined, since its location depends on the experimental time and on the sample size [10]. Supercritical liquid–gas boundaries (Widom line) are lines in the pressure–temperature (pT) diagram that delimit more liquid-like and more gas-like states of a supercritical fluid. Thus, the critical anomalies on the Widom lines demarcates two regions, the liquid-like and the gas-like being a hallmark of a supercritical fluid.

A review [12] described a phase diagram (Fig. 5.5), in which the liquid–liquid phase transition line exists only at high pressures—hundreds of atmospheres. In the diagram, the Widom line is an extension of the phase boundary past the critical point and into the supercritical regime. In the supercooled regime, water was hypothesized to exist in two distinct liquid phases: high-density liquid (HDL) and low-density liquid (LDL).

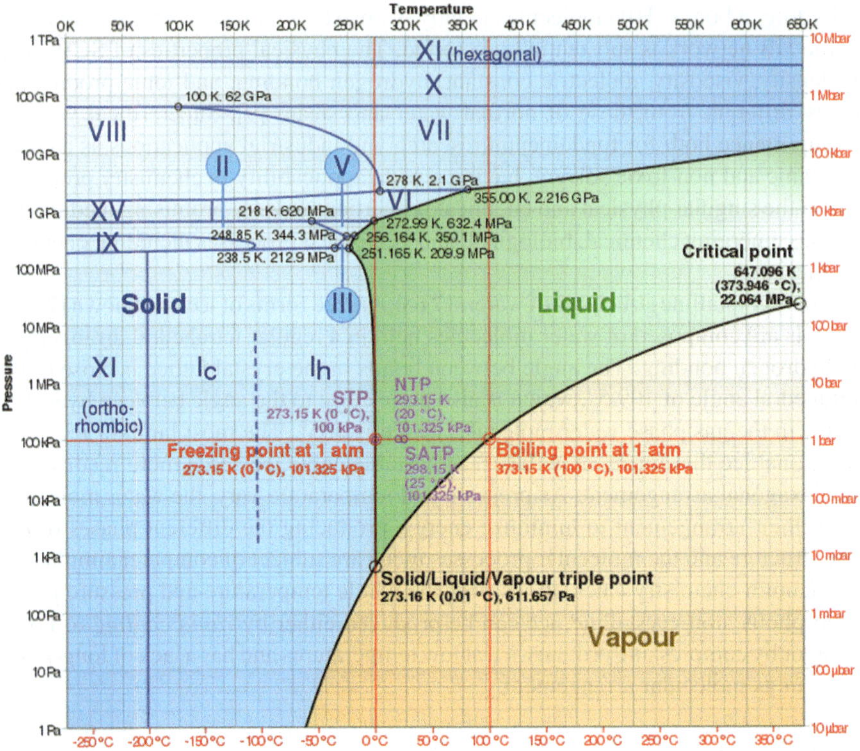

Fig. 5.3 Water phase diagram. https://www.google.com/search?q=Water+phase+diagram.&sxsrf=ALeKk01x-Ynz5zwmGau-PkNlWymKIJh_vA:1622217173777&tbm=isch&source=iu&ictx=1&fir=CHoqv9OxwxkzbM%252CgNswwOMUqqr4pM%252C_&vet=1&usg=AI4_-kRxR2o8BcieQR3dpuVQclsNMXS5SQ&sa=X&ved=2ahUKEwiho5Dz3ezwAhXi8bsIHdVNADwQ9QF6BAgQEAE&biw=1283&bih=650#imgrc=CHoqv9OxwxkzbM

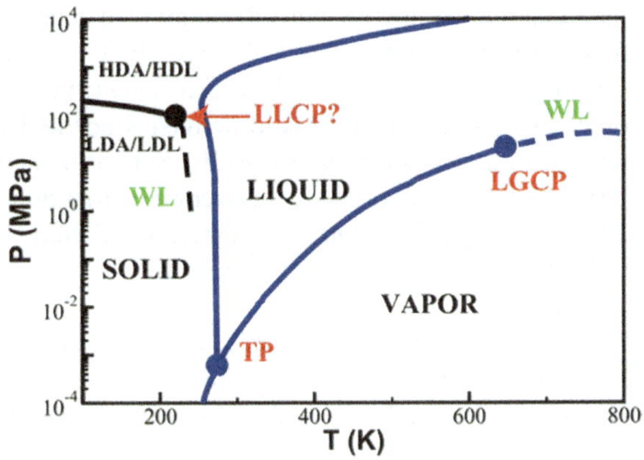

Fig. 5.4 Phase diagram suggested in [11]. With permission from American Chemical Society

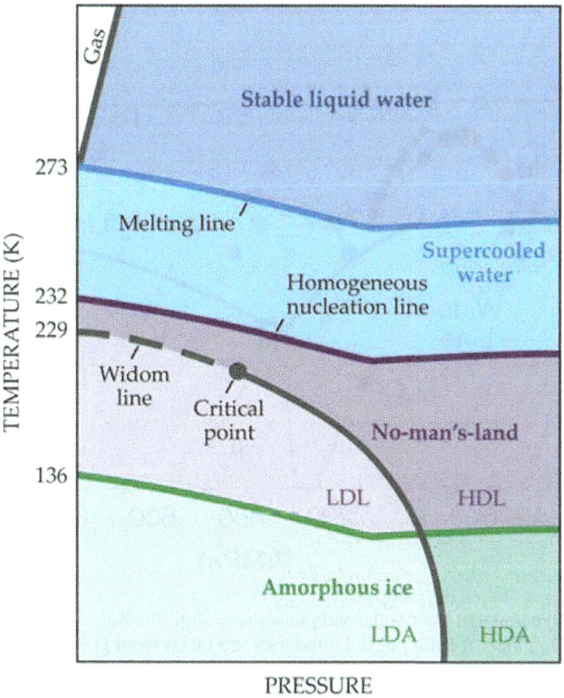

Fig. 5.5 A PHASE DIAGRAM of water and its metastable states. Deep in the supercooled regime, water is hypothesized to exist in two distinct liquid phases: high-density liquid (HDL) and low-density liquid (LDL), in analogy to the well-known high-density amorphous (HDA) and low-density amorphous (LDA) ices. New experiments suggest that the liquid–liquid phase boundary ends at a critical point. The Widom line that extends past the critical point is characterized by maxima in the correlation length and several thermo-dynamic response functions [12]. With permission from American Institute of Physics

To show the structurally anomalous regime in water that encloses the region of diffusional anomaly, which in turn encloses the region of density anomaly, the TIP4P/2005 water model was used for simulation (Fig. 5.6) [13]. This nested structure gives rise a cascade of anomalies, where progressive enhancement of the degree of structural anomaly gives rise to various transport and thermodynamic anomalies.

Figure 5.7 demonstrates sophistry relationship between water in the liquid, supercooled, supercritical, and various ice states.

5.2.2 *Supercooled Water Structure*

Data on research on supercooled and glassy water published by 2017 were reviewed focusing on the possible origins of its complex behavior [4]. It was stressed that

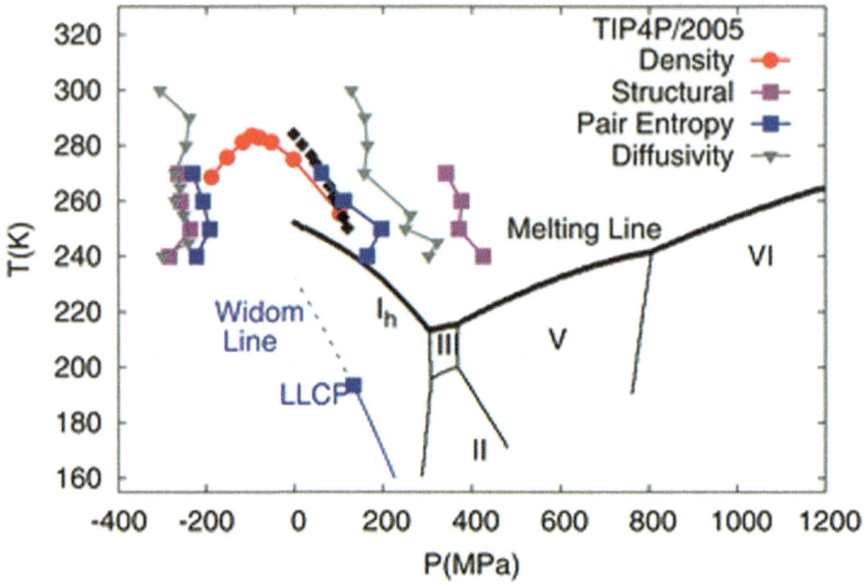

Fig. 5.6 Thermodynamics of the condensed phases of water, illustrated for the TIP4P/2005 rigid-body model of water. Data for the phase boundaries are taken from [15]. The experimental TMD line is shown in filled black diamonds [13]. With permission of Royal Society of Chemistry

the strong directionality of the water–water interaction and the competition between local energy, local entropy, and local density play the central role. The explicitly on bulk water in metastable conditions, that is, water below its equilibrium melting temperature, both as a supercooled liquid and as an amorphous solid was focused. In a separate section, the possibility of linking polyamorphism with a liquid–liquid transition that could take place in the no-man's land, the temperature–pressure window, in which homogeneous nucleation prevents the investigation of water in its metastable liquid form, were debated. Experiments showed that metastable liquid water can transiently exist on the millisecond timescale down to $T = 227$ K and that tetrahedral ordering increases in supercooled water.

Results on oxygen–oxygen radial distribution functions of liquid water at ambient and supercooling conditions and comparison between the experimental and calculated temperature dependence of the density of liquid water derived from neutron diffraction measurements are presented in Fig. 5.8 [14]. From these measurements, the position and height of the nearest neighbor maximum in $g_{OO}(r)$ were found to be 2.80(1) Å and 2.57(5), respectively. The resulting elastically scattered X-ray intensity, $IX(Q)$, from liquid water under ambient conditions from the different datasets, is compared in Fig. 5.8.

The Lorch function, defined as $\sin(Q\Delta)/Q\Delta$, where $\Delta = \pi/Q_{\max}$ was employed. Using the *APS* dataset as an example, the total intermolecular PDF, $g\text{mol}X(r)gX\text{mol}r$, produced with and without the Lorch modification functions is plotted in Fig. 5.9.

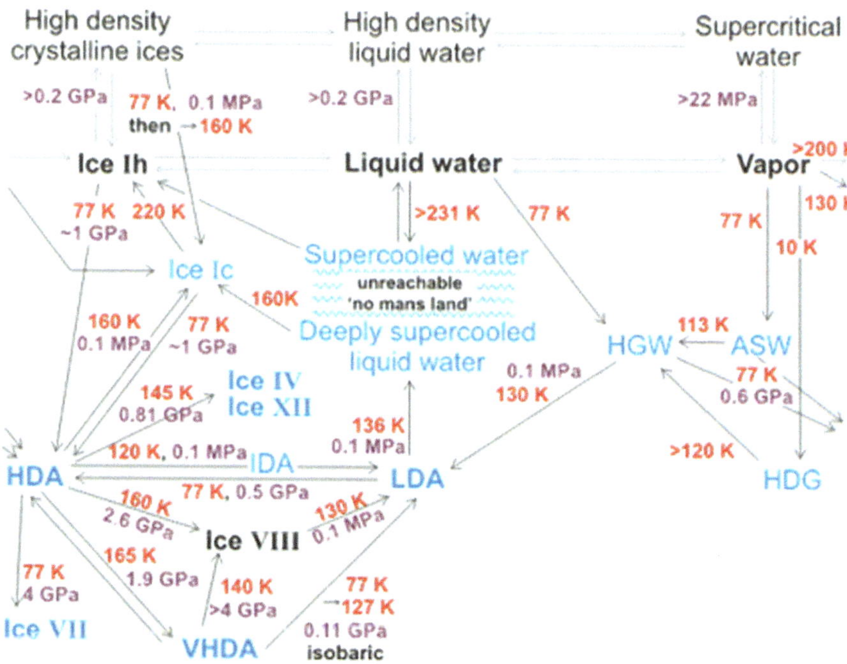

Fig. 5.7 Various types of water LDA is the low-density amorphous ice, HAD is the high-density amorphous ice, and VHDA is the very high-density amorphous ice. This page was established in 2004 and last updated by Martin Chaplin on January 2020. This work is licensed under a Creative Commons Attribution-Noncommercial-No Derivative Works 2.0 UK: England & Wales License. With permission from Dr. Martin Chaplin

The structure and energetics of supercooled water were studied by means of X-ray Raman and Compton scattering [16]. Under supercooled conditions down to 255 K, the oxygen K-edge measured by X-ray Raman scattering suggests an increase of tetrahedral order similar to the conventional temperature effect observed in non-supercooled water. Compton profile differences indicate contributions beyond the theoretically predicted temperature effect, suggesting a decrease of the electron mean kinetic energy by 3.3 ± 0.7 kJ (mol K)$^{-1}$.

X-ray scattering was utilized to study the structure of bulk liquid water below temperature limit of homogeneous nucleation (T_H) of ~232 K using oxygen K-edge X-ray emission spectroscopy (XES) [17]. Non-resonant O K-edge XES was performed at the Soft X-ray Materials Science instrument at the Linac Coherent Light Source (LCLS) employing a nominal photon energy of 550 and a nominal pulse duration of 100 fs. Figure 5.10 displays O K-edge X-ray emission (XE) spectra measured at LCLS at non-resonant conditions (550 eV) on the water droplets at three different temperatures corresponding to ~290, 226, and 222 K.

A number of alternative hypotheses to explain the results were discussed. The interpretation that the line shape of each component changes when approaching lower

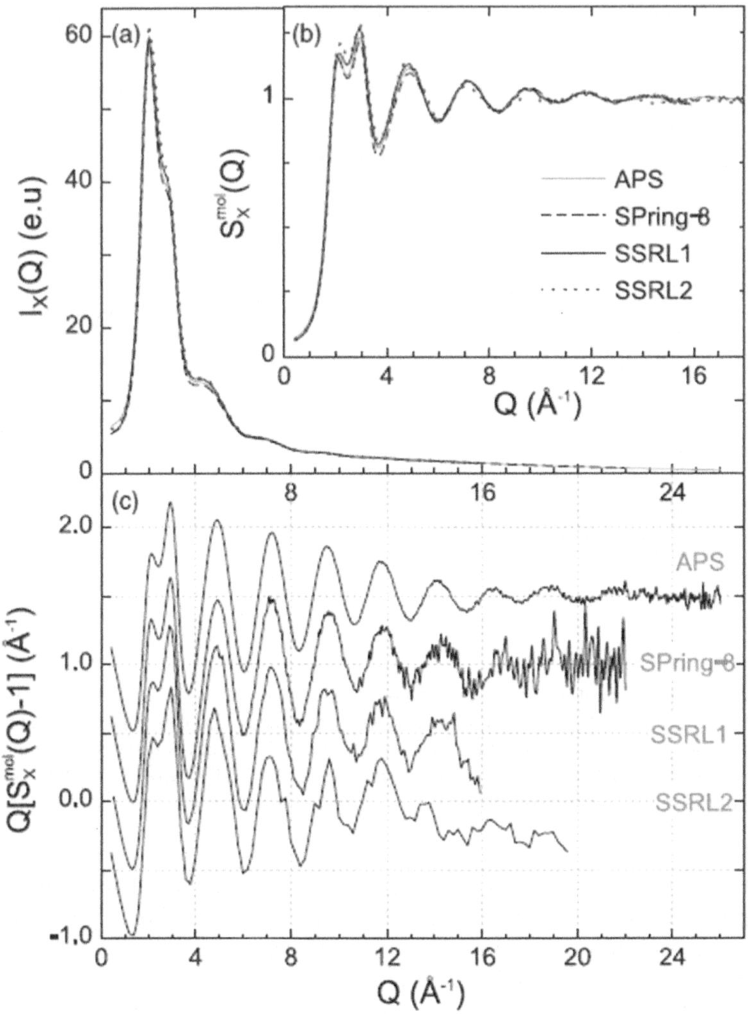

temperatures, where, in particular, the peak assigned to the proposed disordered component would become more symmetrical as vibrational interference becomes more important was preferred. According to suggested mechanism: (1) water behaves as a simple liquid at temperatures close to the boiling point, where most interactions are isotropic and dominated by non-directional van der Waals interactions; (2) as the liquid cools down, the water molecules start to stick to each other through directional H-bonds; (3) as a result, two classes of configurations, tetrahedral structures with four strong H-bonds (LDL-like regions) and asymmetric structures with distorted H-bonds (HDL-like regions) appeared; (4) there are fluctuations between the LDL-like

◄**Fig. 5.8** Four recent synchrotron X-ray diffraction measurements of liquid H_2O at ambient temperatures. They are: *APS* dataset (22 °C) measured using 114.76 keV photons and an amorphous silicon area detector (*APS*: light blue line), *SPring-8* dataset (26 °C) measured at beamline 04B2 using 61.62 keV photons, and a Ge detector (*SPring-8*: black dashed line), and *SSRL1* (25 °C) measured using 17 keV photons, a Ge analyzer and a photo-multiplier tube detector (*SSRL1*: black line), and *SSRL2* data (23 °C) measured using 19.6 keV photons, a curved graphite analyzer and position-sensitive detector (*SSRL2*: red dotted line 3). **a** Scattering intensity $IX(Q)$ in electron units, normalized to the molecular form factor. **b** The X-ray weighted structure factors, $S\text{mol}\,X(Q)$. This is plotted again in **c** as $Q[S\text{mol}\,X(Q) - 1]$ to emphasize the structure at large Q-values (the datasets in **c** are shifted vertically by 0.5 from each other for clarity). Also note that the datasets differ in their Q-step. The *SSRL2* data have a Q-step of 0.2 $Å^{-1}$, the *SSRL1* data have a Q-step of 0.1 $Å^{-1}$ while both the *SPring8* and *APS* data are plotted with a 0.025 $Å^{-1}$ Q-step. **a** Comparison between oxygen–oxygen radial distribution functions of liquid water at ambient conditions derived from X-ray diffraction measurements (black) and calculated from molecular dynamics simulations performed with the coarse-grained mW model (blue) and the empirical E3B models (green and orange). **b** Comparison between the experimental and calculated temperature dependence of the density of liquid water at ambient pressure. The experimental values are in black [14]. With permission from American Institute of Physics

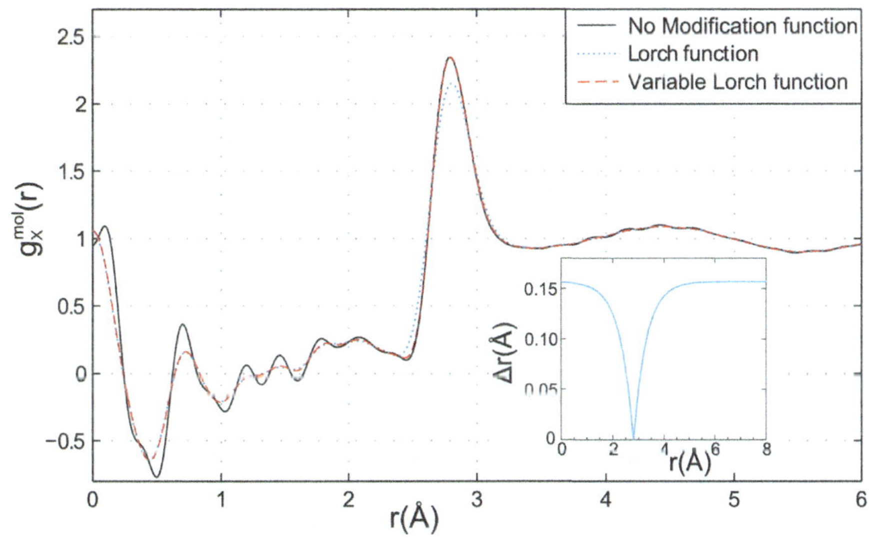

Fig. 5.9 Comparison of the total molecular PDF $g\text{mol}X(r)gX\text{mol}r$ with different modification functions in the Fourier transform of $S\text{mol}X(Q)SX\text{mol}Q$. (1) without a modification function (black solid line), (2) a Lorch modification function (light blue dotted line), and (3) a variable Lorch function (red dashed line). The variable Lorch function, shown in the inset, is tuned so that it does not broaden the $g\text{mol}X(r)gX\text{mol}r$ around $r = 2.8$ Å. Using the *APS* dataset as a demonstration, the variable Lorch function reduces the Fourier transform oscillations, which have a frequency of about $\pi/Q_{\max}$ in $g\text{mol}X(r)gX\text{mol}r$, while leaving the first neighbor O–O peak intact. See detail in [14]. With permission from American Institute of Physics

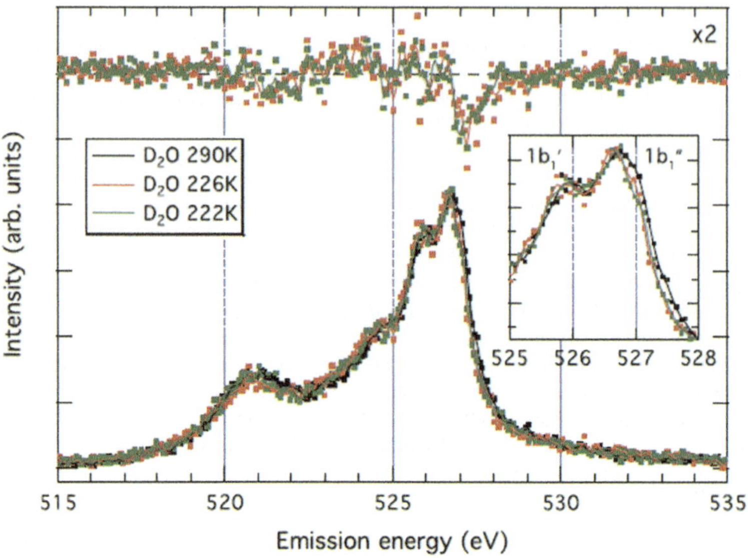

Fig. 5.10 O K-edge X-ray emission (XE) spectra of liquid D2O measured at ~290 K (black curve), 226 K (red curve), and 222 K (green curve). The XE spectra were recorded at LCLS using an incident photon energy of 550 eV, a nominal pulse length of 100 fs, and a beam focus of 75 × 520 μm^2 (horizontal × vertical) at the interaction point, corresponding to a peak fluence of 0.27 J/cm^2. The top shows the difference spectra (magnified by a factor of 2) with respect to the XE spectrum of D_2O at 290 K. The inset shows an enlarged view of the lone-pair (1b1) region of the XE spectra. All XE spectra were background subtracted with an 8th order polynomial and area normalized between 500 and 550 eV [17]. With permission from American Institute of Physics

and HDL-like structures; (5) as water is supercooled, the correlation length increases indicating that the LDL-like regions become significantly larger; (6) at cooling as the HDL-like structures become more symmetrical, similar to HAD ice, the two OH groups in the water molecule encounter more similar environments; and (7) the latter effect can lead to increase vibrational lifetime interference due to both OH groups contributing in the core-excited electronic state, rather than only one. Thus, the study observations provided direct evidence for the existence of a Widom line in supercooled water, support the liquid–liquid critical point LLCP scenario, and demonstrated the importance of nuclear quantum effects [17].

The structure factor of H_2O in the low-q SAXS (small-angle X-ray scattering) region averaged at a given temperature, which in the low-q region, increased upon cooling was established and the integrated structure factor between 0.12 and 0.20 Å^{-1} for both H_2O and D_2O data, where a maximum in the integrated low-q structure factor at 229.2 ± 1 K for H_2O and 233.7 ± 1 K for D_2O was observed [18]. The integrated structure factor at the maxima was found to be higher for H_2O than for D_2O. The peak position as a function of temperature, and temperature at which the liquid underwent the most rapid increase toward tetrahedral and coordination is pointed out in a figure.

The spectra maximum was seen for both H_2O and D_2O at 229.2 ± 1 and 234.5 ± 1 K, respectively. These observations confirmed the existence of a Widom line.

Supercooling at freezing is associated with the shorter H–O bond in water molecules with fewer than four neighbors and also with the longer H–O bond being in contact with hydrophilic surface or with the longer H–O bond being compressed [19]. In this case, averaging the background long-range interactions due to other H_2O molecules or protons and omitting the nucleus quantum effect on fluctuations can take place. The initially longer H–O bonds in the colder water react more slowly to the relaxation at the freezing point than the bonds in the warmer water, because of the lower momentum of relaxation–memory effect and hot water freezes faster than its cold (the Mpemba paradox). Forces and relaxation dynamics of the segmented O:H O bond with H^+ at heating, freezing, salting, clustering, compressing, and liquid and solid cooling were schematically shown in Fig. 3.21. Figure 5.11 confirms the universality of the O:H O bond cooperative relaxation under various stimuli obtained from MD calculations using the force field code. Results indicate that O:H serves as the master and H O as the slave of water in (a) under mechanical compression and in (c) cooling liquid in the region of $T > T_m$; the H O bond serves as the master in (b) the liquid–solid transition phase ($T < T_m$) and in (d) upon molecular under coordination. The arrows next to the "master" segment point in the direction of density gain ((a) and (c)) or loss ((b) and (d)).

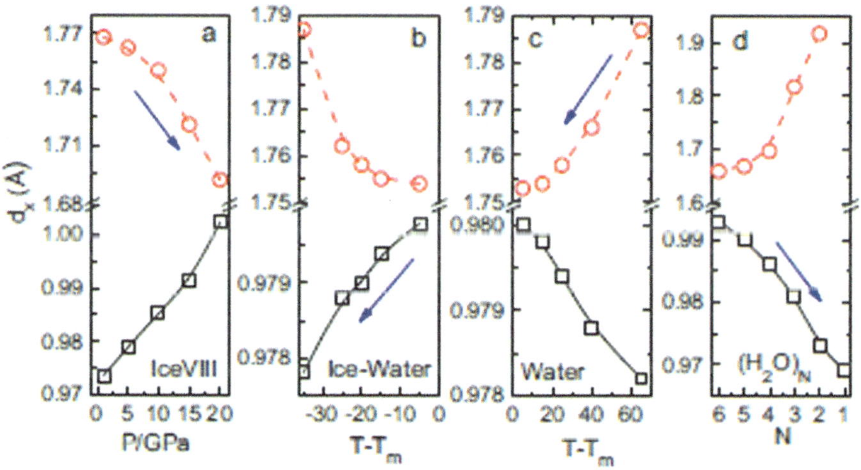

Fig. 5.11 Universality of the O:H O bond cooperative relaxation. Changes of length of O:H (broken red lines) and H O (solid black lines) are shown for stimulation by: **a** mechanical compression; **b** cooling excitation in the quasi-solid phase; **c** cooling in the liquid phase; and **d** cluster size reduction (molecular under coordination). The arrows indicate the master segments and their relaxation directions under excitation. If either of the O:H or the H O shrinks, the other one expands, regardless of the applied stimulus or structural phases, because of the inter-oxygen Coulomb coupling. The arrows in **a** and **c** show the direction of density gain as dO O shortens; the arrows in **b** and **d** show the direction of density loss. All processes are reversible [19]. With permission from Elsevier

5.2.3 Two Liquid Forms of Supercooled Water

In work [18], the evidence in support of the two-liquid theory was presented. Femtosecond X-ray pulses off a stream of water microdroplets that were evaporatively cooled under vacuum to as low as 227.7 K, or − 45.4 °C were scattered. From different portions of the scattering curves, four quantities and correlation length a measure of the local molecular order were extracted. All four quantities for H_2O peaked at similar temperatures, above 229 K. The correlation length of H_2O does not diverge; instead, it peaks at a finite value near 229 K, then decreases at lower temperatures.

Ising-like model to estimate the ratio between thermal expansivity and specific heat (the Grüneisen parameter Γ_s) in supercooled water was proposed [20]. Generally, the Ising model consists of discrete variables that can be in one of two states (+1 or − 1) allowing each variable to interact with its neighbors. Near the critical pressure and temperature, Grüneisen parameter (Γ_s), which describes the effect that changing the volume of an object has on its vibrational properties, becomes significantly sensitive to thermal fluctuations of the order parameter, indicating that two energy scales are governing the system. In this condition, high- and low-density liquids are coexisted and the coexistence becomes indistinguishable at the critical point in the supercooled phase in the no-man's land regime. The temperature dependence of the compressibility, sound velocity, and pseudo-Grüneisen parameter Γ_w was also reported. It was concluded that the behavior of the Grüneisen parameter is enhanced near pressure and temperature values that display anomalous behavior and thus support the presence of a liquid–liquid critical point is governed by two distinct energy scales.

By computer simulations of three popular water models and the analysis of recent X-ray scattering experimental data, it was shown that there are two overlapped peaks hidden in the apparent "first diffraction peak" of the structure factor (Fig. 5.12) [21]. Ordinary peak was proposed to correspond to the neighboring O–O distance as in ordinary liquids, and the other anomalous peak corresponds to a longer distance.

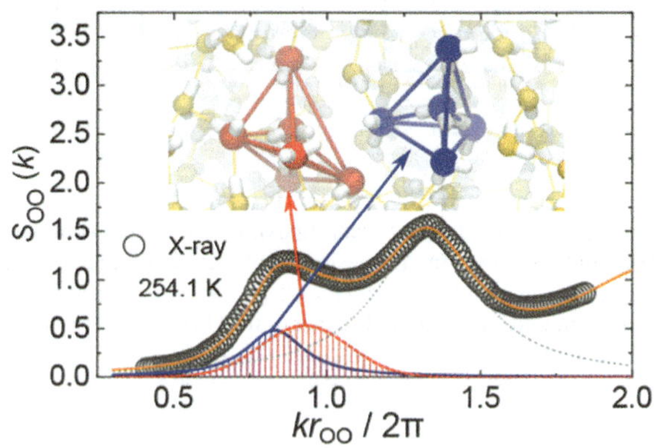

Fig. 5.12 Schematic presentations of coexistence of two types of local structures in the coordination number distribution [21]. With permission from American Chemical Society

This anomalous peak arises from the most extended period of density wave and is associated with a tetrahedral water structure and can be identified as the first sharp diffraction peak. Evidence that liquid water is indeed a mixture of two types of local structures, from simulations of three popular water models paves a way for detailed analysis of recent scattering experiments. Additional evidence for the coexistence of two types of local structures in the scattering function was also given in Sect. 4.5.

An image of two different phases of liquid water—a high-density liquid and a low-density liquid, as a function of temperature in the supercooled regime is shown in Fig. 5.13.

The microscopic origin of water's anomalies by inspecting the hydrogen-bond network (HBN) and the spatial organization of low-density-liquid (LDL) like

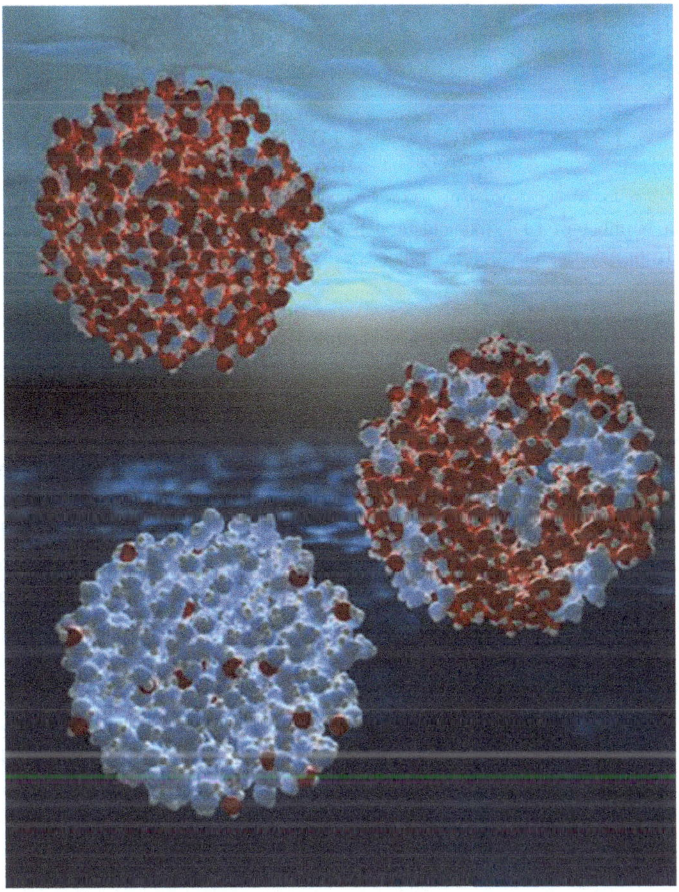

Fig. 5.13 Experiments using femtosecond X-ray free electron lasers illuminated fluctuations between two different phases of liquid water—a high-density liquid (red) and a low-density liquid (blue)—as a function of temperature in the supercooled regime. https://www.osa-opn.org/home/newsroom/2017/december/x-ray_studies_probe_water_s_elusive_properties/

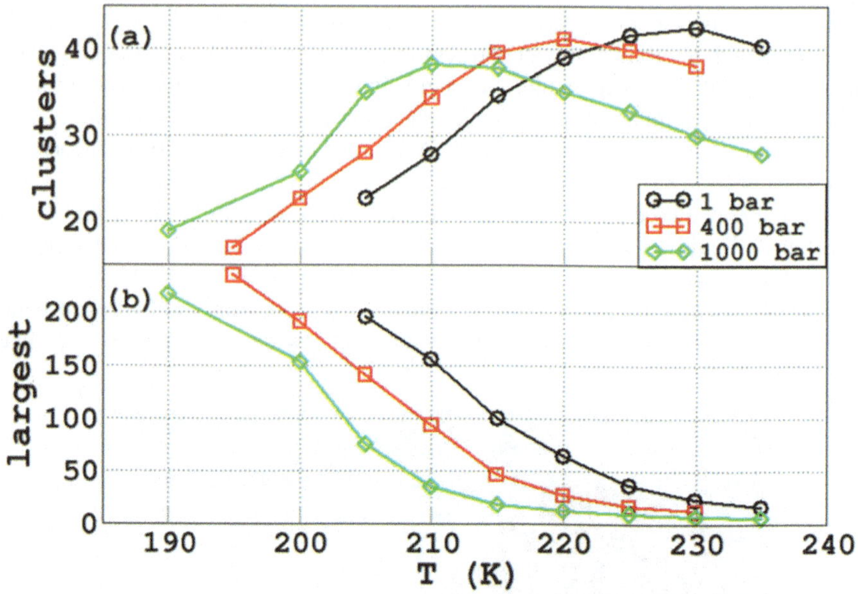

Fig. 5.14 **a** Number of LDL-like clusters. **b** Size of the largest LDL-like cluster [22]. With permission from American Institute of Physics

and high-density-liquid (HDL)-like environments were investigated using classical molecular dynamics simulations [22]. Isobaric cooling of a sample composed of 512 water molecules from ambient to deeply undercooled conditions at three pressures, namely 1, 400, and 1000 bars, was in focus of the simulation. Data on temperature-dependent isobaric heat capacity and isothermal compressibility the three pressures were reported (Fig. 5.14). It was found that the Widom line temperature (T_W) is 220–230 K for 1 bar, 210–220 K for 400 bars, and 200–205 for 1000 bars. The calculation revealed the number of LDL-like clusters and demonstrated a non-monotonic behavior with maxima at temperatures slightly higher than the WL as well as decreasing trends starting, upon cooling, in correspondence with the T_{WL}. This observation evidences that, in correspondence with the T_{WL}, the LDL-like clusters start forming an extended network further expanding upon cooling. It was also found that the pace of growth of the largest cluster shows two distinct regimes: a slow growth at temperatures above T_{WL} and a faster growth starting in correspondence with T_{WL} upon cooling the sample.

Figure 5.15 reports the number of HDL-like clusters and the size of the largest HDL-like cluster. At temperatures above the T_{WL}, the system is composed of one large HDL-like aggregate spanning the entire simulation box and enclosing smaller LDL-like clusters. Upon cooling, the largest HDL-like cluster gets continuously depleted at the expense of LDL-like environments.

Figure 5.16 displays a two-dimensional 4 Å-thick slice of our sample at 400 bars and at three temperatures: $T = 230$ K (a), $T \sim T_{WL} = 215$ K (b), and $T = 195$ K (c). At

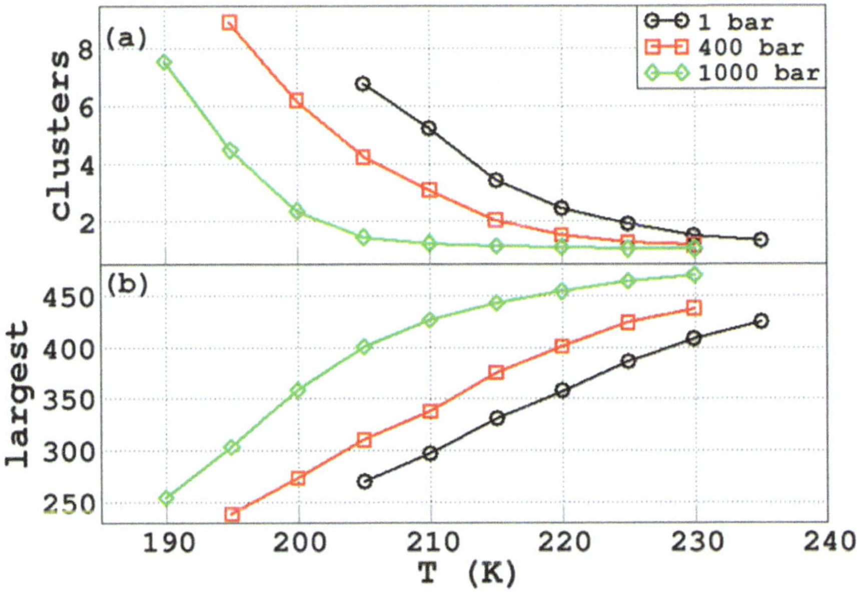

Fig. 5.15 Number of HDL-like clusters and the size of the largest HDL-like cluster [22]. With permission from American Institute of Physics

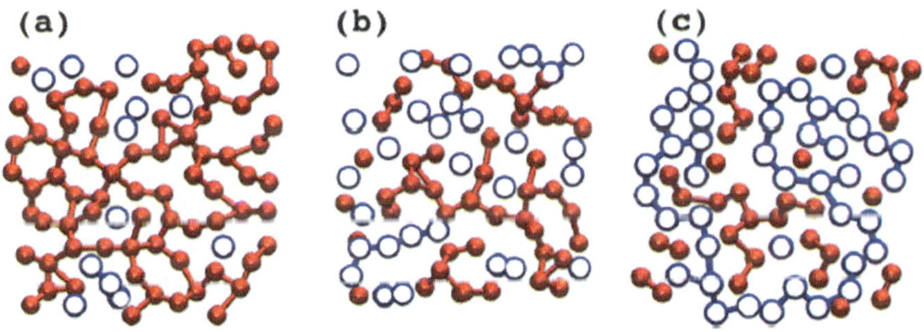

Fig. 5.16 Two-dimensional slices of the LDL-like network (blue, open spheres) and of the HDL-like network (red, filled spheres) at 400 bars and three temperatures [22]. With permission from American Institute of Physics

$T = 230$ K, the sample is permeated by the large HDL-like cluster and its underlying HBN (red sticks), embracing LDL-like molecules and clusters. Upon cooling the sample to $T \_ T_{WL} = 215$ K is now fragmented and alternated with the likewise fragmented LDL-like network. In addition, at $T = 195$ K, it was observed that the HDL-like and the LDL-like networks are now almost independent and separated.

The work main results can be summarized as follows: (1) the HDL-like dominating cluster undergoes fragmentation caused by the percolation of LDL-like aggregates following a spinodal-like kinetics which occurs at a "critical" concentration of _20–30% in LDL; (2) the hydrogen-bond network (HBN) within LDL-like environments is characterized by an equal number of pentagonal and hexagonal rings that create a state of maximal frustration between a configuration that promotes crystallization (hexagonal ring) and a configuration that hinders it (pentagonal ring); (3) the spatial organization of HDL-like environments underwent a marked variation becoming more pronounced upon increasing the pressure; and (4) the latter supports the hypothesis of a liquid–liquid critical point. Thus, these results revealed that water's anomalies occur under specific conditions of composition and connectivity and enlighten the role of the underlying HBN and rationalize the complex microscopic origin of water's anomalies as the cooperative effect of several factors acting synergistically.

5.2.4 Miscellaneous Properties

A molecular dynamics study of the self-dynamics of water molecules in deeply supercooled liquid states suggested that the decay of single-particle dynamics correlation functions is characterized by a fast initial relaxation toward a plateau and by a region of self-similar dynamics, followed at late times by a stretched exponential decay [23]. The results were interpreted in the framework of the mode-coupling theory for supercooled liquids. In a frame of the interpretation, the apparent anomalies of the transport coefficients in water on lowering the temperature were related to the formation of cages and to the associated slow dynamics resulting from the presence of long-lived molecular cages. The thermodynamic response functions in the region where water exhibited anomalous behavior were investigated [24]. Using molecular dynamics simulations of the proposed model TIP4P/2005, both the maximum in density and the minimum in the isothermal compressibility of water along a number of isobars were revealed The calculation showed that: (1) the model correctly describes the decrease in the temperature of the density maximum with increasing pressure; (2) at atmospheric pressure, the model exhibits an additional minimum in density at a temperature of about 200 K, in good agreement with recent experimental work on supercooled confined water; (3) the model also presents a minimum in the isothermal compressibility close to 310 K; (4) the expansion coefficient (kT) showed a minimum at 46.5 °C at $P = 1$ bar; and (5) the model described the decrease in the temperature of the density maximum; and the shift toward slightly higher temperatures of the minimum in _kT as the pressure increases.

The review [11] described simulations and experiments on anomalous water. Several scenarios have been proposed to explain the anomalous properties that become strongly enhanced in the supercooled region (Fig. 5.17).

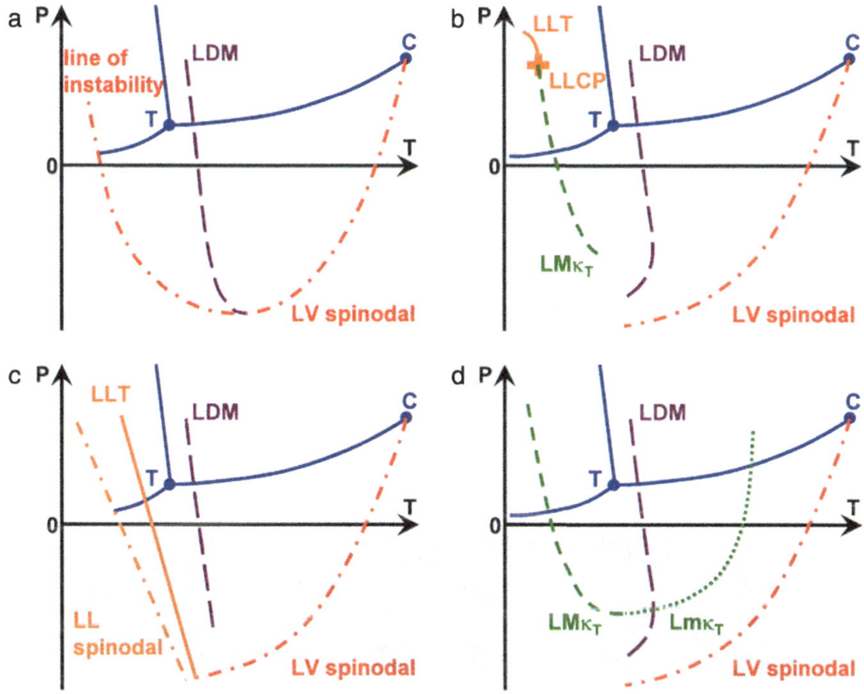

Fig. 5.17 Scenarios that might account: **a** Speedy's stability limit conjecture 33, **b** Poole et al.'s second critical point 23, **c** Poole et al.'s "weak bond"-modified van der Waals model, now the critical-point-free scenario 46, and **d** Sastry et al.'s singularity-free scenario 31. Continuous blue curves show the known equilibrium coexistence lines between liquid, solid, and vapor with the triple point marked as *T*. Liquid–vapor equilibrium terminates at the critical point *C*. The long-dashed purple line shows the line of density maxima (LDM), and the short-dashed and dotted green lines are the lines of isothermal compressibility maxima ($\mathrm{LM}\kappa\mathrm{T}$) and minima ($\mathrm{Lm}\kappa\mathrm{T}$), respectively. Dash–dotted lines indicate lines of instability. In scenarios **a** and **c**, the LDM keeps a negative slope and ends at a line of instability. In scenarios **b** and **d**, the LDM reaches a maximum temperature and changes its slope, eventually merging with a line of density minima (not shown for clarity). When the scenario comprises a liquid–liquid transition, it is displayed with a continuous orange line (LLT), and the liquid–liquid critical point is shown as an orange plus [25]. Published with permission of PNAS article containing the requested material published under a CC BY-NC-ND license

In a scenario (Fig. 5.19a), the boundary of the liquid state at high temperatures is reversing its temperature dependence. Figure 5.19b describes the second critical-point scenario wherein a second critical point exists at positive pressure, where it terminates a line of liquid liquid transition. The critical-point-free scenario by which the liquid–liquid transition exists but the moves sufficiently to negative pressures were shown in Fig. 5.19c. The singularity-free scenario occurs at different temperatures but without a liquid–liquid transition and with a critical point (Fig. 5.19c, d). The following results were reported: (1) the second critical-point scenario has been investigated extensively; (2) a coexistence line between a high-density amorphous phase and a low-density amorphous phase would continue in a coexistence line between a high-density and a low-density liquid phase terminating in a liquid–liquid critical point, LLCP; (3) an approaching this LLCP from the one-phase region, a crossover in the term dynamics and dynamics was found. Based on data of a temperature-dependent balance between a high-density liquid and a low-density liquid favored by, respectively, entropy and enthalpy, a consistent picture of the thermodynamics of bulk water, were composed and discussed.

P–T and *T–ρ* phase diagrams of different thermodynamic scenarios for anomalous liquids were discussed in details [4]. A singularity-free scenario (Fig. 5.18) is consistent with experimental and numerical observations that supercooled water approaches the structure of a fully bonded random tetrahedral network on cooling. The crossover from a weakly tetrahedral structure at ambient *T* to a fully developed tetrahedral structure at very low *T* suggests a sigmoidal *T* dependence of energy

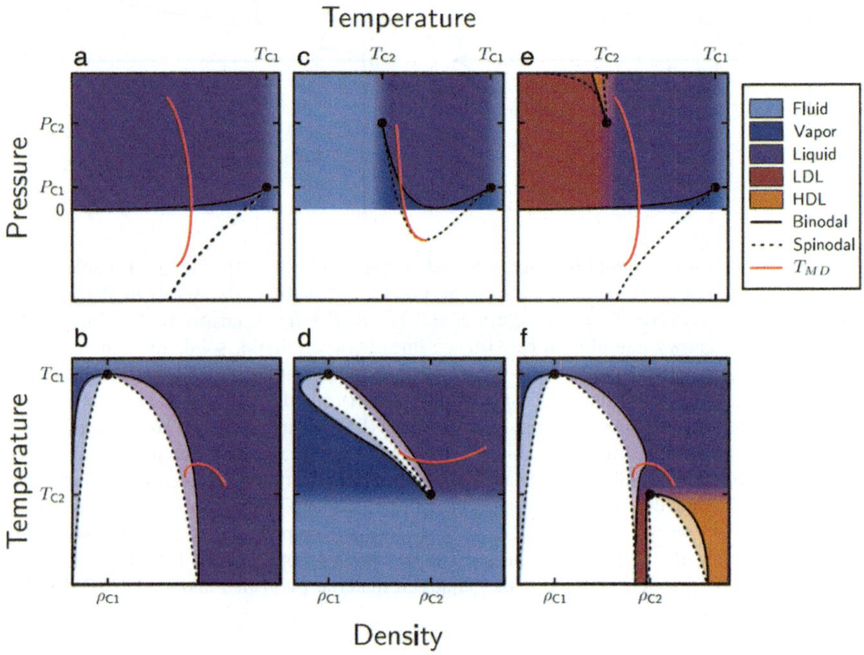

Fig. 5.18 *P–T* and *T–ρ* phase diagrams of different thermodynamic scenarios for anomalous liquids. Binodals are drawn as solid lines, (mean-field) spinodals as dashed lines, and the TMDTMD line as a red solid line. **a**, **b** Show the behavior of a simple system with only a gas–liquid CP. **c**, **d** Show the behavior of a system with a reentrant spinodal. **e**, **f** Show the behavior of a system with a gas–liquid CP and a liquid–liquid (LL) CP (based on the ST2 model). The light regions in **b**, **d**, and **f** between the binodals and the spinodals are the regions where the corresponding phases are metastable [4]. Published with permission of PNAS article containing the requested material published under a CC BY-NC-ND license Published with permission of PNAS article containing the requested material published under a CC BY-NC-ND license

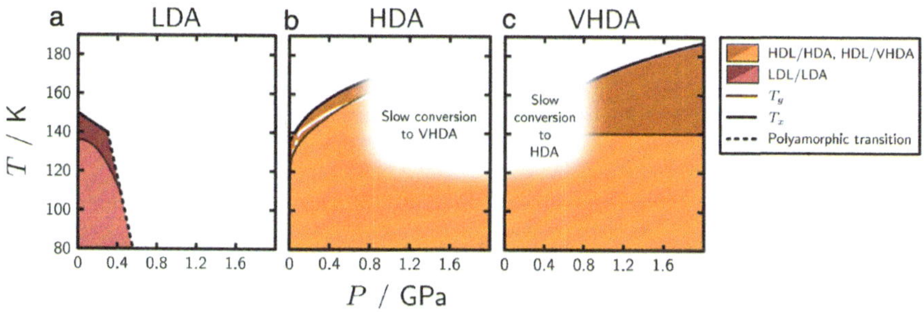

Fig. 5.19 *P–T* regions of existence of LDA (**a**), HDA (**b**), and VHDA (**c**) including the temperatures of crystallization (Tx) and polyamorphic transition as well as the temperature of the glass-to-liquid transition (T_g). The T_g and polyamorphic transition lines are based on available experimental data. The T_g lines are based on the known calorimetric data in combination with the expected pressure dependence based on numerical and non-calorimetric results [4]. Published with permission of PNAS article containing the requested material published under a CC BY-NC-ND license

and volume and a maximum in their T derivatives. In thermodynamically consistent speedy scenario [15], the reentrance is provided by closed-loop phase diagrams (Fig. 5.18), where the spinodal starts in the upper critical point (CP) and terminates at the lower one.

Phase diagram of non-crystalline water (Fig. 5.19) provides guidance to summarize the *P–T* behavior of LDA, HDA, and VHDA. The *P–T* window where these glasses can be observed, bounded by the temperatures of crystallization Tx and polyamorphic transition was in focus of consideration. The calorimetric glass-to-liquid transition as characterized by its onset temperature T_g. LDA was also revealed.

Analysis of available experimental data on such properties of water as heat capacity, speed of sound, dynamic viscosity, and thermal conductivity showed the anomalous behavior of water in the vicinity of critical points, namely the liquid–gas critical point and the second hypothetical critical point in supercooled region [26]. Importantly, that these properties demonstrate anomalous maxima and minima in a region which is far from both critical points and, therefore, a novel region of anomalous properties of water which cannot be related to critical fluctuations was found. The investigation of anomalous behavior of water is restricted to two relatively narrow regions.

The purpose of the study [27] was to generate an expression for the supercooled liquid water on the base of a model for thermodynamic properties of water from subzero temperatures up to 373 K at ambient pressure. The obtained properties for supercooled water such as the mass density r and specific heat under constant pressure Cp, thermal conductivity k, (c) sound velocity v, and (d) surface tension g_s of bulk water were found to be in agreement with literature data (Fig. 5.20). A simple model for heat capacity, density, and thermal expansion of ice was also derived from 170 K up to melting point.

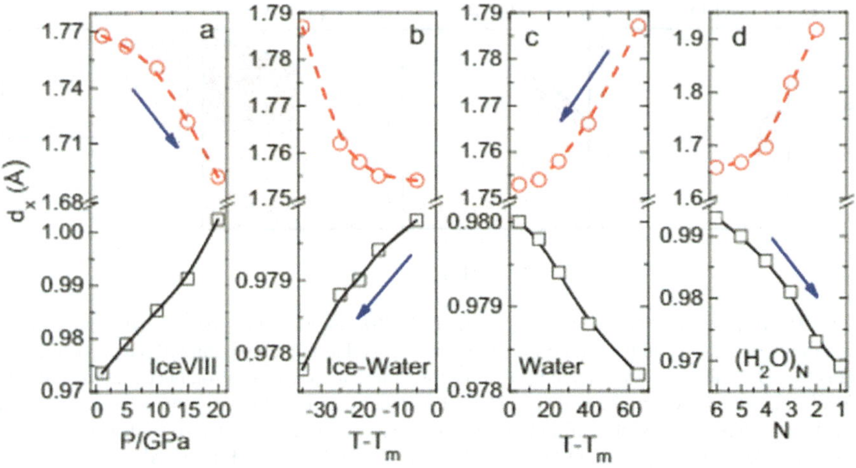

Fig. 5.20 Temperature dependence of **a** the mass density *r* and specific heat under constant pressure Cp; **b** thermal conductivity *k*; **c** sound velocity *v* and **d** surface tension g_s of bulk water. Reproduction of **d** the measured temperature-dependent g_s results in an estimation of the Debye temperature $\theta_{DL} = 198$ K and molecular cohesive energy $E_S(0) = 0.38$ eV in the skin, which gives rise to $E_L = E_b(0)/4 = 0.095$ eV/non-bond for the skin O:H non-bond dissociation. Reprinted with permission from [28]. With permission from American Physical Society

Femtosecond X-ray laser pulses were used to probe micrometer-sized water droplets that were cooled down to 227 K in vacuum in the experiment performed at PAL-XFEL [18]. Micrometer-sized water droplets undergoing evaporative cooling at ≈ 0 bar were probed with ultrafast X-ray scattering (pulse duration < 50 fs), photon energy 5.5 keV. The large-area detector used at PAL-XFEL has the capability to cover an extended *q* range and thereby simultaneously record the wide-angle X-ray scattering (WAXS) region near the first diffraction peak and the small-angle X-ray scattering (SAXS) region, which is sensitive to density fluctuations. Employing this technique, the temperature-dependent scattering structure factor of H_2O was determined. The difference in the maximum value of the isothermal compressibility between the two isotopes shows the importance of nuclear quantum effects. A continuous series **of droplets**, each with a uniform diameter of ~14 μm, was generated by a droplet dispenser in vacuum, and the scattering patterns were generated by X-ray pulses with a photon energy of 5.5 keV. Various temperature points were measured by changing the distance between the dispenser and the measurement point. *T*he wide dynamic range of the detector (MX225-HS, Rayonix) allowed to simultaneously measure the SAXS and WAXS regions for each X-ray shot. Data from three different temperature points (227.7, 231.0, and 237.2 K) were shown.

Data on isothermal compressibility as a function of the pressure difference compared to the LLCP estimated by MD simulations performed with the iAMOEBA water model and correlation length extracted from X-ray scattering at the low-momentum transfer region were presented [18]. A connection of the experimental data to the liquid–liquid critical point (LLCP) was demonstrated in a figure. The maximum of the compressibility as a function of the pressure difference compared to the LLCP estimated by MD simulations performed with the iAMOEBA water model. A magnified view from 450 to 1750 bar. H_2O and D_2O have compressibility maxima of ~105×10^{-6} and ~95×10^{-6} bar^{-1}, which correspond to pressure differences with respect to the LLCP of ~800 and ~950 bar, respectively. Wisdom lines with the inflection point of population conversion and maxima in length scale of fluctuations involving high- and low-density liquids together with the estimated relative positions of the LLCP for H_2O and D_2O were indicated.

For water simulated with two different potentials, TIP5P and ST2, it was found that the temperature of a breakdown of the Stokes–Einstein (SE) relation between the diffusion constant D and the alpha relaxation time t_a. Correlates with the temperature of the maximum in specific heat that corresponds to crossing the Widom line [29]. Specifically, results for $D\ t_a/T$ collapse onto a single "master curve" if temperature is replaced by $T - T_W(P)$. The size of the mobile molecule clusters (dynamical heterogeneities) increased sharply near $T_W(P)$, and mobile particle cluster size also collapses onto a single master curve if T is replaced by $T - T_W(P)$. The crossover to a more locally structured low-density liquid (LDL) as $T \rightarrow T_W(P)$ appears to be well correlated both with the breakdown of the Stokes–Einstein relation and with the growth of dynamic heterogeneities. It was hypothesized that in these conditions liquid–liquid phase transition can occur. It was expected that calculated the alpha relaxation time t_a would behave similar to viscosity, and hence, according to the Stokes–Einstein (SE) relation, $D\ t_a/T =$ constant. Both t_a and the diffusion constant D for the TIP5P display rapid changes at low T.

The authors summarized that: (1) supercooled water exhibits a breakdown of the Stokes–Einstein relation between the diffusion constant D and the alpha relaxation time; (2) the temperature of the decoupling of diffusion and alpha relaxation correlates with the temperature of the maximum in specific heat that corresponds to crossing the Widom line $T_W(P)$; (3) the size of the mobile molecule clusters (dynamical heterogeneities) increases sharply near $T_W(P)$; (4) mobile particle cluster size $n(t^*)$ for different pressures, where t^* is the time for which the mobile particle cluster size is largest, collapse onto a single master curve if T is replaced by $T\ T_W(P)$; (5) the crossover to a more locally structured low-density liquid (LDL) as $T_W(P)$ correlated both with the breakdown of the Stokes–Einstein relation and with the growth of dynamic heterogeneities; and (6) the results are consistent with the possibility that the breakdown of the SE relation in supercooled water at low pressures is associated with the hypothesized liquid–liquid phase transition.

Data on research on supercooled and glassy water published by 2017 were reviewed focusing on the possible origins of its complex behavior [4]. Experiments showed that metastable liquid water can transiently exist on the millisecond timescale down to $T = 227$ K and that tetrahedral ordering increases in supercooled water. It was stressed that the strong directionality of the water–water interaction, and the competition between local energy, local entropy, and local density plays the central role. The explicitly on bulk water in metastable conditions, that is, water below its equilibrium melting temperature, both as a supercooled liquid and as an amorphous solid was also be focused. In a separate section, the possibility of linking polyamorphism with a liquid–liquid transition that could take place in the no-man's land, the temperature–pressure window in which homogeneous nucleation prevents the investigation of water in its metastable liquid form were debated.

5.3 Supercritical Water

Molecular dynamics of supercritical water (SCW), the Widom line as the crossover between liquid-like and gas-like behavior in supercritical fluids, the pseudo-critical regions for supercritical water, local structural fluctuations, and hydrogen bonding and structural transitions in supercritical water, etc., were in focus of attention in last decades [30–34].

Two differences between pressurized liquid water and supercritical water were revealed by means of the supercritical water-cooled reactor [33]. The first is that, while in normal pressurized water one can cause phase transition (boiling) by increasing the temperature or by decreasing the pressure, this is not possible in supercritical water. Some quantities (like density, compressibility, heat capacity) exhibit anomalous behavior in the supercritical region. The location of these anomalies was represented by lines in pressure–temperature space, connecting the extrema or inflection points of these quantities. According to the suggested fluid phase diagram of water, at critical point (647 $\pm$ 2 K; 22.064 $\pm$ 0.001 MPa (NIST)), above and below the vapor pressure line, one can find stable liquid water or steam, respectively.

The contribution of hydrogen-bonding interactions to the formation of local density inhomogeneities in supercritical water at near-critical conditions has been studied employing molecular dynamics simulations [34]. The results revealed the strong effect of water molecules forming one and two hydrogen bonds on the determination of the local density augmentation in the fluid. The correlation between local orientational order, derived from analysis of the trigonal and tetrahedral order parameters, and hydrogen bonding was obtained. Significant evidence about the existence of a liquid-like to gas-like structural transition in supercritical water at the density range close to 0.2 ρ_c was also presented.

Classical MD simulations of supercritical water using polarized and polarizable models have been performed in order to study the anomalous behavior of SCW for a wide range of thermodynamic conditions [35]. The simulations at different thermodynamic conditions (state points) were carried out using GROMACS v.2016–4 code55. The supercell was consisting of 2048 water molecules. The simulation revealed an anomalous behavior due to nanoscale density fluctuations (inhomogeneity) which is demarked through the Widom line (maxima in response factions) and affects the properties was were examined. Developed theory predicted that liquid-like (LL) and gas-like (GL) supercritical water are metastable phases, and that the Widom line zone is marking the crossover. Besides the Widom line, in fluids, can be one more line as a dynamical crossover which is termed the Frenkel line. This line demarcates two regions in which the fluid behaves as non-rigid (dense gas-like behavior) and rigid-liquid (solid-like behavior). A fluid was characterized supercritical when the temperature and pressure are higher than the critical point (in the case of water $T_c > 647.096$ K and $P_c > 220.640$ bar). Beyond the critical point, no physically observable difference between a liquid and a gas exists and hence only a single fluid-phase region was considered. The following data were presented in separated figures: average number of hydrogen bonds (⟨nHB⟩) per water molecule as a function of density, static dielectric constant along the studied isobars as a function of temperature, self-diffusion coefficient of water at different isobars in respect to temperature, and supercritical isobars using different interatomic potentials. The distribution of mass density along four isobars at different temperatures is shown in Fig. 5.21.

The presence of inhomogeneous patterns in the density is evident mainly in the gas-like phase (Fig. 5.22) which is in consensus with the absence of the 2nd peak in the radial distribution function suggesting the loss of the tridimensional tetrahedral structure. The experimental Widom line of water was in agreement with the current theory of supercritical fluid pseudo-boiling. With the increase of temperature, the height of the first peak increases due to the low-density cluster formation which starts during the destruction of the hydrogen-bond networks.

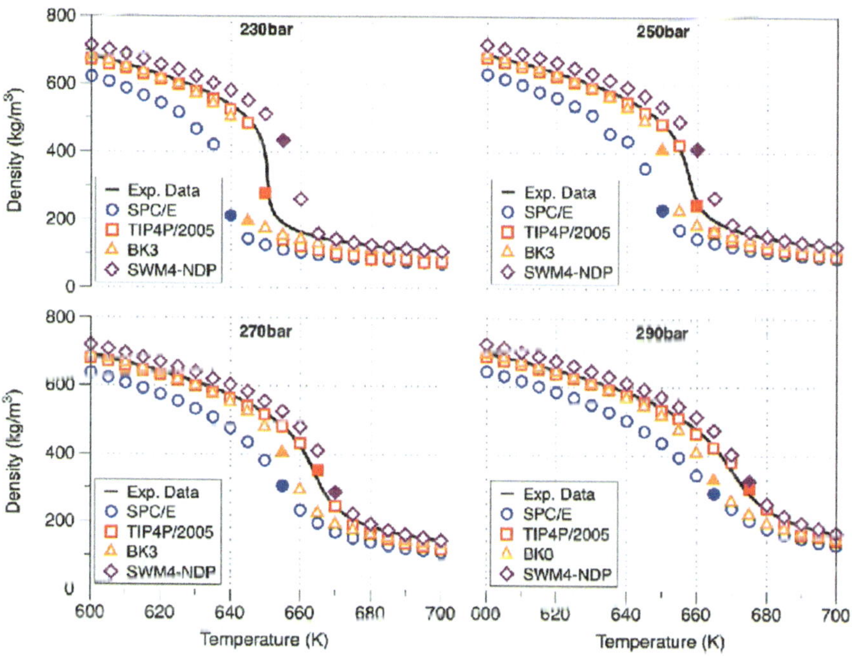

Fig. 5.21 Supercritical isobars using different interatomic potentials (SPC/E, TIP4P/2005, BKE, and SWM4-NDP) in comparison with experimental results. The filled symbols describe the temperature at which the maximum heat capacity line (point in the Widom line) is crossed (Supplementary Information). The Widom point signs the change of water behavior from liquid-like to gas-like. The black line refers to experimental data obtained using water EoS36 [35]

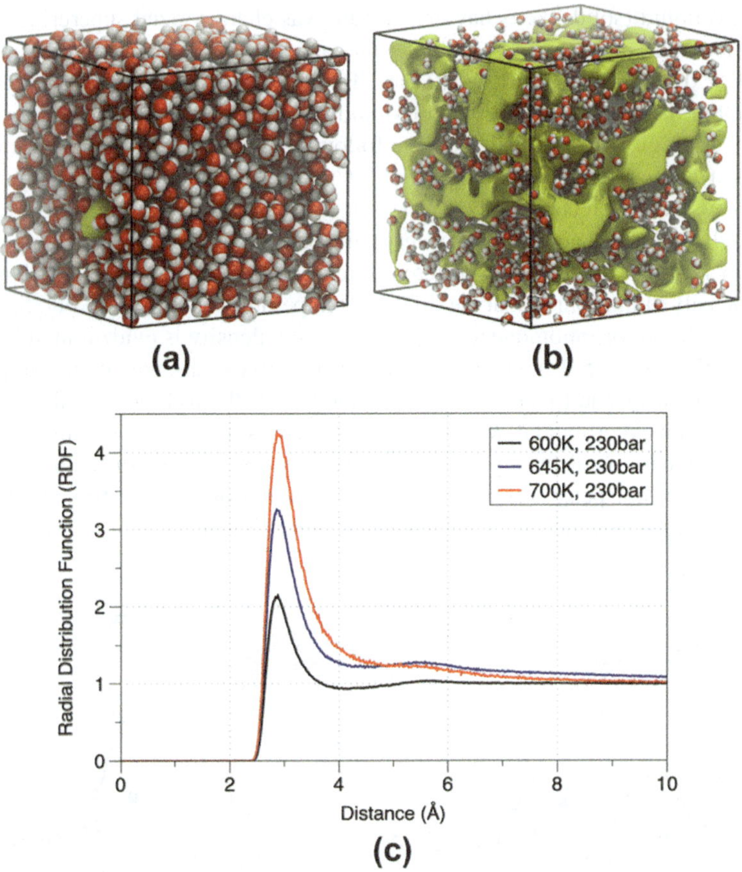

Fig. 5.22 Water molecules and void isosurfaces represent those points in space in which all water molecules exhibit a distance of more than 3 Å portraying the inhomogeneity in **a** the liquid-like phase and **b** the gas-like phase. The red color presents the oxygen atoms, the gray the hydrogen atoms, and the yellow isosurfaces the voids inside the supercell. For the visualization, the VMD code was used. **c** Radial distribution function (RDF) of oxygen–oxygen interactions in the liquid-like and gas-like phases in respect to different pressures using the TIP4P/2005 mode [4]

The average number of hydrogen bonds per water molecule (⟨nHB⟩) as a function of the temperature for the studied pressures was determined [35]. With increasing temperature, a progressive decrease of the ⟨nHB⟩ occurs asymptotically approaching zero at higher temperatures. The ⟨nHB⟩ in temperatures above T_c (647.17 K) is always below the percolation threshold indicating the absence of a continuous network of hydrogen bonds.

The following conclusions were formulated: (1) the absence of the second peak in the oxygen–oxygen RDF indicated the SCW inhomogeneity which was intensified in the gas-like phase, (2) the temperature increase leads to the increase of the first peak in the oxygen–oxygen RDF suggesting the cluster formation due to the destruction of hydrogen-bond network; (3) in higher temperature regimes, an increase of the monomers fraction indicated a sheet-like structure which yields to an increased ability of dissolving nonpolar compounds; (4) the collapse into a single curve of the number of hydrogen bonds and self-diffusion coefficients in respect to density indicated that the main factor governing diffusion is the density; and (5) only the polarizable models (BK3 and SWM4-NDP) was estimated with high-accuracy the dielectric constant due to use of multiple polarization sites. Overall, the polarized TIP4P/2005 model accurately predicted the properties of water in both liquid-like and gas-like regions. All interatomic potentials captured the anomalous behavior providing a direct evidence of molecular-scale inhomogeneity [35].

Neutron imaging techniques were utilized to monitor density fluctuations of supercritical water while the system evolves rapidly from liquid-like (LL) to gas-like (GL), as the Widom line is crossed during isobaric heating [36]. The neutron imaging experiments were performed using a dedicated setup for in situ measurements under supercritical water conditions (designed and developed by SITEC-Sieber Engineering AG, Switzerland, and Computer Power SRL, Romania). Utilizing this technique, the density fluctuations of water flow passing through a porous medium made from activated carbon fibers and the conversion from LL to GL supercritical water at Widom line crossover were examined. The experimental Widom line of water was found to be in agreement with the current theory of supercritical fluid pseudo-boiling which is the state where density changes and the heat capacity are maximal.

5.4 Ice

5.4.1 General

One of the most intriguing anomalies of water is its ability to exist as distinct amorphous ice forms, glass polymorphism, or polyamorphism (Fig. 5.29) [37–41]. In addition to ordered crystals, ice also can come in amorphous and other metastable forms. Amorphous ice has loosely tetrahedral in molecular structure and kinetically stable in the pressure–temperature stability space. There is a rich variety of metastable states reported **for** glassy water which is defined as solids that lack long-range order and are regarded as vitrified liquids, in which the molecules are kinetically immobilized. Ice molecules can exhibit more than eighteen or different *phases* packing geometries of which that depend on temperature and pressure. "Ordinary" *ice I_h* has a *hexagonal crystalline structure* with minute traces of cubic ice denoted as ice I_c. The ice I_h is characterized by *dielectric constant* (ε_r) 3.15 and refractive index (n) 1.309. The most common phase transition to ice I_h occurs when liquid water is cooled below 0 °C (273.15 K, 32 °F) at standard atmospheric pressure. When water is cooled rapidly (quenching), up to three different types of amorphous ice can form depending on the history of its pressure and temperature. Figure 5.23 displays types of high-density crystalized, high-density water, supercooled water, and supercritical water.

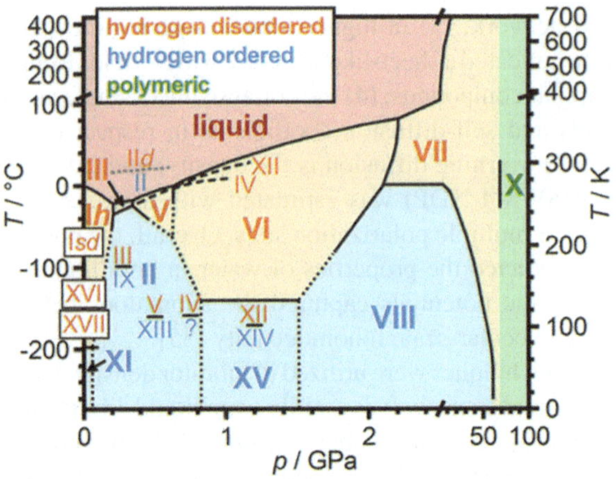

Fig. 5.23 Phase diagram of ice including the regions of thermodynamic stability of liquid water (red), the hydrogen-disordered (orange), and hydrogen-ordered (blue) phases of ice as well as the "polymeric" ice X (green). Stable phases of ice are highlighted by large bold Roman numerals, whereas metastable states are indicated by a smaller font size. Black dotted lines are extrapolated phase boundaries at lower temperatures. Dashed lines indicate the metastable melting lines of ices IV and XII. The metastable phase boundary between ice II and IId (gray dotted line) has been predicted computationally. The melting line of a metastable phase discovered by Chou et al. has been assigned to ice XII [7]. With permission from American Institute of Physics

5.4.2 Ice I Structures

In liquid water, each molecule is hydrogen-bonded to approximately 3.4 other water molecules, while in ice each molecule is hydrogen-bonded to 4 other molecules (Figs. 5.24 and 5.25) https://www.nyu.edu/pages/mathmol/textbook/info_water.htm.

The close similarity between the equilibrium H–O–H angle in water of 104.52 and the tetrahedral angle of 109.46 _ suggests that water tends to form tetrahedral H-bonds upon crystallization. The most stable ice phase at atmospheric pressure below the melting temperature (T_m) and above 72 K is hexagonal ice with a tetrahedral arrangement of the oxygen atoms [37].

Fig. 5.24 A comparison structures of water (**a**) and ice (**b**). Notice the empty spaces within the ice structure. https://www.nyu.edu/pages/mathmol/textbook/info_water.htm

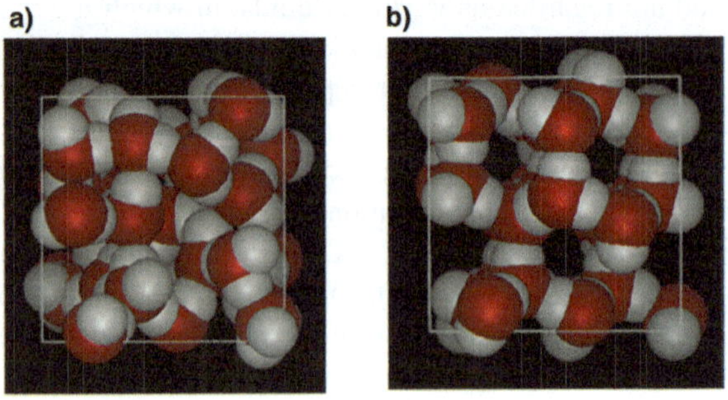

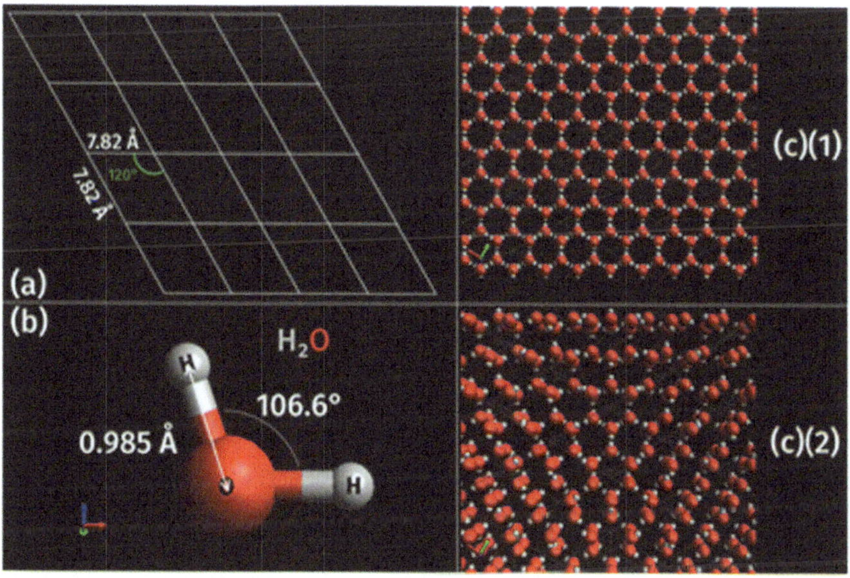

Fig. 5.25 Three-dimensional crystal structure of H_2O ice I_h (**c**) is composed of bases of H_2O ice molecules (**b**) located on lattice points within the two-dimensional hexagonal space lattice (**a**). https://en.wikipedia.org/wiki/Ice

The behavior of structures of H_2O crystalline ices Ih, Ic, XI, VII, VIII, VI (Fig. 5.26) was studied in molecular dynamics experiment [39]. Based on the simulation results of the above-mentioned ices and previous simulation of ices II, III, IX, IV, and XII, the general regularities of dynamic properties of water molecules in crystalline water ices were formulated. The calculations showed that ice spontaneous nucleation creates once a sufficient number of relatively long lived hydrogen bonds to form a compact initial nucleus, and initial nucleus then slowly changes shape and size until it reaches a stage that allows rapid expansion, resulting in crystallization of the entire system.

A molecular dynamics trajectory that captures the molecular processes involved in the freezing of pure water was calculated [40]. For a calculation, the TIP4P water model was used. It was found that ice nucleation occurs involving a sufficient number of relatively long-lived hydrogen bonds which develop spontaneously at the same location to form a fairly compact initial nucleus. Then, the initial nucleus slowly changes shape and size until it reaches a stage that allows rapid expansion, resulting in crystallization of the entire system. Figure 5.27 shows the hydrogen-bond structure of the system at different times after quenching. In the quiescent period (Fig. 5.30a), the hydrogen-bond network contains mainly five-, six-, and seven-membered rings. Hydrogen bonds have a relatively long lifetime (τ_{life}) of more than 2 ns and long-lasting hydrogen bonds appear occasionally and randomly at various locations. At $t \approx 256$ ns, a polyhedral structure composed of long-lasting hydrogen bonds forms spontaneously (Fig. 5.27) by altering hydrogen bonds with

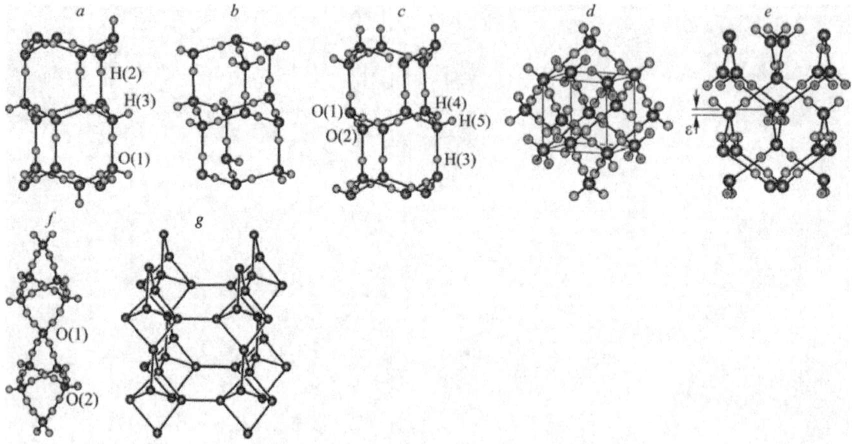

Fig. 5.26 Structures of ices Ih, Ic, XI, VII, VIII, and VI. Structure fragments of proton-disordered ices Ih (**a**) and Ic (**b**) with a possible arrangement of H atoms; a fragment of the proton-ordered ice XI structure (**c**). A fragment of the ice VII structure (**d**); all possible positions of H atoms are displayed, their population is 0.5; the unit cell is designated. A fragment of the ice VIII structure (**e**); the shift of independent frameworks relative to each other is shown. In **d** and **e** H atoms of one of the two identical independent Ic water frameworks forming structures VII and VIII are blacked out. A basic element of the ice VI structure is a chain of water molecules stretched along the *c* axis (**f**); all possible positions of H atoms are shown. Association of chains into one of the two identical independent water frameworks forming the ice VI structure (**g**); the chains forming the second framework pass through the channels of this framework; positions of H atoms are not designated [39]

surrounding water molecules, and finally "anchors" at the position it occupies at $t \approx 290$ ns (Fig. 5.30c). At the end of period of rapid growth, a "stacked honeycomb" structure, consisting of six-membered rings, is established (Fig. 5.30d). During the final freezing stage ($t > 320$ ns), gradual formation of a perfect honeycomb structure (Fig. 5.30e) is completed.

Figure 5.28 indicates the number of long-lasting hydrogen bonds ($\tau_{\mathrm{life}} > 2$ ns) as a function of time which changes intermittently in the quiescent period, fluctuates in the initial nucleation period, and then rapidly increases after that.

At $t \approx 256$ ns, several long-lasting bonds appear by chance in the same location, where they form a polyhedral structure (Fig. 5.29) which evolves into a stable initial ice nucleus. The observation that the initial ice nucleus forms by chance as a result of random hydrogen-bond network rearrangement dynamics implies that the length of the quiescent period is randomly distributed.

Experimental diffraction data on pair distribution functions of stacking-disordered ice I were used to determine local information on the fractions of cubic and hexagonal stacking [41]. The ice Ih consists of layers of hydrogen-bonded water molecules that form annulated six-membered rings in the armchair conformation. The thermodynamically stable form of ice I is hexagonal (ice Ih) (Fig. 5.30a) while cubic stacking

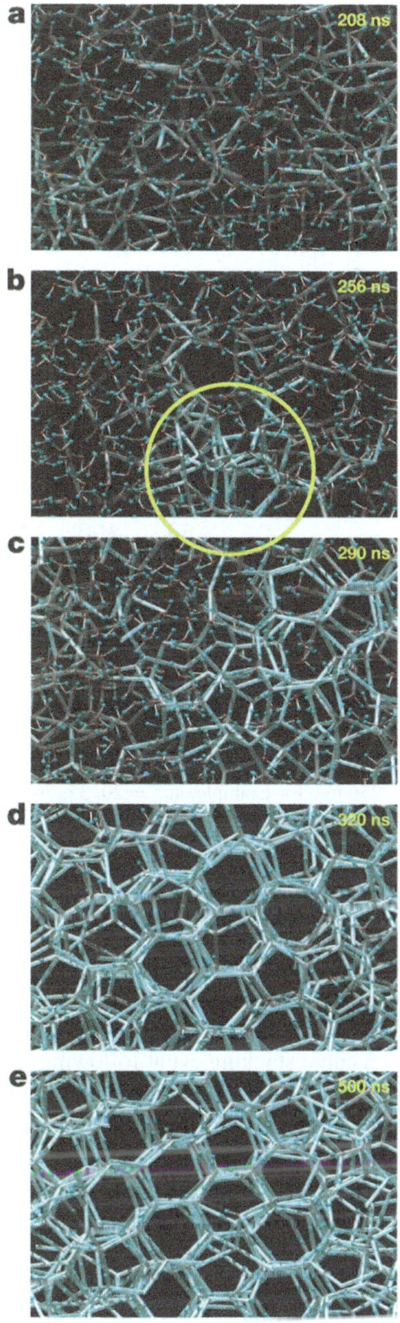

Fig. 5.27 Hydrogen-bond network structure of water at a given time in inherent structures. **a** At time $t =$ 208 ns; **b** at $t = 256$ ns; **c** at $t = 290$ ns; **d** at $t = 320$ ns; and **e** at $t = 500$ ns. Lines indicate hydrogen bonds among water molecules, and the intermolecular bonds of water participating in such hydrogen bonds. Bright blue lines indicate "long-lasting" hydrogen bonds, which have persisted longer than a specified threshold value before a given time t. The brightest blue lines are those with a lifetime longer than 2 ns ($\tau_{\text{life}} > 2$ ns). An initial nucleus is formed in the region circled in **b** [40]

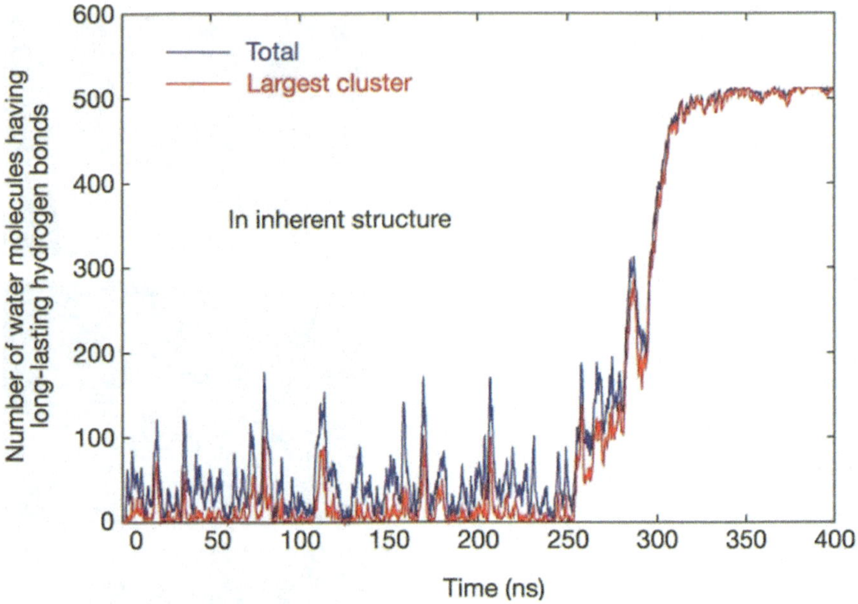

Fig. 5.28 Number of water molecules having long-lasting hydrogen bonds ($\tau_{life} > 2$ ns) and the size of the largest cluster in inherent structures. The number of water molecules having "long-lasting" hydrogen bonds is indicated by a dark-blue line. See Methods and Fig. 2 legend for the definition of "long-lasting" hydrogen bonds. The red line indicates the number of water molecules belonging to the largest cluster. A cluster here is defined as a group of water molecules connected by long-lasting hydrogen bonds. We note that the number fluctuates intermittently in the quiescent period (t = 0–256 ns), for example, at t = 20, 35, 70, 80, 110, 160, 170, and 207 ns. A Fourier transform of this number fluctuation during the quiescent period yields a $1/f$ type power spectrum [40]

of the layers is also possible (Fig. 5.30b). Consequently, a perfect cubic ice I structure (Fig. 5.30c) consists of only six-membered rings in the armchair conformation whereas ice Ih contains a 50:50 mixture of boat and armchair rings.

The four second-order stacking probabilities for cubic hexagonal stacking were considered, namely Φccc, Φhcc, Φchc, and Φhhc. Φhcc and describe the probability of a cubic stacking event following (hc) stacking in a stackogram (Fig. 5.31). In the figure, the first-order stacking probabilities are plotted against each other. The bottom left and corner correspond to fully hexagonal ice Ih, and the fully cubic ice Ic state is located at the top right-hand corner. The diagonal line connecting the ice Ih and ice Ic corners is the line of random stacking since Φcc = _Φhc.

The experimental diffraction data collected at 80 K for ice Ih and the corresponding experimental $D(r)$ total distribution functions are shown in Fig. 5.32a, b. The ice Isd sample, obtained after the phase transition from ice VI (T_A = 156 K), contains significant amounts of cubic stacking. The various low-r peaks (Fig. 5.35b) were assigned starting from the O–O distance at 1.0 Å to the hydrogen-bonded OO

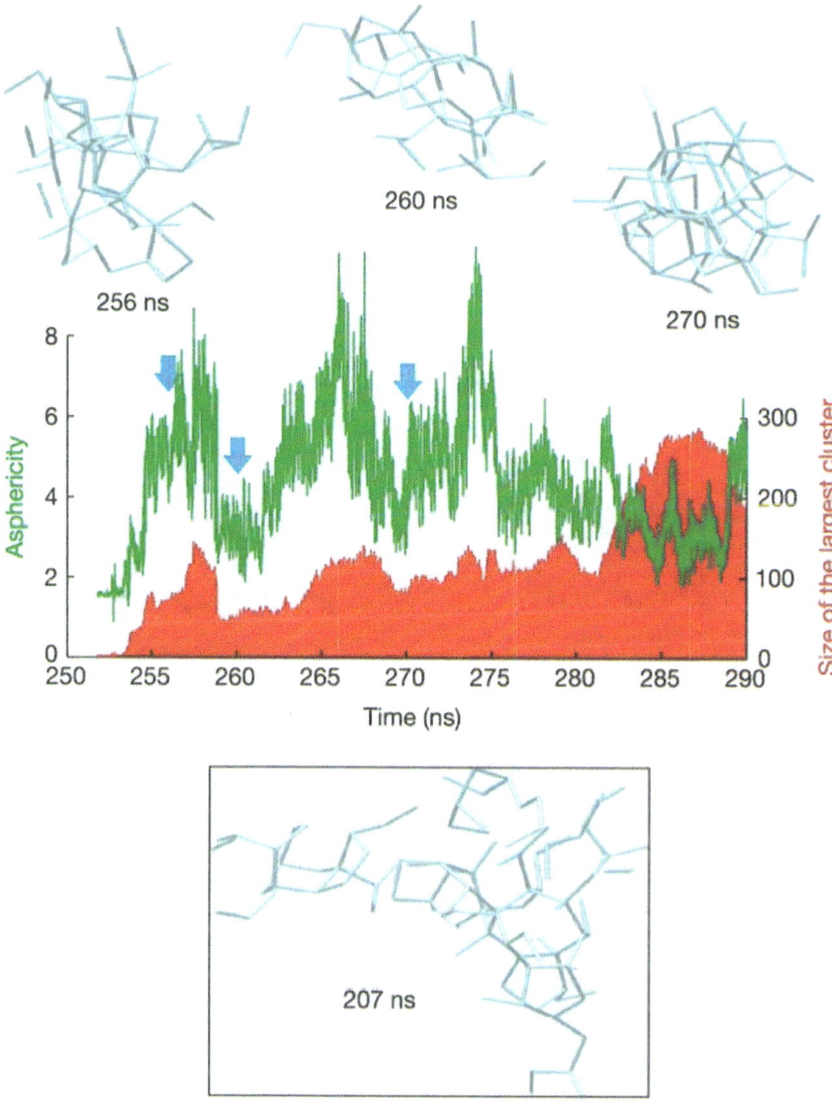

Fig. 5.29 Asphericity and size of the largest cluster. Asphericity (green line) is calculated at 10-ps intervals for inherent structures. A cluster is defined as a group of water molecules connected by long-lasting hydrogen bonds. We assign Voronoi polyhedra to individual water molecules belonging to this nucleus. The shape of the nucleus is represented by a polyhedron which consists of these Voronoi polyhedra. Asphericity of the nucleus is defined as the cube of the surface area divided by the square of the volume of this polyhedron, renormalized to yield the asphericity of a sphere to be 1. Asphericity increases as the shape deviates from the sphere; for example, it is 1.9 for a cube. The size of the largest cluster (red bar lines) is defined as the number of water molecules belonging to the nucleus. The structures of the largest cluster (nucleus) for $t = 256$, 260, and 270 ns are also shown (blue arrows from left to right), as well as that at $t = 207$ ns (whose cluster size is 100) in the inset [40]

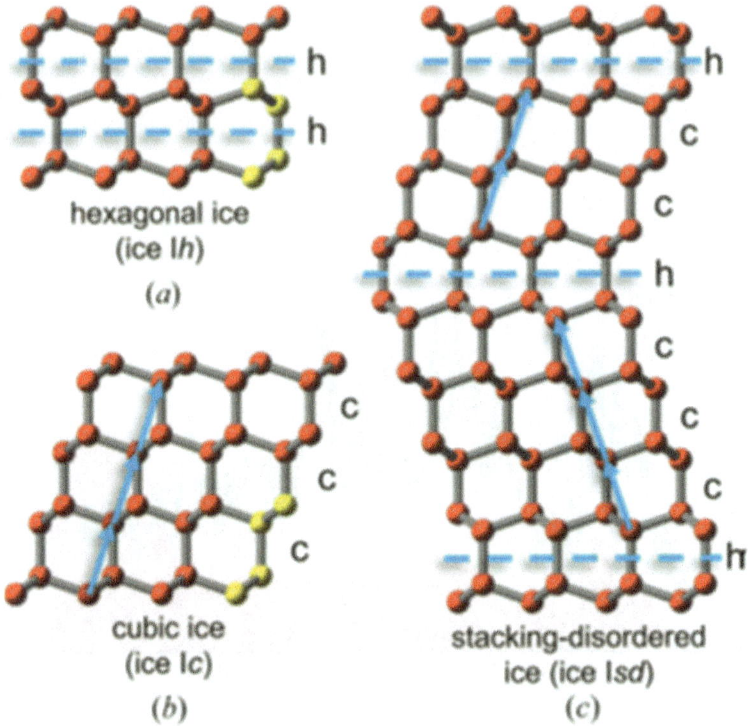

Fig. 5.30 Stacking of layers in **a** hexagonal, **b** cubic, and **c** stacking-disordered ice I. Red spheres indicate oxygen atoms. Six-membered rings in the boat and armchair conformations are highlighted in yellow in **a** and **b**, respectively. Hydrogen atoms have been omitted for clarity. Hexagonal or cubic stacking of layers is indicated by "h" and "c." respectively. Dashed lines show the mirror planes associated with hexagonal stacking, whereas cubic stacking is highlighted by displacement vectors [41]. With permission from Journal of Applied Crystallography IUCr Journals

distance at 2.75 Å. Differences arising from cubic and hexagonal stacking manifests at distances above the average layer separation of 3.7 Å.

Figure 5.33 summarizes data on the structural diversity of the ice family.

5.4.3 *Physical Properties*

As can be seen from Figs. 5.34, 5.35, and 5.36, the assessed density, thermal expansion coefficient, and vapor pressure of ice derived from it, agree well with the literature data and the obtained equations have good extrapolation capabilities especially to lower temperatures [27].

Calculated **thermal expansion** coefficient of liquid water compared with the literature values [27].

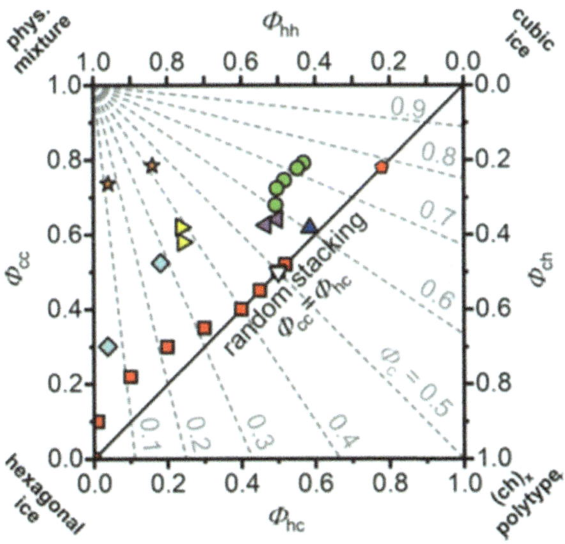

Fig. 5.31 "Stackogram" used to describe the structure of stacking-disordered ice including first-order memory effects. The diagonal solid line indicates random stacking where $\Phi cc = \Phi hc$. Lines of constant cubicity (Φc) are shown as dashed gray lines. All stacking probabilities shown in this plot were determined from the analysis of diffraction data. The **ice Isd** samples were obtained by D_2O vapor deposition at low temperatures (cyan diamonds) as well as ambient pressure heating of D_2O ice IX (yellow right-pointing triangles), D_2O ice V (magenta left-pointing triangles), D_2O CO_2 clathrate hydrate (orange stars) H_2O ice II (green circles) and H_2O ice VIII (blue up-pointing triangle). H_2O ice Isd obtained from homogenous and heterogeneous freezing of water is indicated by the down-pointing open triangle and red squares, respectively. The red pentagon corresponds to _20 nm ice Isd particles for which random stacking is assumed [41]

Using Raman spectroscopy, new insights into the spectroscopic properties of ice were gained [42]. Stacking disorder in a semiquantitative fashion (Fig. 5.37) was detected and quantified by comparing the Ramon spectra of stacking disordered ice with spectra of the hexagonal ice Ih. The results suggested that the spectral signature of stacking disorder arises from a greater structural diversity on the local length scale, vibrational modes that appear due to the lower-symmetry environments, and a strengthening of the covalent bonds.

Figure 5.38 presents the normalized Raman spectra of H_2O ice Isd samples from different parent materials (spectra 2–6) and of H_2O ice Ih (spectrum 1) in the coupled ν(O–H) spectral region. Similar spectroscopic trends were seen in the coupled ν(O–D) spectral region of D_2O ice Isd samples and D_2O ice I.

In Fig. 5.39, it can be seen that the μ(O–D) of ice Isd from ice V is centered at a higher wavenumber and is broader compared to the μ(O–D) feature of ice Ih. The larger half-width of the feature can be explained by the presence of more diverse local environments in ice Isd.

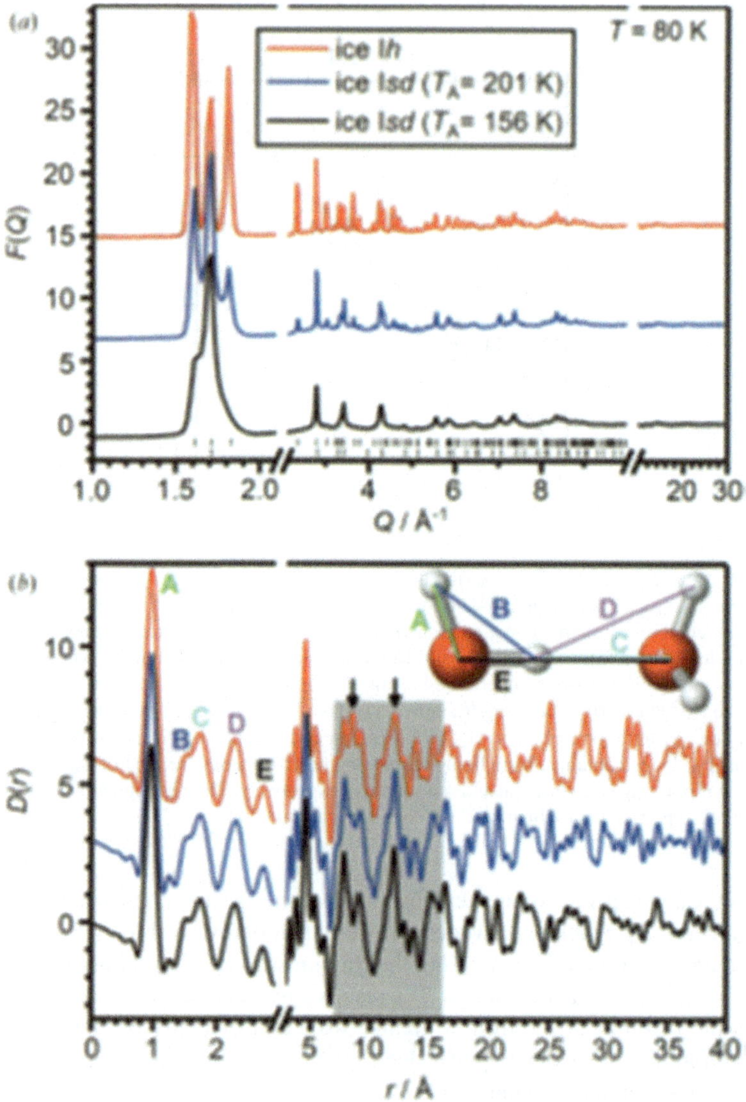

Fig. 5.32 **a** Experimental $F(Q)$ diffraction data collected at 80 K for ice Ih as well as two ice Isd samples annealed at 156 and 201 K, respectively. Tick marks indicate the expected positions of Bragg peaks for ice Ih (top) and ice Ic (bottom) for $Q < 10$ Å^{-1}. **b** Corresponding $D(r)$ data, with the features in the low-r region assigned to the intra- and intermolecular distances as indicated. The r range used for fitting is indicated by a gray-shaded area, and the arrows indicate the major differences in this range. The red and blue curves in **a** and **b** are shifted vertically for clarity [41]. With permission from Journal of Applied Crystallography IUCr Journals

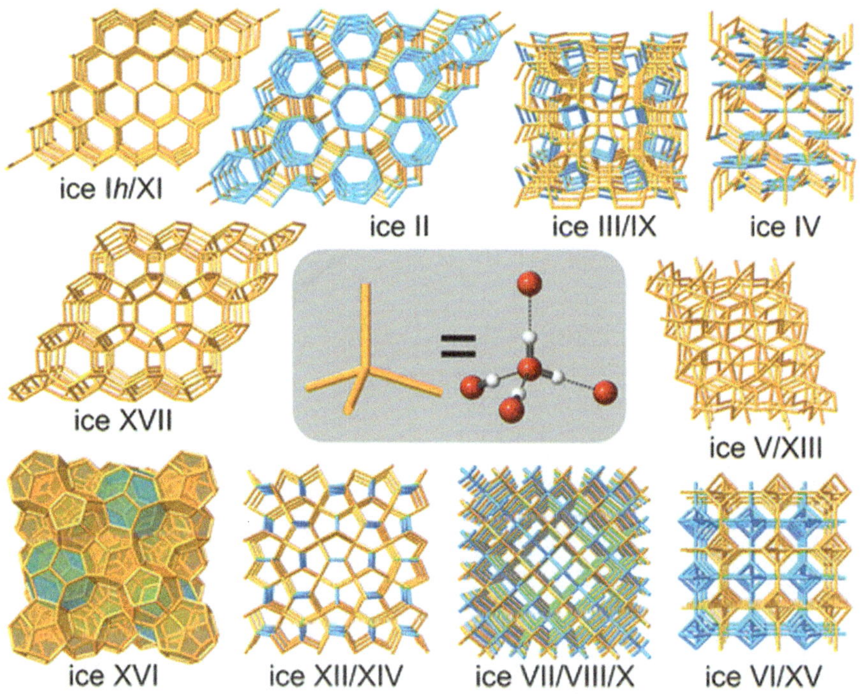

Fig. 5.33 Hydrogen-bonded networks of the various polymorphs of ice. The inset shows how the structure of a local hydrogen-bonded environment of a water molecule corresponds to the orange stick-representation used in this figure. Some of the network topologies are shared between the hydrogen disordered phases and their hydrogen-ordered counterparts. Noteworthy structural features are highlighted in cyan. These include the six-membered rings of the "ice nanotubes" in ice II, the fourfold spirals in ice III/IX, the interpenetrated six-membered rings in ice IV, the second independent hydrogen-bonded networks in ice VI/XV and ice VII/VIII/X, and the "zigzag" chains of ice XII/XIV. The 512 and 5126 4 cages of ice XVI are shown as orange and cyan polyhedra, respect [7]. With permission from American Institute of Physics

Obtain findings were compared to results from diffraction and calorimetry. Advantages and disadvantages of the techniques with respect to detecting stacking disorder in ice I were discussed.

The dynamics of ice I were studied via dielectric spectroscopy and nuclear magnetic resonance relaxometry on protonated and deuterated samples obtained by heating high-density amorphous ices as well as crystalline ice XII [43]. The corresponding dielectric response in the range from 0.1 Hz to 100 kHz and nuclear magnetic resonance (NMR) relaxometry effects of the Ic(h) → Ih transition, which probe the dynamics typically on the 10–100 MHz scale were examined. Results of dynamic investigations of cubic ice produced from different initial phases, from unrelaxed high-density amorphous ice (eHDA and uHDA, respectively), and from crystalline ice XII were compared. As an example, Fig. 5.40 presents the results obtained for a sample of ice I that was prepared from deuterated eHDA and heated

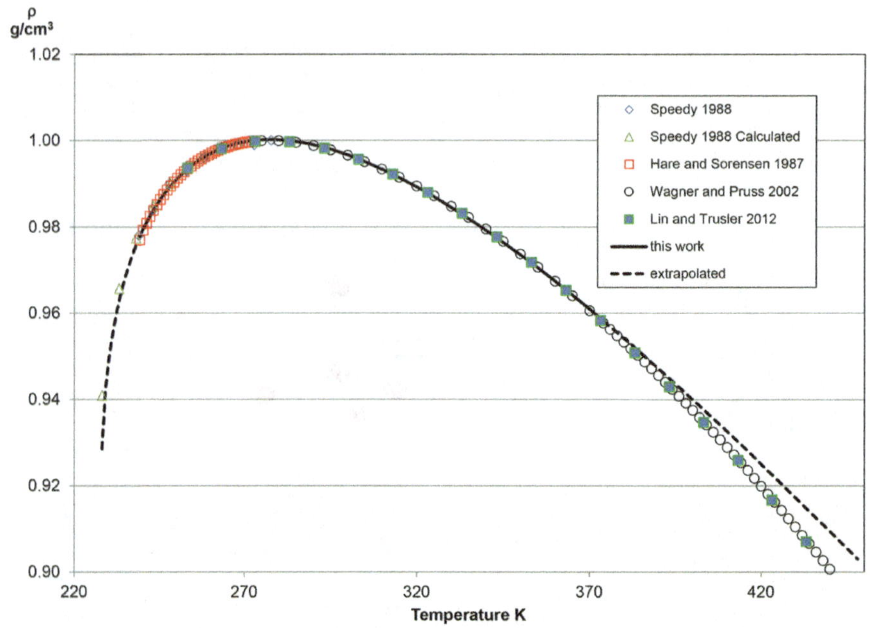

Fig. 5.34 Assessed density ρ of liquid water compared with experimental data. Dotted lines indicate extrapolated values [27]. With permission from American Chemical Society

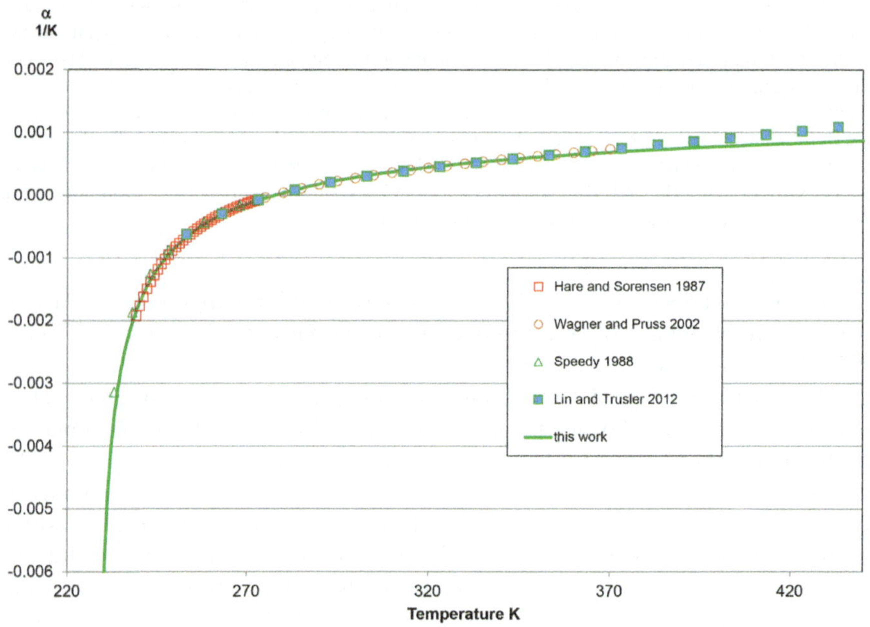

Fig. 5.35 Calculated thermal expansion coefficient of liquid water compared with the literature values [27]. With permission from American Chemical Society

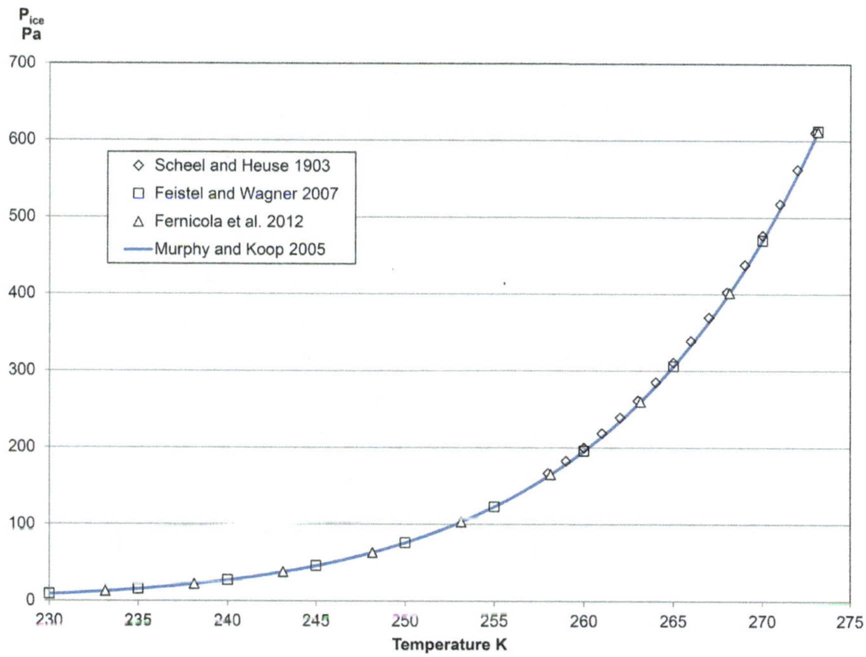

Fig. 5.36 Vapor pressure of ice p_{ice} according to various authors [27]. With permission from American Chemical Society

to 200 K at a rate $|q| = 0.5$ K/min. The low-density amorphous ice (LDA) ice Ic transformation occurs near 160 K. The consequences of the temperature cycling and hallmarks for the cubic to hexagonal transition and the ice sample cycled down after reaching 270 K were demonstrated in the figure.

In nuclear spin relaxometry experiments, the deuteron spin–lattice relaxation times, T_1, of ice I generated from uHDA and from eHDA were derived from longitudinal magnetization recovery curves, $M(t)$, recorded at 200 K (Fig. 5.41).

For comparison, T_1 measurements at Tp on ice I made from uHDA and from eHDA, as well as from hexagonal ice produced by freezing water were summarized in Fig. 5.42.

Thus, despite the fact that the NMR and the dielectric data probe the dynamics of ice I on different frequency scales, they show very similar effects. The rearrangements of the hydrogen matrix of "cubic" ice to hexagonal ice were found to be sensitive to dynamical changes initiated by stacking disorder.

Recent study [44] provided measurements for improving the high-pressure and low-temperature representation of liquid water. Sound speeds of pure liquid water were measured between 0.1 and 700 MPa, from 353 K down to the melting curves of the ice phases. A new equation of state was derived extending from 240 to 500 K and from 0.1 to 2300 MPa, covering much of the subcritical domain of water up to the ice VI–ice VII transition.

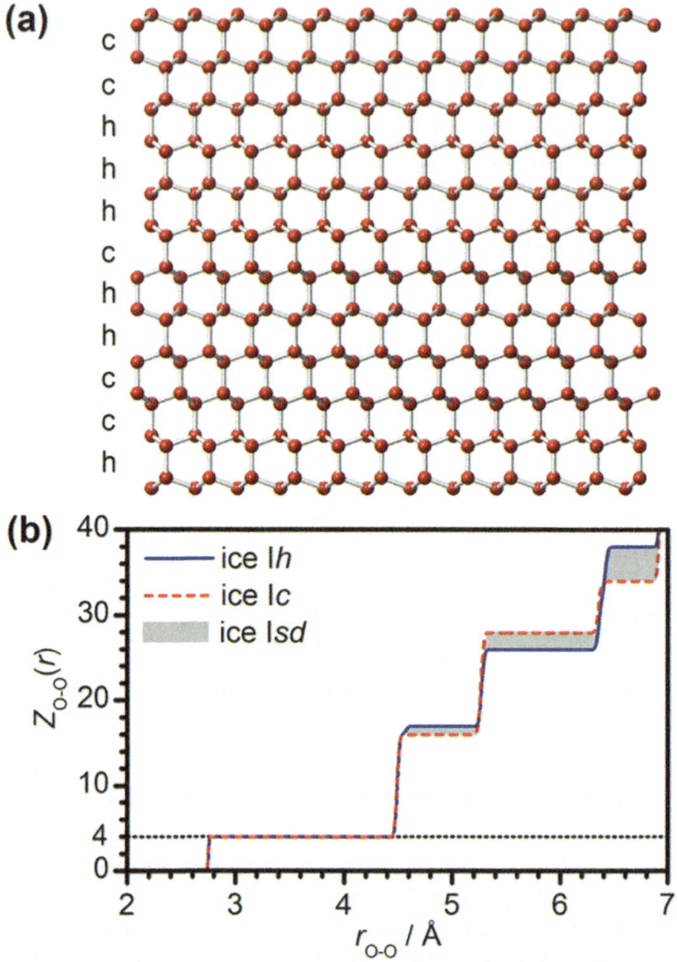

Fig. 5.37 **a** Possible stacking sequence of layers in ice I*sd*. Hexagonal or cubic stacking is indicated by "h" or "c," respectively. Hydrogen atoms have been omitted for clarity. **b** Running oxygen–oxygen coordination number of the pure ice I*h* and I*c* structures. Gray-shaded areas indicate the range of possible values for ice I*sd*. The running coordination number is the average number of oxygen atoms contained within a sphere with oxygen at its center and radius r_{O-O} [42]. With permission from American Chemical Society

5.4.4 Amorphous Ice

Low-density amorphous ice can be formed by cooling the liquid rapidly below T_g or by vapor deposition onto a cold plate, kept below T_g at near-atmospheric pressure [1]. Alternatively, low-density amorphous (LDA) ice has been formed by temperature- or pressure-induced transformations [45]. Specifically, the first-order transition between the low- and high-density amorphous H_2O can be observed with a piston-cylinder

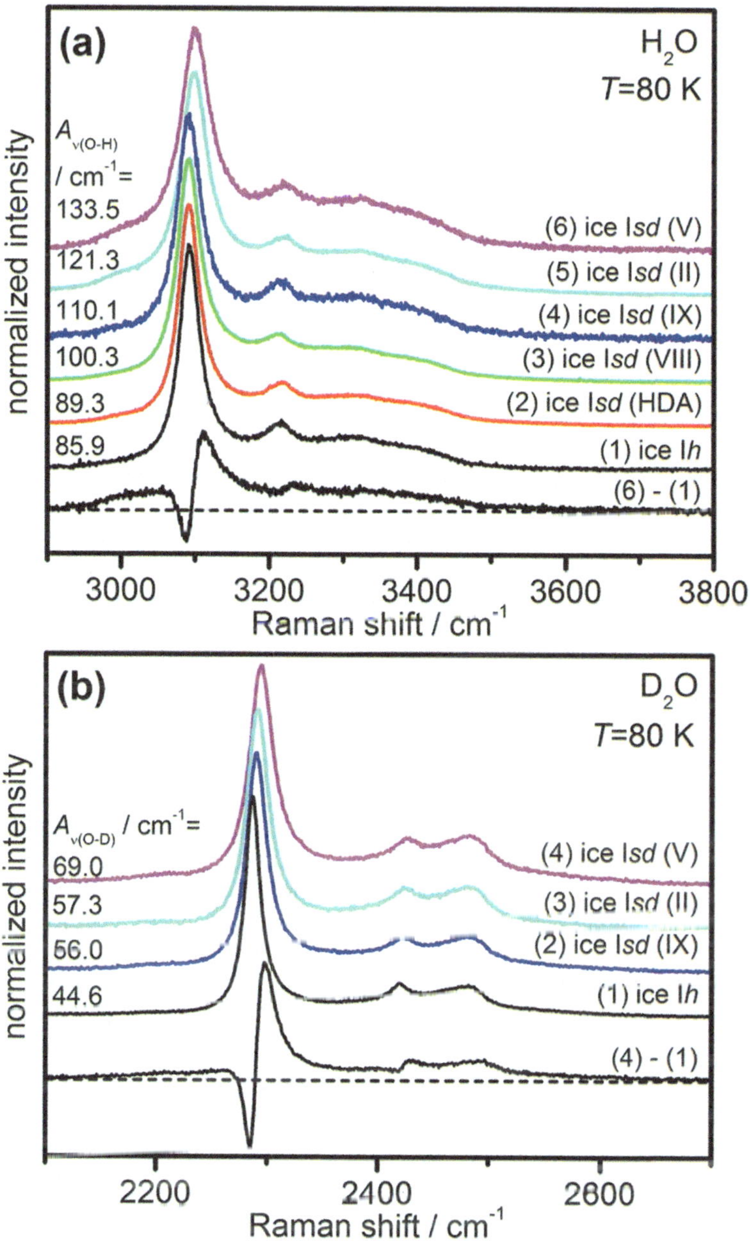

Fig. 5.38 Raman spectra of the coupled **a** ν(O–H) and **b** ν(O–D) stretching regions of H_2O and D_2O ice Ih and ice Isd samples, respectively. The spectra were normalized to the maximum intensity of the most intense feature and shifted vertically for clarity. The parent materials of ice Isd are given in parentheses. The bottom spectra in both panels are difference spectra, as indicated. The listed integrated peak areas, Aν(O–H) and Aν(O–D), are average values of the spectra shown and typically four additional spectra [42]. With permission from American Chemical Society

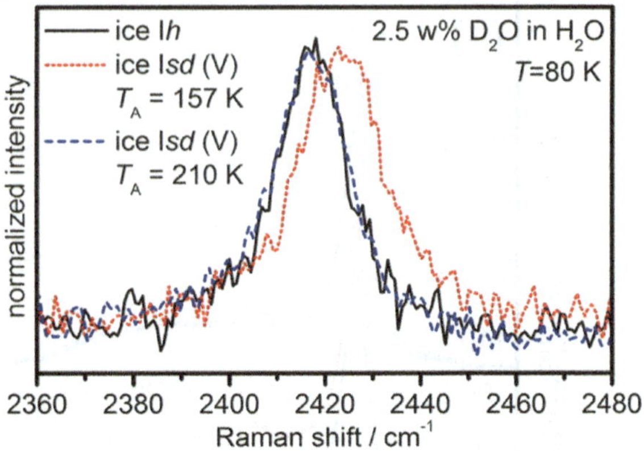

Fig. 5.39 Normalized Raman spectra of the decoupled μ(O–D) region of H_2O ice Isd samples from ice V and ice Ih, both containing 2.5 w% D_2O [42]. With permission from American Chemical Society

apparatus from 77 to 140 K. The transition was found to occur reversibly and abruptly at ~135 K and ~0.2 GPa with a volume change of 0.20 ± ~0.01 cm^3/g and some hysteresis. In pioneering works [45, 46], it was found that when ice Ih is pressurized above 10 kbar at 77 K, it undergoes a pressure-induced amorphization to high-density amorphous (HDA) ice with a density of 1.17 g/cm^3 at zero pressure which is 24% higher than that of low-density amorphous (LDA). The high-density amorphous ice exhibits a collapse of the second hydration shell compared to LDA ice of 3.7 Å, with an interstitial molecule inside of 3.3 Å that is not directly H-bonded to the center molecule [47]. Very high-density amorphous (VHDA) ice of 1.25 g/cm^3 can be formed when HDA is annealed at temperatures above 130 K and pressures above 0.8 GPa [48].

Evidence from molecular simulations and from X-ray diffractogram and Raman spectrum, that the nucleation of ice proceeds through the formation of a new metastable phase, Ice 0, was provided [49]. It was found that Ice 0 is similar to the supercooled liquid, and that on growth it gradually converts into a stacking of cubic Ice I_c and hexagonal I_h. X-ray diffractogram and Raman spectrum of VHDA is differed from that of HDA, and the hydrogen-bonded O–O distance increases from 2.82 Å in HDA to 2.85 Å in VHDA. Suggested crystal planes of above-mentioned ices and distribution of the average angle $\langle \cos\theta \rangle$ between the dipole moment of a molecule and its hydrogen-bonded neighbors are presented in Fig. 5.43.

Figure 5.44 indicates that (1) the nucleation events happen inside the precursor regions, and that the nucleus embedded in a precursor consists of Ice 0, as evidenced by bond-orientational parameters; (2) crystal nucleation is triggered by pre-ordered regions in the supercooled liquid whose local symmetry is consistent with that of the crystal to be formed; and (3) the critical nucleus consists of an Ice I_c core surrounded by a shell with Ice 0 character, which is retained in the post-critical nucleus.

The focus of the review [50] was the glass transition in high-density amorphous ice (HDA). At ambient pressure, HDA is metastable against both ice I and LDA, whereas at >0.2 GPa HDA is no longer metastable against low-density amorphous ice (LDA). A glass transition temperature T_g of 160 K at 0.40 GPa, 122 K at 1.0 GPa,

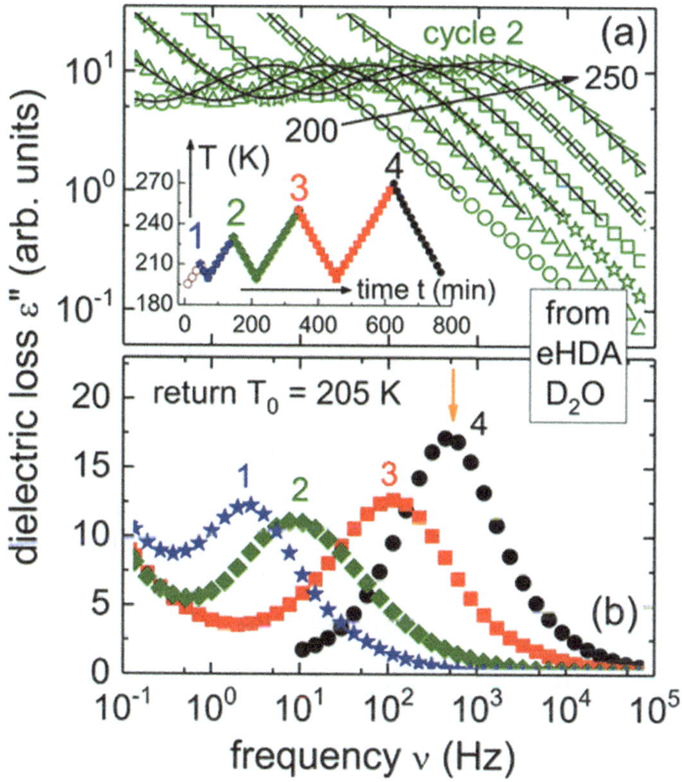

Fig. 5.40 **a** Dielectric loss spectra of ice I obtained from deuterated eHDA. The measurements were taken from 200 to 250 K in steps of 10 K. The solid lines are fits using (1) [43]; see the text for details. The inset shows the employed temperature cycling scheme. Spectra are shown for the second cycle. The scan rate was $|q| = 0.4$ K/min. For each cycle, the minimum and the maximum temperatures are the return temperature $T0$ and the peak temperature Tp, respectively. The numbers 1, 2, 3, and 4 mark the beginnings of each temperature cycle, and for eHDA/D2O they correspond to $Tp = 210$, 230, 250, and 270 K, respectively. **b** Loss measurements carried out at $T0 = 205$ K after each cycle, i.e., after cooling down ice I obtained from deuterated eHDA from Tp. An increase in Tp leads to a faster dielectric relaxation at $T0$. The arrow reflects the loss maximum of hexagonal ice-cooled from the liquid state and agrees well with the measurement carried out for $Tp = 270$ K. With permission from American Institute of Physics

and 116 at ambient pressure was taken place. LDA and HDA underwent two distinct transitions, separated by about 20 K. The glass transition temperature for HDA increases with pressure. Data on glass transition temperatures for very (V)HDA were interpreted within a coherent picture of distinct glass transition temperatures for HDA and VHDA.

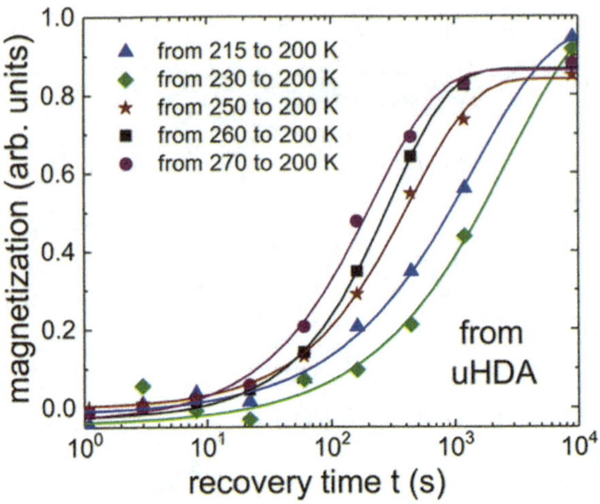

Fig. 5.41 Deuteron magnetization recoveries of ice I recorded after heating uHDA. All measurements were performed at 200 K after returning from higher temperatures *T*p [43]. With permission from American Institute of Physics

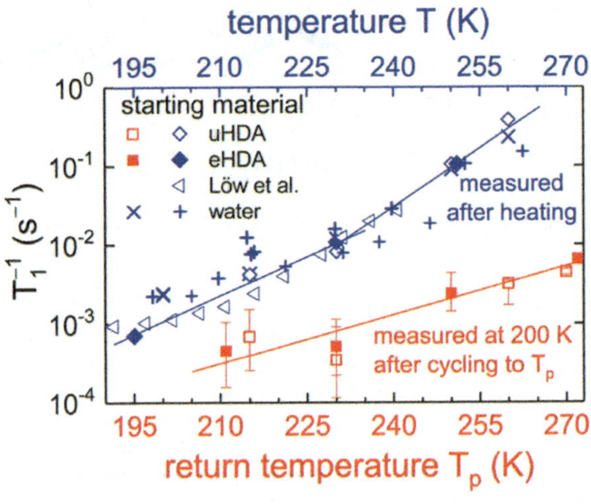

Fig. 5.42 Spin–lattice relaxation rates of ice I as obtained from various starting materials. The diamonds refer to measurements carried out after stepping the temperatures up to various *T*p (upper scale) and the squares refer to experiments performed after cycling the temperature down to *T*0 = 200 K (lower scale). The lines are drawn to guide the eye [43]. With permission from American Institute of Physics

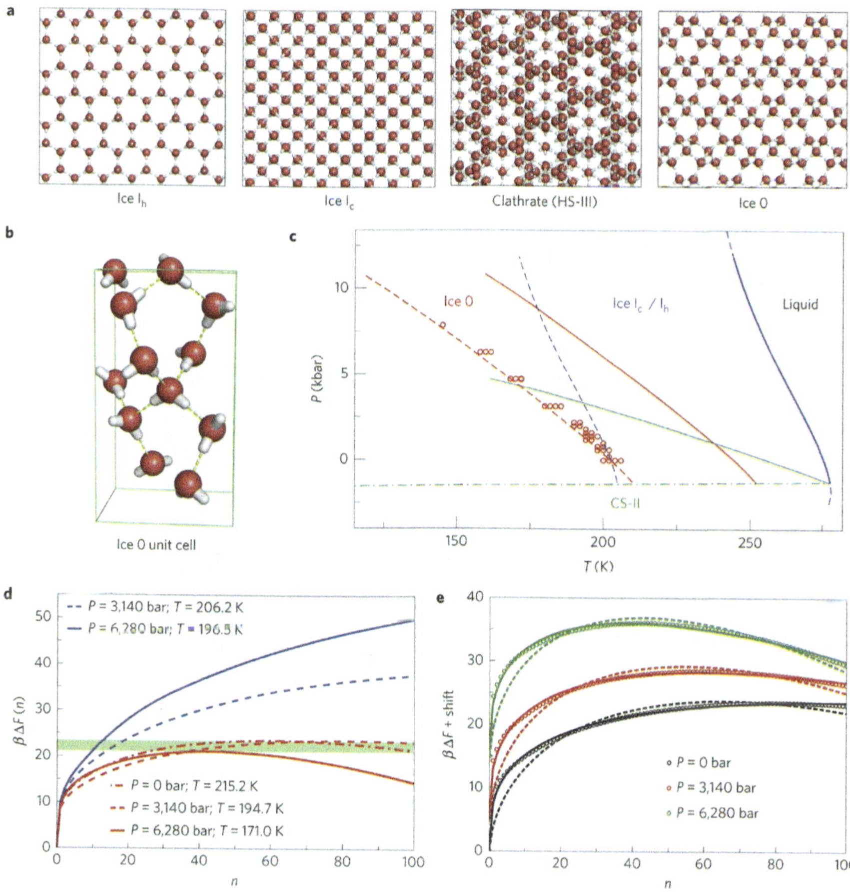

Fig. 5.43 (Top) Crystal planes for the stable phases and ice 0. (Bottom left) Distribution of the average angle $\langle \cos\theta \rangle$ between the dipole moment of a molecule and its hydrogen-bonded neighbors for TIP4P/2005 water at $T = 200$ K and $P = 1$ bar in the liquid phase and the ice Ic, Ih, and ice 0 phases. (Bottom right) The same as in the left panel but for second-nearest neighbors [49]

Review [51] described the advances in X-ray and neutron scattering studies of water structure, from ambient temperature to the deeply supercooled and amorphous states. Water diffusive and collective dynamics, in disparate thermodynamic conditions and environments were also considered. The following aspects of the area were comprehensively discussed: (1) the ability to measure X-ray and neutron diffraction of water with high accuracy in an extended range of momentum transfers which allowed the derivation of detailed O–O pair correlation functions; (2) a panorama of the diffusive dynamics of water in a wide range of temperatures (from 400 K down to supercooled water) and pressures (from ambient up to multiple GPa); (3) the comparison of results obtained by quasi-elastic neutron scattering under high pressure with

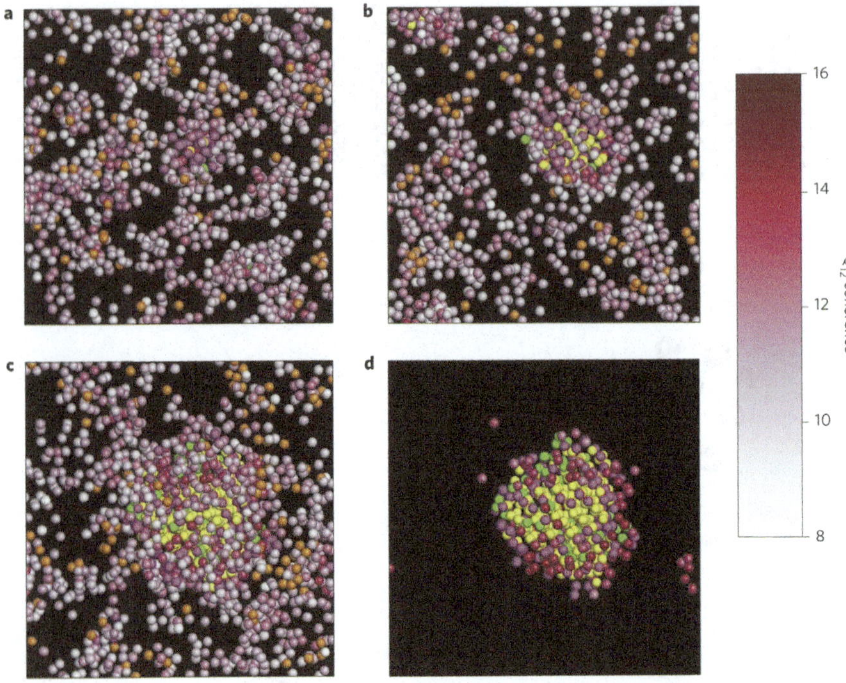

Fig. 5.44 Snapshots from the water nucleation. **a–d** The color code is as follows: Ice I_c (yellow), Ice I_h (green), clathrate (orange), and Ice 0 (magenta). In **a–c,** we also plot particles for which the number of connections is equal or bigger than 6, and color them according to their total coherence. **a** Precritical nuclei, composed predominantly of Ice 0 and embedded in a cloud of high Q_{12} coherency. **b** A nucleus of critical size ($n_c = 80$) where an Ice I_c core is formed. **c** Post-critical nucleus. **d** Same as **c**, but showing only particles with 12 or more connections [49]

the existing data from nuclear magnetic resonance, dielectric and infrared measurements, and modeling; (4) a detailed description of the vibrational dynamics of water as measured by inelastic neutron scattering is presented; (5) the dependence of the water vibrational density of states on temperature and pressure, and in the presence of biological molecules; and (6) results about the collective dynamics of water and its dispersion curves as measured by coherent inelastic neutron scattering and inelastic X-ray scattering in different thermodynamic conditions. The radial distribution functions of different amorphous ices observed by neutron diffraction as measured at the spallation source ISIS at the SANDALS spectrometer are depicted in Fig. 5.45.

At 0.07 GPa from a transformation of high- (HDA) to low-density amorphous ice (LDA) at 140 (H2O) and 143 K (D2O) was investigated employing powder X-ray diffractograms [52]. A double halo peak on the diffractograms the indicative of the simultaneous presence of LDA and HAD was detected. Growth of the low-density form takes place randomly inside of the high-density matrix. The thermal stability of HDA against transformation to LDA at ambient pressure significantly increases with decreasing recovery pressure and reaches its maximum at 0.07 GPa. An equilibrated

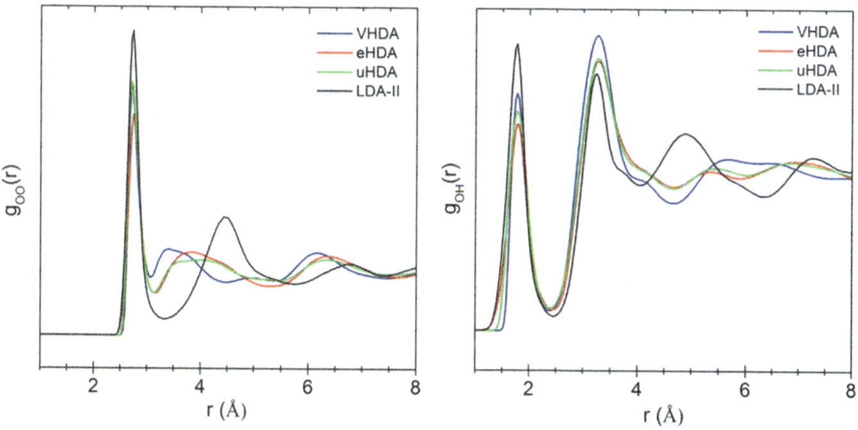

Fig. 5.45 Comparison of O–O and O–H pair distribution functions of different **amorphous ices**: VHDA (blue), eHDA (red), uHDA (green), and LDA-II (black). Data have been obtained from neutron scattering measurements using EPSR structure refinement [51]. With permission from American Chemical Society

nature of the amorphous sample in the pressure range of $p \lesssim 0.2$ GPa was proved. All these findings support the scenario of a first-order liquid–liquid transition in bulk water.

A phase diagram for amorphous solids and liquid supercooled water (Fig. 5.46) and was derived using large-deviation theory [53]. The theory predicted that: (1) along with non-equilibrium transitions between the ergodic liquid and two distinct amorphous solids, coexistence between these two amorphous solids takes place; (2) a non-equilibrium triple point where two amorphous phases and the liquid coexist; (3) whereas the amorphous solids are long-lived and slowly aging glasses, their melting can lead quickly to the formation of crystalline ice; (4) melting of the higher density amorphous solid at low pressures takes place in steps, transitioning to the lower-density glass before accessing a non-equilibrium liquid from which ice coarsens; and (5) the amorphous solids of water exist in several different forms. It is partitioned with the onset temperature, $T_o(p)$, which is the crossover temperature below which liquid-phase dynamics are spatially heterogeneous.

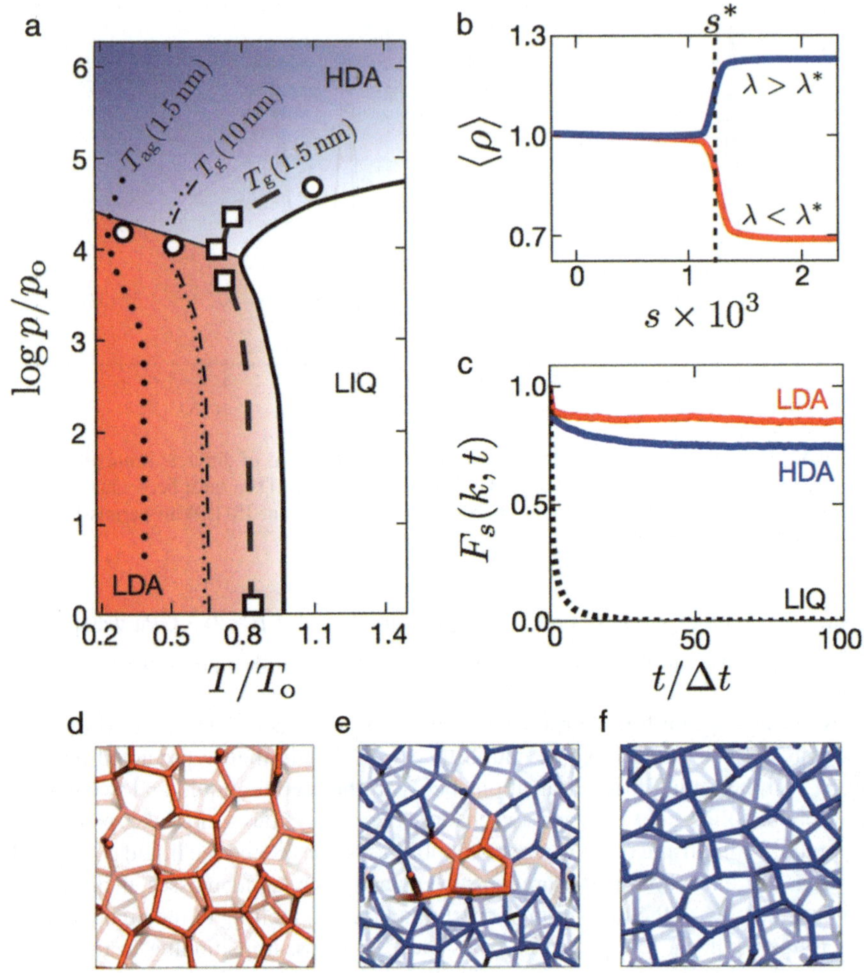

Fig. 5.46 Liquid, LIQ, and non-equilibrium HDA and LDA solids. **a** Pressure–temperature phase diagram for water, with the liquid onset temperature line (solid), glass transition lines (dashed), and apparent glass transition lines (dotted). Squares locate points where non-equilibrium *s*-ensemble calculations locate coexistence between non-equilibrium phases in the mW model. Circles locate transitions observed in experiments and in non-equilibrium relaxation simulations of the mW model. **b** The mean reduced density $\langle\rho\rangle$ as a function of non-equilibrium control parameters computed for the mW model with the *s* ensemble near the non-equilibrium triple point. **c** Van Hove self-correlation functions for the three phases at the liquid's principal wave vector *k*, all computed for the mW model at conditions near the non-equilibrium triple point. **d–f** Snapshots from simulations, where bond connecting molecular centers *i* and *j* are colored according to the value of $\eta ijk\eta ijk$ averaged over a second neighbor *k* (the quantity $\eta ijk\eta ijk$ being defined below in *Distinct Phases and Coexistence*). The bond is red if this value is less than 0.1; otherwise, it is blue. A typical configuration of LDA ice is pictured in **d**, that of a domain of LDA ice in coexistence with HDA ice is in **e**, and that of HDA is in **f** [53]. Published with permission of PNAS article containing the requested material published under a CC BY-NC-ND license

References

1. T. Loerting, M. Bauer, I. Kohl, K. Watschinger, K. Winkel, E. Mayer, Cryoflotation: densities of amorphous and crystalline ices. J. Phys. Chem. B **115**, 14167 (2011)
2. P. Gallo, H.E. Stanley, Supercooled water reveals its secrets. Science **358**(6370), 1543–1544 (2017)
3. P.G. Debenedetti, H.E. Stanley, Supercooled and glassy water. Phys. Today **56**, 40–46 (2003)
4. P.H. Handle, T. Loerting, F. Sciortino, Supercooled and glassy water: metastable liquid(s), amorphous solid(s), and a no-man's land. PNAS **114**, 13336–13344 (2017)
5. A.H. Harvey, Properties of ice and supercooled water, in *CRC Handbook of Chemistry and Physics*, ed. by W.M. Haynes, D.R. Lide, T.J. Bruno, 97th edn. (CRC Press, Boca Raton, FL, 2017)
6. M. Chaplin, *Water Structure and Science* (London South Bank University, 2012)
7. C.G. Salzmann, Advances in the experimental exploration of water's phase diagram. J. Chem. Phys. **150**, 060901 (2019)
8. G. Galli, D. Pan, A closer look at supercritical water. PNAS **110**, 6250–6251 (2013)
9. Y. Marcus, *SUPERCRITICAL WATER. A Green Solvent: Properties and Uses* (Wiley, 2012)
10. O. Mishima, H. Stanley, The relationship between liquid, supercooled and glassy water. Nature **396**, 329–335 (1998)
11. P. Gallo, K. Amann-Winkel, C.A. Angell, M.A. Anisimov, F. Caupin, C. Chakravarty, E. Lascaris, T. Loerting, A.Z. Panagiotopoulos, J. Russo, J.A. Sellberg, H.E. Stanley, H. Tanaka, C. Vega, L. Xu, L.G.M. Pettersson, Water: a tale of two liquids. Chem. Rev. **116**, 7463–7500 (2016)
12. J.L. Miller, Super cooled water goes supercritical. Phys. Today **71**, 18 (2018)
13. D. Nayar, C. Chakravarty, Water and water-like liquids: relationships between structure, entropy and mobility. Phys. Chem. Chem. Phys. **15**, 14162–14177 (2013)
14. L.B. Skinner, C.C. Huang, D. Schlesinger, L.G.M. Pettersson, A. Nilsson, A. Benmore, C.J. Benchmark, Oxygen-oxygen pair-distribution function of ambient water from X-ray diffraction measurements with a wide Q-range. J. Chem. Phys. **138**, 074506 (2013)
15. L. Rovigatti, V. Bianco, J.M. Tavares, F. Sciortino, Communication: re-entrant limits of stability of the liquid phase and the speedy scenario in colloidal model systems. J. Chem. Phys. **146**, 041103 (2017)
16. F. Lehmkühler, Y. Forov, T. Büning, C.J. Sahle, I. Steinke, K. Julius, T. Buslaps, M. Tolan, M. Hakala, C. Sternemann, Intramolecular structure and energetics in supercooled water down to 255 K. Phys. Chem. Chem. Phys. **18**, 6925–6930 (2016)
17. J.A. Sellberg, T. McQueen, H. Laksmono, S. Schreck, M. Beye, D.P. DePonte, B. Kennedy, D. Nordlund, R.G. Sierra, D. Schlesinger, T. Tokushima, I. Zhovtobriukh, S. Eckert, V.H. Segtnan, H. Ogasawara, K. Kubicek, S. Techert, U. Bergmann, G.L. Dakovski, W.F. Schlotter, Y. Harada, M.J. Bogan, P. Wernet, A. Föhlisch, L.G.M. Pettersson, A. Nilsson, X-ray emission spectroscopy of bulk liquid water in "no-man's land". J. Chem. Phys. **142**, 044505 (2015)
18. K.H. Kim, A. Späh, H. Pathak, F. Perakis, D. Mariedahl, K. Amann-Winkel, J.A. Sellberg, J.H. Lee, S. Kim, J. Park, K.H. Nam, T. Katayama, A. Nilsson, Maxima in the thermodynamic response and correlation functions of deeply supercooled water. Science **358**, 1589 (2017)
19. Y. Huang, X. Zhang, Z. Ma, Y. Zhou, C.Q. Sun, Hydrogen-bond relaxation dynamics: resolving mysteries of water ice. Coord. Chem. Rev. **285**, 109–165 (2015)
20. G.O. Gomes, H.E. Stanley, M. de Souza, Enhanced Grüneisen parameter in supercooled water. Sci. Rep. **9**, 12006 (2019)
21. R. Shi, H. Tanaka, Direct evidence in the scattering function for the coexistence of two types of local structures in liquid water. J. Am. Chem. Soc. **142**, 2868–2875 (2020)
22. F. Martelli, Unravelling the contribution of local structures to the anomalies of water: the synergistic action of several factors. J. Chem. Phys. **150**, 094506 (2019)
23. F. Sciortino, P. Gallo, P. Tartaglia, S.-H. Chen, Supercooled water and the kinetic glass transition. Phys. Rev. E Stat. Phys. Plasmas Fluids Relat. Interdiscip. Topics **54**, 6331–6343 (1996)

24. H.L. Pi, J.L. Aragones, C. Vega, E.G. Noya, J.L.F. Abascal, M.A. Gonzalez, C. McBride, Anomalies in water as obtained from computer simulations of the TIP4P/2005 model: density maxima, and density, isothermal compressibility and heat capacity minima. Mol. Phys. **107**, 365–374 (2009)
25. D.T. Limmer. D. Chandler, Theory of amorphous ices. PNAS **111**, 9413–9418
26. Y.D. Fomin, E.N. Tsiok, V.N. Ryzhov, V.V. Brazhkin, A novel anomalous region of water. arXiv:1706.02923 (2017). https://www.researchgate.net/publication/317543209
27. H. Sippola, P. Taskinen, Activity of supercooled water on the ice curve and other thermodynamic properties of liquid water up to the boiling point at standard pressure. J. Chem. Eng. **63**, 2986–2998 (2018)
28. M. Zhao, W.T. Zheng, J.C. Li, Z. Wen, M.X. Gu, C.Q. Sun, Atomistic origin, temperature dependence, and responsibilities of surface energetics: an extended broken-bond rule. Phys. Rev. B **75**, 085427 (2007)
29. P. Kumar, S.V. Buldyrev, S.R. Becker, P.H. Poole, F.W. Starr, H.E. Stanley, Relation between the Widom line and the breakdown of the Stokes-Einstein relation in supercooled water. PNAS **104**, 9575–9579 (2007)
30. T.L. Mizan, P.E. Savage, R.M. Ziff, Molecular dynamics of supercritical water using a flexible SPC model. J. Phys. Chem. **98**, 13067–13076 (1994)
31. L. Xu, P. Kumar, S.V. Buldyrev, S.-H. Chen, P.H. Poole, F. Sciortino, H.E. Stanley, Relation between the Widom line and the dynamic crossover in systems with a liquid–liquid phase transition. Proc. Natl. Acad. Sci. U.S.A. **102**, 16558–16562 (2005)
32. G.G. Simeoni, T. Bryk, F.A. Gorelli, M. Krisch, G. Ruocco, M. Santoro, T. Scopigno, The Widom line as the crossover between liquid-like and gas-like behaviour in supercritical fluids. Nat. Phys. **6**, 503–507 (2010)
33. A.R. Imre, U.K. Deiters, T. Kraska, I. Tiselj, The pseudocritical regions for supercritical water. Nucl. Eng. Des. **252**, 183 (2012)
34. I. Skarmoutsos, E. Guardia, J. Samios, Local structural fluctuations, hydrogen bonding and structural transitions in supercritical water. J. Supercrit. Fluids **130**, 156164 (2017). https://doi.org/10.1016/j.supflu.2017.08.004 and references therein
35. K. Karalis, C. Ludwig, B. Niceno, Supercritical water anomalies in the vicinity of the Widom line. Sci. Rep. **9**, 15731 (2019)
36. F. Maxim, C. Contescu, P. Boillat, B. Niceno, K. Karalis, A. Testino, C. Ludwig, Visualization of supercritical water pseudo-boiling at Widom line crossover. Nat. Commun. (2019). https://doi.org/10.1038/s41467-019-12117-5
37. V.F. Petrenko, R.W. Whitworth, *Physics of Ice* (Oxford University Press, Oxford, 1999)
38. M. Chaplin, *Water Structure and Science* (London South Bank University). Archived from the original on 23 Mar 2017
39. E.A. Zheligovskaya, Molecular dynamics study of crystalline water ices. J. Struct. Chem. **49**, 459–471 (2008)
40. M. Matsumoto, S. Saito, I. Ohmine, Molecular dynamics simulation of the ice nucleation and growth process leading to water freezing. Nature **416**, 409–413 (2002)
41. H.Y. Playford, T.F. Whale, B.J. Murray, M.G. Tuckerc, C.G. Salzmann, Analysis of stacking disorder in ice I using pair distribution functions. J. Appl. Cryst. **51**, 1211–1220 (2018)
42. T.H.G. Carr, J.J. Shephard, C.G. Salzmann, Spectroscopic signature of stacking disorder in ice I. J. Phys. Chem. Lett. **5**, 2469–2473 (2014)
43. C. Gainaru, E. Vynokur, K.W. Köster, V. Fuentes-Landete, N. Spettel, J. Zollner, T. Loerting, R. Böhmer, Dynamic signatures of the transition from stacking disordered to hexagonal ice: dielectric and nuclear magnetic resonance studies. J. Chem. Phys. **148**, 134502 (2018)
44. O. Bollengier, J.M. Brown, G.H. Shaw, Thermodynamics of pure liquid water: sound speed measurements to 700 MPa down to the freezing point, and an equation of state to 2300 MPa from 240 to 500 K. J. Chem. Phys. **151**, 054501 (2019)
45. O. Mishima, Reversible first-order transition between two H_2O amorphs at ~0.2 GPa and ~135 K. J. Chem. Phys. **100**, 5910 (1994)

46. G. Pallares, M. El Mekki Azouzi, M.A. González, J.L. Aragones, J.L.F. Abascal, C. Valeriani, F. Caupin, Anomalies of water at negative pressure. Proc. Natl. Acad. Sci. U.S.A. **111**, 7936–7941 (2014)
47. J.L. Finney, A. Hallbrucker, I. Kohl, A.K. Soper, D.T. Bowron, Phys. Rev. Lett. **88**, 225503 (2002)
48. T. Loerting, C. Salzmann, I. Kohl, E. Mayer, A. Hallbrucker, A second distinct structural "state" of high-density amorphous ice at 77 K and 1 bar. Phys. Chem. Chem. Phys. **3**, 535 (2001)
49. J. Russo, F. Romano, H. Tanaka, New metastable form of ice and its role in the homogeneous crystallization of water. Nat. Mater. **13**, 733–739 (2014)
50. T. Loerting, V. Fuentes-Landete, P.H. Handle, M. Seidl, K. Amann-Winkel, C. Gainaru, R. Böhmer, The glass transition in high-density amorphous ice. J. Non Cryst. Solids **407**, 423–430 (2015)
51. K. Amann-Winkel, M.-C. Bellissent-Funel, L.E. Bove, T. Loerting, A. Nilsson, A. Paciaroni, D. Schlesinger, L. Skinner, X-ray and neutron scattering of water. Chem. Rev. **116**, 7570–7589 (2016)
52. K. Winkel, E. Mayer, T. Loerting, Equilibrated high-density amorphous ice and its first-order transition to the low-density form. J. Phys. Chem. B **115**(48), 14141–14148 (2011)
53. D.T. Limmer, D. Chandler, Theory of amorphous ices. PNAS **111**, 9413–9418 (2014)

Chapter 6
Hydration of Simple Biological Molecules and Ions

Abstract The main subject of this chapter is the critical analysis of data on the thermodynamic properties of hydrated amino acids, DNA, RNA, proton, hydroxide, and ions of biological importance and especially on hydration enthalpy–entropy compensation effect. These data are important for understanding the molecular processes of formation and denaturation and recognition of biological molecules, enzyme catalysis as well as in drug discovery.

6.1 Introduction

The hydration state of small biological molecules and ions in aqueous solution is a subject of keen interest, with both practical and scientific importance [1–14]. Quantitative data on the hydration of these objects can form the basis for the corresponding study of complex biological systems, such as proteins, nucleic acids, and biomembranes. For example, protein folding, dynamics, and ligand binding are all closely linked to this solvation energy. Differences in solvation energy between states are required for physics-based studies of a range of interest in drug discovery including ionization states, tautomer ratios, bioactive conformations, solubility, catalysis, loop modeling, phase transfer, and aggregation and fold prediction. In this regard, experimental investigations utilizing a whole arsenal of physical methods such as calorimetric, viscosity and ultrasound techniques, X-ray and neutron scattering, nuclear and electron magnetic resonances, and dielectric spectroscopy in the microwave region, as well as calculations using advanced theoretical methods in the area are of particular importance.

The free energy of transferring a molecule from vacuum to water plays a central role in aqueous systems. The hydration free energy is described by hydrogen bond and the van der Waals energy, the solute cavity formation energy in the water phase, and the change in electrostatic solute–solvent interaction energy. Solvation free energies can now be calculated precisely from molecular simulations, providing a valuable test of the energy functions underlying these simulations [1]; for example, approaches for calculating the solvation free energies of small, neutral organic molecules from molecular simulations were briefly reviewed and illustrated by applying them to

G. I. Likhtenshtein, *Biological Water*, Soft and Biological Matter,
https://doi.org/10.1007/978-3-030-82503-4_6

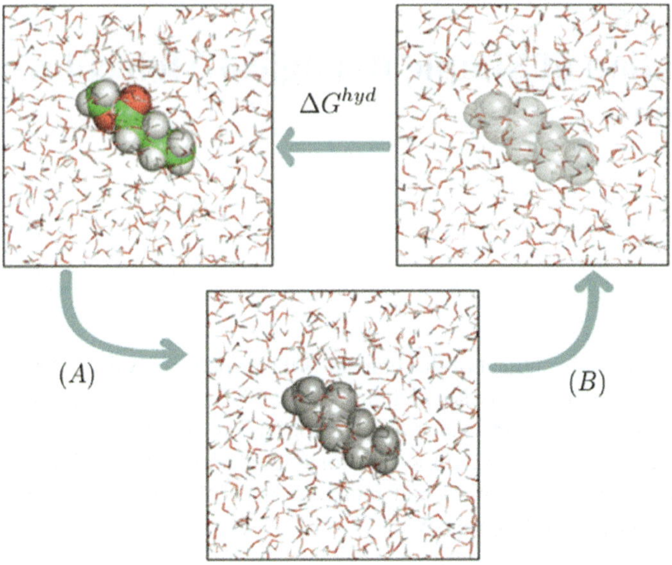

Fig. 6.1 Chematic presentation of a small molecular hydration [1]. With permission from American Chemical Society

calculate hydration free energies (Fig. 6.1). These approaches allow to compute free energy differences from a simulation or set of simulations and appear to be a particularly robust and general-purpose approach for this task. An update version to FreeSolv database of experimental and calculated hydration free energies, enthalpies and entropies of hydration, as well as electrostatic and non-polar components of solvation free energies of neutral compounds was presented. A statistical-mechanical framework for estimation of solvation entropies and enthalpies was proposed [2]. Structural modifications to interacting systems led to changes in both the enthalpy (heat) and entropy of the process that compensate each other [3].

6.2 Small Organic Molecules

6.2.1 Theoretical Aspects. Computation

A nonlinear solvation model was used to study the hydration of 181 small organic molecules [9]. This modified Langevin-Debye model takes in consideration that in strong electric fields, the solvent polarization becomes saturated and the dielectric no longer responds linearly in the applied field. Atomic charges and radii of the solute molecule were described by a standard classical force field, and the optimized potentials for liquid simulation all atom (OPLS-AA) force field were applied. The nonlinear solvation model was improved by introducing a few charge-scaling parameters for some functional groups. Theoretical work [8] was addressed to the performance of different atomic charges to reproduce experimental hydration free energies in the FreeSolv database in combination with the GAFF force field. Atomic charges were calculated by two atoms-in-molecules approaches, Hirshfeld-I and Minimal Basis Iterative Stockholder (MBIS). To account for polarization effects, the charges were derived from the solute's electron density computed with a solvent model and the energy required to polarize the solute was added to the free energy cycle.

Two different methods for estimating solvation free energies: an implicit solvent approach based on the Poisson–Boltzmann equation (PBE) and an explicit solvent approach using "alchemical" free energy calculations were developed [10]. PBE describes the distribution of the electric potential in solution in the direction normal to a charged surface. Alchemical free energy calculations employ unphysical ("alchemical") intermediates to estimate the free energies of various physical processes. The hallmark of this approach is the use of potential energy functions representing alchemical intermediate states that cannot exist as real chemical species. The Poisson–Boltzmann method treats water as a uniform, high dielectric (ε), responding to the electrostatic fields emanating from a charge distribution (F) embedded in low-dielectric solute molecules. "Alchemical" free energy calculations provided a rigorous way to calculate free energy differences between thermodynamic state with a solute in water versus a state with water plus a separate solute in vacuum within classical statistical mechanics, assuming the underlying force field accurately describes the physics. For a set of 17 small molecules (Table 6.1; Fig. 6.2), the application of these methods showed good agreement of experimental and theoretical values.

6.2.2 Amino Acids and Simple Peptides

Using molecular dynamics-based free energy simulations, relative solvation free energies for pairs of N-acetyl-methylamide amino acids (Ala Ser, Val Thr, Phe-Tyr, Val-Ala, Thr-Ser, Phe-Ala, and Tyr-Ser) and the corresponding pairs of side chain analogs were computed [14]. Differences in relative solvent affinity $\Delta\Delta\Delta A$ between amino acids and side chain analogs of up to 66% or, in absolute numbers, 4.9 kcal/mol (Ala-Ser) were obtained. Contributions to the free energy from solvent exclusion and self-solvation estimated separately were separately estimated. In solvent simulations 243 TIP3P water molecules, the values of relative solvation free energy differences $\Delta\Delta A_{\mathrm{solv}}$ for selected pairs of N-acetyl-X-methylamide amino acids ($\Delta\Delta A_{\mathrm{solv}}$ AA of Ala-Ser, Val-Thr, Phe-Tyr, Val-Ala, Thr-Ser, Phe-Ala, and Tyr-Ser) and for pairs of the corresponding side chain analogs ($\Delta\Delta A_{\mathrm{solv}}{}^{\mathrm{SC}}$ of methane(Me)-methanol(MeOH), propane(Pr)-ethanol(EtOH), Toluene (Tol)-p-cresol(p-Cre), Pr-Me, EtOH-MeOH, Tol-Me and p-Cre-MeOH) were summarized in a table. Analysis of the results indicated that the determining contribution for small polar amino acids arises from interactions between the polar backbone and the polar functional group

Table 6.1 Experimental free energies of solvation [10]. With permission from American Chemical Society

No.	Compound	Exptl value
1	Glycerol triacetate	– 8.84[10]
2	Benzyl bromide	– 2.38[10,76]
3	Benzyl chloride	– 1.93[10,76]
4	*m*-bis(trifluoromethyl)benzene	1.07[10]
5	*N,N*-dimethyl-*p*-methoxybenzamide	– 11.01[77]
6	*N,N*-4-trimethylbenzamide	– 9.76[77]
7	Bis-2-chloroethyl ether	– 4.23[10,76]
8	1,1-diacetoxyethane	– 4.97[10]
9	1,1-diethoxyethane	– 3.28[8,10]
10	1,4-dioxane	– 5.05[K,10]
11	Diethyl propanedioate	– 6.00[10]
12	Dimethoxymethane	– 2.93[K,10]
13	Ethylene glycol diacetate	– 6.34[10]
14	1,2-diethoxyethane	– 3.54[K,10]
15	Diethyl sulfide	– 1.43[5](–1.55)[78]
16	Phenyl formate	– 3.82[79](–4.08)[80]
17	Imidazole	– 9.81[K1]

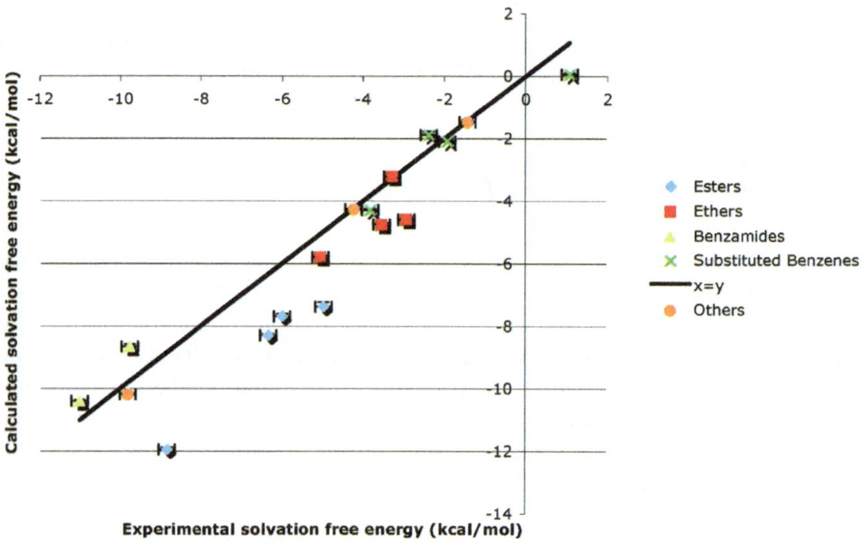

Fig. 6.2 Solvation free energies in explicit solvent, calculated with alchemical free energy calculations and AM1-BCC v2.6A24 charges, versus experiment. Molecules are color coded by group as described in the text. Vertical (computed) error bars are one standard deviation; horizontal (estimated experimental) error bars are 0.2 kcal/mol [10]. With permission from American Chemical Society

of respective side chain in the gas phase and, consequently, the solvent affinity of small polar amino acids depends strongly on the backbone conformation.

Because of experimental restrictions, absolute solvation free energies of proteins or amino acids are not available and solvation free energies were estimated based on side chain analog data [15]. Estimates of the solvation free energy were relied on the additivity assumption by first measuring solvation free energies of small model compounds considered representative of parts of the system of interest, followed by adding the contributions of the model compounds. For example, full amino acids were split into *N*-methylacetamide, representing the backbone, and side chain analogs, e.g., methanol for Ser, etc. These data allowed to estimate the solvent affinity of individual amino acids. Molecular dynamics-based free energy simulations to calculate absolute hydration free energies for 15 *N*-acetyl-methylamide amino acids with neutral side chains were carried out and compared with solvation free energies for side chain analogs. Calculation of the solvation free energies of all canonical neutral *N*-acetyl-*X*-methylamide amino acids was performed. The process of calculation of the solvation free energies of all canonical neutral *N*-acetyl-*X*-methylamide amino acids began by calculating relative solvation free energy differences of all amino acids with respect to an intermediate state, pseudo-glycine (PG), which resembles Gly, except for the atom type of the Ca carbon, followed by calculation of the absolute solvation free energy of PG. The absolute solvation free energy differences for 15 blocked amino acids (Gly, Ala, Val, Leu, Ile, Ser, Thr, Cys, Met, Asn, Gln, Phe, Tyr, Trp, and the two uncharged His tautomers Hid and Hie;), using the CHARMM27 force field, were summarized in a table.

In Fig. 6.3, the solvation free energy results for *N*-acetyl-methylamide amino acids (+) from this study with the results for the side chain analog results (×) by other works are compared.

Two effects were discussed. The first is a steric effect, "solvent exclusion" accounts for the fact that atoms in the interior of a large molecule do not interact with water and, therefore, will not contribute to the solvent affinity of the molecule. The second, larger contribution, "self-solvation," was identified after "solvent exclusion" was found insufficient for explaining the observed differences.

In work [16], using the AMBER ff03 charge model and water models (3-site TIP3P and for 4 siteTIP4P-Ew), calculated hydration free energies of natural and unnatural amino acids were compared to experimental hydration free energies of corresponding side chain analogs. The data on calculation performed on 19 natural amino acids and previously derived for 17 unnatural amino acids together with experimental hydration free energies of corresponding side chain analog were discussed in details. It was found that the ff03 charge model is correlated with experimental hydration free energies but underestimates the solubilities of several polar natural and unnatural amino acids. The experimental hydration free energy was assessed via measurement of the partial pressure P_s of a solute in the vapor phase and via the number density concentration of solute in aqueous solution. When this two-phase system is at equilibrium, the hydration free energy is defined as

$$(1_{\text{Ghydration}}) = k_B T \ln(P_s \text{ls} k_B T). \tag{6.1}$$

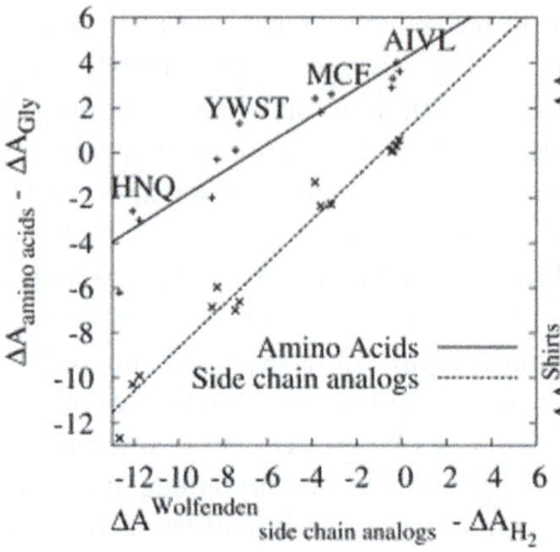

Fig. 6.3 Comparison of blocked amino acid solvation free energies with their corresponding side chain analog data (kcal/mol). (*Amino acids*) Results for the blocked amino acids relative to Glycine was reported in [15]. (*Side chain analogs*) Side chain analog results of Shirts et al. relative to he side chain analog of Glycine (H_2). The approximate horizontal position of each amino acid is indicated by its one-letter code. The computational results relative to Gly ($\Delta A_{\text{amino acids}} - \Delta A_{\text{Gly}}$) or its side chain analog H_2 ($\Delta A_{\text{Shirtssidechain analogs}} - \Delta A_{H_2}$; y axis) are plotted against the experimental solvation free energy differences of the side chain analogs reported by Wolfenden et al. also relative to H_2 ($\Delta A_{\text{Wolfendensidechain analogs}} - \Delta A_{H_2}$; x axis). If the contributions of side chain and backbone to the solvation free energy were purely additive, both lines should be identical [15]. With permission from Elsevier

This potential described the interactions between the solvent water and blocked peptide solute. The hydration free energy was calculated as G(side chain hydration) $= -G$(Acetyl-Xaa-NH_2 hydration) $-$ G(Acetyl-Gly-NH_2 hydration).

A soft-core potential was used to describe the interactions between the solvent water and a blocked peptide solute. A visual depiction of the starting and end states involved in a one-step transformation are shown in Fig. 6.4.

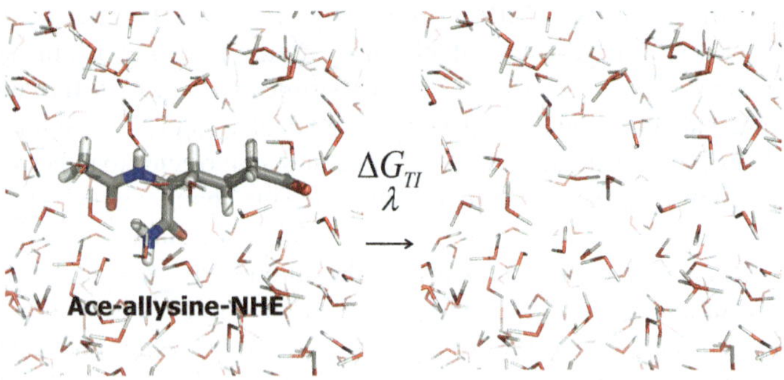

Fig. 6.4 Visual depiction of one-step transformation used for hydration free energy calculation. The blocked peptide representing allysine (ALK) is shown [16]. With permission from Elsevier

6.2.3 Nucleotides

Hydration properties of adenine nucleotides and orthophosphate (Pi) in aqueous solutions adjusted to pH = 8 with NaOH were studied by high-resolution microwave dielectric relaxation (DR) spectroscopy at 20 °C [17]. DR spectrum measurements were carried out utilizing a microwave network analyzer (PNA8364B-85070E, Agilent, Santa Clara, CA). DR spectra were recorded for the frequency range of 0.2–26 GHz at 20 ± 1 °C. DR spectra of aqueous solutions of ATP, ADP, and AMP and Pi at pH 8.0, where the major ionic species were ATP^{4+}, ADP^{3-}, AMP^{2-}, and $HPO_4{}^{2-}$ with Na^+ as the counter ion in respective solutions were measured and analyzed using a mixture theory combined with a least-squares Debye decomposition method. Solutions of Pi and adenine nucleotides showed qualitatively similar dielectric properties described by two Debye components. One component was characterized by a relaxation frequency (υ_c = 18.8–19.7 GHz) significantly higher than that of bulk water (17 GHz) and the other by a much lower υ_c (6.4–7.6 GHz). The components are referred to here as hyper-mobile water and constrained water, respectively. By contrast, a hydration shell of only the latter type was found for adenosine (fc ~ 6.7 GHz). In Fig. 6.5, the difference of static electrical conductivity between

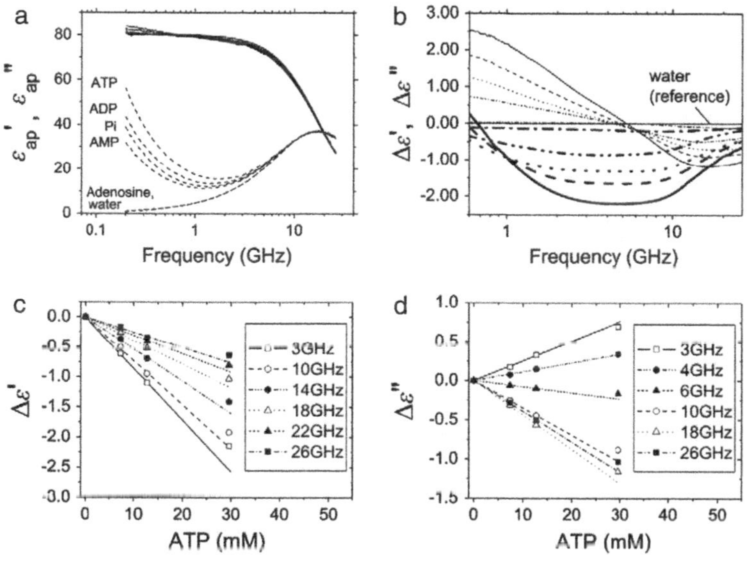

Fig. 6.5 Dielectric spectra of aqueous solutions of ATP, ADP, AMP, adenosine, and Pi at 20.00 °C. **a** Dielectric spectra $\varepsilon a * p(f)$. Solid and dashed lines indicate $\varepsilon a'p$ and $\varepsilon a''p$, respectively. **b** Difference dielectric spectra Difference dielectric spectra ($\Delta\varepsilon' = \varepsilon'ap - \varepsilon'w$, thick line; $\Delta\varepsilon'' = \varepsilon''ap - \varepsilon''w$, thin line) were averaged over n experiments and smoothed for ATP ($n = 4$, solid line), ADP ($n = 6$, dashed line), AMP ($n = 4$, dotted line), adenosine ($n = 5$, chain line), and Pi ($n = 8$, two-dot chain line). In all cases, error bars are much smaller than the symbol sizes of $\Delta\varepsilon'$ and $\Delta\varepsilon''$ [17]. With permission from Elsevier

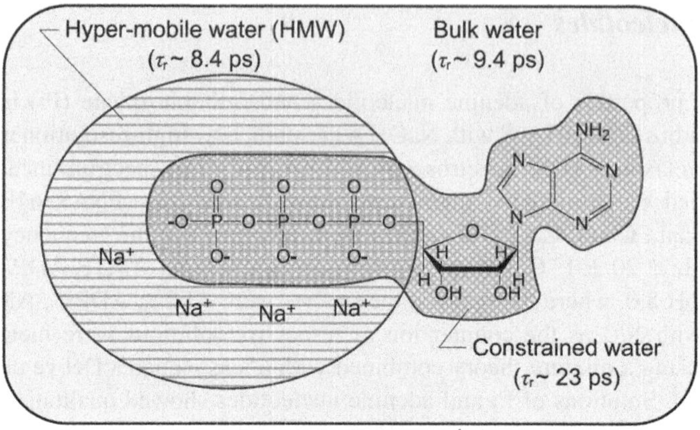

Fig. 6.6 An image of hydration structure of ATP revealed by the present study. The ATP anion (ATP^{4-}) coexists with charge-balancing Na^+ ions. In embracing bulk water with a dielectric relaxation time τ_r ~ 9.4 ps (=1/(2πfcw)) at 20 °C, the adenosine portion has a constrained water shell with τ_r ~ 23 ps (=1/(2πfc2)), whereas the phosphoryl groups are surrounded by a constrained water layer with τ_r ~ 23 ps together with a three to four layers of hyper-mobile water with τ_r ~ 8.4 ps (=1/(2πfc1)) [17]. With permission from Elsevier

a given sample at a specified concentration and pure water ($\Delta\varepsilon'$ and $\Delta\varepsilon''$) of the ATP solution at several sampled frequencies are plotted as a function of the solute concentration.

An image of hydration structure of ATP is depicted in Fig. 6.6. The present results indicated that phosphoryl groups are mostly responsible for affecting the structure of the water surrounding the adenine nucleotides by forming one constrained water layer and an additional three or four layers of hyper-mobile water.

To determine the complex frequency-dependent dielectric response of four nucleotides at different dilute concentrations the terahertz time domain spectroscopy (THz-TDS) was used [18].

Water solution of protonated and deprotonated forms of the four mononucleotides dAMP, dCMP, dGMP, and dTMP was studied experimentally by equilibrium measurements using an electrospray mass spectrometer equipped with a drift cell and theoretically by computational methods including molecular modeling and density functional theory calculations [19]. Experiments carried out in positive and negative ion mode, and calculations of the protonated and deprotonated forms of the four nucleotides revealed that: (1) for deprotonated anionic nucleotides, the experimental enthalpies of hydration (ΔH°) are similar for all four systems and varied between −10.1 and −11.5 kcal mol^{-1} for the first water molecule (n) (1) and −8.3 and −9.6 kcal mol^{-1} for additional water molecules (n) (2–4); (2) the first water molecule binds to the charge-carrying phosphate group; (3) simulations of deprotonated mononucleotides with four water molecules yielded a large number of structures with similar energies; (4) in some of the structures, all four water molecules cluster around the phosphate group, and in other structures, the four water molecules each hydrate a different functional group of the nucleotide; (5) experimental ΔH° values for the protonated cationic mononucleotides ranged from -10.5 to -13.5 kcal/mol with more negative values (-12 kcal/mol) for dCMP, dGMP, and dTMP; (6) the least negative value for dAMP. For $n = 2$–4, the ΔH° values varied from -6.9 to -9.7 kcal/mol and were similar in value to the deprotonated nucleotides except for dAMP; (7) the first water molecule binds to the charge-carrying group for dCMP, dGMP, and dTMP; and (8) the charge-carrying N3 group is self-solvated by the phosphate group.

Experimental $\Delta H^{\circ}n$ and $\Delta S^{\circ}n$ values for the sequential hydration of deprotonated and protonated mononucleotides were listened in a table. Figure 6.7 showed the temperature dependence of the experimental data in terms of ΔG°_n values for the example of dCMP. In all cases, the first water molecule is bound more strongly

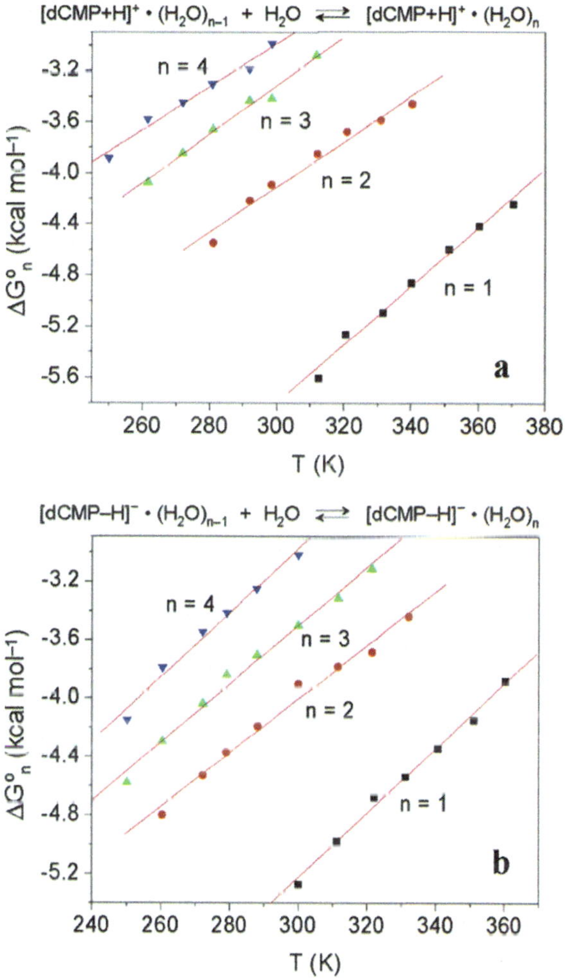

Fig. 6.7 Plot of the free energy of hydration versus temperature measured for the **a** protonated and **b** deprotonated nucleotide dCMP [19]. With permission from American Chemical Society

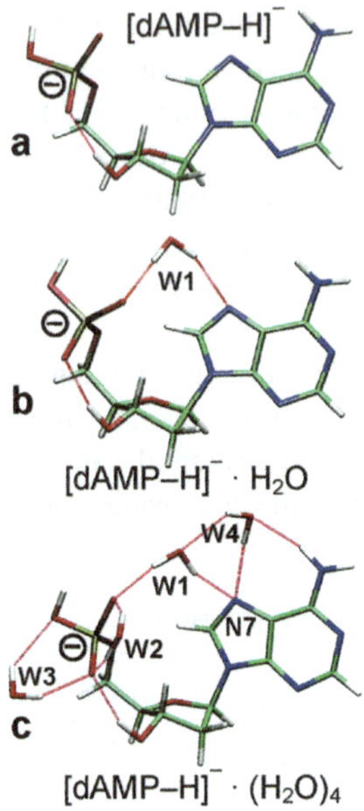

Fig. 6.8 Low-energy structures of deprotonated dCMP hydrated by four water molecules obtained by molecular mechanics (AMBER) simulations. **a** Lowest energy most typical structure, **b** isomer showing extensive phosphate hydration, **c** isomer with a water molecule in the second solvation shell, **d** isomer with water molecules bound to phosphate, sugar, and base [19]

$\Delta H°_1 > 10$ kcal/mol) than the following water molecules ($n = 2, 3, 4$). This effect was found to be most pronounced for the positive ions $[dCMP^{+}H]^{+}$, $[dGMP^{+}H]^{+}$, and $[dTMP^{+}H]^{+}$. The $\Delta H°_n$ values for $n = 2$, 3, and 4 are constant at −91 kcal/mol for all positive and negative ions with the exception of $[dAMP^{+}H]^{+}$. The $\Delta S°_n$ values are strongly correlated with the corresponding $\Delta H°_n$ values.

Some examples of AMBER simulations of deprotonated mononucleotides hydrated by four water molecules are shown in Fig. 6.8. The figure indicates structures and the lowest energy isomers of [dAMP-H]-â$(H_2O)_4$ and [dCMP-H]-â$(H_2O)_4$, respectively, with three water molecules solvating the phosphate group examples of a common hydration pattern. Calculated geometry optimized structure of protonated dehydrated dAMP is depicted in Fig. 6.9.

The authors concluded that: (1) the first water molecule binds more strongly to all of the protonated and deprotonated mononucleotide ions than the second through fourth water molecules; (2) the first water molecule adds to the charged site in all cases except for $[dAMP^{+}H]^{+}$ where the charged site is inaccessible to the water molecule; (3) fourth water molecules have about the same binding energy (9 kcal/mol^{-}); and (4) the second through fourth water molecules was predicted to bind equally to a number of functional groups on the mononucleotides including charge-proximate and charge remote sites or to already bound water molecules in a second solvation shell (Fig. 6.10).

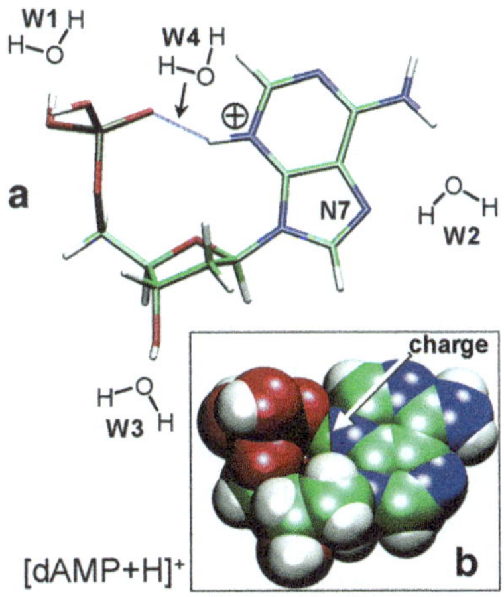

Fig. 6.9 Geometry optimized structure of protonated dAMP (dehydrated) calculated at the B3LYP/6-311++G** level. **a** The primary hydration sites W1 through W4 with binding energies listed in Table 3 are schematically shown. The water molecule at site W4 inserts into the blue hydrogen bond between the proton on the base and the phosphate group thereby significantly changing the dAMP conformation (see text). **b** Spacefilling representation of the same structure showing the efficient shielding of the charge. The proton involved in the hydrogen bond from the base to the phosphate (location of the charge) is highlighted in green [19]

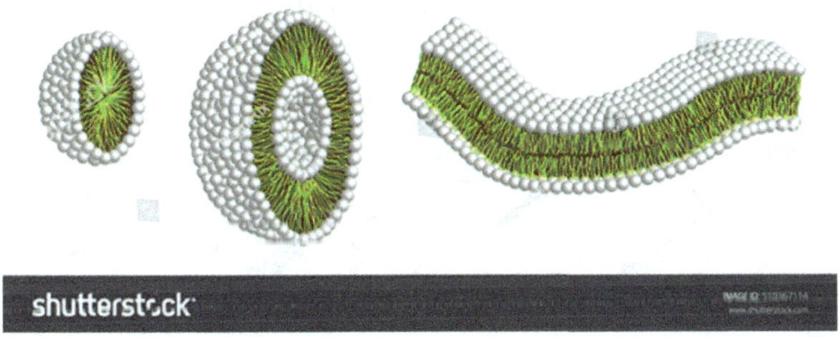

Fig. 6.10 Micelle, liposome and lipid bilayer https://www.shutterstock.com/search/liposome

6.2.4 Lipids. Liposomes

Liposomes composed of a lipid bilayer separating an aqueous internal compartment from the bulk aqueous phase. Micelles are closed lipid monolayers with a fatty acid core and polar surface or polar core with fatty acids on the surface (inverted micelle).

These structures are gaining attention as simple models of biological membranes [20–23]. Biological membranes are the basic component of all biological cells. Lipid bilayer membranes are primary building blocks of organisms. These membranes exhibit a diverse composition in order to separate functional compartments, to provide chemical and physical processes, and to control signaling processes. In plasma membranes, the lipid composition between the two leaflets is highly asymmetric and changes dynamically to trigger environmental responses. Numerous studies have paid attention to the contribution of water to determine the thickness and area of lipid membranes, membrane structure, dynamic and stability, and its influence on permeability properties. Due to the complex structure of the real biological membrane, different simplified bilayer lipid membranes were developed to allow experiments that cannot be performed with natural membranes, including supported bilayer lipid membranes (sBLMs), tethered bilayer lipid membranes (tBLMs), and floating bilayer lipid membranes (fBLMs) [24].

In work [23], monolayers and bilayers composed by 1,2-dimiristoyl-sn-glycero-3-phosphocholine (DMPC) and 1,2-di-O-tetradecyl-sn-glycero-3-phosphocholine (14:0 diether PC) (Figs. 6.11 and 6.12) were studied by steady-state fluorescence spectroscopy, FTIR spectroscopy, and surface pressure isotherms. Specifically, experimental part of the study included time-resolved fluorescence measurements, attenuated total reflection (ATR)-FTIR spectroscopy, surface pressure and measurements of dipole potential and compressibility. The excitation generalized polarization (GP_{ex}) function was calculated from the emission intensities using Eq. 6.1:

$$GP_{ex} = \frac{I_{440} - I_{480}}{I_{440} + I_{480}} \tag{6.2}$$

where I_{440} and I_{480} correspond to the emission maximum in the gel and the liquid crystalline state, respectively.

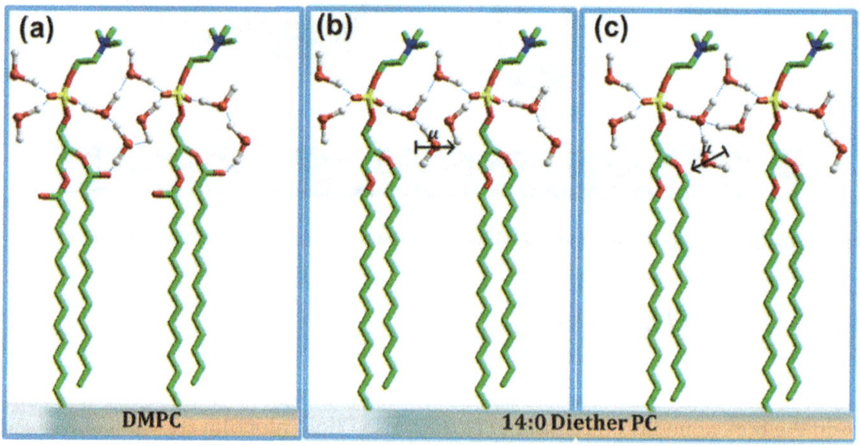

Fig. 6.11 **a** DMPC molecule shows the formation of an intramolecular water bridge between a PO and a CO group; **b** the water bridge is broken due to the absence of a CO Group; and **c** the water bound to the PO can rotate around the H-bond as a result of reorientation of the water dipole of 90° with respect to the membrane plane [23]. With permission from American Chemical Society

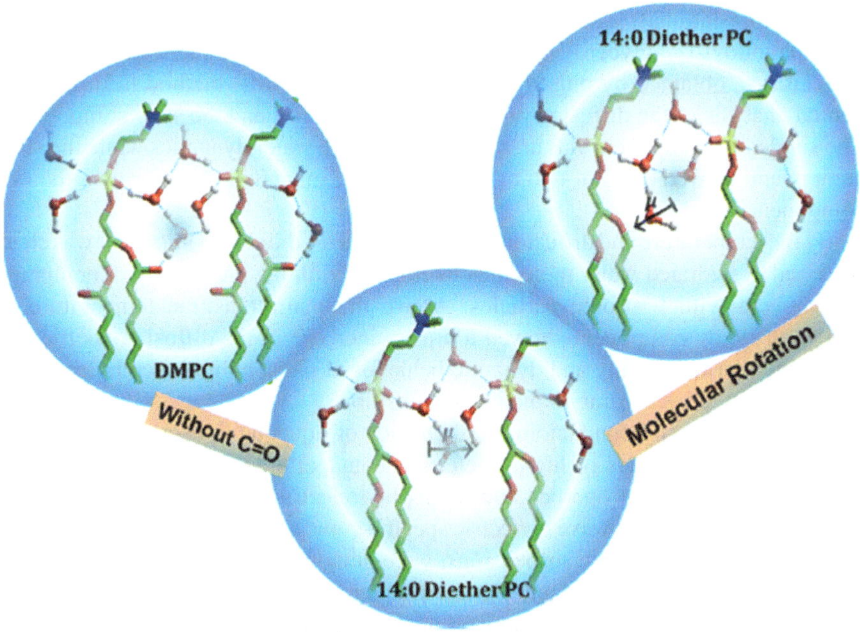

Fig. 6.12 Schematic presentation artificial membrane investigated in [23]. With permission from American Chemical Society

Generalized polarization of 1,2-di-O-tetradecyl-sn-glycero-3-phosphocholine in correlation with Fourier transform infrared (FTIR) analysis showed a higher level of polarizability of water molecules in the membrane phase around the phosphate groups both below and above transition temperature T_m. This promotes a different response in compressibility and dipole moment of the interphase, relating to different H bonding of water molecules with phosphates (PO) and carbonyl (CO) groups. The compressibility and dipole moment data indicated that: (1) the formation of an intramolecular water bridge between a PO and a CO group in DMPC molecule, (2) the water bridge is broken due to the absence of a CO Group; and (3) the bounded to the PO water rotates around the H-bond as a result of reorientation of the water dipole of 90° with respect to the membrane plane H.

The results also indicated that the lack of carbonyl groups gives lipids different surface physicochemical properties when compared with those of the diacyl/ester PC. The main differences are: (1) a change in the hydration state visualized as a stronger interaction of water with the phosphates together with an increase of more polarizable water; (2) an increase in the cohesion between lipids, indicating more propensity to form a condensed phase at lower pressures; and (3) water with different relaxation properties around the head groups can be modulated by the presence of carbonyl groups with consequences on dipole surface potential and compressibility properties. Confined water molecules in the membrane interphase could play a key role in the structure, function, and dynamics of many biological systems.

The translational diffusion coefficient (TDC) of water molecules in 1,2-dipalmitoyl-*rac*-glycero-3-phosphocholine (DPPC) liposome/water dispersion systems was obtained by pulsed field gradient (PFG) NMR method [25]. Biexponential analysis of the diffusion process revealed the fast and the slow components. The single fast component was associated to the pure water dynamics while the slow component showed TDC of two or three orders of magnitude smaller than the fast component.

By the temperature and diffusion time dependences, the slow component was characterized as hydrated water molecules interacting with the phospholipid membrane and moving with liposome particles. Above the main-transition temperature of DPPC, the slow component showed a drastic increase in the diffusion coefficient, indicating that the fluidity and dynamic behavior of phospholipid membranes were abruptly changed according to the phase transition.

The dielectric and conductometric behavior of aqueous liposome suspensions, the interfacial properties of highly heterogeneous colloidal systems, the interplay between lipid bilayer properties and an applied electric field were reviewed [26]. Applications of the radiowave dielectric spectroscopy, addressed to the investigation of DPPC aqueous liposome suspensions, were demonstrated. It was evidenced how dielectric spectroscopy, in an appropriate frequency range, is capable of furnishing information on the structural organization of the lipid bilayer on the changes in the lipid chain conformation at the pre- and main-transition temperature, and on counterion accumulation effects, in the neighboring bilayer surface. In addition, the effects of the surface charge density on the spontaneous vesicle formation and the influence of an external electric field on the double-layer structure were described and discussed in details.

Solutions of liposomes composed of binary mixtures of anionic dioleoylphosphatidylserine (DOPS) and zwitterionic dioleoylphosphatidylcholine (DOPC) were investigated with label-free angle-resolved (AR) second harmonic scattering (SHS) and electrophoretic mobility measurements [27]. The membrane surface potential was extracted from the AR-SHS response. As the DOPS membrane content is varied from 0 to 100%, the surface potential changes from -10 to 145 mV at a few percent of DOPS in the membrane after which it remains constant. In addition, the formation of a condensed layer of Na^+ counterions around the outer leaflet of the liposome was predicted and confirmed.

Nuclear relaxation measurements of ^1H and ^{17}O of water have been applied to study the kinetics of water diffusion across vesicular lipid membranes [28]. In addition, in the lipid segmental motion was detected in ^{13}C magnetic resonance T_1 measurements. Differentiation between the intra- and extravesicular media was achieved by entrapping Mn^{2+} inside the vesicles. The following results were reported: (1) the water permeability of egg phosphatidylcholine vesicles was 2.9 $\times$ 10^{-3} cm/s through the temperature range 0–65° with an activation energy of 10.5 kcal/mol; (2) the water permeability across vesicular bilayers of l-α-dipalmitoylphosphatidylcholine exhibits a sharp change through the lipid phase transition; (3) the permeability in the lipid crystal phase (45°) was 7.2 $\times$ 10^{-3} cm/s with an activation energy of 7.2 kcal/mol; and (4) below the transition at the gel phase

(35°), a permeability is 1.0×10^{-3} cm/s. The change in water permeability through the phase transition was correlated with the changes observed in the lipid segmental motion detected from the ^{13}C magnetic resonance T_1 measurements.

Molecular dynamics of Dipalmitoyl Phosphatidylcholine liposome–water dispersion was observed by dielectric spectroscopy measurement in higher frequency range from 1 to 30 GHz for time-domain reflectometry method [29]. The experiment indicated that the relaxation time distribution parameter β decreased around the gel-liquid crystalline phase transition temperature with increasing temperature from 36 to 48 °C. It was suggested that surrounding water of lipid membrane may form the particular water structure in the LC phase different from bulky water in the gel phase with a change in the mobility for hydrophobic regions of the lipid bilayer.

The above-mentioned basic fundamental properties of model lipid membranes underlie the structure and molecular dynamics of various types of liposomes.

Lipophilicity is defined as the ability of a chemical compound to a lipophilic environment, that is, to be dissolved in fats, oils, lipids, and non-polar solvents. Lipophilicity is quantitatively characterized by the partition coefficient (*P*) which describes the partition of the compound between aquatic and non-polar solvent. Review [30] describes lipophilicity parameters currently used in drug design and Quantitative Structure–Activity Relationship (QSAR) studies. Factors affected on the dual nature of lipophilicity as the outcome of non-polar and polar interactions were discussed. Non-polar interactions were equated with hydrophobicity and being expressed by steric terms like molecular volume, polarizability, molar refractivity contribute positively to lipophilicity. The polar interactions which include ion–dipole, dipole–dipole, or hydrogen-bond interactions were expressed by electronic constants, dipole moments, or hydrogen-bond parameters. A global polarity parameter, designated as *L*, has been proposed.

Figure 6.13 provides a rough summary of the most important information discussed in the review. The figure focuses on the distinction between lipophilicity and phospholipophilicity, as well as on the impact and common characteristics of chromatographic techniques in the assessment of both properties.

Free energy profiles in micellar systems, as well as mixed lipid bilayers, were investigated, taking advantage of both methods, that is, MD simulations and COSMO-RS, referred to as COSMOmic [31]. Free energy profiles of more than 200 solutes were predicted and validated by means of experimental partition coefficients. The partitioning in lipid bilayer systems containing different hydrophobic tail groups (DOPC (1,2-dioleoyl-sn-glycero-3-phosphocholine), SOPC (stearoyl-oleoylphosphatidylcholine), DMPC (1,2-dimyristoyl-sn-glycero-3-phosphocholine), and POPC (1-palmitoyl-2-oleoylsn-lycero-3-phosphocholine)) as well as mixed bilayers was calculated.

To predict the absorption of a drug, the octanol/water and the liposome/water models were compared [32]. The parameters used to relate the two models were the distribution coefficients between liposomes and water and octanol and water and the fraction of drug orally absorbed. For this study, 66 drugs were collected from literature sources and divided into four groups according to charge and ionization degree: neutral; positively charged; negatively charged; and partially ionized/zwitterionic.

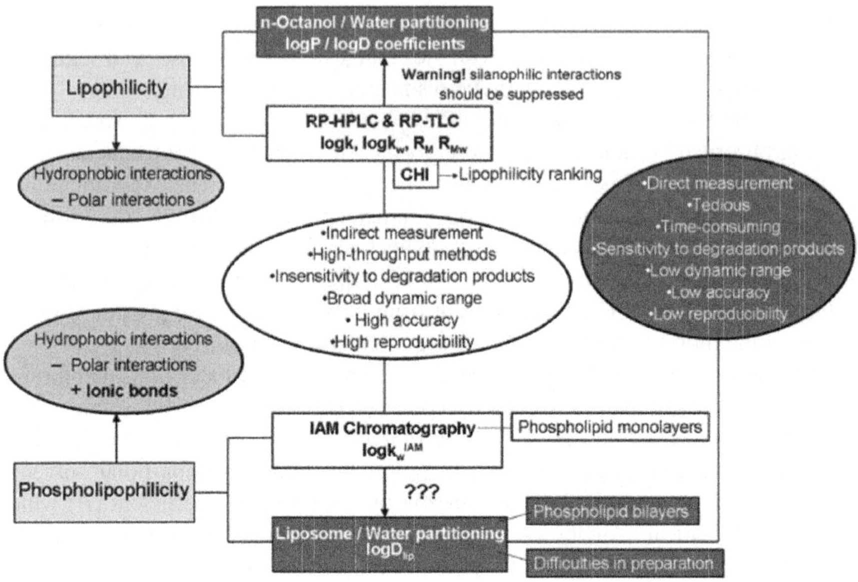

Fig. 6.13 Schematic representation of the basic features of lipophilicity and phospholipophilicity, the related indices and the impact of chromatographic techniques in lipophilicity/phospholipophilicity assessment [30]. With permission from Elsevier.

These results showed that the neutral compounds only interact with membranes through hydrophobic bonds, whereas charged drugs favor electrostatic interactions established with the liposomes.

Three different modeling approaches to describe the liposome–water partition coefficient (K_{lipw}) of organic ions, including 36 cations, 56 anions, 2 divalent cations and 2 zwitterions (plus 207 neutral compounds for ensuring model consistency) were compared [33]. The result indicated that: (1) the empirical correlation with the octanol–water partition coefficient of the corresponding neutral species yielded better results for the prediction of anions than for cations; (2) the polyparameter linear free energy relationship (pp-LFER) model performs worse for anions/cations; and (3) a program COSMOmic based on quantum chemistry and fluid phase thermodynamics has the widest applicability domain and is proved to the only model applicable for multiply charged ions.

Membrane/water partition coefficient ($K_{\mathrm{m/w}}$) is a parameter used to characterize the membrane permeability of compounds. Quantitative structure property relationship (QSPR) models for logarithm of liposome/water partition coefficient $K_{\mathrm{lip/w}}$ (log $K_{\mathrm{lip/w}}$) of the neutral organic chemicals and the neutral form of ionogenic organic chemicals (IOCs) (log K-neutral), ionic form of IOCs (log $K_{\mathrm{lip/w}}$-ionic), the speciation corrected liposome–water distribution ratios at a pH = 7.40 (log $D_{\mathrm{lip/w}}$-(pH = 7.40)) were examined [34]. Two modeling methods (multiple linear regressions (MLR) and k-nearest neighbor (kNN)) were used to predict employed variables which could be calculated from the molecular structure directly. The determination coefficient (R^2), leave-one-out cross-validation $Q^2(Q^2_{LOO})$ and bootstrapping coefficient (Q^2_{BOOT}), the external validation coefficient (Q^2_{EXT}) of all the models met the acceptable criteria ($Q^2 > 0.600$, $R^2 > 0.700$).

An approach for predicting bioaccumulation factors (BAFs) was proposed [35]. Liposome—water partitioning coefficients as determined by using solid-phase microextraction (SPME) were evaluated for the cause of assessing bioaccumulation potential of hydrophobic organic chemicals. The method was mapped in terms of mass balance, mode of spiking, kinetics, and reproducibility. In addition, the robustness of liposomes as partitioning phase was investigated in terms of chemical loading, and pH and ionic strength of the medium.

An important result of the work [36] was the proposal of a common protocol for constructing computational HII systems that correspond to the experimental data. Direct force measurements on a broad class of hydrophobic and hydrophilic surfaces have enabled the development of a general interaction potential per unit area

$$W(D) = -2\gamma_i \mathrm{Hy} \exp(-D/\mathrm{DH}) \tag{6.3}$$

where Hy is a non-dimensional hydraparameter, that applies to both hydrophobic and hydrophilic interactions between extended surfaces of the separation distance, D is the separation distance between two hydrophobic interfaces, DH $\approx$ 0.3–2 nm is the decay length for both hydrophobic and hydrophilic interactions, γ_i is the interfacial tension. This potential allows one to quantitatively account for additional attractions. For Hy > 0, the interaction potential describes interactions between partially hydrophobic surfaces, with the maximum hydrophobic interaction while for the potential becomes repulsive for Hy < 0, corresponding to partially hydrophilic (hydrated) interfaces. The following values of the parameters of Eq. 6.3 were found to be typical: $\gamma_i = 50$ mJ/m^2 for hydrocarbon interfaces in water, the decay length DH $\approx$ 1 nm and Hy $= 1 - a_0/a$, where a_0 is the hydrophilic area and a is the hydrophobic area at a given interface.

6.2.5 *Saccharides*

Carbohydrates are among the most important naturally occurring biomolecules, with the premier role in metabolism of all living species. These molecules also have a myriad of other functions, working as bioprotective agents against dehydration and other environmental pressures and are important to the food industry as well as being a potential source of fuels. The hydration state of saccharides in aqueous solution is a subject of keen interest, with both practical and scientific importance, for it is involved in the reception of sweetness and environmental stress tolerance including dehydra tion protection [5–7, 13, 37, 38]. The hydration of various saccharide molecules has been experimentally estimated using calorimetric, viscosity and ultrasound techniques, X-ray scattering and neutron scattering, nuclear magnetic resonance (NMR), and dielectric spectroscopy in the microwave region. This investigation has shed light on structure and dynamics of the saccharides and water molecules strongly bonded to the solute [13] and references therein). In addition, the hydration of a simple sugar is an essential model for understanding interactions between hydrophilic groups and interfacial water molecules [13]. One of the more intriguing aspects of carbohydrate chemistry is that despite having very similar molecular structures, sugars have very different properties [37]. For instance, there is a sensible difference in sweet taste between glucose and trehalose, even though trehalose is a disaccharide that comprised two glucose units, suggesting a different ability of these two carbohydrates to bind to sweet receptors.

To investigate how individual hydroxyl groups in a glucose–water system are locally hydrated, first-principles molecular dynamics, a localized orbital approach, was employed for study a glucose–water system (Fig. 6.14) [38]. The main reported results are: (1) the local hydration of the hydroxyl groups is sensitively affected by seemingly small variations in the local electronic structure and bond polarity of the groups; (2) the local hydration of the hydroxyl groups is roughly two-coordinated; (3) the hydroxyl groups enhance the dipole moment of water molecules in the first hydration shell, but only when the groups form donor H-bonds; and (4) the chemical bonding of a hydroxyl group that is mainly relevant for H bonding and can be described by three localized orbitals: two lone-pair orbitals (the acceptor side) and one orbital associated with the covalent OH-bond (the donor side).

To provide an atomic-level insight of unique biochemical role of the alpha, alpha-$(1 \rightarrow 1)$-linked glucopyranoside dimer over the other glycosidically linked sugars, systematic computational work for a series of 13 disaccharides was performed [7]. The stable hydrogen-bond network is derived from the formation of long-lived water bridges at the expense of decreasing the dynamics of the water molecules. The molecular dynamic computation revealed that: (1) superior osmotic and cryoprotective abilities of trehalose can be explained on the basis of conformational and hydration characteristics of the trehalose molecule; (2) there was very little hydration adjacent to the glycosidic oxygen of trehalose; (3) the dynamic conformation of trehalose was less flexible than any of the other sugars due to this anisotropic hydration; (4) to act as a sugar template, conformational rigidity was required for stable interactions with hydrogen-bonded water molecules; (5) trehalose made an average of 2.8 long-lived hydrogen bonds per each MD step, which was much larger than the average of 2.1 for the other sugars; and (6) dynamic reduction of water by trehalose was basing on each of the lowest translational diffusion coefficients and the lowest intermolecular coulombic energy of the water molecules around trehalose was taken place. Thus, trehalose functions as a "dynamic reducer" for solvent water molecules based on its anisotropic hydration and conformational rigidity have been established.

In order to precisely evaluate dynamics in the global hydration state of monosaccharides and disaccharides, the complex refractive index of these saccharides in solutions was determined, using THz time-domain spectroscopy (THz-TDS). between 0.2 and 3.0 THz [13]. Since THz waves have much stronger absorption in bulk water with picosecond dynamics than water molecules in the global hydration state that have dynamics in the order of nanoseconds or subnanoseconds, the hydration

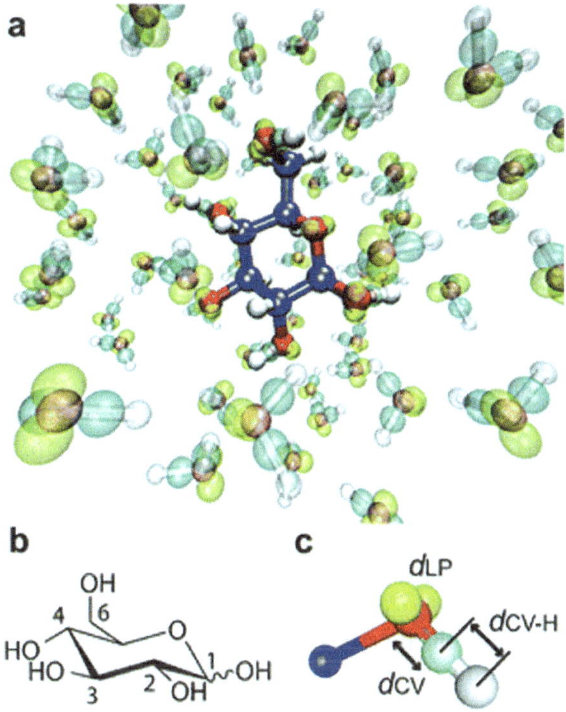

Fig. 6.14 The glucose–water system. **a** A snapshot from an ab initio MD simulation on glucopyranose in aqueous solution. Carbon, oxygen and hydrogen atoms are represented by blue, red and white respectively. Transparent yellow and light blue represent the centers of localized Wannier functions associated with lone pairs and covalent OH-bonds respectively. The molecular graphic was created by VMD.57. **b** The drawing of glucopyranose with the numbering scheme of its carbon atoms. The wavy bond at the anomeric site (C1) indicates either the a or b anomer. **c** A localized orbital description of the electronic properties of a hydroxyl group. The color representation is the same as in (**a**). To evaluate the electronic properties, I used three distances: dLP, dCV and dCV–H. The same definitions hold for water molecules. The distance dLP is the average distance from the oxygen atom to the two centers of the lone-pair Wannier orbitals. The distances dCV and dCV–H are the distances from the center of the covalent OH-bond orbital to the oxygen atom and to the hydrogen atom, respectively. The molecular graphic was created by gOpenMol [38]. With permission from Royal Society of Chemistry

number was calculated using absorption in the THz region. The influence of the strong absorption of bulk water was reduced by applying THz time-domain attenuated total reflection (THz TD-ATR) spectroscopy which allows to obtain the complex refractive index of the saccharide solutions. Typical powder samples of monosaccharides and disaccharides as shown in Table 6.2 were dissolved in distilled water at room temperature.

As it is seen from Table 6.2, the lowest concentrations (0.292 M), sucrose and maltose have almost double numbers of hydration water compared with galactose and mannose and hydration number decreases as the concentration of solution increases.

Table 6.2 Sample solutions of monosaccharides and disaccharides with mole concentrations (M). Mass percent (wt%) converted is shown in the bracket [13]. With permission from Elsevier

Monosaccharides: $C_6H_{12}O_6$		
D(+)-Glucose	0.292–0.877 M	(4.83–22.16 wt%)
D(+)-Galactose	0.292–0.877 M	(4.83–22.16 wt%)
D(+)-Mannose	0.292–0.877 M	(4.83–22.16 wt%)
Disaccharides: $C_{12}H_{22}O_{11}$		
Sucrose	0.292–1.754 M	(9.58–49.26 wt%)
Maltose	0.292–0.877 M	(9.58–27.15 wt%)

The calculated number of hydration water around different saccharide solutes is shown in Fig. 6.14. Hydration number decreases as the concentration of solution increases, for each of the different saccharides measured. In Fig. 6.15, the hydration number for glucose and the other monosaccharides measured, differed by about 3, while those of sucrose and maltose were almost the same. Thus, results indicated that the global hydration state is closely related to the number of hydrophilic groups and steric configuration of hydroxyl groups in saccharide molecules.

The hydration of specific sites and at the three-dimensional configuration of water molecules around three carbohydrates (glucose, cellobiose, and trehalose), combining neutron diffraction data with computer modeling, was examined [37]. The neutron diffraction experiments have been augmented by two different tools: Empirical potential structure refinement (EPSR) and ANGULA. A set of EPSR configurations has been analyzed using the program ANGULA. EPSR is a computational technique which has been explicitly designed to provide atomic scale detail

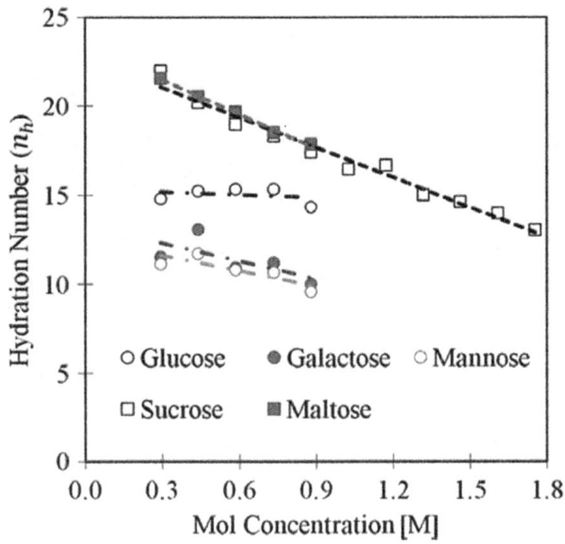

Fig. 6.15 Hydration number of monosaccharides and disaccharides. At the diluent concentration (0.292 M), the hydration numbers are 15.2 (glucose), 11.2 (galactose), 10.5 (mannose), 21.9 (sucrose) and 21.1 (maltose). The broken and chain lines are drawn only to guide the eye [13]. With permission from Elsevier

for disordered systems, using a set of neutron diffraction data to constrain the three-dimensional model of the solution in question. The program ANGULA can be used to characterize the position of neighboring water molecules around the solute molecules—glucose, trehalose, or cellobiose. It was suggested that the possibility that the different properties of carbohydrates, namely their ability to interact with receptors and/or other proteins, might be related to their different hydration shells. On this path, the hydration shell of glucose in solution has been determined using neutron diffraction experiments, enhanced by isotopic substitution (NDIS). Results were compared with the hydration shell of two disaccharides, namely trehalose and cellobiose, in three different situations: for glucose as a monomer, and when it is in two glucose-containing disaccharides with a different glycosidic bond between the glucose monomers, namely a glycosidic bond for trehalose 4-glycosidic bond for cellobiose.

The nearest neighbor radial distribution functions for the O_SO_W correlations for the selected hydroxyl and the hydroxymethyl groups for the special density maps (SDMs) for each sugar were examined (Fig. 6.16). O_S is the oxygen atom on the solute molecule, either that of the hydroxyl or the hydroxymethyl group, and O_W

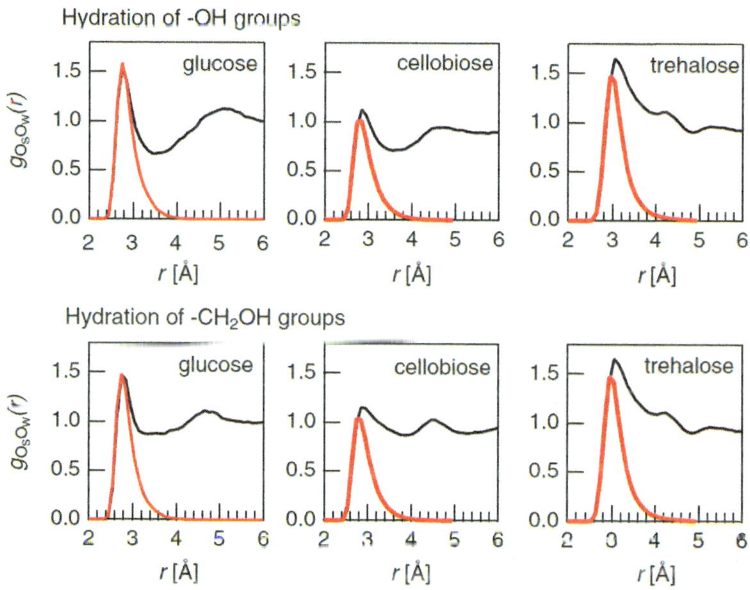

Fig. 6.16 Site–site correlation between the nearest neighbor water oxygen, O_W, and the oxygen atom of the solute, O_S, hydroxyl (top row, red line) and hydroxymethyl group (bottom row, red line). Radial distribution functions for the O_S–O_W interactions for the hydroxyl and hydroxymethyl groups for the spatial density maps SDMs are shown as a black line. The closest water oxygens for glucose and cellobiose are located about 2.8 Å away from the solute hydroxyl and hydroxymethyl sites, a distance compatible with strong hydrogen bonding. The closest water oxygens around trehalose are located instead at about 3.0 Å away from both –OH and $–CH_2OH$ groups, suggesting relatively weaker hydrogen bonding [37]. With permission from Elsevier

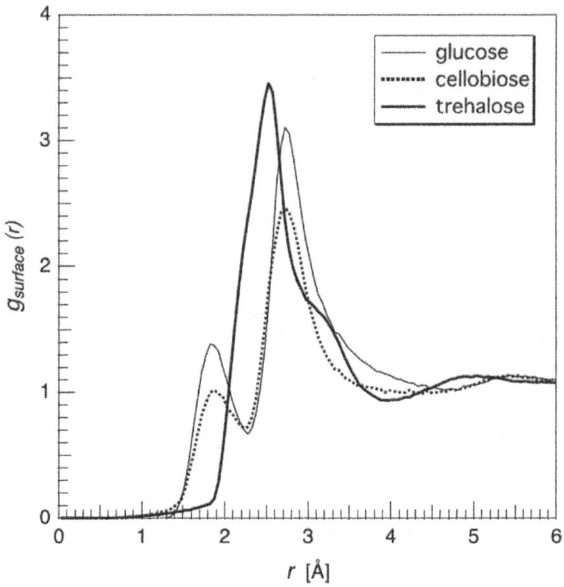

Fig. 6.17 The distance between water molecules and the central solute molecule. This plot reports the distribution function of water molecules around the "surface" of a solute. It reveals that the closest water molecules for glucose and cellobiose are located about 1.8 away from the solute surface, a distance compatible with hydrogen bonding between hydroxymethyl and hydroxyl groups and the solvent. Conversely, the closest water molecules around trehalose are located at about 2.5 away from the solute surface, a distance too large to suggest strong hydrogen bonding between the solute and the solvent [37]. With permission from Elsevier

is the water oxygen. As it is seen from the figure, nearest neighbor water oxygen is located at about 2.8 Å from O_S for both glucose and cellobiose –OH and –CH_2OH groups, while the distance between O_S and O_W for trehalose is slightly larger (about 3 Å), for both its hydroxyl and hydroxymethyl sites. These water molecules are likely those located between the two trehalose rings (Fig. 6.17). This finding confirms the presence of relatively weaker hydrogen bonded water molecules around these specific groups of trehalose.

The calculated three-dimensional arrangement of water molecules around the entire solute molecule obtained with ANGULA is displayed in Fig. 6.18. The figure shows the possibility of the B site (–OH) to act both as a H-bond acceptor (through its O atom) with a receptor H atom (for instance that of a –NH group) and as a donor through its H atom with a nearest neighbor water molecule the atomic length scale. The results stress difference between the SDM of water around glucose (or cellobiose) or around trehalose.

Based on the above data and theoretical calculations, the following conclusions were made: (1) identical chemical groups can have radically different hydration patterns depending on their location on a given molecule; (2) the same glucose unit, present in all carbohydrates investigated, has a different overall hydration if it is in

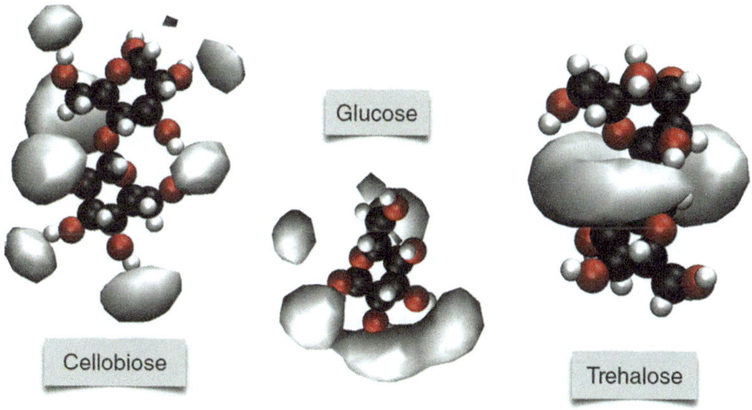

Fig. 6.18 The three-dimensional arrangement of water molecules around the entire solute molecule. These spatial density maps have been calculated for glucose, cellobiose, and trehalose over the distance range 0–3.5 Å at 37% probability [37]. With permission from Elsevier

solution as a single molecule or paired with other glucose units but with different glycosidic bond; (3) the observed difference between the overall hydration of the carbohydrates investigated might explain their different sweetness; (4) the possibility that these interactions are mediated by water molecules, bridging between the ligand and the receptor, could be an important feature in solvent recognition processes; and (5) the global hydration state is closely related to the number of hydrophilic groups and steric configuration of hydroxyl groups in saccharide molecules.

All in all, the results indicated that identical chemical groups can have radically different hydration patterns depending on their location on a given molecule. These differences can be linked with the specific activity of glucose, cellobiose, and trehalose as a sweet substance, as building block of cellulose fiber, and as a bioprotective agent, respectively.

Although sucrose and trehalose are both considered as stabilizing molecules for the purpose of preserving biological materials during, for example, lyophilization or cryo-preservation, trehalose has been shown to be more efficient than other sugar molecules in preserving different biological molecules against stress [39]. In addition, there are several important differences in their properties when they interact with water, such as differences in solubility, viscosity, and glass transition temperature. To investigate how these two disaccharides differ in their water interaction, the structure of aqueous solutions of these disaccharides was studied by using neutron and X-ray diffraction in combination with empirical potential structure refinement (EPSR) modelling [39]. The effect the two sugars have on their aqueous environment and how the sugars interact with themselves and with water was compared. As it is schematically illustrated in Fig. 6.19, the presence of trehalose appears to have a stronger distractive effect on the structure of bulk water than sucrose.

On the basis of the isotope subtraction method to highlight interactions with the carbon-bound hydrogen atoms (M atoms), as shown in Fig. 6.20, the M atoms of

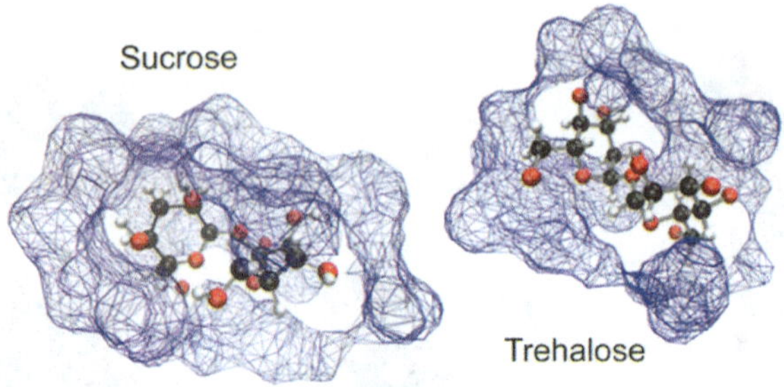

Fig. 6.19 Schematic representation different effect of sucrose and trehalose on the aqueous environments [39]. With permission from American Chemical Society

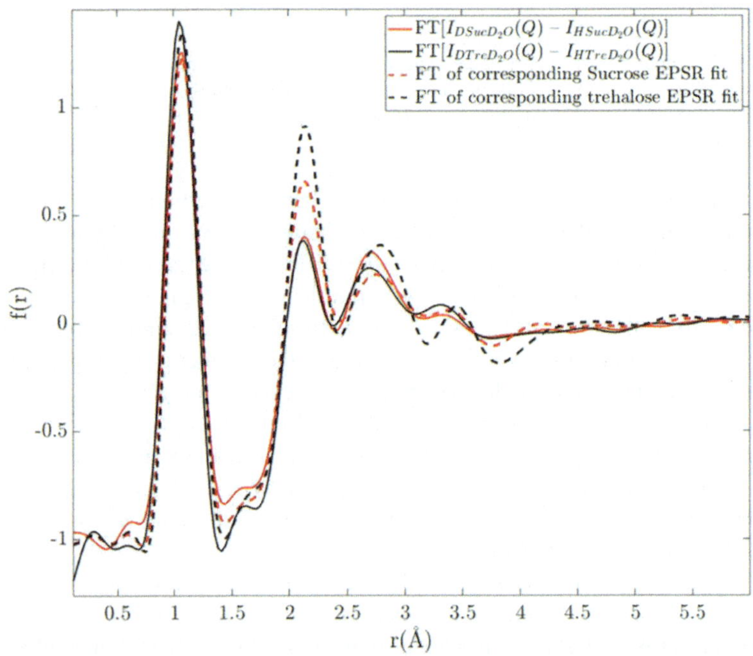

Fig. 6.20 Correlation functions obtained from the Fourier transform of the difference between the differential scattering cross sections from D-Sug in D_2O and H-Sug in D_2O (red solid line for sucrose and black solid line for trehalose). Red and black dashed lines represent the Fourier transform of the corresponding EPSR fits for sucrose and trehalose, respectively. The first peak in $f(r)$ at about 1.1 Å is entirely due to intramolecular C–M correlations and consequently not related to the hydration of the disaccharides [39]. With permission from American Chemical Society

the two sugars bind to similar numbers of water molecules in the two solutions. In addition, trehalose exhibits a stronger dynamical coupling (i.e., longer residence times of a water–trehalose hydrogen bond and slower diffusion of water for trehalose than for other disaccharides). Thus, there are indications for that trehalose perturbs the water structure more than sucrose.

6.3 Ion Hydration. Water as a Ligand

6.3.1 Ion Hydration Properties

After coming apart from crystal, the individual ions are then surrounded by solvent particles in a process called solvation. For example, the individual Na^+ ions are surrounded by water molecules with the oxygen atom oriented near the positive ion while chloride ions are surrounded by water molecules with the opposite orientation (Fig. 6.21). Hydration helps to stabilize aqueous solutions by preventing the positive and negative ions from coming back together and forming a precipitate. Hydration of ions is of pivotal importance for many fundamental biological processes. Various spectroscopic and theoretical methods are used to study of the ion hydration structure and the retardation of the hydration bond dynamics in the vicinity of anions and cations [3, 40–72].

A sodium cation is solvated by water molecules with their partially negative charged lone pairs pointing inwards toward the positively charged sodium ion https://en.wikipedia.org/wiki/Hydration_number.

The fundamental experimental information on ion hydration thermodynamics comes typically from calorimetric measurements performed on charge-neutral salts. The enthalpies, entropies, and Gibbs free energies of formation of salts in aqueous solutions have been measured with high accuracy and tabulated [34].

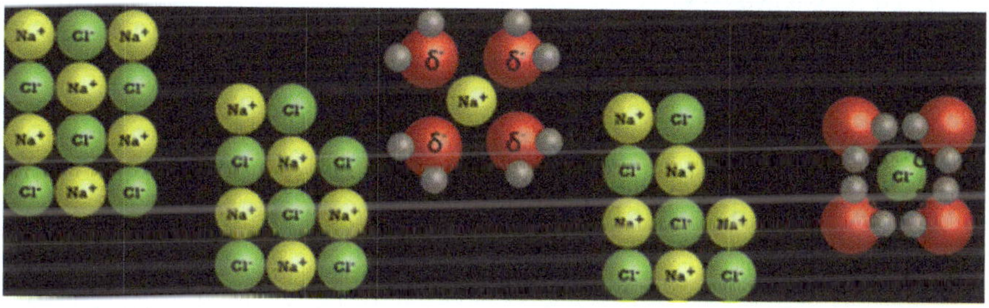

Fig. 6.21 The action of the polar water molecules takes the crystal lattice apart/t. The positively charged sodium ions in the crystal attract the oxygen end of the water molecules because they are partially negative. The negatively charged chloride ions in the crystal attract the hydrogen end of the water molecules because they are partially positive https://courses.lumenlearning.com/chemintro/chapter/dissolving-process/

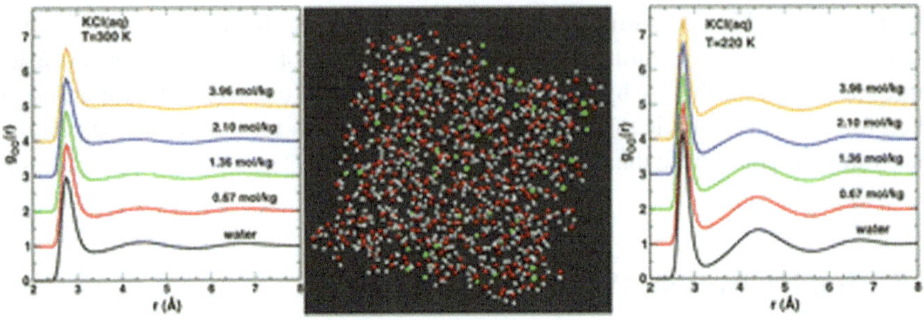

Fig. 6.22 The effect of salt on the structure of water [40]. With permission from Royal Society of Chemistry

The structural properties of aqueous solutions of NaCl, KCl, and KF salts at ambient conditions and upon supercooling at constant pressure were examined by means the method of molecular dynamics simulation at increasing concentration of the salt starting from $c = 0.67\ \text{mol kg}^{-1}$ up to $3.96\ \text{mol kg}^{-1}$ [40]. The modifications of the hydration shells and the changes in the water structure induced by the presence of the anion indicated that the oxygen–oxygen structure is strongly dependent on the ionic concentration. These changes are almost independent from the cation and the hydrogen bonding is preserved at all concentrations and temperatures. Main effect of increasing the ionic concentration is the tendency of the water structure to assume the high-density liquid form predicted for pure water upon supercooling. The effect of salt on the structure of water is illustrated in Fig. 6.22 where the radial distribution functions (RDF), $g_{\mathrm{OO}}(r)$, for NaCl (aq) at different concentrations and temperatures are indicated [40].

Figure 6.23 describes the behavior of the cation–water radial RDF $g_{\mathrm{ZO}}(r)$ and $g_{\mathrm{ZH}}(r)$ for the Na or the K cation in the NaCl (aq) and KCl (aq) systems at increasing ion concentrations and temperatures $T = 300$ K and $T = 220$ K. It is seen in the figure that the position of the first peak does not change with increasing ion concentration or decreasing temperature.

From the results, it is evident that in the cation hydrations shells, the arrangement of water around the K ions has a weaker order with respect to water molecules in the hydration shells of Na ions. Nevertheless, a strong distortion of the HB network upon increasing the concentration in both NaCl (aq) and KCl (aq) with not many differences between the two cases was found. Thus, the data on the effects of the perturbation of the HB network both in NaCl (aq) and KCl (aq) support the hypothesis of an analogy between concentration of ions in solution and enhancement of external pressure applied to pure water.

The hydration of the alkali metal ions in aqueous solution has been studied by large angle X-ray scattering (LAXS) and double difference infrared spectroscopy (DDIR) [42]. The results of the LAXS and DDIR measurements showed that the sodium, potassium, rubidium, and cesium ions are weakly hydrated with only a single shell of water molecules and the smaller lithium ion is more strongly hydrated, most

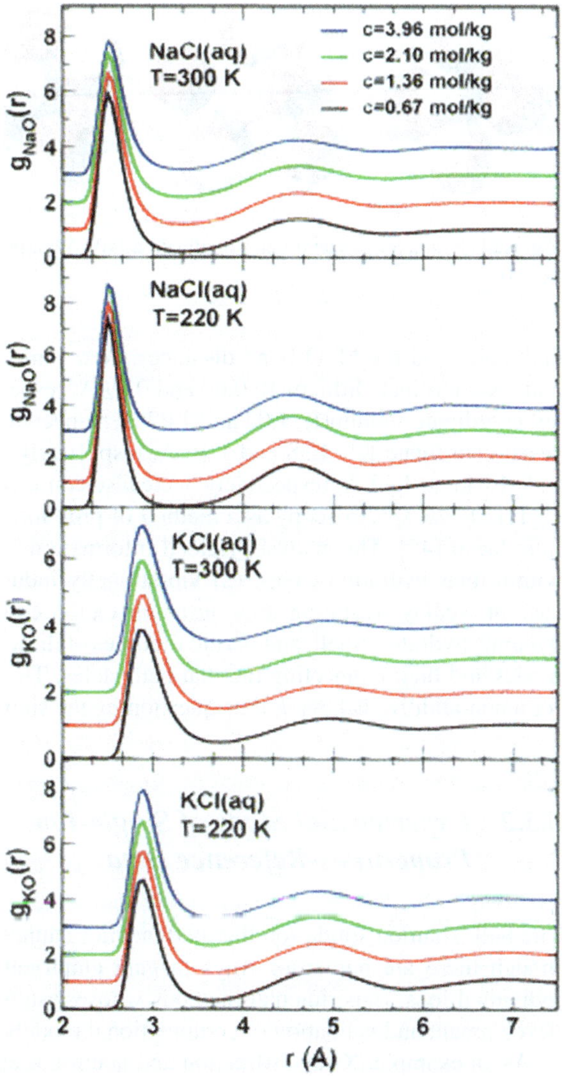

Fig. 6.23 Cation–oxygen RDF $g_{NaO}(r)$ on the two top panels and $g_{KO}(r)$ on the two bottom panels for increasing concentration and temperatures $T = 300$ K and $T = 220$ K [40]. With permission from Royal Society of Chemistry

probably with a second hydration shell present. The influence of the rubidium and cesium ions on the water structure was found to be very weak. For the alkali metal ions even though the M–O interactions are weak and the number of complexes of potassium, rubidium and cesium with well-defined coordination geometry are very small. Nevertheless, relatively strong correlation between M–O bond distances and coordination numbers exists. The mean M–O bond distance in the hydrated sodium, potassium, rubidium, and cesium ions in aqueous solution has been determined to be 2.43(2), 2.81(1), 2.98(1), and 3.07(1) Å, which corresponds to six-, seven-, and eight-coordination (Fig. 6.24) which supported by the linear relationship of the hydration

Fig. 6.24 Examples of metal ions coordination [40]. With permission from American Chemical Society

enthalpies and the M–O bond distances. New ionic radii were proposed for four- and six-coordinate lithium(I), 0.60 and 0.79 Å, respectively, as well as for five- and six-coordinate sodium(I), 1.02 and 1.07 Å, respectively. The ionic radii for six- and seven-coordinate K^+, 1.38 and 1.46 Å, respectively, and eight-coordinate Rb^+ and Cs^+, 1.64 and 1.73 Å, respectively, were also confirmed.

THz-FTIR spectroscopy as a method of prob ion hydration and ion pairing was introduced [45]. The analysis yielded information on: (1) strongly hydrating ions within their hydration cages, (2) vibrationally induced charge fluctuations in the case of weakly hydrating ions on anion–cation cooperativity, (3) the size of the dynamic hydration shell, and (4) the lifetimes of these collective ion-hydration water modes and their connecting thermal bath states. The study also provided evidence for a non-additive behavior, thus questioning the simplified Hofmeister model.

6.3.2 Experimental Study of Single-Ion Properties—Reference Data

The bulk solution single-ion thermodynamics cannot be experimentally measured, though there are numerous ways to gain empirical insight into the behavior of individual ions. Experimental methods whose results are most frequently used for development and validation of computational models were listed in [50].

As an example, X-ray diffraction and neutron scattering techniques can provide structural information in the form of ion–water pair distribution functions or estimates of the effective ion radius and coordination number. Effective ionic radii have been used to correlate hydration thermodynamic properties and o serve as a reference for molecular model development.

The fundamental experimental information on ion hydration thermodynamics comes typically from calorimetric measurements performed on charge-neutral salts. The enthalpies, entropies, and Gibbs free energies of formation of salts in aqueous solutions have been measured with high accuracy and tabulated [54]. A characteristic quasi-linear compensation relationship between the entropy and enthalpy was revealed. When these quantities were plotted \, the separation between cation and anion lines reflects the charge asymmetry of water molecules because any charge-symmetric solvent would solvate the positive and negative ions identically, and the corresponding trend lines would coincide.

In the work [55], the solvation free energy in conjunction with the solvation entropy was used to construct thermodynamically sound force fields for the alkali and halide ions for the simulation of ion-specific effects in aqueous environment. Several problems during optimization of the effective ionic radius from the first peak of the radial ion–water distribution function were discussed in detail. Various possibilities to deal with these problems were considered and an optimized force field for the halide anions that reproduces the free energy and the entropy of solvation was presented. Three different parameter sets for every alkali metal cations with a small, intermediate, or large Lennard-Jones interaction strength, where the Lennard-Jones diameters were optimized to reproduce the solvation free energy, were suggested. The ionic radius was reproduced with acceptable accuracy by this optimization strategy, suggesting that the proposed force fields are reliable for calculation energy and entropy of solvation.

A combined terahertz and femtosecond infrared spectroscopic study of water dynamics around magnesium, lithium, sodium, and cesium cations, as well as sulfate, chloride, iodide, and perchlorate anions revealed that the effect of ions and counterions on water can be strongly interdependent and non-additive [56]. In certain cases, the effect can expand beyond the first solvation shell of water molecules directly surrounding the ion.

Atomistic molecular dynamics simulations of double strand DNA (dsDNA) with four alkaline earth metal ions (Mg^{2+}, Ca^{2+}, Sr^{2+}, and Ba^{2+}) (Fig. 6.25) to elucidate the physical interactions that govern DNA–ion binding were performed to identity of metal ions surrounding DNA [41]; force field parameterized to reproduce experimental ion hydration data; (2) this force field was validated for modeling DNA–ion interactions by reproducing experimental ion counting data and ion–phosphate distances; (3) the spatial distribution of the alkaline earth metal ions surrounding a DNA duplex was analyzed using RDFs (Fig. 6.25) and occupancy values; and (4) radial distances were used to identify direct and indirect binding and occupancy values provided quantitative comparisons for the number and type of contacts direct and indirect contacts formed between the alkaline earth metal ions and dsDNA.

The radial distribution functions (RDFs) between each alkaline earth metal ion and water oxygen atoms in the presence of dsDNA are shown in Fig. 6.26.

The main results of the investigation are: (1) alkaline earth metal ions prefer to bind at the phosphate backbone compared to the major groove and negligible binding occurs in the minor groove; (2) larger alkaline earth metal ions with variable first solvation shells (Ca^{2+}, Sr^{2+}, and Ba^{2+}) involve in both direct and indirect binding, where indirect binding increases with ion size; (3) the strength of its first solvation shell of Mg^{2+} predicts indirect binding only; (4) ions bound to the phosphate backbone form fewer contacts per ion compared to the major groove; (5) within the major groove, metal ions preferentially bind to guanine–cytosine base pairs and form simultaneous contacts with the N7 and O6 atoms of guanine; (6) RDFs and occupancy values demonstrate that Mg^{2+} ions bind to dsDNA through indirect binding

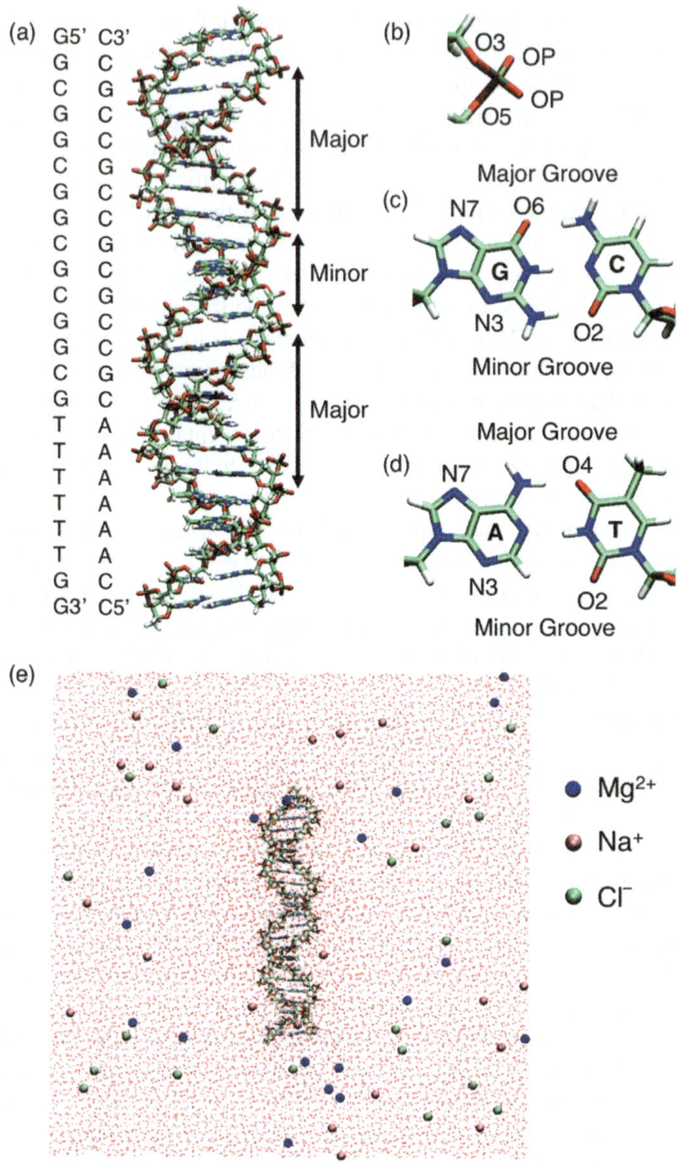

Fig. 6.25 **a** Nucleotide sequence and molecular structure of the 23-mer DNA duplex. Double headed arrows signify the major and minor grooves. Molecular structure of the **b** phosphate group and the **c** guanine (G)—cysteine (C) and **d** adenine (A)—thymine (T) base pairs. Labels show the electronegative atoms on the phosphate backbone and major and minor grooves. **d** Initial configuration of dsDNA, ions, and water oxygen atoms in the simulation box [41]. With permission from Royal Society of Chemistry

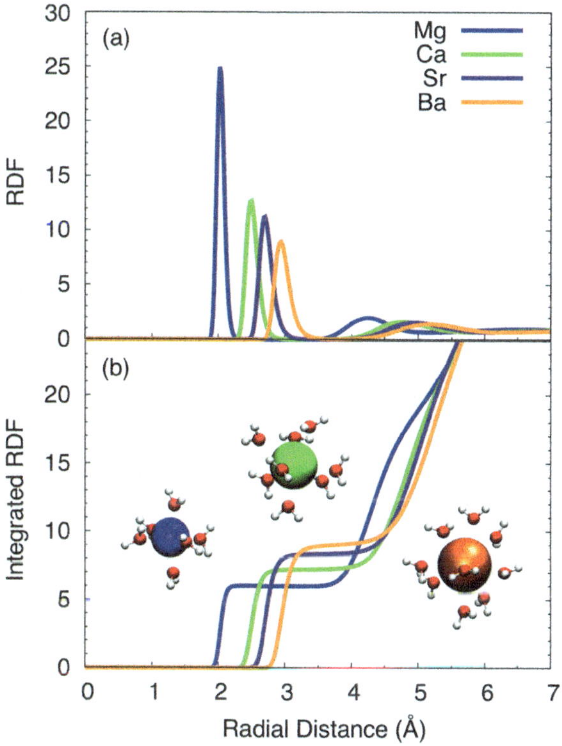

Fig. 6.26 **a** The RDF and **b** integrated RDF between each alkaline earth metal ion and water oxygen atoms. Insets show Mg^{2+} (blue), Ca^{2+} (green), and Ba^{2+} (orange) solvated by 6, 8, and 9 water molecules, respectively. Ions are shown with unscaled van der Waals radii [41]. With permission from Royal Society of Chemistry

due to its tightly held first solvation shell and Ca^{2+}, Sr^{2+}, and Ba^{2+} ions can lose one or more water molecules in their first solvation shell and bind to dsDNA through direct and/or indirect binding; and (7) the total occupancy of dsDNA decreases in the following order: $Ca^{2+} > Sr^{2+} > Ba^{2+} > Mg^{2+}$. Thus, the interplay among ion size, DNA–ion interaction, and the size and flexibility of the first solvation shell are key to predicting how alkaline earth metal ions interact with DNA.

6.3.3 Water as a Ligand

Metal aquo-complexesarecoordination compounds containing metal ions with only water as a ligand. The majority of aquo-complexes, such as $[Fe(H_2O)_6]^{3+}$ (Fig. 6.27), have the general formula $[M(H_2O)_6]^{n+}$, with $n = 2$ or 3 and an octahedral structure. Some physical properties hydrated cations in biological systems are listed in Table 6.3, https://en.wikipedia.org/wiki/Metal_aquo_complex.

The results on the water exchange reaction between coordination shells around metal ions in aqueous solution which is a fundamental process in understanding the

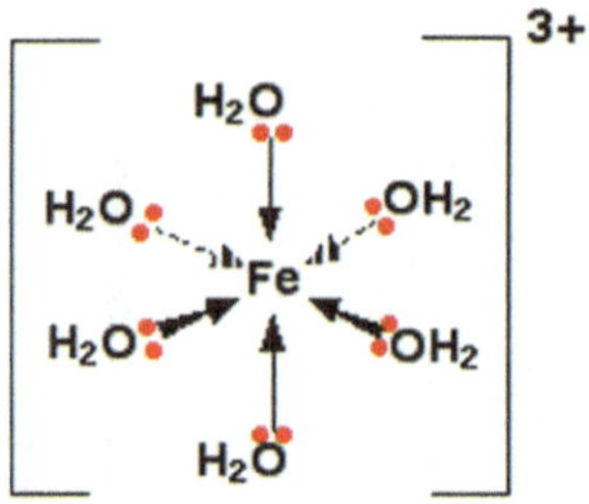

Fig. 6.27 Structure of the octahedral $[Fe(H_2O)_6]^{3+}$·aquo-complex https://www.chemguide.co.uk/inorganic/complexions/whatis.html, https://en.wikipedia.org/wiki/Metal_aquo_complex

Table 6.3 Physical properties of transition metal aquo-complexes. https://en.wikipedia.org/wiki/Metal_aquo_complex

Complex	Colour	Electron config.	M–O distance (Å) [2]	Water exchange rate (s^{-1}, 25 °C) [3]	$M^{2+/3+}$ self-exchange rate (M^{-1} s^{-1}, 25 °C)
$[Ti(H_2O)_6]^{3}$	Violet	$(t_{2g})^1$	2.025	1.8 x 10^5	N/A
$[V(H_2O)_6]^{2+}$	Violet	$(t_{2g})^3$	2.12	8.7 x 10^1	Fast
$[V(H_2O)_6]^{3+}$	Yellow	$(t_{2g})^2$	1.991 [4]	5.0 x 10^2	Fast
$[Cr(H_2O)_6]^{2+}$	Blue	$(t_{2g})^3(e_g)^1$	2.06 and 2.33	1.2 x 10^8	Slow
$[Cr(H_2O)_6]^{3+}$	Violet	$(t_{2g})^3$	1.961	2.4 x 10^{-6}	Slow
$[Mn(H_2O)_6]^{2+}$	Pale pink	$(t_{2g})^3(e_g)^2$	2.177	2.1 x 10^7	N/A
$[Fe(H_2O)_6]^{2+}$	Pale blue-green	$(t_{2g})^4(e_g)^2$	2.095	4.4 x 10^6	Fast
$[Fe(H_2O)_6]^{3+}$	Pale violet	$(t_{2g})^3(e_g)^2$	1.990	1.6 x 10^2	Fast
$[Co(H_2O)_6]^{2+}$	Pink	$(t_{2g})^5(e_g)^2$	2.08	3.2 x 10^6	N/A
$[Ni(H_2O)_6]^{2+}$	Green	$(t_{2g})^6(e_g)^2$	2.05	3.2 x 10^4	N/A
$[Cu(H_2O)_6]^{2+}$	Blue	$(t_{2g})^6(e_g)^3$	1.97 and 2.30	5.7 x 10^9	N/A

reactivity of these ions in chemical and biological systems were reviewed [57]. The paper demonstrated the complementary of experimental studies and computer simulations or quantum chemical calculations performed on such systems. A variety of experimental and different computer techniques applied the large range of exchange rate constants was described. Very fast exchange reactions between first and second coordination shell and between second shell and bulk solvent were simulated by classical molecular dynamics technique. Reaction pathways for water exchange on metal ions with a less labile first coordination shell and corresponding transition states were calculated. A simple model accounted for the thermodynamics of hydration of ions with charges varying from ±1 to ±4, depending only on their radii, when taken as spheres was suggested [58]. The model specifies the width of the hydration shell, Δr, the number of water molecules immobilized and electrostricted in it, proportional to $|z|/r$, and the bulk and high-field dielectric properties of the water [58].

6.4 Hydration Enthalpy–Entropy. Compensation Effect

The relationship, typically linear one, between the standard enthalpy and entropy and the energy and entropy of activation (the compensation effect, CE) is widespread in chemical and biological reactions [60–80]. This effect is also observed in hydration processes and received explanations within the framework of the model assuming cooperative rearrangement of a cluster of water molecules accompanying a chemical or physical act, for example, a change in the configuration of a reaction complex in a chemical reaction or a conformational transition [60–67].

These pioneering works presented evidences for the existence of a specific linear relationship between the entropy change and the enthalpy change in a variety of processes of small solutes in water solution; the processes include solvation of ions and non-electrolytes, hydrolysis, oxidation–reduction, ionization of weak electrolytes, and quenching of indole fluorescence among others as well as in enzyme catalysis. Below are examples of compensation effects during ion hydration (Figs. 6.28, 6.29 and 6.30). Thermodynamic and kinetic compensation effects in enzymatic reactions are described in Chap. 8.

The hydration enthalpy and Gibbs free energy of proton and hydroxide were calculated by means of a combination of ab initio density functional theory and a polarizable continuum model within the self-consistent reaction field method [59]. The ion–water cluster models included up to 13 water molecules solvating the ions allowed to describe explicitly the first and second solvation shells. Vibrational contributions the enthalpy and entropy were taken into account. A proposed model of the hydrated proton included three molecules in the first hydration shell and nine molecules in the second shell was suggested. The calculated proton hydration enthalpy of ≈ -1150 kJ/mol and hydroxide hydration enthalpy, ≈ -520 kJ/mol, and hydration free energy, ≈ -400 kJ/mol, are in good agreement with available results from cluster–ion solvation data.

The study [80] was based the cluster-pair-based approximation (CPA) technique and assumptions that the hydration properties of single cations and anions in growing clusters rapidly converge to each other following a monotonous trend. Using a combination of simulation techniques to study the transition between alkali halide ions in small clusters and bulk solution, it was found that this convergence is slow and involves a change in trends, which can result in significant errors in the original estimated single-ion properties. The resulting hydration thermodynamic properties of ions (Fig. 6.28) [79] were obtained by the simulation methodology described in publication [80] and revealed enthalpy–entropy compensation relationship.

In work [55], data on the solvation free energy in conjunction with the solvation entropy was used to construct thermodynamically sound force fields for the alkali

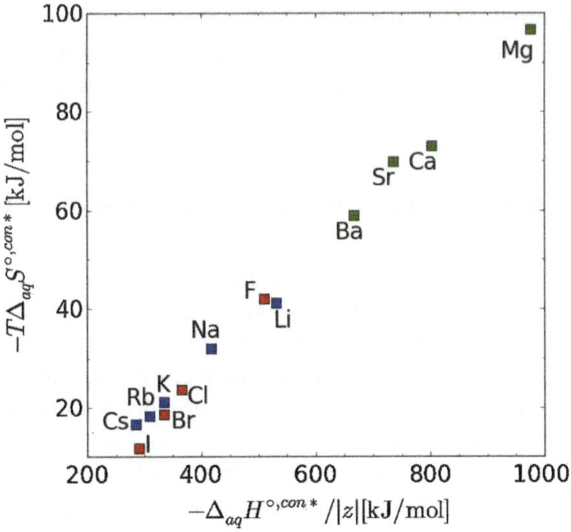

Fig. 6.28 Conventional single-ion hydration entropy, as a function of hydration enthalpy. The plotted quantities are conventional values shifted so that proton hydration enthalpy and Gibbs free energy are equal to those recommended by Conway: 5 $\Delta aqH^{\circ,con}* = -1104$ kJ/mol, $\Delta aqG^{\circ,con}* = -1066$ kJ/mol, and $T\Delta aqS^{\circ,con}* = \Delta aqH^{\circ,con}* - \Delta aqG^{\circ,con}*$. Monovalent cations are shown in blue, divalent cations in green, and halide anions in red [79]. With permission from Elsevier

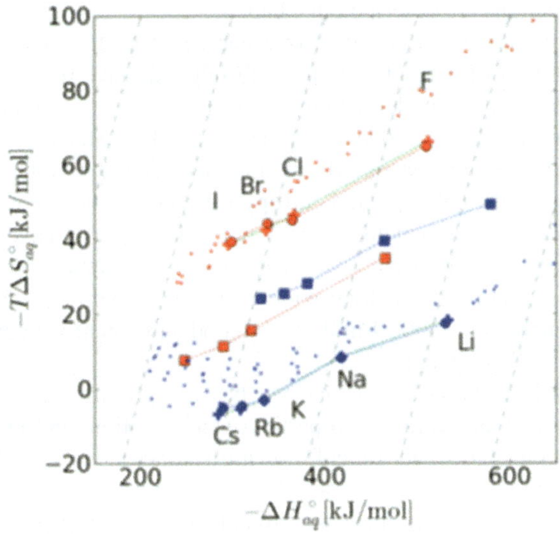

Fig. 6.29 Real bulk single-ion hydration entropy ($-T\Delta S_{aq}$) versus enthalpy ($-\Delta H_{aq}$) from molecular simulations (dots: data from Horinek and coworkers 25 with our values of φX sp, circles: results obtained with modified models including anion-hydrogen LJ potential), the original CPA values (squares), and the corrected CPA values (crosses). Cations are shown in blue, anions in red, and the gray dashed lines denote free energy isolines, $\Delta G_{aq} = \Delta H_{aq} - T\Delta S_{aq}$ [55]. With permission from American Institute of Physics

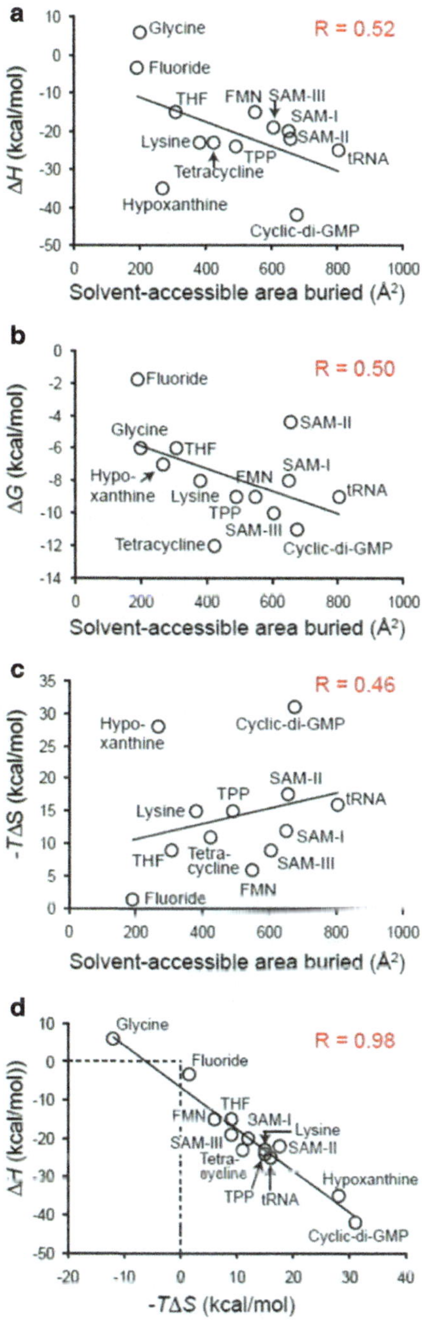

Fig. 6.30 Calorimetric analysis of riboswitches. **a–c** Solvent-accessible ligand surface area buried (A^2) by riboswitch binding does not correlate strongly with the binding enthalpy (**a**), free energy (**b**), or entropy (**c**). This contrasts with the strong correlation ($R = 0.89$) between binding free energy and solvent-accessible area buried by in vitro selected aptamers [51]. **d** A wide spectrum of entropy–enthalpy distributions demonstrate significant compensation ($R = 0.98$), resulting in a relatively narrow range of ligand binding affinities for riboswitches [74]. With permission from Elsevier

and halide ions to simulate ion-specific effects in aqueous environment. The value of effective ionic radius derived from the first peak of the radial ion–water distribution function and force field for the halide anions that reproduces the free energy and the entropy of solvation were optimized. Three different parameter sets for every alkali metal cations with a small, intermediate, or large Lennard-Jones interaction strength to reproduce the solvation free energy were given. The ionic radius was reproduced with acceptable accuracy by this optimization strategy, suggesting that the proposed force fields are reliable for calculation energy and entropy of solvation. Plot of the calculated hydration entropy ($-T\Delta S_{aq}$) versus enthalpy ($-\Delta H_{aq}$) demonstrates their compensation behavior (Fig. 6.29).

Riboswitches are phylogenetically widespread non-coding mRNA domains that directly bind cellular metabolites. A wide spectrum of enthalpy–entropy compensation regimes across riboswitches that bind a diverse set of ligands, giving rise to a relatively narrow range of physiologically relevant free energies and ligand affinities was demonstrated (Fig. 6.30) [74].

A statistical-mechanical framework for estimation of solvation entropies and enthalpies based on the analysis of water as a mixture of correlated water oxygens and water hydrogens, employing for grid-based analyses of molecular dynamics simulations, was proposed [2]. Entropic and enthalpic contributions in hydrated solutes (Figs. 6.31 and 6.32) were computed using Mutual Information Expansion. This approach promoted to evaluate to pairwise interactions and calculate directly from a distance-based hydrogen bonding energy algorithm, correspondingly. The per-voxel structures of the total entropy, enthalpy, and free energy are depicted in Fig. 6.32a–c,

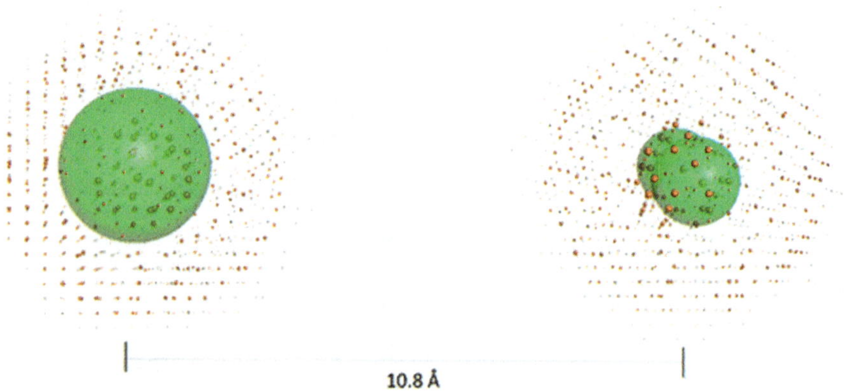

Fig. 6.31 Per-voxel nonbulk entropy (solid orange) and enthalpy (translucent green) for a water simulation with two restrained water oxygens that are ~10.8 Å apart from each other. Each voxel is represented by a sphere whose size correlates with the magnitude of the thermodynamic quantity estimated for that volume element. For better visualization, both enthalpies and entropies have been independently rescaled based on the range of their magnitudes. We note that the volume of the spheres representing the enthalpy of the voxels harboring the restrained oxygens is larger for the left field because the tethered particle remains in a single voxel, whereas the one on the right-hand side visits more than one voxel [2]. With permission from American Chemical Society.

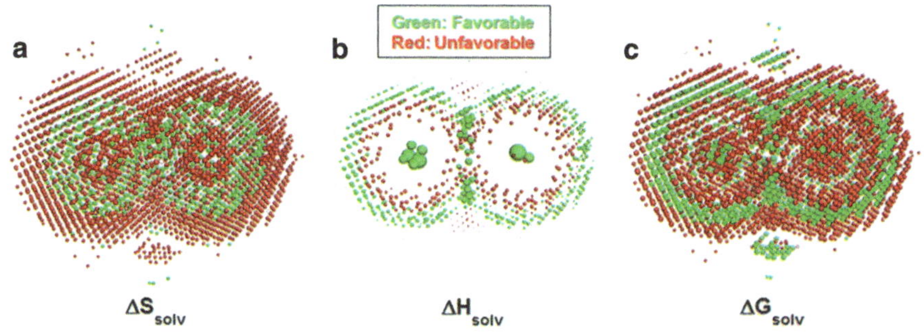

Fig. 6.32 Per-voxel nonbulk structure of **a** ΔS_{solv}, **b** ΔH_{solv}, **c** ΔG_{solv}, and **d** higher-than-bulk occupancy of those voxels with nonzero ΔG_{solv}, for a water system with two water oxygens ~4.3 Å apart. For each structure, all quantities were normalized with respect to the minimum (making the magnitude of all values greater than or equal to 1). The natural logarithm of the resulting quantities was then taken. Each field was subsequently rescaled and clipped to obtain the visually tractable fields shown [2]. With permission from American Chemical Society

respectively. The higher-than-bulk occupancy of those voxels with non-negligible free energy is indicated in Fig. 6.32d.

The excess enthalpy, HE, was defined as to be in turn equal to

$$\text{HE} = H - \text{Hideal} \tag{6.4}$$

The computation gave the following entropic and enthalpic contributions: $T\Delta S_{\text{solv}} = 2.07$, * SE = −5.32 kcal/mol, $\Delta H_{\text{solv}} = 2.02$ * HE = −26.18 kcal/mol, and $\Delta G_{\text{solv}} = -20.86$ kcal/mol, HE = −10.78 kcal/mol, that $T\Delta S_{\text{solv}} = 2.07$ * SE = −5.32 kcal/mol, $\Delta H_{\text{solv}} = 2.02$ * HE = −26.18 kcal/mol, and $\Delta G_{\text{solv}} = -$ 20.86 kcal/mol. The computed excess free energy (GE) is GE = HE − TSE = −10.78 + 3.76 = −7.02 kcal/mol at 10 million frames, whereas the corresponding experimental value is −6.33 kcal/mol.

Approaches for calculating the solvation free energies of small, neutral organic molecules from molecular simulations and by applying them calculations of aqueous solvation free energies (hydration free energies) were reviewed [1]. An update to FreeSolv database of experimental and calculated hydration free energies of neutral compounds provided input files in formats for several simulation packages with a single protocol. Using this approach, theoretical enthalpies and entropies of hydration were calculated and compared with some experimental enthalpies and entropies as well as electrostatic and non-polar components of solvation free energies.

A general statistical-mechanical Gaussian density of states model of a complex system was analyzed to explore whether and under what conditions extra-thermodynamic compensation can occur and what it reveals about the system [81].

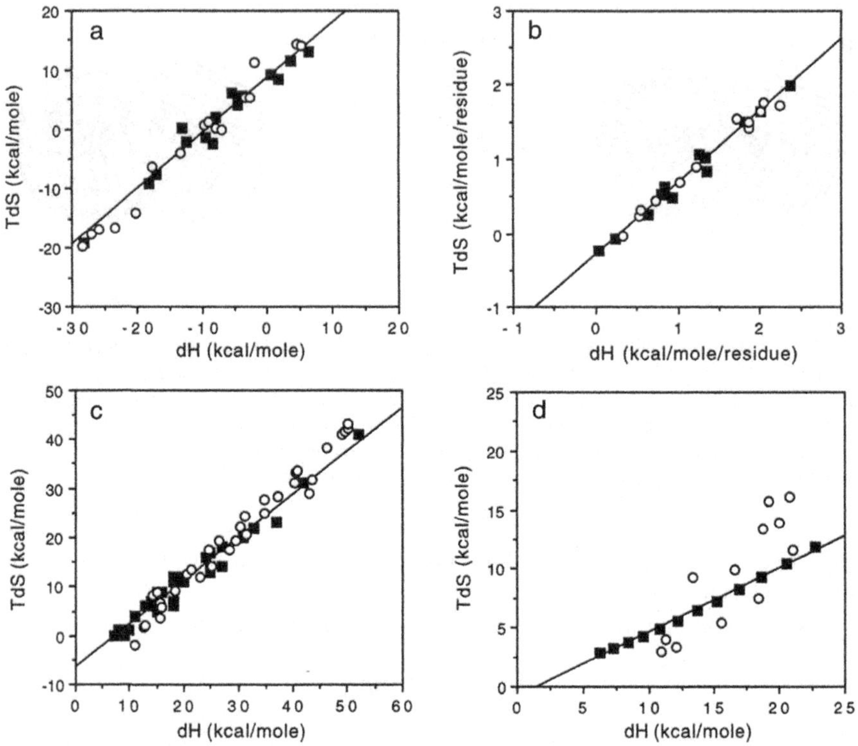

Fig. 6.33 Entropy (expressed as TS at 298 K)–enthalpy compensation plots for **a** calcium binding to proteins, **b** protein unfolding, **c** hydrogen exchange in cytochrome c, and **d** alkane solvation. Using randomly generated ΔH values with experimental ΔG's (white circle) level spreads (SD, σ) of 0.5 (x), 1 (white square), 2 (black diamond), and 3 (black square) kcal/mol, respectively [81]. With permission from Wiley and Son

This model shows that the most likely behavior to be seen is linear S–H compensation over a rather limited range of perturbations with a compensation temperature $T_c = d\Delta H/d\Delta S$ within 20% of the experimental temperature. Experimental data on entropy–enthalpy compensation plots with a high correlation coefficient for calcium binding to proteins, protein unfolding, hydrogen exchange in cytochrome c, and alkane solvation (Fig. 6.33) from point of view of the theory were discussed. Using suggested probability distribution, a plot of DS versus DE was generated for a systematic series of perturbations. The resulting plot was found to be ellipsoidal with the major axis aligned along the $T_c \rightarrow T$ direction (Figs. 6.33 and 6.34).

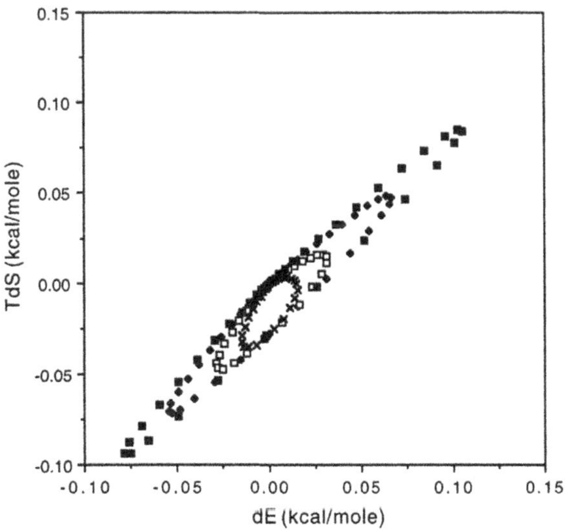

Fig. 6.34 Entropy (TS) versus energy (*E*) plot for the Gaussian density of states model at 298 K, with energy level spreads (SD, σ) of 0.5 (*x*), 1 (white square), 2 (black diamond), and 3 (black square) kcal/mol, respectively [81]. With permission from Wiley and Son

References

1. G.D.R. Matos, D.Y. Kyu, H.H. Loeffler, J.D. Chodera, M.R. Shirts, D.L. Mobley, Approaches for calculating solvation free energies and enthalpies demonstrated with an update of the FreeSolv database. Chem. Eng. Data **62**, 1559–1569 (2017)
2. C. Velez Vega, D.J.J. McKay, T. Kurtzman, V. Aravamuthan, R.A. Pearlstein, J. Duca, Estimation of solvation entropy and enthalpy via analysis of water oxygen–hydrogen correlations. J. Chem. Theory Comput. **11**, 5090–5102 (2015)
3. A.I. Dragan, C.M. Read, C. Crane-Robinson, Enthalpy–entropy compensation: the role of solvation. Eur. Biophys. J. **46**, 301–308 (2017)
4. V. Molinero, T. Cagin, I.I.I. Goddard, W.A. Sugar, Water and free volume networks in concentrated sucrose solutions. Chem. Phys. Lett. **377**, 469–474 (2003)
5. P. Bordat, A. Lerbret, J.P. Demaret, F. Affouard, M. Descamps, Comparative study of trehalose, sucrose and maltose in water solutions by molecular modelling. Europhys. Lett. **65**, 41 (2004)
6. A. Lerbret, P. Bordat, F. Affouard, M. Descamps, F. Migliardo, How homogeneous are the trehalose, maltose, and sucrose water solutions? An insight from molecular dynamics simulations. J. Phys. Chem. B **109**, 11046–11057 (2005)
7. Y. Choi, K.W. Cho, K. Jeong, S. Jung, Molecular dynamics simulations of trehalose as a "dynamic reducer" for solvent water molecules in the hydration shell. Carbohydr. Res. **341**, 1020–1028 (2006)
8. M. Riquelme, A. Lara, D.L. Mobley Toon Vestraelen, A.R. Matamala, E. Vöhringer-Martinez, Hydration free energies of organic molecules in the FreeSolv database calculated with polarized atom in molecules atomic charges and the GAFF force field. Preprint submitted on 13, 21 (2018)
9. L. Sandberg, R. Casemyr, O. Edholm, Calculated hydration free energies of small organic molecules using a nonlinear dielectric continuum model. J. Phys. Chem. B **106**(32), 7889–7897 (2002)
10. A. Nicholls, D.L. Mobley, J.P. Guthrie, J.D. Chodera, C.I. Bayly, M.D. Cooper, V.S. Pande, Predicting small-molecule solvation free energies: an informal blind test for computational chemistry. J. Med. Chem. **51**, 769–779 (2008)
11. J.P. Guthrie, A blind challenge for computational solvation free energies: introduction and overview. J. Phys. Chem. B **113**, 4501–4507 (2009). ([PubMed: 19338360])

12. P.V. Klimovich, D.L. Mobley, Predicting hydration free energies using all-atom molecular dynamics simulations and multiple starting conformations. J. Comput. Aided Mol. Des. **24**, 307–316 (2010)
13. K. Shiraga, Y. Ogawa, N. Kondo, A. Irisawa, M. Imamura, Evaluation of the hydration state of saccharides using terahertz time-domain attenuated total reflection spectroscopy. Food Chem. **140**, 315–320 (2013)
14. G. Konig, S. Boresch, Hydration free energies of amino acids: why side chain analog data are not enough. J. Phys. Chem. B **113**, 8967–8974 (2009)
15. K.M. Biswas, D.R. DeVido, J.G. Dorsey, Evaluation of methods for measuring amino acid hydrophobicities and interactions. J. Chromatogr. A **1000**, 637–655 (2003)
16. G.A. Khoury, N. Bhatia, C.A. Floudas, Hydration free energies calculated using the AMBER ff03 charge model for natural and unnatural amino acids and multiple water models. Comput. Chem. Eng. **71**, 745–752 (2014)
17. G. Mogami, T. Wazawa, N. Morimoto, T. Kodama, M. Suzuki, Hydration properties of adenosine phosphate series as studied by microwave dielectric spectroscopy. Biophys. Chem. **154**, 1–7 (2011)
18. P. Glancy, W.P. Beyermann, Dielectric properties of fully hydrated nucleotides in the terahertz frequency range. J. Chem. Phys. **132**, 245102 (2010)
19. D. Liu, T. Wyttenbach, M.T. Bowers, Hydration of mononucleotides. J. Am. Chem. Soc. **128**, 15155–15163 (2006)
20. M. Luckey, *Membrane Structural Biology: With Biochemical and Biophysical Foundations* (Cambridge University Press, Cambridge, 2014)
21. A. Del Regno, R. Notman, Permeation pathways through lateral domains in model membranes of skin lipids. Phys. Chem. Chem. Phys. **20**, 2162–2174 (2018)
22. Z.F. Su, J. Juhaniewicz-Debinska, S. Sek, J. Lipkowski, Water structure in the submembrane region of a floating lipid bilayer: the effect of an ion channel formation and the channel blocker. Langmuir **36**, 409−418 (2020)
23. H.A. Pérez, J.P. Cejas, A.S. Rosa, R.E. Giménez, E.A. Disalvo, M.A. Frías, Modulation of interfacial hydration by carbonyl groups in lipid membranes. Langmuir **36**, 2644−2653 (2020)
24. G. Fragneto, T. Charitat, J. Daillant, Floating lipid bilayers: models for physics and biology. Eur. Biophys. J. **41**, 863–874 (2012)
25. C. Lutgebaucks, C. Macias-Romero, S. Roke, The bi-modality diffusion of water molecules in liposome/water dispersion systems analyzed by pulsed field gradient spin echo NMR method. Trans. Mater. Res. Soc. Jpn **41**, 359–362 (2016)
26. F. Bordi, C. Cametti, S. Sennato, Electrical properties of aqueous liposome suspensions. Adv. Planar Lipid Bilayers Liposomes **4**, 281–320 (2006)
27. C. Lutgebaucks, C. Macias-Romero, S. Roke, Characterization of the interface of binary mixed DOPC:DOPS liposomes in water: the impact of charge condensation. J. Chem. Phys. **146**, 044701/1–044701/7 (2017)
28. C. Lipschitz-Farber, Y. Degani, Kinetics of water diffusion across phospholipid membranes. Proton and oxygen-17-NMR relaxation studies. Biochim. Biophys. Acta Biomembr. **600**, 291–300 (1980)
29. Y. Hosoi, R. Kita, N. Shinyashiki, S. Yagihara, Dielectric study of liposome dispersion using statistical analysis of jitter (2010). https://doi.org/10.14723/tmrsj.35.861
30. C. Giaginis, A. Tsantili-Kakoulidou, Alternative measures of lipophilicity: from octanol-water partitioning to immobilized artificial membrane retention. J. Pharm. Sci. **97**, 2984–3004 (2008)
31. T. Ingram, S. Storm, L. Kloss, T. Mehling, S. Jakobtorweihen, I. Smirnova, Prediction of micelle/water and liposome/water partition coefficients based on molecular dynamics simulations, COSMO-RS, and COSMOmic. Langmuir **29**, 3527–3537 (2013)
32. F. Esteves, C. Moutinho, C. Matos Correlation between octanol/water and liposome/water distribution coefficients and drug absorption of a set of pharmacologically active compounds. J. Liposome Res. **23**, 83–93 (2013)
33. K. Bittermann, S. Spycher, K.-U. Goss, Comparison of different models predicting the phospholipid-membrane water partition coefficients of charged compounds. Chemosphere **144**, 382–391 (2016)

34. S. Lin, X. Yang, H. Liu, Development of liposome/water 1 partition coefficients predictive models for neutral and ionogenic organic chemicals. Ecotoxicol Environ. Saf. **179**, 40–49 (2019)
35. S.A. van der Heijden, M.T.O. Jonker, Evaluation of liposome−water partitioning for predicting bioaccumulation potential of hydrophobic organic chemicals. Environ. Sci. Technol. **43**(23), 8854–8859 (2009)
36. S.H. Donaldson, A. Røyne, K. Kristiansen, M.V. Rapp, S. Das, M.A. Gebbie, D.W. Lee, P. Stock, M. Valtiner, J. Israelachvili, Developing a general interaction potential for hydrophobic and hydrophilic interactions. Langmuir **31**, 2051−2064 (2015)
37. L. Maugeria, S. Buschb, S.E. McLainc, L.C. Pardod, F. Brunia, M.A. Riccia, Structure-activity relationships in carbohydrates revealed by their hydration. Biochem. Biophys. Acta **1861**, 1486–1493 (2017)
38. T. Suzuki, The hydration of glucose: the local configurations in sugar–water hydrogen bonds. Phys. Chem. Chem. Phys. **10**, 96–105 (2008)
39. C. Olsson, J. Swenson, Structural comparison between sucrose and trehalose in aqueous solution. J. Phys. Chem. B **124**, 3074–3082 (2020)
40. P. Gallo, D. Corradini, M. Rovere, Ion hydration and structural properties of water in aqueous solutions at normal and supercooled conditions: a test of the structure making and breaking concept. Phys. Chem. Chem. Phys. **13**, 19814–19822 (2011)
41. M.P. Long, S. Alland, M.E. Martin, C.M. Isborn, Molecular dynamics simulations of alkaline earth metal ions binding to DNA reveal ion size and hydration effects. Phys. Chem. Chem. Phys. **22**, 5584–5596 (2020)
42. J. Mähler, I. Persson, A study of the hydration of the alkali metal ions in aqueous solution. Inorg. Chem. **51**, 425–438 (2012)
43. M.T. Record Jr., E. Guinn, L. Pegram, M. Capp, Introductory lecture: interpreting and predicting Hofmeister salt ion and solute effects on biopolymer and model processes using the solute partitioning model. Faraday Discus. **160**, 9–44 (2013)
44. A. Salis, B.W. Ninham, Models and mechanisms of Hofmeister effects in electrolyte solutions, and colloid and protein systems revisited. Chem. Soc. Rev. **43**, 7358–7377 (2014)
45. G. Schwaab, F. Sebastiani, M. Havenith, Ion hydration and ion pairing as probed by THz spectroscopy. Angewante Chem **58**, 3000–3013 (2019)
46. T.L. Beck, The influence of water interfacial potentials on ion hydration in bulk water and near interfaces. Chem. Phys. Lett. **561**, 1–13 (2013)
47. P. Li, L.E. Song, K.M. Merz, Systematic parameterization of monovalent ions employing the nonbonded model. J. Chem. Theory Comput. (2015). https://doi.org/10.1021/ct500918t
48. M.M. Reif, P.H. Hunenberger, Computation of methodology-independent single-ion solvation properties from molecular simulations. III. Correction terms for the solvation free energies, enthalpies, entropies, heat capacities, volumes, compressibilities, and expansivities of solvated ions. J. Chem. Phys. **134**, 144103 (2011)
49. S. Habershon, Zero-point energy effects in anion solvation shells. Phys. Chem. Chem. Phys. **16**, 9154–9160 (2014)
50. L. Vlcek, A.A. Chialvo, Clusters to bulk solutions: recent insights from molecular modeling. Fluid Phase Equilib. **407**, 58–75 (2016)
51. I.S. Joung, T.E. Cheatham, Determination of alkali and halide monovalent ion parameters for use in explicitly solvated biomolecular simulations. J. Phys. Chem. B **112**, 9020–9041 (2008)
52. A.A. Chialvo, L. Vlcek, Toward the understanding of hydration phenomena in aqueous electrolytes from the interplay of theory, molecular simulation, and experiment. Fluid Phase Equilib. https://doi.org/10.1016/j.fluid.2015.05.014
53. Y. Marcus, Effect of ions on the structure of water: structure making and breaking. Chem. Rev. **109**, 1346–1370 (2009)
54. D.D. Wagman, W.H. Evans, V.B. Parker, R.H. Schumm, L. Halow, S.M. Bailey, K.L. Churney, R.L. Nuttall, The NBS tables of chemical thermodynamic properties—selected values for inorganic and C-1 and C-2 organic substances in SI units. J. Phys. Chem. Ref. Data **11**, 1 (1982)

55. D. Horinek, S.I. Mamatkulov, R.R. Netz, Rational design of ion force fields based on thermodynamic solvation properties. J. Chem. Phys. **130**, 124507 (2009)
56. K.-J. Tielrooij, N. Garcia-Araez, M. Bonn, H.J. Bakker, Cooperativity in ion hydration. Science **328**(5981), 1006–1009 (2010)
57. L. Helm, A.E. Merbach, Water exchange on metal ions: experiments and simulations. Coord. Chem. Rev. **187**, 151–181 (1999)
58. Y. Marcus, A simple empirical model describing the thermodynamics of hydration of ions of widely varying charges, sizes, and shapes. Biophys. Chem. **51**, 111–112 (1994)
59. J.A. Mejías, S. Lago, Calculation of the absolute hydration enthalpy and free energy of H^+ and OH^-. J. Chem. Phys. **113**, 7306 (2000)
60. G.I. Likhtenshtein, Regularities in the energy and entropy properties of enzymatic processes. Biofizika **11**, 24–32 (1966)
61. G.I. Likhtenshtein, On the nature of compensation phenomena in liquid phase reactions. Russ. J. Phys. Chem. **44**, 1980–1990 (1970)
62. G.I. Likhtenshtein, *Chemical Physics of Redox Metalloenzymes* (Springer, Heidelberg, 1988)
63. G.I. Likhtenshtein, *New Trends in Enzyme Catalysis and Mimicking Chemical Reactions* (Kluwer Academic/Plenum Publishers, New York, 2003)
64. E.T. Denisov, O.M. Sarkisov, G.I. Likhtenshtein, *Chemical Kinetics. Fundamentals and Recent Developments* (Elsevier Science, Amsterdam, 2003)
65. R. Lumry, S. Rajender, Enthalpy-entropy compensation phenomenon in water solutions of proteins and small molecules: ubiquitous property of water. Biopolymers **9**, 1125–1227 (1970)
66. R. Lumry, R.B. Gregory, Free energy management in protein reactions: concepts, complications, and compensation, in *The Fluctuating Enzymes*, ed. by G.R. Welch (Wiley, New York, 1986)
67. R. Lumry, The new paradigm for protein research, in *Protein-Solvent Interactions* ed. by R.B. Gregory (Marcel Dekker Inc., New York, 1995)
68. A. Cornish-Bowden, Enthalpy–entropy compensation and the isokinetic temperature in enzyme catalysis. J. Biosci. **42**, 665–670 (2017)
69. M.B. Bushuev, D.P. Pishchur, E.B. Nikolaenkovac, V.P. Krivopalov, Compensation effects and relation between the activation energy of spin transition and the hysteresis loop width for an iron(II) complex. Phys. Chem. Chem. Phys. **18**, 16690–16699 (2016)
70. R. Schmid, A.M. Miah, V.N. Sapunov, New table of the thermodynamic quantities of ionic hydration: values and some applications (enthalpy-entropy compensation and born radii). Phys. Chem. Chem. Phys. **2**, 97–102 (2000)
71. M. Schauperl, M. Podewitz, T.S. Ortner, F. Waibl, A. Thoeny, T. Loerting, K.R. Liedl, Balance between hydration enthalpy and entropy is important for ice binding surfaces in antifreeze proteins. Sci. Rep. **7**, Article number: 11901 (2017)
72. A.I. Dragan, C.M. Read, C. Crane-Robinson, Enthalpy–entropy compensation: the role of solvation. Eur. Biophys. J. **46**, 301–308 (2017)
73. A.A. Edwards, J.M. Mason, K. Clinch, P.C. Tyler, G.B. Evans, V.L. Schramm, Altered enthalpy-entropy compensation in picomolar transition state analogues of human purine nucleoside phosphorylase. Biochemistry **48**, 5226–5238 (2009)
74. J. Zhang, C.P. Jones, A.R. Ferré-D'Amaré, Global analysis of riboswitches by small-angle X-ray scattering and calorimetry. Biochim. Biophys. Acta **1839**, 1020–1029 (2014)
75. L. Jen-Jacobson, E. Engler, L.A. Jen-Jacobson, Structural and thermodynamic strategies for site-specific DNA binding proteins. Structure **8**, 1015–1023 (2000)
76. B. Breiten, M.R. Lockett, W. Sherman, S. Fujita, M. Al-Sayah, H. Lange, C.M. Bowers, A. Heroux, G. Krilov, G.M. Whitesides, Water networks contribute to enthalpy/entropy compensation in protein–ligand binding. J. Am. Chem. Soc. **135**, 15579–15584 (2013)
77. K.K. Frederick, M.S. Marlow, K.G. Valentine, A.J. Wand, Conformational entropy in molecular recognition by proteins. Nature **448**, 325–329 (2007)
78. S.-R. Tzeng, C.G. Kalodimos, Protein activity regulation by conformational entropy. Nature **488**, 236–240 (2012)

79. L. Vlcek, A.A. Chialvo, Single-ion hydration thermodynamics from clusters to bulk solution. Fluid Phase Equilib. **407**, 58–75 (2016)
80. L. Vlcek, A.A. Chialvo, J.M. Simonson, Correspondence between cluster-ion and bulk solution thermodynamic properties: on the validity of the cluster-pair-based approximation. J. Phys. Chem. A **117**, 11328–11338 (2013)
81. K. Sharp, Entropy-enthalpy compensation: fact or artifact? Protein Sci. **10**, 661–667 (2001)

Chapter 7
Protein Hydration

Abstract Knowledge of the molecular structure of hydrated proteins is absolutely necessary for understanding their properties such as functional activity, stability, and interaction with other structures in biological systems. The following are typical examples of research in this area. Proteins in solution affect the structural and dynamic properties of the bulk water at the protein–water interface, contributing to the order of the hydration water. This chapter on the quantitative parameters of the mobility of hydrated proteins employing such modern physical methods revealed vibrational, rotational, and translational motions and chemical exchange covering a broad range in space and time, from individual water molecules to larger pools and from femtosecond to microsecond timescales. Systematic studies of the intramolecular mobility of a number hydrated proteins over a wide temperature range (20–300 K) by the combined physical labeling methods allow the motion to be measured at a correlation frequency of $\upsilon_c = 10^{-2}$–10^{10} s^{-1}. Four independent physical methods, that is, spin, Mossbauer, fluorescence labeling and ESR revealed appearance anharmonic motion at temperature about 220 K.

7.1 Introduction

The hydration layer surrounding a protein (Figs. 7.1 and 7.2) plays an essential role in its biochemical function and consists of a heterogeneous ensemble of water molecules with different local environments and different dynamics. Structural fluctuations and molecular excitations of hydrating water molecules cover a broad range in space and time, from individual water molecules to larger pools and from femtosecond to microsecond timescales [1–37]. The protein backbone motions can be energetically coupled with protein-bound water molecules. Water motion in the hydration layer is central to protein fluctuation, an essential determinant to its structural stability, dynamics, recognition, allosteric signals, and enzymatic activity. A whole arsenal experimental and computational techniques have been employed to study hydrated protein dynamics, hierarchy of different processes, and the coupling between protein and hydration water dynamics [4–40].

G. I. Likhtenshtein, *Biological Water*, Soft and Biological Matter,
https://doi.org/10.1007/978-3-030-82503-4_7

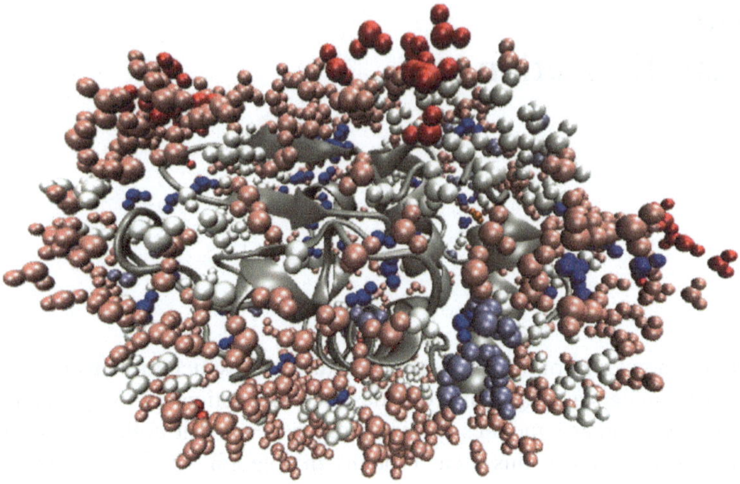

Fig. 7.1 Cartoon presentation of a protein hydration shells [35]. With permission from American Chemical Society

Fig. 7.2 Illustration to emphasize the need for structural changes in the hydration layer and hence the necessity of water molecules as "essential assistants" in the vicinity of a protein for its activity [30]. With permission from American Chemical Society

The importance of protein dynamics for their biological activity is now well recognized. Figure 7.2 illustrates schematically the need for structural changes in the hydration layer and hence the necessity of water molecules in the vicinity of a protein for its activity [30].

The present conception of the intramolecular dynamics of proteins is based on a hypothesis put forward in the 1950s and 1960s. First, Lumry and Eyring [6] assumed that substrate–enzyme interaction was accompanied by a certain structural distortion of both substrate and enzyme. Another basis for the modern concept of protein molecular dynamics was the Linderstrom-Lang hypothesis (1959) regarding the structural fluctuation of the protein macromolecule, which is visible in the ability of inner peptide groups to exchange hydrogen atoms for water protons [28]. According to the Koshland-induced fit theory [29], the conformational structure of the enzyme can be induced to fit the structure of the substrate during their interaction.

In the 1960s and 1970s, much indirect evidenced were obtained in favor of protein intramolecular mobility, by an analysis of data on the entropy and energy specificity of enzyme catalysis [1, 2, 7–9]. The first direct experimental evidence of the intramolecular mobility of a protein matrix in nanosecond temporal region was obtained using spin and Mössbauer labels and probes in the 1973s by Likhtenshtein group [11–15], fluorescence technique by Lakowicz group [19], and theoretically by Karplus et al. [24]. The main fundamental conclusions about the nanosecond dynamics of hydrated proteins were independently confirmed in the following decades experimentally [10, 17, 18, 20, 21, 38] and by excellent theoretical calculations [24–26].

In particular, biophysical labeling methods provide a unique possibility for monitoring local intermolecular dynamics properties in a wide range of correlation times ($\tau_c = 10^2$–10^{-10} s). They enable the investigation of the dynamics of different parts of protein globules in the vicinity of the spin, fluorescence, phosphorescence, and Mössbauer labels under various conditions (temperatures 30–330 K, Fig. 7.3), water content, viscosity, substrates and inhibitor additions, etc. Theoretical calculations using modern computer technology have also made a significant contribution to the field [22–26].

The comparative analysis of available data [12, 13] revealed an apparent discrepancy between the physical labeling approach and certain other physical methods.

The above-mentioned discrepancy was easily explained in the frame of a model [12–14] according to which nanosecond intermolecular mobility consists of movement of the relatively large and rigid parts of the protein macromolecules (domains). Such hinge-like oscillation of the tightly packed polypeptide blocks does not give a measurable contribution to the overall heat capacity and the helicity degree of polypeptide chains, but it strongly affects the mobility of the spin and Mössbauer labels firmly bound to the protein blocks, the mobility of spin and fluorescence labels and native chromophores located in cavities between the blocks, etc, At a later stage, the mechanism proposed was confirmed by numerous independent experimental investigations and theoretical calculations [4] and references therein.

In 1970s, physical labeling experiments showed that the anharmonic nanosecond dynamics of the proteins appear only if the water content of the sample being studied exceeds a critical value by about 10–25% [4, 11–16]. See for example Fig. 7.4. Later, the depending of critical degree of hydration depends on the protein's nature and temperature and on the nature of the solvent, in the presence of cryoprotectors,

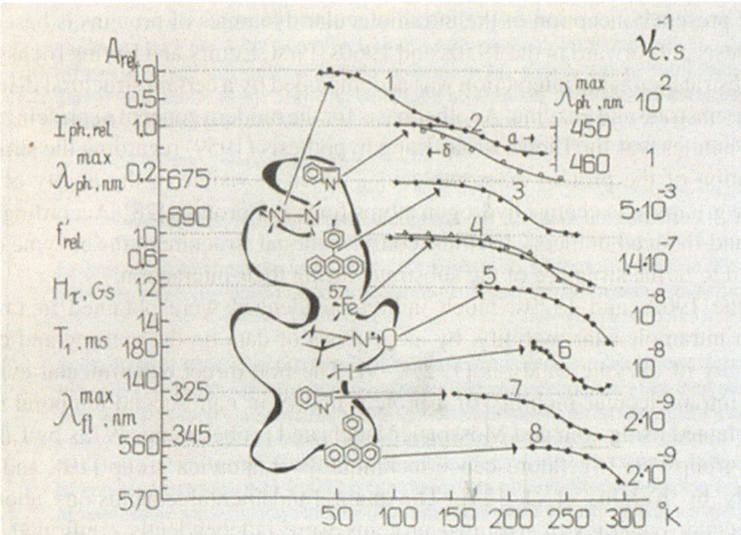

Fig. 7.3 Temperature dependences of dynamic parameters of physical labels tethered or incorporated in serum albumins. Left: Schematic diagram of the label location; right: characteristic time of appropriate method. 1 Radical pair recombination, 2 tryptophane phosphorescence, 3 eosin label phosphorescence, 4 Mossbauer label, 5 nitroxide spin label, 6 NMR, 7 tryptophan fluorescent, and 8 eosin label fluorescence [9]

in particular, was confirmed [10–18]. For example, a study of neutron scattering within the molecular dynamics of solid-state lysozyme indicated a significant effect of glycerol and water on dynamic transition temperatures (T_d) [20]. The transition was not detected in the dehydrated protein, but the transition temperature was found to be equal to 210 and 270 K in samples containing 30% D_2O 50% glycerol, respectively. At subzero temperatures, the intramolecular dynamics of enzyme in the picosecond timescale studied by neutron scattering and computer simulation methods [21] confirmed that normalized integrated elastic intensity dropped drastically at 220 K in rigid solution (D_2O and mixtures CD_3OD/D_2O, and $DMSO/D_2O$) but not in dry samples.

At the next stages, the researchers' interests were focused on measuring the quantitative parameters of the mobility of hydrated proteins employing such modern methods as neutron scattering spectroscopy small-angle X-ray and neutron scattering, quasi-elastic incoherent neutron scattering (QENS), elastic incoherent neutron scattering (EINS), Fourier transform infrared spectroscopy, time-resolved spectroscopic measurements, the time-dependent Stokes shift, dielectric spectroscopy light scattering techniques, X-ray and neutron crystallography, NMR, terahertz spectroscopy, mass spectroscopy, thermodynamics, physical labeling, and computer simulations.

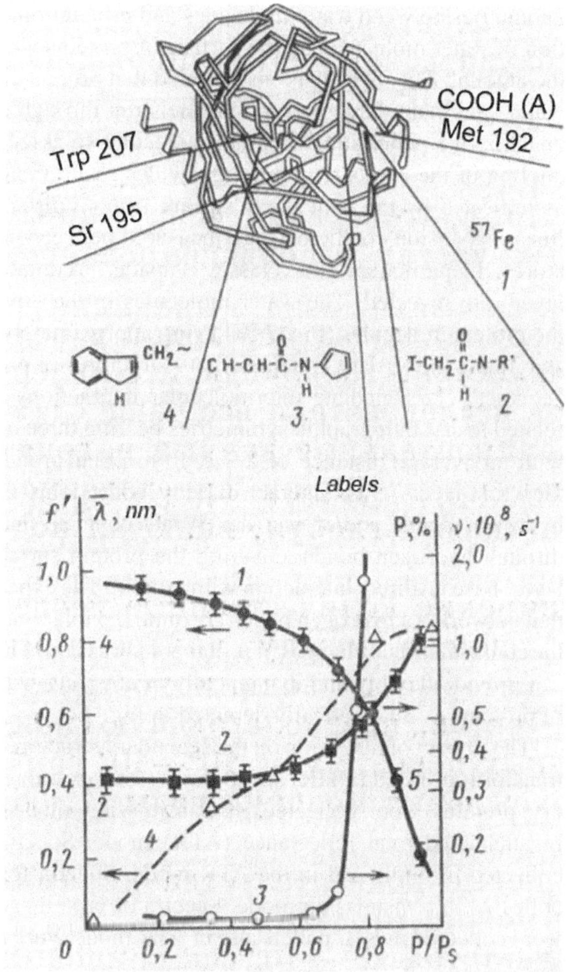

Fig. 7.4 Dependence of the dynamic parameters of physical labels attached to a-chymotrypsin and the depth of hydrolysis of cynamoyl a-chymotrypsin on relative humidity. 1 The amplitude of the mean displacement of the Mossbauer label ^{57}Fe; 2 the rotation frequency of the nitroxide spin label; 3 P(%) the depth of the cynamoyl a-chymotrypsin hydrolysis; 4 the relaxation shift of tryptophan fluorescence [14]

7.2 Hydrated Protein–Water Structure and Properties

Proteins in solution affect the structural and dynamic properties of the bulk water at the protein–water interface, resulting in a contribution to the order of the hydration water. Knowledge of the molecular structure of hydrated proteins is absolutely necessary for understanding their properties such as functional activity, stability, and interaction with other structures in biological systems. The following are typical examples of research in this area [36–52].

To understand the role of water at molecular and atomic levels, structures and interactions at the protein water interface have been investigated by cryogenic X-ray crystallography [38]. Using the structural models of proteins, including several hydration water molecules, the characteristics in hydration structures, the interaction geometries between water molecules and proteins, and the local and global distribution of water molecules on the surface of proteins were systematically analyzed for the amount. The investigations revealed that protein crystals contain more than 27% water, and many large solvent channels run through crystals, retaining fluidity and enabling the penetration of small molecules to associate with proteins. After flash cooling of the crystals, protein molecules retain their three-dimensional structures as indicated by the **root** mean **square** (r.m.s.) differences of less than 0.5 Å, and linear expansion coefficients are measured between cryogenic and ambient temperatures. In particular, four classes: "inside," "contact," "first layer," and "second layer" are revealed. The water molecules in the "inside" class occupy cavities in the protein molecules (Fig. 7.5) having approximately four hydrogen-bond partners stabilizing the tertiary and quaternary structures of proteins. Water molecules in the "contact" class mediate intermolecular interactions between adjoining molecules related to crystallographic symmetries bearing three or four hydrogen-bond partners with an average distance of 2.9 Å from polar protein atoms. Water molecules in the "first-layer" class interact directly with atoms of the protein surface through hydrogen bonds and/or van der Waals contacts and form aggregates connected through hydrogen bonds, covering the protein surface. Molecules in the "second layer' have no direct interaction with protein. The experimental results also suggested that networks of hydrogen bonds are quite flexible to accommodate and/or to regulate the conformational changes of proteins such as domain motions. Hydration structures can provide flexibly and dynamically in coupling with the conformational changes of proteins as schematically depicted in Fig. 7.6

The impact of hydration on the secondary structure of proteins employing Fourier transform infrared (FTIR) spectroscopy was examined [44]. Spectra of lyophilized dry proteins were collected in transmission mode by palletizing, mulling, and in attenuated total reflectance (ATR) mode. Spectra for hydrated proteins were collected in mulls and in the ATR mode. Spectra for reconstituted solutions were collected in transmission mode. Spectra of protein–sucrose colyophilized mixtures were collected in KBr pellets and in ATR mode. Pure proteins underwent significant

(*a*)

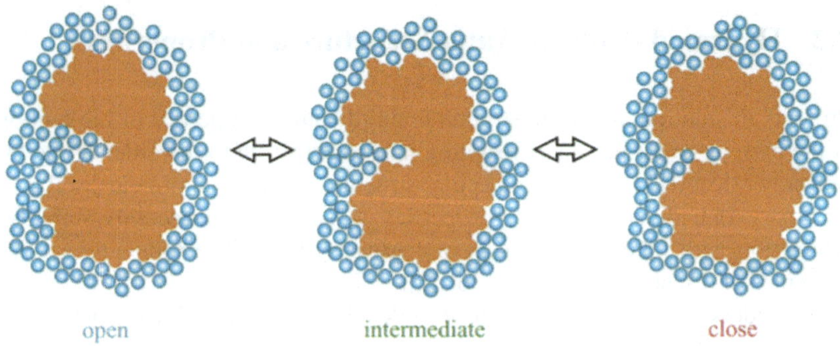

Fig. 7.5 (*Overleaf.*) **a** Schematic illustration of the hydration structural changes accompanying domain motions of a protein [38]. With permission from Royal Society of Chemistry

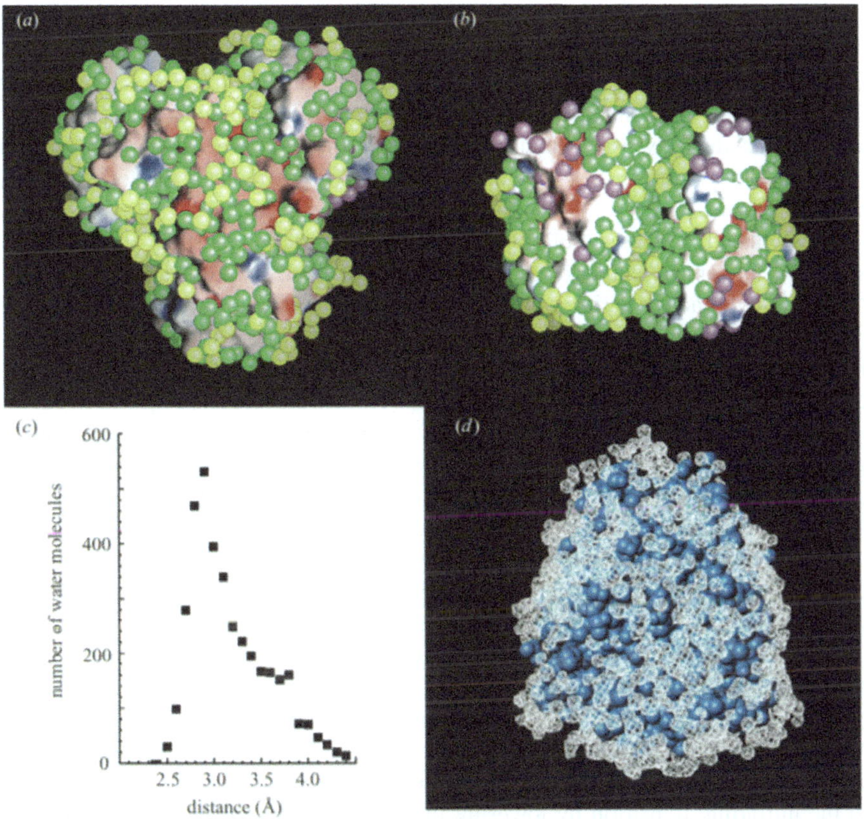

Fig. 7.6 Illustrations showing the distribution of hydration water molecules around **a** scytalone dehydratase at 37 K and **b** the killer toxin at 110 K The green, yellow, and purple spheres (diameter of 3 A°) are the hydration water molecules in the "first-layer," the "second-layer," and the "contact" classes, respectively. The proteins are illustrated by their molecular surface with the electrostatic potentials visualized by the program. **c** A plot showing the distance distribution of hydration water molecules from atoms on the protein surface. The data were prepared from 15 crystal structures refined using diffraction data collected below 110 K. **d** A distribution of mass densities of hydration water molecules (white fishnets) around trypsin shown by a blue-colored space-filling model. The density is reconstructed by superimposing the "first-layer" hydration water molecules identified in three crystal forms with different molecular packing of trypsin. The method for the density calculation was described previously. The data shown in the four panels were prepared using the FESTKOP program suite for hydration structure analyses developed by the author [38]. With permission from Royal Society of Chemistry

change in structure upon lyophilization, reforming upon reconstitution. ATR spectra differed from transmission spectra in peak intensity and position, suggesting a more native-like structure. The effect of hydration on the structure was found to be protein dependent, ranging from loss of native structure (IgG) to partial reformation of native structure (BSA). Importantly, the secondary structure of proteins in a hydrated state was not equivalent to that in solution.

Terahertz (far-infrared) absorption spectra of the 80 residue viral protein, a five helix bundle, in the 2.1–2.8 THz region at pH 7 were measured [37]. It was found that the wild type at pH 7 has a much more pronounced effect on long-distance solvation water than mutants replacing a single polar glutamine side chain with aromatic residues (tyrosine, histidine). Bringing the wild type and mutants closer to the unfolding transition by lowering the pH likewise reduces the long-distance solvation effect.

Molecular dynamics simulations and analytic modeling to study the hydration shell structure and dynamics of a typical globular protein, ubiquitin, and of the spruce budworm hyperactive antifreeze protein (Fig. 7.6) over the 230–300 K temperature range were combined [36]. The results were summarized as follows: (1) The dynamical heterogeneity arises mostly from the distribution of protein surface topographies and is little affected by temperature; (2) the ice-binding face of the antifreeze protein induces a short-ranged enhancement of water structure and a greater slowdown of water reorientation dynamics than the non-ice-binding faces; (3) the hydration shell of the ice-binding face remains less tetrahedral than the bulk and is not "ice-like"; (4) the hydrogen bonds between water and the ice-binding threonine residues are particularly strong due to a steric confinement effect, thereby contributing to the strong binding of the antifreeze protein on ice crystals; the mapping of the *Cf*AFP ice-binding surface sites is composed (Fig. 7.7).

Terahertz (far-infrared) absorption spectra of the 80 residue viral protein, a five helix bundle, in the 2.1–2.8 THz region at pH 7 were measured [37]. It was found that the wild type at pH 7 has a much more pronounced effect on long-distance solvation water than mutants replacing a single polar glutamine side chain with aromatic residues (tyrosine, histidine). Bringing the wild type and mutants closer to the unfolding transition by lowering the pH likewise reduces the long-distance solvation effect.

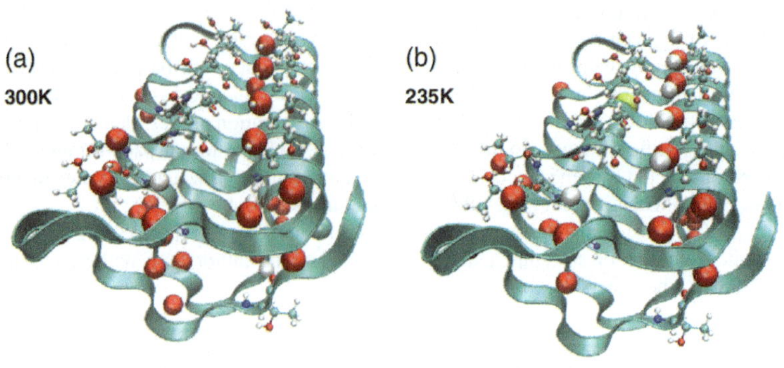

Fig. 7.7 Mapping of the *Cf*AFP ice-binding surface sites where the 5% slowest water molecules are located at 300 K [panel (**a**)] and at 235 K [panel (**b**)]. The color code is the same as shown in Fig. 7.4. The residues are represented in small balls and sticks [36]. With permission from American Institute of Physics

Some results of MD simulations that show the natural neighborhood of selected natural probes, W7 in myoglobin, W123 in lysozyme, and W3 in sweet protein monellin are displayed in Fig. 7.8. It was concluded that that the slow component approximately in the 20–80 ps range arises from slow water molecules present in the hydration layer, while the more than 100 ps component has multiple origins, namely adjacent charges in amino acid side chains, hydrogen–bonded water molecules, and a dynamically coupled motion between side chain and water. The results also suggested that the hydration layer water molecules are largely responsible for the slow component, whereas the initial ultrafast decay arises predominantly (approximately 80%) due to the bulk.

Binding affinities of three different peptides for an oncoprotein MDM2 were evaluated using the MD simulation which provided microscopic information on biomolecule–intramolecular and biomolecule–water hydrogen bonds, water structure near a biomolecule, and structural fluctuations of a biomolecule [40]. The solute hydration of three different peptides for an oncoprotein MDM2 was decomposed into processes 1 and 2 (Fig. 7.9). A cavity matching the geometric characteristics of the solute at the atomic level is created in process 1. Solute-water van der Waals and electrostatic interaction potentials are incorporated in process 2. The angle-dependent integral equation theory combined with our morphometric approach was applied to process 1, and the three-dimensional reference interaction site model theory was employed for process 2. Process 1 corresponds to the hydrophobic hydration, while solute-water van der Waals and electrostatic interaction potentials are incorporated in process 2.

Evaluation of changes in energetic (a) and entropic (b) components of thermodynamic quantities upon binding *of* three different peptides for an oncoprotein MDM2 is schematically depicted in Fig. 7.10.

The local hydration environment of self-assembled peptide nanostructures (SPNs) based on β-sheet-forming peptides was examined in detail to understand the driving

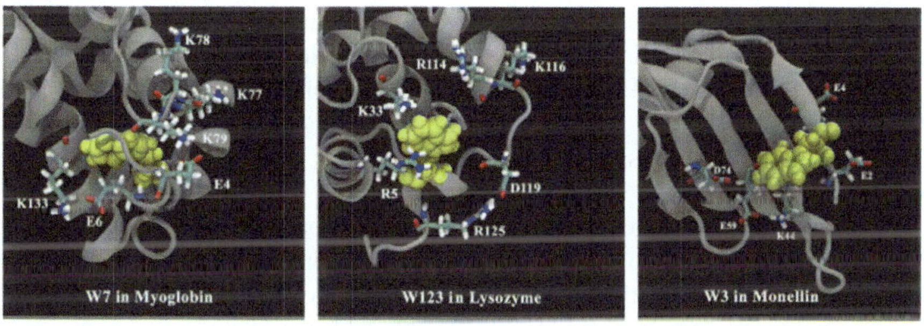

Fig. 7.8 Snapshots from MD simulations (initial frame at $t = 0$) that show the natural neighborhood of selected natural probes, W7 in myoglobin (left), W123 in lysozyme (middle), and W3 in sweet protein monellin (right). All the probes are embedded in a polar environment provided by the nearby amino acid side chain residues in their wild form. Tryptophan residues (used as a natural probe) are shown in yellow. The figures are prepared using VMD [39]. With permission from American Institute of Physics

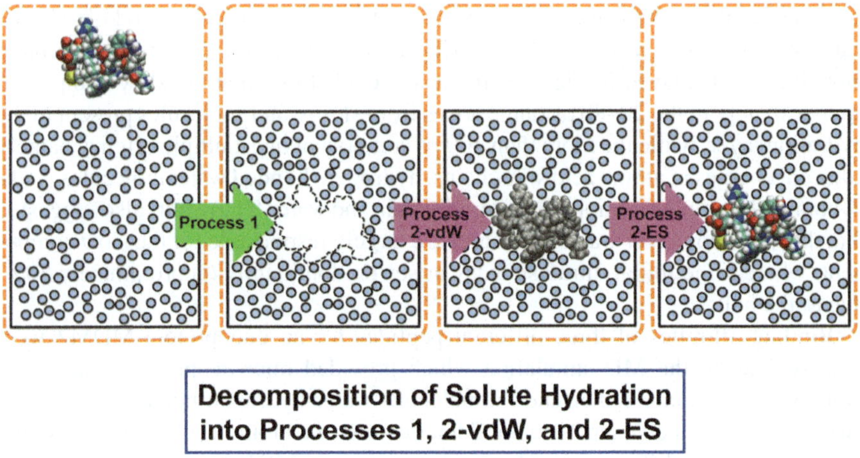

Fig. 7.9 Decomposition of solute hydration into two processes. In process 1, a cavity whose size and polyatomic structure correspond to those of the solute is created in water. In processes 2-vdW and 2-ES, respectively, solute-water van Der Waals (vdW) and electrostatic (ES) interaction potentials are incorporated. Before process 2-ES, the partial charges of the solute toms are all set at zero. After process 2-ES, the full partial charges are assigned to the solute atoms [40]. With permission from Elsevier

force of the discrete geometric structural self-assembling phenomena for the peptide structures [50]. For this purpose, W-Band CW-EPR spectroscopy, *Q*-band Mims electron nuclear double resonance (ENDOR) spectroscopy, and *Q*-band ESEEM spectroscopy are employed. Hydrogen-bond formation and geometry, and hydrophobicity of the local environments at various spin-labeled sites in SPNs were in focus of the investigation. The block copolypeptides used in this study have two functional domains: a self-assembling domain and a hydrophilic/bioactive domain (Fig. 7.11a). To make the peptides EPR active, site-directed spin labeling with a 2,2,6,6-tetramethylpiperidine-1-oxyl (TEMPO) maleimide (TEMPO-MI) at the C-terminus (peptides 1, 2, and 4) and N-terminus (peptides 3 and 5), located in the exterior and interior of SPNs, respectively, was performed. Figure 7.11b exhibits the W-band (94 GHz) CW-EPR spectrum of peptide 3 (KF 150 mM) taken at 70 K. Via EPR simulations, the spin Hamiltonian parameters of g- and A-tensor principal values were obtained.

To investigate the hydrogen-bonding geometry, *Q*-band (34 GHz) pulse *Q*-band Mims electron nuclear double resonance ENDOR) on samples prepared in D_2O was carried out. This method elucidates the hyperfine interaction of an unpaired electron with surrounding nuclei. Using the obtained spin Hamiltonian parameters, the hydrogen–bonding geometry between the nitroxide radical and the hydrogen-bonded deuterium was drown. The result indicated that three atoms (D···O•–N) lie on the same plane.

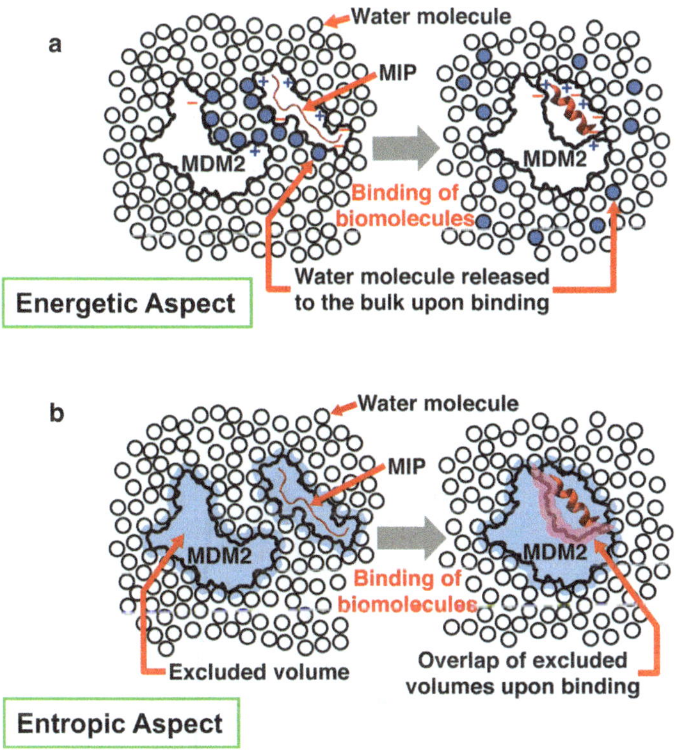

Fig. 7.10 Evaluation of binding affinities of three different peptides for an oncoprotein MDM2 [40]

The observation modulation depth in electron spin-echo envelop modulation (ESEEM) spectra represented the amount of D_2O molecules adjoining the nitroxide radical, and the hydration environment near the nitroxide radical is attached to self-assembled peptide nanostructures (SPNs). Figure 7.12 displays a schematic representation of the proposed SPN structure and its local hydration based on obtained EPR results.

A model expressing the dipolar response of solutions in terms of a single parameter related to the interface dipole moment induced in the interfacial water by electromagnetic radiation was proposed [51]. The model was applied to experimental THz absorption of hydrated sugars, amino acids, and the lambda repressor protein, a transcription inhibitor of bacteriophage Lambda (Fig. 7.13). To extract the interfacial dipole moment and corresponding frequency-dependent interfacial susceptibilities and to calculate the absorption coefficient, molecular dynamics (MD) simulations were used. Hydrophobic amino acids and sugars gave rise to near zero interface dipoles, while strongly hydrophilic ones are best described by a positive projection of the interface dipole on the external field. The calculations elaborated that the induced interface dipole is altered significantly depending on the solute identity and

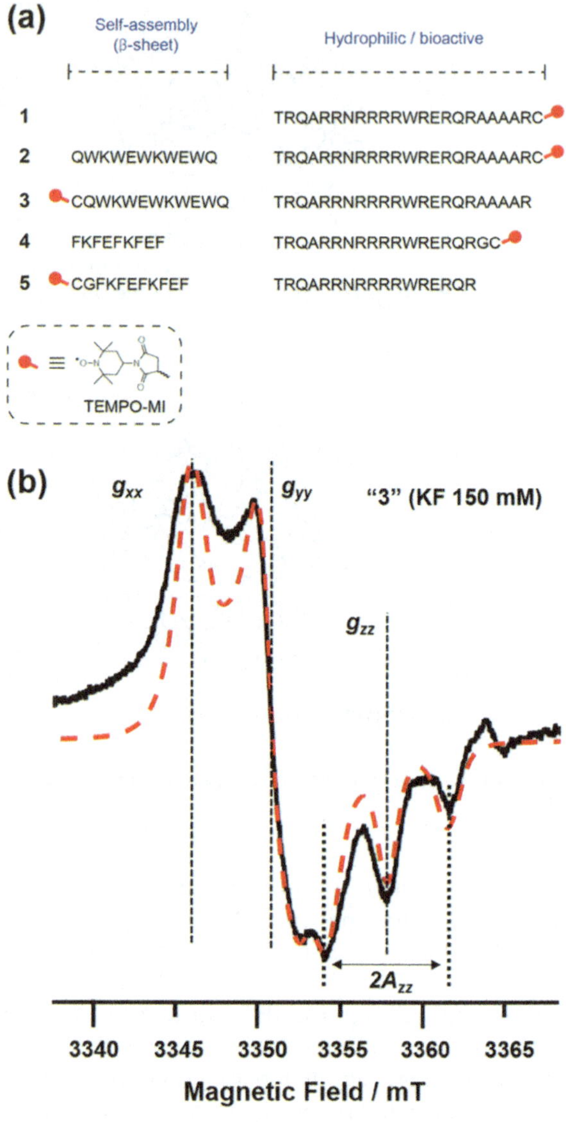

Fig. 7.11 **a** Self-assembling peptides labeled with TEMPO-MI radical groups. **b** W-band CW-EPR (black) and its simulation (red) of peptide 3 (KF 150 mM). Simulation parameters; $g =$ [2.00720, 2.00410, 2.00020], $A =$ [12.6, 21.0, 102.2] MHz [50]. With permission from American Chemical Society

is nearly zero for weakly hydrophilic sugars and increases for polar, and zwitterionic amino acids and for proteins.

The purpose of the study [41] was to systematically investigate the solvation properties of lysozyme, myoglobin, and bovine serum albumin (BSA). In water solution using a vector network, analyzer-driven quasi-optical bench covers the WR-3 waveguide band (0.220–0.325 THz). The concentration-dependent absorption of protein molecules ranging from 2 to 100 mg/ml was determined by treating

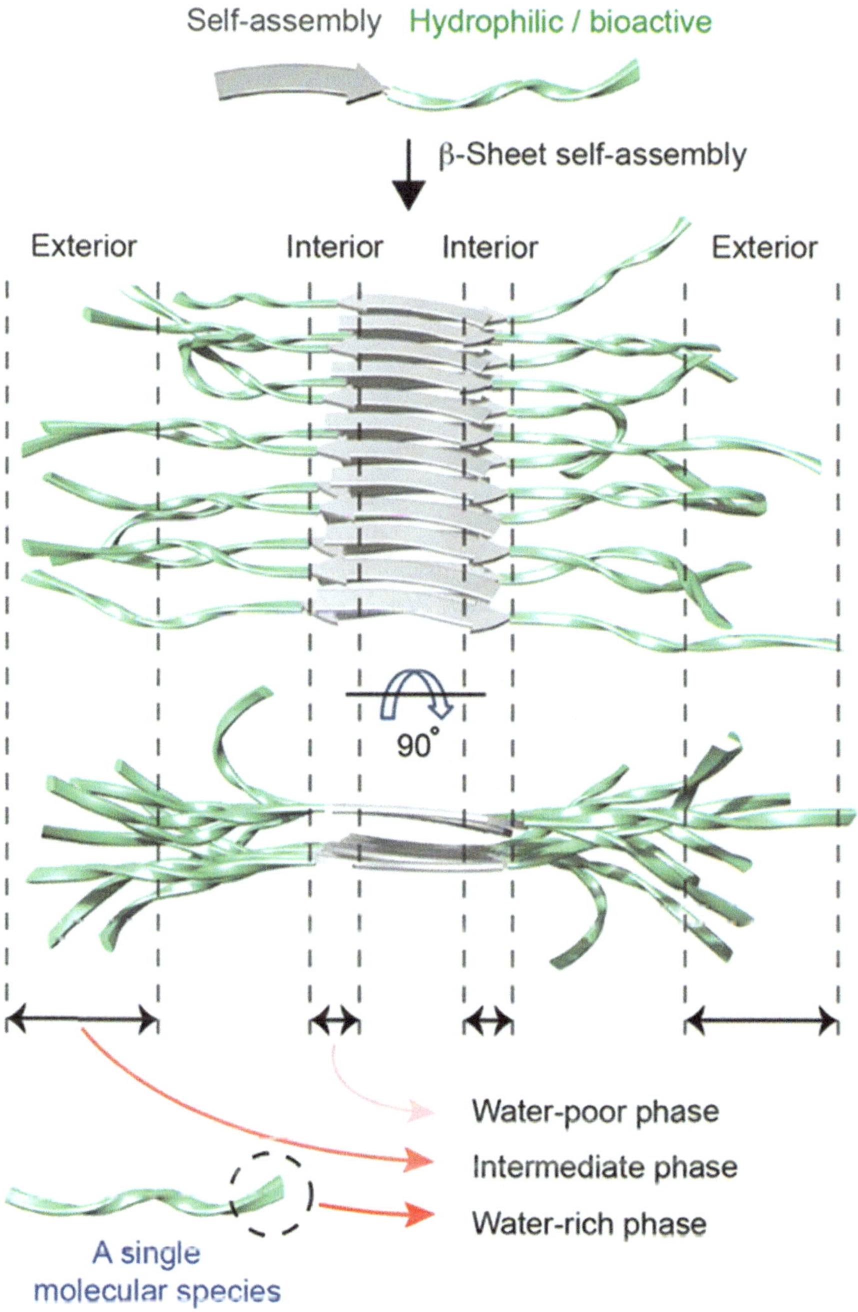

Fig. 7.12 Suggested SPN structure and its local hydration states [50]. With permission from American Chemical Society

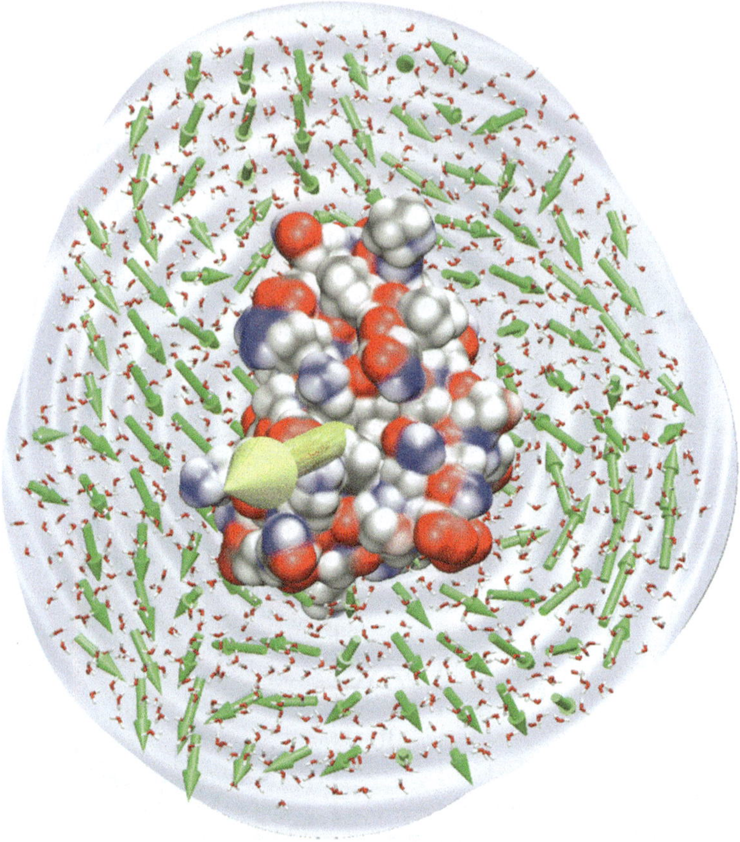

Fig. 7.13 Cartoon of fluctuating water dipolar polarization, coarse grained to the size of the green arrows, in the hydration shell of the protein. The hydration dipoles make up a fast subsystem absorbing THz radiation, while protein dipole (indicated by a large arrow), even though modulated by fast vibrations, makes a vanishingly small contribution to the absorption coefficient. Since the structure of the hydration shell is mostly dictated by the local hydrogen-bonding and electrostatic interactions of hydration waters with the surface residues, the cross-correlation between the protein and hydration shell dipoles builds up in a region 2–4 nm thick, as indicated by the area rendered blue in the cartoon [51]. With permission from American Institute of Physics

the solution as a two-component model first; then, based on protein absorptivity, the extent of the hydration shell is estimated. Absorption curves of all three proteins sharply peak toward a dilution limit that was attributed to the enhanced flexibility of protein and amino acid side chains. The derived estimate of hydration shell thickness for each protein indicated that protein–water interaction dynamics extends beyond 2–3 water solvation layers as was predicted by molecular dynamics simulations and other techniques such as NMR, X-ray scattering, and neutron scattering. According to the estimations, the radius of the dynamic hydration shell is 16, 19, and 25 Å, respectively, for lysozyme, myoglobin, and BSA proteins and correlates with the dipole moment of the protein. It was suggested that the existence of a broader hydration shell will have an important influence on molecular recognition events such as substrate binding and the formation of macromolecular complexes as the effective capture radius may be extended.

Details of the interaction between human lysozyme proteins, both native and fibrils, and their water environment by intense terahertz time–domain spectroscopy were revealed in study [42]. The amplitude and the phase of the THz field transmitted through 0.5-mm-thick lysozyme—water solutions at room temperature were measured. From the time-domain THz experiments, the full dielectric properties of the solutions, both the index of refraction and the absorption coefficient, were calculated. A suggested rigorous dielectric model is allowed to determine the amplitude and phase of the oscillating dipole induced by the THz field in the volume containing native human lysozyme or fibrils and its hydration water. Separating the amplitude and phase of this response made it possible to reveal the amplitude of the induced dipole. The observed changes in the THz response indicate protein–protein interactions mediated by extended hydration layers.

The impact of hydration on the secondary structure of proteins immunoglobulin G (IgG) and bovin serum albumin (BSA) using FTIR spectroscopy was investigated [44]. Spectra of lyophilized dry, hydrated, reconstituted proteins, and protein–sucrose colyophilized mixtures were collected in different modes. Pure proteins underwent significant change in structure upon lyophilization, reforming upon reconstitution. The effect of hydration on the structure was protein-dependent, ranging from loss of native structure (IgG) to partial reformation of BSA. It was concluded that spectra collected in different modes are not directly comparable, and the secondary structure of proteins in a hydrated state was not equivalent to that in solution.

The coupling of water with the hydrophilic moieties of hydrated lysozyme ($h =$ 0.3) was studied by means of Fourier transform infrared and NMR with the high-resolution magic angle spinning probe spectra in the temperature interval 180–350 K [45]. The intensity of the amide vibrational modes of hydrated lysozyme reflects thermal evolution of the amide vibrational modes of hydrated lysozyme. Above 225 K, the flexibility of hydrated lysozyme sharply increases due to the softening of the hydrogen-bond network that the hydration water develops on its surface. Above 320 K, water behaves as a normal liquid since the HB lifetime becomes less than picoseconds. The peaks belonging to the protons of lysozyme begin to appear at $\approx$ 320 K. Thus, the presented experimental data suggest that the main mechanisms that define the thermal limits of the enzymatic activity of proteins are given by the hydrogen–bond coupling between hydration water and protein hydrophilic groups.

Measurements on the ultrafast vibrational dynamics of amide I vibrations of proteins at the lipid membrane/H_2O interface using femtosecond time-resolved sum frequency generation vibrational spectroscopy were performed [47]. The experiments revealed that the relaxation time of the amide I mode shows a very strong dependence on the H_2O exposure, indicating the exposure of amide I bond to H_2O, opening up a resonant relaxation channel and facilitating direct resonant vibrational energy transfer from the amide I mode to the H_2O bending mode. In particular, measurements on the ultrafast vibrational dynamics of amide I vibrations of proteins at the lipid membrane/H_2O interface using several peptides, including melittin, LKα14, mastoparan (MP), influenza M2 proteins (AM2 and BM2), and KALP23 as the models were carry out. The relative amount of peptide bonds that are exposed to H_2O obtained by measuring the ratio of the hydrogen–deuterium exchange (HDX) of the amide proton was determined. Figure 7.14 indicates the amide A spectra of peptides before and after 5-h HD, the amide I frequency of melittin (1651 cm^{-1}), and the amide I frequency of α-helices shifts from 1665 to 1648 cm^{-1} upon hydration.

Studies of hydrated films of an antiparallel β-sheet peptide (LK$_7\beta$) adsorbed on glass slides using chiral (SFG) in conjunction with isotopic labeling, isotopic dilution, and kinetic measurements of H_2O–D_2O exchange were reported [53]. Experiments of isotopic labeling, isotopic dilution of water, and H_2O–D_2O exchange kinetics corroborate the assignments of the chiral SFG response to water stretching modes. Because individual water molecules are achiral, the chiral SFG response indicates the formation of chiral superstructures of water around the antiparallel β-sheet. Therefore, the LK$_7\beta$ secondary structure can imprint its chirality onto the surrounding water, that is, protein secondary structures can orient the surrounding water.

Several techniques based on the combination of pulsed electron paramagnetic resonance (EPR) with site-directed spin labeling were used to quantify water accessibility of plant light-harvesting complex Iib (LHCIIb) [52]. Several mutant versions of the Lhcb1*2 (AB80) gene from pea (*Pisum sativum*) that contain a single cysteine were constructed by replacing serine or valine for cysteine. The mutation positions, as well as a schematic picture of the LHCIIb protein, are shown in Fig. 7.15. In all mutants, the native single cysteine at position 79 was replaced by serine. The modified protein was spin labeled on cysteine by adding 3-(2-iodoacetamido)-2,2,5,5-tetramethyl-1-pyrrolidinyloxyl (IAA-PROXYL). Accessibility parameters for different residues in LHCIIb were determined by electron spin-echo envelope modulation spectroscopy in the presence of deuterated water. Deuterium contrast in transversal relaxation rates, analysis of longitudinal relaxation rates, and line shape analysis of electron spin-echo-detected EPR spectra as well as by the conventional techniques of measuring the maximum hyperfine splitting and progressive saturation in continuous-wave EPR were carried out. A semiquantitative estimate of the polarity of the spin label environment for the different residues in LHCIIb was obtained by comparison with A_{zz} of IAA-PROXYL in solvents with different polarity. Analysis data on the maximum extrema splitting $2A_{zz}$ unveiled that mutants S3r, S52r, and S59r have environments less polar than this mixture but more polar than methanol ($\varepsilon = 32.6$ at 25 °C), V90r and S123r have an environment with a polarity comparable to that of methanol, and mutants S160r and V196r have an environment with a polarity similar to that of isopropanol ($\varepsilon = 18$ at 25 °C).

The average density of exchangeable protons at distances in the range of $\sim$ 6–20 Å from the label was characterized by experimental water accessibility parameter, $\Pi(D_2O)$.

The dependence of the intensity of the central nitroxide peak on microwave power for the different residues in LHCIIb was studied in the absence and presence of relaxation agents 20% O_2 and $N_2/(Cr(ox)_3)^{3-}$ within spin labelspin probe approach [23]. Differences in water accessibility between different mutants were clearly seen

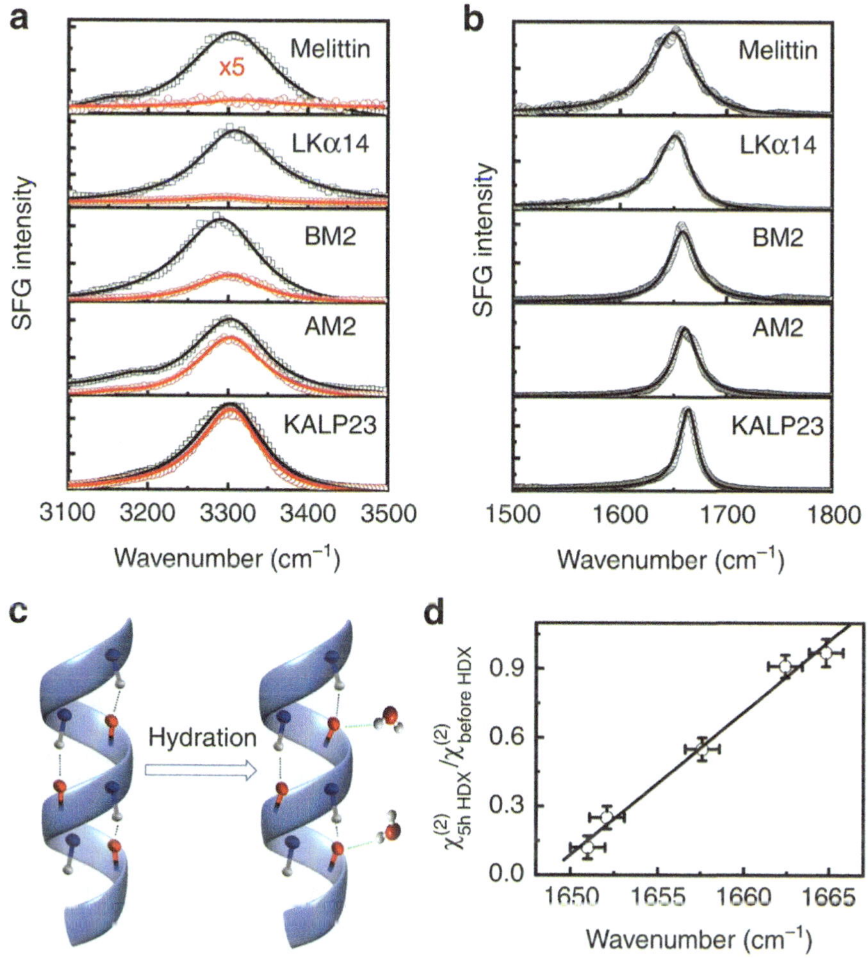

Fig. 7.14 Sum-frequency generation (SFG) spectra. The ssp (denoting s-, s-, and p-polarized sum-frequency output, visible input, and infrared input, respectively) SFG spectra of the peptides of amide A and amide I modes, sketch for the hydration of amide I mode, and the relationship between hydrogen–deuterium exchange (HDX) ratio and amide I frequency. **a** The ssp SFG spectra of the peptides in the N–H region. The spectra in black curves are measured at lipid bilayer/H_2O interface while the ones in red curves are measured at lipid bilayer/D_2O interface after 5-h HDX experiments. **b** The ssp amide I SFG spectra of peptides at lipid/H_2O interface. The solid curves in **a**, **b** are the fitting curves using supplementary (7.1). **c** Sketch for the carbonyl groups forming a bifurcated H bonding to water molecules upon hydration. **d** Fitting amplitude ratio (χ5hHDX(2)/χbeforeHDX(2)) of the ssp SFG spectra in the N–H region after 5-h HDX and before HDX against the amide I frequency. The vertical error bars represent experimental error in determining the amplitude ratio from the fitting procedure (see supplementary Note 2). The horizontal error bars indicate the standard error of the mean of the fitting center frequency from five independent spectra [47]

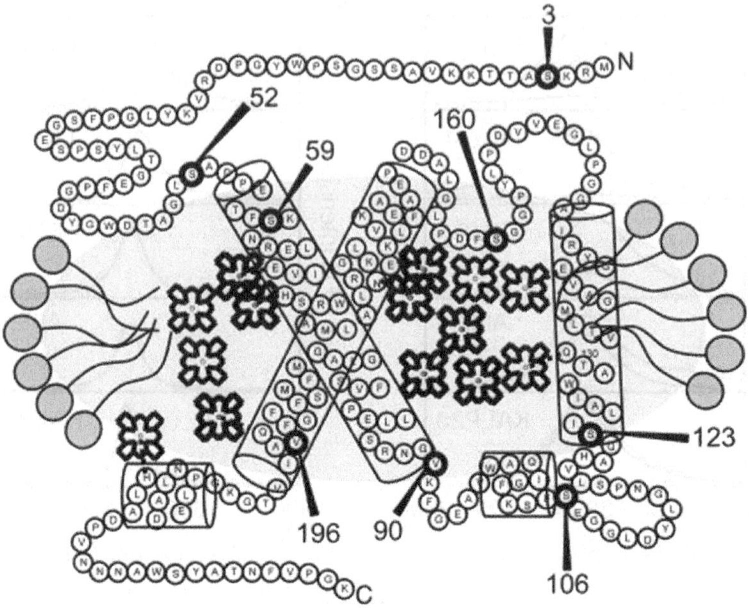

Fig. 7.15 Schematic structure of LHCIIb. Each circle with a letter represents an amino acid residue. Circles labeled with numbers show mutation positions where spin labels were attached in this work. Cylinders represent the helices, and chlorophylls are symbolized by schematic porphyrin formulas. Other pigments and lipids are not shown. The gray oval around the protein represents the micelle, as expected from analyzing the crystal structure. The top corresponds to the stromal side and the bottom to the lumenal side of the thylakoid membrane [52]. With permission from Elsevier.

in primary three-pulse ESEEM data, and main data analysis method was based on modulation depth. The various water accessibilities determined from ESEEM data are presented in Fig. 7.16. As can be seen, $\Pi(D_2O)$, the deuterium peak intensity, I, and the average number, n, of deuterium nuclei correlate very well with each other.

Thus, simultaneous determination of the polarity-dependent ^{14}N hyperfine splitting A_{zz}, the dynamics-sensitive longitudinal relaxation time and intensity ratio h_1/h_2 in ESE-detected EPR, and the ESEEM water accessibility parameter $\Pi(D_2O)$ that quantifies the average number of water molecules in van der Waals contact with the label provided complementary information on water accessibility of a spin label in a membrane protein The authors concluded that the techniques presented in this work can be applicable to protein that can be spin labeled, regardless of its size or degree of structural order.

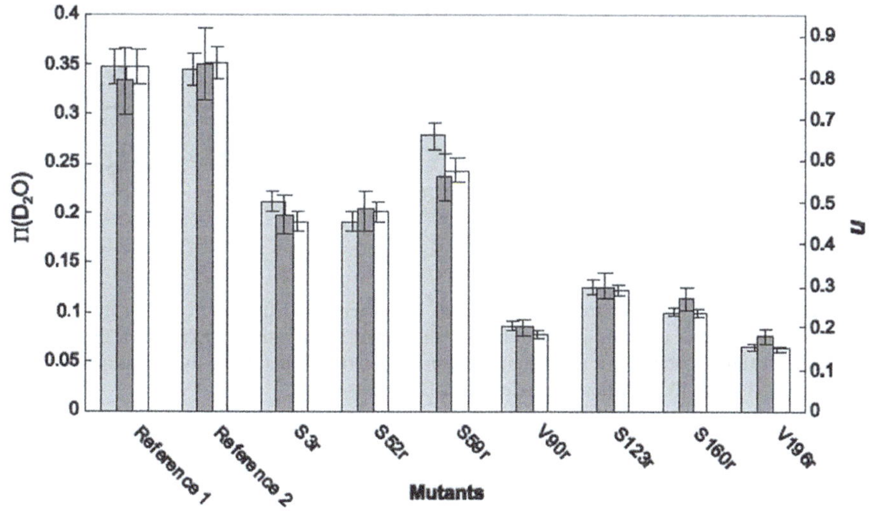

Fig. 7.16 Water accessibility parameters obtained by different analysis procedures from deuterium three-pulse ESEEM data influence the 1/*e* time τ_1 for longitudinal relaxation and transversal relaxation time, τ_2, in the solid state, and mechanisms are not quantitatively understood for protein samples in glassy frozen solutions, interpretation of relaxation data is not as straightforward as interpretation of CW EPR and ESEEM data, the effects that govern electron spin relaxation of nitroxide labels in LHCIIb were considered on the basis of empirical relaxation parameters. The difference in relaxation rates between protonated and deuterated samples is sensitive to proton concentration on a different length scale than ESEEM modulation depth. Due to its r^{-6} dependence, the modulation depth is strongly biased toward distances of closest approach, i.e., to water molecules in van der Waals contact with the label. For example, concentrations of exchangeable protons in this distance range are computed range between 10 nm^{-3} for mutant S123r and 30 nm^{-3} for mutant S3r. These values are compared to a proton concentration in pure water of 67 nm^{-3}. The relaxation anisotropy was quantified in terms of an anisotropy ratio, h_1/h_2, of spectral intensities at two positions in the spectrum that roughly correspond to the orientations in the nitroxide molecular frame along the orbital lobes of the nitrogen p_z orbital and perpendicular to it [52]. With permission from Elsevier

7.3 Hydration Water Dynamics

The structure and function of biomolecules are strongly influenced by their hydration shells. Structural fluctuations and molecular excitations of hydrating water molecules cover a broad range in space and time, from individual water molecules to larger pools and from femtosecond to microsecond timescales [53–81].

Figure 7.17 illustrates timescale of the structural dynamics in proteins and hydration water. The intrinsic timescales of elementary molecular processes here are ultrafast. They cover a time range from sub 100 fs for the fastest structural fluctuations in the water shell up to tens of picoseconds for the escape of some water molecules from the shell into the bulk and for energy transport within biomolecules and from them into the hydration shell.

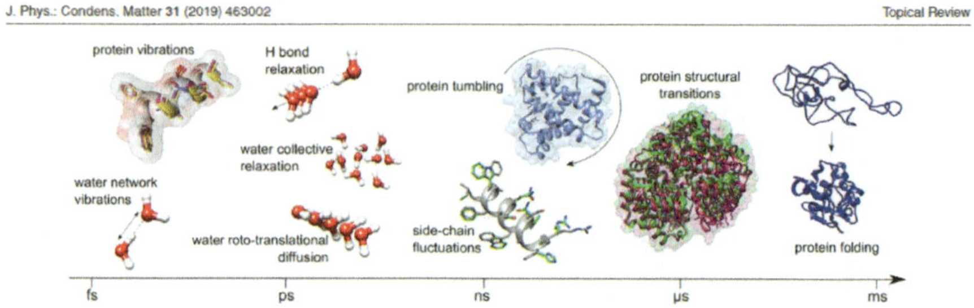

Fig. 7.17 Timescale of the structural dynamics in proteins and hydration water, from protein and water vibrational motions in the femtosecond range to diffusive relaxation in the pico- to nanosecond range, up to large-scale protein structural transitions and folding dynamics in the micro- to millisecond range [31]. Published under a CC-BY license

7.3.1 Vibration

The role of vibrational processes and especially of resonant vibrational energy transfer through dipole coupling in protein and hydrogen bonds between H_2O and the biomolecule on the entire energy transfer pathways is extremely important for understanding detailed molecular mechanism of the functional activity of proteins.

To monitor functionally relevant, global, and subglobal collective modes in the terahertz (THz) frequency range, the absorption spectrum of solvated bovine serum albumin (BSA) from 0.3 to 3.72 THz (10–124 cm^{-1}) was experimentally determined [43]. The terahertz absorption spectrum of both our buffer blank and buffered BSA solutions was measured, and the amount of buffered water displaced by the BSA solute was calculated. The hydration effect of the solvated protein via two assumptions with (s) and without correction to account for hydration layer surrounding each BSA molecule in solution was examined and hydration layer of 300 water molecules was estimated (Fig. 7.18).

Recent progress in theory and molecular dynamics simulations as well as in ultrafast vibrational spectroscopy has led to detailed insight into processes of vibrational relaxation of water signal between 1674 and 1676 cm^{-1} with the protein signal between 1649 and 1655 cm^{-1} and energy dissipation [47]. As an example, Fig. 7.19 depicted coupling scheme and energy diagram of interaction between amide carbonyl and energy level of the vibrational relaxation of the amide I mode. The protein backbone motions can thus be energetically coupled with protein-bound water molecules.

To understand the low-frequency intermolecular vibrational spectrum of water present in the hydration layer of the protein villin headpiece subdomain or HP-36, molecular dynamics simulation has been carried out [59]. The calculations revealed that compared to bulk water, these bands are non-uniformly blue shifted for water near the helices, while the extent of shifts being more pronounced for water molecules hydrogen bonded to the protein residues.

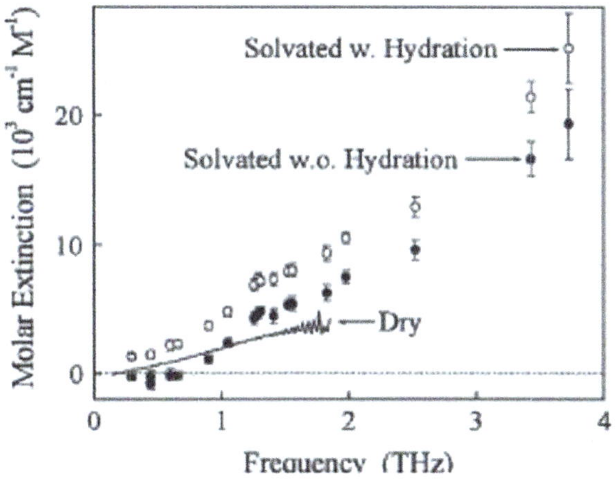

Fig. 7.18 Hydration effect of the solvated protein via two assumptions with (s) and without (d) correction to account for hydration layer surrounding each BSA molecule in solution. Uncertainty in the actual hydration suggests that we should use the region between these two extremes, with (s) and without (d) hydration shell correction, as the absorption spectrum of solvated BSA. The monotonic frequency-dependent increase previously reported below 2 THz in dry samples (solid line) continues to at least 3.72 THz in solution phase (this study). The apparent sharp resonances in dry BSA measurements are artifacts arising from the large noise-to-signal ration at the higher frequencies, whereas below 1 THz, the spectrum of solvated BSA is within error of that reported for the dry protein pellets (solid line), significant deviations are observed at higher frequencies where solvation may have served to enhance the terahertz dynamics of BSA. It is perhaps worth noting that the correspondence observed below 1 THz is quantitative; that is, the absolute absorbance of dry and solvated BSA are within error of one another, and no scaling was required in order to produce these effectively overlapping spectra [43]. With permission from John Wiley and Sons

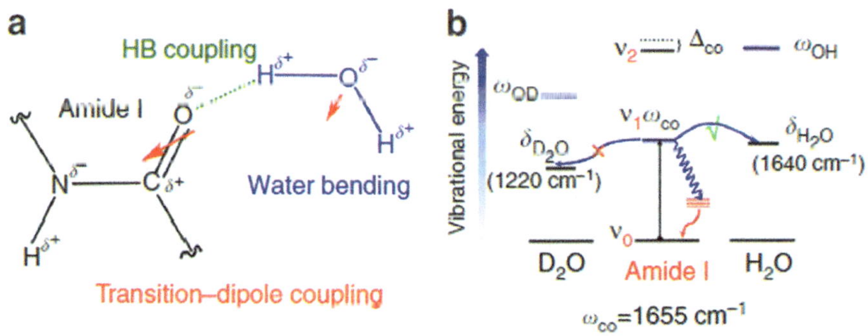

Fig. 7.19 Coupling scheme and energy diagram. **a** Interaction between amide carbonyl and water through hydrogen bonding (HB) and transition–dipole coupling. **b** Energy-level depiction of the vibrational relaxation of the amide I mode to the water bending mode [47]

Inelastic neutron scattering experiments and molecular dynamics simulations have been used to investigate the low-frequency modes hydration water in biomolecules including peptides in the region between 0 and 100 meV [56]. Interestingly, at 200 K, the measured low-frequency density of states of hydration water molecules of hydrophilic peptides is similar to that of high-density amorphous ice, whereas for hydrophobic biomolecules, it is comparable to that of low-density amorphous ice behavior.

A method that uses the magnitude of the associated thermal response of H_2O as an internal normalization standard for 2D-IR spectra and its application was reported [46]. IR pump–probe and 2D-IR spectra of stock solutions of bovine serum albumin (BSA) in H_2O-based buffer (pH = 7.5) were recorded using the ultra B (10 kHz repetition rate) and lifetime (100 kHz) laser systems at the STFC Rutherford Appleton laboratory. The ability to measure 2DIR signals of proteins in water provided a route to normalization of spectra via the magnitude of the thermal response of the water that follows protein amide I excitation. Several different methods were used to extract the protein peak intensity at a Tw value of 250 fs and the thermal water signal at $T_w = 5$ ps. An example is demonstrated for 2D-IR spectra obtained with both the lifetime and ultra instruments (Fig. 7.20). Projection of the lifetime 2D-IR spectra onto the probe frequency axis of the 2D-IR spectrum allowed the BSA response to be quantified by averaging the signal in the $T_w = 250$ fs spectrum. The value of the thermal water response was measured using the $T_w = 5$ ps spectrum near 1675 cm^{-1}. The correlation of the signals obtained for 30 and 50 mg/mL samples and comparable data obtained with the ultra spectrometer are indicated in the figure.

It was concluded that using the water response, which is temporally separated from the protein signal, to normalize the spectra allowed significant reduction of the impact of measurement-to-measurement fluctuations on the data can be advanced. This normalization method enables creation of calibration curves for measurement of absolute protein concentrations and facilitates reproducible difference spectroscopy.

7.3.2 *Rotational and Translational Motions*

The early studies [3] revealed three fractions of the nanosecond rotational molecular dynamics of water with corresponding correlation frequencies υ_c, namely a layer closely associated with the protein III ($\upsilon_c = 10^8 - 10^7$ s^{-1}), a layer at a distance 3–5 Å from the protein surface II with $\upsilon_c = 10^8 - 10^9$ s^{-1}), and a more mobile layer with I $\upsilon_c = 10^9 - 10^{10}$ s^{-1}) extending at a distance in some cases at the distance up 20 Å. However, the problem of the coupling between the collective dynamics of proteins and their hydration water is of particular concern. A nanosecond intermolecular mobility was thought to consist of movement of the relatively large and rigid parts of the protein macromolecules [3] (Fig. 7.21).

As shown in Fig. 7.22, the observation of surface hydration dynamics on two timescales and a series of correlations provide a molecular picture of water motions and their coupling with protein fluctuations in the layer [57]. The dynamic exchange

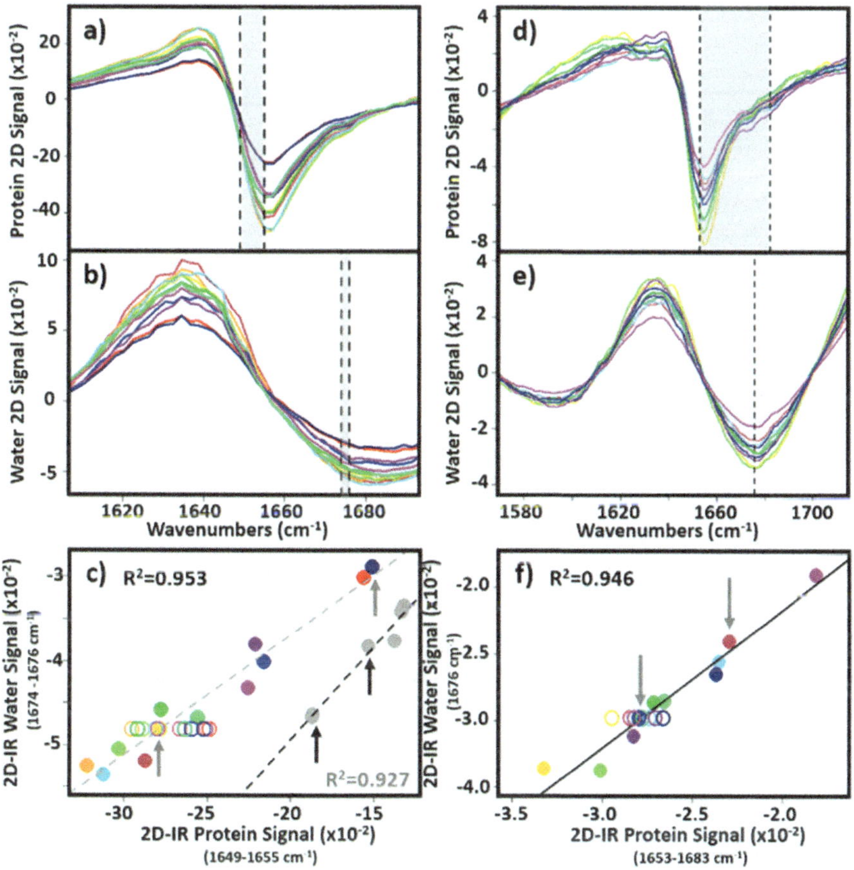

Fig. 7.20 **a–c** Projection of 2D-IR spectra of 50 mg/mL BSA in H_2O buffer over pump frequencies between 1580 and 1720 cm^{-1} onto the probe frequency axis at waiting times of **a** 250 fs showing the protein response and **b** 5 ps showing the water response. **c** Correlation of the water signal between 1674 and 1676 cm^{-1} with the protein signal between 1649 and 1655 cm^{-1} (shaded areas in parts **b** and **a**, respectively). Each solid-colored circle indicates an individual measurement. Open circles show the spread of the spectra after scaling to the water response of the gold spectrum (see text). Gray circles show the results obtained using a 30 mg/mL BSA solution. **d–f** Using the ultra instrument, the 2D-IR spectra of 30 mg/mL BSA in H^2O buffer are projected onto the probe axis at waiting times of **d** 250 fs and **e** 5 ps using the pump frequency range between 1400 and 1700 cm^{-1}. **f** Correlation of the water signal at 1676 cm^{-1} with the protein signal between 1653 and 1683 cm^{-1} (shaded area in part d). Open circles show the spread of the spectra after scaling to the water response of the blue spectrum [46]. With permission from American Chemical Society

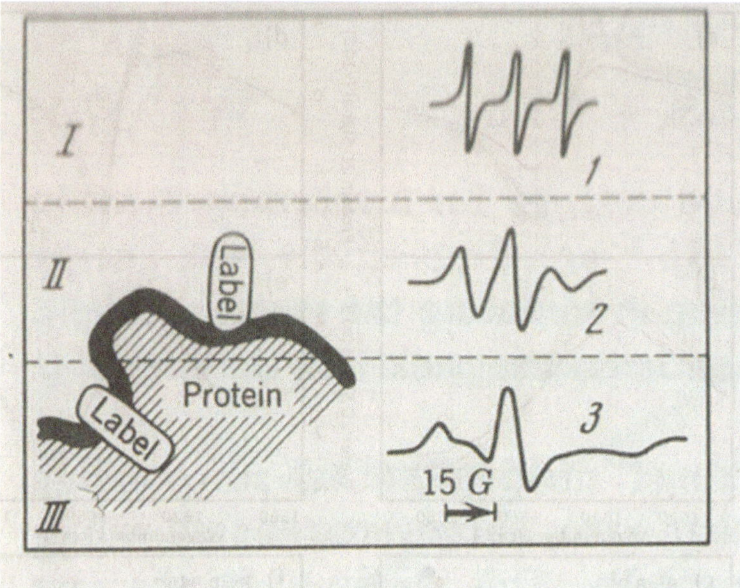

Fig. 7.21 Mobility of nitroxide spin labels in various layers of a hydrated protein matrix: (1–3) Typical ESR spectra of nitroxide radicals submerged in the respective matrix layers; (I) outer aqueous layer ($v = 10^{10}$–10^{11} s^{-1}), (II) glycerol-like layer ($v = 10^{9}$–10^{10} s^{-1}), and protein globule ($v = 10^{7}$–10^{11} s^{-1}); v is the rotational correlation time [3]

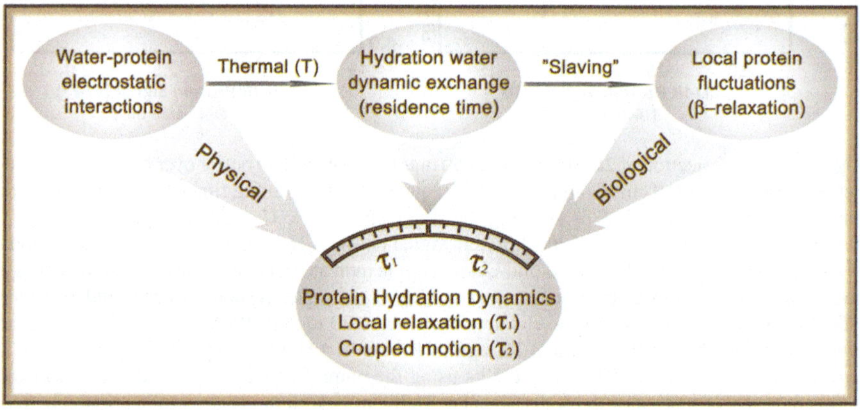

Fig. 7.22 Schematic representation of the mechanism of protein surface hydration. The fundamental electrostatic interaction between water and protein is the key for hydration layer formation. Under thermal fluctuation, the hydration layer water is in dynamic exchange with bulk water (residence time). Such water fluctuations in and out control local protein motions. The hydration dynamics reported here represent two types of collective water fluctuations inside the layer: initial fundamental physical motion of libration and rotation and subsequent network restructuring coupled with slaved protein motions, a dynamic process of biologically relevant water–protein interactions [57]. Published with permission of PNAS article containing the requested material published under a CC BY-NC-ND license

of hydration layer water with outside bulk water occurs on the picosecond timescales (residence time). The collective water network rearrangements in the layer strongly couple with protein fluctuations, all facilitated by bulk water dynamic exchange, occur on tens to hundreds of picoseconds. Thus, the dynamics of protein surface hydration bridge ultrafast bulk water motions, govern protein fluctuations, maintain protein's intact structure and flexibility, lubricate its various recognitions, and mediate a variety of catalytic reactions in enzymatic function.

Based on the overview of neutron scattering, molecular dynamic simulations, NMR and dielectric spectroscopy results, a general picture of protein dynamics covering timescales from faster than ps to microseconds and the influence of hydration water on different relaxation processes were presented [32]. As it is seen in Fig. 7.23, a significant difference in dynamics of hydration water and protein takes place, and dynamics of protein hydration water are slowed down relative to bulk water by a factor of 2–4. In contrast to hydration water, characteristic relaxation time of the coupled protein hydration water relaxation is on 10–100 ps timescale. The following major conclusions were formulated: (1) Internal protein dynamics spread over a wide time range from faster than picosecond to longer than microseconds; (2) the structural relaxation in hydrated proteins appears on the microsecond timescale, while faster processes present mostly motion of side groups and some domains; (3) hydration water plays a crucial role in protein dynamics on all timescales and controls the coupled protein hydration water relaxation on 10–100 ps timescale; and (4) changes in amount of hydration water affect not only general friction, but also influence significantly the protein's energy landscape.

In study [67], both dynamic nuclear polarization (DNP) and field cycling relaxometry (FCR) were employed to determine the coupling factor and translational correlation time given by [68]

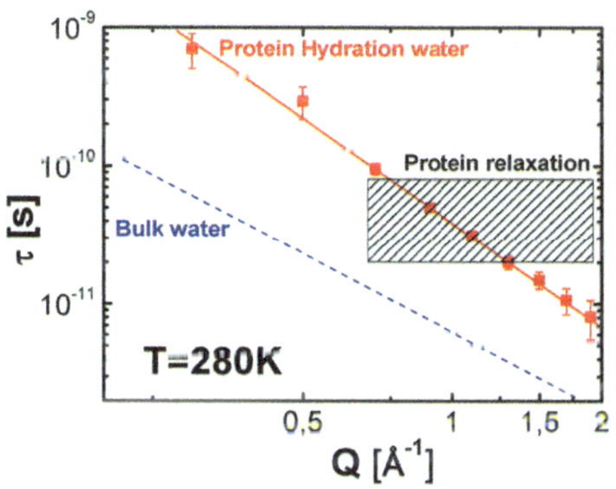

Fig. 7.23 Wave vector, Q, dependence of relaxation time at $T = 280$ K in protein hydration water (symbols) and bulk water (dashed line) [66] and characteristic relaxation of the protein dynamics for lysozyme, GFP, and myoglobin. Bulk water shows normal diffusion behavior with $\tau \sim Q^{-2}$, while hydration water shows stronger Q dependence and is ~ 4 times slower than bulk water on local length scale ($Q \sim 1.5$–2 Å^{-1}). In contrast, proteins exhibit very broad and Q-independent relaxation spectra (shaded area), indicating localized nature of the relaxation process [32]. With permission from Elsevier

$$t = d^2/D_\mathrm{I} + D_\mathrm{S} \tag{7.1}$$

where d is the distance of closest approach of the two spins, D_I is the diffusion coefficient of the solvent molecules, and D_S is the diffusion coefficient of the electron spin-bearing molecules. The key parameter of interest when determining water dynamics on the surface of and inside biomolecular or other soft matter systems is the water diffusion coefficient. DNP and FCR experiments were performed with the radical 4-oxo-2,2,6,6-tetramethyl-1-piperidinyloxy (4-oxo-TEMPO) dissolved in water at room temperature. In DNP experiments, the applied power driving the electron spin transition was varied, and the ^{1}H NMR signal enhancements were measured for several different radical concentrations. The obtained value of D_I 1.6 $\times$ 10^{-9} m^2 s^{-1} represents an average diffusion coefficient for water that is weighted toward water molecules very close to the radical, with 80% of the relaxation coming within 5 Å and 90% within 10 Å. This value for the hydration layer water on the surface of an unfolded protein is in good agreement with the value of 1.65 $\times$ 10^{-9} m^2 s^{-1} for the hydration layer water on the surface of a model protein, as determined by quasi-elastic neutron scattering (QENS). Results of an FCR experiment for a sample of 5 mM 4-oxo-TEMPO dissolved in water showed the following values: correlation time $\tau = 24$ ps, $d = 2.4$ Å, and $\rho = 0.36$.

To study protein–water interactions, ^{15}N-resolved NOESY and ROESY spectra of the 76 amino acid protein ubiquitin encapsulated in reverse micelles dissolved in pentane were detected [60]. Consequent use of ^{13}C-resolved NOESY and ROESY experiments offered by reverse micelle encapsulation allowed significantly expanding the characterization of the hydration dynamics at the surface of ubiquitin. By manipulating the molecular reorientation time of the reverse micelle particle through use of low viscosity liquid propane, the $T_{1\rho}$ relaxation time constants of ^{1}H bonded to ^{13}C were sufficiently lengthened. This allows to obtain high-quality rotating frame nuclear Overhauser effects. The main features of the ubiquitin hydration dynamics shown in the map in Fig. 7.24 are as follows: (1) the obvious clustering of the fast hydration dynamic sites; (2) grouping of sites where interactions with water are quite long-lived; (3) clustering sites along the outer surface of the mixed β-sheet and along the interface of the β-sheet and R-helix indicates regions of protein surface with greatly slowed hydration dynamics; and (4) intermediate hydration dynamic clusters particularly around the 3_{10} helix.

Hydration dynamics are at the surface of ubiquitin. A ribbon representation of the structure encapsulated human ubiquitin (PDBcode 1G6J, conformer 25) 28 is shown. Hydration probes sites are represented as spheres. Probe sites with unique hydrogen chemical shifts are shown as small spheres at the location of the probe hydrogen. Probe sites that are degenerated, such as the hydrogens of a methyl group whose chemical shifts are averaged, are shown as large spheres at the location of the bonded carbon. The spheres are colored according to the probe σNOE/σROE values. σNOE/σROE ratios near – 0.5 indicate the slowest hydration dynamics, while values near 0 are the regions of fast hydration dynamics. Backbone amide hydrogens which are solvent exposed but showed no cross-peaks to solvent are colored orange. These

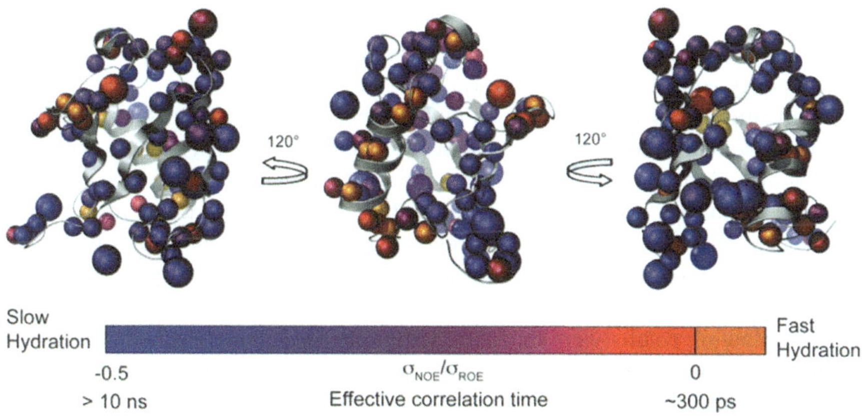

Fig. 7.24 Hydration dynamics at the surface of ubiquitin. A ribbon representation of the structure encapsulated human ubiquitin (PDBcode 1G6J, conformer 25) 28 is shown. Hydration probes sites are represented as spheres. Probe sites with unique hydrogen chemical shifts are shown as small spheres at the location of the probe hydrogen. Probe sites that are degenerate, such as the hydrogens of a methyl group whose chemical shifts are averaged, are shown as large spheres at the location of the bonded carbon. The spheres are colored according to the probe σNOE/σROE values. σNOE/σROE ratios near −0.5 indicate the slowest hydration dynamics, while values near 0 are the regions of fast hydration dynamics. Backbone amide hydrogens which are solvent exposed but showed no cross-peaks to solvent are colored orange. These sites are interpreted as the locations of fastest hydration dynamics. The carbon-resolved hydration dynamic data correspond well with the ^{15}N-resolved measurements. Sites in the rigid limit have residence times and surface dynamics slower than 10 ns. Sites showing no NOE but a detectable ROE have effective correlation times on the order of 300 ps. Intermediate σNOE/σROE values potentially arise from a complicated scaling due to both the timescale of motion and its geometric details [60]. With permission from American Chemical Society

sites are interpreted as the locations of fastest hydration dynamics. The carbon-resolved hydration dynamic data correspond well with the ^{15}N-resolved measurements. Sites in the rigid limit have residence times and surface dynamics slower than 10 ns. Sites showing no NOE but a detectable ROE have effective correlation times on the order of 300 ps. Intermediate σNOE/σROE values potentially arise from a complicated scaling due to both the timescale of motion and its geometric details [60]. It was concluded that regions of long-lived hydration water correspond with regions of the protein that participate in molecular recognition of binding partners suggesting that the contribution of the solvent entropy to the entropy of binding has been maximized through evolution.

Molecular dynamics simulations and analytical models were used to access site-resolved information on the reorientation dynamics of individual water molecules within the hydration shell of acetylcholinesterase, subtilisin Carlsberg, lysozyme, and ubiquitin and to elucidate the molecular origins of the dynamical perturbation of hydration shell water around relative to bulk water [35]. The simulations showed that all four proteins have very similar hydration shell dynamics, despite their wide range of sizes and functions, and differing secondary structures. It was suggested that this arises from the similar local surface topology and surface chemical composition of the four proteins and that protein conformational fluctuations induce a dynamical heterogeneity within the hydration layer.

Polarization-resolved femtosecond pump–probe infrared spectroscopy has been used to measure the reorientation dynamics of water molecules in aqueous bovine α-lactalbumin solutions [48]. Following an initial vibrational excitation by a polarized pulse, the anisotropy decay was measured with a femtosecond time resolution. Since this decay is the sum of the rotational relaxations of all water molecules within the sample, it was possible to determine simultaneously the number of perturbed water molecules [48].

The following results were reported: (1) Slowly reorienting surface waters was identified; (2) their hydrogen-bond lifetime and reorientation dynamics were established and compared to the experimentally measured anisotropy decay; (3) surface waters form fewer hydrogen bonds than the bulk and within the hydration layer water is slower when donating more hydrogen bonds; (4) at concave sites, the protein–water hydrogen bonds break preferably via translational diffusion rather than via a hydrogen-bond jump mechanism; (5) water molecules reorient slower near these sites than at convex water-exposed sites; (6) protein misfolding leads to an increased exposure of hydrophobic groups, inducing relatively faster surface water dynamics; (7) the larger exposed surface slows down a larger amount of water. The authors concluded that while for native proteins hydrating water is slower near hydrophobic than near hydrophilic residues, misfolding causes hydrophobic water to reorient relatively faster because exposure of hydrophobic groups destroys concave protein cavities with a large excluded volume.

The hierarchy in the protein conformational energy landscape and dynamics based on a series of temperature-dependent magic angle spinning multinuclear nuclear magnetic resonance relaxation measurements in a hydrated nanocrystalline globular protein GB1 was established [61]. Simultaneous measurement of the temperature dependence of 13 different spin relaxation probes reporting on dynamic modes distributed throughout the solvated microcrystalline protein allowed for a unified description of the essential conformational energy surface, relating the amplitude and activation of solvent, sidechain, and backbone motions. Changes in picosecond–nanosecond and nanosecond–millisecond motions were identified by the analysis of longitudinal (R1) and transverse (R1r and R2′) relaxation rates.

The measurement of different ^{1}H, ^{13}C, and ^{15}N relaxation rates (14) from a sample allows to determine different dominant rotational motional modes that affect bulk solvent, hydration water, protein side chains, and protein backbone in a site-specific manner.

The coherent picture for the dynamic transitions in protein behavior based on the comparison of the observed dynamical transitions at different locations of the protein–solvent system and the associated changes in the average activation energies for the dominant motional is summarized in the figure. In the region below 160 K, translations and rotations of solvent molecules are arrested, and protein motions are decoupled from the dynamic of the solvent. The amplitude of motions within these states increases gradually, concomitant with dynamics getting faster, as temperature increases from 110 to 160 K. Solvent rotations, which become active at ~ 160 K and dominate solvent relaxation rates above TI (~195 K), allow side chains to access different rotameric states. At TII (~220 K), larger-amplitude, nanosecond timescale anisotropic backbone peptide motions, with substantially increased activation energies, become the dominant source of longitudinal relaxation.

Summary of hierarchical dynamic behavior of the protein–solvent system as observed by solid-state NMR in a microcrystalline globular protein GB1. The approximate temperature for the transitions between dominant dynamic modes is indicated on the blue axis. The image in the top right corner represents an ensemble extracted from a 200-ns molecular dynamics simulation in a crystalline environment. The left panel presents a simplified representation of the link between small- and larger-amplitude backbone motional modes. At low temperatures, the protein backbone is constrained to small-amplitude modes separated by low-energy barriers, within substrates separated by high barriers. A temperature increases, these modes are dynamically sampled, enabling larger-amplitude anisotropic modes [61]. All these results support strong coupling between protein and solvent dynamics above 160 K, with fast solvent motions, slow protein side chain motions, and fast protein backbone motions being activated consecutively. Low activation energy and small-amplitude local motions dominate at low temperatures, while larger-amplitude, anisotropic, and functionally relevant motions involving entire peptide units becoming dominant at temperatures above 220 K.

The main features of water structure and dynamics around biomolecules summarized in [55] are as follows: (1) The arrangement and local interactions of water molecules in the first layer around the biomolecule are strongly influenced by the heterogeneous geometric and chemical character of the biomolecular surface; (2) starting from the interfacial water layer, the hydration shell assumes the structure of bulk-like water within a few layers, typically less than five layers; (3) the fastest fluctuations of water structure in the first few water layers occur on a timescale of a few hundred femtoseconds and are mainly related to water's librational (hindered rotational) degrees of freedom; (4) the biomolecular shell hydrogen-bond network rearranges via the large-amplitude jumps executed by water molecules when they exchange hydrogen-bond acceptors; (5) the water shell around biomolecules serves as a heat sink for excess energy released in the radiationless decay of electronically excited states and/or vibrational relaxation; and (6) electric fields at the interface between a biomolecule and its water shell reach values up to 100 MV/cm.

Data on probability of rotational time constants of water molecules in bulk and proteins shells are summarized in Fig. 7.25 which indicates distribution of first-rank rotational time constants of water molecules that have residence times more than 100 ps inside of the PHL for four proteins, namely (a) lysozyme, (b) myoglobin, (c) monellin, and (d) protein-G, and the bulk (blue).

The water molecules whose mobility is impacted by protein–surface interactions collectively compose a hydration layer (Fig. 7.26) [58]. This layer typically consists of the nearest two–four layers of water molecules from the protein surface due to the magnitude of motional retardation near the surface. Majority water molecules in the hydration layer have been measured to move ~ 2–10 times slower on average

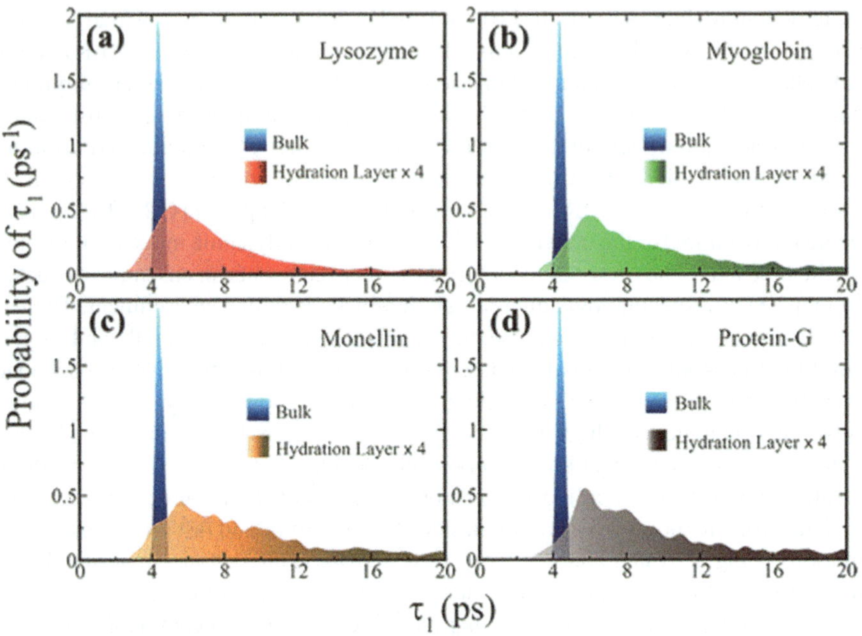

Fig. 7.25 Distribution of first-rank rotational time constants of water molecules that have residence times more than 100 ps inside of the PHL for four proteins, namely **a** lysozyme, **b** myoglobin, **c** monellin, and **d** protein-G, and the bulk (blue). $[C1i(t) = \prec P1(\hat{\mu}0i, \hat{\mu}ti)$, where $P1(x) = x$ and $\hat{\mu}ti$ are the unit vectors along any one O–H bond vector of the ith water molecule at time t. Hence, $\tau 1(i) = \int 0 \infty C1i(t)dt$ is the first-rank rotational time constant of the ith water molecule [30]. With permission from American Chemical Society

than bulk water, while a minority move the nanosecond range. Bound waters that are buried within the protein structure exhibit exchange with bulk water on the order of tens of nanoseconds or longer.

The average rotational correlation time of water molecules in the protein hydration shell was determined for three protein systems of different size: ribonuclease A, human serum albumin, and fibrinogen, employing theoretical and experimental NMR relaxation methods [49]. Specifically, ^{1}H NMR spectra were detected, and the spin–lattice relaxation rates were measured using the $(180° - \tau - 90° - t)n$ sequence. The field-dependent relaxation or nuclear magnetic resonance (NMRD) profiles were measured on a Stelar SPINMASTER FFC field cycling spectrometer (operating in the field range from 2.4×10^{-4} to 8.2×10^{-1} T). The magnetic relaxation dispersion studies gave information about the rotational mobility of a protein and the buried waters, as well as on the number of buried water molecules that are associated with the protein. The fast chemical exchange between the bulk (A) and the water molecules in the layers around a protein surface (B), on the basis of the equilibrium both $R_1{}^{NS}$ and $R_1{}^{SE}$ can be given as follows:

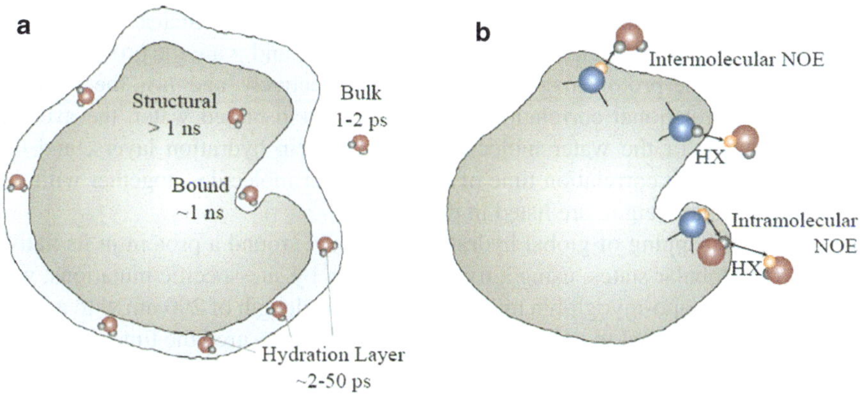

Fig. 7.26 **a** Water mobility within and near a protein (tan) is illustrated for water molecules integrated into the protein structure (structural), surface waters exhibiting long-lived interactions with the protein surface (bound), hydration waters, and waters in bulk solution. **b** Mechanisms of magnetization exchange between protein hydrogens (orange) and water hydrogens (gray) are illustrated schematically. Water dynamics may be evaluated by measurement of dipolar exchange via intermolecular NOEs (top), but such measurements are complicated by hydrogen exchange (HX)-mediated interactions (middle and bottom). For interpretation of the references to color in this figure legend, the reader is referred to the web version of this article [58]. With permission from Elsevier

$$wR_{1\,\exp} = \chi_b R_{1b} + \chi_f R_{1f} \tag{7.2}$$

where $wR_{1\exp}$ is the experimental relaxation rate of water in the presence of the protein and R_{Ib} and R_{1f} the water relaxation rates of the pure bound and free environments, respectively. The values of χ_b and χ_f are the molar fractions of water in bound and bulk conditions and χ_b can be defined as follows:

$$\chi_b = \frac{n[M(\mathrm{H_2O})_n^{\wedge}]}{[\mathrm{H_2O}] + [M(\mathrm{H_2O})_n^{\wedge}]} \approx \frac{n[M(\mathrm{H_2O})_n^{\wedge}]}{[M(\mathrm{H_2O})]} \tag{7.3}$$

where χ_f is the free water molar fraction: $\chi_f = 1 - \chi_b \approx 1$, considering that $[\mathrm{H_2O}] \gg [M(\mathrm{H_2O})nA]$. Dipolar non-selective $R_1{}^{\mathrm{NS}}$ and selective $R_1{}^{\mathrm{SE}}$ spin–lattice relaxation rates can be described as [50, 51]. For any (*i*, *j*) dipolar coupling, $R_1{}^{\mathrm{NS}}$ and $R_1{}^{\mathrm{SE}}$ assume the explicit form

$$R_1^{\mathrm{NS}} = \frac{3}{10}\frac{\gamma_H^4 \hbar^2}{r_{ij}^6}\left[\frac{4\tau_c}{1+4\omega_H^2\tau_c^2} + \frac{\tau_c}{1+\omega_H^2\tau_c^2}\right]$$
$$R_1^{\mathrm{SE}} = \frac{1}{10}\frac{\gamma_H^4 \hbar^2}{r_{ij}^6}\left[\frac{3\tau_c}{1+\omega_H^2\tau_c^2} + \frac{6\tau_c}{1+4\omega_H^2\tau_c^2} + \tau_c\right] \tag{7.4}$$

where τ_c is the rotational correlation time modulating the internuclear interactions, r_{ij} is the internuclear distance between i and j atoms, ω and γ are the proton Larmor frequency and the proton magnetogyric ratio, respectively, and $\hbar$ is the Planck's constant. The rotational correlation time of the protein buried water, the average correlation time of the water molecules in the protein hydration layers, and the tumbling rotational correlation time of the bulk water molecules, together with the protein molecular weight, are listed in the table.

The direct mapping of global hydration dynamics around a protein in its native and molten globular states, using a tryptophan scan by site-specific mutations, was reported [57]. In apo-myoglobin mutants, a laser wavelength of 290 nm with a pulse width of 90 fs was used to excite tryptophan and then measured the time-dependent emission spectrum by fluorescence up-conversion, a nonlinear gating method with femtosecond time resolution. Sixteen mutants were found to be appropriated for mapping global hydration, as depicted in Fig. 7.27. With 16 tryptophan mutants and in 29 different positions and states, two distinct water dynamics in the hydration layer on a few ($\approx$ 1–8 ps) and tens to hundreds of picoseconds ($\approx$ 20–200 ps) were observed. They represent the initial local relaxation and subsequent collective network restructuring, respectively, which strongly correlated with protein's structural and chemical properties.

The experimental decay times of fluorescence frequency shift of protein-attached fluorescence probes ranged from 10 ps to several nanoseconds were attributed to hydration water motions several orders of magnitude slower than those in the hydration shell of small solutes [62]. It was argued that the slow decay in the fluorescence

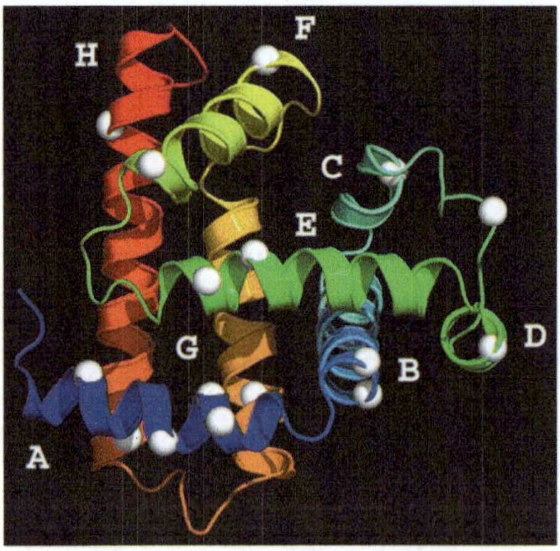

Fig. 7.27 Global mapping of surface hydration dynamics of apoMb. The x-ray crystal structure of *sperm whale* myoglobin (Protein Data Bank ID code 1MBD) in the holo form with eight helices A–H colored from blue to red is shown. The 16 white balls indicate positions of mutation with tryptophan one at a time. In the apo form, the F-helix melts to a random coil [57]. Published with permission of PNAS article containing the requested material published under a CC BY-NC-ND license

shift can be explained by a ubiquitous solvent polarization mechanism, which can be qualitatively understood with the aid of a dielectric continuum model.

Staphylococcal nuclease (SNase) is a protein with a single tryptophan (W140) residue that is partially buried in the protein with a fluorescence emission peak at 332 nm. This partially buried tryptophan reports only two relaxation timescales of a few to tens of picoseconds, probing only the inner-layer interfacial water molecules in the hydration shell [63, 64]. With systematic mutations of the three neighboring charged residues around W140 using an alanine scan, it shows the dominant relaxation on the picosecond timescales from interfacial hydration water molecules [64]. In study [63], four proteins of wild type (WT), single mutation of K110A (K), double mutation of K110AE129A (KE), and triple mutation of K110AE129AK133A (KEK) (Fig. 7.31) were prepared by an alanine scan for the fixed probe (W140). Simultaneously, the relaxations of hydration water and local tryptophan with systematic temperature changes were measured, employing the femtosecond-resolved fluorescence up-conversion method. On the top of local snapshots of four proteins at 2 ns from a 20 ns with two inner layers of about 60–70 water molecules within a distance of 10 Å to the indole-ring center of tryptophan are indicated in Fig. 7.28a. More than 90% of those water molecules are within 7 Å to the protein surface. Figure 7.28 shows the plots of the pairs of two relaxations (τ_{2W}, τ_{2S}) in Fig. 7.28b and (τ_{3W},

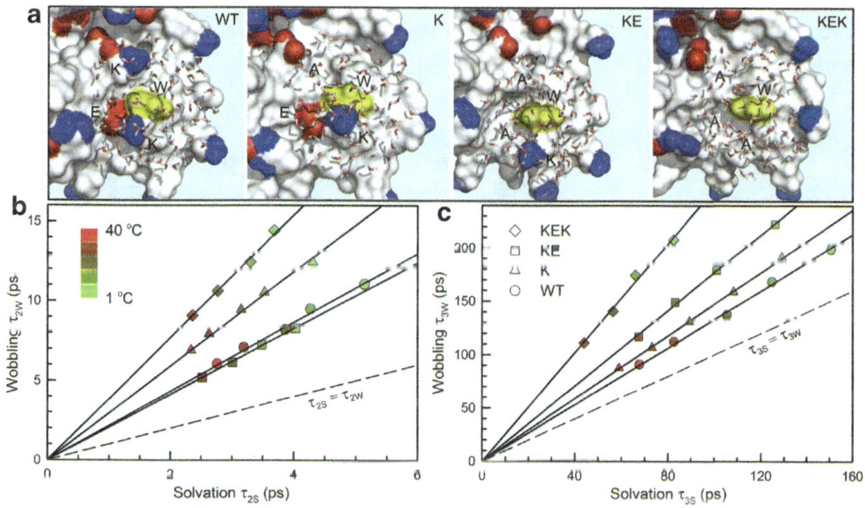

Fig. 7.28 Correlations between hydration water motions and tryptophan side chain relaxations. **a** Snapshots of MD simulations for the four proteins indicating different local protein surface topology and hydration water molecules within 10 Å from the indole-ring center of the tryptophan probe. **b** and **c** Various linear time correlations between tryptophan side chain relaxations and hydration water motions on two distinct timescales with temperature changes. Scattered points show the experimental data, and solid lines are the linear fits. Dashed lines represent the equal line of $\tau_{iW} = \tau_{iS}$. Clearly, all protein side chain relaxations are slower than hydration water motions [63]. With permission from American Chemical Society

τ_{3S}) in Fig. 7.28c with temperature dependence. Thus, the water–protein relaxations are strongly coupled, and the hydration water molecules govern such fluctuations on the picosecond timescales.

To reveal the effect of different protein secondary structures on hydration water, hydration dynamics around a β-barrel protein, rat liver fatty acid-binding protein (rLFABP), were systematically characterized [65]. A tryptophan scan to the protein surface one at a time was employed, and a total of 17 different sites was examined by the femtosecond-resolved fluorescence spectroscopy. The global mapping of inner layers hydration dynamics and protein side chain motions around rLFABP is given in Fig. 7.32 with the solvation speeds, S2 and S3, and tryptophan angular speeds, ω_2 and ω_3 (Fig. 7.29).

The experiments revealed that: (1) The hydration dynamics are highly heterogeneous around the protein surface of both α-helical and β-sheet proteins; (2) the outer layer water of the hydration shell behave like a bulk and relaxes in hundreds of femtoseconds; (3) The inner layer water collectively relaxes in two timescales, reorientation motions in a few picoseconds and network restructuring in tens to a hundred picoseconds; (4) the hydration dynamics are always faster than local protein relaxations and drive the protein fluctuations on the picosecond timescale; (5) the hydration dynamics in general are more retarded around β-sheet structures than α-helical motifs; and (6) thicker hydration shell and a more rigid interfacial hydration network are observed in the β-sheet protein. Overall, these findings elucidate the intimate relationship between water–protein interactions and dynamics on the ultrafast timescale [65].

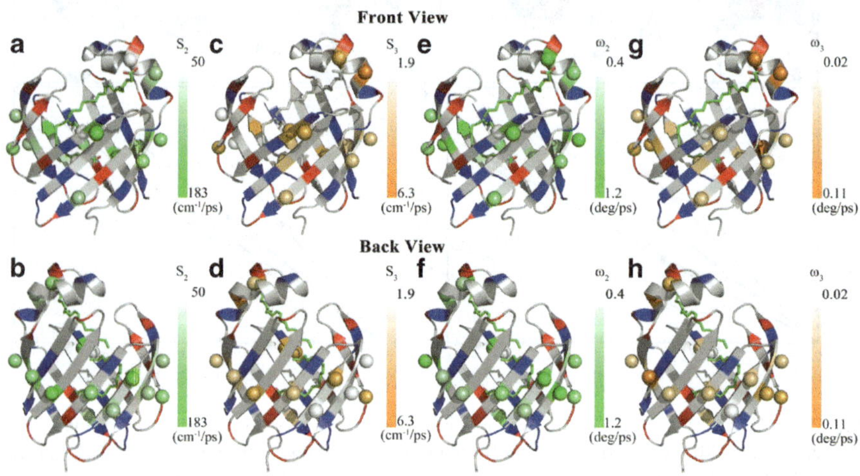

Fig. 7.29 Color scale representation **of** hydration dynamics (S2 and S3) **and** local protein side chain relaxations (ω_2 and ω_3) at 17 positions of rLFABP. The mutant sites are shown in balls, with the colors scaled to represent fast (dark) and slow dynamics (light) [65]. With permission from American Chemical Society

To examine the hydration-level-dependent diffusion and the hydrogen-bond dynamics of the hydration water, MD simulations of an all-atom model were performed with hydrated crystalline staphylococcal nuclease (SNase) [76]. From the MD simulation of two hydration levels, $h = 0.30$ and 0.40, at 300 K, the time-dependent MSDs of hydration water at two hydration levels and bulk water at 300 K were estimated (Fig. 7.30). Lifetime of hydrogen bond between hydration water and protein was found to be picosecond timescale and amplitude of the fluctuation is an order of a few Å. It was suggested that the hydration water molecules sometimes move to the different hydration sites or the outer hydration layers, and they are not always oscillating in the harmonic potential. Thus, the dynamics of hydration water can be described as Brownian motion rather than harmonic motion.

The full study of both the longitudinal and transverse coherent collective motions in the maltose-binding protein (MBP)–solvent system was performed [77]. The maltose-binding protein (MBP) in its ligand-free conformation in a crystal-like packing at$h = 0.42$ at 150 K was examined. The MD simulations were carried out using NAMD2. Following more than 10 ns of MD simulation at constant pressure, a 150 ps run was performed at constant volume, and the atomic positions and velocities, saved every femtosecond, were collected for the analyses. The dynamical coherent structure factor $S(Q, E)$ longitudinal $C_L(Q, E)$ and transverse $C_T(Q, E)$ spectra of water tot, water-2.5, and protein carbon atoms have been calculated at Q values ranging from 0.2 to 2 Å^{-1}. The current spectra for the different water selections calculated at $Q = 0.2$ Å^{-1} were found to be similar those of bulk liquid water. In contrast, the spectra corresponding to the selections of water molecules that are closer to the protein surface (water-4, water-3, and water-2.5) reveal different characteristics.

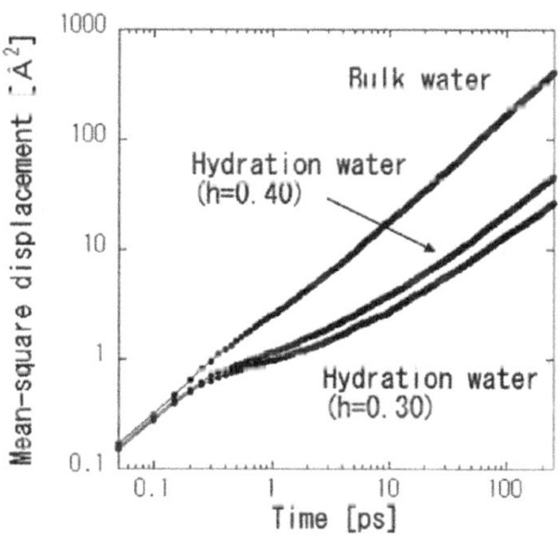

Fig. 7.30 Time-dependent MSDs of hydration water at two hydration levels and bulk water at 300 K obtained by MD simulations, averaged over all water molecules in the system at time zero proteins [76]

Focusing on study of the dynamical properties of water-2.5 indicated that two of the three modes in each of the $C\alpha(Q, E)$ spectra have an acoustic-like character while the third one is associated with an optical branch characterized by an almost constant energy. The high-energy acoustic mode is the main propagating excitation in the longitudinal current spectra, and the low-energy acoustic mode is the dominant propagating excitation in the transverse current spectra. Finally, the authors concluded that: (1) Fluctuations of protein hydration water (PHW) largely influence protein motions; (2) the water dynamics of collective modes belong to the protein; (3) the properties of these modes, in particular, their propagation velocities and amplitudes, indicate a strengthening of the interactions and a higher rigidity of the network of solvent molecules close to the protein surface; (4) the coherent collective dynamics of PHW significantly differs from that of liquid water; and (5) a very strong dynamical coupling between a protein and its hydration water takes place.

7.4 Chemical Exchange

The fourth after the above-described processes of vibration, reorientation, and translation of water molecules in proteins is a chemical exchange (Fig. 7.31).

Applications of water polarization transfer and water-edited SSNMR and NOE experiments to water interactions with proteins, lipid membranes, and carbohydrates were reviewed in [69]. Three polarization transfer mechanisms, chemical exchange, spin diffusion, and nuclear Overhauser effects (NOE)**,** involving in water polarization transfer to biomolecules, were manifest themselves at different temperatures, magic angle-spinning frequencies, and pulse irradiations [69]. In particular, the water polarization transfer technique has been used to **study the** water–protein interactions and to derive atomic-resolution details of the kinetics and mechanism of ion conduction in channels and pumps (Fig. 7.32). Using this approach, the water polarization transfer to the transmembrane domain of the influenza M2 protein to obtain information on the structure of this tetrameric proton channel was measured. Solid-state NMR (SSNMR) method was employed to evaluate the water interaction with the four-helix bundle formed by the influenza M2 transmembrane peptide (M2TM). The

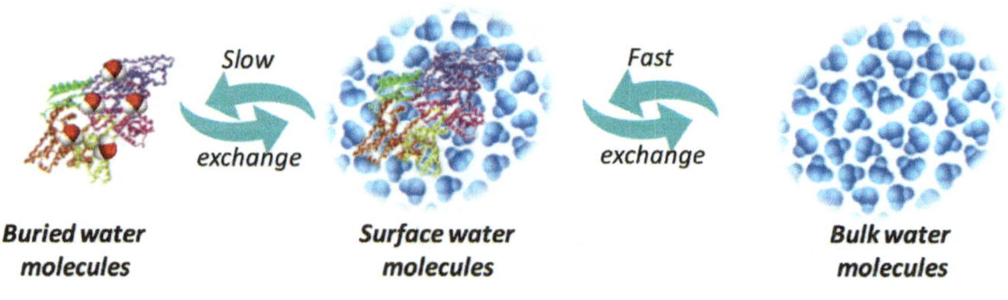

Fig. 7.31 Schematic presentation of chemical exchange in apoprotein [49]. With permission from Elsevier

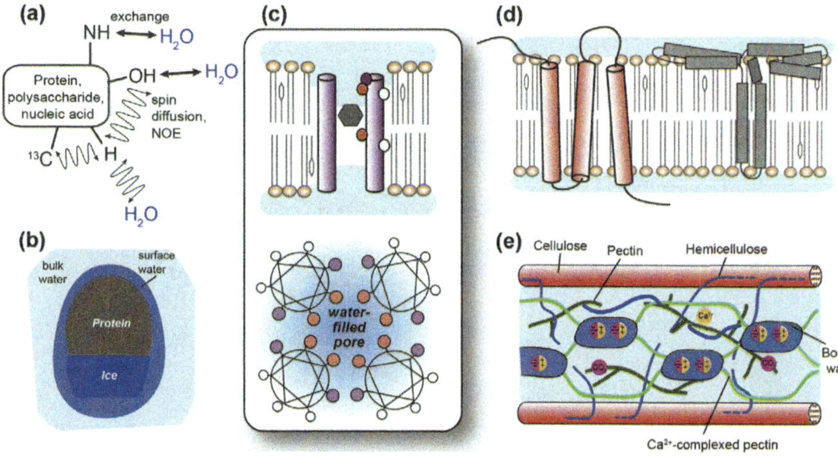

Fig. 7.32 Main mechanisms of water–biomolecule polarization transfer (**a**) and applications of solid-state NMR to investigations of water–protein and water–polysaccharide interactions (**b**–**e**). a Water protons can transfer polarization to biomolecules via chemical exchange, dipolar spin diffusion, and NOE. Distinguishing liquid water and ice by solid-state NMR provides structural information on ice-binding proteins. **c** Water polarization transfer to ion channels has been used to investigate ion conduction kinetics, oligomeric structure of the channel, and to identify lipid-facing versus pore-facing residues. **d** Water-edited solid-state NMR experiments have been used to determine the topology of membrane proteins in terms of surface-exposed loops (left) and helices (right) versus membrane-embedded residues. **e** Distinct water interactions with different polysaccharides of the plant cell wall provide information on the three-dimensional structure of the cell wall [69]. With permission from Elsevier

experiments indicated that at neutral pH, chemical exchange rates range from 100 s^{-1} for amide protons in protein backbones to 1000 s^{-1} for amino protons of Lys, His and Arg, to 10,000 s^{-1} for hydroxy protons of Tyr.

Solution NMR offers a route to a site-resolved view of the dynamics of protein–water interactions through the nuclear Overhauser effects between hydration water and the protein in the laboratory **NOE** and rotating (ROE) frames of [70]. In the comprehensive review, the following general aspects in the area were considered: (1) foundation theory; (2) overcoming artifacts and limitations; (3) two-dimensional versus three-dimensional experiments; (3) the NOE spectroscopy (NOESY) and ROE spectroscopy (ROESY) experimental setup; (4) identification of hydrogen exchange; (5) quantification of hydrogen exchange relayed magnetization; (6) data collection general fitting strategy, and (7) simplified analysis in the absence of hydrogen exchange. Importance of the hydrogen exchange in hydrated proteins was stressed. Figure 7.33 illustrates the positions of molecules of water, surfactant, and low viscosity solvent around a protein and examples of magnetization that can contribute to the measured NOE2.

The methodological advantages of the H_2O/D_2O exchange technique in incoherent neutron scattering (INS) and the compatibility of INS and MD simulation as tools for studying protein dynamics and hydration water were demonstrated [76]. The

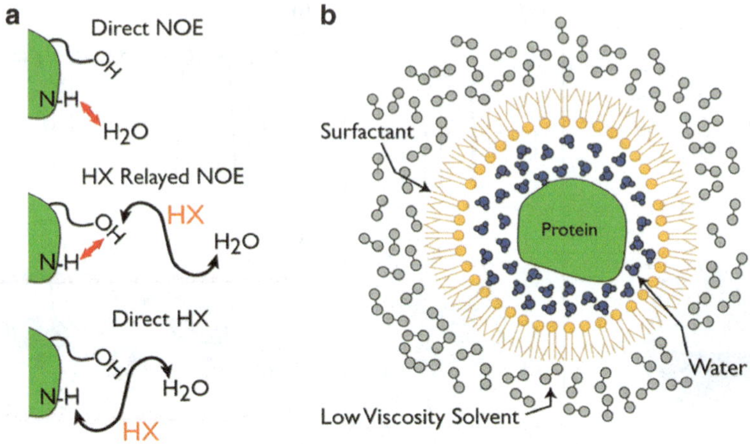

Fig. 7.33 **a** Examples of magnetization that can contribute to the measured NOE; **b** 2 encapsulation of single protein molecules within the water core of a reverse micelle overcomes these limitations, main advantages of which are the suppression hydrogen exchange and elimination of bulk water [70]. With permission from Elsevier

dynamics of hydration water in hydration-induced PDT of staphylococcal nuclease (SNase) by performing an INS experiment with H_2O/D_2O exchange of hydration water and MD simulation were examined. Neutron scattering experiments were performed with the triple axis spectrometer LTAS in the JRR-3 M reactor, which has an energy resolution of ~ 106 μeV, in Tokai, Japan. The q-range covered was 0.4–2.0 Å^{-1}. Neutron scattering measurements were carried out at intervals of 10 K between 80 and 300 K. As shown in Fig. 7.34, the INS due to the hydration water can be obtained by the difference in the scattering intensities of an H_2O-hydrated sample and a D_2O-hydrated sample.

In recent times, comprehensive review [70] suitable for characterization by NOE and ROE spectroscopy was presented. The contribution of detected NOE intensity arising from magnetization relayed by hydrogen exchange and various aspects of fitting obtained NOE, selectively decoupled NOE, and ROE time courses were discussed.

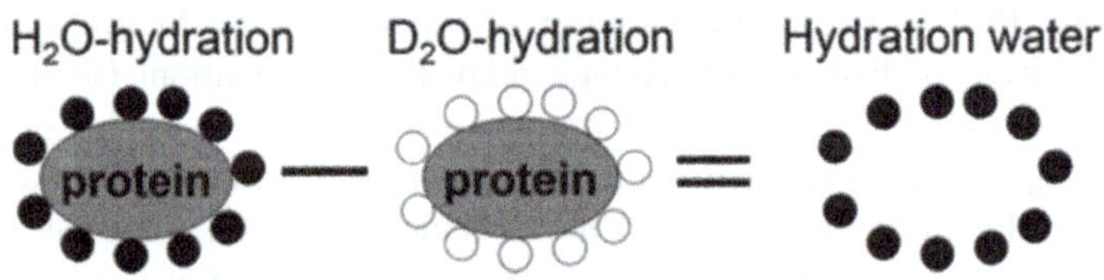

Fig. 7.34 Concept of H_2O/D_2O exchange for estimating scattering intensity of hydration water based on difference in scattering intensity between H_2O- and D_2O-hydrated proteins [76]

7.5 Hydrated Proteins Low-Temperature Transitions

Systematic studies of the intramolecular mobility of a number hydrated proteins such as bovine and human serum albumins (Fig. 7.3), lysozyme, myoglobin, ferredoxin, a-chymotrypsin, and bacterial photosynthetic center over a wide temperature range (20–300 K) by the combine physical labeling methods, allowing the motion to be measured at a correlation frequency of $\upsilon_c = 10^{-2}–10^{10}\ s^{-1}$, were carried out [3, 4, 9–18]. Four independent physical methods, that is, spin, Mossbauer, fluorescence labeling, and ESR revealed appearance anharmonic motion at temperature about 220 K.

All in all, the experiments revealed the following: (1) Appropriate physical methods detect the mobility of labels starting from the temperature of liquid nitrogen; (2) the lower the value of the characteristic frequency of the method, the lower the temperature at which the label mobility can be recorded; (3) the mobility recorded results from a glass-like gradual softening of the protein–water matrix, rather than from an individual phase transition; and (4) at physiological ambient conditions, all labeling methods indicate the mobility of labels and surrounding media in the nanosecond temporal region.

An analysis of the picosecond-to-nanosecond motions of green fluorescent protein (GFP) and its hydration water, using neutron scattering spectroscopy and hydrogen/deuterium contrast, revealed that hydration water suppresses protein motions at lower temperatures ($<\sim$200 K) and facilitates protein dynamics at high [81]. The separation of protein and hydration water dynamics was made by using hydrogenated GFP protein, dry (dry h-GFP), and in D_2O (h-GFP/D_2O) in a combination of the above methods. Samples of green fluorescent protein were hydrated to a ratio of 0.4 (mass solvent/mass protein) of the appropriate solvent by placing the sample in a hydration chamber under nitrogen and subsequently sealed in aluminum sample holders. Neutron scattering measurements at several temperatures were performed on three spectrometers: The high flux backscattering spectrometer (HFBS) NG2 at the National Institute of Standards and Technology, the backscattering spectrometer (BASIS), and the cold neutron chopper spectrometer (CNCS).

As an example, structural relaxation in the hydration water was detectable at T ~ 220 K as a broadening of QENS spectra becomes more significant at higher T = 280 K. Analysis of the QENS spectra at 280 K showed a strong Q-dependence of the observed broadening. Temperature variation of mean squared displacements ($\prec r^2 \succ$) in the GFP three samples d-GFP/H_2O in d-GFP/H_2O reflected motions of hydration water is indicated in Fig. 7.35.

Measurement of $\prec r^2 \succ$ as a function of temperature for the in d-GFP/H_2O showed no increase in slope up to T ~ 180–190 K. Above this temperature, $\prec r^2 \succ$ increases sharply with T. In contrast to d-GFP/H_2O sample, $\prec \mathbf{r}^2 \succ$ of h-GFP in both the dry and hydrated states exhibits a change of slope at T ~ 120 K. This change can be ascribed to methyl group rotations that reach the nanosecond time window already at low

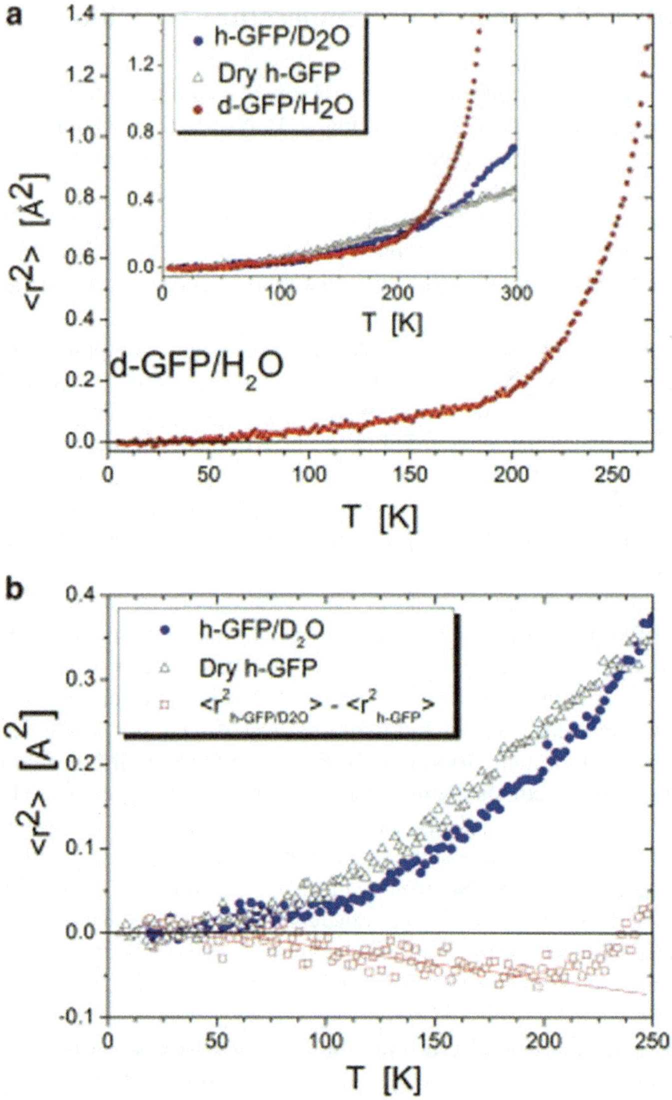

Fig. 7.35 Temperature variation of mean squared displacements. **a** $\prec r^2 \succ$ in d-GFP/H_2O reflects motions of hydration water. (Inset) $\prec r^2 \succ$ in dry h-GFP, h-GFP/D_2O, and d-GFP/H_2O. **b** $\prec r^2 \succ$ in h-GFP/D_2O (solid circles) and dry h-GFP (open triangles). The difference (open squares) monotonically decreases with temperature up to T ~ 200 [81]. With permission from Elsevier

temperature. The motion of water molecules in hydrated shell of the d-GFP/H_2O occurs up to ten times slower than that in bulk water

Experimental data demonstrated that the hydration water is harmonic at temperatures < ~180–190 K. The dynamics of the hydration water exhibit changes at ~180–190 K which indicated the glass transition in the hydrated protein. The results confirmed significant differences in the dynamics of protein (nanosecond timescale) and its hydration water picosecond at high temperatures. The hydration water exhibits diffusive dynamics, while the protein motions are localized to < ~3 Å. On the basis of QENS and NMR data on hydrated lysozyme powders, Chen and coworkers suggested that a structural transition from a predominantly low density (less fluid) to a predominantly high-density (more fluid) form of hydration water takes places at the dynamical transition [74]. A switch in dynamic behavior of the hydration water on lysozyme occurred at 220 K was described as a fragile-to-strong dynamic crossover [74]

The dynamics of hydration water in the protein lysozyme in the temperature range 180 K < T < 360 K were studied using Fourier transform infrared and nuclear magnetic resonance (NMR) spectroscopies [75]. By analyzing the thermal evolution of spectra of the OH stretching vibration modes and the NMR self-diffusion (DS) and spin–lattice relaxation time (T_1), the existence of two dynamical transitions in the protein hydration water was revealed. Below the first transition, at about 220 K, the hydration water fragile-to-strong dynamic crossover resulting in the loss of the protein conformational flexibility occurs. Above the second transition, at about 346 K, the dynamics of the hydration water dominate by the non-hydrogen-bonded fraction of water molecules, as the result of the protein unfolding.

The coupling between maltose-binding protein dynamics and hydration water dynamics was examined by perdeuteration, temperature-dependent neutron scattering, and molecular dynamics simulations [72]. As it is seen in Fig. 7.36, mean square displacements of water (red) and protein motions (blue) both show a coupled broad transition at 220 K. Thus, the protein dynamical transition appears to be driven by the onset of hydration water translational motion.

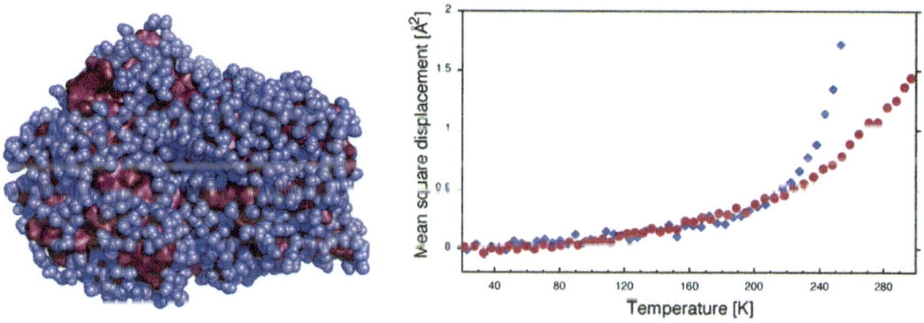

Fig. 7.36 Dynamical transition of protein hydration water. Mean square displacements of maltose-binding protein H atoms (blue) and of its hydration water measured by neutron scattering (red) [72]. With permission from American Chemical Society

Neutron scattering and calorimetry in the study of a low hydration sample of a 15-residue hybrid peptide from cecropin and mellitin CA(1–7)M(2–9) of antimicrobial activity were performed [71]. MQuasi-elastic and low-frequency inelastic neutron spectra were measured at defined hydration levels—a sample with specific residual hydration ($h = 0.060$ g/g), a H_2O-hydrated ($h = 0.49$), and a D_2O-hydrated one ($h = 0.51$). Averaged mean square proton mobilities were derived over a large temperature range (50–300 K), and the vibrational density of states (VDOS) was evaluated for the hydrated samples. The heat capacity of the H_2O-hydrated CA(1–7)M(2–9) peptide was measured by adiabatic calorimetry in the temperature range 5–300 K, for different hydration levels. The revealed different internal motions of protein are as follows: (1) main structural (segmental) relaxation reaching $\tau \sim 100$ s at glass transition temperature Tg of hydrated proteins; (2) domain and side group motions, conformational jumps with nanosecond relaxation times at room temperature; (3) coupled protein–solvent motions present localized diffusion of residues and side group motions with relaxation time of tens of picosecond at room temperature; (4) methyl group rotation with Arrhenius temperature dependence; and (5) fast picosecond relaxation with a rather weak temperature dependence of its characteristic relaxation time. These results indicated that both hydration water and the peptide are involved in the experimentally observed transitions.

The human protein DP tau which is involved in the pathogenic processes accompanying Alzheimer disease was examined, combining neutron scattering and protein perdeuteration [73]. Similar atomic mean square displacements over a large temperature range for the tau protein and its hydration water were found, indicating coupling between them. The extracted mean square displacements also revealed a greater motional flexibility of intrinsically disordered proteins (IDP). The results provide evidence that protein and hydration water motions mutually affect and shape each other. In particular, the degree of coupling between protein and hydration water dynamics by comparing the temperature dependences of their mean square displacement (MSDs) was assessed (Fig. 7.37). For the disordered tau, the protein and water MSDs follow each other up to 250 K (Fig. 7.37a), whereas they diverge above 220 K for the folded protein MBP (Fig. 7.40b).

In the review [54], the main experimental and computational results that provided evidence for the dynamical transition, with a focus on the role of hydration water dynamics in sustaining functional protein dynamics, the coupling, and mutual influence of hydration water dynamics were discussed. As an example, when studied as a function of temperature, functionally relevant dynamics were observed at protein dynamical transition (~ 240 K) in hydrated proteins in lysozyme (Fig. 7.38) [54].

To distinguish between protein dynamics and interfacial water dynamics, experiments on elastic incoherent neutron scattering (EINS) of perdeuterated C-phycocyanin (CPC) were carried out [78]. Perdeuterated samples were obtained by growing the cells in a medium prepared with 90% heavy water. Powders of hydrogenated (hCPC) and perdeuterated (dCPC) CPC protein have been hydrated, respectively, with D_2O or H_2O and measured by EINS to separately probe protein dynamics (hCPC/D_2O) and water dynamics (dCPC/H_2O) at different time- and length scales. The time- and length scales of the probed atomic motions were determined by the

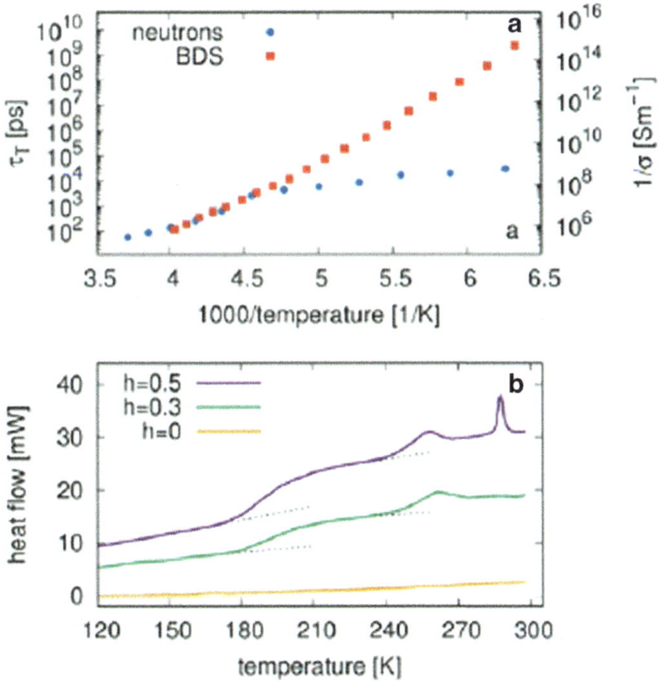

Fig. 7.37 Dynamical coupling of protein dynamics with hydration water motions for different protein classes. Temperature-dependent atomic MSDs of proteins (orange data points) are compared with those of hydration water (blue data points) **a** IDP tau. **b** Folded, globular MBP. An ensemble of 100 tau structures was generated using a coil library and a structure with a calculated radius of gyration (63 Å) identical to that measured by SAXS (63 Å) was chosen for the upper half of panel A. Approximately, 1000 water molecules forming the first hydration shell are shown in panel A, corresponding to the experimental hydration level of 0.4 g water/g protein. Data were obtained on the IN16 spectrometer at the ILL, Grenoble, France. Arrows indicate the temperature at which water and protein MSD diverge (250 K (**a**) and 220 K (**b**) for tau and MBP, respectively). MSDs > 260 K were omitted for the protein (tau and MBP) experiments for the sake of comparison with hydration water MSDs (tau and MBP), which could be determined only up to 260 K [73]. With permission from Elsevier

energy resolution and the Q-range accessible employing the IN13 backscattering neutron spectrometer at the Institute Laue-Langevin (ILL, Grenoble, France) and the Mibémol neutron time-of-flight spectrometer at the Laboratoire Leon-Brillouin (LLB, France). Evolution of protein dynamics and interfacial water dynamics as a function of temperature is reported in Fig. 7.39. The dynamical behavior remains harmonic until 220 K. At that temperature, a deviation from the linearity of total "fast" (< 20 ps) MSD is observed indicating that flexibility of both protein and hydration water drastically becomes anharmonic because of the onset of diffusive, large-amplitude motions observed on that ps timescale. A perfect coincident behavior

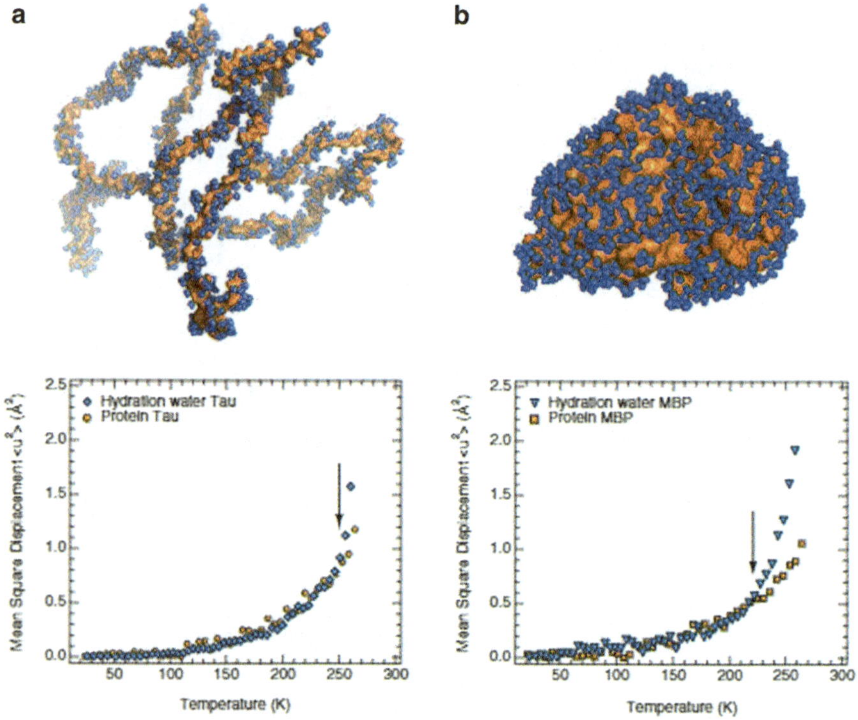

Fig. 7.38 Thermodynamic origin of the dynamical transition of hydration water. **a** Quasi-elastic neutron scattering on lysozyme hydration water reveals an Arrhenius to super-Arrhenius crossover for the translational relaxation time (τ_T, left *y* axis) at the dynamical transition temperature, while the conductivity ($1/\sigma$, right *y* axis) measured on the same protein by BDS does not detect any crossover. **b** Differential scanning calorimetry reveals the existence of a glass transition at ~ 180 K and a first-order-like endothermic peak at the dynamical transition temperature in hydrated myoglobin [h = [water mass]/[protein mass]] defines the hydration level [54].

between the interfacial water and the **CPC protein** with a dynamical transition at 220 K for both was noted [78].

For large-amplitude motions, related to the small Q-region, the dynamics remain harmonic for both protein and hydration water until roughly TE 240 K, where a deviation from the linear behavior of total MSD is observed for both samples corresponding to the hydration-dependent dynamical transition. On higher resolution (< 400 ps), two different types of motions were separated: (i) localized motions with the same amplitude for CPC and hydration water and two transitions at ~ 170 and ~ 240 K for both; (ii) large-scale fluctuations exhibiting for both water molecules and CPC protein a single transition at ~ 240 K, with a significantly higher amplitude for the interfacial water than for CPC. **No coupling** between bulk water dynamics and protein dynamics all along the temperature range was shown.

The "dynamic" or "glass" transition in hydrated proteins can be triggered by the strong hydrogen-bond coupling in the hydration water [79]. To investigate dynamics

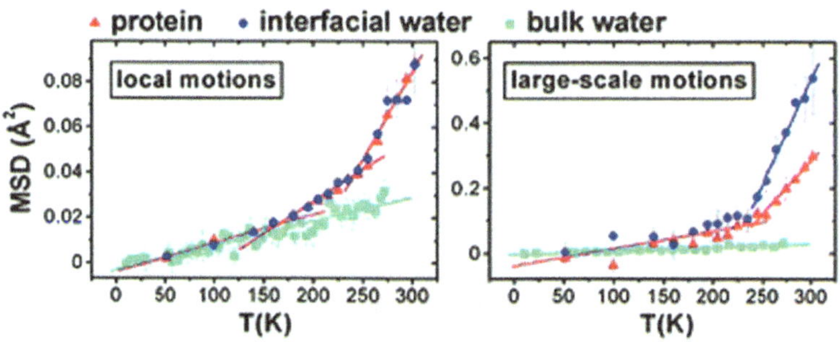

Fig. 7.39 Temperature dependence of MSD of protein (hCPC–D_2O, filled triangles) and interfacial water (dCPC–H_2O, open circles) both hydrated at B0.38 g water g_1 dry powder. Solid lines are linear regressions of hu2i below and above the "dynamical transition" temperature TE 220 K. The slope change has been attributed to the onset of anharmonic dynamics. No difference is observed between protein dynamics and hydration water dynamics, probed in hCPC–D_2O and dCPC–H_2O samples, respectively. Measurements have been performed on Mibémol spectrometer (LLB) with B140 meV (FWHM), a wavelength of 5.2 Å, and a Q-range of 0.6oQo2.1 Å1. Only processes faster than B20 ps contribute to the displacements [78]. With permission from Royal Society of Chemistry

behavior in the globular protein lysozyme hydrated with a single monolayer of water at different temperatures, the vibrational bending modes of water and C=O and N–H peptide groups of lysozyme were measured employing Fourier transform infrared spectroscopy. FTIR absorption measurements were performed at ambient pressure by using a Perkin Elmer spectrum GX Fourier transform spectrometer in the attenuated total reflection (ATR) geometry. The vibrational bending mode of the pure bulk water and the peptide modes falling in this bending spectral region ($1300 < \nu < 1800\ \text{cm}^{-1}$): amide I, amide II, and amide III (lower panel) were identified. The peptide groups were described in IR and Raman spectroscopy with nine characteristic bands named amide A, B, and I–VII in order of decreasing frequency. Figure 7.40 reports the obtained spectra in the water lysozyme system with a hydration level of $h = 0.3$, at three different temperatures: in the native state below [$T = 190$ K, panel (a)] and above [$T = 300$ K, panel (b)] the protein dynamical crossover and in the irreversible denatured state [$T = 350$ K, panel (c)], the protein denatures at TD $\approx$ 345 K. Thus, the temperature of the "dynamic" protein transition is the same as the fragile-to-strong dynamic transition in confined water. Therefore, the fragile-to-strong dynamic transition in water governs the nature of the H-bonds between water and peptides.

To obtain the dynamic protein transition and the dynamic crossover in water, neutron and NMR spectroscopy were utilized. The mean squared displacement MSD was evaluated from the elastic neutron scattering data of the Lyso/D_2O and averaged over all the hydrogen atoms. The MSD behavior indicates the "softness" of the protein. The upper panel of Fig. 7.41 presents MSD as a function of T for hydrated lysozyme with H_2O (circles, upper curve) and D_2O at $h = 0.3$. The inset indicated the inverse of the water self-diffusion coefficient D, obtained using NMR spectroscopy. From these two curves, it was concluded that the water fragile-to-strong dynamic crossover (FSDC) temperature of the hydration water and the dynamic transition

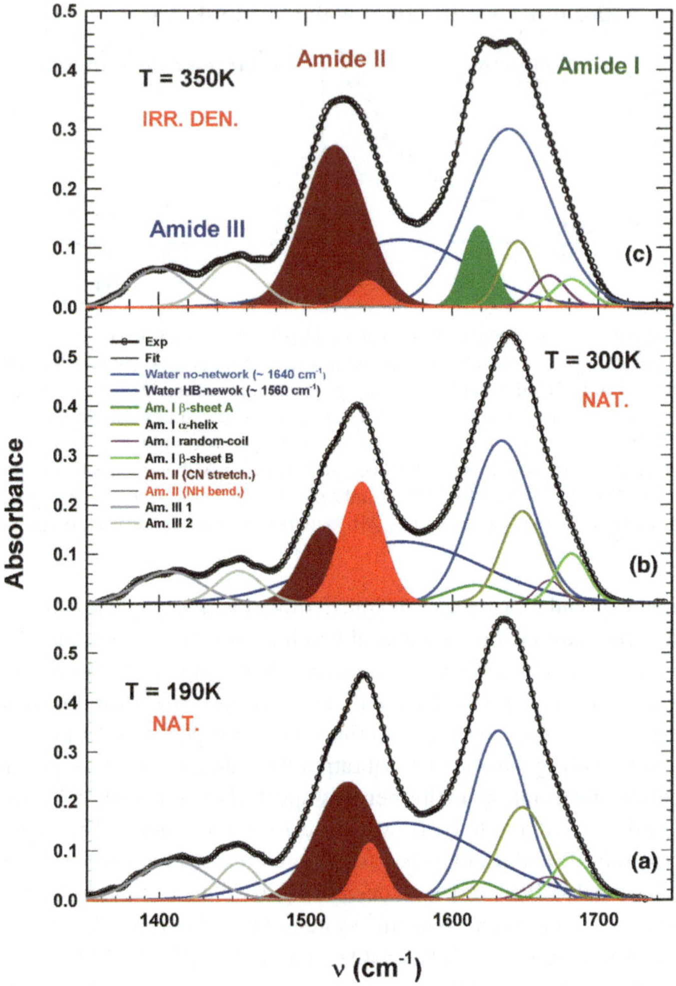

Fig. 7.40 Measured FTIR spectra in the water lysozyme system with the hydration level $h = 0.3$, at three different temperatures, respectively, in the native state below [$T = 190$ K (**a**)] and above [$T = 300$ K (**b**)] the protein dynamical crossover and finally in the irreversible denatured state [$T = 350$ K (**c**)]. The figure shows the water and the peptide (amide I, II, and III) components [79]. With permission from American Institute of Physics

temperature of the protein are approximately the same. The fragile-to-strong dynamic transition in water governs the nature of the H-bonds between water and peptides Thus, protein flexibility, essential to their enzymatic catalysis and their other biological functions, results from conformational disorder in the protein. The obtained

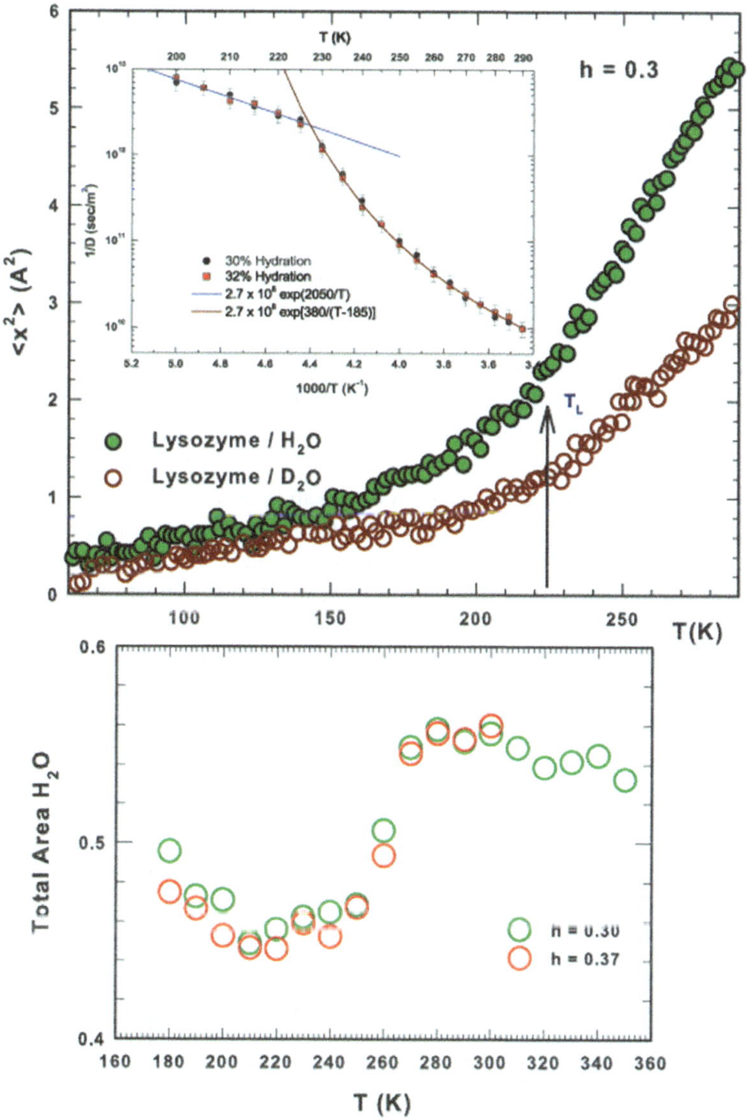

Fig. 7.41 Evidence for the protein dynamic transition. (Top panel) The mean squared atomic displacement MSD averaged over all the hydrogen atoms, $\langle X^2 \rangle$, as a function of temperature for the H_2O and D_2O-hydrated lysozyme samples at the hydration level $h = 0.30$ and 0.32. The arrow indicates the crossover temperature TC. In the inset is reported the behavior of the protein–water self-diffusion coefficient $D(T)$, at the same hydration, that shows the strong-fragile crossover at TL TC. (Bottom panel) The spectral area, relative to the total FTIR measured area, of the water bending components for $h = 0.3$ and 0.37 [79]. With permission from American Institute of Physics

results indicate that above the water FSDC all protein–water recovers the hydrogen-bond dynamics that had been frozen in the harmonic "glass" state and protein flexibility, essential to their enzymatic catalysis and their other biological functions, results from conformational disorder in the protein.

Picosecond to and RNase A powders using dielectric spectroscopy and hydrogen atom single particle dynamics simulations focused on tens of picoseconds and nanosecond relaxation processes were studied [80].

The enzymes were lyophilized, and the dry protein powder was hydrated using isopiestic equilibration to obtain 0.4 g of water per gram of protein. Hydration levels were determined by thermogravimetric analysis. Dielectric spectra in the frequency range from 102 to 107 Hz were measured using a Concept 80 system from Novocontrol that allowed to observe temperature dependence of the relaxation times observed in hydrated proteins (Fig. 7.42). Detailed analysis of the MD simulations and comparison to dielectric data indicated that: (1) fast (tens of picoseconds) and slow (nanosecond) relaxation processes were observed in dielectric spectra of hydrated protein samples; (2) the observed relaxation process in the nanosecond time range of hydrated protein spectra is mainly due to protein atoms in the protein backbone, side chains; (3) both surface and buried protein atoms contribute to the slow processes; (4) surface atoms demonstrate slightly faster relaxation dynamics; (5) analysis of the water molecule residence and dipolar relaxation correlation behavior indicates that the hydration water relaxes at much shorter timescales; (6) protein hydration water exchanges with the bulk water on a timescale faster than 100 ps; and (7) there are very few hydration water molecules that do not exchange with bulk water on a timescale of 1 ns.

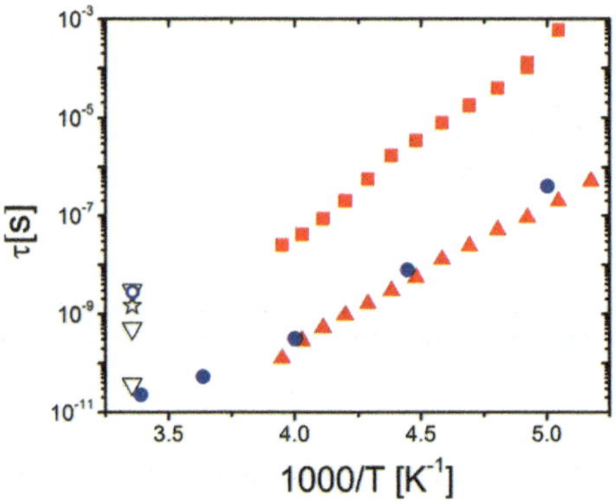

Fig. 7.42 Temperature dependence of the relaxation times observed in hydrated proteins by dielectric spectroscopy. Hydrated lysozyme ($h = 0.4$) powder 2, faster relaxation process; 9, slower relaxation process. Protein solutions at $T = 298$ K: r, RNase A; g, RNase A; and O, lysozyme b, neutron scattering data for lysozyme relaxation in hydrated powder ($h = 0.4$). Error bars for dielectric data are the size of symbols and represent the standard deviation [80]. With permission from American Institute of Physics

Various aspects of the molecular dynamics of hydrated proteins are described and discussed in a series of monographs [82–97].

References

1. G.I. Likhtenshtein, Regularities in the energy and entropy properties of enzymatic processes. Biofizika **11**, 24–32 (1966)
2. G.I. Likhtenshtein, On the nature of compensation phenomena in liquid phase reactions. Russian. J. Phys. Chem. **44**, 1980–1990 (1970)
3. G.I. Likhtenshtein, *Spin Labeling Method in Molecular Biology* (N.Y. Wiley Interscience, 1976)
4. G.I. Likhtenshtein, *New Trends In Enzyme Catalysis and Mimicking Chemical Reactions* (Kluwer Academic/Plenum Publishers, N.Y., 2003)
5. E.T. Denisov, O.M. Sarkisov, G.I. Likhtenshtein, Chemical Kinetics, in *Fundamentals and Recent Developments* (Elsevier Science, 2003)
6. R. Lumry, H. Eyring, Conformational changes in proteins. J. Phys. Chem. **58**, 110–120 (1954)
7. R. Lumry, S. Rajender, Enthalpy-entropy compensation phenomenon in water solutions of proteins and small molecules: ubiquitous property of water. Biopolymers **9**, 1125–1227 (1970)
8. R. Lumry, R.B. Gregory, Free energy management in protein reactions: concepts, complications, and compensation, in *The Fluctuating Enzymes*, ed. by G.R. Welch (New York, Wiley, 1986)
9. G.I. Likhtenshtein, *Biophysical Labeling Methods in Molecular Biology* (Cambridge University Press, Cambridge, N.Y., 1993)
10. G.I. Likhtenshtein, Water and dynamics of proteins and membranes. Studia Biophysica **111**, 89–100 (1986)
11. G.I. Likhtenshtein, Study of protein dynamics by spin-labeling, Mösbauer spectroscopy, and NMR, in *Special Collogue Amper on Dynamic Processes in Molecular Systems*. ed. by A. Losche (Karl-Marx University, Leipzig, 1979), pp. 100–107
12. G.I. Likhtenshtein, Water and protein dynamics, in *L'eau et les Systemes Biologoque*, eds. by A. Alfsen, A.J. Bertran (CNR, Paris, 1976), pp. 45–43
13. G.I. Likhtenshtein, *Spin Labeling Method in Molecular Biology* (Wiley Interscience, N.Y., 1976)
14. E.N. Frolov, O.V. Belongova, G.I. Likhtenshtein, Investigation of the mobility of spin and Mössbauer labels bound to macromolecules, in *Equilibrium Dynamics of the Native Structure of Protein*, ed. by E.A. Burshtein (Izdatelstvo Akademii Nauk SSSR, Pushchino, 1977), pp. 99–142
15. E.N. Frolov, A.N. Mokrushin, G.I. Likhtenshtein, V.A. Trukhtanov, V.I. Goldansky, Investigation of dynamic structure of proteins using gamma-resonance labels. Dokl. Akad. Nauk SSSR **212**, 165–168 (1973)
16. O.V. Belonogova, G.I. Likhtenshtein, V.I. Krinichnyi, Electron transfer in donor-acceptor pairs in modified a-chymotrypsin. Effect of microviscosity, macroviscosity and local polarity. Photochem. Photobiol. A Chem. **107**, 35–42 (1997)
17. G.I. Likhtenshtein, *Spin Labeling Method in Molecular Biology* (Moscow Nauka, 1974). (In Russian); O.V. Belonogova, G.I Likhtenshtein, G.I. Levashov, A.V. Khmelnitskij, N.L. Klyachko, K. Martinek, A spin-lavel study of the state of the active center microenvironment of a-chymotrypsin solubilized in octan, using the surfactant aerosol OT. Biokhimia **48**, 379 (1983)
18. G.I. Likhtenshtein, A.I. Kotelnikov, study of the functional intermolecular mobility of proteins by the physical technique. Moleculyarnaya Biologiya **17**, 505–518 (1983)
19. G.I. Likhtenshtein, Amorphous-crystaline model of intermolecular dynamics in global proteins. Biofizika **30**, 30–37 (1985)

20. J. Lakowicz, G. Weber, Quenching of protein fluorescence by oxygen. Detection of structural fluctuations in proteins on the nanosecond time scale. Biochemistry **12**, 4171–4179 (1973)
21. V. Reat, R. Dunn, M. Ferrand, J.L. Finney, R.M. Daniel, J.C. Smith, Solvent dependene of dynamic transition in protein solution. Proc. Natl. Acad. Sci. USA **97**, 9961–9966 (2000)
22. A.M. Tsai, D.A. Neumann, L.N. Bell, Molecular dynamics of solid-state lysozyme as affected by glycerol and water: a neutron scattering study. Biophys. J. **79**, 2728–2732 (2000)
23. K. Wütrich, *NMR of Proteins and Nucleic Acids* (A Wiley-Interscience Publication, N. Y., 1986)
24. G.I. Likhtenstein, Y.B. Grebentchikov, P.K. Bobodzhanov, Y.V. Kokhanov, Spin label-spin probe methods. Mol. Biol. (Moscow) **4**, 782–789 (1970)
25. M. Karplus, M. Cammon, Dynamics of protein structure. Sci. Am. **251**, 4–12 (1986)
26. M. Karplus, G.A. Petsko, Molecular dynamics simulations in biology. Nature **347**, 631–639 (1990)
27. D. Vitkup, D. Ringe, G.A. Petsko, M. Karplus, Solvent mobility and the protein 'glass' transition. Nature Struct. Biol. **7**, 34–35 (2000)
28. K.H. Linderstrom-Lang, J.A. Schellmann, Protein structure and enzyme activity, in *The Enzyme V.1*, eds. by P.D. Boyer, M. Lardy, Myrback (Academic Press, N.Y., 1959), pp. 443–510
29. D.J. Koshland, Mechanism of transfer of enzymes, in *The Enzyme V.1*, eds. by P.D. Boyer, M. Lardy, Myrback (Academic Press, N.Y., 1959), pp. 305–346
30. S. Mondal, S. Mukherjee, B. Bagchi, Protein hydration dynamics: much ado about nothing? J. Phys. Chem. Lett. **8**, 4878–4884 (2017)
31. G. Schirò, M. Weik, Role of hydration water in the onset of protein structural dynamics. J. Phys. Condens. Matter **31**, 463 (2019)
32. S. Khodadadi, A.P. Sokolov, Atomistic details of protein dynamics and the role of hydration water. Biochem. Biophys. Acta. **1861**, 3546–3552 (2017)
33. B. Bagchi, Water dynamics in the hydration layer around proteins and micelles. Chem. Rev. **105**, 3199 (2005)
34. B. Bagchi, *Molecular Relaxation in Liquids* (Oxford University Press, USA, 2012)
35. A.C. Fogarty, D. Laage, Water dynamics in protein hydration shells: the molecular origins of the dynamical perturbation. J. Phys. Chem. B **118**, 7715–7729 (2014)
36. E. Duboué-Dijon, D. Laage, Comparative study of hydration shell dynamics around a hyperactive antifreeze protein and around ubiquitin. J. Chem. Phys. **141**, 22D529 (2014). https://doi.org/10.1063/1.4902822
37. S. Ebbinghaus, S.J. Kim, M. Heyden, X. Yu, M. Gruebele, D.M. Leitner, M. Havenith, Protein Sequence- and pH-Dependent Hydration Probed by Terahertz Spectroscopy. J. Am. Chem. Soc. **130**, 2374–2375 (2008)
38. M. Nakasako, Water–protein interactions from high-resolution protein crystallography. Trans. R. Soc. Lond. **B 359**, 1191–12 06 (2004)
39. S. Mondal, S. Mukherjee, B. Bagchic, Origin of diverse time scales in the protein hydration layersolvation dynamics: a simulation study. J. Chem. Phys. **147**, 154901 (2017)
40. M. Kinoshita, T. Hayashi, Accurate and rapid calculation of hydration free energy and its physical implication for biomolecular functions. Biophys. Rev. **12**, 469–480 (2020)
41. O. Sushko, R. Dubrovka, R.S. Donnan, Sub-terahertz spectroscopy reveals that proteins influence the properties of water at greater distances than previously detected. J. Chem. Phys. **142**, 055101 (2015)
42. F. Novelli, S.O. Pour, J. Tollerud, A. Roozbeh, D.R.T. Appadoo, E.W. Blanch, J.A. Davis, Time-Domain, THz spectroscopy reveals coupled protein–hydration dielectric response in solutions of native and fibrils of human lysozyme. J. Phys. Chem. **B 121**, 4810–4816 (2017)
43. J. Xu, K.W. Plaxco, S.J. Allen, Probing the collective vibrational dynamics of a protein in liquid water by terahertz absorption spectroscopy. Protein Sci. **15**, 1175–1181 (2006)
44. S. Luthra, D.S. Kalonia, M.J. Pikal, Effect of hydration on the secondary structure of lyophilized proteins as measured by fourier transform infrared (FTIR) spectroscopy. J. Pharm. Sci. **96**, 2910–2921 (2007)

45. F. Mallamace, C. Corsaro, D. Mallamace, S. Vasi, C. Vasi, G. Dugo, The role of water in protein's behavior: the two dynamical crossovers studied by NMR and FTIR techniques. Comput. Struct. Biotechnol. J. **13**, 33–37 (2015)
46. S. Hume, G.M. Greetham, P.M. Donaldson, M. Towrie, A.W. Parker, M.J. Baker, N.T. Hunt, 2D-infrared spectroscopy of proteins in water: using the solvent thermal response as an internal standard. Anal. Chem. **92**, 3463–3469 (2020)
47. J. Tan, J. Zhang, C. Li, Y. Luo, S. Ye, Ultrafast energy relaxation dynamics of amide I vibrations coupled with protein-bound water molecules Ultrafast energy relaxation dynamics of amide I vibrations coupled with protein-bound water molecules. Nat. Commun. **10**, 1010 (2019)
48. Z.F. Brotzakis, C.C.M. Groot, W.H. Brandeburgo, H.J. Bakker, P.G. Bolhuis, Dynamics of Hydration Water Around Native and Misfolded α-Lactalbumin. J. Phys. Chem. B **120**, 4756−4766 (2016)
49. C. Bonechia, G. Tamasia, A. Pardinia, A. Donatia, V. Volpia, G. Leonea, M. Consumia, A. Magnania, C. Rossia, Ordering effect of protein surfaces on water dynamics: NMR relaxation study. Biophys. Chem. **249**, 106149 (2019)
50. D. Jeong, S. Han, Y.-B. Lim, S.H. Kim, Investigation of the hydration state of self-assembled peptide nanostructures with advanced electron paramagnetic resonance spectroscopy. ACS Omega **4**, 114–120 (2019)
51. M. Heyden, D.J. Tobias, D.V. Matyushov, Terahertz absorption of dilute aqueous solutions. J. Chem. Phys. **137**, 235103 (2012)
52. A. Volkov, C. Dockter, C., Bund, H. Paulsen, G. Jeschke, Pulsed EPR determination of water accessibility to spin-labeled amino acid residues in LHCIIb. Biophys. J. **96**, 1124−1141 (2009)
53. E.A. Perets, E.C.Y. Yan, Chiral water superstructures around antiparallel β sheets observed by chiral vibrational sum frequency generation spectroscopy. J. Phys. Chem. Lett. **10**, 3395–3401 (2019)
54. G. Schirò, M. Weik, Role of hydration water in the onset of protein structural dynamics. J. Phys. Condens. Matter **31**, 463002 (2019)
55. D. Laage, T. Elsaesser, J.T. Hynes, Water dynamics in the hydration shells of biomolecules. Chem. Rev. **117**, 10694–10725 (2017)
56. D. Russo, J. Teixeira, L. Kneller, J.R.D. Copley, J. Ollivier, S. Perticaroli, E. Pellegrini, M.A. Gonzalez, Vibrational density of states of hydration water at biomolecular sites: hydrophobicity promotes low density amorphous ice behavior. J. Am. Chem. Soc. **133**, 4882–4888 (2011)
57. L. Zhang, L. Wang, Y.-T. Kao, W. Qiu, Y. Yang, O. Okobiah, D. Zhong, Mapping hydration dynamics around a protein surface. Proc. Natl. Acad. Sci. USA **104**, 18461–18466 (2007)
58. P.N. Gallo, J.C. Iovine, N.V. Nucci, Toward comprehensive measurement of protein hydration dynamics: facilitation of NMR-based methods by reverse micelle encapsulation. Methods **148**, 146–153 (2018)
59. S. Chakraborty, S.K. Sinha, S. Bandyopadhyay, Low-frequency vibrational spectrum of water in the hydration layer of a protein: a molecular dynamics simulation study. J. Phys. Chem. B **111**, 13626–13631(2007)
60. N.V. Nucci, M.S. Pometun, A.J. Wand, Mapping the hydration dynamics of ubiquitin. J. Am. Chem. Soc. **133**, 12326–12329 (2011)
61. J.R. Lewandowski, M.E. Halse, M. Blackledge, L. Emsley, Direct observation of hierarchical protein dynamics. Science **348**, 578 (2015)
62. B. Halle, L. Nilsson, Does the dynamic Stokes shift report on slow protein hydration dynamics? J. Phys. Chem. B **113**, 8210–8213 (2009)
63. Y. Qin, M. Jia, J. Yang, D. Wang, L. Wang, J. Xu, D. Zhong, Molecular origin of ultrafast water−protein coupled interactions. J. Phys. Chem. Lett. **7**, 4171–4177 (2016)
64. M. Jia, J. Yang, Y. Qin, D. Wang, H. Pan, L. Wang, J. Xu, D. Zhong, Determination of protein surface hydration by systematic charge mutations. J. Phys. Chem. Lett. **6**, 5100–5105 (2015)
65. J. Yang, Y. Wang, L. Wang, D. Zhong, Mapping hydration dynamics around a β-barrel protein. J. Am. Chem. Soc. **139,** 4399−4408 (2017)
66. C.-Y. Chenga, J. Varkey, M.R. Ambroso, R. Langen, S. Hana, Hydration dynamics as an intrinsic ruler for refining protein structure at lipid membrane interfaces. PNAS **110**, 16839 (2013)

67. B.D. Armstrong, S. Han, Overhauser dynamic nuclear polarization to study local water dynamics. J Am Chem Soc **131**, 4641–4647 (2009)
68. J.H. Freed, Dynamic effects of pair correlation functions on spin relaxation by translational diffusion in liquids. II. Finite jumps and independent T_1 processes. J. Chem. Phys. **68**, 4034–4037 (1978)
69. J.K. Williams, M. Hong, Probing membrane protein structure using water polarization transfer solid-state NMR. J. Magn. Reson. **247**, 118–127 (2014)
70. C. Jorge, B.S. Marques, K.G. Valentine, A. J. Wand, Characterizing protein hydration dynamics using solution NMR spectroscopy. Methods Enzymol. **615**, 77–101 (2019)
71. M. Bastos, N. Alves, S. Maia, P. Gomes, A. Inaba, Y. Miyazaki, J.-M. Zanotti, Hydration water and peptide dynamics—two sides of a coin. A neutron scattering and adiabatic calorimetry study at low hydration and cryogenic temperatures. Phys. Chem. Chem. Phys. **15**, 16693–16703 (2013)
72. K. Wood, A. Frolich, A. Paciaroni, Moulin M, Hartlein, G. Zaccai, D.J. Tobias, M. Weik, Coincidence of dynamical transitions in a soluble protein and its hydration water: direct measurements by neutron scattering and MD simulations. J. Am. Chem. Soc. **130**, 4586–4587 (2008)
73. F.-X. Gallat, A. Laganowsky, K. Wood, F. Gabel, L. van Eijck, J. Wuttke, M. Moulin, M. Härtlein, D. Eisenberg, J.-P. Colletier, G. Zaccai, M. Weik, Dynamical coupling of intrinsically disordered proteins and their hydration water: comparison with folded soluble and membrane proteins. Biophys. J. **103**, 129–136 (2012)
74. S.-H. Chen, L. Liu, E. Fratini, P.A. Baglioni, A. Faraone, E. Mamontov, Observation of fragile-to-strong dynamic crossover in protein hydration water. Proc. Natl. Acad. Sci. USA **103**, 9012–9016 (2006)
75. F. Mallamace, Role of the solvent in the dynamical transitions of proteins: the case of the lysozyme-water system. J. Chem. Phys. **127**, 045104 (2007)
76. H. Nakagawa, M. Kataoka, How can we derive hydration water dynamics with incoherent neutron scattering and molecular dynamics simulation? Biophys. Physicobiol. **16**, 213–219 (2019)
77. V. Conti Nibali, G. D'Angelo, A. Paciaroni, D.J. Tobias, M. Tarek, On the coupling between the collective dynamics of proteins and their hydration water. J. Phys. Chem. Lett. **5**, 1181–1186 (2014)
78. S. Combet, J.M. Zanotti, Further evidence that interfacial water is the main "driving force" of protein dynamics: a neutron scattering study on perdeuterated C-phycocyanin. Phys. Chem. Chem. Phys. **14**, 4927–4934 (2012)
79. F. Mallamac, P. Baglioni, C. Corsaro, S.H. Chen, D. Mallamace, C. Vasi, H.E. Stanley, the influence of water on protein properties. J Chem Phys. **141** 165104 (2014)
80. S. Khodadadi, J.E. Curtis, A.P. Sokolov, Nanosecond relaxation dynamics of hydrated proteins: water versus protein contributions. J. Phys. Chem. B **115**, 6222–6226 (2011)
81. J.D. Nickels, H. O'Neill, L. Hong, M. Tyagi, G. Ehlers, K.L. Weiss, Q. Zhang, Z. Yi, E. Mamontov, J.C. Smith, A.P. Sokolov, Dynamics of protein and its hydration water: neutron scattering studies on fully deuterated GFP. Biophys J. **103**, 1566–1575 (2012)
82. D.R. Livesay, *Protein Dynamics. Methods and Protocols* (Springer Science, New York, 2014)
83. R. Zhou, *Molecular Modeling at the Atomic Scale: Methods and Applications in Quantitative Biology (Series in Computational Biophysics)* (CRC Press, 2015)
84. C.L. Brooks, M. Karplus, B. Pettitt, Proteins: *A Theoretical Perspective of Dynamics, Structure, and Thermodynamics* (Wiley-Interscience, 1988)
85. H. Frauenfelder, *Pressure Studies of Protein Dynamics* (Enero, 1992)
86. K. Han, X. Zhang, M. Yang, Protein conformational dynamics, in *Advances in Experimental Medicine and Biology*, 805 (Springer, 2014)
87. J. Zhang, *Molecular Dynamics Analyses of Prion Protein Structures: The Resistance to Prion Diseases Down Under* (Springer, 2018)
88. A. Kessel, N. Ben-Tal, *Introduction to Proteins: Structure, Function, and Motion*, 2nd edn. (Chapman & Hall/CRC Computational Biology Series, CRC, 2018)

89. N. Salvi, *Intrinsically Disordered Proteins Dynamics Binding and Function* (Elsevier Inc., 2019)
90. M. Fuxreiter, *Computational Approaches to Protein Dynamics: From Quantum to Coarse-Grained Methods* (Series in Computational Biophysics) (CRC Press, 2015)
91. C.L. Meyerkord, H. Fu, *Protein-Protein Interactions: Methods and Applications* (Humana Press, 2015)
92. N. Salvi, *Intrinsically Disordered Proteins: Dynamics, Binding, and Function* (Elsevier, 2019)
93. N. Rama Krishna, L. Berliner, *Structure Computation and Dynamics in Protein NMR* (Springer, 2013)
94. D.J. Rigden, *From Protein Structure to Function with Bioinformatics* (Springer, 2017)
95. M.N.V. Prasad Gajula, *A Novel Approach to Study Protein Dynamics by EPR and MD Simulations* (Scholar's Press, 2014)
96. A.K. Srivastava, *Investigation of Protein Dynamics Using the Genetic Code Expansion* (Lambert, 2017)
97. I. Bahar, R.L. Jernigan, K.A. Dill, *Protein Actions Principles and Modeling* (Taylor and Frances, 2017)

Chapter 8
Water in Enzyme Catalysis as a Promoter and Chemical Reagent

Abstract The concepts of the dynamic structure of globular proteins and its role in the enzyme catalysis are supported by a large body of experimental and theoretical data. This chapter provides multiple evidence of the influence of the environment, primarily water, on the molecular dynamics of enzyme active sites and its chemical behavior. The experiments carried out on dry and moistened powders and at the presence of trehalose and sucrose, which excluded any motion by the macromolecule as a whole revealed nanosecond intermolecular mobility consisting of movement of the relatively large and rigid parts of the protein macromolecules modulated by mobility of the surrounding water. Many kinetic and thermodynamic studies of enzyme reaction series reveal the existence of an activation enthalpy–entropy linear correlation, the enthalpy–entropy compensation effect (EECE). The EECE theoretical model and the effect experimental examples are described.

8.1 Introduction

The concepts of the dynamic structure of globular proteins put forward in the 1950s and 1960s [1–3] have found numerous experimental and theoretical support. The changes in proteins confirmation upon binding to the substrate predicted by Lumry and Eyring and Koshland have also been numerously confirmed [1, 3]. According to Kramers theory, solvent viscosity results in friction against proteins in solution, and this should result in decreased motion, inhibiting catalysis in motile enzymes. Coupling between the collective dynamics of proteins and their hydration water was unambiguously established in 1970th–1980th [4–13]. Nevertheless, a connection between the enzymes molecular dynamics, its hydration, and their functional activity remained the most challenging problem.

The protein dynamics concept was developed into the theory of "complementarity" of the free energy of the chemical reaction in the enzyme active center and the enzyme conformational free energy. In this theory, energy redistribution occurs in such a way that the energy profile of the process as a whole is eventually flattened [14–16]. Regarding to enzyme reactions, the Lumry concepts of the enzyme reactions were supplemented by the principle of the dynamic adaptation of the enzyme

G. I. Likhtenshtein, *Biological Water*, Soft and Biological Matter,
https://doi.org/10.1007/978-3-030-82503-4_8

conformational structure to the substrate configuration in consecutive reactions to the enzyme put forward by Likhtenshtein [17, 18]. Such an adaptation promotes both the first step of precise orientation and the subsequent chemical steps, without allowing cleavage of the contacts needed for the chemical mechanism.

8.2 Role of Water in the Dynamics and Chemical Reactions of Enzymes Active Sites

There are numerous evidences of the influence of the environment, primarily water, on the molecular dynamics of enzymes active sites and its chemical behavior [14–27] (see also Section). Below are some typical examples.

The first direct evidence of the nanosecond intramolecular mobility of heme-protein globules at ambient condition was first obtained in 1978 by the spin and Mössbauer labeling methods in the Likhteshtein and Goldansky groups [19]. Later this finding was repeatedly confirmed [7–9, 20–26].

In study [28], the experiments were carried out on dry and moistened powders, which excluded any motion by the macromolecule as a whole. The ^{57}Fe atoms were incorporated into the heme group in myoglobin and hemoglobin. Given that the rigid heme ring is bound to the protein by numerous contacts, it is evident that anharmonic motion of heme above 200 K is related to the intramolecular mobility of the protein globule. This mobility appears only at a critical degree value of hydration samples (P/Ps > 0.65). The increase of mobility recorded by Mössbauer spectroscopy correlates with the data for isotopic H–D whose relatively higher amplitude must be accompanied by the displacement of the helical polypeptide chain. Such an anharmonic nanosecond motion with $\prec \Delta x \succ nh > 0.2$ Å was also recorded at temperature $T > 210$ K in myoglobin using spin and fluorescence labeling methods [28, 29] and references therein. The flexibility of the cavity of the myoglobin active site is evidenced by the mobility of a spin probe, a derivative of isocyanate attached to the heme group in the single crystal. At room temperature, the mobility parameters were found as follows: Correlation frequency τ_c is about 10^8, $E_{app} = 2.8$ kcal/mole, $\Delta S_{app} = 14$ e.u.

Nuclear forward scattering of synchrotron radiation was employed used to determine the quadrupole splitting and the mean square displacement of the iron atom in deoxymyoglobin in the temperature range between 50 K and 243 [22]. The results strongly support the picture that protein dynamics seen at the position of the iron can be understood by harmonic motions in the low-temperature regime, while above 200 K, diffusive motions in limited space are present.

By varying the temperature (−15 to 65 °C) and humidity for samples of carbonmonoxy myoglobin embedded in trehalose-glass, it was possible to observe a hierarchy of distinct geminate recombination phases that extend from nanosecond to almost seconds that can be directly associated with rebinding from specific hydrophobic cavities [23]. The use of mutant forms of Mb reveals the role of key residues in modulating ligand access between these cavities and the distal heme pocket. A room-temperature molecular dynamics (MD) simulation on a system containing carboxy-myoglobin (MbCO) molecule in a sucrose–water matrix of identical composition (89% [sucrose/(sucrose + water)] w/w) was performed [24].

The following results were reported: (1) as for trehalose and sucrose, the amplitude of protein atomic mean square fluctuations, on the nanosecond timescale, is reduced with respect to aqueous solutions; (2) different heme pocket structures are observed in the two systems; (3) the joint distribution of the magnitude of the electric field at the CO oxygen atom and of the angle between the field and the CO unit vector indicates a secondary maximum in sucrose, absent in trehalose; (4) the CO stretching band profile differences are evidenced by infrared spectroscopy in sucrose- and trehalose-coated MbCO and in particular the appearance of a further substate in sucrose; (5) analysis of hydrogen bonds at the protein–solvent interface shows that the fraction of water molecules shared between the protein and the sugar is lower in sucrose than in trehalose, in spite of a larger number of water molecules bound to the protein in the former system; and (6) the latter indicates a lower protein-matrix coupling, as observed by Fourier transform infrared (FTIR) experiments.

Viscosity and humidity strongly affect photochemical reactions in proteins. In work [26], spin-labeled α-chymotrypsin with the methionine-192 group modified with 4-iodacetamide-2,2,6,6-tetramethylpiperidine-1-oxyl was prepared. The effects of temperature (T = 200–300 K), viscosity, and humidity on the rate constant of photoreduction of the label nitroxide fragment (acceptor) by the protein tryptophan residue in the excited singlet state (donor) were investigated employing X-band ESR spectroscopy. In parallel, the dynamics of the microenvironment in the vicinity of the nitroxide and tryptophan were studied by high-resolution 2 mm ESR and fluorescence techniques, respectively. The experiments revealed that the changes in the rate constant of photoreduction correlate with the changes in the parameters of the local dynamics of the donor–acceptor pair microenvironment (correlation time $\tau_c \le 10^{-7}$ s) recorded by ESR spectroscopy and fluorescence.

Work [27] provided a full analysis and a detailed study of slow relaxation following photoexcitation of Trp-7 (W7) in myoglobin and the dynamics of its hydration layer employing molecular dynamics methods (Fig. 8.1). To calculate the time-dependent Stokes shift following photoexcitation of Trp-7 (W7) in myoglobin, molecular dynamics simulations were used. Simulations for the 154-residue protein myoglobin were conducted using a double precision version of the GROMACS package23–25 and GROMOS96 force field. Solvent dynamics were examined using the SPC/E water model because the dynamical properties like the diffusion constant are in good agreement with experiment for this model.

Examination of the structures before and after the transition point indicated that: (1) the structure transition at 10 ns is associated with the loop region between R-helices E and F, residues 77–86 (Fig. 8.2) (2) the loop is closer to the indole during the first 10 ns of the molecular dynamics run; and (3) after the transition, the loop residues fluctuate more, and R-helix E is more disordered near the loop. As it is seen in Fig. 8.2, for isomer 1, the protein response shows life time dependence beyond ~ 20 ps, while the water response is responsible for the slowest $\tau_2 = 56$ ps component present in

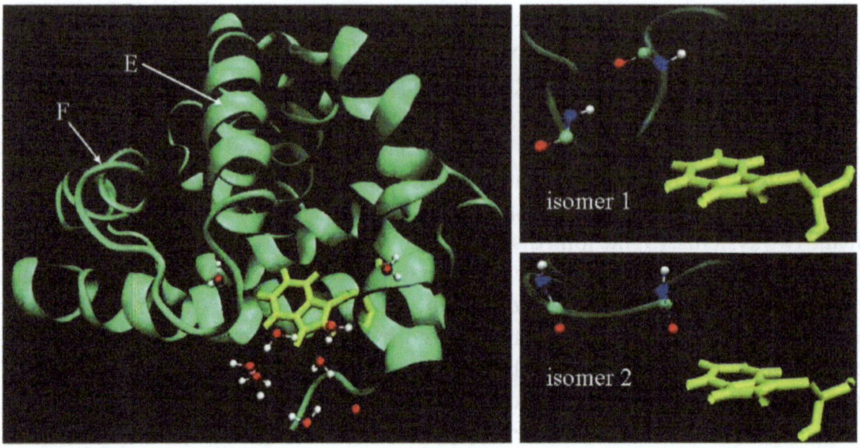

Fig. 8.1 Representative snapshots of tryptophan W7 and the loop region consisting of residues 77–86 that portray the structural transition. On the left, the E and F helices, the tryptophan W7, and some waters of hydration are shown. The proximity of the loop joining helices E and F is evident. On the right, snapshots from the isomer 1 and 2 substates are shown. The atoms of the peptide bonds connecting residues 78 and 79 and 80 and 81 are explicitly shown, reflecting the shift in angles relative to W7 [27]. With permission from American Chemical Society

the overall response. These features are seen in the linear response approximation, in the response calculated from non-equilibrium molecular dynamics trajectories, and freezing both isomers 1 and 2 at the instant of photoexcitation which removes all long time dynamics Therefore, for the case of W7 in myoglobin, protein motion is essential for observation of slow dynamics of the Stokes shift, and the latter is not an inherent feature of water trapped.

In conclusion, the experiments revealed that (1) the slow ($\sim$ 60 ps) timescale component of the Stokes shift of photoexcited W7 in myoglobin arises from coupled protein–water motion; (2) protein flexibility is required to observe the slow Stokes shift dynamics; (3) the water molecules can be classified into a more labile population that is typically further from the protein and a tightly bound, nearest neighbor layer within $\sim$ 3.3 Å of the protein; (4) orientational and translational motion is slowed considerably within the hydration layer, particularly for the tightly bound subpopulation; and (5) diffusion in the hydration layer is in the orders of magnitude slower than in bulk water.

In the search for a correlation between an enzyme functional activity and its molecular dynamics agitation by humidity, α-chymotrypsin was modified by a spin label attached to the methionine-92 group in the region of the active site and by a Mössbauer label in the enzyme globule periphery (Fig. 7.4). The protein tryptophane group also served as an intrinsic fluorescence label. Agitation of the label mobility in the nanosecond region at relative humidity about $P_{rel} = 0.6$ correlates with a sharp increase in the hydrolysis degree of the cyanomoyl α-cymotrypsin

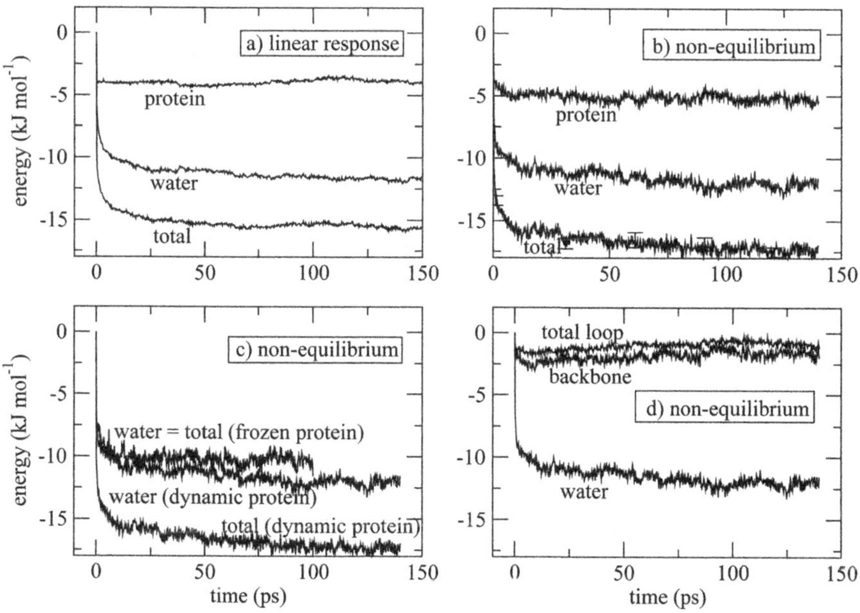

Fig. 8.2 Fluorescence stokes shift for isomer 1. **a** Linear response theory estimate for the total Stokes shift and the contributions from indole-water and indole-protein interactions. The time interval used to calculate the linear response correlation functions, (3) and (4), and sampled for initial conditions of the non-equilibrium trajectories. **b** Total Stokes shift and the contributions from indole-water and indole-protein interactions calculated from 360 non-equilibrium molecular dynamics trajectories. Error bars represent one standard deviation as estimated from block averages obtained by breaking the total dataset into three parts 69. **c** Comparison of the Stokes shift calculated with full protein dynamics, as in (**b**), or from 120 trajectories propagated with the protein coordinates frozen following photoexcitation. When the protein is frozen, the water response is the total response. **d** Contribution of the loop region (residues 77–86) to the Stokes shift and the contribution that arises from the backbone atoms of the loop [27]. With permission from American Chemical Society

covalent compound (Fig 7.4) [30–32]. This result first experimentally highlighted the significance of nanosecond dynamics in the chemical step of the enzyme reaction.

Quasi-elastic neutron and light scattering techniques along with molecular dynamics simulations were employed to study the influence of hydration on the internal dynamics of lysozyme [32]. The protein powder was dissolved in D_2O to replace exchangeable hydrogen atoms in lysozyme molecules with deuterium atoms. After lyophilization, hydration was achieved by isopiestic equilibration of the deuterium-exchanged protein with saturated solutions of LiCl, NaCl, and K_2SO_4 in D_2O which resulted in hydration levels of 0.05, 0.18, and 0.30, respectively. Samples with $h = 0.45$, 0.50, and 0.80 were prepared by adding D_2O. Two spectrometers, the high flux backscattering spectrometer (HFBS) NG2, and the disk-chopper time-of-flight spectrometer (TOF) NG4 were employed to cover a sufficiently broad energy range (1 meV, 240 MHz)–(100 meV, 24 THz) for analysis of the complex relaxation spectra of proteins. A solution of lysozyme was prepared with equimolar amounts of the hexasaccharide of N-acetylglucosamine at pH values between 8 and 9 and then was lyophilized to give the dry enzyme–substrate complex, which was subsequently rehydrated to the required hydration levels, and catalytic activity has been measured as a function of hydration level. Data on effect humidity on the protein dynamics were compared with correspondent dependence of its enzymatic activity (enzymatic hydrolysis rate, v_0, of the lysozyme to hexasaccharide of N-acetylglucosamine ((GlcNAc)6)(q) on the humidity [33].

Three major relaxation processes that contribute to the observed internal dynamics of lysozyme in the picosecond to nanosecond time range fluctuations of methyl groups, fast picosecond relaxation, and a slow relaxation process were identified. As an example, the observed hydration dependence of the slow relaxation process (mobile fraction of hydrogen atoms was found to be correlated with the hydration dependence of catalytic activity (Fig. 8.3): The mobile fraction remains very small up to $h = 0.18$, and it increases strongly between $h = 0.18$ and $h = 0.50$; a further weak increase is observed above $h = 0.50$ [33]. These results suggest that the slow relaxation process includes motions important for enzymatic activity of lysozyme. As it is seen in Fig. 8.3, the fast process is activated at much lower hydration levels than enzymatic activity and is significant even in the dry protein. The dependence of fast conformational fluctuations on hydration level is similar to the hydration dependence of hydrogen exchange.

The experiments revealed that:: (1) low-temperature onset of anharmonicity at $T = 100$ K ascribed to methyl-group dynamics is not sensitive to hydration level; (2) the

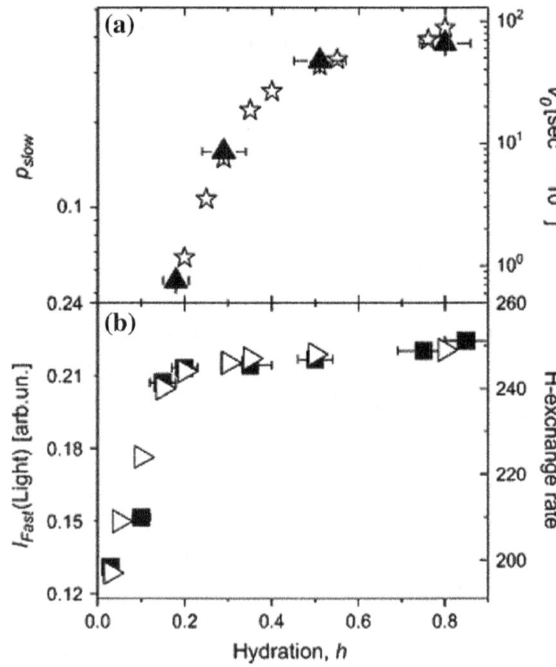

Fig. 8.3 **a** Parallel comparison of hydration dependences of the mobile fraction of hydrogen atoms involved in the slow relaxation process (:) and enzymatic reaction rate, v_0, of the lysozyme to hexasaccharide of N-acetylglucosamine ((GlcNAc)6)(q) as estimated by Rupley and colleagues (5.74). **b** Parallel comparison of hydration dependences of the integrated QES intensity of the fast process (n) and hydrogen exchange rate in units of moles of exchanged H atoms/1 mol of lysozyme/24 h ($T = 295$ K) [27]. With permission from Elsevier

increase of hydration level animates the fast relaxation process and then activate the slow relaxation process at $h = 0.2$; (3) the quasi-elastic scattering intensity associated with the slow process increases sharply with an increase of hydration to above $h = 0.2$; (4) activation of the slow process is responsible for the dynamical transition at $T = 200$ K; (5) an increase in the hydration level strongly increases the QES intensity of lysozyme; (6) behavior upon hydration appears to be different at low and high frequency; (7) a step-like increase in QES intensity is observed between $h = 0.2$ and $h = 0.5$ for the slow dynamics, whereas a monotonic increase of the QES intensity with h is observed at high frequency; (8) the temperature dependence of $\prec r^2 \succ$ indicates that an additional relaxation process enters the studied frequency window at temperature above $T_D = 220$ K, but only a hydration levels 0.2 h; and (9) The dependence of the slow process on hydration correlates with the hydration dependence of the enzymatic activity of lysozyme, whereas the dependence of the fast process seems to correlate with the hydration dependence of hydrogen exchange of lysozyme.

8.3 Effect Viscosity on Mechanism of Enzyme Reaction

The study of enzyme reaction detailed mechanisms is fundamentally important to understanding of biochemistry, cellular metabolism, and drug development.

According to the Kramer's, Lumry's conception [8, 15, 16, 34] and the principle of dynamic adaptation [8], each chemical stage of the enzymatic reaction requires a corresponding change in protein conformation, and a different effect of the solvent viscosity on the rate of different stage should be expected. This makes it possible by changing the viscosity to change the rates of individual stages and even reveal "hidden" stages. The Kramers' theory predicted that enzymes alternating between widely different conformations during catalysis should be inhibited by viscosity.

The overall activity of animal fatty acid synthetase at the saturation level of substrate concentration decreased when the solvent viscosity, η, of the reaction mixture was increased with as glycerol, sucrose, and polyethylene glycol [35]. The activity of the enzyme changed roughly proportional to η^{-p}, where $p = 1.0$ for glycerol, $p = 0.66$ for sucrose, and $p < 0.6$ for polyethylene glycol with different molecular sizes. The use of kinetic solvent viscosity effects (KSVEs) for steady-state kinetic analyses to probe whether substrate binding is diffusion-controlled and whether product release is the rate-limiting step in the catalytic cycle of an enzyme reaction was reviewed [36]. KSVEs identifying the enzyme reaction steps were demonstrated. Figure 8.4 schematically illustrates the different effect of viscosity depending on which stage (conformational or chemical one) limits the process. The use of KSVEs in combination with other techniques, such as kinetic isotope effects, pH effects, and site-directed mutagenesis, can provide a detailed view of the mechanism of enzyme action. For example, such an approach can measure a ratio of rate constants that provides an indication of how either the ES complex proceeds forward through k_3 rather than dissociating through k_2 (i.e., k_3/k_2 or forward commitment

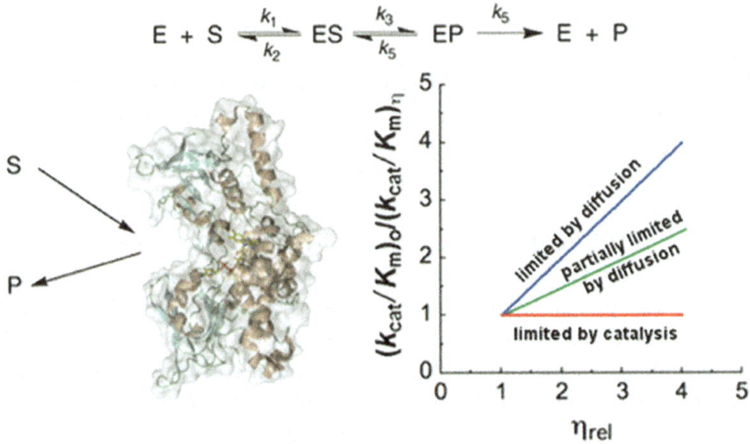

Fig. 8.4 Schematic illustration of different effect of viscosity on the kinetic parameters of an enzyme [35]. With permission from American Chemical Society

to catalysis) or the EP complex reverts backward through k_4 rather than proceeding forward through k_5 (i.e., k_4/k_5 or reverse commitment to catalysis).

The basic theory of solution viscosity and its effect on enzyme activity, important experimental considerations, and potential outcomes and brief discussion of some examples were presented [37]. As an example, an irreversible reaction for which there is no reverse flux of intermediates from EP to ES ($k_4 \approx 0$) and one for which $k_5 > k_4$ are practically indistinguishable. Nevertheless, they are not the same. Instead, when $k_5 > k_4$, it follows that the EP complex partitions forward to release the product rather than reverting back to the ES complex, making the flux through the reverse kinetic step k_4 negligible. Despite the reaction behaving as if it were irreversible, it is not. Indeed, one can envision altering environmental conditions by increasing viscosity to decrease the magnitude of the k_5 rate constant, thereby making k_5 comparable to k_4; this would result in the reaction behaving as if it were reversible. Consistent with Kramer's theory, the reaction rate of ATP hydrolysis by the H^+-ATPase from the plasma membrane of *Kluyveromyces lactis decreases by* adding increasing concentrations of glycerol, sucrose, and trehalos. The determination of enzyme kinetics and the application of Kramers' equation to evaluate the effect of viscosity on the rate of ATP hydrolysis by the H^+-ATPase were evaluated. At 20 °C, trehalose, sucrose, or glycerol increased solvent viscosity while inhibiting the H^+-ATPase, mainly through a decrease in V_{max}. In addition, increasing the temperature resulted in diminished effects viscosity on the V_{max} [36].

The flavoprotein nitroalkane oxidase catalyzes the oxidative denitrification of a broad range of primary and secondary nitroalkanes to yield the respective aldehydes or ketones, hydrogen peroxide, and nitrite [38]. Increasing the solvent viscosity did not affect the k_{cat} or k_{cat}/K_M value significantly, establishing that nitroethane binding is at equilibrium.

8.4 Mechanism of Protein Dynamics Related to Its Activity

The comparative analysis of the data obtained [7–13, 17, 18] revealed an apparent discrepancy between the physical labeling approach and certain other physical methods. Thus, the measurements of the temperature dependence of the heat capacity (c_p) of proteins, a-chymotrypsin, lysozyme, myoglobin, collagen, etc., at wide range of temperature at various degrees of hydration indicated only a monotonic increase of c_p and did not detect pronounced phase transitions before thermal denaturation [39–41]. However, spin, fluorescence and Mössbauer labeling, H-D exchange, non-elastic neutron scattering, and absorption spectra of proteins detected sharp transitions within this temperature range [7–13, 18, 28–30]. (See for example Fig. 7.3).

The above-mentioned apparent discrepancy was easily explained by proposal [7, 8, 17, 18, 30] that nanosecond intermolecular mobility consists of movement of the relatively large and rigid parts of the protein macromolecules modulated by mobility of the surrounding water. Such hinge-like oscillation of the tightly packed polypeptide blocks does not give a measurable contribution to the overall heat capacity and the helicity degree of polypeptide chains, but it strongly affects the mobility of the Mössbauer labels firmly bound to the protein blocks, the mobility of spin and fluorescence labels, native chromophores located in cavities between the blocks, etc. At a later stage, the mechanism later was confirmed by independent experimental investigation and theoretical calculations. Similar concepts of hinge-bending or domain motion introduced later were cited in [18].

The dynamic behavior of the protein molecules was described in terms of a "crystalline-amorphous" model [42]. According to model, the protein globule is made up of rigid blocks a-helical portions, β-structures, closely packed side groups, and softer, amorphous regions (unordered, loose portions, and water molecules). The blocks are bound by relatively flexible hinges and fairly weak interactions in the contacts between the blocks and the amorphous regions. Thus, the block system is embedded in a viscous amorphous medium. The globule is surrounded by a glycerol-like sheath composed of exposed surface groups and water molecules. In enzyme reactions, a rearrangement of the water assembly ("melting" or "crystallization"), which is accompanied elementary act chemical reaction, is the main contributor in experimental activation energy and entropy.

Recent data on proteins containing multiple folded domains separated by flexible linkers were reviewed [43]. As an example, the structural and dynamical properties of a multidomain protein, the three-domain protein TIA-1 with known high-resolution structures (Fig. 8.5a), were investigated combining coarse-grained molecular dynamics simulations with small-angle scattering data. The three-folded domains of TIA-1, RNA recognition motifs 1, 2, and 3 (RRM1, RRM2, and RRM3), are connected by linkers that provide a high degree of structural flexibility to the complex. To analyze measured SAXS small angle X-ray scattering (SAXS) and small-angle neutron scattering (SANS) data (Fig. 8.5b, c), the coarse grained force field Martini version 3.0.beta.4.17 [20, 21] in combination with GROMACS 5.1.4

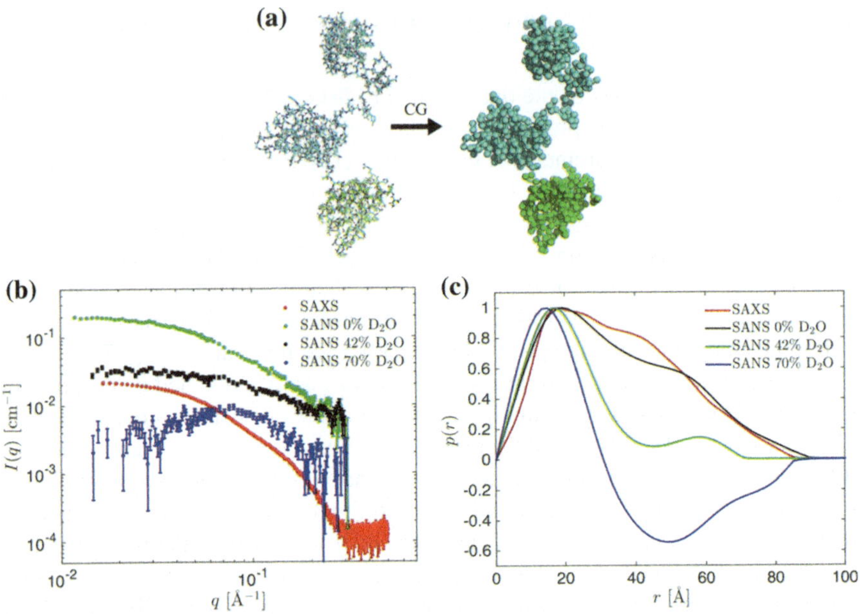

Fig. 8.5 TIA-1 structure and experimental SAXS and SANS data. **a** All-atom model of TIA-1 and the corresponding coarse-grained Martini model. **b** Experimental SAXS (black) and SANS data (green: 0% D_2O, red: 42% D_2O, blue: 70% D_2O). **c** Corresponding pair distance distribution functions, $p(r)$ [43]. Publish Content Under a CC BY License

or 2016.5 was used. The calculation revealed increasing the strength of protein–water interactions results in ensembles that are in good agreement with experiments. A general strategy for studying the conformation of m formation of multidomain proteins in solution that combines coarse-grained simulations with small-angle X-ray scattering data was suggested.

8.5 Enthalpy–Entropy Compensation

8.5.1 General

Many kinetic and thermodynamical studies concerning homologous reaction series report the existence of an activation enthalpy–entropy linear correlation (compensation plot), its slope being the temperature at which all the members of the series have the same rate constant (isokinetic temperature). Constable (1925) [44] found a straight line relation between the logarithm of the frequency factor (corresponding approximately to the entropy of activation $\Delta S^{\#}$) and the energy of activation (corresponding approximately to the enthalpy of activation $\Delta H^{\#}$). The next principle step in the area was made by Leffler [45] and Exner [46].

The enthalpy–entropy compensation effect in enzymatic catalysis was first described by Likhnenshtein in [4] and then confirmed by Lumry [5]. Both authors explained this effect by the rearrangement of the water ensemble accompanying the elementary chemical act. The activation parameters of this act are included in the experimental values of the energy and entropy of the activation [5, 6]. The energy (E_a) and entropy ($\Delta S^{\#}$) activation of an enzymatic reactions are found to be sensitive to the structure of substrate, temperature regions, pH, solvent composition, etc.

Recently, various aspects of the enthalpy–entropy compensation effect (EECE) were widely discussed [47–65].

For numerous enzyme reactions, the following linear dependences are satisfied [4, 5]:

$$E_a = \beta \Delta S_a^{\ddagger} + \Delta H_0^{\ddagger} \tag{8.1}$$

where $\Delta H^{\ddagger}{}_0$ is constant regardless of substituents, and $\Delta S^{\dagger}$ is different for each substituent. Inserting the Gibbs free energy equation and combining like terms produces the following equation:

$$\Delta G_a^{\ddagger} = \Delta H_0^{\ddagger} - (T - \beta_c)\Delta S^{\ddagger} \tag{8.2}$$

β_c has the dimension of temperature and is referred to as the isokinetic (or isoequilibrium) temperature because at $\beta_c = T$, $\Delta G_a{}^{\ddagger} = \Delta H^{\ddagger}{}_0$, that is, all reactants in the series having different $\Delta S^{\ddagger}$ ran with the same rate.

The difference between the E_a for reactants can be related to the difference between $\Delta S^{\ddagger}$ for the same reactants:

$$\delta E_a = \boldsymbol{\beta}_c \delta \Delta S^{\ddagger} \tag{8.3}$$

For a chemical reaction in active sites of enzymes, the experimental values of E_a and $\Delta S^{\#}$ are apparent parameters, which jointly reflect the entire complexity of the process (presence of intermediate microsteps, involving accompanying conformational transitions and solvent molecules reorganization).

The kinetic enthalpy–entropy compensation effect in protein denaturation is noted to be enormous ranges up to $E_a = 120$ kcal/mole and $\Delta S^{\#}$ up to 400 eu [54]. This quantity has been found to be highly sensitive to external conditional (pH, moisture content, ions present, etc.). The finding pointed out above may be accounted for the general bases of the assumption that the chemical elementary step on the enzyme is accompanied by the rearrangement of the conformational structure of protein globules and surrounding water molecules. The thermodynamic parameters of such a rearrangement should make key contribution to measured activation energy and entropy. At the same time, it is expected that the contribution of the movement of the solid domain fragments of the globule to the change in entropy will be insignificant in comparison with the contribution of water.

Considering the above arguments, we can represent the apparent activation parameters in the form of [6].

$$E_a = E_r + n\mathrm{DH}_0 \quad \text{and} \quad \mathrm{DS}^{\#} = \mathrm{DS}_r^{\#} + n\mathrm{DS}_0 + A \tag{8.4}$$

where E_r and $\Delta S_r{}^{\#}$ are attributed to energy and entropy activation of the chemical step, respectively; n is the number of particles in the surrounding cooperative assemble; ΔH_0 and ΔS_0 are the standard enthalpy and entropy of the rearrangement of one elementary unit (primary water), and the A value depends on the manner in which the particles are packed. Combination of 8.3 and 8.4 gives

$$\delta(E_r + n\mathrm{DH}_0) = \boldsymbol{\beta}_c\delta(\mathrm{DS}_r^{\#} + n\mathrm{DS}_0) \tag{8.5}$$

Suggestion that all series reactants have the same activation parameters of the chemical step and only are different by the number of molecules in the ensemble ($\delta E_r = 0$, $\delta\Delta S_r{}^{\#} = 0$) leads to

$$b_c = \mathrm{DH}_0/\mathrm{DS}_0 = T_m \tag{8.6}$$

where T_m is the "melting" temperature of the cooperative assemble which is expected to be close to average temperature of the experiment. In the general case, the chemical reaction and accompanying rearrangement cannot be regarded as independent events. Suggesting linear function of E_r and $\Delta S_r{}^{\#}$ of n with the coefficients being, respectively, equal to a_1 and b_1, and taking in consideration 8.5 we have

$$\delta(E_r + a_1 n\mathrm{DH}_0) = \beta_c\delta(\mathrm{DS}_r^{\#} + b_1 n\mathrm{DS}_0) \tag{8.7}$$

The relation

$$a = (a_1/b_1) = \beta_c/T_m \tag{8.8}$$

The **a** can be useful as a quantitative parameter of the influence of the environment on the activation energy of the enzymatic reaction which can be measured using experimental values of β_c and T_m. Equation 8.4 allows to roughly estimate number of water molecules in an enzyme active site. For example taking $E_r = 20$ kJmol^{-1} for an enzyme fast chemical step and the molar enthalpy of fusion for ice $\Delta H_0 =$ 6.00 kJmol^{-1} (https://en.wikipedia.org/wiki/Enthalpy) typical values $E_a = 60$–80 kJmol^{-1}, 8.4 gives reasonable number of the water molecules in the ensemble in vicinity of the active cite $n = 7$–10. These values are in reasonable agreement with the corresponding experimental data (Chap. 7).

8.5.2 Examples

An analysis of ten site-specific protein–DNA complexes led to the following important conclusions [55] (1) for complexes with relatively undistorted DNA, favorable enthalpy change drives unfavorable entropy change; (2) for complexes with highly distorted DNA, unfavorable ΔHo is driven by favorable ΔSo; (3) protein–DNA associations have isothermal enthalpy–entropy compensation, distinct from temperature-dependent compensation; (4) all complexes have favorable ΔHo from direct protein–DNA recognition interactions and favorable ΔSo from water release; (5) systems that strongly distort the DNA nevertheless have net unfavorable, and systems with little DNA distortion have net favorable ΔHo. Thus, isothermal enthalpy–entropy compensation implies that a structurally optimal, unstrained fit is achieved at the cost of entropically unfavorable immobilization, whereas an enthalpically weaker, strained interface entails smaller entropic contribution.

Calmodulin as a model system to investigate the role for changes in protein conformational entropy in the high affinity association of proteins was employed utilizing isothermal titration calorimetry and NMR methods [59]. Peptides representing the calmodulin-binding domains of the smooth muscle myosin light chain kinase (smMLCKp), the neuronal and endothelial nitric oxide synthases (nNOSp and eNOSp), the calmodulin kinase kinase (CaMKKαp), the calmodulin kinase I (CaMKIp)1, and the phosphodiesterase (PDEp) (Fig. 8.6) were synthetized and examined. NMR order parameters Q_{axis} were determined from T_1 and $T_{1\rho}$ deuterium relaxation measured at two magnetic fields. Rotational correlation times and Q NH were determined by ^{15}N relaxation measurements1 obtained at two magnetic fields. $Q_{\text{CO–C}\alpha}$ parameters were determined by transverse cross-correlated relaxation between ^{13}CO chemical shift anisotropy and the ^{13}CO–^{13}Cα dipolar interactions. Titration of CaM with each of the peptides revealed that all six of the resulting complexes have a 1:1 stoichiometry and are in slow exchange with their dissociated components on the NMR 1H chemical shift timescale. All peptides have a basic amphiphillic character and form alpha-helical structure when bound to calmodulin (Fig. 8.6a). The Gibbs free energy (ΔG), enthalpy (ΔH), and entropy ($- T\Delta S$) for the formation of the six calcium-saturated CaM-peptide complexes at 35 °C are shown in Fig. 8.6b.

As it is seen from Fig. 8.7, the changes in the residual entropy of calmodulin upon binding to the six peptides, obtained by simple summation of the individual local entropies, indicate a linear correlation ($R^2 = 0.78$) with the corresponding entropy of binding.

Human purine nucleoside phosphorylase (PNP) belongs to the trimeric class of PNPs and is essential for catabolism of deoxyguanosine [47]. Binding 17 different low molecular weight Immucillin, transition state analog inhibitor, to the first subunit of human purine nucleoside phosphorylase was characterized by the Gibbs energy remained fairly constant at – 40 to – 50 kJ/mol, the enthalpy changed from – 92 to – 33 kJ/mol, and the entropy factor from + 35 to – 10 kJ/mol [47]. Thus, substantial enthalpy–entropy compensation in a process is largely enthalpy-driven

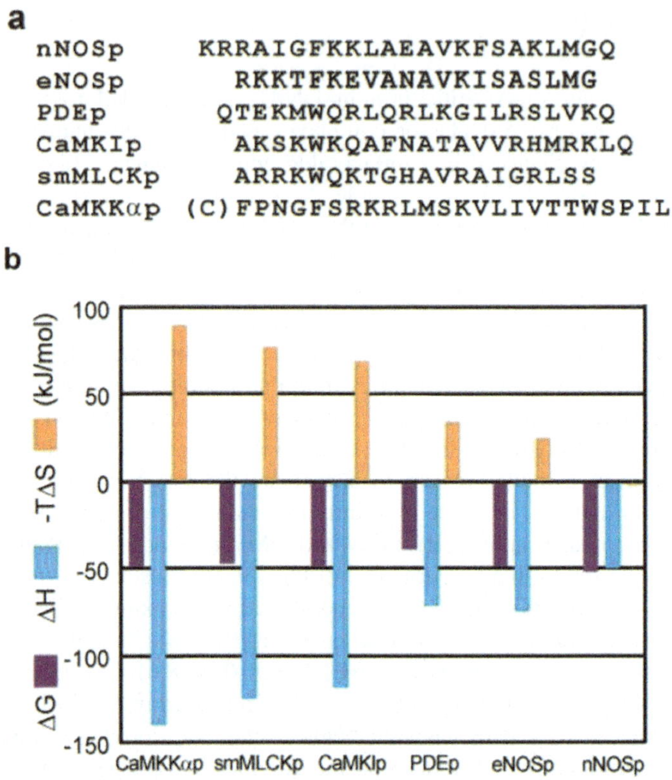

Fig. 8.6 Calmodulin-binding peptides: **a** Sequence alignment of the six calmodulin-binding peptides. Note that CaMKKαp binds in opposite orientation to CaM relative to the other five peptides. For PDEp, a C15S mutation has been used to avoid complications with oxidation. **b** Thermodynamic parameters. The Gibbs free energy (ΔG), enthalpy (ΔH), and entropy ($-T\Delta S$) for the formation of the six calcium-saturated CaM-peptide complexes at 35 °C. Values are tabulated in the supplementary information [59]

(Fig. 8.8). Titrations of human PNP using isothermal calorimetry indicate that binding of a structurally rigid first-generation Immucillin ($K_d = 56$ pM) is driven by large negative enthalpy values ($\Delta H = -21.2$ kcal/mol) with a substantial entropic ($-T\Delta S$) penalty. Flexible inhibitors bind with high affinity because of reduced entropic penalties. Tight binding of the 17 Immucillins is characterized by large enthalpic contributions, emphasizing their similarity to the transition state. The tightest-binding inhibitors ($K_d = 5 - 9$ pM) have increased conformational flexibility, while flexible inhibitors bind with high affinity because of reduced entropic penalties. Entropic penalties are proposed to arise from conformational freezing of the PNP inhibitor complex with the entropy term dominated by protein dynamics.

The mechanism of enthalpy–entropy (H/S) compensation in protein–ligand binding in a system of protein and ligands—human carbonic anhydrase (HCA) and a series of benzothiazole sulfonamide ligands with different patterns of fluorination

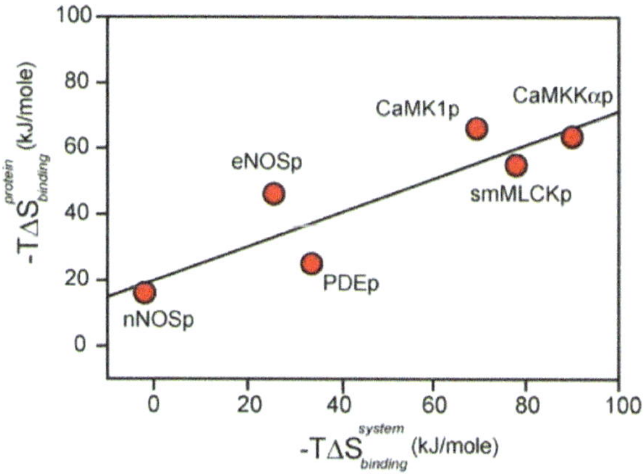

Fig. 8.7 Correlation of the change in residual conformational entropy of calmodulin with the change in the total entropy of binding of a target domain. The change in conformational entropy was estimated as described in the methods and supplementary information. Propagation of measurement error in fitted order parameters results in uncertainties in conformational entropy less than the size of the symbols used. The fitted linear correlation coefficient (R2) of conformational entropy versus the entropy of binding is 0.78 with a slope of 0.51 [59].

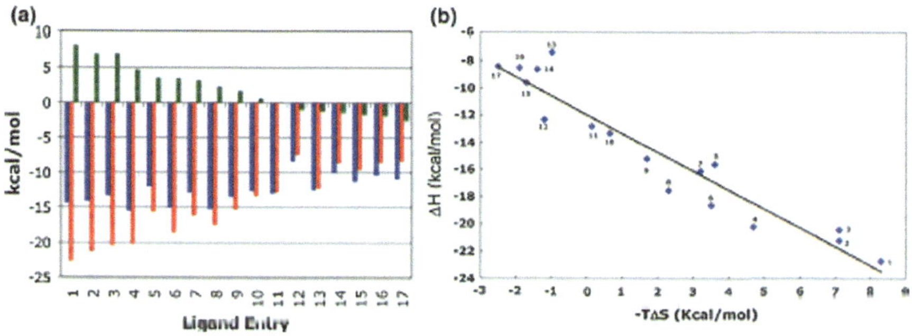

Fig. 8.8 **a** Thermodynamic signatures of a variety of Immucillins binding to the first subunit of human PNP at 300 K: ΔG (*blue*), ΔH (*red*), and $-T\Delta S$ (*green*), exhibiting EEC compensation. **b** Plot of the data showing that as the enthalpy, ΔH, becomes more positive by about 60 kJ/mol, the entropy factor $-T\Delta S$ compensates by becoming more negative by about 45 kJ/mol [47]. With permission from American Chemical Society

was examined [56]. A system of protein and ligands—human carbonic anhydrase (HCA) and a series of benzothiazole sulfonamide ligands with different patterns of fluorination was used to define enthalpy/entropy (H/S) compensation in this system thermodynamically and structurally. Analysis reveals that differences in the structure and thermodynamic properties of the waters surrounding the bound ligands are an important contributor to the observed H/S compensation. Therefore, the contact interactions between the protein and the ligand for biomolecular recognition and in determining the thermodynamics of binding are related to the molecules of water filling the active site of a protein and surrounding the ligand.

Changes in conformational entropy were determined when protein–ligand interactions occur among the catabolite activator protein (CAP) complexes with **DNA-binding** domain employing NMR spectroscopy [57]. NMR experiments were performed on Varian 800 and 600 MHz spectrometers and Bruker 700 and 600 MHz spectrometers at 32 °C. Results showed that changes in conformational entropy can dominate the total entropy of binding (Fig. 8.9a, b) and conformational entropy determines the magnitude of the total binding entropy and whether the total entropy of binding is positive (unfavorable) or negative (favorable). In some CAP variants, despite considerable changes in conformational entropy and thus in the total entropy of binding, the effect on the expected binding affinity is negligible owning to enthalpy–entropy compensation. This effect was explained by contribution of "melting" water in the thermodynamic parameters. Other interesting results are as follows: (1) in certain CAP complexes, the compensation is strong, resulting in little or no difference in the free energy of binding and (2) in several complexes, the observed marked alterations in entropy resulted in weaker enthalpy compensation, giving rise to differences in affinity of as much as two orders of magnitude.

Figure 8.9 indicates correlation but not identity between the total entropy of binding ($-\ T\Delta S$) as measured by isothermal titration calorimetry (ITC) and the conformational entropy ($-\ T\Delta S_{\mathrm{conf}}$) as determined by NMR for DNA binding to each of the eleven CAP variants.

By compiling the work of many laboratories investigating riboswitches (regulatory segment of a messenger RNA molecule) and using small-angle X-ray scattering (SAXS) and isothermal titration calorimetry (ITC),global analysis of segments of riboswitches, segments of noncoding mRNA, was performed [58]. A wide spectrum of enthalpy–entropy compensation regimes across riboswitches that bind a diverse set of ligands, giving rise to a relatively narrow range of physiologically relevant free energies and ligand affinities, was demonstrated. Enthalpy and entropy changes that accompany ligand binding to riboswitches reported local and global structural changes, formation of new tertiary interactions, and changes in quaternary structure, in addition to changes in solvation and counterion distribution, and vibrational modes. The available data analysis revealed the strongly entropy-driven binding of glycine and the predominantly enthalpy-driven binding of c-di-GMP to their respective riboswitches. Comparison of the thermodynamic profiles of twelve representative riboswitches and one in vitro selected aptamer indicated that ligand binding to these RNAs is primary driven by favorable enthalpy changes (mean $\Delta H = -\ 22$ kcal/mol) and is impeded by unfavorable entropy changes (mean $-\ T\Delta S = 12$ kcal/mol). Interestingly, natural riboswitches exhibit only weak correlation between the ligand solvent-accessible surface buried and the binding free energy ΔG, enthalpy ΔH, or entropy $-\ T\Delta S$.

Well-established cases of enthalpy–entropy compensation (EEC) published by 2017 were reviewed [50]. It was concluded that the large fluctuations in enthalpy and entropy observed come from variations in the amounts of water immobilized or

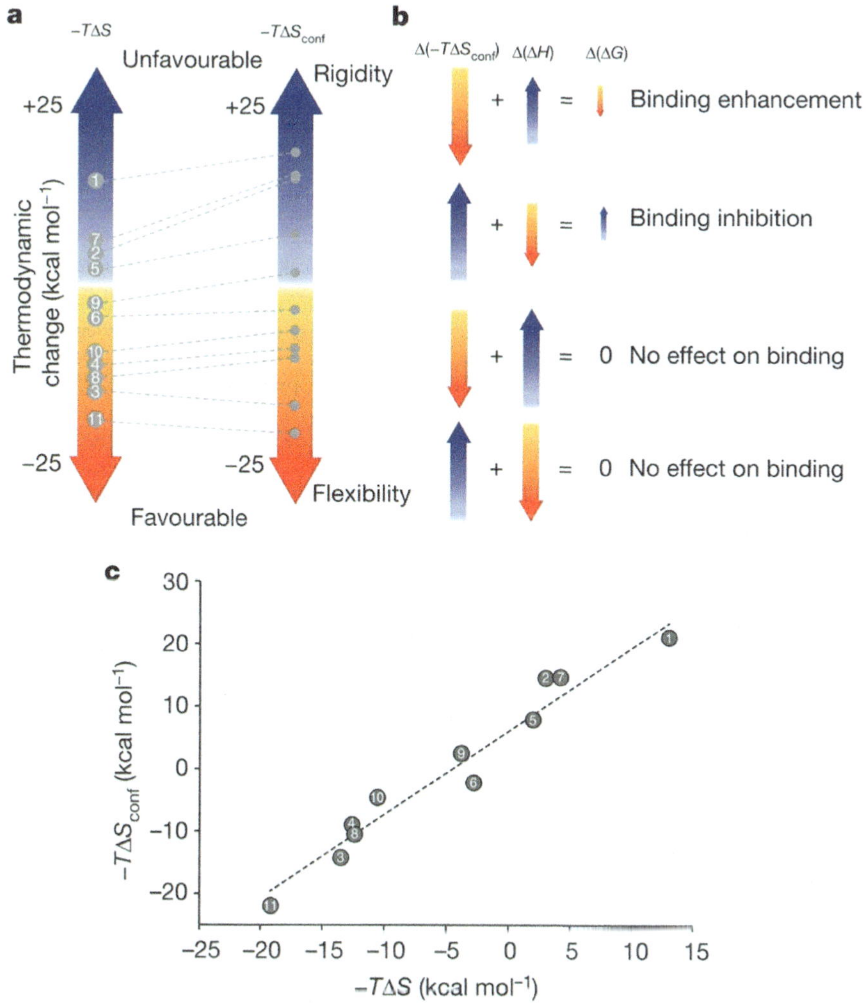

Fig. 8.9 **a** Total entropy of binding ($-T\Delta S$) as measured by ITC and the conformational entropy ($-T\Delta S_{conf}$) as determined by NMR for DNA binding to each of the eleven CAP variants. The numbering of the CAP variants is as in Fig. 1a. **b** The change in conformational entropy can result in binding enhancement, binding inhibition or no effect in the expected affinity, depending on the tendency for enthalpy–entropy compensation. **c** Relationship between the total binding entropy and the conformational entropy of binding for all eleven CAP variants [57]

released on forming complexes. Two systems exhibiting enthalpy–entropy compensation (EEC) showed a correlation between calorimetric entropies and local mobilities and were interpreted to mean conformational control of the binding entropy/free energy. Large compensating changes in the enthalpies and entropies of binding were explained in the frame of approach to engineering elevated affinities through the addition of ionic links, as they generate increased entropy without affecting the

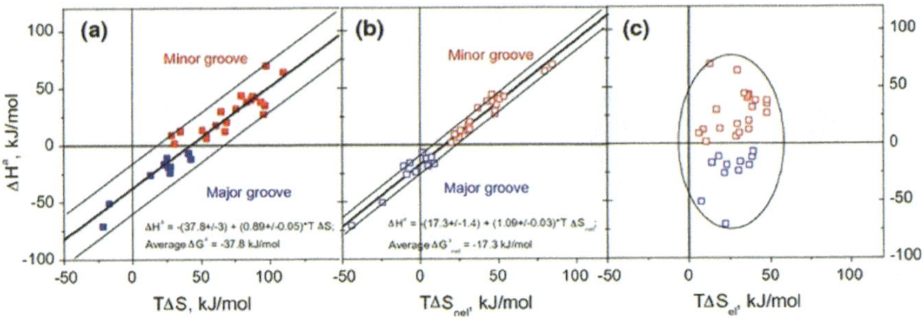

Fig. 8.10 Energetics of DNA-binding domain (*DBD*) binding to target DNAs: dependence of the entropy factor, $T\Delta S$, on the enthalpy, ΔH. Minor groove binders, *red symbols*; major groove binders, *blue symbols*. **a** Total entropy factors, **b** non-electrostatic entropy factors, **c** electrostatic entropy factors. Data, all at 20 °C [50]

enthalpy. Effects of the entropy factor, $T\Delta S$, on the enthalpy, ΔH on energetics of DNA-binding domain binding to target DNAs is illustrated in Fig. 8.10.

By compiling the work of many laboratories investigating riboswitches using small-angle X-ray scattering (SAXS) and isothermal titration calorimetry (ITC), uncover general trends, and common themes were uncovered [58]. A wide spectrum of enthalpy–entropy compensation regimes across riboswitches that bind a diverse set of ligands, giving rise to a relatively narrow range of physiologically relevant free energies and ligand affinities, was demonstrated. A meta-analysis of ligand binding to riboswitches showed a wide level of enthalpy–entropy compensation (Fig. 8.11). This covered a range of ~ 200 kJ/mol from the entropically driven glycine riboswitch to the strongly enthalpy-driven cyclic-di-GMP riboswitch.

Thermodynamic measurements on processes involving biological macromolecules in aqueous solution, which illustrate the ubiquitous phenomenon of entropy–enthalpy compensation in the processes including protein–ligand binding/unbinding, with compensation temperatures vary by about 50 K around an

Fig. 8.11 Enthalpy and entropy factors of riboswitches activated by a variety of effectors at temperatures between 288 and 310 K [58]. With permission from Elsevier

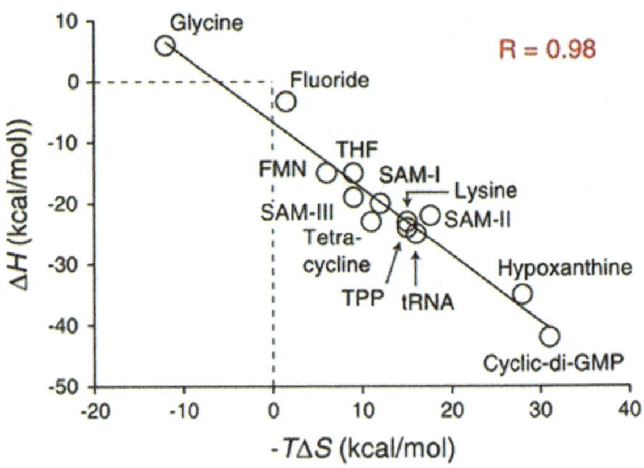

average near 293 K [60]. Results showed that incorporating both near-exact entropy–enthalpy compensation (due to solvent relaxation) and multi-excitation entropy (from vibrational quanta) leads to a compensation temperature in water of about 230 K. Using molecular simulations, it was shown that, for solvation of six different HIV-1 protease inhibitors, CCs in the solute cause enthalpy–entropy compensation (EEC) as large as 10–30 kcal/mol [61]. Due to conformational changes, CCs in HIV-1 protease EEC are shown to vary significantly upon modifying its bound ligand.

Recent review [65] summarized known sources of enthalpy/entropy compensation and lays out a conceptual framework for understanding and avoiding or exploiting this phenomenon in biophysical systems. The suggested association of a protein and ligand in buffered aqueous solution was presented as the sum of nine processes indicating rearrangement water molecules in an enzyme active cite. The scheme does not take into account the change in the conformation of the protein during the binding of the substrate, which along with the water molecules in the cavity of the active center in turn causes a rearrangement in the structure of the surrounding water and makes a decisive contribution to the thermodynamic and kinetic parameters of the process. Schematic illustration of a model process is depicted in Fig. 8.12.

There is a possibility that some of enthalpy–entropy compensation might be caused by accidental **errors**. A computer program has been developed to evaluate the probability that experimental errors might lead to a linear compensation plot parting from an initial randomly scattered set of activation parameters (*p*-test) [64] (2016). Application of this program to kinetic data for 100 homologous reaction series extracted from bibliographic sources has allowed concluding that most of the reported compensation plots can hardly be explained by the accumulation of experimental errors.

Paper [66] explored one of the mathematical explanations for the compensation effect and for the isokinetic. Both theoretical and numerical simulations led to conclusion that random errors in kinetic data generate an apparent compensation effect (the statistical compensation effect) when the true Arrhenius parameters are constant. Expressions for the gradient of data points on a plot of $\ln A$ against E derived when experimental kinetic data were analyzed by linear regression, by nonlinear regression, and by weighted linear regression shown that the most appropriate analysis technique depends critically on the error structure of the kinetic data. When this plot is a straight line with a gradient close to 1/RT, then confidence ellipses can be calculated for each data point to investigate whether the apparent compensation effect arises from random errors in the kinetic measurements or has some other origin.

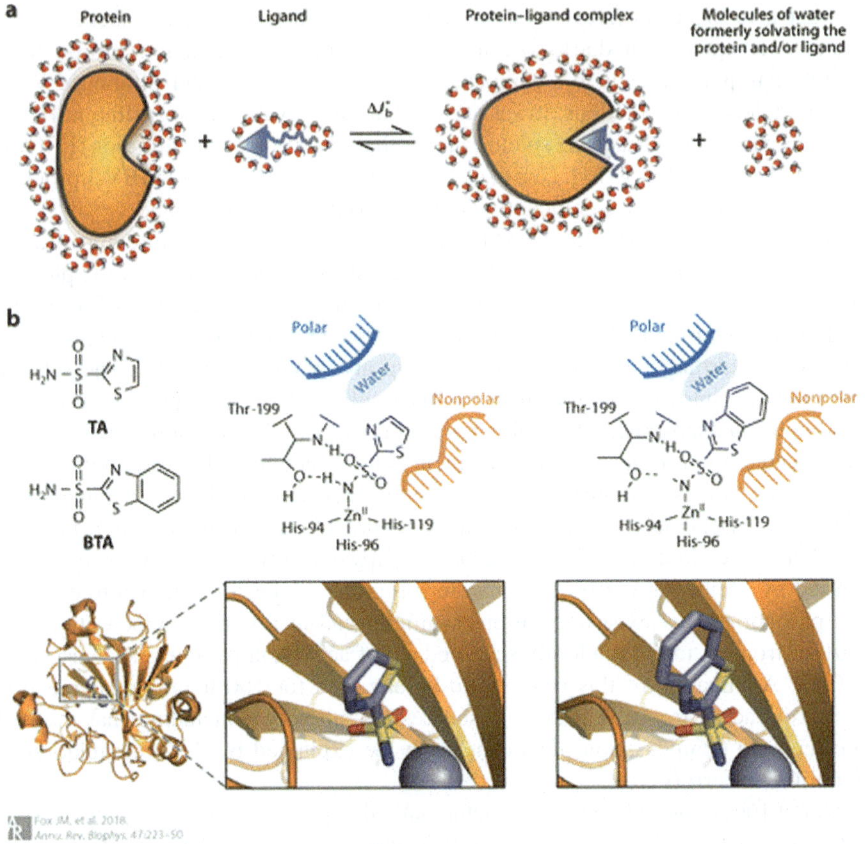

Fig. 8.12 Model systems. **a** Biomolecular recognition can be broken down into nine processes: i the formation of protein–ligand contacts, ii the rearrangement of water initially solvating the protein, iii the rearrangement of water initially solvating the ligand, iv the formation of a hydration structure around the protein–ligand complex, v changes in the conformation of the protein, vi changes in the conformation of the ligand, vii changes in the dynamics of the protein, viii changes in the dynamics of the ligand, and ix changes in the organization of—and interactions associated with—buffer ions. This schematic illustrates processes i–viii. **b** A demonstration of the use of human carbonic anhydrase II to examine the influence of differences in ligand structure—in the absence of differences in protein conformation—on binding. The ligands 1,3-thiazole-2-sulfonamide (TA) and benzo[*d*]thiazole-2-sulfonamide (BTA) differ by a benzene ring but not in binding geometry and, thus, reveal the thermodynamic contribution of the benzene ring to binding. This specific comparison came from our benzo-extension study. The illustration of bound ligands was adapted with permission from [36]. Additional abbreviation: _*J* ◦ *b*, the change in a thermodynamic property *J* upon binding [65]. With permission from American Chemical Society

8.6 Water as a Chemical Reagent in Enzyme Catalysis

8.6.1 Hydrolase

Water is an attacking nucleophilic reagent in enzymatic hydrolytic reactions activated in the enzyme active cite [67, 68]. Hydrolases is a class of more than 200 enzymes that catalyze the hydrolysis of several types of compounds. Esterases include (1) lipases, which break ester bonds (between a carboxylic acid and an alcohol) in lipids; (2)phosphatases, which act analogously upon phosphates; (3) glycosidases which sever bonds between sugar molecules in carbohydrates; and (4) peptidases which hydrolyze peptide bonds (between the carboxylic acid group of one amino acid and the amino group of another) within protein molecules. Some typical examples of enzymatic hydrolase reactions are given below.

The principal biological role of acetylcholinesterase (AChEis) in termination of impulse transmission at cholinergic synapses by rapid hydrolysis of the neurotransmitter, acetylcholine (ACh) was investigated (Figs. 8.13, 8.14, 8.15 and 8.16) [69]. The acetylcholine esterase assists in transforming the neuron impulse. The hydrolase breaks the acetylcholine into choline and acetic acid.

As can be seen from Fig. 8.14, the hydrolysis mechanism includes a nucleophilic attack of an oxygen atom on the carboxyl group of serine, accompanied by synchronic proton transfer to histidine.

The three-dimensional structure of acetylcholinesterase from Torpedo californica electric organ has been determined by x-ray analysis to 2.8 angstrom resolution. The form crystallized is the glycolipid-anchored homodimer that was purified subsequent to solubilization with a bacterial phosphatidylinositol-specific phospholipase C [70].

The proposed model of the enzyme active center confirmed by the determination of the 3D structure of *Tc*AChE complexed with a transition state analog, *m*-(*N,N,N*-trimethylammonio)-2,2,2-trifluoroacetophenone (TMTFA) is depicted in Fig. 8.16. The participation of water in the enzymatic hydrolysis of acetylcholine is illustrated in the figure [62]. In addition to the two subsites of the catalytic center, AChE was shown to possess one or more additional binding sites for ACh and for other quaternary ligands.

Cysteine proteases, thiol proteases, share a common catalytic mechanism that involves a nucleophilic cysteine thiol in a catalytic triad or dyad [71]. Structure of cysteine peptidase gingipain K (Kgp), which is a major virulence factor of *Porphyromonas gingivalis* in periodontitis, and the scheme depicting relevant interactions involved three water molecules are displayed in Figs. 8.17 and 8.18.

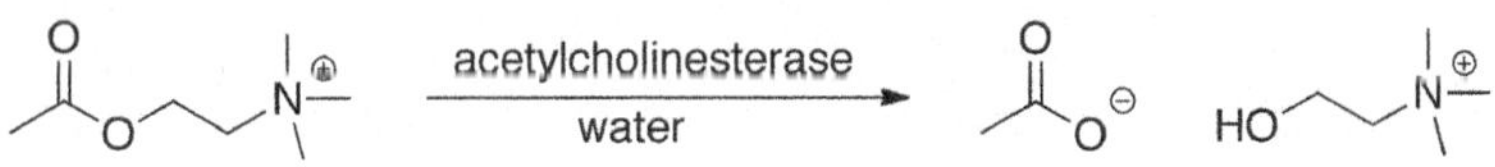

Fig. 8.13 Acetylcholinesterase catalyzes the hydrolysis of acetylcholine to acetate ion and choline. https://en.wikipedia.org/wiki/Acetylcholinesterase

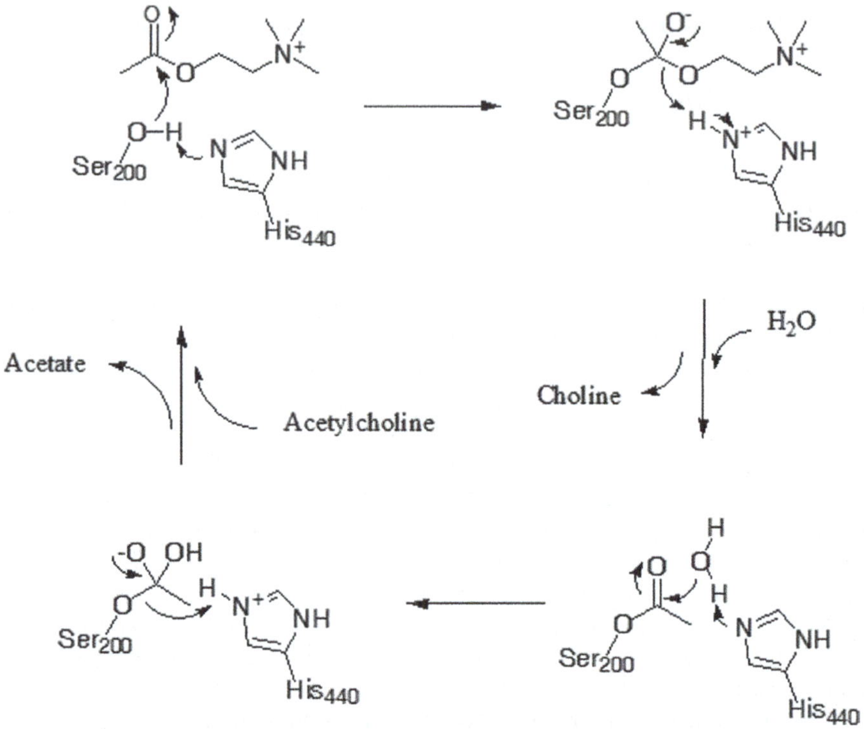

Fig. 8.14 Mechanism for hydrolysis of acetylcholine, https://en.wikibooks.org/wiki/Structural_Biochemistry/Enzyme/Acetylcholinesterase

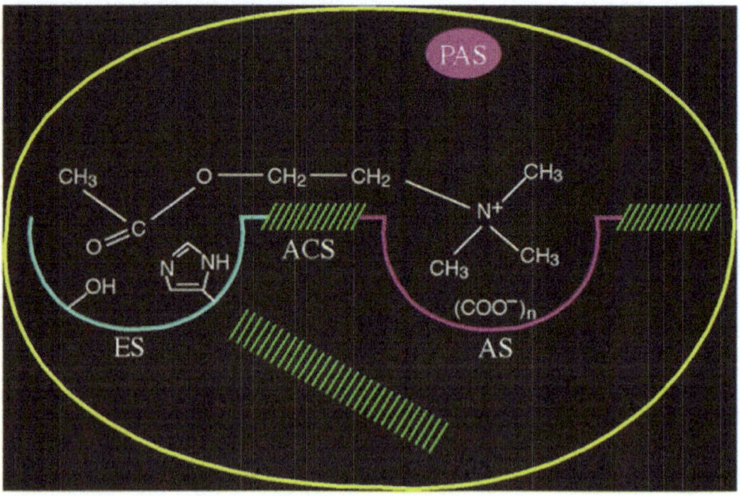

Fig. 8.15 Proposed model of the enzyme active center of *Tc*AChE [69]. With permission from Elsevier

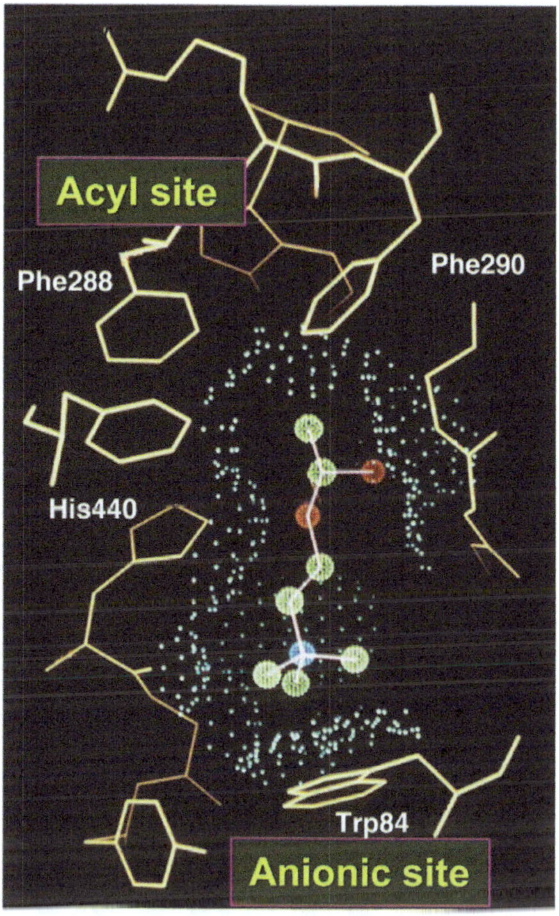

Fig. 8.16 Active site of AChE [69]. With permission from Elsevier

The cysteine protease mediates cleavage of a peptide bond including a key stage of the water nucleophilic attack (Fig. 8.19) [73].

Phosphatases catalyze the hydrolysis of a phosphomonoester, removing a phosphate moiety from the substrate (Fig. 8.20).

Eukaryotic protein phosphatases are structurally and functionally diverse enzymes that are represented by three distinct gene families [74]. Two of these, the PPP and PPM families, dephosphorylate phosphoserine and phosphothreonine residues, whereas the protein tyrosine phosphatases (PTPs) dephosphorylate phosphotyrosine amino acids. Tyrosine specific protein phosphatases catalyze the removal of a phosphate group attached to a tyrosine residue, via a cysteinyl-phosphate enzyme intermediate [75]. The protein tyrosine-specific protein phosphatases are metalloenzymes and dephosphorylate their substrates in a single reaction step using a metal-activated nucleophilic water molecule. In the catalytic subunit of Ser/Thr PP1 metal ions, identified as Mn^{2+} and Fe^{2+} are located in the active site at the three-way joint of the β

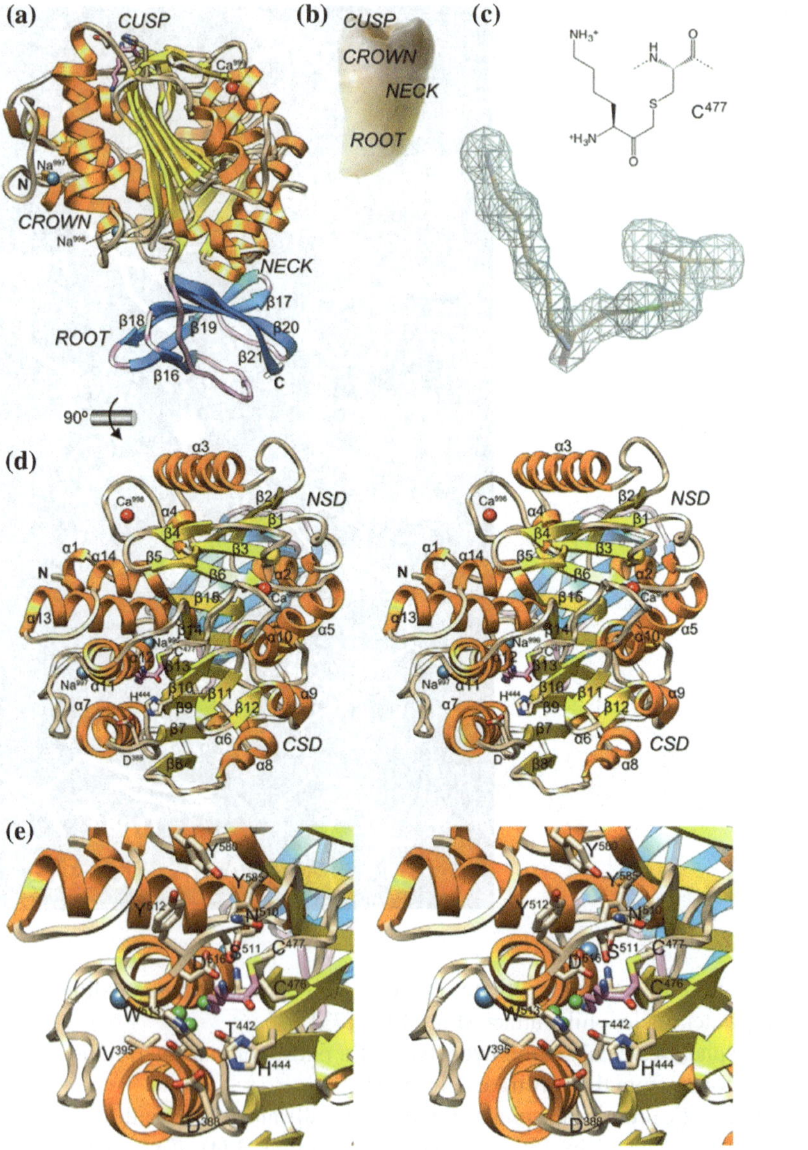

sandwich and the two helical domains (Fig. 8.21). Coordination of these two metal ions is provided by three histidines, two aspartic acids, and one asparagine. The two metal ions bind and activate a water molecule, which initiates a nucleophilic attack on the phosphorous atom.

◀**Fig. 8.17** General architecture of KgpB. **a** ribbon-type plot of Kgp(CD + IgSF) showing the regular secondary structure elements (CD, α-helices as *orange ribbons*, β-strand as *yellow arrows*, and coils in *tan*; IgSF, β-strand as *blue arrows* labeled β16–β21 and coils in *lilac*). The N and C termini and the four structural cations (calciums Ca^{999} and Ca^{998} in *red* and sodiums Na^{997} and Na^{996} in *blue*) are depicted. The (putative) catalytic triad, Cys^{477} (covalently modified at Sγ with an L-lysinylmethyl group; carbons in *lilac*), His^{444}, and Asp^{388}, is further shown as *sticks*. *Black arrows* point to the solvent channel. **b** Picture of a tooth with its parts labeled. **c** ($2mF_{obs} - DF_{calc}$)-type Fourier map of the region around the catalytic cysteine Cys^{477} obtained with diffraction data to 1.75 Å resolution and contoured at 1σ above threshold. **d** Structure of Kgp(CD + IgSF) in cross-eye stereo in standard orientation [27, 72], which corresponds to a horizontal 90° rotation of the view in **a**, i.e., viewing the CD cusp region. Regular secondary structure elements of the CD (helices α1–α14 and strands β1–β15) are labeled. The NSD is on *top*, and the CSD is at the *bottom*. **e** Close-up of *D* in stereo centered on the non-primed side of the active site. Residues framing the specificity pocket S_1 and pocket S_2 are labeled. *Small green spheres* represent solvent molecules I [71]. Published under the CC-BY license

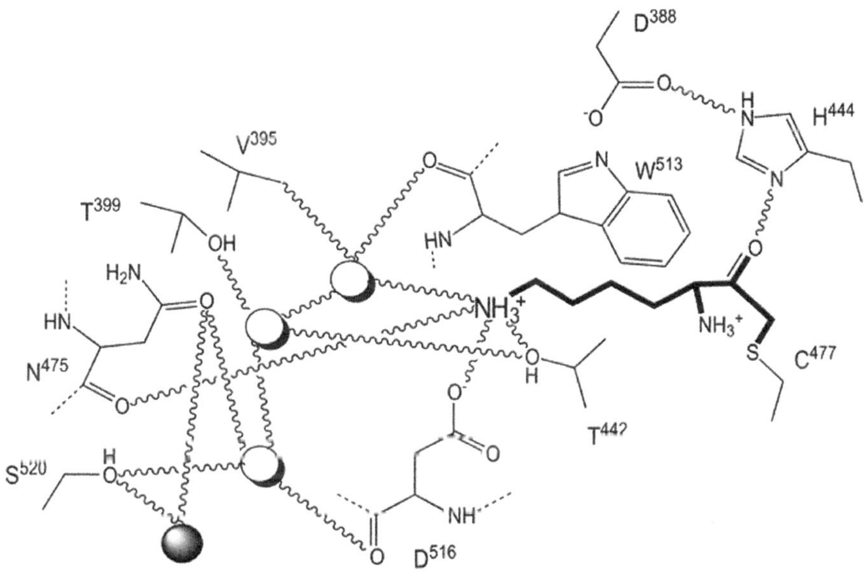

Fig. 8.18 Interactions in the S_1 pocket. The scheme depicts relevant interactions (as *curved lines*) of the L-lysinylmethyl group (*bold trace*) in the specificity pocket. The three solvent molecules in the pocket are shown as *white spheres*, and the first solvent of the large inner solvent channel is shown as a *gray sphere* molecules [71]. Published under the CC-BY license

Exopeptidases, such as aminopeptidases and carboxypeptidase, detach the terminal amino acids from the protein chain, while endopeptidases, such as trypsin, chymotrypsin, pepsin, papain, and elastase, attack internal peptide bonds of a protein [76]. Aspartic, glutamic, and metallo-proteases activate a water molecule which performs a nucleophilic attack on the peptide bond to hydrolyze. Serine, threonine, and cysteine proteases using a nucleophilic catalytic triad perform a nucleophilic

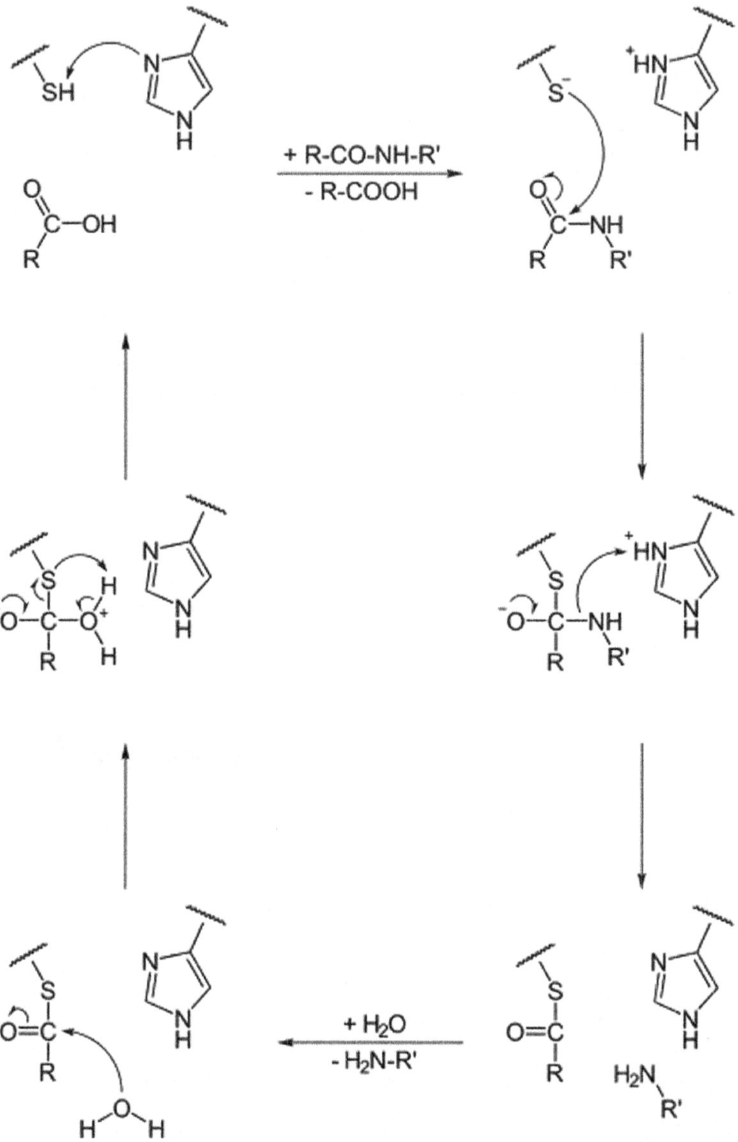

Fig. 8.19 Reaction mechanism of the cysteine protease mediated cleavage of a peptide bond https://en.wikipedia.org/wiki/Cysteine_protease

attack to covalently link the protease to the substrate protein, releasing the first half of the product. This covalent acyl-enzyme intermediate is then hydrolyzed by activated water to complete catalysis by releasing the second half of the product and regenerating the free enzyme. The background of the findings and crystal structures,

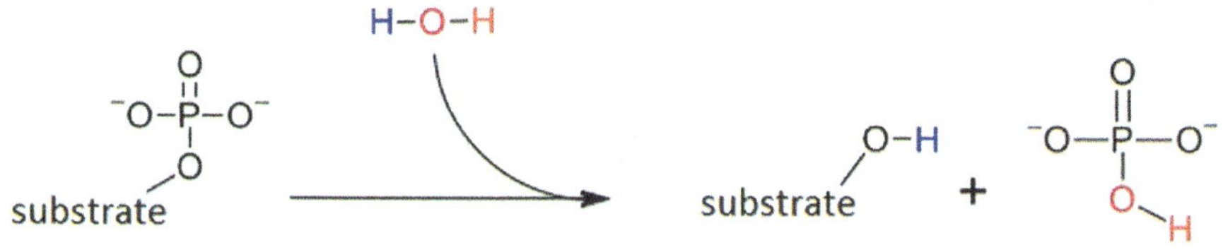

Fig. 8.20 Phosphatases reaction https://en.wikipedia.org/wiki/Phosphatase

catalytic mechanisms, substrates specificities, and distribution of the new peptidase including sedolisin and kumamolisin were reviewed 77].

According to the catalytic mechanism for sedolisin proposed in [77], Glu80 acts as a general base to accept a proton from Ser287 during the nucleophilic attack and then as a general acid to protonate the leaving group (N–H of P10-Phe) during the cleavage of the scissile peptide bond (Fig. 8.22). Another acidic residue forming the oxyanion hole, Asp170, acts as a general acid catalyst to protonate the carbonyl oxygen of P1-Glu during the formation of the tetrahedral intermediate and as a general base for the formation of the acyl-enzyme. Proton transfer from Asp170 to the carbonyl of P1-Glu was suggested to be important in the stabilization of the tetrahedral intermediate and the formation of a low-barrier hydrogen bond between the carboxyl group of P1-Glu and Asp170 in the lowering of the free energy barrier for the cleavage of the peptide bond.

A protease (peptidase proteinase) is an enzyme that catalyzes proteolysis, the breakdown of proteins into smaller polypeptides or single amino acids by hydrolysis [72]. They do this by cleaving the peptide bonds within proteins by hydrolysis. In eqolisin, scytalidoglutamic peptidase, the active site of which contains a catalytic dyad, glutamic acid (E), and glutamine (Q), the suggested hydrolytic mechanism involves nucleophilic attack of a general base (Glu136) activated water (OH^-) on the *si* face of the scissile peptide carbonyl carbon atom to form a tetrahedral intermediate (Fig. 8.23). Electrophilic assistance and oxyanion stabilization is provided by the side chain amide of Gln53 and protonation of the leaving-group nitrogen can be accomplished by the general acid function of the protonated carboxyl group of Glu136.

8.6.2 Oxygen-Evolving Complex

The most essential for the formation of the biomass on earth is the biochemical process of splitting of two water molecules into molecular oxygen with the formation of energy reached pigment P_{680} in the reduced state and release of four protons. This unique process occurs in oxygen-evolving complex located in the photosystem II (PSII) complex in the thylakoid membrane creating the present day oxygen level of ~ 21% in our atmosphere [78–89].

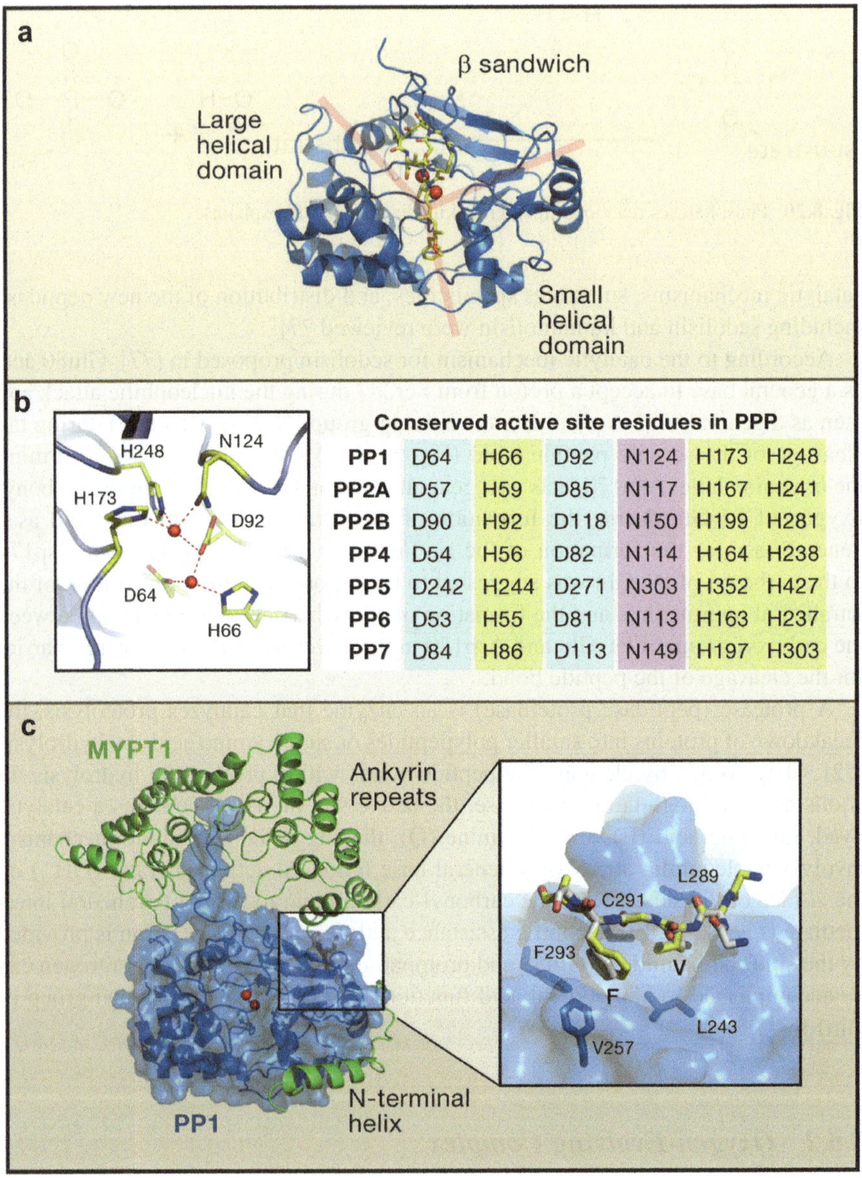

Conserved active site residues in PPP

PP1	D64	H66	D92	N124	H173	H248
PP2A	D57	H59	D85	N117	H167	H241
PP2B	D90	H92	D118	N150	H199	H281
PP4	D54	H56	D82	N114	H164	H238
PP5	D242	H244	D271	N303	H352	H427
PP6	D53	H55	D81	N113	H163	H237
PP7	D84	H86	D113	N149	H197	H303

◂**Fig. 8.21** Structure and mechanism of protein phosphatase 1 **a** Structure of the catalytic subunit (blue) of protein phosphatase 1 (PP1) bound to okadaic acid (OA, yellow ball and stick). A *Y*-shaped surface groove (pink) is defined by the three domains of PP1. The two metal ions (red spheres) are Mn^{2+} (manganese) and Fe^{2+} (iron). **b** Conserved coordination of the binuclear metal center in the PPP family phosphatases. The metal-binding scheme of PP1 where six highly conserved amino acids bind to the two metal ions is indicated on the left. The corresponding metal-binding residues from other PPP family members are shown on the right. **c** Structure of the PP1 catalytic domain (blue) bound to the regulatory subunit myosin phosphatase targeting subunit (MYPT1) protein (green). Three elements of MYPT1 are involved in the interactions: a VxF (V, valin; x, any amino acid; F, phenylalanine) motif that binds to the conserved hydrophobic surface of PP1, an ankyrin repeat domain that caps the C terminus of PP1, and a hydrophobic N-terminal helix that docks onto the surface of PP1 [75]. With permission from Elsevier

In PSII, the four-electron oxidation of two water molecules is catalyzed by a functional unit, which is known as the oxygen-evolving complex (OEC) or water-oxidizing complex (WOC). The OEC consists of an inorganic core comprising four manganese, one calcium, and at least five oxygen bridges (Mn_4O_xCa cluster) (Fig. 8.24).

The Kok cycle includes a sequence of proton and electron transfer events at each S-state transition, along with the most probable oxidation states of the cluster. In the photocycle involving for intermediate states: $S_0 \rightarrow S_1 \rightarrow S_2 \rightarrow S_3 \rightarrow S_4$ oxidized cluster (Mn_4O_5Ca) is reduced by P_{680} and $P_{680}{}^+$, in turn generating a neutral tyrosine radical which acts as an oxidant for water splitting. Protonation states of water ligands and oxo bridges are directly involved in tuning the electronic structures and oxidation potentials of OEC. It was suggested that protonation states of water ligands and oxo bridges are directly involved in tuning the electronic structures and oxidation potentials of OEC.

Structural and spin parameters of the PS II Mn_4O_5Ca cluster were established in the pulse ^{55}Mn-ENDOR and EPR experiments [81–89] The following mechanism of the water photooxidation was proposed: (1) in the $S_0 \rightarrow S_1$ transition, central Mn_B(III) is oxidized to Mn_B(IV) followed by deprotonation of the μ_2-hydroxo in the $S_1 \rightarrow S_2$ transition, and (2) Mn_D is oxidized from Mn_D(III) to Mn_D(IV), and water molecules are either not deprotonated or transfer the proton within an H-bridge to a nearby residue.

In a framework of conception of multi-electron processes suggested by Likhtenshtein and Shilov in 1970 [84–89] into inner sphere of transition metal complexes, a four-electron mechanism of two water oxidation to O_2 was proposed. According to thermodynamic estimation, the formation HO_2 and OH intermediate complexes is energetically strongly unfavorable. Therefore, this reaction can occur as four-electron elementary step with the Mn cluster reduction and oxidation of two water molecules partly deprotonated

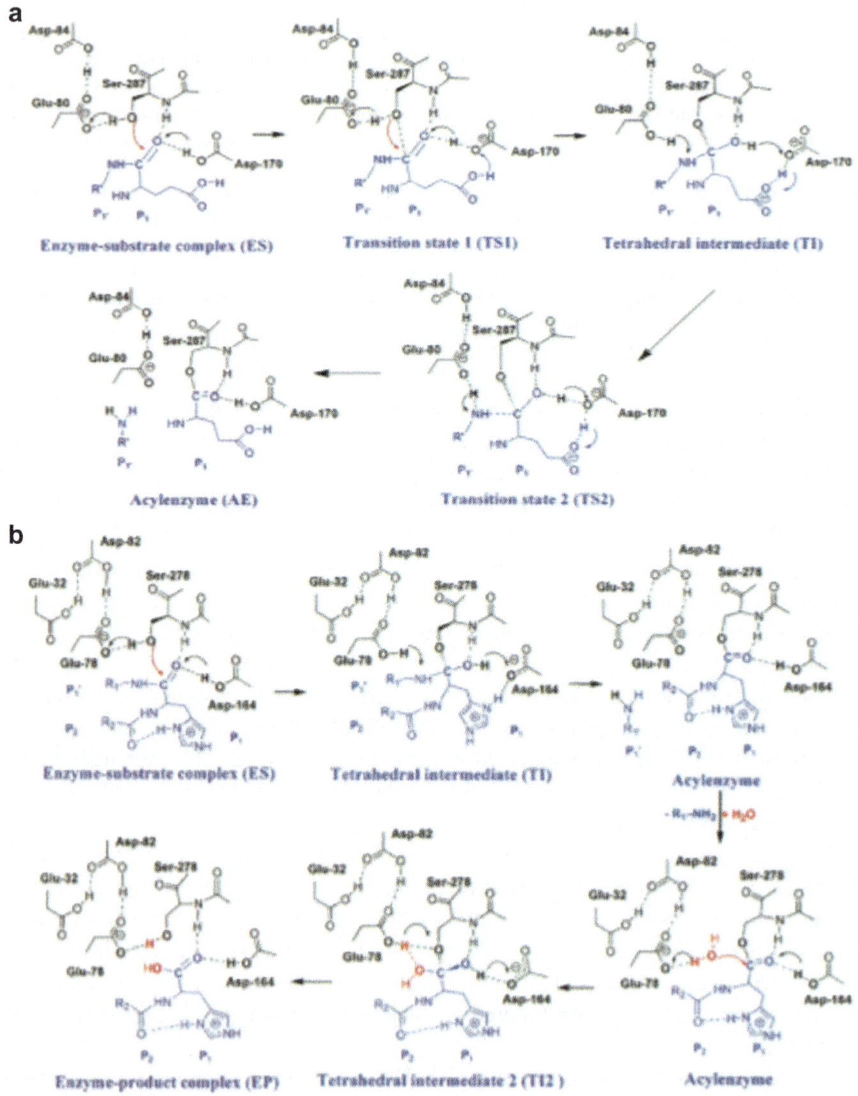

Fig. 8.22 Catalytic mechanism of sedolisin and kumamolisin-As through QM/MM free energy simulations. **a** Catalytic mechanism of the acylation reaction of a peptide substrate (Leu-Leu-Glu#Phe-Leu) catalyzed by sedolisin that is consistent with QM/MM free energy simulations. The black bend arrows indicate the directions of the proton transfers, and red bend arrows the nucleophilic attacks. **b** Catalytic mechanism and the role of the key active site residues for both acylation and deacylation reactions of kumamolisin-As based on computer simulations. The black arrows indicate the directions of the proton transfers, and red arrows the nucleophilic attacks. The atoms from the enzyme, substrate, and water are colored in black, blue, and red, respectively [76]. With permission from Oxford University Press

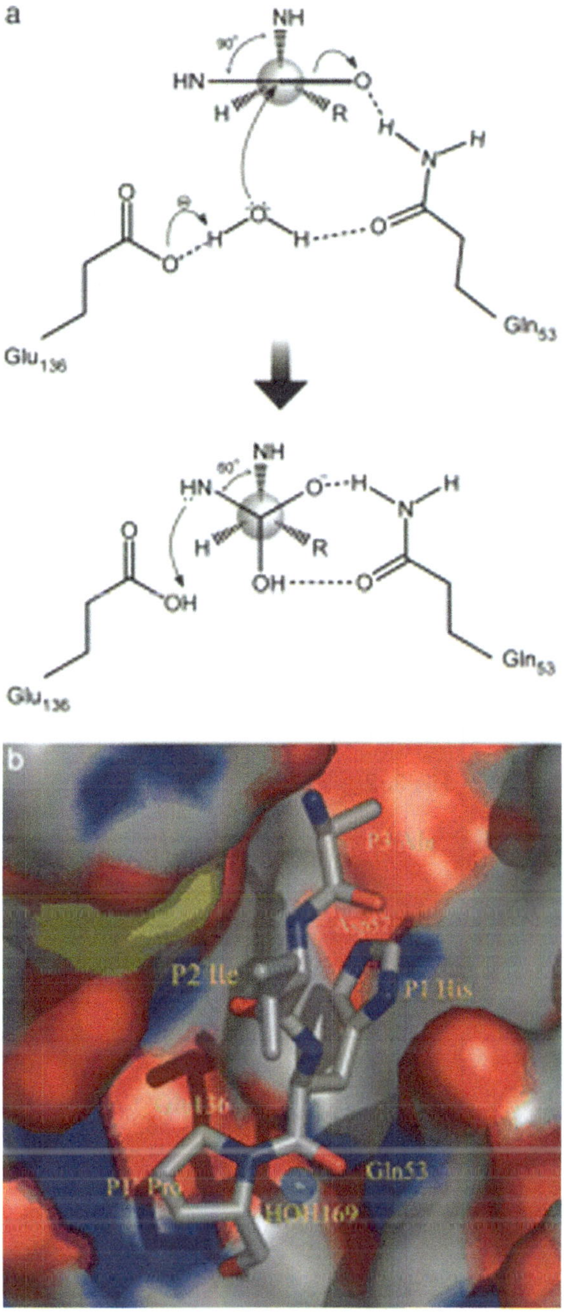

Fig. 8.23 Catalytic mechanism of eqolisin. **a** The proposed catalytic mechanism of eqolisin. The water molecule hydrogen bonded to both Glu136 and Gln53 is the nucleophile. The general base is the carboxylate of Glu136. The side chain amide of Gln53 assists in the nucleophilic attack and stabilizes the tetrahedral intermediate by hydrogen bonding. **b** A model of the product Ala-Ile-His-COO^- bound in the active site of eqolisin. The nucleophilic attack by the activated OH^- ion would be on the *si*-face of the scissile peptide; its position coincides with the oxygen atom of the carboxyl group closest to the enzyme surface in the above figure. The surface of eqolisin is represented and colored according to the underlying atoms (slate, carbon; blue, nitrogen; red, oxygen) [72]. Published with permission of PNAS article containing the requested material published under a CC BY-NC-ND license

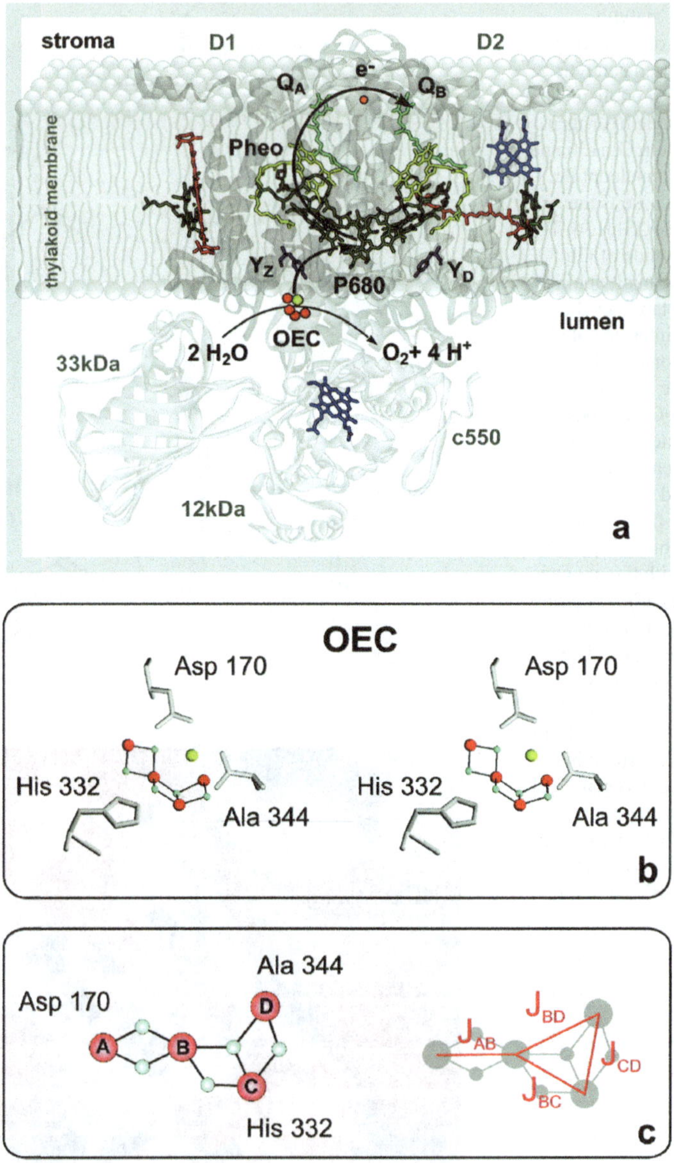
stroma
D1
D2
e-
Q_A
Q_B
thylakoid membrane
Pheo
Y_Z
P680
Y_D
lumen
2 H_2O
OEC
O_2+ 4 H^+
33kDa
c550
12kDa
a
OEC
Asp 170
Asp 170
His 332
Ala 344
His 332
Ala 344
b
Ala 344
Asp 170
A
B
C
D
His 332
J_{AB}
J_{BD}
J_{CD}
J_{BC}
c

◂**Fig. 8.24** **Panel a** shows a schematic view of the photosystem II (PSII) complex in the thylakoid membrane that is based on the 3.0 Å crystal structure. For clarity, the inner antenna proteins CP43 and CP47, which are involved in harvesting the light energy, and the cytochrome b559 subunits are not shown. The other core proteins of PSII are shown in gray with corresponding labels. The cofactors of the D1, D2, cyt c550, and cyt b559 proteins are placed in color on top of the proteins (see text for more details). **Panel b** presents a stereoview of the Mn_4O_xCa cluster together with selected ligands that are referred to in this work. The structure shown was derived by Yano et al. on the basis of polarized EXAFS spectroscopy on PSII single crystals (Model III). The view is also approximately along the membrane plane and from a similar angle as that in **panel a**. Mn ions are shown in red, oxygen in cyan, and Ca in yellow. **Panel c** displays a schematic top view (approximately along the membrane normal) of the Mn_4O_5 core of the Mn_4O_xCa cluster using the same color code as that above. This schematic representation allows a clearer visualization of the bridging motifs between the four Mn ions, which are labeled A, B, C, and D. On the right side of **panel** c, the general coupling scheme is shown that is employed in this study to derive the electronic structure of the S_0 and S_2 states [81]. With permission from American Chemical Society

8.7 Concerted Enzymatic Reactions

In a concerted elementary reaction, a substrate is simultaneously attacked by different active reagent with acid and basic groups, nucleophile and electrophile, reducing and oxidizing agents or two radicals, providing by such a way lowing of energy activation. Nevertheless, certain kinetic limitation exists on the realization of reaction which are accompanied by a change in the configuration of a large number of nuclei because of high energy reorganization [90–93] and low possibility of concerted reactions involving four or more bonds (principle of non-perfect synchronization or "minimal motion"). In order to reconcile these seemingly contradictory trends, the principle of "optimum motion" was formulated by Likhtenshtein [17, 18, 89, 94–96]. According to this principle, the number of nuclei whose configuration is change in the elementary act of a chemical reaction must be sufficiently large to provide favorable energetics for the step and, at the same time, sufficiently small for the maintenance of high value of the synchronization probability during motion along the reaction pathway to the reaction product.

According a simplified model of Alexandrov [97], dependence of the rate of the concerted chemical reactions on the number of degree of freedom involved in the transition in a concert reaction occurs as a result of the simultaneous transition (taken approximately 10^{-13} s) of a system of independent n oscillators, with the mean displacement of nuclei φ_0, from the ground state, to the activated state in which this displacement exceeded a certain critical value φ_{cr}. If the activation energy of the concerted process $E_{syn} > n$RT, the theory gives the following expression for the synchronization factor which is the ration of the pre-exponential factor (8.9).

$$a_{syn} = \frac{n}{2^{n-1}}\left(\frac{nRT}{\pi E_{syn}}\right)^{\frac{n-1}{2}} \quad (8.9)$$

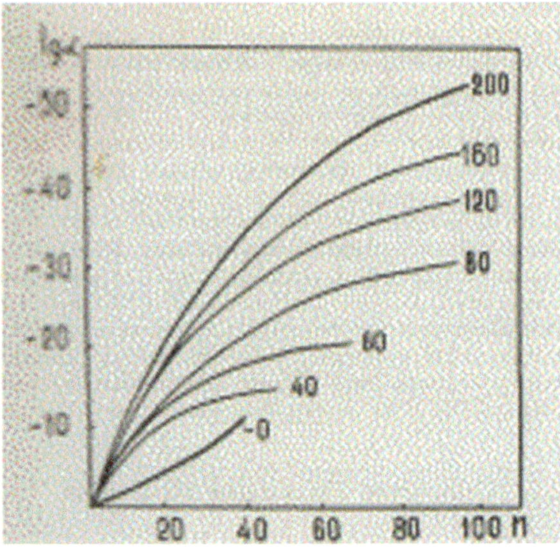

Fig. 8.25 Theoretical dependence of the synchronization factor (a_{syn}) on number of degree of freedom of nuclei involving in the concerted reaction. The curves have been constructed in accordance with 8.9 [18]

Here, n is the number of vibrational degree of freedom of the nuclei participating in the concert transition. Analysis of 2.44 provides a clear idea of the scale of the synchronization factor, depending on the number of broking and newly forming chemical bonds and the energy activation (Fig. 8.25).

For example, at moderate energy activation, typical for chemical step of enzyme reactions (20–40 kJ/mol), the incorporation of each new nucleus into the transition state can lead to about ten-fold decrease in the rate state of the process. The condition preferring the concerted reaction as opposite to the direct one with a rate constant k_{dir} and energy activation E_{dir} is the inequality (8.10) which can be considered as quantitative expression of the principle of "optimal motion".

$$\frac{k_{syn}}{k_{dir}} > a_{syn} \exp\left(\frac{E_{dir} - E_{syn}}{RT}\right) \tag{8.10}$$

It is obvious that according to the principle of optimum motion, the participation of water molecules in synchronous chemical elementary act should decrease the act energy activation and, in turn, decrease the value of synchronization factor.

References

1. R. Lumry, H. Eyring, Conformational changes in proteins. J. Phys. Chem. **58**, 110–120 (1954)
2. K. H. Linderstrom-Lang, X Schellmann, Protein structure and enzyme activity, in *The Enzyme*, V.1, eds. by P.D. Boyer, M. Lardy, Myrback (Academic Press, N.Y., 1959), pp. 443–510
3. D.J. Koshland, Mechanism of transfer of enzymes, in *The Enzyme V.1*, eds. by P.D. Boyer, M. Lardy, Myrback (Academic Press, N.Y., 1959), pp. 305–346

4. G.I. Likhtenshtein, Regularities in the energy and entropy properties of enzymatic processes. Biofizika **11**, 24–32 (1966)
5. R. Lumry, S. Rajender, Enthalpy-entropy compensation phenomenon in water solutions of proteins and small molecules: ubiquitous property of water. Biopolymers **9**, 1125–1227 (1970)
6. G.I. Likhtenshtein, On the nature of compensation phenomena in liquid phase reactions. Russian. J. Phys. Chem. **44**, 1980–1990 (1970)
7. G.I. Likhtenshtein, Water and protein dynamics, in *L'eau et les Systemes Biologoque*, eds. by A. Alfsen, A.J. Bertran (CNR, Paris, 1976), pp. 45–43
8. G.I. Likhtenshtein, *Spin Labeling Method in Molecular Biology* (Wiley Interscience, N.Y., 1976)
9. G.I. Likhtenshtein, Study of protein dynamics by spin-labeling, Mösbauer spectroscopy, and NMR, in *Special Collogue Amper on Dynamic Processes in Molecular Systems*. ed. by A. Losche (Karl-Marx University, Leipzig, 1979), pp. 100–107
10. G.I. Likhtenshtein, F. Febbrario, R. Nucci, Intramolecular dynamics and conformational transitions in proteins studied by biophysical labeling methods. Common and specific features of proteins from thermophylic microorganisms. Spectrochemica Acta Part A. Biomol. Spectroscopy **56**, 2011–2031 (2000); G.I. Likhtenshtein, Water and protein dynamics, in *L'eau et les Systemes Biologoque*, eds. by A. Alfsen, A.J. Bertran (CNR, Paris, 1976), pp. 45–43
11. G.I. Likhtenshtein, Structure and dynamics of metalloenzymes studied by biophysical labeling methods. J. Molecul. Catal. **48**, 129–138 (1988)
12. E.N. Frolov, O.V. Belongova, G.I. Likhtenshtein, Investigation of the mobility of spin and Mössbauer labels bound to macromolecules, in Equilibrium Dynamics of the Native Structure of Protein, ed. by E.A. Burshtein (Izdatelstvo Akademii Nauk SSSR, Pushchino, 1977), pp. 99–142
13. G.I. Likhtenshtein, Water and dynamics of proteins and membranes. Studia Biophysica **111**, 89–100 (1986)
14. R. Lumry, R.B. Gregory, Free energy management in protein reactions: concepts, complications, and compensation, in *The Fluctuating Enzyme*. ed. by G.R. Welch (Wiley, New York, 1986), pp. 3–185
15. R. Lumry, On the interpretation of data from isothermal processes. Methods Enzymol. **259**, 628–720 (1995)
16. R. Lumry, The new paradigm for protein research, in *Protein-Solvent Interactions*, ed. by R.B. Gregory (Marcel Dekker Inc., New York, 1995)
17. E.T. Denisov, O.M. Sarkisov, G.I. Likhtenshtein, *Chemical Kinetics Fundamentals and Recent Developments* (Elsevier Science, 2003)
18. G.I. Likhtenshtein, *New Trends In Enzyme Catalysis and Mimicking Chemical Reactions* (Kluwer Academic/ Plenum Publishers, N.Y., 2003)
19. O.V. Belonogova, E.N. Frolov, S.A. Krasnopolskaya, B.B. Atanasov, V.K. Gins, E.N. Mukhin, A.A. Levina, A.B. Andreeva, G.I. Likhtenstein, V.I. Goldansky, Study of the effect of the degree hydration on the mobility of mossbauer atoms at the active sites of metalloenzymes and metelcarriers. Dokl. Akad. Nauk SSSR **241**, 219–222 (1978)
20. F. Parak, E.W. Knapp, D. Kucheida, Protein dynamics. Mössbauer spectroscopy on deoxymyoglobin crystal. J. Mol. Biol. **161**, 177–194 (1982)
21. E.R. Bauminger, S.G. Cohen, I. Nowik, S. Ofer, J. Yariv, Dynamics of heme iron in crystals of metmyoglobin and deoxymyoglobin. PNAS **80**, 736–740 (1983)
22. K.C. Keppler, K. Achterhold, A. Ostermann, U. van Bürck, A.I. Chumakov, R. Rüffer, W. Sturhahn, E.E. Alp, F.G. Parak, Nuclear forward scattering of synchrotron radiation by deoxymyoglobin. Eur. Biophys. **29**, 146–152 (2000)
23. D. Dantsker, U. Samuni, A.J. Friedman, M. Yang, A. Ray, J.M. Friedman, Geminate rebinding in trehalose-glass embedded myoglobins reveals residue-specific control of intramolecular trajectories. J. Mol. Biol. **315**, 239–251 (2002)
24. G. Cottone, S. Giuffrida, G. Ciccotti, L. Cordone, Molecular dynamics simulation of sucrose- and trehalose-coated carboxy-myoglobin. Proteins **1**(59), 291–302 (2005)

25. K.D. Rector, J. Jiang, M.A. Berg, M.D. Fayer, Effect of solvent viscosity on protein dynamics: infrared vibrational echo experiment and theory. J. Phys. Chem. **105**, 1081–1082 (2001)
26. O.V. Belonogova, G.I. Likhtenshtein, V.I. Krinichnyi, Electron transfer in donor-acceptor pairs in modified α-chymotrypsin. Effect of microviscosity, macroviscosity and local polarity. Photochem. Photobiol. A Chem. **107**, 35–42 (1997)
27. T. Li, A.A. Hassan, S.J. Singer, Origin of slow relaxation following photoexcitation of W7 in myoglobin and the dynamics of its hydration layer. J. Phys. Chem. B **112**, 16121–16134 (2008)
28. G.I. Likhtenshtein, *Labeling, Biophysical*, ed. by R. Meyers, R. (ed.), Encyclopedia of Molecular Biology and Molecular Medecine, Vol. 7 (VCH, New-York, 2008), pp. 157–178
29. G.I. Likhtenshtein, *Chemical Physics of Redox Metalloenzyme Catalysis* (Springer-Verlag, Berlin, 1988)
30. G.I. Likhtenshtein, *Biophysical Labeling Methods in Molecular Biology* (Cambridge University Press, Cambridge, N.Y., 1993)
31. B.Y. Roslyakov, Y.I. Khurgin, Study of cyanamoyl-a chymotrypsin hydrolysis in the solid state. Biokhimiya **37**, 447–493 (1972)
32. J.H. Roh, J.E. Curtis, S. Azzam, V.N. Novikov, I. Peral, Z. Chowdhuri, R.B. Gregory, A.P. Sokolov, Influence of hydration on the dynamics of lysozyme. Biophys. J **91**, 2573–2588 (2006)
33. J.A. Rupley, G. Careri, Protein hydration and function. Adv. Protein Chem. **41**, 37–172 (1991)
34. H.A. Kramers, Brownian motion in a field of force and the diffusion model of chemical reactions. Physica **7**, 284–304 (1940)
35. H. Kyushiki, A. Ikai, The effect of solvent viscosity on the rate-determining step of fatty acid synthetase. Proteins **8**, 287–293 (1990)
36. G. Gadda, P. Sobrado kinetic solvent viscosity effects as probes for studying the mechanisms of enzyme action. Biochemistry **57**, 3445–3453 (2018*)*
37. S. Uribe, G. Sampedro, Measuring solution viscosity and its effect on enzyme activity. Biol. Proc. Online **5**, 108–115 (2003)
38. J. Gadda, P.F. Fitzpatrick, Solvent isotope and viscosity effects on the steady-state kinetics of the flavoprotein nitroalkane oxidase. FEBS Lett. **587**, 2785–2789 (2013)
39. P.L. Privalov, J.G. Gill, Stability of protein structure and hydrophobic interaction. Adv. Prot. Chem. **39**, 191–234 (1998)
40. G. Rialdi, E. Battistel, Thermodynamic properties of immobilized enzymes. Calorim. Anal. Therm. **24**, 401–404 (1993)
41. E. Battistel, F. Attanasio, G. Rialdi, Thermal stability of immobilized α-chymotrypsinogen. J. Therm. Anal. Calorim. **61**, 513–525 (2000)
42. G.I. Likhtenshtein, Amorphous-crystaline model of intermolecular dynamics in global proteins. Biofizika **30**, 3–37 (1985)
43. A.H. Larsen, Y. Wang, S. Bottaro, S. Grudinin, L. Arleth, K. Lindorff-Larsen, Combining molecular dynamics simulations with small-angle X-ray and neutron scattering data to study multi-domain proteins in solution. PLoS Comput. Biol. **16**, e1007870 (2020)
44. The mechanism of catalytic decomposition. Proc. Roy. Soc. Lond. A **108**, 355–378 (1925)
45. J.E. Leffler, The enthalpy-entropy relationship and its implications for organic chemistry. J. Am. Chem. Soc. **20**, 1202–1231 (1955)
46. O. Exner, On the enthalpy–entropy relationship. Coll. Czech. Chem. Comm. **26**, 1094–1113 (1964)
47. A.A. Edwards, J.M. Mason, K. Clinch, P.C. Tyler, G.B. Evans, V.L. Schramm, Altered enthalpy-entropy compensation in picomolar transition state analogues of human purine nucleoside phosphorylase. Biochemistry **48**, 5226–5238 (2009)
48. A. Cornish-Bowden, Enthalpy–entropy compensation and the isokinetic temperature in enzyme catalysis. J. Biosci. **42**, 665–670 (2017)
49. K. Sharp, Entropy-enthalpy compensation: fact or artifact? Protein Sci. **10**, 661–667 (2001)
50. A.I. Dragan, C.M. Read, C. Crane-Robinson, Enthalpy–entropy compensation: the role of solvation. Eur Biophys J **46**, 301–308 (2017)

51. J.F. Perez-Benito, M. Mulero-Raichs, Enthalpy-Entropy compensation effect in chemical kinetics and experimental errors: a numerical simulation approach. J. Phys. Chem. A **120**, 7598–7609 (2016)
52. J.D. Chodera, D.L. Mobley, Entropy and enthalpy compensation: role and ramification in biomolecule ligand recognition and design. Annu. Rev. Biophys. **42**, 121–142 (2013)
53. A. Pan, A. Pan, A.K. Rakshit, S.P. Moulik, Enthalpy-entropy compensation (EEC) effect: a revisit. Phys. Chem. B **119**, 15876–15884 (2015)
54. G.I. Likhtenshtein, T.V. Troshkina, On the kinetics of the cooperative processes in proteins. Molecul. Biol. (Moscow) **2**, 654–663 (1968)
55. L. Jen-Jacobson, E. Engler, L.A. Jen-Jacobson, Structural and thermodynamic strategies for site-specific DNA binding proteins. Structure **8**, 1015–1023 (2000)
56. B. Breiten, M.R. Lockett, W. Sherman, S. Fujita , M. Al-Sayah, H. Lange, C.M. Bowers, A. Heroux, G. Krilov, G.M. Whitesides, Water networks contribute to enthalpy/entropy compensation in protein– ligand binding. J Am Chem Soc **135**, 15579–15584 (2013)
57. S-R. Tzeng, C.G. Kalodimos Protein activity regulation by conformational entropy. Nature **488**, 236–240 (2012)
58. J. Zhang, C.P. Jones, A.R. Ferré-D'Amaré, Global analysis of riboswitches by small-angle X-ray scattering and calorimetry. Biochim Biophys Acta **1839**, 1020–1029 (2014)
59. K.K. Frederick, M.S. Marlow, K.G.A.J, Valentine, J. Wand, Conformational entropy in molecular recognition by proteins. Nature **448**, 325–329 (2007)
60. L. Movileanu, E.A. Schiff, Entropy-Enthalpy compensation of biomolecular systems in aqueous phase: a dry perspective. Monatsh. Chem. **144**, 59–65 (2013)
61. M. Ahmad, V. Helms, T. Lengauer, O.V. Kalinina, Enthalpy–Entropy compensation upon molecular conformational changes. J. Chem. Theory Comput. **11**, 1410–1418 (2015)
62. J. Zhang, J.S. Liu, On side-chain conformational entropy of proteins. PLoS Comput. Biol. **2**(12), e168 (2006). https://doi.org/10.1371/journal.pcbi.0020168
63. J.A. Caro, K.W. Harpole, V. Kasinath, J. Lim, J. Granja, K.G. Valentine, K.A. Sharp, A. Joshua Wand, Entropy in molecular recognition by proteins. PNAS **114**(25), 6563–6568 (2017)
64. Y. Matsuura, M. Takehira, Y. Joti, K. Ogasahara, T. Tanaka, N. Ono, N. Kunishima, K. Yutani, Thermodynamics of protein denaturation at temperatures over 100 °C: CutA1 mutant proteins substituted with hydrophobic and charged residues. Sci. Rep. 5, 15545 (2015)
65. J.F. Perez-Benito, M. Mulero-Raichs, Enthalpy–entropy compensation effect in chemical kinetics and experimental errors: a numerical simulation approach. J. Phys. Chem. A **120**, 7598–7609 (2016)
66. P.J. Barrie, The mathematical origins of the kinetic compensation effect: 1. The effect of random experimental errors. Phys. Chem. Chem. Phys.**14**, 318–326 (2012)
67. Hydrolase—Chemistry Encyclopedia—Water, Examples, Molecule. www.chemistryexplained.com. Retrieved 2018-04-29
68. Hydrolase. Britannica Academic, Encyclopedia Britannica (2018). https://academic-eb-com.proxy.wexler.hunter.cuny.edu/levels/collegiate/article/hydrolase/41737
69. H. Dvira, I. Silman, M. Harela, T.L. Rosenberryc, J.L. Sussman, Acetylcholinesterase: from 3D structure to function. Chem. Biol. Interact. **6**(187), 10–22 (2010)
70. J. Sussman, J.L. Sussman, M. Harel, F. Frolow, C. Oefner, A. Goldman, L. Toker, I Silman structure of acetylcholinesterase from torpedo Californica: a prototypic acetylcholine-binding protein. Science **253**(5022), 872–879 (1991)
71. I. de Diego, F. Veillard, M.N. Sztukowska, T. Guevara, B. Potempa, A. Pomowski, J.A. Huntington, J. Potempa, F. Xavier Gomis-Rüth, Structure and mechanism of cysteine peptidase gingipain K (Kgp), a major virulence factor of Porphyromonas gingivalis in periodontitis. J. Biol. Chem. **289**, 32291–32302 (2014)
72. M. Fujinaga, M.M. Cherney, H. Oyama, K. Oda, M.N. James, The molecular structure and catalytic mechanism of a novel carboxyl peptidase from *Scytalidium lignicolum*. Proc. Natl. Acad. Sci. USA **101**, 3364–3369 (2004)

73. J.A. Huntington , J. Potempa , F. Xavier Gomis-Rüth, Structure and mechanism of cysteine peptidase gingipain K (Kgp), a major virulence factor of Porphyromonas gingivalis in periodontitis. J. Biol. Chem. **289**, 32291–302 (2014)
74. D. Barford, A.K. Das, M.P. Egloff, The structure and mechanism of protein phosphatases: insights into catalysis and regulation. Annu. Rev. Biophys Biomol. Struct. **27**, 133–164 (1998)
75. Y. Shi, Serine/threonine phosphatases: mechanism through structure. Cell **139**, 468–483 (2009)
76. K. Oda, New families of carboxyl peptidases: serine-carboxyl peptidases and glutamic peptidases. J. Biochem. **151**, 13–25 (2012)
77. Q. Xu, J. Yao, A. Wlodawer, H. Guo, Clarification of the mechanism of acylation reaction and origin of substrate specificity of the serine-carboxyl peptidase sedolisin through OM/MM free energy simulations. J. Phys Chem. B. **115**, 2470–2476 (2011)
78. S.I. Allakhverdiev, *Photosynthesis: New Approaches to the Molecular, Cellular, and Organismal Levels* (Scrivener Publishing LLC, 2016)
79. H.J.M. Hou, M.M. Najafpour, G.F. Moore, S.I. Allakhverdiev, *Photosynthesis: Structures, Mechanisms, and Applications* (Springer, 2017)
80. J. Raymond, R. Blankenship, The origin of the oxygen-evolving complex. Coord. Chem. Rev. **252**, 377–383 (2008)
81. L.V. Kulik, B. Epel, W. Lubitz, J. Messinger, Electronic Structure of the Mn_4OxCa cluster in the S_0 and S_2 states of the oxygen-evolving complex of photosystem II based on pulse 55Mn-ENDOR and EPR spectroscopy. J. Am. Chem. Soc. **129**, 13421–13435 (2007)
82. P. Chernev, S. Fischer, H. Dau, Light-driven formation of manganese oxide by today's photosystem II supports evolutionarily ancient manganese-oxidizing photosynthesis. Nat. Commun. **11**, 6110 (2020)
83. G.I. Likhtenshtein, *Chemical Physics of Redox Metalloenzymes* (Springer-Verlag, Heidelberg, 1988)
84. G.I. Likhtenshtein, *New Trends in Enzyme Catalysis and Mimicking Chemical Reactions* (Kluwer Academic/ Plenum Publishers, N.Y., 2003), p. 19
85. E.T., Denisov, O.M. Sarkisov, G.I. Likhtenshtein,*Chemical Kinetics. Fundamentals and Recent Developments* (Elsevier Science, 2003), p. 18
86. G.I. Likhtenshtein, *Solar Energy Conversion. Chemical Aspects* (WILEY-VCH, 2012)
87. G.I. Likhtenshtein, *Electron Spin in Chemistry and Biology: Fundamentals, Methods, Reactions Mechanisms, Magnetic Phenomena, Structure Investigation* (Weinhem Springer, 2016)
88. G.I. Likhtenshtein, A.E. Shilov, On the thermodynamics of fixation of molecular nitrogen. Russian J. Phys. Chem. **44**, 849–856 (1970)
89. N.N. Semenov, A.E. Shilov, G.I. Likhtenshtein, Multi-electron mechanisms in chemistry and biology. Dokl. Akad. Nauk SSSR **221**, 374–1377 (1975)
90. R.A Marcus, Electron transfer past and future, in *Advances in Chemical Physics*, 107, Part 1, eds. by J. Jortner, M. Bixon (John Wiley & Sons, 1999), pp. 1–6
91. R.A. Marcus, On the theory of oxidation-reduction reactions involving electron transfer. J. Chem. Phys. **24**, 966–978 (1956)
92. F.G. Bordwell, Are nucleophylic bimolecular concerted reactions involving four or more bonds a myth? Accounts Chem. Res.
93. C.F. Bernasconi, The principle of nonperfect synchronization: more then qualitative concept. Acc. Chem. Res. **25**, 9–16 (1992)
94. G.I. Likhtenshtein, On the principle of "optimum motion' in elementary acts of chemical and biological processes, I. Estimation of the synchronization factor for model processes. Kinetika I Kataliz (Kinetics Catal.) **28**, 878–882 (1977)
95. G.I. Likhtenshtein, E.A. Mullokandov, On the principle of "optimum motion' in elementary acts of chemical and biological processes, II. Mechanisms of the reduction of molecular nitrogen and liberation of molecular oxygen from water, Kinetika I Kataliz (Kinetics Catal.) **28**, 883–888 (1977)

96. G.I. Likhtenshtein, On the principle of "optimum motion' in elementary acts of chemical and biological processes. III. Analysis of probabilistic mechanisms of some chemical and boichemical processes, Kinetika I Kataliz (Kinetics Catal.) **28**, 1255–1260 (1977)
97. I.V. Alexandrov, Dependence of the rate of the concerted chemical reactions on the number of degree of freedom involved in the transition, Teor. Exper. Khim. (Theor. Experim. Chem.) **12**, 299–306 (1976)

Chapter 9
Nucleic Acids Hydration

Abstract Molecular properties of deoxyribonucleic acid (DNA) govern important biological processes such as DNA replication and DNA repair. Ribonucleic acids (RNAs) play the critical role in various cellular processes. Structure and molecular dynamics of water in hydrated DNA and RNA studied by advanced physical methods are the focus of this chapter. The results showed that electric forces at the DNA surface mainly originate from water molecules in the first two hydration layers and the water-bridge lifetime in the DNA and RNA groove varies from 1 to 300 ps, depending on the sequence. Effect of ions residing in the hydration shells and pockets/grooves of DNA/RNA is also described.

9.1 Introduction

The Swiss scientist Friedrich Miescher in 1868 discovered nucleic acids (DNA) and raised the idea that they could be involved in heredity [1, 2]. The Miescher's prophecy came true after 83 years when Watson and Crick presented the structure of DNA as the primary information-carrying molecules in cells [3].

Deoxyribonucleic acid (DNA) is directly linked to important biological processes such as DNA replication and DNA repair. Ribonucleic acid (RNA) molecules play an active role within cells by catalyzing biological reactions, controlling gene expression, or sensing and communicating responses to cellular signals. Water and ions control chemical and physical processes of nucleic acids, directly and/or indirectly facilitating the functions of biomolecules such as nucleic acids (DNA and RNA), ions residing in the hydration shells and pockets/grooves of DNA/RNA. (See for example Fig. 9.1). Hydration water serves as a microscopic manifestation of structural stability and functions of nuclear acids [3–34]. The helical structure of DNA is stabilized by the solvation environment, where a change in the hydration state can lead to dramatic changes to the DNA structure. Dehydration can modify the structure from the classic right-handed B-form to the shorter and wider A-form and the left-handed Z-form. In order for nucleic acids to associate, many ordered water molecules surrounding them must be released from their interaction surfaces; this entropically affects the

G. I. Likhtenshtein, *Biological Water*, Soft and Biological Matter,
https://doi.org/10.1007/978-3-030-82503-4_9

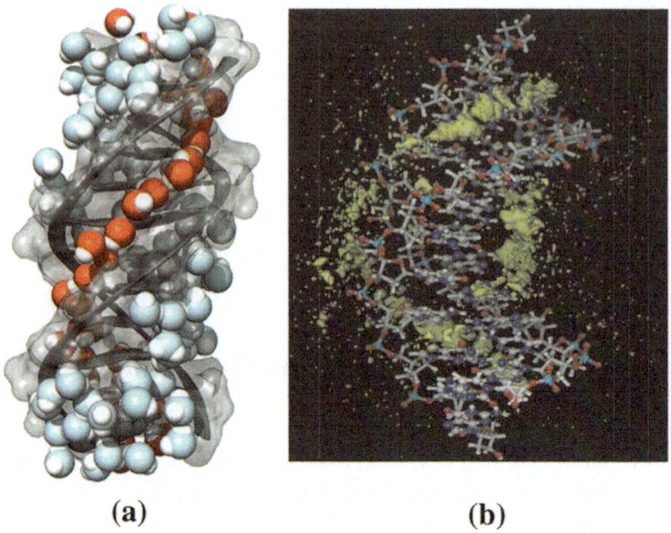

Fig. 9.1 Regions of high Na^+ density around d(CGCGAATTCGCG)$_2$ obtained as a superposition of 6007 snapshots corresponding to 60-ns MD [14]. Published with permission of PNAS article containing the requested material published under a CC BY-NC-ND license

molecular association [1, 2]. Interactions of DNA strands with small molecules and proteins are also mediated by the solvation structure [3–34].

Electric forces at the DNA surface have been shown to mainly originate from water molecules in the first two hydration layers (Fig. 9.2). These layers of water around DNA are critical to its structure and biological function. Such water and ions in the hydration/condensation layer show hydrogen-bond (vibrational) and diffusive (rotational and translational) dynamics in ultrafast timescales ranging from tens of femtoseconds (10^{-15} s) to several nanoseconds 10^{-9} s. Water forms a hydration shell which interacts with nucleic acids through local hydrogen bonds, short- and intermediate-range Coulomb forces from the dipolar water molecules, and hydrophobic forces. In the case of DNA and RNA, the hydration pattern at the surface of the double helix is highly inhomogeneous with the rather rigid so-called spine of water molecules in the minor groove and a preferred hydration of the ionic phosphate groups of the backbone. In addition to such structural and thermodynamic effects, the kinetics of water molecules is also of importance in the hydration of biomolecules.

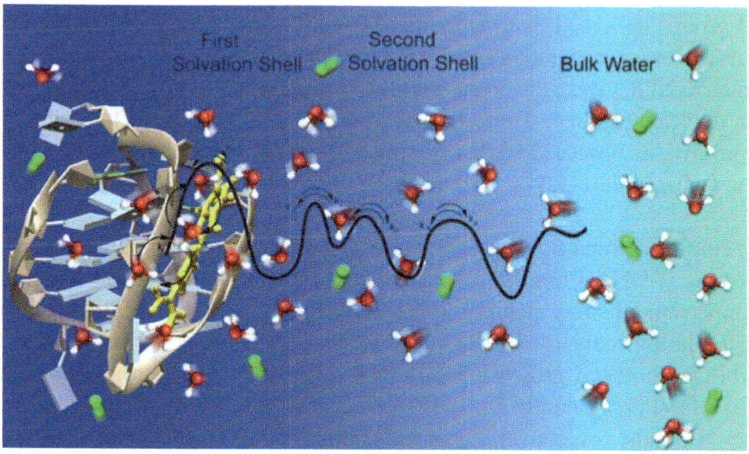

Fig. 9.2 Schematic representation depicting the situation of perturbed water molecules on a rugged potential surface near GqDNA. Water with broad distribution of residence times can have different trapping times at local minima and thus different exchange rates (ki's) between minima on the potential surface. Such distributions of exchange of subdiffusive water can ultimately lead to the observed dispersed (power law) solvation dynamics in GqDNA [13]. With permission from American Chemical Society

9.2 DNA

9.2.1 Structure of Water in Hydrated DNA

The x-ray crystal structure of the double-helical *B*-DNA dodecamer of sequence C-G-C-G-A-A-T-T-C-G-C-G has been solved and refined *independently* in three forms: 1 the parent sequence at room temperature; 2 the same sequence at 16 K, and 3 the 9-bromo variant C-G-C-G-A-A-T-T^{Br}C-G-C-G at 7 °C in 60% (v/v) 2-methyl-2.4-pentanediol [12]. The latter two structures showed extensive hydration along the phosphate backbone.

The x-ray analysis revealed that (1) sixty-five solvent peaks were associated with the phosphate backbone or an average of three per phosphate group; (2) nineteen other molecules form a first shell of hydration to base edge N and O atoms within the major groove; (3) 36 more water molecules are found in upper hydration layers and tend to occur in strings or clusters spanning the major groove from one phosphate group to another; (4) in the minor groove, the zigzag spine of hydration being responsible for stabilizing the *B*-form of DNA was found; and (5) upper level hydration in the minor groove consists mainly of strings of water molecules extending across the groove, with few contacts to the spine below.

The experimental observations of nuclear Overhauser effects (**NOE**) between water protons and the protons of the d-$(CGCGAATTCGCG)_2$ and d-$(AAAAATTTTT)_2$ were reported[13]. The positive sign of NOEs with the thymine methyl groups indicated that the residence times of the hydration water molecules near these protons in the major groove of the DNA must be shorter than about 500 ps. Negative NOEs were observed with the hydrogen atoms in position 2 of adenine in both duplexes studied. These findings proved that a "spine of hydration" in the minor groove, as observed by X-ray diffraction in DNA crystals, is present also in solution and is an integral part of the molecular architecture both in solution and in crystals.

The conformational transitional of substates B(I) and B(II) of the phosphodiester backbone in B-DNA, contributed to DNA flexibility and recognition of indolicidin, an antimicrobial 13-mer peptide, were studied by rapid scan FTIR spectroscopy [16]. Correlation analysis of IR absorption changes occurring within seconds after a reversible incremental growth of the DNA hydration shell identified water populations w(1) (PO_2^--bound) and w(2) (non-(PO_2^--bound) exhibiting weaker and stronger H-bonds, respectively, than those dominating in bulk water. The analysis revealed that the B(II) substate is stabilized by w(2) and a contiguous water network (at 12–14 H_2O molecules per DNA phosphate) of reduced ν(OH) bandwidth is formed. In this state, hydration water cooperatively stabilizes the B(I) conformer via the entropically favored replacement of w(2)-DNA interactions by additional w(2)-water contacts. It was suggested that the water rearrangements contribute to the recognition of DNA by indolicidin, preferentially binding to the B(I) conformer in a water-mediated induced fit. In addition, the FTIR spectra resolve sequential steps leading from PO_2^{--}-solvation to substate transition and eventually to base stacking changes in the complex.

X-ray diffraction experiments at cryogenic temperatures have shown crystallographic water molecules in the minor groove of DNA, which has led to the notion of a spine of hydration of DNA [17]. Chiral nonlinear vibrational spectroscopy of two DNA sequences confirmed that such structural water molecules exist in solution at ambient conditions having a form of chiral superstructure: a chiral spine of hydration (Fig. 9.3). Thus, a chiral water superstructure is templated by a biomolecule.

Achiral signature the CGCGAATTCGCG helical structure was obtained by taking the difference between the + 45°and −45° signals, analogously to circular dichroism [18]). The strong difference between the + 45°pp and − 45°pp spectra across the water OH stretch vibrational frequency region demonstrates the existence of a chiral water superstructure templated by the helical structure. Presented data confirm the existence of a DNA minor groove spine of hydration at room temperature and shows that the chiral structure of biomolecules can be imprinted on the surrounding solvation structure. In addition, the chiral water superstructure exhibits a broad range of hydrogen-bonding strengths reflecting the large heterogeneity of the water structure and dynamics in the solvation shell of DNA.

Hydration forces between DNA molecules of the A- and B-form in oriented calf-thymus DNA fibers were studied utilizing a newly developed small- and wide-angle X-ray scattering (SAXS and WAXS) enabling simultaneous in situ control of temperature and relative humidity (RH) range of 98–70% [19]. In this condition, DNA undergoes the B- to A-form transition. The derivation of a force–distance curve with a characteristic exponential decay constant of ~2 Å for both A and B-DNA indicated

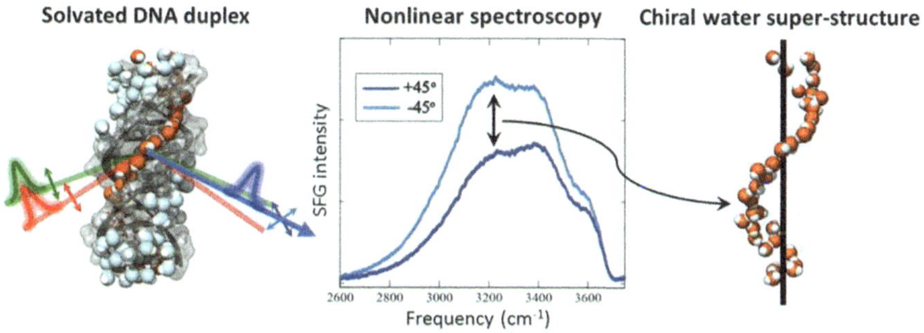

Fig. 9.3 Observation of a chiral water superstructure surrounding DNA by chiral sum-frequency generation (SFG) spectroscopy. Water molecules form helical superstructures following the contours of the DNA in the minor groove (red) and major groove (light blue). DNA was bound at the glass/bulk water interface and probed by chiral SFG. Chiral SFG is a nonlinear chiral optical spectroscopic method, relying on a visible beam at frequency ωvis and an infrared beam at frequency ωIR to generate an SFG signal at frequency ωvis + ωIR. Figure modified with permission from ref 1. Copyright 2017 American Chemical Society [17]. With permission from American Chemical Society

the change in DNA separation in response to variation in humidity. The investigation confirmed that the hydration force remains dominant even in the dry A-DNA state and at surface separation down to ~1.5 Å, within the first hydration shell. It was also found that the observed chemical potential difference between the A and B states could be attributed to the water layer inside the major and minor grooves of the A-DNA double helices and surface separation range is ~1.5 Å–6 Å and for DNA in either A- or B-form. The contribution of water within the major and minor grooves of A-DNA at separations < 2.5 Å was considered. The transition from B-form to A-form (Fig. 9.4) was examined by plotting the change of the double helix pitch as a function of relative humidity and showed that (1) the onset of the transition from B to A-form occurs at ~87% RH within a narrow range (~0.3% RH) and is discontinuous; (2) At RH between 82.7% and 86.7% (vertical dashed lines), there is evidence of coexistence of B- and A-forms in the diffraction data; (3) At RH < 80%, the signatures of the B-form are no longer detectable; and (4) at RH = 88.6%, the sample is in the B phase in which all DNA are in the B-form. At RH = 82.7%, the majority of the DNA in the sample is in the A-form (A phase).

Ability of generation of 3D hydration maps of single DNA molecules under near-physiological conditions frequency-modulation atomic force microscopy (FM-AFM) was demonstrated [20]. K FM-AFM utilizes a sharp tip located at the edge of a cantilever, which is oscillated at its resonance frequency, defined by a π/2 phase lag of the tip position relative to the driving force. A 3D-resonance frequency mapping proceeds in the following steps: (1) tip—sample interactions shift the cantilever resonance frequency during scanning; (2) imaging is accomplished by enabling an addi tional feedback loop (z-loop), which controls the tip—sample distance by driving the

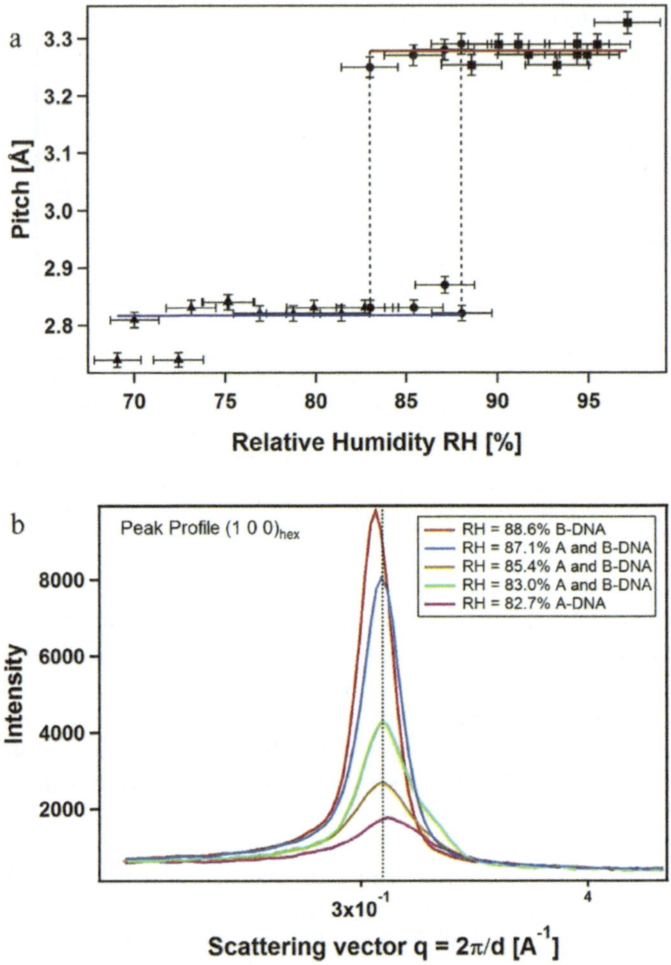

Fig. 9.4 Conversion of B-DNA to A-DNA driven by reduced humidity. **a** Pitch of DNA double helix as a function of humidity showing the transition from B- to A-form starting at RH ~87%. Between RH ~82 and 87% (dashed vertical lines), there is coexistence of A- and B-forms. Error bars for RH were determined by propagating thermistor uncertainties through the Claussius–Clapeyron equation, while uncertainties in pitch were determined by calculating the detector's qz resolution at the first and second layer lines of B- and ADNA, respectively. **b** Intensity profiles of the first equatorial peak in the humidity range of pure B-DNA, to a coexistence of B- and A-DNA, and finally to pure A-DNA. In the coexistence range (between RH 83% and 87%), the peak position (hence interhelical separation) remains unchanged (vertical line) form [19]. With permission from Elsevier

sample stage so that the resonance frequency remains constant while the tip scans the surface; (3) a 2D topography image is generated by displaying the tip–sample separation needed to maintain a preset resonance frequency across the image; and (4) the 3D-resonance frequency mapping is generated by adding another step to the 2D topography scan.

Figure 9.5 depicts four consecutive 3D scans of the same section of a single DNA molecule. The minor groove at the bottom appears open (Fig. 9.5 a, d) but partially covered by the phosphate backbone in (Fig. 9.5b, c). Panels e and f of the figure indicate frequency shift cross sections taken along the solid and dashed lines in Fig. 9.5d, respectively. The cross section of Fig. 9.5e, taken along the minor groove, shows clear hydration oscillations, while the cross section of Fig. 9.5f indicated no hydration oscillations along the phosphate backbone.

Relationship between the heat capacity change, ΔCp, and the nature of the surface dehydrated cannot be modeled for DNA by the use of small molecules, and therefore, the contiguous surfaces of the grooves cannot be treated as the sum of small component molecules such as nucleotides. To overcome this problem, alternative approach in which the heat capacity changes, ΔCp, was measured for the formation of several protein/DNA complexes, and the calculated contribution from protein dehydration subtracted to yield the heat capacity change attributable to dehydration of the DNA was developed. The polar and apolar surface areas of the DNA dehydrated on complex formation were calculated from the known structures of the complexes. The calculation indicated that dehydration of apolar surfaces in both grooves is essentially identical and accompanied by a reduction in ΔCp by about 3 J K^{-1} mol^{-1} $(Å^2)^{-1}$, a value of somewhat greater magnitude than observed for

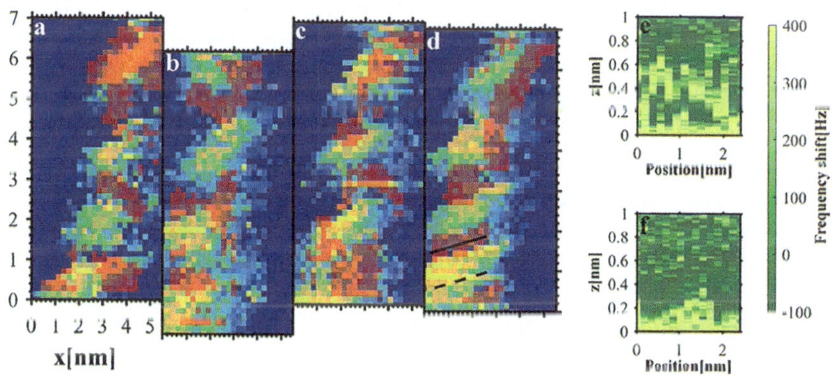

Fig. 9.5 **a–d** Hydration of the same section of a DNA molecule, measured in four consecutive 3D scans. Panels **b–d** were shifted vertically with respect to panel a to align identical grooves. Pixels shaded red indicate high hydration density. Shaded pixels located to the sides of the molecule (dark blue region) have been removed for clarity. **e** Cross section of frequency shift data along the solid line in panel d that passes through a hydrated domain in the groove showing clear hydration oscillations. **f** Cross section of frequency shift data along the dashed line in panel d that passes through the sugar–phosphate backbone and does not show hydration oscillations [20]. With permission from American Chemical Society.

proteins {$\Delta Cp = -1.79$ J K^{-1} mol^{-1} $(Å^2)^{-1}$}. In contrast, dehydration of polar surfaces is very different in the two grooves: In the minor groove, ΔCp increases by 2.7 J K^{-1} mol^{-1} $(Å^2)^{-1}$, but in the major groove, although ΔCp is also positive, it is low in value: +0.4 J K^{-1} mol^{-1} $(Å^2)^{-1}$.

The experimental observations of nuclear Overhauser effects (NOE) between water protons and the protons of the d-$(CGCGAATTCGCG)_2$ and d-$(AAAAATTTTT)_2$ were reported [13]. The positive sign of NOEs with the thymine methyl groups shows that the residence times of the hydration water molecules near these protons in the major groove of the DNA must be shorter than about 500 ps. Negative NOEs were observed with the hydrogen atoms in position 2 of adenine in both duplexes studied. These findings unambiguously indicated that a "spine of hydration" in the minor groove, as observed by X-ray diffraction in DNA crystals, is present also in solution and is an integral part of the molecular architecture both in solution and in crystals.

9.2.2 *Water Dynamics in Hydrated DNA*

Time-resolved emission spectra (TRES) from **fluorescence** decays provide insights about the dynamics of solute—time-dependent fluorescence Stokes shift (TDFSS), and solvation correlation function can be represented in terms the time-resolved fluorescence Stokes shift (TRFSS) experiments (Figs. 9.6 and 9.7) [13, 21, 24]. In the experiments, fluorescent probes, embedded inside DNA, were directly excited by a short laser pulse, and the subsequent time-dependent fluorescence Stokes shift (TDFSS) was monitored. In DNA, the fluorescent probe in the electronic excited state is stabilized by the dynamic water molecules, the motional counterions, and charged/ dipolar parts of DNA molecule which reside within *10–15 A° from the probe-solute. Thus, the measured fluorescence Stokes shift dynamics of probe reports the collective dynamic response of local water, counterions, and DNA parts at the probe site.

Comparison of dynamics over broad time range obtained in different works using TRFSS indicated that the power-law dynamics is very similar within the initial three decades from 100 fs to 100 ps (Fig. 9.7).

A hypothesis that repairs enzymes senses the changes of local collective dynamics of water, counterions, and DNA parts (primarily electrostatic), rather than only the structural changes during the recognition of the mismatches were tested [13].

Specifically, the TRFSS of two minor groove binders, Hoechst 33,258 and DAPI, near a TT mismatch base pair in the AT-rich minor groove of DNA was monitored (Fig. 9.8). The TRFSS results were compared with the dynamics measured by same ligands in the minor groove of normal DNA over five decades of time from 100 fs to 10 ns. Results showed that the fluorescence probe depicts different solvation dynamics in the minor groove of normal DNA. In addition, it was found that introduction of T-T mismatch in the minor groove gives rise to very similar solvation dynamics probed by both DAPI and Hoechst, which follow a single power-law

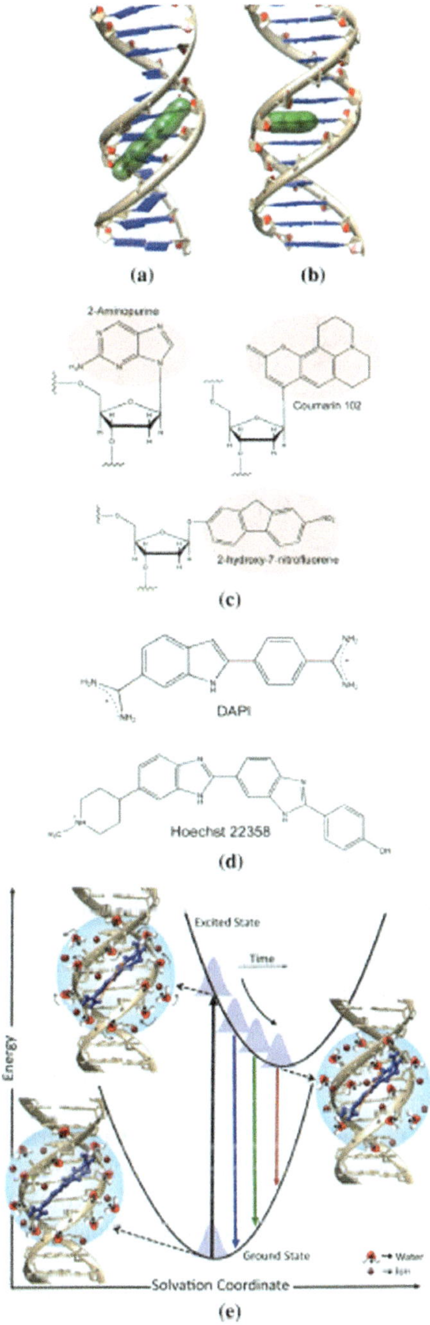

Fig. 9.6 Fluorescent probes used for DNA solvation dynamics: **a** non-covalently attached groove-bound and **b** covalently attached base-stacked probes. Examples of **c** base-stacked and **d** minor groove-bound probe structures. **e** Schematic of solvation dynamics process in DNA where the excited probe energy is stabilized by dynamic water, counterions, and parts of DNA which stay within 10–15 A° of probe-solute [13]

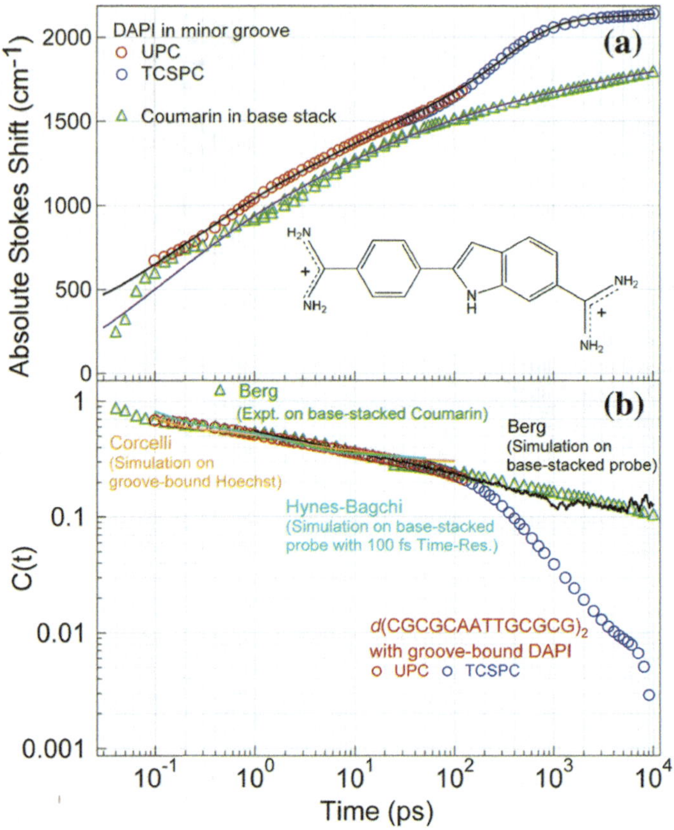

Fig. 9.7 **a** Comparison of (absolute) Stokes shift of DAPI in 14-mer Watson–Crick DNA and base-stacked Coumarin-102. DAPI data (circles: red-UPC; blue-TCSPC) and Coumarin data (green triangles). Solid lines fit to data with single power law (Coumarin-102 data) and power law multiplied with bi-exponential decay (DAPI data). **b** Comparison of solvation correlation function CE(t) of groove-bound DAPI (circle) and base-stacked Coumarin (triangle) [24]. With permission from American Chemical Society

relaxation (with exponent 0.23–0.24) over entire five decades of time from 100 fs to 10 ns.

G-quadruplex DNA (GqDNA) structures act as promising anticancer targets for small molecules. Solvation dynamics of a ligand (DAPI: 4′,6-diamidino-2-phenylindole) inside antiparallel-GqDNA was studied through direct comparison of time-resolved experiments to molecular dynamics (MD) simulation [22]. Dynamic Stokes shifts of DAPI in GqDNA prepared in H_2O buffer and D_2O were compared to find the effect of water on ligand solvation, and experimental dynamics (in H_2O) was in turn directly compared with the dynamics computed from 65 ns simulation on the same DAPI-GqDNA complex. Ligand solvation follows power-law relaxation which was found to be from ~100 fs to 10 ns. Simulation results showed that

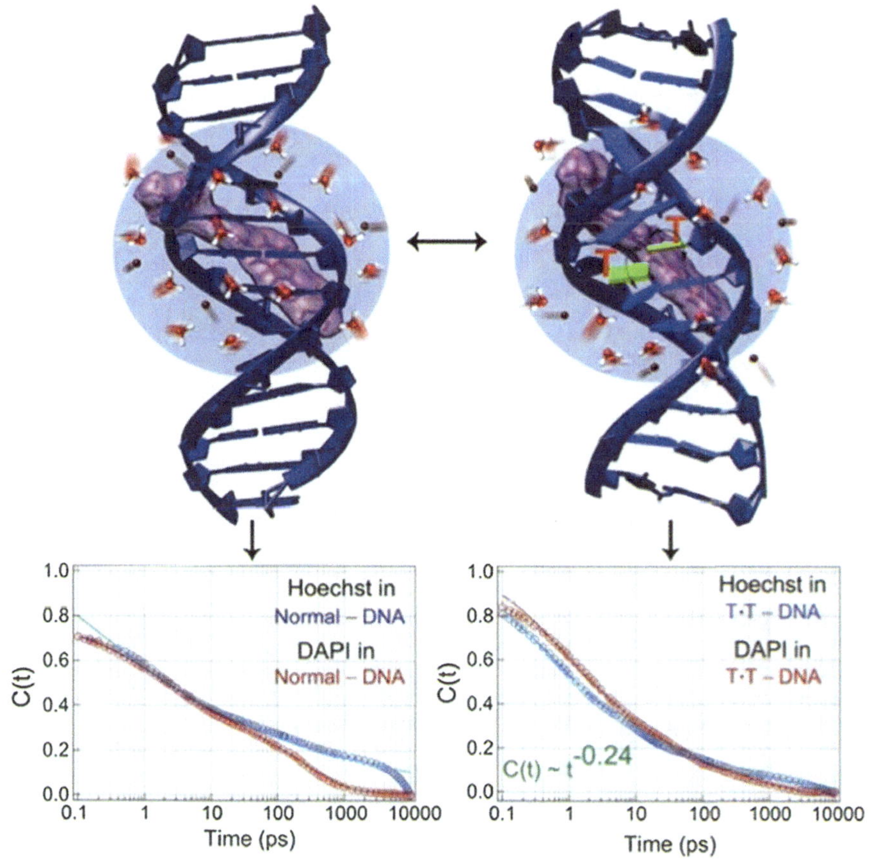

Fig. 9.8 Comparison of solvation correlation functions of DAPI and Hoechst in normal and T-T DNA from 100 fs to 10 ns in DNA. Both probes show faster power-law dynamics in T-TDNA suggesting the unique T_T- effect on the dynamics [13]. With permission from American Chemical Society

relaxation below ~5 ps is dominated by water motion, while both water and DNA contribute comparably to dictate long-time power-law dynamics. Simulation results also suggest that anomalous solvation dynamics may have origin in subdiffusive motion of perturbed water near GqDNA.

The lifetime for a particular hydration pattern in the DNA minor groove, the water bridge, for various DNA sequences showing that the water-bridge lifetime varies from 1 to ~ 300 ps was calculated using MD trajectories [25]. The considered double-stranded B-form DNA was composed of a 12-mer, CGCGN1N2N3N4CGCG (Fig. 9.9), where for N1N2N3N4, all possible arrangements of A, T, G, and C were took into consideration, resulting in 136 systems with different tetrameric patterns. The force fields for DNA and water were obtained from AMBER ff99 and TIP3P,

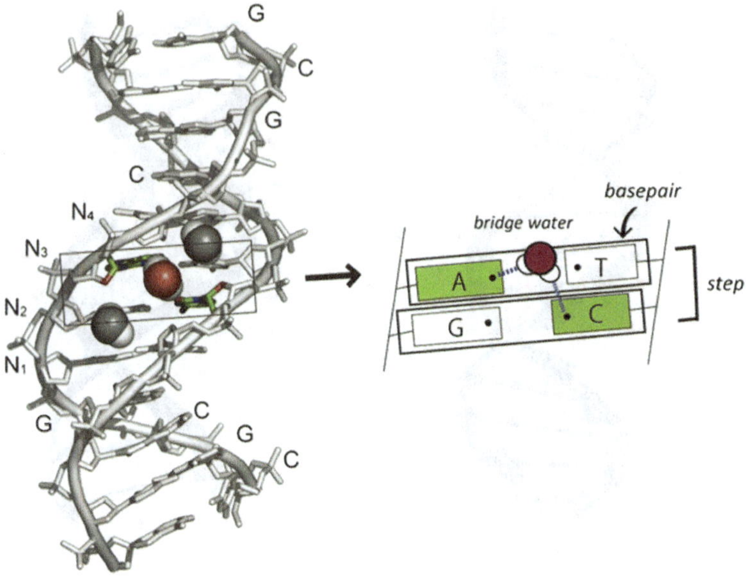

Fig. 9.9 Hydration of 12-mer DNA observed in the minor groove (a snapshot from the MD trajectories). The lifetime was calculated for the water molecule (shown by red and white spheres) that appears at the site of the N2N3 base pair step. This water molecule forms a bridge through hydrogen bonding between the bases of distinct strands [see the schematic view showing a GA step (N2 = G and N3 = A) on the right-hand side [25]. With permission from Elsevier

respectively. About 5000 water molecules were placed around each DNA, and a truncated octahedral periodic boundary condition was imposed. The calculation revealed that the process under study followed 1/k(V(step))P(m), where P(m) and V(step) are two crucial factors, namely the probability of forming a specific hydrogen bond in which more than one donor atom participates and the structural fluctuation of DNA, respectively.

The proposed mechanism by which a water molecule might dissociate from the water-bridge-forming site is sketched in Fig. 9.10 [26]. According to the mechanism, breaking of the associated hydrogen bond occurs when multiple hydrogen bonds are formed and the residence time of a water molecule is $\tau = 1/kPm$, where k is the rate constant of the hydrogen-bond breaking in the M state s [27] **D. Laage,** J.T. Hynes, Reorientational dynamics of water molecules in anionic hydration shells, Proceedings of the National Academy of Sciences of the United States of America 104, 11,167–11,172 (2007). This picture is consistent with the model which suggested that the water oxygen usually receives two hydrogen atoms from the neighboring waters and can sometimes receive three or one. When such over- and under-coordinated water molecules encounter each other, one excess hydrogen flips from the former to the latter.

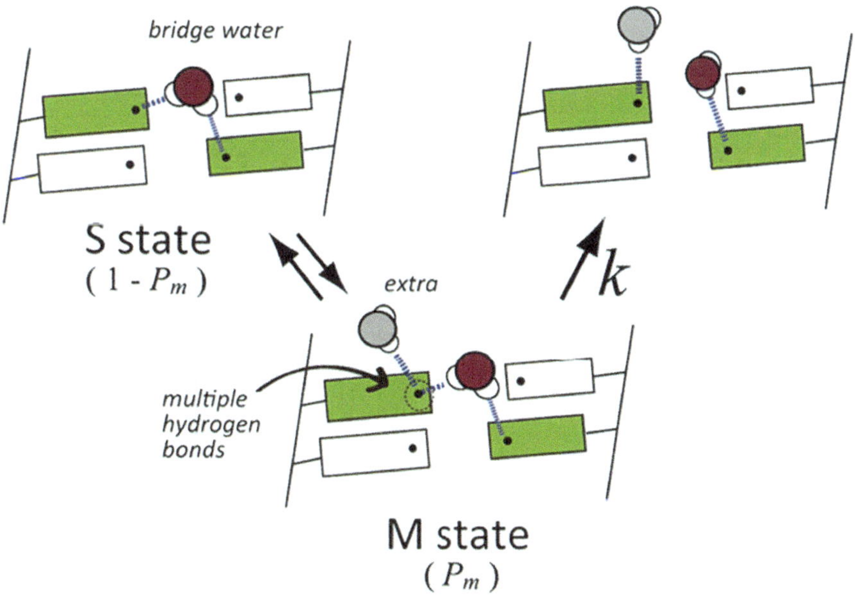

Fig. 9.10 Multiple hydrogen bonds (center) to an acceptor and the proposed mechanism for water dissociation from the bridge-forming site. The water-bridge-forming states can be classified into two cases, those with and without multiple hydrogen bonds, which are called M and S states, respectively. Pm and 1 − Pm denote the probabilities of being in the M and S states, respectively. The parameter k denotes the rate constant in the M state at which the hydrogen bond between DNA and the bridge water breaks [25]. With permission from Elsevier

The results showed that the water-bridge lifetime in the DNA minor groove varies from 1 to 300 ps, depending on the sequence, and the variety can be described by equation $\tau = 1/k(V_{step})$Pm, where V_{step} is called the base pair step deformability. Thus, water dissociation occurs when a particular hydrogen-bonding pattern appears and the DNA structural fluctuation V_{step} contributes to the rate constant k.

To investigate the dynamics and escape process associated with the free energy change of water molecules close to G-quadruplex DNA (GDNA), molecular dynamics simulations combined with the steered molecular dynamics and umbrella sampling were performed [27]. All MD simulations were carry out utilizing the program 121 and the CHARMM27 force field. The stochastic motion of a water molecule steered MD (SMD) and umbrella sampling simulations were used to construct the H_2O and D_2O and to pull the water molecule completely from the hydration shell to the bulk water. The established locations of various water molecules GDNA and the corresponding free energy change (or PMF) are shown in Fig. 9.11. Owing to the structural analysis, the H-bonds between the target water molecule (wt) and biomolecular residue (i.e., S, S1, and S2) were able to be classified into

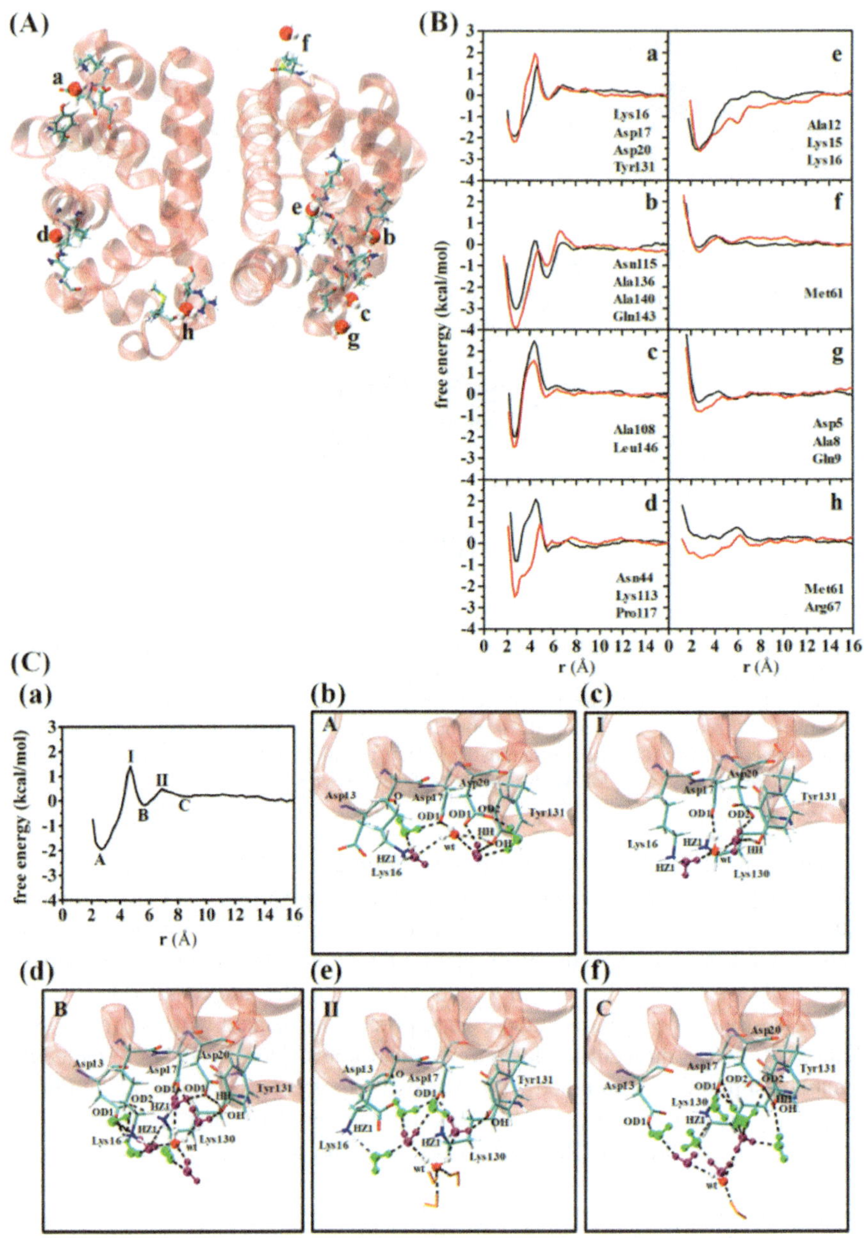

Fig. 9.11 Free energy profile and snapshot of the molecular structure in the GDNA system. (A) The local position of wt. The atoms of the GDNA nucleotide residues are drawn using a stick model and are colored using atoms, and wt is in a ball-and-stick model. (B) Free energy profile of wt escape from the GDNA surface along dN. For H2O (D2O) molecules, the profile is plotted using a black (red) line. (C) The free energy profile with respect to the local structure of wt. In part C, **a** the free energy profile is extracted from part a in part B for the H_2O molecule and **b**–**f** the detailed local structure corresponds to the A, I, B, II, and C states, respectively [27]. With permission from American Chemical Society

(1) direct H-bond, wt-S, (2) one indirect H-bond, wt-w1-S, and (3) two indirect H-bonds, wt-w1-w2-S, where w1 is the nearest neighbor water molecule, and w2 is the second nearest neighbor water molecule. Several special hydration structures, such as a caged H-bond network and edge bridge structures, were also found.

Among the results of the work, we note those related to the dynamic properties of water.

The distribution of residence times of water molecules within each specific hydration layer around the GDNA was analyzed. A drastic change in both mean residence time τ_{res} and long-time tail power α-value with respect to the surface normal distance d_N, which reflects the hydration shell thickness d_N of ~3–5 Å, was observed. The water isotope effect prolongs the dynamic timescales of libration motion in the inner hydration shell, and the dipole relaxation processes in the bulk strengthen the DNA spine water structure. The dipole correlation function with respect to the d_N shows non-Debye relaxation processes with three timescales. The solvent isotope effect was seen to explicitly prolong the residence time and dipole relaxation times of the hydration waters. The potential of the mean force (PMF) of water molecules escaping from the biomolecular surface into the bulk state depending on H-bond rupture processes and ranges from 0.2 to 4.2 kcal/mol was calculated.

Using molecular dynamics simulations dynamical properties of the water confined in hydrophilic protein–protein and protein–DNA interfaces were investigated [28]. It was found that the interfacial water exhibits glassy slow relaxations even at 300 K. Rotational dynamics showed a logarithmic decay that was observed in glass-forming liquids at deeply supercooled states. Such slow water dynamics was argued to be induced by the hydrophilic binding surfaces, which is in opposition to the picture that the hydration water slaves protein motions. An example is depicted in Fig. 9.12 showing the MSD for the hydration water of the barnase–barstar complex in a log–log scale. All the curves display an initial t^2 region describing the following ballistic behavior: (1) for the bulk water, this motion is followed by the diffusion dynamics proportional to t; (2) for the surface and interfacial water, the MSD is reduced compared to the bulk behavior; and (3) for the bridging water, which is a special type of the interfacial water forming concurrent hydrogen bonds with two proteins.

The reorientation and hydrogen-bond dynamics of water molecules within the hydration shell of a B-DNA dodecamer were investigated via molecular dynamics simulations and an analytic jump model [35]. Pronounced spatial heterogeneity was found at the DNA interface and was explained via the jump model by the diversity in local DNA interfacial topographies and DNA–water H-bond interactions. Most of the hydration shell is moderately retarded with respect to the bulk, while some water molecules confined in the narrow minor groove exhibiting very slow dynamics. The DNA conformational fluctuations modulate the water dynamics caused an additional source of heterogeneity which is especially strong in the minor groove. The groove width fluctuations occur on the same timescale as the water H-bond rearrangements, leading to a strong dynamical disorder.

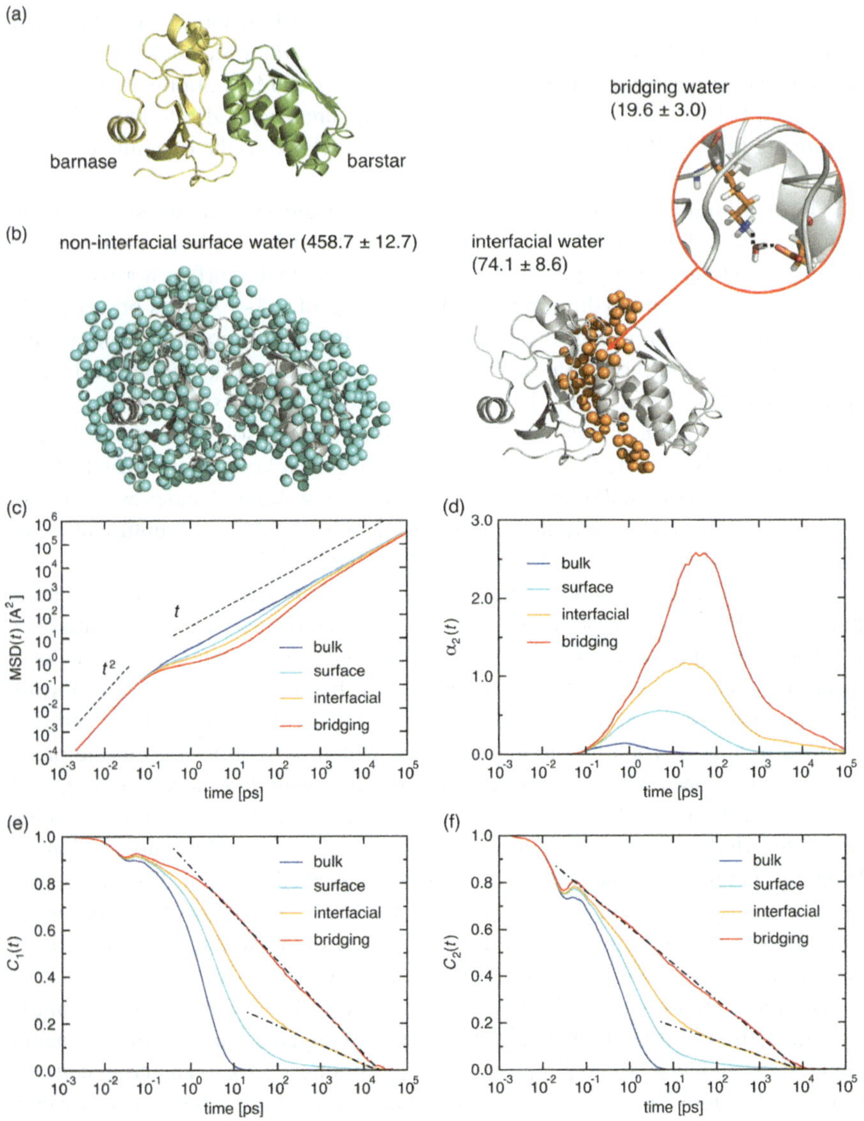

Fig. 9.12 **a** Structure of the barnase–barstar complex. **b** Illustration of the non-interfacial surface water (spheres-colored cyan represent water oxygen positions), interfacial water (orange), and bridging water forming concurrent hydrogen bonds (dotted lines) with two proteins. The average number of those waters ± standard deviation is reported in parentheses. **c**–**f** MSD of the center-of-mass position (**c**), non-Gaussian parameter $\alpha_2(t)$ **d**, rotational correlators $C_1(t)$ (**e**), and $C_2(t)$ (**f**) versus logarithmic time axis. In these panels, the curves for the bulk water (blue), non-interfacial surface water (cyan), interfacial water (orange), and bridging water (red) are displayed with different colors. The dashed lines in panel (c) represent a square (t^2) and linear (t) time variation, whereas the dotted-dashed lines in **e** and **f** a logarithmic decay [28]. Song-Ho Chong, Sihyun Ham, Anomalous Dynamics of Water Confined in Protein–Protein and Protein–DNA Interfaces J. Phys. Chem. Lett. 7, 19, 3967–3972 (2016) PERM NO FIGURE

The water structure within the minor and major grooves of the double-stranded CGCAAATTTGCG oligonucleotide was suggested from a MD simulation [29] and the X-ray experiments at cryogenic temperatures [15]. The water OH stretch **vibrational** frequency probed was proved to be very sensitive to the local hydrogen-bonded network with lower vibrational frequencies corresponding to stronger hydrogen-bonded configurations. The OH stretch of the water–air surface spans the frequency range of 3000–3800 cm^{-1}, and strongly hydrogen-bonded water exhibited lower vibrational frequencies. The NH stretch vibrations between DNA base pairs appeared as two narrower features at 3200 and 3350 cm^{-1}. Pure chiral signals can be obtained in SFG spectroscopy employing the light fields linearly polarized at an odd angle, e.g., ±45°, containing both p and s polarizations.

The dielectric relaxation study of extended hydration of the of dilute aqueous solution of two DNA samples, extracted from salmon sperm (SS DNA) and calf thymus (CT DNA) (Fig. 9.13) at different concentrations in the frequency range of 0.3–2.1 THz, was reported [30]. The frequency-dependent complex dielectric response has been fitted according to a Debye relaxation model assuming three relaxation modes having the timescales of the order of 8 ps, 200 fs, and 80 fs. The relaxation dynamics of water was obtained from the frequency-dependent complex dielectric constant, and the frequency-dependent real (ε') and imaginary (e'') dielectric constants of the samples were calculated (Fig. 9.14).

The observed relaxation time constants do not deviate much from those of bulk water and do not follow any particular trend with increasing DNA concentration indicating a feeble alteration of the hydrogen-bonded network in the studied dilute range.

Native salmon testes DNA containing 2000 base pairs was studied in femtosecond pump–probe experiments under conditions of full hydration and at a reduced hydration level with two water layers around the double helix [31]. Femtosecond pump and probe pulses independently tunable in the mid-infrared were generated in two homebuilt optical parametric amplifiers (OPA) pumped by the output of an amplified Ti:sapphire laser system. Molecular couplings between DNA and water together

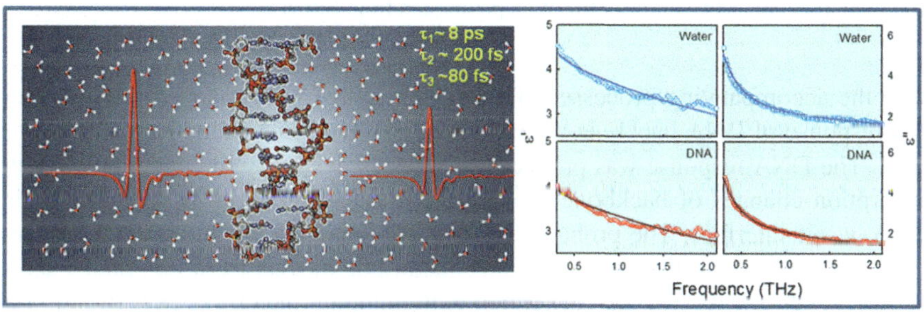

Fig. 9.13 Schematic presentation of DNA structure, three relaxation modes timescales and the frequency dependence of imaginary (e'') dielectric constants [30]. With permission from Elsevier

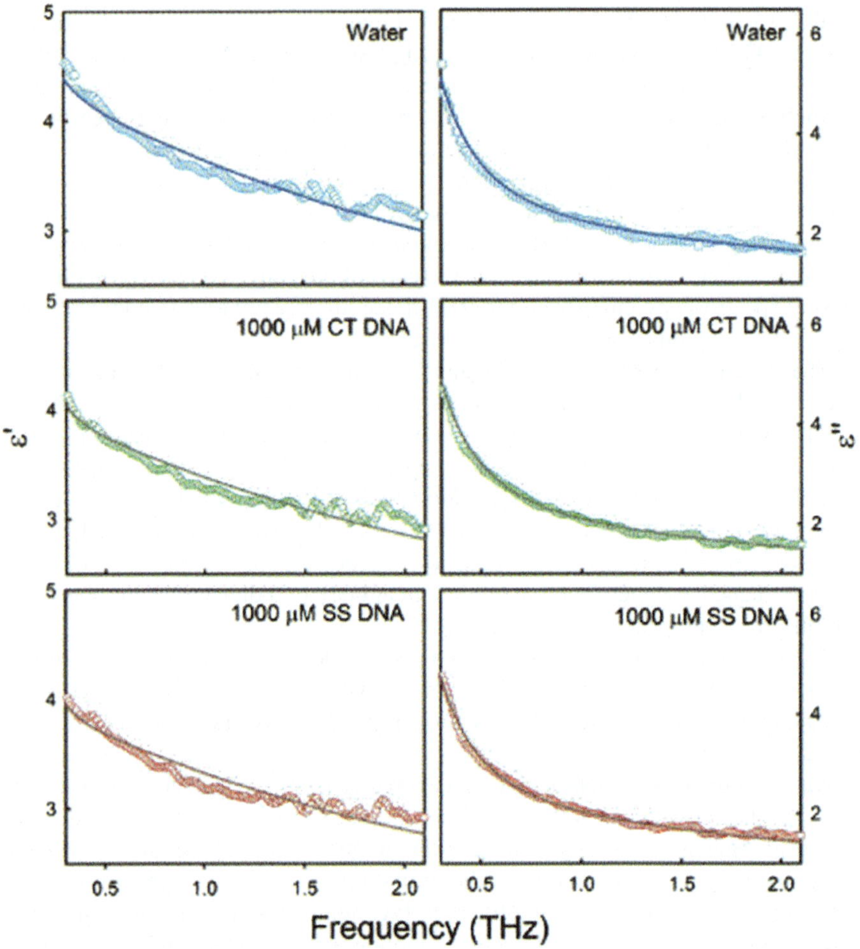

Fig. 9.14 Measured complex dielectric function of water, 1000 lM CTDNA, and 1000 lM SSDNA. The solid lines are Debye fittings [30]. With permission from Elsevier

with the accompanying processes of energy exchange were mapped via the ultrafast response of DNA backbone vibrations after OH stretch excitation of the water shell. The DNA response was probed in a time-resolved way by mapping transient absorption changes of backbone modes which are sensitive probes located at the DNA–water interface. The probe pulses cover a frequency range between 940 and 1120 cm^{-1} and more than two orders of magnitude in time from less than 1 up to 100 ps. A set of data was also generated with pumping the OH stretch vibration of water molecules by a pulse centered at 3450 cm^{-1} and probing the response of DNA backbone vibrations.

The fast changes of DNA absorption after OH stretch excitation to structural changes in the water shell which couple to DNA through the local electric fields were assigned, while the second slower process was attributed to a flow of excess energy from the water shell into DNA. Figure 9.15 presents transient spectra for (a) DNA at 92% r.h (relative humidity) corresponding to roughly two closed water layers around the double helices, (b) DNA in solution, and (c) neat water. As it is seen in the figure, the transient spectra for 92% r.h. (Fig. 9.14 a) are dominated by the DNA response, while the spectra of the solution sample consist of contributions from DNA and from the water environment. The transient spectra of Figs. 9.14a and b) display a response of all backbone modes occurring simultaneously with the excitation of the water shell.

Recent results from 2D-IR spectroscopy of native and artificial DNA giving insight into local hydration geometries, elementary molecular dynamics, and the mechanisms behind them were presented, together with theoretical calculations of molecular couplings and molecular dynamics simulations [33]. The results revealed (1) the femtosecond fluctuation dynamics of the water shell, (2) the short-range character of Coulomb interactions, (3) the strength and fluctuation amplitudes of interfacial electric fields, and (4) backbone vibrations of DNA and RNA which sensitive noninvasive probes of the complex behavior of hydrated helices.

Range and magnitude of electric fields at biomolecular interfaces and their fluctuations in a time window down to the subpicosecond regime calling for electric field mapping in space and time were investigated employing femtosecond two-dimensional infrared spectroscopy and ab initio-based theory [34]. By building the water shell layer by layer, fluctuating electric fields at the surface of native salmon DNA were established via their interactions with backbone vibrations in a wide range of hydration levels. The experiment also revealed water molecules in the first two layers as the predominant source of interfacial electric fields, which fluctuate on a 300 fs timescale with an amplitude of 25 MV/cm due to thermally excited water motions (Fig. 9.16).

The dimethylphosphate anion (DMP) is a fundamental building block of the phosphodiester backbone in DNA [36]. Electric fields at the phosphate water interface, which considered to be solvated in bulk water, and the fluctuating electric forces exerted on the $(PO_2)^-$ moiety were calculated by combining the ab initio-based effective fragment potential approach. The approach, accounting for electric fields due to static multipoles and polarization contributions due to induced dynamic dipoles, demonstrated that the total time-averaged electric field generated by water molecules arises to a large extent from the first water layer contributing 82%. The second layer contributes some 18%. The calculation also showed that the solvent electric field experienced by the phosphate group is the dominant contribution to the pronounced solvatochromism of the asymmetric $(PO_2)^-$ stretch vibration.

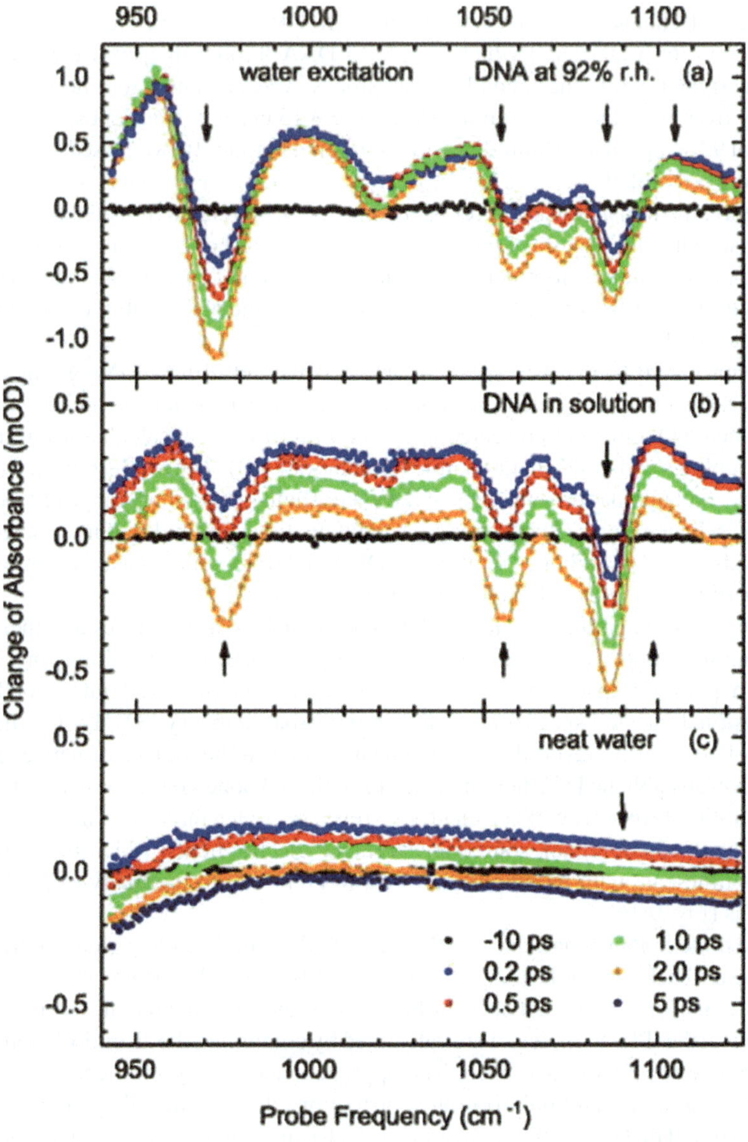

Fig. 9.15 **a** Transient pump–probe spectra of (**a**) DNA at 92% r.h., **b** DNA in solution, and **c** neat water after excitation of the water OH stretch vibration by pulses centered at 3450 cm^{-1}. The change of absorbance $\Delta A = \_\log(T/T0)$ in mOD is plotted as a function of probe frequency for different delay times (T, T0: transmission of the sample with and without excitation). The arrows mark the spectral positions at which the time-resolved transients were measured [31]. With permission from American Institute of Physics

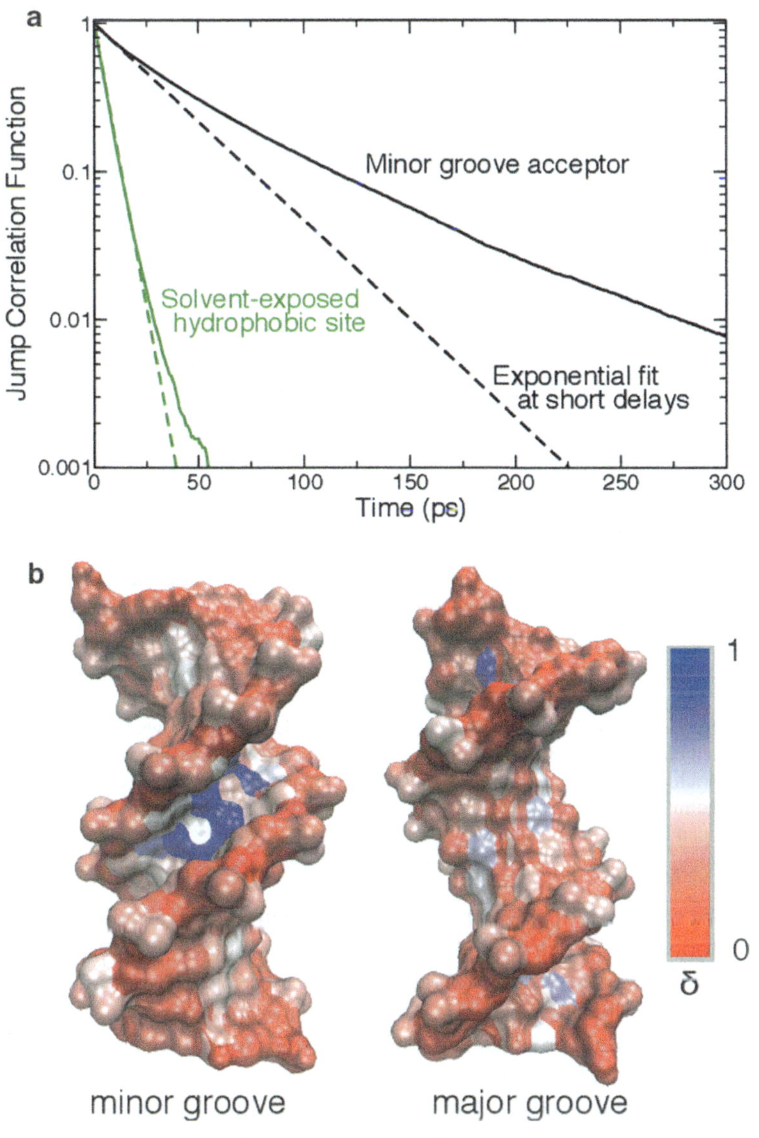

Fig. 9.16 DNA grooves [35]. With permission from American Chemical Society

9.3 RNA

Ribonucleic acids (RNAs) play critical roles in various cellular processes. RNAs as an elementary constituent of biological cells exhibit complex biochemical functionality, including the transmission of information in the form of mRNA, RNA-mediated catalytic function in ribosomes, and the encoding of genetic information in viruses.

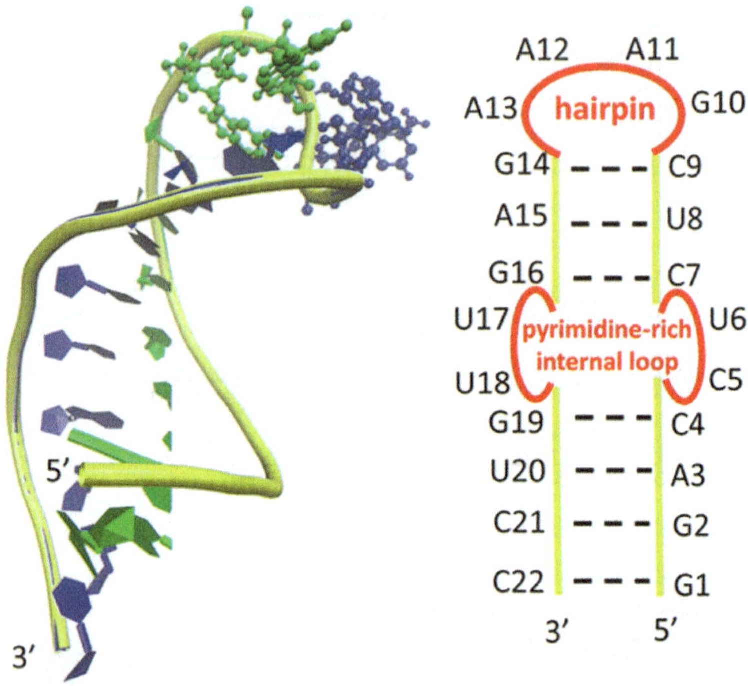

Fig. 9.17 Structure of poliovirus RNA (PDB ID: 1N66). Left panel shows the RNA hairpin in a typical folded configuration. Right panel presents the RNA base pair arrangements containing the pyrimidine-rich internal loop with a 50CU30/50UU30 non-canonical tandem base pair, known to be conserved in the 30-untranslated region of poliovirus-like enteroviruses [45]. With permission from American Institute of Physics

Interactions of RNA with the surrounding water shell, as well as counterions, determine the formation of such three-dimensional RNA structures (Fig. 9.17) Numerous combined experimental–theoretical investigation of structure and ultrafast hydration dynamics of an A-form RNA double helix in water revealed an ordered arrangement of water molecules and provides boundary conditions for the ion atmosphere around the polyanionic RNA [37–47].

It is known that heavy water stabilizes attenuated viral RNA against thermo degradation. Using a combination of parallel tempering and well-tempered metadynamics **simulation** in light water and in D_2O, the free energy surface associated with the folding/unfolding of a RNA hairpin containing a non-canonical base pair motif which is conserved within the 30-untranslated region of poliovirus-like enteroviruses was calculated [37]. The coordinates of the 22-nucleotide polioviral RNA hairpin were obtained from PDB data bank. Microcanonical (NVE) simulations were used for the evaluation of RNA hydration, namely, RNA-water HB time correlation function, basewater-basewater bridge (WB) time correlation function, water-water HB autocorrelation functions at specified distances around RNA surface, and cross correlation of RNA-water collective motions. Utilizing this approach, monitoring through an intramolecular hydrogen bond (HB), size, shape, and flexibility of RNA structures was performed. To investigate folding/unfolding of the hairpin, two approaches, namely parallel **tempering (PT)31** and metadynamics (MetaD) simultaneously into a combined method PTMetaD, were utilized. Simulations revealed a considerable increase of the stability of the folded basin in heavy water which is translated into a higher melting temperature in D_2O by 41 K compared with light water. D_2O strengthens the HB network in the solvent, lengthens inter-residue water-bridge lifetime, and weakens dynamical coupling of the hairpin to its solvation environment.

To elucidate the hydration dynamics, a geometric criterion for hydrogen bonding was formulated. According the criterion, a HB donor (D) and acceptor (A) are considered hydrogen bonded if the D-A distance is less than 3.5 Å, and the D-H-A angle is less than 35°.

Suggested two-state model consists of a fully flexible unfolded (U) and one-folded (F) native structure (Fig. 9.18). The regions covered by white lines in the figure are lowest in energy (0–5 kJ/mol), one referring to the folded state (left panel) and the other to the unfolded state (right panel).

The role of water molecules in 89 protein–tRNA and single-stranded RNA complexes taken from the **Protein Data Bank** was discussed [38]. **A. Barik,** Ranjit Prasad Bahadur, Hydration of protein–RNA recognition sites. Nucleic Acids Research, **42,** 10,148–10,160 (2014). Main research findings are as follows: (1) protein–RNA interfaces are hydrated less than protein–DNA interfaces, but more

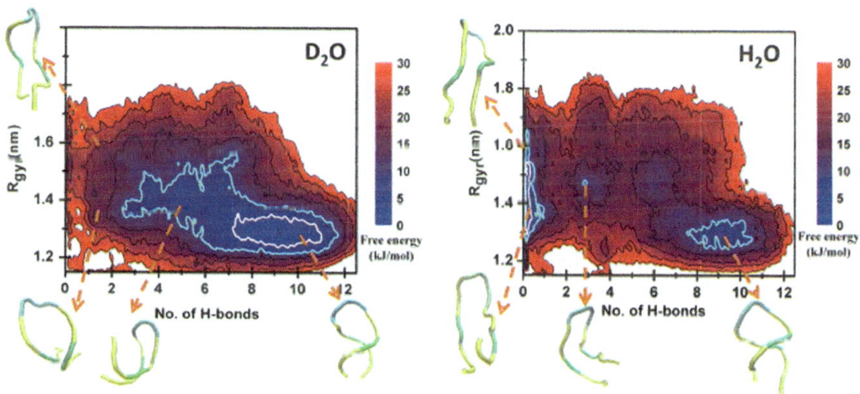

Fig. 9.18 Free energy surface at an interval of 5 kJ/mol and as a function of inter-base HB and radius of gyration of RNA heavy atoms at 300 K is shown. The representative RNA structures (the GAAA tetraloop in cyan and the pyrimidine-rich internal loop in green) of the most relevant minima in the two water solvents are highlighted. Propensity toward folded configuration, characterized by compact structures and larger number of inter-base HB, can be clearly seen in the D_2O solvent. For easy guidance, the regions in the energy range 0–5 kJ/mol are covered by white lines while that between 5 and 10 kJ/mol are covered by cyan lines [37]. With permission from American Institute of Physics

than protein–protein interfaces; (2) majority of the waters at protein–RNA interfaces forms multiple H-bonds, making H-bonds have preferences for the polar groups of RNA than its partner protein; (3) The spatial distribution of waters makes interfaces with ribosomal proteins and single-stranded RNA relatively "dry" than interfaces with tRNA and duplex RNA; (4) in contrast to protein–DNA interfaces, mainly due to the presence of the 2_OH, the ribose in protein–RNA interfaces is hydrated more than the phosphate or the bases; (5) The minor groove in protein–RNA interfaces is hydrated more than the major groove; (6) the strands make the highest number of water-mediated H-bonds per unit interface area followed by the helices and the non-regular structures; and (7) the preserved waters at protein–RNA interfaces make higher number of H-bonds than the other waters and contribute toward the affinity in protein–RNA 480 structures with polypeptide chains of minimum 30 amino acid residues and polyribonucleotides of minimum five nucleotides.

Specifically, 2440 waters bound at the interface, contributing to the recognition process identified, and an average protein RNA complex contains about 217 bound waters, of which 27 (equivalent to 11 per 1000 A°2 of B) are at the interface. The frequency distribution of interface waters making up to four H-bonds about 60% of which make one or two H-bonds with the protein polar groups was shown. The percentage increases to 67% with the RNA polar groups. About 42% of B is contributed by the helices involving in 43% of all water–protein H-bonds. An analysis of available datasets also revealed that (1) the interface of arginyl-tRNA synthetase and its cognate **tRNA** (PDB id: 1F7U) containing 116 waters are distributed throughout the interface, making it "wet" (*d*r = 0.84; (Fig. 9.19a); (2) all the tRNA interfaces in this dataset are "wet" with *d*r between 0.67 and 1.00 with ribosomal proteins are "dry" with *d*r > 1.0; (3) in the ribosomal protein L1–mRNA interface (PDB id: 2HW8), the waters are distributed along the periphery of the interface, making it "dry" *dr* = 1.18 Fig. 9.19b); (4) majority of the interfaces involving duplex RNA are "wet" with an average *dr* of 0.98; and (5) in the Fab–RNA interface (PDB id: 2R8S), the waters are distributed throughout the interface, making it "wet" (*d*r = 0.95; Fig. 9.19 5C).

24 interface waters in the ribosomal protein L1–mRNA complex (PDB id: 2HW8), of which four are preserved, were identified (Fig. 9.20) [39]. They make H-bonds with the same polar groups in the bound and the unbound form of the protein. In the entire dataset, 266 preserved waters, which are equivalent to 5.2 per 1000 °A2 of B, were found.

Pseudo-uridine (Ψ), an isomer of the nucleoside uridine in which the uracil is attached to the sugar via the C5-C1′ instead of the N1-C1′ glycosidic bond, is the most common chemical modification present in RNA. To explain experimentally observed sequence dependence of the effect of Ψ on the thermodynamic stability of RNA duplexes, the structure, dynamics, and hydration of RNA duplexes with an internal Ψ-A base pair in different nearest neighbor sequence contexts were investigated [39]. To examine whether structural factors could be responsible for the differences observed in the increment of thermodynamic stabilities, the NMR structures of two 9-mer RNA duplexes with 5′-GΨC/3′-CAG and 5′-CΨG/3′-GAC motifs were determined. To elucidate how the replacement of the middle U-A base pair with

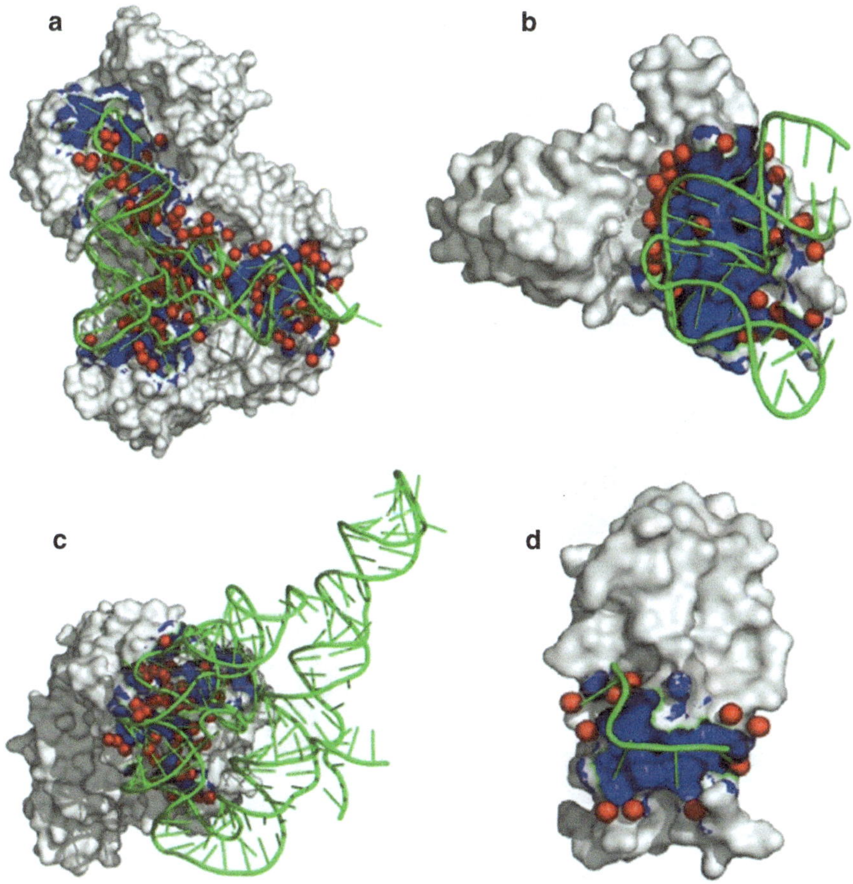

Fig. 9.19 Hydration pattern in protein–RNA interfaces. In each diagram, the protein chain is shown as molecular surface, and the RNA chain is shown as ribbon. The interface region is colored blue, and the interface water molecules are represented by red sphere. **a** A "wet" interface between arginyl-tRNA synthetase and its cognate tRNA (1F7U; $dr = 0.84$). **b** A "dry" interface between ribosomal protein L1 and mRNA (2HW8; $dr = 1.18$). **c** A "wet" interface between FAB and duplex RNA (2R8S; $dr = 0.95$). **D** A "dry" interface between CspB and single-stranded RNA (3PF4; $dr = 1.46$) [38]. With permission of Oxford University Press

Ψ-A in duplexes with 5′-GΨC/3′-CAG, 5′-CΨG/3′-GAC, 5′-AΨU/3′-UAA, and 5′ UΨA/3′-AAU motifs a systematic study using molecular dynamics (MD) simulations was also performed. The two-dimensional NOESY experiments involving non-exchangeable 1H resonances were recorded at 25 °C. 2D NOESY spectra at 30 and 35 °C were acquired for duplex-GΨC and -CΨG, respectively, with the broad spectral width of ~ 7002 Hz. High-resolution DQF-COSY spectra were acquired with narrow spectral width of 2520 Hz. All molecular mechanics (MM) calculations and 500 ns molecular dynamic (MD) simulations were performed with AMBER 14

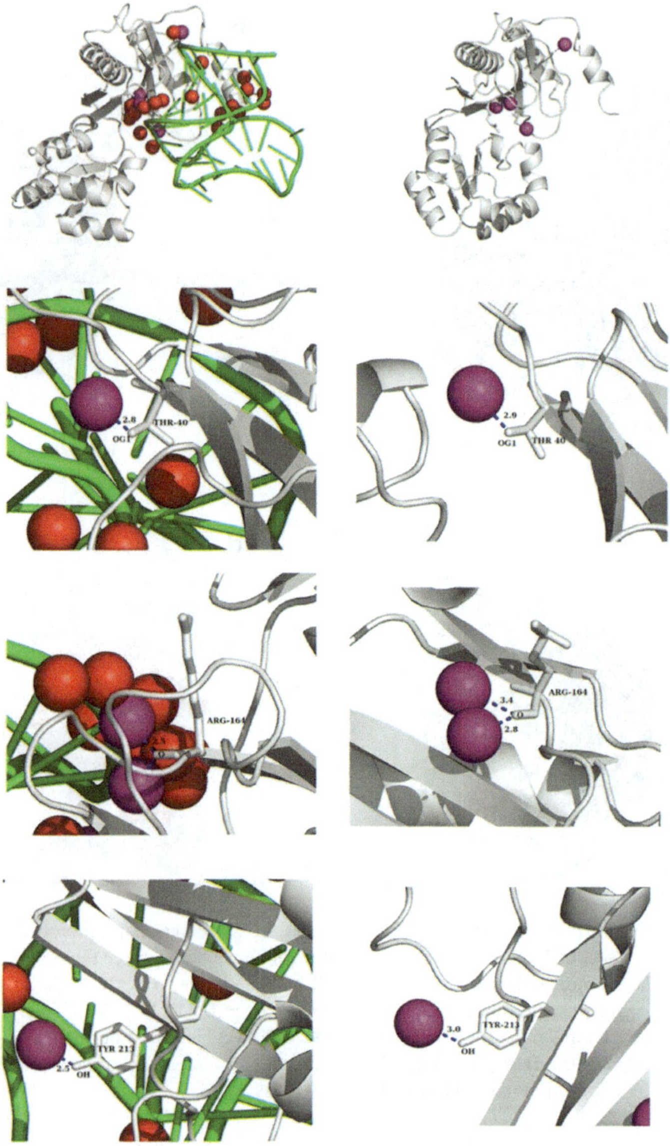

Fig. 9.20 Preserved interface water molecules in a protein–RNA recognition site. In the top left panel, the bound complex between ribosomal protein L1 and mRNA (2HW8) with interface waters (in red) and preserved waters (in magenta) are shown. In the top right panel, the unbound protein and the four preserved waters are shown. Protein and RNA chains are shown in gray and green color, respectively. The panels below show the conservation of four interface waters. These preserved waters make same H-bond in the bound and unbound forms of the protein. Downloaded from https://academic.oup.com/nar/article/42/15/10148/2435300 by Ben Gurion University of the Negev-Medical Library user on [38]

simulation package and AmberTools1557. In addition, high-level ab initio methods were applied, and calculations for eight unique base pair steps for the RNA structures derived from MD simulations were performed.

The radial distribution functions (RDF) plot of the water oxygen atoms with respect to the HN1-Ψ5 atom (Fig. 9.21) for the duplex-CΨG indicates the formation of a well-defined first hydration shell between 1.5 Å and 2.5 Å with a maximum at 2.50 Å.

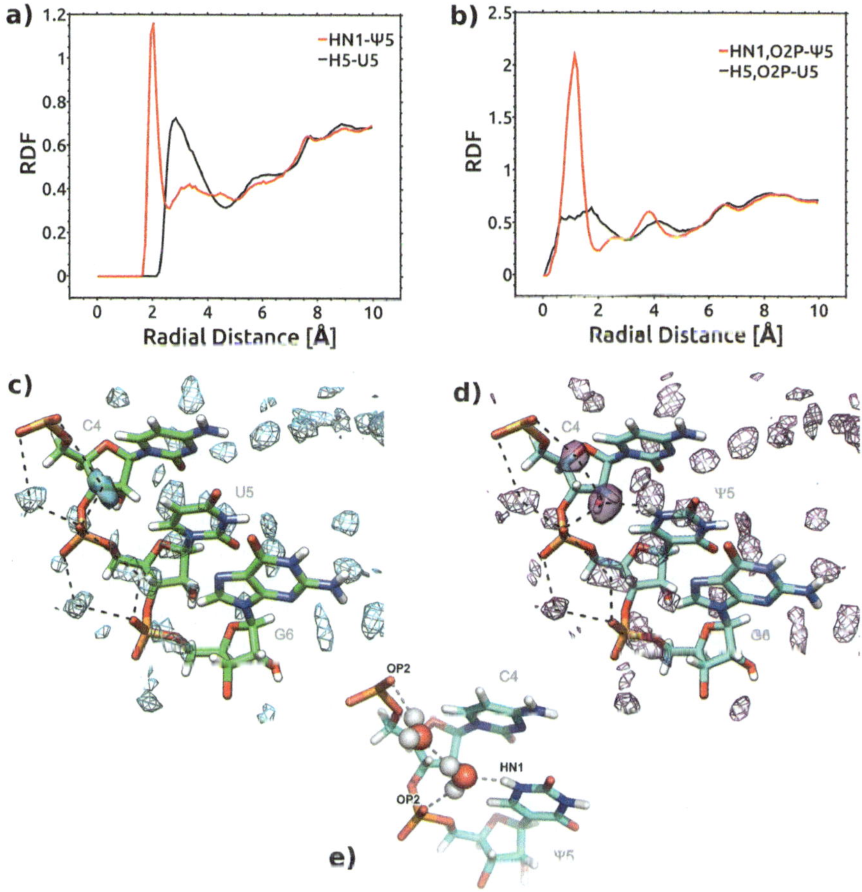

Fig. 9.21 Change in the hydration pattern upon U to Ψ modification for duplex-CΨG and duplex-CUG. RDFs of water oxygen atoms **a** around the HN1-Ψ5 and H5-U5 atoms, respectively; **b** around the geometric center of the OP2-Ψ5 and HN1-Ψ5 atoms or OP2-U5 and H5-U5 atoms. Red—Ψ-modified duplexes; Black—unmodified duplexes. Water occupancy contoured at equivalent levels and hydrogen bonding patterns (in dotted lines) viewed from the major groove. **c** Reference duplex-CUG; **d** Ψ-modified duplex-CΨG; differences in hydration patterns are shown in solid. (**e**) Snapshot of two water molecules making contact between HN1, OP2 atoms of Ψ and OP2 atom of the preceding residue [40]

High-level quantum mechanical (QM) method was used to study stacking interaction energies for eight base pair steps containing the Ψ-A and the reference U-A base pairs. The calculation showed the dependence of the change in the stacking energies on the sequence context, ranging from −1.59 to 0.23 kcal/mol. The relative binding free energies calculated for the Ψ-modified and unmodified duplexes were found to be between 0.61 and −1.47 kcal/mol. The RDF plot of the water oxygen atoms with respect to the HN1-Ψ5 atom for the duplex-CΨG indicates the formation of a well-defined first hydration shell between 1.5 Å and 2.5 Å with a maximum at 2.05 Å. (Fig. 9.21a).

To highlight that sufficient hydration of non-crystallized RNA could provide high-resolution solid-state NMR (SSNMR) spectra, with similar spectral quality to the crystallized RNA, two different RNAs varying in molecular sizes were prepared, including a genomic from the dimerization initiation site of HIV-1 (DIS-HIV-1) and an adenine riboswitch (riboA71) [40]. Sufficient hydration greatly improved the SSNMR spectral quality of RNAs in solids. In this condition, the 2D 1H-detected 1H–15 N SSNMR exhibited signals characteristic of both the canonical and the non-canonical base pairs, indicating the presence of secondary structures and native-like tertiary interactions of RNA. Obtained results suggested that preservation of hydration in RNA solids ensures high-quality SSNMR spectra without the need for forming RNA crystals.

Review [41] discussed the methods for RNA production and purification as well as numerous approaches for isotope labeling of RNA. A few examples that emphasize the instrumental role of isotope labeling and ssNMR for studying RNA as part of large ribonucleoprotein complexes were given. It was stressed that isotope labeling, either as nucleotide-specific, atom-specific, or segmental labeling, can resolve resonance overlaps and reduce the line width, thus allowing ssNMR studies of RNA domains as part of large biomolecules or complexes. The Minireview by Yang and Wang provided a concise overview of recent progress in magic-angel-spinning solid-state NMR (MAS-SSNMR) strategies toward structural and dynamic characterization on RNA in the solid state) [42]. The topics include the procedures of RNA sample preparation, SSNMR spectroscopy for RNA characterization, and recent successful examples of RNA studied by SSNMR spectroscopy.

The first ultrafast pump–probe and 2D-IR study of an RNA double helix in a water environment augmented by in-depth theoretical modeling of the RNA vibrational response were reported [44]. To monitor solvation dynamics, RNA backbone vibrational modes of phosphate and sugar groups in the frequency range 900–1300 cm^{-1} were examined reported. Pump–probe and 2D-IR spectra were recorded in the range of RNA backbone modes where prominent peaks appear for the asymmetric (PO_2^-) stretch vibration P1 (doublet), the symmetric PO_2^- stretch vibration P2, and backbone linker modes L1/2. Characteristic RNA bands at 1120 cm^{-1} with a shoulder at 1135 cm^{-1} arise from the OH group in the 2′ position with contributions from $C4' - O4' - C1'$ stretching modes located on the ribose units (Fig. 9.23a). Vibrational $\nu = 1$ lifetimes were found to be in the 0.3–1.1 ps range with the fastest population decay for P1 modes (Fig. 9.24b). A 2D-IR spectrum of RNA was plotted as a function of excitation (ν1) and detection frequency (ν3) in Fig. 9.23c. The findings correlate with X-ray diffraction data where individual water molecules link the 2′–OH group with ribose oxygen atoms and oxygen atoms of the phosphate group and neighboring phosphate groups are bridged by individual water molecules, forming a more ordered hydration shell around RNA. Among other results of the work are as follows (1) the presence of an OH group attached to the 2′ position of the ribose units has a strong impact on the vibrational spectrum and the hydration pattern in which the phosphate and sugar groups serve as distinct interaction sites for water molecules; (2) the action of the fluctuating electric force is generated by low frequency, e.g., librational motion of solvating water molecules; and (3) this action directly manifested in the vibrational line shape of the backbone modes that report on local field fluctuations in the 2–3 neighboring water layers (Fig. 9.22).

Two-dimensional infrared (2D-IR) spectroscopy in a range from 950 to 1300 cm^{-1} was employed to elucidate the character, dynamics, and couplings of backbone modes of a double-stranded RNA (Fig. 9.24) [46]. Double-stranded RNA oligomers containing 23 alternating adenine-uracil (A-U) base pairs in Watson–Crick geometry were studied in an aqueous environment with more than 190 water molecules per base pair. The 2D-IR spectra were analyzed by theoretical simulations combining a density matrix approach for calculating the nonlinear vibrational response functions in third order and the frequency fluctuation correlation function (FFCF) of the different vibrations. The 2D-IR spectra for different waiting times T were measured in 3-pulse photon-echo experiments with a heterodyne detection of the nonlinear signal. Mid-infrared pump–probe 2D-IR measurements were complemented by temporally and spectrally resolved pump–probe measurements in order to determine the decay times of the $v = 1$ states of the different backbone vibrations. Temporally and spectrally resolved pump–probe measurements were performed in parallel to the 2D experiments. In molecular dynamics (MD) simulations, an initial model structure of a AU 23-mer RNA double strand (A-form, sequence: 5′-UUAUAUAUAUAUAUAUAUAUAUU-3′ and its complementing strand) was generated with the nucleic acid builder (NAB) of the AmberTools program suite as canonical Watson–Crick A-RNA double helix. Quantum mechanical molecular mechanics (QM/MM) normal mode analysis was also performed. Suggested solvation structure of base step 7–8 with counterions is shown in Fig. 9.23.

The experiments revealed that (1) interactions with the fluctuating water shell gave rise to spectral diffusion on a 300 fs timescale, leading to a quasi-homogeneous line shape of the symmetric PO_2^- stretching mode of the strongly hydrated phosphate groups; (2) the decay time of the fast FFCF component of 300 fs is shorter than the picosecond hydrogen bond lifetimes at the water–RNA interface slowing fluctuations which are moderate compared to bulk water; (3) the fast decay can reflect fluctuations of short-range electric fields which originate from the dipolar water molecules in the first and second hydration layer undergoing subpicosecond librational motions within their steric constraints set by the surface, and/or hydrogen bonds between water molecules in a geometry distinctly different from bulk water; and (4) the fast component gives rise to spectral diffusion of vibrational frequencies, whereas

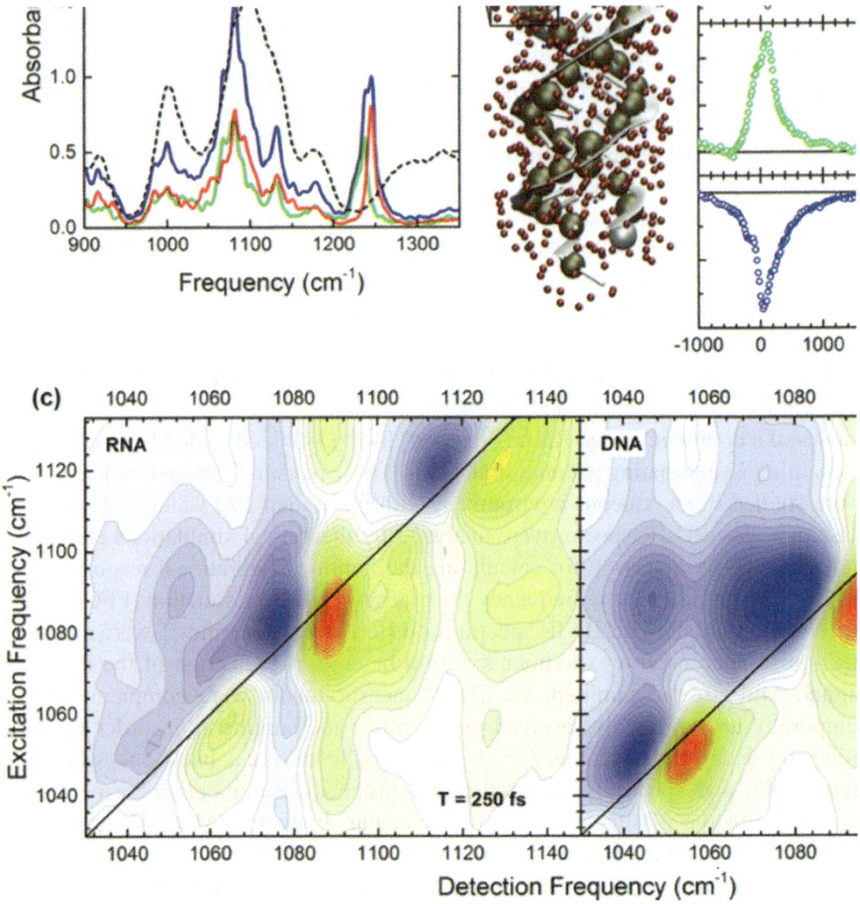

Fig. 9.22 **a** Experimental (top) and simulated (bottom) linear absorption spectrum, obtained by QM/MM normal mode analysis along molecular dynamics trajectories (3• 200 ns = 0.6 μs total simulation time) of an (A-U)23 A-RNA double helix; inset: sugar–phosphate backbone. **b** Pump–probe transients recorded at the frequency position of symmetric P2 and asymmetric P1 phosphate stretch vibrations. **c** 2D-IR spectra of RNA (left) and DNA (right) revealing profound differences in mode number, coupling pattern, and line shape (normalized to the maximum positive signal [44]

the slow component with a decay time much longer than the 0.36 – 1.1 ps vibrational lifetimes of the backbone modes causes quasi-static inhomogeneous broadening.

To systematically investigate motions in HIV-1 TAR RNA as a function of hydration, deuterium solid-state NMR was utilized [47]. Dynamics in solid-state samples of TAR RNA was monitored using ^{13}C solution relaxation methods under a wide range of hydration conditions from nearly dry to fully hydrated

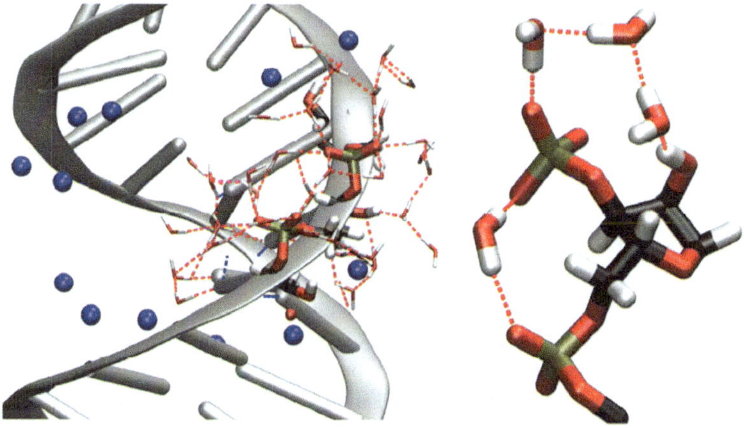

Fig. 9.23 Solvation structure of base step # 7–8) with counterions shown in blue. The magnification (right) highlights a $(PO_2)^-\ldots(PO_2)^-$ bridging water molecule and a hydrogen-bond network ring structure mediated by the C2′-OH group as discussed in the main text [46]. With permission from American Chemical Society

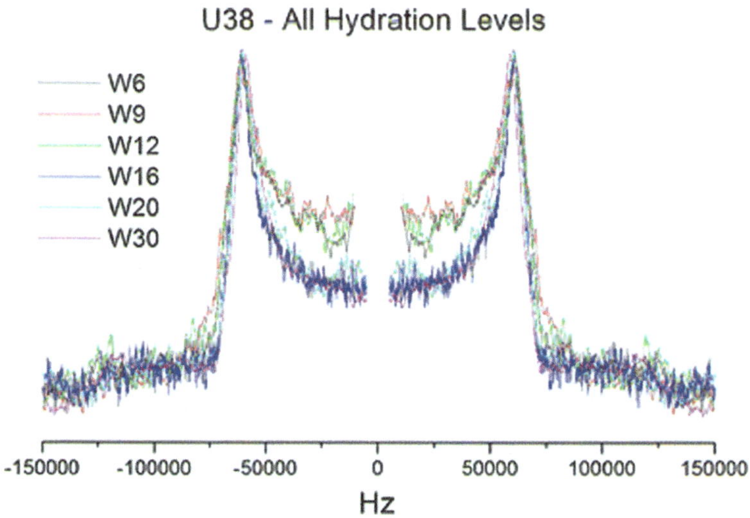

Fig. 9.24 Overlay of solid -tate line shapes for U38 at different levels of hydration from W = 6 to W = 30. The signal in the center of the spectrum is dominated by the sharp peak corresponding to residual 2H in the deuterium-depleted water and has been removed to allow closer inspection of the data [47].

[47] G.L. Olsen, M.F. Barda J, D.C. Echod, G.P. Drobn, G. Varani, Hydration dependent dynamics in RNA, J Biomol NMR 45, 133–142 (2009) The following TAR RNA 29-mer constructs were used: 50-GGC CAGA-20FU-CUG(pS)AGCGAAAGC*UCUCUGGCC-30 ("U38"), 50-GGCCAGA*U-20FC-UG(pS)AGCGAAAGCU CUCUGGCC-30 ("U23"), and 50-GGCCAGA-20FU-C*UG (pS)AGCGAAAGCUCUCUGGCC-30 ("U25"), where pS indicates a phosphorothioate label. All NMR experiments were performed on a homebuilt spectrometer at 11.7 T field using a homebuilt single-channel static deuterium probe. T_{1z} relaxation times at 76.76 MHz and T_{1Q} relaxation times were determined using saturation recovery experiments and the broadband Jeener–Broekaert pulse sequence, respectively. Powder-averaged T1Z and T1Q relaxation times were determined using nonlinear least-squares fits of integrated powder spectra to a single exponential. Deuterium line shape data were obtained using a quadrupole echo pulse sequence. All measurements were made at room temperature (26 °C), but with continuous application of 20 °C cooling air.

Dynamics at three uridine residues in different structural environments ranging from helical to completely unrestrained was probed. Distinct changes in ^{2}H solid-state relaxation times and line shapes at each site as hydration levels increase were observed. By measuring solid-state and solution state ^{13}C relaxation measurements, it was established that ns–fs motions, being indicative of collective dynamics, arise in the RNA as hydration reaches a critical point coincident with the onset of bulk hydration. The visible transition of samples to a gel-like appearance is not evident until 30 waters/nucleotide. It was also found that all three sites show changes in the line shapes at intermediate hydration levels coincident with initiation of bulk hydration, indicating the onset of ns–ls motions.

The NMR experiments revealed a number specific effects of water content on structure and dynamics of the RNA fragments (Fig. 9.25): (1) at low hydration levels waters per nucleotide, $W = 6$ and $W = 9$), hydration of the backbone is substantial; (2) in this condition, hydration of the bases and sugars remains incomplete, and the line shapes for all three sites are consistent with the presence of only small amplitude, fast limit, or "rigid" dynamics; (3) moderate hydration levels were reached at $W = 12$; (4) hydration to about $W = 12$–16, local hydration is effectively completed; (5) addition of water beyond this point contributes to bulk hydration, but does not contribute further to local hydration of the bases; (6) hydration beyond W = 16 typically results in swelling of DNA samples; (7) ns–fs motions are activated by W = 16 and then do not change significantly; (8) for the TAR RNA samples, the visible transition of samples to a gel-like appearance is not evident until 30 waters/nucleotide; (9) for U25 spectral density functions (ps/radian) changes from 4 ps up 8 ps for W = 6 and w = 30, respectively.

References

1. R. Dahm, Discovering DNA: Friedrich Miescher and the early years of nucleic acid research. Hum. Genet. **122**, 565–581 (2008)
2. B. Bryson, *A Short History of Nearly Everything* (Broadway Books, 2015)
3. J.D. Watson, F.H. Crick, molecular structure of nucleic acids: A structure for Deoxyribose nucleic acid. Nature **171**, 37 (1953)
4. W. Saenger, Water and Nucleic Acids. in *Principles of Nucleic Acid Structure. Springer Advanced Texts in Chemistry*. (Springer, New York, NY 1984).
5. E. Westhof, D. L. Beveridge, in F. Franks (ed) *Hydration of Nucleic Acids* (Cambridge University Press, Cambridge, 1989)
6. G.A Jeffrey, W. Saenger, Hydration of nucleic acids. in *Hydrogen Bonding in Biological Structures* (Berlin, Heidelberg, Springer 1994)
7. V. Makarov, B. Pettitt, M. Feig, Solvation and hydration of proteins and nucleic acids: a theoretical view of simulation and experiment. Chem. Res. **35**, 376–384 (2002)
8. M. Kabeláč, P. Hobza, Hydration and stability of nucleic acid bases and base pairs. Phys. Chem. Chem. Phys. **9**, 903–917 (2007)
9. J. Barciszewski, J. Jurczak, S. Porowski, T. Specht, V.A. Erdmann, The role of water structure in conformational changes of nucleic acids in ambient and high-pressure conditions. Eur. J. Biochem. **260**, 293–307 (1999)
10. S.D. Verma, N. Pal, M.K. Singh, S. Sen, Sequence-dependent solvation dynamics of minor-groove bound ligand inside duplex-DNA. J. Phys. Chem. B **119**, 11019–11029 (2015)
11. A. Pascal, E. Westhof, Water and ion binding around RNA and DNA (C, G) oligomers. J. Mol. Biol. **300**, 1113–1131 (2000)
12. M.L. Kopka, A.V. Fratini, H.R. Drew, R.E. Dickerson, Ordered water structure around a B-DNA Dodecamer: a quantitative study. J. Mol. Biol. **163**, 129–146 (1983)
13. H.I.M. Shweta, SOBHAN SEN, dynamics of water and ions around DNA: What is so special about them? J. Biosci. **43**, 499–518 (2018)
14. S.Y. Ponomarev, K.M. Thayer, D.L. Beveridge, Ion Motions in molecular dynamics simulations on DNA. Proc. Natl. Acad. Sci. USA **101**, 14771–14775 (2004)
15. E. Liepinsh, G. Otting, K. Wüthrich, NMR observation of individual molecules of hydration water bound to DNA duplexes: direct evidence for a spine of hydration water present in aqueous solution. Nucleic Acids Res. **20,** 6549–6553 (1992)
16. H. Khesbak, O. Savchuk, S. Tsushima, K. Fahmy, The role of water H-bond imbalances in B-DNA substate transitions and peptide recognition revealed by time-resolved FTIR spectroscopy. J. Am. Chem. Soc. **133**, 5834–5842 (2011)
17. M.L. McDermott, H. Vanselous, S.A. Corcelli, P.B. Petersen, DNA's chiral spine of hydration. ACS Cent. Sci. **3**, 708–714 (2017)
18. E.A. Perets, E.C.Y. Yan, The H_2O Helix: the chiral water superstructure surrounding DNA. ACS Cent. Sci. **3**, 683–685 (2017)
19. R. Casea, H. Schollmeyera, P. Kohlb, E.B. Sirotac, R. Pynnd, K.E. Ewerta, C.R. Safinyaa, Y. Lib, Hydration forces between aligned DNA helices undergoing B to A conformational change: In-situ X-ray fiber diffraction studies in a humidity and temperature controlled environment. J. Struct. Biol. **200**, 283–292 (2017)
20. K. Kuchuk, U. Sivan, Hydration structure of a single DNA molecule revealed by frequency-modulation atomic force microscopy. Nano Lett. **18**, 2733–2737 (2018)
21. A.I. Dragan, C.M. Read, C. Crane-Robinson, Hydration differences between the major and minor grooves of DNA were revealed from heat capacity measurements. Eur Biophys J. **48**, 131–138 (2019)
22. N. Pal, H. Shweta, M.K. Singh, S.D. Verma, S. Sen, Power-Law solvation dynamics in G-quadruplex DNA: role of hydration dynamics on ligand solvation inside DNA. J. Phys. Chem. Lett. **6**, 1754–1760 (2015)

23. M.A. Berg, R.S. Coleman, C.J. Murphy, Nanoscale structure and dynamics of DNA. Phys. Chem. Chem. Phys. **10,** 1229–1242 (2008); Zewail A Biological water: a critique. Chem. Phys. Lett. 503 1–11 (2011) CHAPTER 12
24. N. Pal, S.D. Verma, S. Sen, Probe position dependence of DNA dynamics: comparison of the time-resolved shift of groove-bound to base-stacked probes. J. Am. Chem. Soc. **132**, 9277–9279 (2010)
25. Y. Yonetani, H. Kono, What determines water-bridge lifetimes at the surface of DNA? Insight from systematic molecular dynamics analysis of water kinetics for various DNA sequences. Biophys. Chem. **160**, 54–61 (2012)
26. D. Laage, J.T. Hynes, Reorientational dynamics of water molecules in anionic hydration shells. Proc. Natl. Acad. Sci. USA **104**, 11167–11172 (2007)
27. Sheh-Yi Sheu, Yu-Cheng Liu, Jia-Kai Zhou, E.W. Schlag, D.-Y. Yang, Surface topography effects of globular biomolecules on hydration water. J. Phys. Chem. B **123,** 6917−6932 (2019)
28. S. Chong, S. Ham, Anomalous dynamics of water confined in protein-protein and protein-DNA interfaces. J. Phys. Chem. Lett. **7**, 3967–3972 (2016)
29. K.E. Furse, S.A. Corcelli, Molecular dynamics simulations of DNA solvation dynamics. J. Phys. Chem. Lett. **1**, 1813–1820 (2010)
30. D. Polley, A. Patra, R.K. Mitra, Dielectric relaxation of the extended hydration sheathe of DNA in the THz frequency region Chem. Phys. Lett. **586,** 143–147 (2013)
31. Y. Liu, B. Guchhait, T. Siebert, B.P. Fingerhut, T. Elsaesser, Molecular couplings and energy exchange between DNA and water mapped by femtosecond infrared spectroscopy of backbone vibrations. Struct. Dyn. **4,** 044015 (2017)
32. E.B. Brauns, M.L. Madaras, R.S. Coleman, C.J. Murphy, M.A, Berg, Complex dynamics in DNA on the picosecond and nanosecond time scales. Phys. Rev. Lett. **88,** 158101–158104 (2002)
33. B.P. Fingerhut, T. Elsaesser, Noncovalent Interactions of hydrated DNA and RNA mapped by 2D-IR Spectroscopy. in M Cho (ed) *Coherent Multidimensional Spectroscopy. Springer Series in Optical Sciences*, vol 226 (Springer, Singapore, 2009). https://doi.org/10.1007/978-981-13-9753-0_8
34. T. Siebert, B. Guchhait, Y. Liu, B.P. Fingerhut, T. Elsaesser, Range, magnitude, and ultrafast dynamics of electric fields at the hydrated DNA surface. J. Phys. Chem. Lett. **7**, 3131 (2016)
35. E. Duboué-Dijon, A.C. Fogarty, D. Laage, Dynamical disorder in the DNA hydration shell. J. Am. Chem. Soc. **138**, 7610–7620 (2016)
36. P. Fingerhut, R. Costard, T. Elsaesser, Predominance of short range Coulomb forces in phosphate-water interactions—a theoretical analysis. J. Chem. Phys. **145,** 115101 (2016)
37. A.K. Pathak, T. Bandyopadhyaya, Water isotope effect on the thermostability of a polio viral RNA hairpin: A metadynamics study. J. Chem. Phys. **146,** 165104 (2017)
38. A. Barik, R.P. Bahadur, Hydration of protein–RNA recognition sites. Nucleic Acids Res. **42,** 10148–10160 (2014)
39. Indrajit Deb, Ł. Popenda, J. Sarzyńska, M. Małgowska, A. Lahiri, Z. Gdaniec, R. Kierzek, Computational and NMR studies of RNA duplexes with an internal pseudouridine-adenosine base pair. Sci. Rep. **9,** 16278 (2019)
40. S. Zhao, Y. Yang, Y. Zhao, X. Li, High-resolution solid-state NMR spectroscopy of hydrated non-crystallized RNA. Chem. Commun. **55**, 13991 (2019)
41. A. Marchanka, C. Kreutz, T. Carlomagno, Isotope labeling for studying RNA by solid-state NMR spectroscopy. J. Biomol. NMR **71**, 151–164 (2018)
42. Y. Yang, S. Wang, Frontispiece: RNA characterization by solid-state NMR spectroscopy. Chemistry **24**, 8698–8707 (2018)
43. Y. Yang, S. Xiang, X. Liu, X. Pei, P. Wu, Q. Gong, N. Li, M. Baldus, S. Wang, Proton-detected solid-state NMR detects the inter-nucleotide correlations and architecture of dimeric RNA in microcrystals. Chem. Commun. **53**, 12886–12889 (2018)
44. B.P. Fingerhut, E.M. Bruening, J. Schauss, T. Siebert, T. Elsaesser, Interactions of RNA and water probed by 2D-IR spectroscopy. EPJ Web Conf. **205**, 10003 (2019)

45. A.K. Pathak. T. Bandyopadhyaya, Water isotope effect on the thermostability of a polio viral RNA hairpin: A metadynamics study, J. Chem. Phys. **146,** 165104 (2017)
46. E.M. Bruening, J. Schauss, T. Siebert, B.P. Fingerhut, T. Elsaesser, Vibrational dynamics and couplings of the hydrated RNA backbone: a two-dimensional infrared study. J. Phys. Chem. Lett. **9**, 583–587 (2018)
47. G.L. Olson, M.F. Barda J, D.C. Echod, G.P. Drobn, G. Varani, Hydration dependent dynamics in RNA. J. Biomol. NMR **45,** 133–142 (2009)

Chapter 10
Biomembranes

Abstract Water structure and dynamics in hydrated biomembranes and its models are the subject of this chapter. In membranes, water shells structure elementary and cohesive water motions (Герц Ильич, я не понимаю, что вы хотите здесь сказать), electric fields at hydrated biointerfaces, and processes of vibrational relaxation and energy dissipation are considered. The water molecules were found to be located inside and outside of the lipid bilayers and in between the protein and the membrane. The total amount of hydration water of 950 molecules per micelle with the number of tightly bound water molecules in the first hydration shell of 310 molecules with slow water dynamics was obtained, and correlated vibrational motion on the subpicosecond timescale and associated collective dynamics of the peripheral membrane protein was experimentally established. The thermodynamic and kinetics properties of the lipid–protein interaction governed by changes in the water activity of monolayers of different lipid composition according to the lateral surface pressure are reviewed.

10.1 Introduction

Biological membranes are universal structural units in living cells and organelles, where they form natural hydrophobic barriers responsible for the selective transport of materials, the generation of energy, recognition, and signaling [1–45]. Biological membranes appear to be composed of a complex mixture of lipids varying in head group size and chain length [5] (Fig. 10.1). Biological membranes are highly organized supramolecular assemblies of lipids and proteins. The phospholipid hydrocarbon chains dominate the bilayer's flexibility, while the lipid head groups govern bilayer packing and the interactions with water and other molecules and ion. According to water species present at the interphase, lipid membrane can act as a water state regulator, determining the interfacial water domains in the surface. The domain is formed by the contact between lipids themselves and between lipids and the water phase, which are needed to trigger adsorption–insertion processes. The water domains are formed by water beyond the hydration shell of the lipid head

G. I. Likhtenshtein, *Biological Water*, Soft and Biological Matter,
https://doi.org/10.1007/978-3-030-82503-4_10

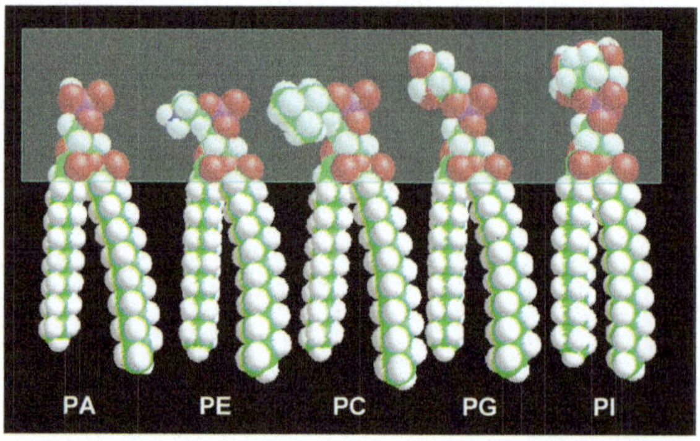

Fig. 10.1 Protrusion of the groups esterified to the phosphate in the surface of the membrane [5]. With permission from Elsevier

groups and are essential to maintain functional dynamical properties of the lipid head groups.

Without a hydrating environment, membranes cannot exist. Water as an abundant commodity stabilizes the structure of biological membranes, controls molecular dynamic and biochemical activities, and regulates interfacial/intermolecular interactions [5–18]. Water molecules form hydrogen bonds with biomolecules and contribute to the adjustment of their properties, such as surface charge, hydrophilicity, structural flexibility, and functional activity perturbating extends through several hydration layers. Retardation of water motions near phospholipid bilayers is extended by the presence of a membrane-bound protein, up to around 10 Å above that protein. Structural fluctuations and molecular excitations of hydrating water molecules cover a broad range in space and time, from individual water molecules to larger pools and from femtosecond to microsecond timescales. Because of water species present at the interphase, the interfacial water domains in the surface can be formed by the contact between lipids themselves and between lipids and the water phase, which are needed to trigger adsorption–insertion processes. The water domains are also formed by water beyond the hydration shell of the lipid head groups and are essential to maintain functional dynamical properties of the lipid head groups.

Water passive transport through biological membranes occurs: (1) by diffusion through the lipid matrix; (2) through transport proteins such as channels and some occlusion transport proteins such as glucose transporters, and (3) through water channel proteins specifically expressed by cells for such purpose. Water-soluble non-electrolytes such as glycerol, erythritol, urea and others can permeate lipid membranes in the liquid crystalline state. Two structural views of the membrane are considered: a solubility model in which mixture of water, lipid, and solute is formed in some degree during the transport process and water and solute diffuse through pores formed in a rigid matrix.

Recent progress in theory, molecular dynamics simulations, and physical methods including magnetic resonance and ultrafast vibrational spectroscopy has led to new and detailed insight into, water shells structure and fluctuations elementary water motions, electric fields at hydrated biointerfaces, and processes of vibrational relaxation and energy dissipation [1–22].

10.2 Biomembranes Models

As first proposed in 1972 [19], the Fluid–Mosaic Membrane Model (F-MMM) depicted biological membranes as a matrix made up of a mostly fluid bilayer of phospholipids with mobile globular integral membrane proteins and glycoproteins that were intercalated into the fluid lipid bilayer. The 1976 elaborations of the basic F-MMM included the interactions of extracellular matrix and membrane-associated cytoskeletal components with cell membranes and their potential influence on the mobility and distribution of trans-membrane glycoproteins [20]. The possibility that less mobile lipid–protein or lipid–lipid domains might exist in membranes as frozen or semifrozen islands of less mobile lipids in a sea of fluid phospholipids was also took in consideration.

A modified F-MMM, taking into account many of the numerous contributions that have been made since the 1970s, is depicted in Fig. 10.2. In this figure,

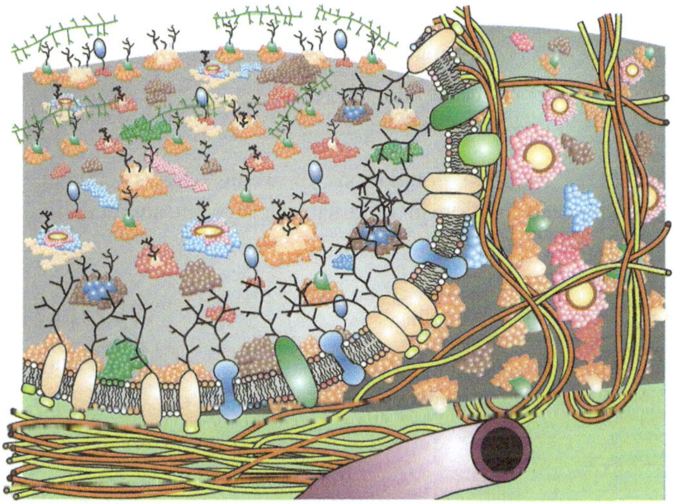

Fig. 10.2 An updated Fluid–Mosaic Membrane Model that contains information on membrane domain structures and membrane-associated cytoskeletal and extracellular structures. Different integral proteins, glycoproteins, lipids, and oligosaccharides are represented by different colors, and where the membrane has been peeled-up to view the inner membrane surface cytoskeletal fencing is apparent that restricts the lateral diffusion of some but not all trans-membrane glycoproteins. Other lateral diffusion restriction mechanisms are also represented, such as lipid domains, integral membrane glycoprotein complex formation (seen in the membrane cut-away), polysaccharide–glycoprotein associations (at the far top left) and direct or indirect attachment of inner surface membrane domains to cytoskeletal elements (at lower left). Although this figure suggests some possible integral membrane protein and lipid mobility restraint mechanisms, it does not accurately present the sizes or structures of integral membrane proteins, cytoskeletal structures, polysaccharides, lipids, submicro- or nano sized domains or membrane-associated cytoskeletal structures or their crowding in the membrane [21]. With permission from Elsevier

different types of interactions occur with integral membrane proteins and glycoproteins, membrane lipids, and membrane-associated cytoskeletal systems and extracellular matrix components. The mobilities, distributions, and aggregation states of membrane components were suggested to be controlled by these non-random interactions. Dynamic changes in the interactions of membrane components which are especially important in determining functional properties were described comprehensive review [21]. ' Thermodynamic considerations, assymmetry of membranes, three classes of membrane proteins, *cis* and *trans*-membrane control, membrane-associated cytoskeletal and extracellular matrix interactions, protein–protein, protein–lipid, lipid–lipid interactions within membranes different forms of mobility restriction in membranes, and hierarchical membrane structures were discussed in details.

The fluid mosaic model describes the dynamic structure of the plasma membrane ranging from 5 to 10 nm in thickness as a mosaic of components including phospholipids, cholesterol, proteins, and carbohydrates (Fig. 10.3). For example, human red blood cells are approximately 8 μm wide and myelin contains 18% protein and 76% lipid and the mitochondrial inner membrane contains 76% protein and 24% lipid.

The membrane interface separates the outer (bulk) aqueous phase from the hydrophobic membrane interior (Fig. 10.4).

Carbonyl exposure to water depends on the water–lipid ratio at each phase state (Fig. 10.5, part A).

The presence of different populations of water species, characterized by its degree of H-bond coordination in fluid saturated or unsaturated lipids, has been also established (Fig. 10.6) [18].

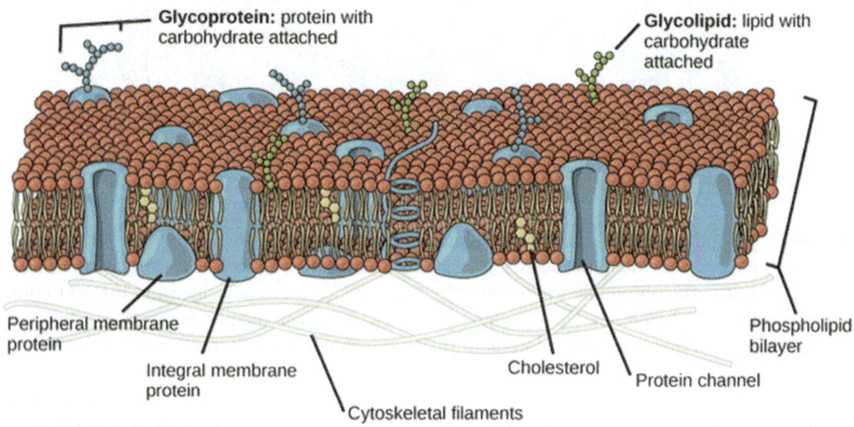

Fig. 10.3 The fluid mosaic model of the plasma membrane: The fluid mosaic model of the plasma membrane describes the plasma membrane as a fluid combination of phospholipids, cholesterol, and proteins. Carbohydrates attached to lipids (glycolipids) and to proteins (glycoproteins) extend from the outward-facing surface of the membrane. https://bio.libretexts.org/Bookshelves/Introductory_and_General_Biology/Book%3A_General_Biology_(Boundless)/5%3A_Structure_and_Function_of_Plasma_Membranes/5.1%

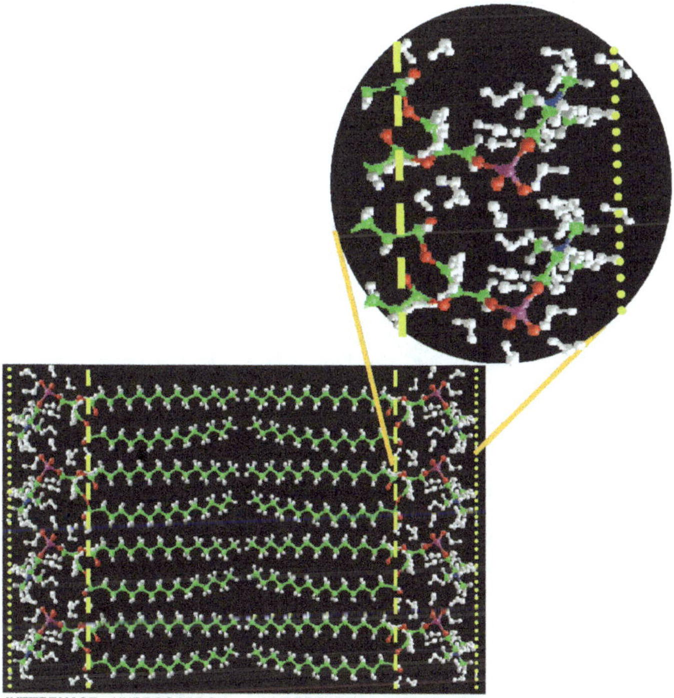

Fig. 10.4 Interphase of a lipid membrane. The interphase region confined between two planes: An inner plane containing the carbonyls (– - –) and an outer plane tangent to the external phosphates. The region between the two planes confines the hydrated groups and water [5]

10.3 Hydrated Biomembranes Properties

10.3.1 Water Structure in Hydrated Membranes

The thermodynamic properties of the lipid–protein interaction governing by changes in the water activity of monolayers of different lipid composition according to the lateral surface pressure were reviewed [27]. Both for structural and thermodynamic purposes were identified as the region confined between two ideal planes: one located at the limit between the hydrocarbon core and the polar head groups, the water–hydrocarbon interface, and an external plane tangent to the hydration shell of the polar head groups. A schematic description of these regions and planes is shown in Fig. 10.7.

Near the phase transition, the lipid membrane expansion causes an increase in capacitance, suggesting water penetration and peptide or protein insertion, and increase of water due to membrane expansion is also denoted. The variation of

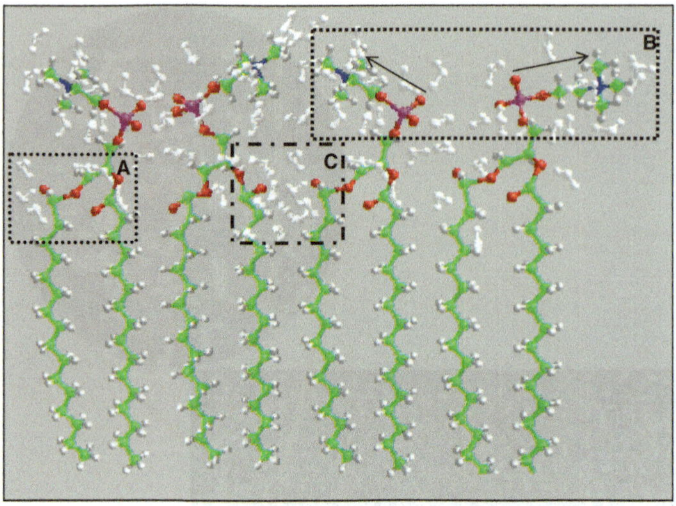

Fig. 10.5 Defects in the lipid surface. Schematic representation of the orientation of the C = O at the water–membrane interface (**a**), phosphocholine or ethanolamine (P–N) alignment (**b**) and CH_2 exposure (**c**) [5]. With permission from Elsevier

Fig. 10.6 Schematic illustration of water present in kinks formed by trans-gauche isomers along the hydrocarbon tails [18]. E.A **Disalvo**, (Ed.), Membrane Hydration

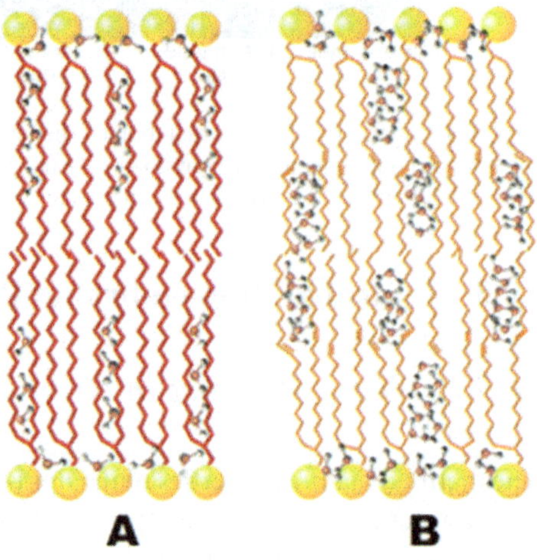

the amount of hydration water and confined water due to membrane expansion is schematically illustrated in Fig. 10.8.

An analysis of the enthalpic vs the entropic changes in the dissolution of different OH bonding solutes in lipid membranes showed that different alcohols can be grouped into three families: one corresponding to compounds in which the hydrophilicity remains constant and hydrophobicity grows with the addition of CH_3 groups in a tetrahedral array, such as in the series ethanol–propanol–isobutanol (Fig. 10.9) [27].

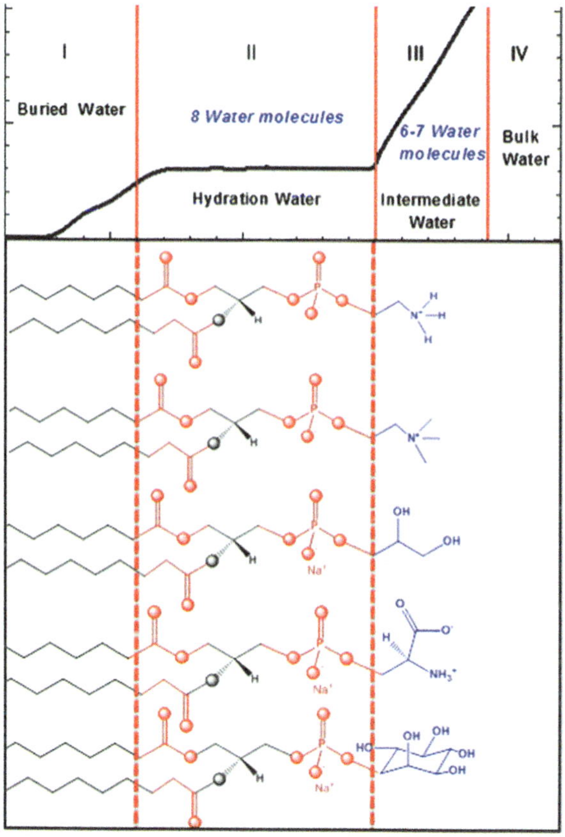

Fig. 10.7 Membrane interphase region. Region I contains only the hydrophobic lipid residues and contains buried water (confined water). This region may vary according to packing and phase state of the lipids Region II begins at the carbonyl groups. In this region, the total system density increases dramatically due to the tight hydration shell of charged groups such as phosphates. It is chemically the most diverse region, containing both hydrophobic and hydrophilic components and hydration water. This region can be considered as a binary solution in which polar head groups with their hydration shells are imbibed in water (second hydration shell). Region III contains the groups bound to the phosphates that protrude into the aqueous phase. In this region, water may preserve its tetrahedral arrangement due to its binding to OH groups or by the adjacency of non-polar moieties such as cholines (clathrates). Region IV is non-perturbed bulk water [27]. With permission from Elsevier

The main results of the work are: (1) head groups protruding toward the water phase organize water in different dynamic and structural patterns; (2) the free energy changes produced by peptide or protein insertion are related to changes in the chemical potential of surface water; (3) lability is related to the domains formed according to cholesterol level and unsaturation when the surface pressure is below that corresponding to the saturation pressure; (4) the excess of water beyond the hydration

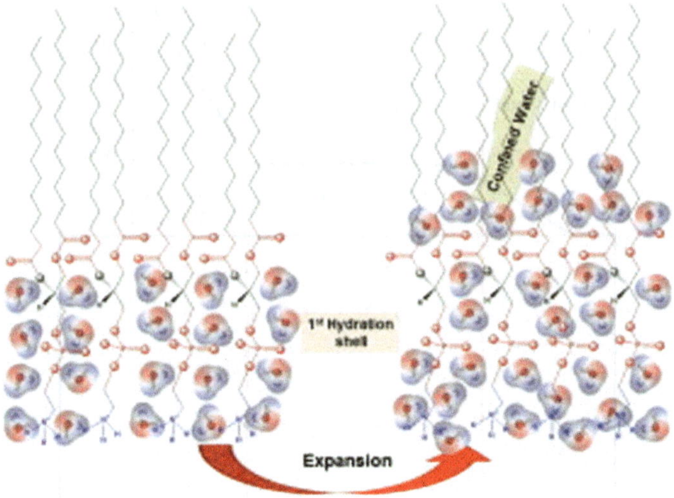

Fig. 10.8 Hydration and confined water in a lipid monolayer upon lateral expansion [27]. With permission from Elsevier

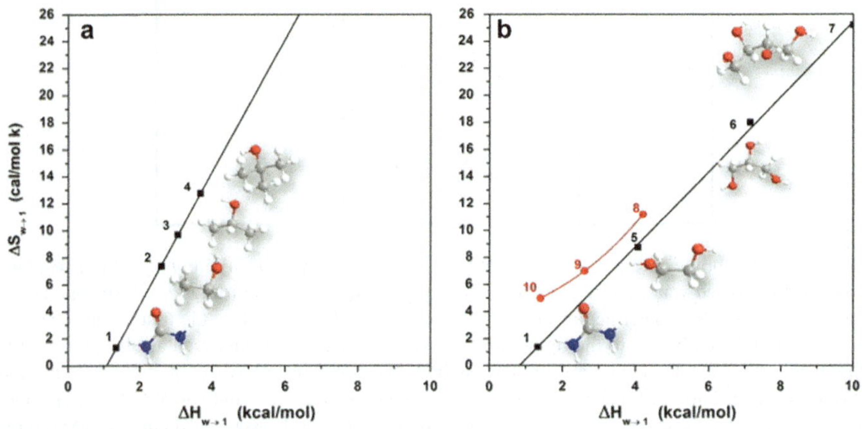

Fig. 10.9 Entropic–enthalpic compensation of the partition of different families of polyalcohols in DMPC bilayers. **a** First family: Urea (1), ethanol (2), propanol (3), isobutanol (4). This family maintains only one H bonding groups (OH) while increasing the hydrophobic character and the molecular volume in a tetrahedral array. Slope: 0.0047; ΔS: −5.4 cal/mol·K. **b** Second family: Urea (1), ethylenglycol (5), glycerol (6), erythritol (7). The hydrophobic–hydrophilic character is maintained along the series, while the size of the molecule increases by adding one HCOH residues. Slope: 0.003; ΔS: − 2.6 cal/mol·K. Third family: Methanol (8), propanol (9), butanol (10). This group maintains only one H bonding group (OH) but the size of the molecule increases by increasing the chain length in one CH_2 in the series. ΔS: +5 cal/mol·K. First and second families would correspond to those obeying a non-classical hydrophobic partition for which the entropic change is negative. Third family would correspond to a classical hydrophobic partition (entropy-driven partition) [27]. With permission from Elsevier

shell is thermodynamically labile and depends on the cholesterol level and the unsaturation saturation ratio; (5) the water domains depend on the phase state of the lipids, specifically near the phase transitions; (6) changes in the kind of water populations are concomitant with the shift of methylene vibrational mode frequencies to higher values; and (7) the increase in isolated populations of methylenes is congruent with the formation of highly ordered water cluster bonded by hydrogen.

A hypothesis, which espouses that cells control activities of membrane-bound enzymes through manipulation of the thermodynamic activity of water in the lipid–water interfacial region, was put forward [22]. The hypothesis is based on the following facts and suggestions: (1) the surface pressure of a lipid monolayer is a direct measure of the thermodynamic activity of water at the lipid–water interface; (2) the surface pressure-dependent activation or inactivation of interfacial enzymes is directly related to changes in the thermodynamic activity of interfacial water; (3) cells may manipulate conformations (and activities) of membrane-bound enzymes by manipulating the (re)activity of interfacial water at various locations in the membrane by localized compression or expansion of the interface; (4) cells may use the membrane-bound hormone receptors, lipid phase transition, and local variations in membrane lipid composition as effectors of local compression and/or expansion of membrane, and thereby local water activity. Several experimental data in the literature were reexamined in the light of this hypothesis.

The structure and electronic properties of phosphatidylcholine (PC) under different degrees of hydration at the single-molecule and monolayer type level were investigated by linear scaling ab initio calculations [24]. The calculation indicated that upon hydration, the phospholipid undergoes long-range conformational rearrangements which lead to a sickle-like ground-state shape. It was also established that the structural unit of the tilted gel-phase PC appears to be a water-bridged PC dimer; and hydration dramatically alters the surface potential, dipole and quadrupole moments of the lipids and consequently guides the interactions of the lipids with other molecules and the communication between cells.

Dual polarization interferometry (DPI) and quartz-crystal microgravimetry (QCM-D) were used to investigate the adsorption of 1,2-Dioleoyl-sn glycero-3-phosphocholine (DOPC) vesicles to a solid hydrophilic surface [25]. The layer of hydration formed between a self-assembled DOPC bilayer and a silica solid support was probed in assemblies constructed using H_2O and D_2O buffers. An AnaLight Flex Bio200 dual-polarization interferometer (DPI, Farfield Scientific, Inc., Cheshire, U.K.) was used to obtain the DPI optical properties of DOPC bilayers. In the frame of QCM-D, real-time frequency and dissipation data were collected utilizing a quartz-crystal microbalance with dissipation. QCM-D was used to measure the mass of the bilayer, including the mass contribution of the coupled solvent that resides between the membrane-solid interface. The thickness of this hydration layer, calculated by relating its mass to the density of the layer, was determined to be 10.46 Å for trapped D_2O and 10.21 Å for trapped H_2O. Thus, this work establishes the feasibility of concurrently using DPI and QCM-D to gauge the extent of hydration in thin films. Changes in frequency (ΔF) and dissipation (ΔD) for the third harmonic after exposing the SiO_2-coated QCM-D sensor to DOPC dispersion in tune gave the sequence H_2O (pure) → DOPC (HEPES) → H_2O (pure) → DOPC (HEPES). *HEPES* (4-(2-hydroxyethyl)-1-piperazineethanesulfonic acid) is a zwitterionic sulfonic acid buffering agent;

In order to determine the atomic scale structure (0–10 Å) of hydration between water and a phosphocholine head group, 1,2-dipropionyl-*sn*-glycero-3-phosphocholine membrane.

(C3–PC) has been investigated in aqueous solution, using a combination of neutron scattering techniques enhanced by neutron diffraction with isotopic substitution, NDIS) [26]. Empirical potential structure refinement (EPSR) modeling was also performed [26]. This lipid molecule was investigated below the critical micelle concentration (405 mM for C3–PC). Neutron diffraction measurements were performed on the small angle neutron diffractometer for liquids and amorphous samples (SANDAL) instrument located at the ISIS Facility (STFC, UK) on C3–PC using a variety of isotopically substituted water solvents, namely, H_2O, HDO, D_2O, and $H_{1.23}$ $D_{0.67}$ O, each at a concentration of 220 mM (1:250 lipid: water ratio) at standard temperature and pressure (298 K, 1 bar). EPSR simulations were performed by building a box of molecules which contained 10 C3–PC molecules and 2500 water molecules. The work main results can be formulated as follows: (1) the MD and EPSR models showed evidence for hydrogen bonding from the water molecules to the P = O oxygen atoms where the coordination number at the first minimum of the distribution functions $gO_2Hw(r)$ (at 2.4 Å) is 2.5; (2) the $gO2O(r)$function gives a coordination number of 2,5 at around 3 Å which indicates that more than one water molecule donates hydrogen bonds to the P = O oxygen atoms; (3) according to the calculation of the spatial density functions SDFs, the single bond P-O-C oxygens (O_{s1}) bisect the—z-axis in the zx plane at 142° and −142° relative to the z-axis, underneath the xy plane and the P = O O_2 oxygens are located along the y-axis at 52° and −52° bisecting the + z-axis in the zy plane, which gives tetrahedral configuration for this group (Fig. 10.10); and (4) for hydration of the onium group, from both the EPSR model of the neutron data and from the MD simulations, the $gC_mO_{w(}r)$ and $g_{NOw}(r)$ functions showed strong peaks at 3.5 and 4.2 Å, respectively.

Figure 10.11 indicates the water orientational correlation functions (OCF) directly above the PO_4PO_4 group in the positive zz-direction, which displays a hydrogen bond as the water hydrogen is oriented toward the PO_4PO_4 group and shows a hydrogen bond being shared between the two $O_4P_4OO_4P_4O$ oxygen atoms.

Phosphate group-water correlations $g_{O2Hw}(r)$ in Fig. 10.12 show two further distinguishable peaks at larger distances, at around 3.5 and 5 Å where the larger distance can related to the atom covalently bound to the water molecule involved in the hydrogen bonding [26].

Molecular dynamics simulations to study the hydration properties of Dipalmitoylphosphatidylcholine (DPPC) bilayers, both in the gel and the liquid crystalline states, were employed [16]. DPPC bilayers composed of 128 DPPC molecules (64 per monolayer in a 8 8′ arrangement) were simulated by means of AMBER12 package. The initial separation between DPPC was 9 Å in a triangular arrangement. The membrane was stabilized at 323 K for the liquid crystalline state and at $T = 298$ K for

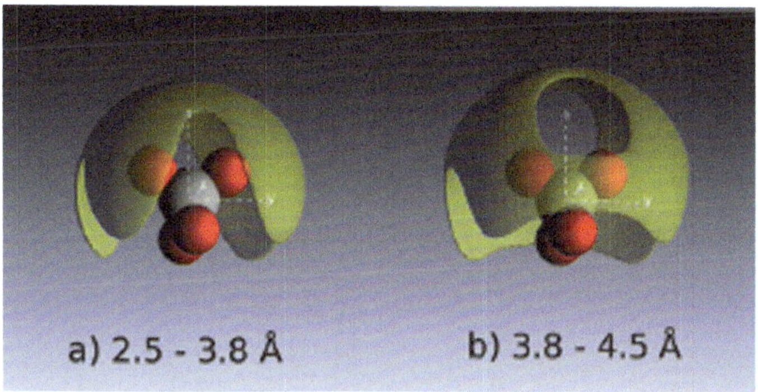

Fig. 10.10 Spatial density functions (SDFs) for water around lipid PO_4 group at the distance range indicated where **a** shows the top 70% of the water molecules in the surrounding shell and **b** shows the top 50% of the water molecules in the surrounding shell [26]. With permission from American Institute of Physics

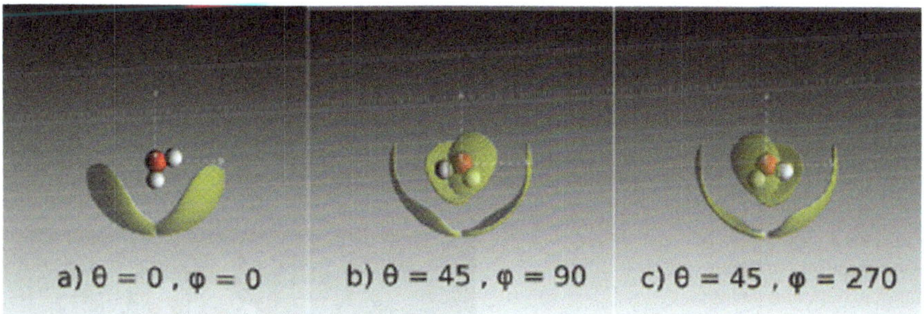

Fig. 10.11 Orientational correlation functions (OCFs) for water around PO_4 group at the distance range of 2.5–3 Å at specified locations relative to the SDF distribution shell shown in Fig. 10.9a. In each case, the top 70% of water molecules are shown, and in each panel, the central water molecule is oriented toward its most likely position [27]. With permission from American Institute of Physics

the gel state and solvated with a total of water molecules around 81 water molecules per lipid). It was obtained that the water species different according to the types of H-bonds they form are distributed in the groups of the membrane both above and below the transition temperature and that most of the water molecules form three H bonds with other water molecules (300 species). A lower number of water molecules form 2 or 4 water–water H-bonds and a much lower amount of water molecules bound to phosphates (110, 210 and 310 species) was found to be. At the tight hydration centers (CO and PO), the H-bond coordination is higher. In addition, as depicted in Fig. 10.13, the existence of less compact (strings or ramified strings) nanoclusters at the buried region within the hydrocarbon chains was found.

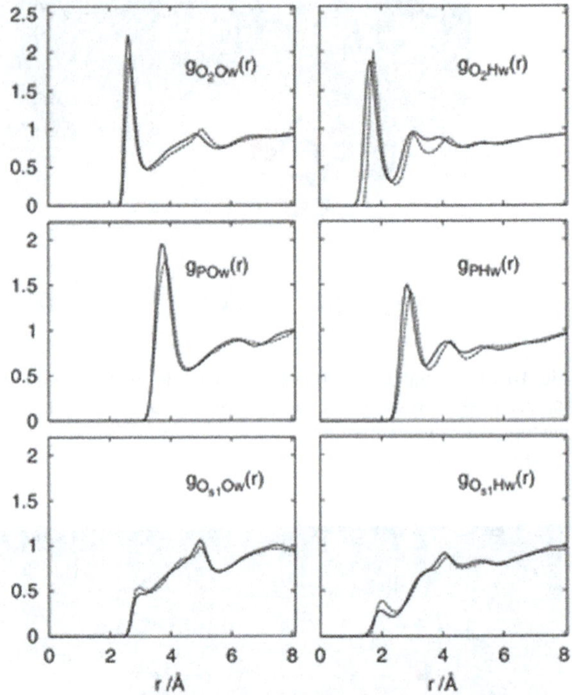

Fig. 10.12 Phosphate group-water correlations where the solid line represents the EPSR data and the dot-dash line represents the MD simulations [26]. With permission from American Institute of Physics

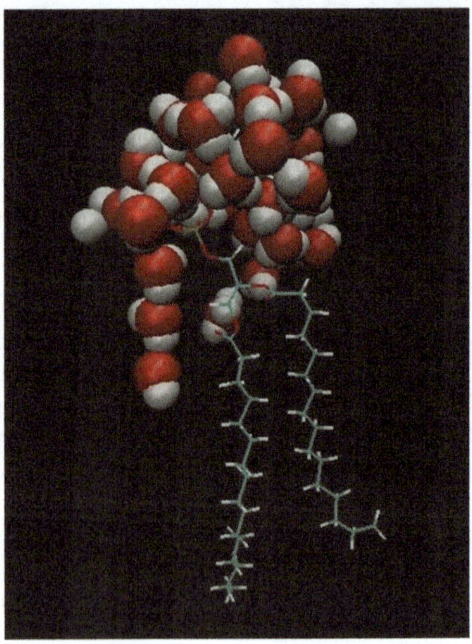

Fig. 10.13 Examples of configurations of water nanoclusters and strings around the lipid chains [16]

Authors concluded that: (1) the gel-fluid transition involves significant changes at the second hydration shell, particularly at the buried region between the hydrocarbon tails; (2) hydrophobic region becomes partially water accessible in the liquid crystalline state, such water molecules present a lower H-bond coordination as compared to the molecules at the primary hydration shell; and (3) the buried water molecules tend to organize themselves in less compact structures, typically strings or branched strings, with a scarce population of isolated molecules. Due to its low light penetration depth, attenuated total reflection Fourier transform infrared ATR-FTIR) spectroscopy is specifically suited to study interlamellar water structure in multibilayers and the long-range water structure disturbance, in particular [29]. Employing this technique, it was established what manner the interlamellar water perturbation is influenced by the hydration level and how it is sequentially modified with a step-by-step dehydration of samples either by water evaporation or by osmotic pressure. The results also indicated that besides different levels of hydration the lipid–water interaction is modulated by different lipid head groups and different lipid phases.

Confocal Raman microscopy has been used to measure depth-dependent profiles of human stratum corneum (SC) in vivo in the high wavenumber (HWN) region [30]. The reported results confirmed that: (1) the hydrogen bound water molecule types contributed differently to the water diffusion process in the SC; (2) the most concentrated double donor–double acceptor (DDAA) and single donor–single acceptor (DA) water molecule types in the SC represent more than 90% of the SC's water and mostly contribute to the water flux in the skin; (3) single donor–double acceptor (DAA) and weakly bound water molecule types represent less than 10% of the SC's water content: (4) the most tightly hydrogen bound water molecule type, DAA, reaches its maximum concentration near the skin surface and does not take part in the water diffusion process via the SC; (5) the hydrogen bonding state of water (DA/DDAA water molecule type ratio) reaches its maximum at the depth of. 30% of the SC thickness; and (6) the natural moisturizing factor (NMF) contribution to the bonding of water in the SC is dominate over that of intercellular lipids (IC).

The structural changes to lipid membrane that ensue from the addition of aliphatic alcohols with various alkyl tail lengths were investigated using small-angle neutron diffraction from flat lipid bilayers. [33] The distance between bilayer thickness and head-to-head distance was found to be constant across the various alcohols (3.7 ± 0.6 Å), while it is about 1.5 Å smaller than observed in the case of neat lipid bilayers. Thus, the addition of an alcohol results in some of the space in the polar head group region being filled with extra water molecules (Figs. 10.14 and 10.15).

The results obtained with AdResS and the comparison with fully atomistic results for the minimal size of the AT region (7 nm) are presented in Fig. 10.16. For AT regions smaller than 7 nm in the x-direction, the discrepancy between the AdResS results and the reference data exceeds 5%. As can be seen in \the figure, the agreement between AdResS and atomistic simulations is satisfactory with the largest discrepancy for the water density (reference bulk value = 980 kgm$^{-3)}$ being about 3% in the Δ region.

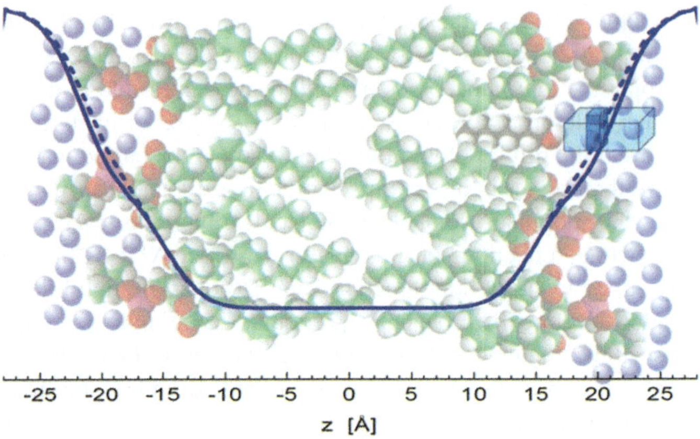

Fig. 10.14 Increase in water hydration at the water-membrane interface upon the addition of an aliphatic alcohol. Water molecules (blue spheres) fill an additional space (the dark portion of a blue rectangle) resulting from increased water penetration and enlarged lateral area above the alcohol molecule [33]. Published with permission MDPI

The novel methodology allowed to suggest a complementary view to the controversial question of the definition of the hydration shell for a membrane by quantifying the minimal size of the region where the atomistic resolution of the molecules is strictly required while the rest of the system plays the role of a generic, structureless thermodynamic reservoir by quantifying the minimal size of the region where the atomistic resolution of the molecules is strictly required while the rest of the system plays the role of a generic, structureless thermodynamic reservoir.

10.3.2 Water Dynamics in Hydrated Membranes

A family of polarity sensitive fluorescent probes (2-(dimethylamino)-6-acylnaphtalenes, i.e., LAURDAN, PRODAN, ACDAN) was used with the aim to monitor solvent relaxation phenomena on protein matrices [36]. The (nanosecond) fluorescent decay of these probes is strongly affected by changes in the local polarity and relaxation dynamics of restricted water molecules existing at the membrane/water interface.

Using LAURDAN spectral imaging and spectral phasor analysis, the growth hydration state of subcellular organelles (lamellar body-like, LB-like) from live A549 lung cancer cells at different post-confluence day was examined [31]. Once incorporated into membranes, the nanosecond fluorescent decay of these probes is strongly affected by changes in the local polarity and relaxation dynamics of restricted water molecules existing at the membrane/water interface. For instance, when glycerophospholipid containing membranes undertake a solid ordered (gel) to liquid disordered phase transition the fluorescence emission maximum of these probes

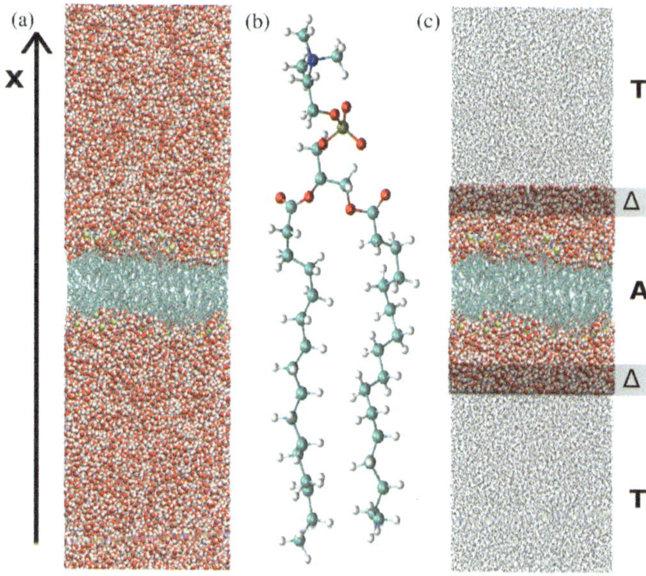

Fig. 10.15 Full atomistic representation of the system. **b** Atomistic representation of the DPPC molecule. ()AdResS simulation setup: the atomistically resolved DPPC bilayer is oriented in the z-y plane and hydrated on top and bottom (along the x-axis) by atomistically resolved water (AT region). Beyond the AT region on both sides is the transition region (Δ) followed by the non-interacting tracer region (TR). DPPC alkane chains are represented in light blue while the negatively charged phosphorus and the positively charged nitrogen atoms in the DPPC head groups are represented in yellow and blue, respectively. **Hydrating water molecules** are shown in a space-filling representation and tracer particles are represented by black beads. Periodic boundary conditions are applied in x, y, and z. [35 J. 10.15 Whittaker, L. Delle Site, Investigation of the hydration shell of a membrane in an [34] open system molecular dynamics simulation. Phys. Rev. Research 1, 033,099 (2019) prresearch@aps.org SEND 20/05/05

shift 50 nm with a significant change in their fluorescence lifetime. The LAURDAN emission observed in LB-like structures was explained by a linear combination of discrete states reflecting the presence of 0, 1, 2, or 3 water molecules. The fluorescence parameters of LAURDAN and PRODAN are also sensitive to cholesterol and other molecules effects, allowing interpretations that correlate changes in membrane packing with membrane hydration. These parameters appeared to be responsive to membrane hydration, which in turn is related to membrane lateral structure. A first step (days 1–7) was characterized by an increase in their size, followed by a second one (days 7–14), where the organelles display a decrease in their global hydration properties. A significant increase in their hydration state upon secretion in these lamellar body-like, LB-like, structures was found. Thus, LAURDAN is a suitable probe to study membrane related phenomena for different reasons: (i) its partition, coefficient from aqueous environments to membranes is relatively high; (ii) it displays an exquisite sensitivity to distinct lipid packing in membranes; and (iii) it generally shows an even partition in membranes displaying [31]. Figure 10.17

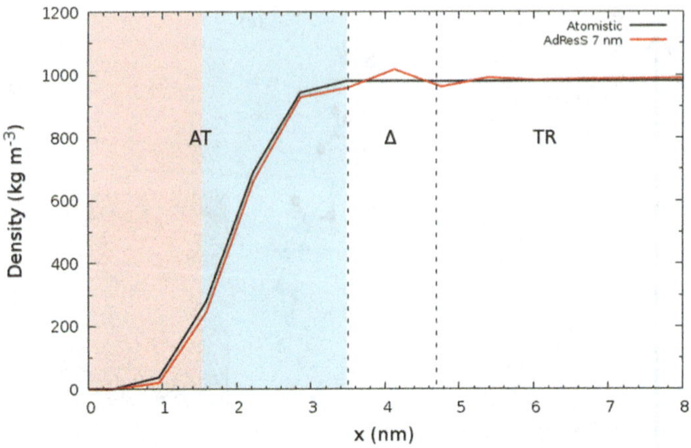

Fig. 10.16. Mass density profile of water in the AdResS $AT = 7$ nm and reference atomistic system. Data from both simulations has been averaged over both sides of the bilayer and symmetrized. Red shading indicates average distance of the phosphorus atoms in each leaflet from the center of the bilayer while blue shading indicates the water-DPPC interface region where density of the bilayer is still nonzero. In each case, shaded regions represent the quantities calculated in the $AT = 7$ nm AdResS simulation [34] 10.16.J. Whittaker, L. Delle Site, Investigation of the hydration shell of a membrane in an open system molecular dynamics simulation. Phys. Rev. Research **1,** 033,099 (2019) prresearch@aps.org **SEND 12/05**

illustrates spectral phasor analysis of the LAURDAN fluorescence in live A549 lung cells to study the hydration and time evolution of intracellular lamellar structures. The membrane cavities model introduced by Parasassi and coworkers [32] suggested that LAURDAN can sense a few molecules of water (i.e., 0 to 5) in their vicinity with a particular relaxation time of the order of nanoseconds.

Hydrated protein motions occur over many decades of time, from femtoseconds to seconds and longer. Fast (ps–ns) protein motion enabled by water dynamics on the surfaces of protein molecules is required for biological function. To understand the coupling between water and protein dynamics for hydrating purple membranes (PM), a molecular dynamics simulation study was performed [35]. The simulation simulations were validated by comparing the temperature and hydration dependence of crystal lattice parameters and water and protein/lipid dynamics with neutron diffraction and spectroscopic data. The temperature dependence of several time-dependent quantities describing different aspects of water dynamics was examined and the correlations between water dynamics and the motion of the protein and lipid components of PM was characterized.

In study [13] by complementary use of the terahertz time-domain spectroscopy (THz-TDS) and the small-angle x-ray scattering (SAXS), the relation between the hydration state and the lamellar structure of hydrated phospholipid bilayers composed of 1,2-dimyristoyl-sn-glycero-3-phosphocholine (DMPC) in the subpicosecond timescale was clarified [13]. Since the faster part of the rotational relaxation dynamics of water molecules lied in the THz frequency range (0.5–2.5 THZ), the

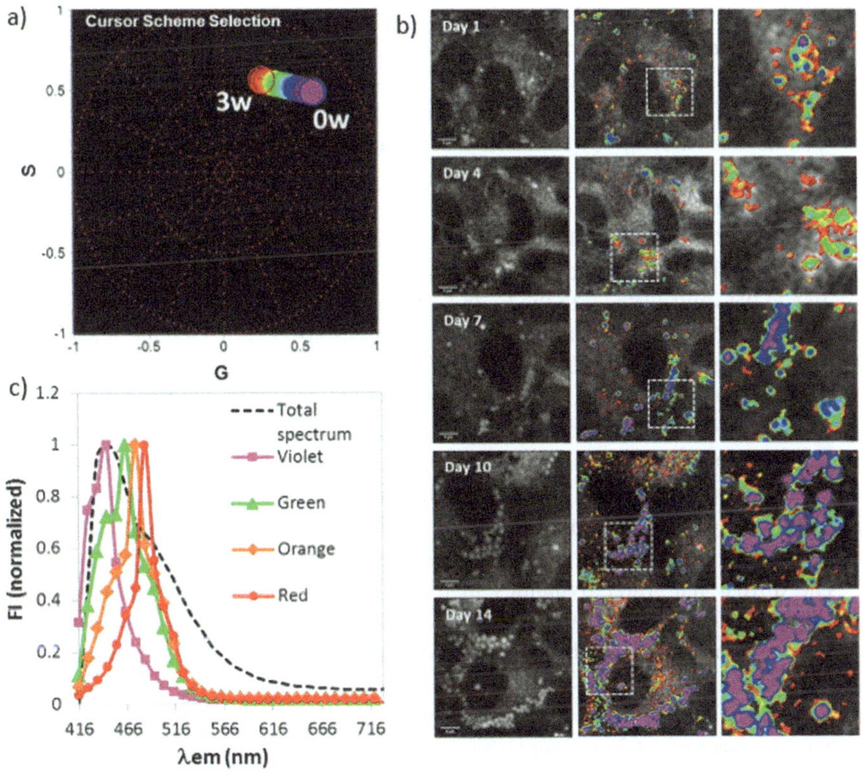

Fig. 10.17 Spectral phasor analysis of LAURDAN emission from intracellular LB-like structures according to the "cavities" model (see text). **a** Cursor design of the pixels of LAURDAN in cell at different stages (3 w and 0 w, referring to the number of water molecules). **b** Fluorescence intensity and pseudo-color reconstruction images of LAURDAN emission in intracellular LB–like structures. The pseudo-colored images were obtained by applying the cursor selection (violet, green, orange and red) in the first harmonic plot shown in (a). **c** Emission spectra of LAURDAN at pixels selected with the different cursors (violet, green, orange and red). Bar scale represents 5 μm [31]. With permission from Elsevier

hydration state in the ultrafast timescale from the change in the complex dielectric constants in the THz region was evaluated. THz-TDS measurements were carried out by using a setup with Er-doped fiber laser (780 nm, 100 fs, 50 MHz) was used as a light source. The SAXS measurements were performed at BL-15 A, Photon Factory, High-Energy Accelerator Research Organization, Japan. Experiments revealed that long-range hydration water includes slightly perturbed water molecules by lipids compared to bulk water, the correlation length of the hydration effect is about 10 Å from the lipid surface, and most of the water molecules in the lamellar phase are regarded as hydration water in the ultrafast timescale. The SAXS results indicated that about 35–40 water molecules per a lipid exist in the 12.5-Å-thick water layer at the fully hydrated condition and 25 water molecules were sufficient for filling the water phase (the cross-sectional area of a lipid is 60 Å^2, and the volume of a bulk water molecule is 30 Å^3. Data on dielectric constant of pure water, DMPC solutions (R = [H_2O]/[DMPC] 190; 113; 78; 60; 35; 19), and DMPC film (R = 4), obtained by THz-TDS measurements (0.4–2.7 THz) were also presented.

A cholesterol derivative, (cholesteryl benzoate) chromium tricarbonyl was synthesized and characterized to probe vibrational properties model lipid membranes, focusing specifically on the membrane–water interface [15]. Cells were prepared from the DMPC/DHPC/chol-BCT molar ratio of 65:24:1. Polarized attenuated total reflectance (ATR) spectroscopy was utilized to measure the orientation of the probe relative to the lipid surface. In the Four-Wave Mixing Experiments, the setups of 2D infrared spectroscopy (2DIR) and RASD were employed to probe the membrane–water interface with chol-BCT as an ultrafast dynamical probe. RASD enables rapid experimental determination of the frequency–frequency correlation function (FFCF). Utilizing FTIR and polarized-ATR spectroscopies, the location of the chromium tricarbonyl motif to be situated at the water–membrane interface with an orientation of $46 \pm 2°$ relative to the vector normal to the membrane surface was established. The interfacial water as sensed by the probe exhibits two to three times slower spectral dynamics than bulk water due to constraints imposed by the interface.

The diffusional water dynamics in the hydration layer of a dipalmitoylphosphatidylcholine (DPPC) bilayer was studied using molecular dynamics simulations [14]. Specifically, the simulation setup consisted of a lipid bilayer formed by 72 DPPC molecules and 2278 SPC/E water molecules (Fig. 10.18). Production runs have a total length of about 6 ms at a temperature of $T = 320$ K in the ensemble with a mean area of 0:65 nm^2 per DPPC molecule. The dynamics of individual water molecules in the vicinity of DPPC bilayer based on the trajectory analysis of extensive MD simulations were examined. By mapping the perpendicular water motion on the ordinary diffusion equation, free energetic and friction effects were disentangled and perpendicular diffusion was shown to be strongly reduced. For example, the lateral water motion exhibits anomalous diffusion up to several nanoseconds and is characterized by further decreased diffusion coefficients.

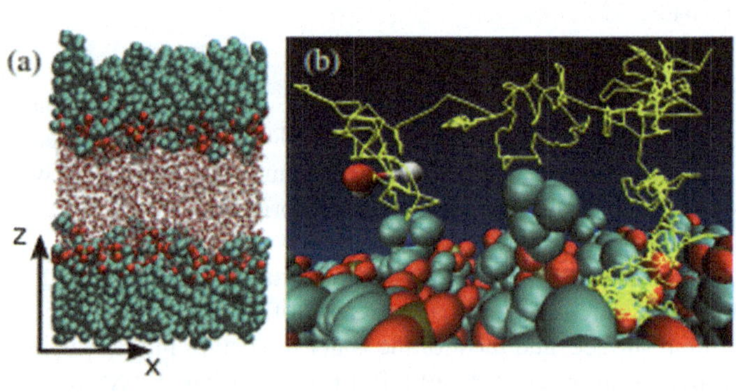

Fig. 10.18 (Color online). Snapshot of the simulation box (top) and trajectory of a water molecule in the vicinity of the membrane: the path of 100 ps duration starts in the lower right (bottom) [14]

In the system under consideration, dynamic behavior was characterized by the mean square displacement (MSD) for water motion parallel to the membrane surface, distances z_0 from the membrane center, lateral diffusivity D_{lat}, the water diffusivity perpendicular to the membrane surface D_Z, and duration time t. The calculation indicated following range above-mentioned parameters: D_{lat} (mm^2/ns) 1–5, Z_0 1.5–2.5 nm. t ps 10^{-1} -10^4 from bulk water. The observed diffusion anisotropy of membrane hydration water was explained by the interplay of two competing effects: (i) Continuum hydrodynamics alone speeds up the lateral diffusion compared to the perpendicular one, implying $D_{lat} = D_z > 1$, and (ii) the lateral diffusion dynamics implying $D_{lat} = D_z < 1$ gives rise to a pronounced anisotropy with $D_{lat} = D_z < 1$. Thus, the findings imply that efficient transport parallel to the membrane surface necessarily also involves fluctuations perpendicular to the lipid leaflet.

Figures 10.19 and 10.20 summarize the following results: (1) the subpicosecond dynamics is independent of z_0, the MSDs differ considerably on larger timescales; (2) water molecules in the center of the water slab ($z_0 = 3.5$ nm) show MSD curves that are indistinguishable from bulk water closer to the membrane; (3) the water

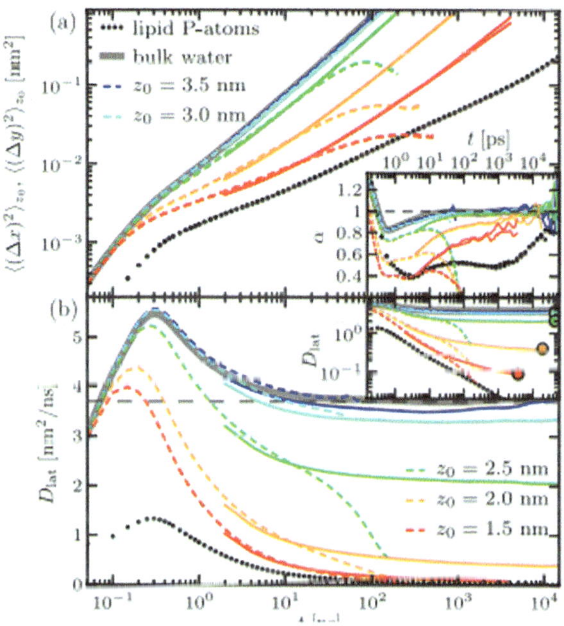

Fig. 10.19 Lateral MSDs] **for water** at different distances z0 from the membrane center (colored lines), for lipid heads (black dots), and for bulk water [25] (gray lines, hidden under the z0 = 3.5 nm water data). Corresponding local exponents α are shown in the inset. **b** Resulting time-dependent diffusion coefficients Dlat [(2)], shown on a double-logarithmic scale in the inset. Filled colored circles denote upper bounds for the long-time diffusion coefficients. In all panels and insets, the same color coding is used, and data for unconstrained and harmonically trapped water molecules are displayed as dashed and solid lines, respectively [14]. With permission from American Physical Society

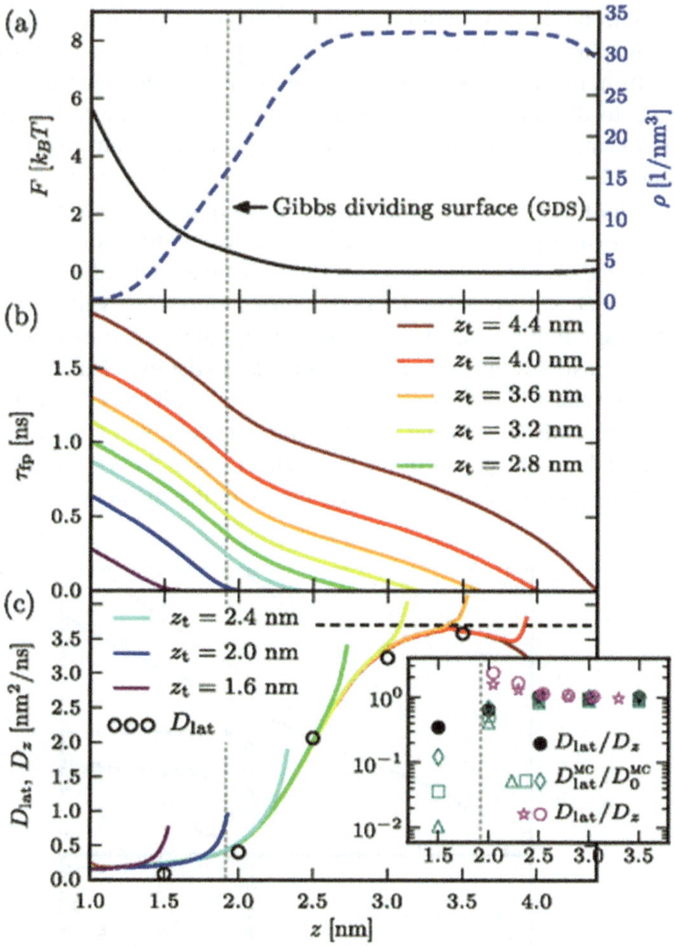

Fig. 10.20 **a** Water density $\rho(z)$ and free energy profile $F(z)$. **b** Water MFPTs to first reach a target distance zt when starting off from z. **c** Perpendicular water diffusivity profiles Dz(z) (solid colored lines) derived via (3). Open circles denote the long-time lateral diffusivities from the inset of Fig. 10.2b. Inset: Water diffusion anisotropy Dlat/Dz at the membrane (filled circles), at solid homogenous hydrophobic (stars) and hydrophilic (open circles) surfaces (Fig. 10.4), and the lateral rescaled diffusion constant DMClat/DMC0 (open triangles, squares, and diamonds) from MC simulations (Fig. 10.5); the z-coordinates of the solid homogenous surfaces have been chosen such that the positions of the respective GDSs agree [14]. With permission from American Physical Society

MSD curves approach the lateral MSDs of the lipids' phosphorus atoms; (4) water density $\rho(z)$, the resulting free energy profile $F(z) \equiv - k_B T \log[\rho(z)/\rho_0]$, and the Gibbs dividing surface (GDS) are indicated; (5) the diffusivity profile reveals bulk-like dynamics with a diffusion coefficient $DH_2O \approx 3.7$ nm^2/ns at $T = 320$ K; and (6) when approaching the membrane, the diffusivity smoothly decreases and levels off at values ~0.2 nm^2/ns in the lipid head group region, about a factor of 20 smaller than in bulk.

To characterize the structure and dynamics and solvation of zwitterionic micelles, high-precision dielectric spectroscopy of dodecylphosphocholine (DPC) micellar solutions in a wide frequency range from 50 MHz to 1.12 THz was utilized [10]. MD simulations were carried down to explain experimental results.

The dielectric response of the samples at high frequency was studied utilizing a GHz-to-THz spectrometer based on the above vector network analyzer together with frequency multipliers from Virginia Diodes, spanning the frequency range from 60 GHz to 1.12 THz. Using two different methods, complex dielectric response spectra from DPC micellar solutions in a large range of frequencies, from 50 MHz to 1.12 THz and from 50 MHz to 50 GHz, a vector network analyzer (PNA N5225A) was combined with a dielectric probe (HP 85070E). In addition, a transmission test set to measure the low-frequency dielectric response (50 MHz to 5 GHz) of solutions.

In the low-frequency range (50 MHz to 50 GHz), the dielectric response of aqueous micellar solutions mainly has contributions from (i) the rotation of micelles; (ii) undulation and motion of head groups)\; (iii) hydration waters in the interfacial region surrounding DPC micelles (dipoles of water molecules in hydration shells); and (iv) the orientational polarization of bulk water molecules (water dipoles). Relaxational processes occur in DPC aqueous solutions, including the motion of DPC surfactant head groups, which have a reorientation time of ~600 ps; hydration waters with times on the order of 30–150 ps; and bulk water with a time of 8 ps. Data on high-frequency response (60 GHz to 1.12 THz) assumed that (i) DPC micelles in solution are spherical with a radius of $R_{micelles}$ and have a spherical hydration shell, (ii) the water molecules in the hydration shell are a part of DPC micelles, and (iii) solvent outside of the DPC micelles has the same dielectric property as that of pure water to gain insight into the experimental observation of the dynamics and structure of hydration waters. The solvation shells of DPC micelles were analyzed by calculating the solvent radial density functions (Fig. 10.21). Radial distribution function (RDF) of water's oxygen around C12 atom, the carbon nearest the head group, the phosphorus atom, and the nitrogen atom was also calculated.

The surface-exposed atoms of DPC micelles having channels are illustrated in Fig. 10.22 which shows waters within the channels. The waters confining and Coulombic interactions at the surface lead to a slowdown in water dynamics.

The total amount of hydration water of 950 molecules per micelle was obtained from the dielectric strengths of bulk water in DPC solutions, which is in agreement with the number of waters in MD simulations found within a distance of 4 Å from the DPC head group atoms. Using the effective-medium approximation for the dielectric measurements at THz frequency, it was able to extract the amount of tightly bound waters in the first hydration shell of 310 molecules with slow water dynamics. Experiments with the THz frequency revealed the collective vibrational modes within DPC micelles, while simulations indicated that the dominant contribution to the peaks comes from large scale motions of the lipids within the confining environment of the micelle (Fig. 10.23).

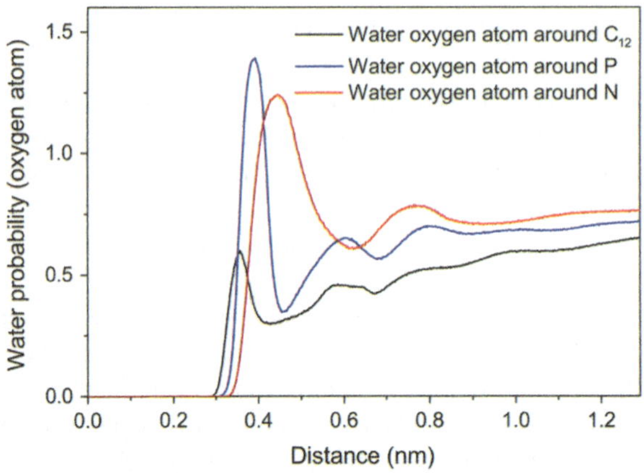

Fig. 10.21 Solvent radial distribution functions (water oxygen atom) around C12 (black line), phosphorus (blue line), and nitrogen (red line) atoms of the DPC molecule [10]. With permission from American Chemical Society

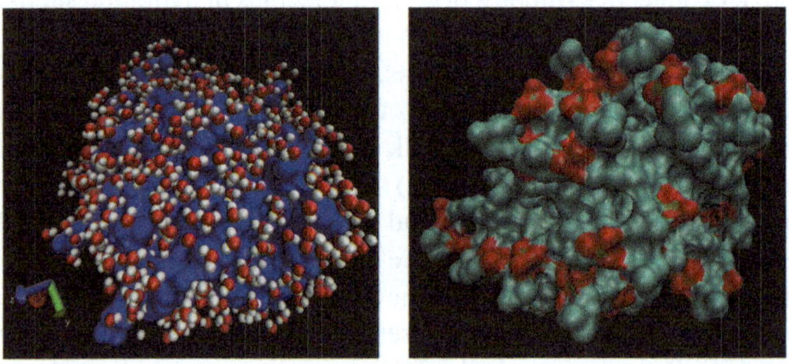

Fig. 10.22 DPC micelle surface rendered (right) with alkyl groups (including trimethyl amine moieties) in aqua, oxygen in red, and phosphorus in gold (no waters are shown) and rendered (left) in dark blue, with solvation shell waters pictured in red [10]

The very high dynamic range of the frequency extenders, together with the precise sample thickness controller, allowed to obtain the highly precise and accurate GHz-to-THz dielectric response spectra (Fig. 10.24).

Dynamics of hydration layers of a dimyristoylphosphatidylcholine (DMPC) bilayer were investigated employing an all-atom molecular dynamics simulation [37]. Continuously, residing interface water molecules which form hydrogen bonds solely among themselves and then concertedly hydrogen bonded to carbonyl, phosphate, and glycerol head groups of DMPC were identified. The calculation revealed the following hierarchy of dynamic processes: (1) the interface water hydrogen bonded to lipids showed slow relaxation rates for translational and rotational dynamics; (2) the mean square displacements and the reorientational autocorrelation functions are slowest for the interfacial waters hydrogen bonded to the buried carbonyl oxygen;

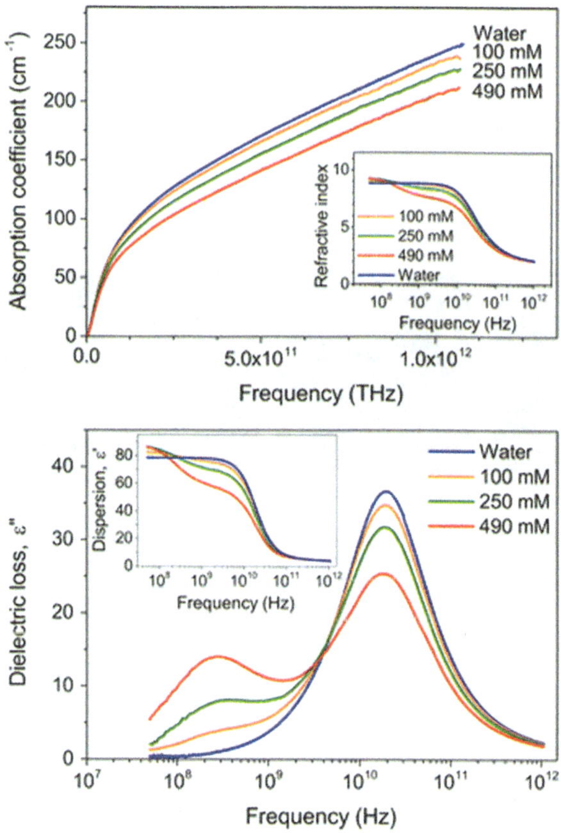

Fig. 10.23 Interaction of DPC micelles with GHz-to-THz radiation providing insight into the liquid's dynamics over picosecond to nanosecond timescales. (Top) the absorption spectra of both DPC micellar solutions and pure water increase monotonically with increasing frequency. The refractive indexes (upper inset) of DPC micelles and water, in contrast, decrease with increasing frequency. (Bottom) the dielectric loss and the dielectric dispersion spectra (lower inset) from DPC aqueous solutions and pure water are obtained from absorption coefficient and refractive index measurements. Data were collected at 25 °C [10]. With permission from American Chemical Society

(3) the autocorrelation functions for interfacial hydrogen bonded networks develop humps during a transition from cage-like motion to eventual power-law behavior of $t^{-3/2}$; (4) the relaxation rates of the networks buried in the hydrophobic core was found to be slower than the networks near the lipid–water interface which is slower than in bulk; (5) since hydrogen-bond breakage follows a translational diffusion dictated mechanism, chemically confined hydrogen-bond networks need an activation energy to diffuse through water depleted hydrophobic environments; (6) the slow relaxation rates of interfacial waters in the vicinity of lipids are originated from the chemical confinement of concerted hydrogen-bond networks. In addition, employing

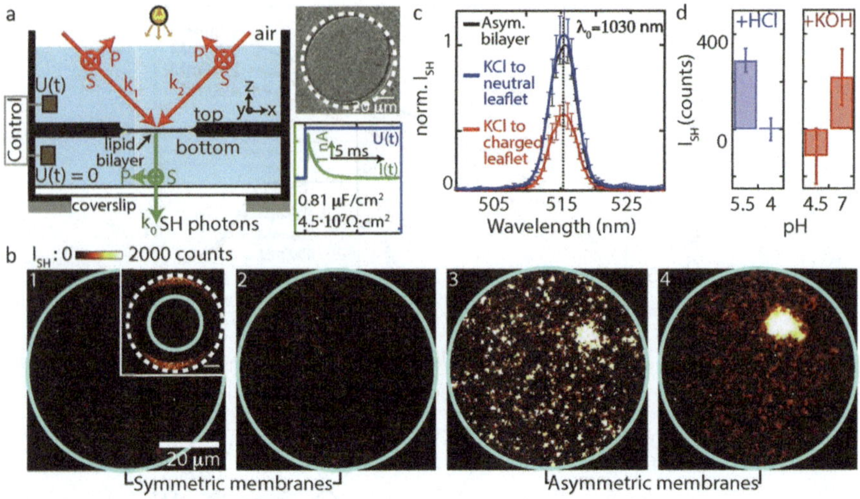

Fig. 10.24 SH imaging of membrane hydration. **a** Two counterpropagating 190-fs beams overlap in space and time to illuminate the membrane. SH photons are collected in the phase-matched direction (see Fig. S3 for details). A white-light image with diffraction rings (indicated by a dashed white circle) and a current response $I(t)$ to a 20-mV bias $[U(t)]$ are also shown. **b** SH images (20 frames, 560 ms each; magnification: 50 ×, NA = 0.65) of a symmetric membrane composed of (1) 75:25 mol % DOPC:Chol, and (2) 37.5:37.5:25 mol % DPPC:DOPC:Chol, and an asymmetric membrane composed of 37.5:37.5:25 mol % DPPS; DOPC:Chol (top leaflet) and 37.5:37.5:25 mol % DPPC:DOPC:Chol (bottom leaflet) (3) single frame, and (4) average of 20 frames. *Inset* in *B* shows the SH response of the Teflon film where the aperture that contains the bilayer is indicated by the white dashed circle [38]. Published with permission of PNAS article containing the requested material published under a CC BY-NC-ND license

reactive flux correlation analysis, Gibbs energy of activation of the hydrogen-bond breaking was calculated. Thus, the networks in the hydration layer of membranes dynamically facilitate the water-mediated lipid–lipid associations.

Imaged membrane-bound oriented water on millisecond timescales was investigated by means of label-free second harmonic (SH) microscopy [38]. As hydrating water molecules are oriented by the ionic groups in the lipid head groups of charged lipids through charge–dipole interactions, they can emit SH photons. Such an approach makes it possible to track, membrane domain diffusion of condensed charged lipid domains, domain structure, and the spatial distribution of charge were tracked on millisecond timescales. The spatiotemporal fluctuation in the membrane potential is large and reveals the importance of charge fluctuations on membrane.

The second-order optical response of water was employed to measure the membrane potential and changes, therein as a function of an external bias. Individual images showed dynamic spatiotemporal fluctuations on the order of 100 mV. The images of membranes composed of (1) 75:25 mol % DOPC:Chol, and (2) 37.5:37.5:25 mol % DPPC:DOPC:hol, and an asymmetric membrane composed of 37.5:37.5:25 mol % DPPS; DOPC:Chol and 37.5:37.5:25 mol % DPPC:DOPC:Chol

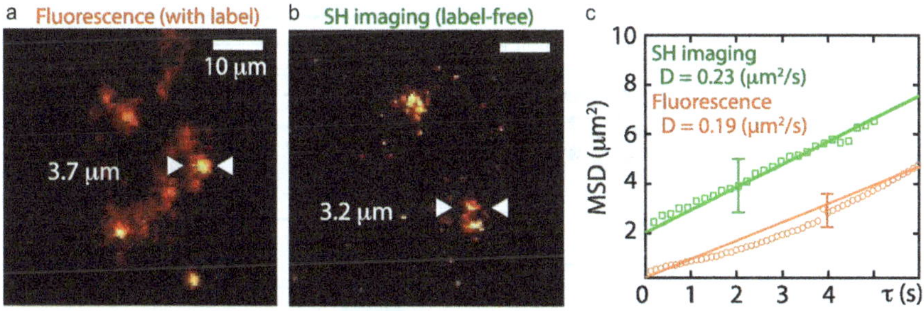

Fig. 10.25 SH imaging of water around liquid-ordered domains of charged lipids. **a** Fluorescence image of an asymmetric membrane composed of 37.5:37.5:25 mol % DPPS/DOPC/Chol + TopFluor (top) and 37.5:37.5:25 mol % DPPC/DOPC/Chol (bottom). **b** SH image of a membrane made with the same composition and asymmetry but now label-free. (Scale bar, 10 μm.) The arrows in *A* and *B* point to the positions where the line profiles were obtained to extract the domain size. Both images show phase separation, leading to DPPS and Chol-rich domains of similar size. **c** MSD as a function of delay time (τ) comparing label-free SH imaging with TopFluor labeled imaging. The calculated diffusion coefficients are comparable. SH scattering images of a 100-μM KCl solution were subtracted from the raw SH images. Symbols are experimental data and lines are linear fits. More data recorded on longer timescales are shown in Fig. S4 [38]

presented in Fig. 10.25 were collected with all beams P-polarized. The presence of a bilayer inside a ~80- to 120-μm-sized circular aperture in a 25-μm-thick Teflon film was confirmed with white-light imaging and electrical recordings (Fig. 10.24a).

The following images are depicted in Fig. 10.24 (1)~3-μm-diameter domains of a liquid-ordered DPPS-rich phase; (2) mean square displacement (MSD) traces, with a diffusion constant of $D = 0.19 \pm 0.01$ μm^2; (3) an SH image recorded with 180-ms acquisition time of a membrane with the same composition, following the hydrating water around the charged head groups of DPPS, label-free; and (4) domains with a similar size and diffusion speed ($D = 0.23 \pm 0.06$ μm^2/s). The spatiotemporal fluctuation in the membrane potential on the order of 100 mV revealed the importance of charge fluctuations on membranes.

The microstructure and collective dynamics of the membrane interfacial hydration shell in zwitterionic, and negatively charged phospholipid membrane bilayers was examined exploring terahertz time-domain spectroscopy (THz-TDS) [9]. Large unilamellar vesicles (LUVs) of ~100 nm diameter of DOPC, DOPC/DOPG (80/20, mol/mol), or DOPC/DOPG (60/40, mol/mol) were utilized. A commercially available THz spectrophotometer, TERA K8 was used for THz-TDS measurements of 0.4–1.2 Thz. A dependence of the critical lipid concentration corresponding to the inflection point on the charge of the lipid head group was observed. The critical lipid concentration corresponding to the inflection point was found to depend on the charge of the lipid head group (Fig. 10.26), thereby implicating membrane electrostatics as a major governing factor in the microstructure and dynamics of water at the membrane interface.

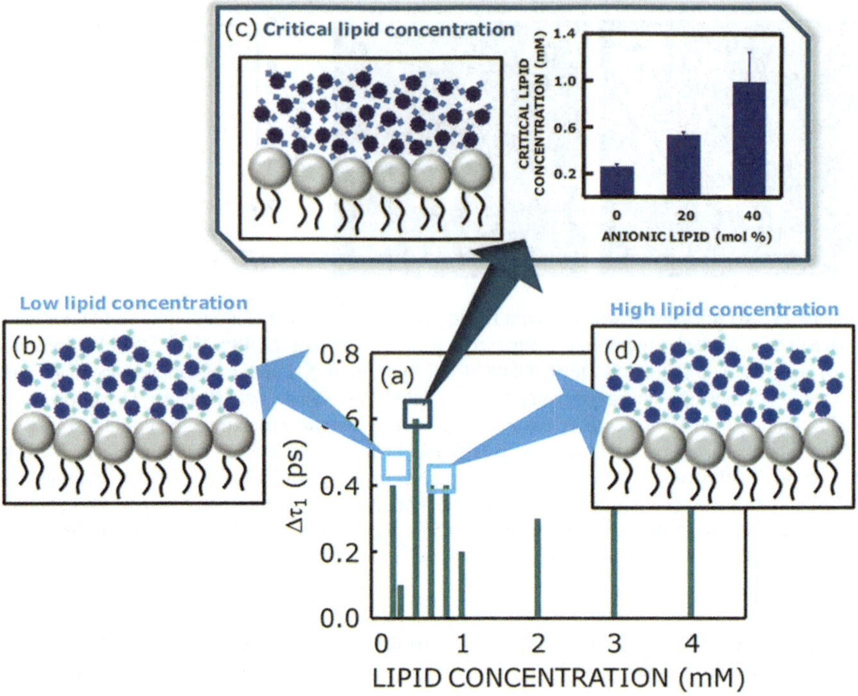

Fig. 10.26 Schematic representation of the effect of phospholipid head group charge on the microstructure and dynamics of water at the membrane interface. Panel a shows a representative "rise and dip" trend of relaxation time constants with lipid concentration. Membrane interfacial water molecules with lower relaxation time constants (and faster hydrogen-bond reorganization) are shown in light blue (panels b and d), and those with higher relaxation time constants (and slower hydrogen-bond reorganization) are shown in dark blue (panel c). The relaxation time constant is maximum at the critical lipid concentration (Ct I) and decreases both before and after Ct [9]. With permission from American Chemical Society

The frequency-dependent absorption coefficient ($\alpha(\nu)$) and refractive index ($n(\nu)$) of the sample, encoding in the changes in amplitude and phase of the transmitted THz pulse, respectively, were determined from the frequency-dependent power and phase of the transmitted pulse. The decrease in the real and imaginary dielectric constants of water molecules in the presence of zwitterionic (DOPC) and negatively charged (DOPC/DOPG) lipid vesicles caused increasing frequency of the incident THz pulse. This effect \was suggested to be associated with the dielectric relaxation response of water, reflecting the reorientation dynamics associated with the permanent dipoles of water molecules. The change in the relaxation time was related to the first (τ_1) and second (τ_2) dielectric relaxation modes with lipid concentration for zwitterionic and negatively charged lipid vesicles. Thus, the cooperative relaxation time constants of the water hydrogen-bond network show a unique "rise and dip" variation with lipid concentration and the modulation of the dielectric relaxation response of membrane interfacial water by lipid composition in a concentration-dependent manner.

Correlated vibrational motion on the subpicosecond timescale and associated collective dynamics of the peripheral membrane protein annexin B12 (Anx) (Figs. 10.27 and 10.28) in an environment was characterized using molecular dynamics simulations [39]. For the membrane, a lipid bilayer consisting of a 7: 3 mixture of **1,2-di-(9Z-octadecenoyl)-*sn*-glycero-3-phosphocholine** (DOPC) and 1,2-dioleoyl-sn-glycero-3-phospho-L-serine (DOPS) lipids with a total of 512 lipids (256 lipids in each leaflet) was utilized. X-ray crystal structure of Anx PDB ID code 1DM5)34 and pre-equilibrated lipid bilayer was placed in a rectangular box with a size of approximately 135 × 135 × 1573 containing ca. 64 000 water molecules. The net charge of the system was neutralized with Na^+ and Cl^- ions.

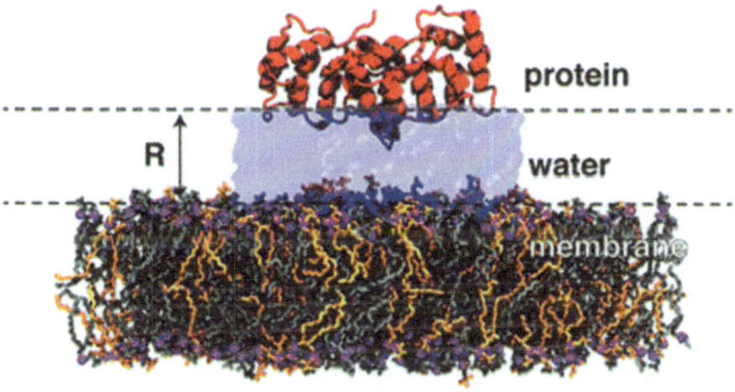

Fig. 10.27 Annexin B12 on a phospholipid bilayer. Starting structures for MD simulations were generated with Anx (red) positioned at various distances R from the bilayer (DOPC: gray, DOPS: orange). **a** Protein–lipid distanceR = 0 (Anx bound to the lipid bilayer). **b** Protein–lipid distance R = 30 (interfacial water between protein and membrane shown in dark blue).??????//////////// **c** and **d** Surface non-hydrogen atoms selected for the cross-correlation analysis. Anx atoms were selected on the membrane-facing side (red) and membrane atoms were selected in a rectangular area of similar size as the protein cross section (white) [39]. With permission from Royal Society of Chemistry

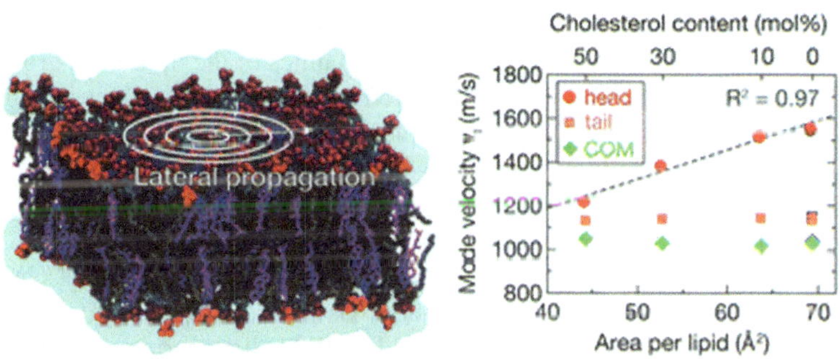

Fig. 10.28 (A **10.27 0** (Pictural illustration of the lateral propagation of the lipid head groups collective dynamics [40]. With permission from Royal Society of Chemistry

Correlated motion of a membrane-associated protein and a lipid bilayer for distinct separation distances was considered. The following subjects were in focus of the computation: (1) protein–membrane cross-correlation spectra I(o.R) for different, separation distances R between the molecular surfaces; (2) cross-correlation spectra of mass-weighted atomic velocities and localized velocity densities for vibrations of non-hydrogen protein or membrane surface atoms with hydration water oxygen atoms; (3) wave propagation velocities of longitudinal, collective protein–water and membrane–water (gray) modes for distinct protein–membrane separation distances R; (4) hydrogen-bond lifetimes tHB of water, varying in the range 3–10 ps at distance 3–20 Å; (5) time-domain representation of the longitudinal mass-weighted velocity cross-correlation functions (TCCF) between protein–membrane, protein–water, and membrane–water atoms for distinct separation distances; and (6) peak intensity of mass-weighted velocity time cross-correlation functions between protein–membrane atoms [39].

The water molecules were found to be located in between the protein and membrane (Fig. 10.27). The observed propagation velocities of the analyzed collective modes v_{mode} are increased significantly in both cases for separation distances R < 25 Å between protein and membrane. The data demonstrated also a sharp modulation of the mode velocities in a narrow distance range (between R = 25–28 Å) corresponding to the thickness of a single hydration shell. The computations revealed that the intensities of membrane–water correlated vibrations are weaker than those for the protein and water because a lipid bilayer can possess a large variety of collective dynamics spanning timescales from 10^{-4}–10^{-10} s. The results also indicated the hydration water-mediated collective motion between biomolecular interfaces over length scales of > 10 Å. Correlated vibrations persist up to distances of 25 Å between both biomolecular surfaces and are mediated by separating layers of water molecules. The simultaneous presence of protein and lipid bilayer interfaces altered these collective properties.

Collective vibrational motion on picosecond timescales in dioleoylphosphatidylcholine lipid bilayers with varying cholesterol content was examined using all-atom molecular dynamics simulation [40]. Cholesterol was found laterally compacted the lipid bilayer and changes the velocity of longitudinal density fluctuations propagating in the plane of the membrane (Fig. 10.28). Interestingly, cholesterol-induced reduction of the area per lipid alters the collective dynamics of the lipid head groups, but not of the lipid tails. In addition, the introduction of cholesterol reduces the number of water molecules interacting with the lipid head groups, leading to a decrease in the velocity of the laterally propagating sound mode. Thus, the stiffening effect of cholesterol is found to be indirect, that is, decreasing the area per lipid weakens the interactions between the lipid head groups and water.

Stratum corneum (SC) lipids serve as a barrier against exogenous chemicals and microbes. In study [41] MD simulation has been performed on a membrane containing an equimolar mixture of ceramide[NP]24, lignoceric acid, and cholesterol. The simulation revealed that the membrane lipids form a hexagonally packed gel phase, hydration of the head group region promoted the lateral motion, and water formed multiple hydrogen bonds with the lipids and the majority of water molecules were clustered together to form pools. In addition, the simulation showed that water molecules outside of the pools were less mobile and stuck in an anhydrous region making it difficult to jump to the head group region of the opposing lipid layer. In this condition, the diffusion of water is slowed down by one or two orders of magnitude compared with the bulk water. Repeat distance and neutron scattering length density calculated from the MD simulation were in good agreement with the experimentally determined values when the bilayer was hydrated by 2 H_2O per lipid.

Various aspects of the water dynamics in hydrated membranes have ben discussed in [42–45].

References

1. M. Shinitzky, Biomembranes: structural and functional aspects. in *Molecular Cell Biology*. 4th edn (VCH Verlagsgesellschaft mbH, Weinheim, 1994)
2. H. Lodish, A. Berk, S.L. Zipursky et al., *Molecular Cell Biology*, 4th edn. (W. H. Freeman, New York, 2000)
3. Q.-X. Jiang, *New Techniques for Studying Biomembranes* (Imprint CRC Press, Location Boca Raton, 2020)
4. M.S.P Sansom, P.C. Biggi, *Molecular Simulations and Biomembranes: From Biophysics to Function* (RSC 2010)
5. E.A. Disalvo, F. Lairion, F. Martini, E. Tymczyszyn, M. Frías, H. Almaleck G.J. Gordillo, Structural and functional properties of hydration and confined water in membrane interfaces. Biochimica et Biophysica Acta (BBA)—Biomemb. **1778,** 2655–267 (2008)
6. E.A. Disalvo, A.M. Bouchet, M.A. Frias, Connected and isolated CH_2 populations in acyl chains and its relation to pockets of confined water in lipid membranes as observed by FTIR spectrometry. Biochimica et Biophysica Acta (BBA)-Biomemb. **1828**, 1683–1689, (2013)
7. C. Päslack, L.V. Schäfer, M. Heyden atomistic characterization of collective protein–water–membrane dynamics. Phys. Chem. Chem. Phys. **21**, 15958–15965 (2019)
8. Z. Zhang, Y. Chen, X. Tang, J. Li, L. Wang, J. Yang, Solid-state NMR shows that dynamically different domains of membrane proteins have different hydration dependence. J. Phys. Chem. B **118**, 9553–9564 (2014)
9. S. Pal, N. Samanta, D. Das Mahanta, R.K. Mitra, A. Chattopadhyay, Effect of phospholipid headgroup charge on the structure and dynamics of water at the membrane interface: a terahertz spectroscopic study. J. Phys. Chem. B **122,** 5066–5074 (2018)
10. D.K. George, A. Charkhesht, O.A. Hull, A. Mishra, D.G.S. Capelluto, K.R. Mitchell-Koch, N.Q. Vinh, New insights into the dynamics of zwitterionic micelles and their hydration waters by gigahertz-to-terahertz dielectric spectroscopy. J. Phys. Chem. B **120**, 10757–10767 (2016)
11. N. Kučerka, J. Gallovám, D. Uhríková, The membrane structure and function affected by water. Chem. Phys. Lipid. **221**, 140–144 (2019)
12. D. Laage, T. Elsaesser, J.T. Hynes, Water dynamics in the hydration shells of biomolecules. Chem. Rev. **117,** 10694–10725 (2017)
13. M. Hishida, K. Tanaka, Long-range hydration effect of lipid membrane studied by terahertz time-domain spectroscopy. Phys. Rev. Lett. **106,** 158102 (2011)
14. Y. von Hansen, S. Gekle, R.R. Netz, Anomalous anisotropic diffusion dynamics of hydration water at lipid membranes. Phys. Rev. Lett. **111,** 118103 (2013)
15. D.G. Osborne, J.A. Dunbar, J.G. Lapping, A.M. White, K.J. Kubarych, Site-specific measurements of lipid membrane interfacial water dynamics with multidimensional infrared spectroscopy. J. Phys. Chem. B **117**, 15407–15414 (2013)

16. L.M. Alarcón, M. de los Angeles Frías, M.A. Morini, M. B. Sierra, G.A Appignanesi, E.A. Disalvo, Water populations in restricted environments of lipid membrane interphases. European Phys. J. E, **39**, 94 (2016)
17. Z. Cao, J. Hutchison, C.R. Sanders, J.U. Bowie, Backbone hydrogen bond strengths can vary widely in transmembrane helices. J. American Chem. Soc. **139**(31) (2017). https://doi.org/10.1021/jacs.7b04819
18. E.A Disalvo, (Ed.), Membrane hydration. in *The Role of Water in the Structure and Function of Biological Membranes* (Springer, 2015)
19. S.J. Singer, G.L. Nicolson, The fluid mosaic model of the structure of cell membranes. Science **175**, 720–731 (1972)
20. G.L. Nicolson, Transmembrane control of the receptors on normal and tumor cells. I. Cytoplasmic influence over cell surface components. Biochim. Biophys. Acta **457,** 57–108 (1976)
21. G. Nicolson, The fluid—mosaic model of membrane structure: still relevant to understanding the structure, function and dynamics of biological membranes after more than 40 years. Biochem. Biophys. Acta. **1838**, 1451–1466 (2014)
22. S. Damodaran, Water at biological phase boundaries: its role in interfacial activation of enzymes and metabolic pathways. Subcell. Biochem. **71**, 233–261 (2015)
23. E. Disalvo, (Eds) *Membrane Hydration. Subcellular Biochemistry*, vol. 71. (Springer, 2015)
24. A. Mashaghi, P. Partovi-Azar, T. Jadidi, N. Nafari, P. Maass, M. Reza Rahimi Tabar, M. Bonn, H.J. Bakker, Hydration strongly affects the molecular and electronic structure of membrane phospholipids. J. Chem. Phys**. 136,** 114709 (2012)
25. T.J. Zwang, W.R. Fletcher, T.J. Lane, M.S. Johal, Quantification of the layer of hydration of a supported lipid bilayer. Langmuir **26**, 4598–4601 (2010)
26. F. Fogila, M.J. Lawrence, C.D. Lorenz, S.E. McLain, On the hydration of the phosphocholine headgroup in aqueous solution. J. Chem. Phy. 133, Article ID 145103, (2010)
27. E.A. Disalvo, M.F. Martini, A.M. Bouchet, A. Hollmann, M.A. Frías, Structural and thermodynamic properties of water-membrane interphases: significance for peptide/membrane interactions. Adv. Coll. Interface. Sci. **211**, 17–33 (2014)
28. R. Tsenkova, Z. Kovacs, Y. Kubota, Aquaphotomics: near infrared spectroscopy and water states in biological systems.in *Membrane Hydration. Subcellular Biochemistry*, ed by Disalvo E. (eds) vol 71 (Springer, Cham 2015), pp. 189–211
29. Z. Arsov, Long-range lipid-water interaction as observed by ATR-FTIR spectroscopy. in *Membrane Hydration. Subcellular Biochemistr*y, ed by Disalvo E. (ed) vol 71 (Springer Cham. 2015), pp. 127–159
30. C. Choe, J. Lademann, M.E. Darvin, Depth profiles of hydrogen bound water molecule types and their relation to lipid and protein interaction in the human stratum corneum in vivo. Analyst **14**, 6329–6337 (2016)
31. L. Malacrida, S. Astrada, A. Briva, M. Bollati-Fogolín, E. Gratton, L.A. Bagatolli, Spectral phasor analysis of LAURDAN fluorescence in live A549 lung cells to study the hydration and time evolution of intracellular lamellar body-like structures. Biochimica et Biophysica Acta (BBA)-Biomemb. **1858,** 2625–2635
32. T. Parasassi, E. Gratton, W.M. Yu, P. Wilson, M. Levi, Two-photon fluorescence microscopy of laurdan generalized polarization domains in model and natural membranes Biophys. J. **72**, 2413–2429 (1997)
33. T. Kondela, J. Gallová, T. Hauß, J. Barnoud, S.J. Marrink, N. Kučerka, Alcohol interactions with lipid bilayers. Molecule **28**, 2078 (2017)
34. J. Whittaker, L. Delle Site, Investigation of the hydration shell of a membrane in an open system molecular dynamics simulation. Phys. Rev. Res. **1,** 033099 (2019)
35. D.J. Tobias, N. Sengupta, Mounir Tarek, Hydration dynamics of purple membranes. Faraday Discuss. **141**, 99–116 (2009)
36. L. Bagatolli, Monitoring Membrane Hydration with 2-(Dimethylamino)-6-Acylnaphtalenes Fluorescent Probes. in *Membrane Hydration. Subcellular Biochemistry*, ed by Disalvo E. (eds) vol 71 (Springer, Cham, 2015)

37. A. Srivastava, A. Debnath, Hydration dynamics of a lipid membrane: Hydrogen bond networks and lipid-lipid associations .J. Chem. Phys. **148,** 094901 (2018)
38. O.B. Tarun, C. Hannesschläger, P. Pohl, S. Roke, Label-free and charge-sensitive dynamic imaging of lipid membrane hydration on millisecond time scales. Proc Natl Acad Sci U S A. **115**, 4081–4086 (2018)
39. C. Paslack, L.V. Schafer, M. Heyden, Atomistic characterization of collective protein–water–membrane dynamics. Phys. Chem. Chem. Phys. **21,** 15958 (2019)
40. C. Päslack, J.C. Smith, M. Heyden, L.V. Schäfer, Hydration-mediated stiffening of collective membrane dynamics by cholesterol. Phys. Chem. Chem. Phys. **21**, 10370–10376 (2019)
41. S. Han, Effect of hydration on a lipid membrane composed of ceramide[NP]24, lignoceric acid, and cholesterol: a molecular dynamics simulation study. Bul. Korean Chem. Soc. **40**, 1215–1221 (2019)
42. M.A. Hemminga, L.J. Berliner, *ESR Spectroscopy in Membrane Biophysics* (Springer, New York, 2007)
43. A.N. Smith, U.T. Twahir, T. Dubroca, G.E. Fanucci, J.R. Long, Molecular rationale for improved dynamic nuclear polarization of biomembranes. J. Phys. Chem. B **120**, 7880–7888 (2016)
44. J.M. Franck, S. Chandrasekaran, B. Dzikovski, C.R. Dunnam, J.H. Freed, Focus: two-dimensional electron–electron double resonance and motions: the challenge of higher frequencies. J. Chem. Phys. **142**, 212302 (2015). (Review)
45. G.I. Likhtenshtein, J. Yamauchi, S. Nakatsuji, A. Smirnov, R. Tamura, *Nitroxides: Application in Chemistry, Biomedicine, and Materials Science* (WILEY-VCH, Weinhem, 2008)

Chapter 11
Hydration of Cells and Tissues

Abstract Biological cells consist of cytoplasm, which contains many biomolecules such as proteins and nucleic acids, enclosed within a membrane. Tissues are composed of anywhere from 20 to 80% water by volume depending on the type of tissue and the location within the tissue. The chapter describes the molecular status of water in cancer, red blood cells, *Escherichia coli*, *Haloarcula marismortui* and miscellaneous cells. Picosecond vibrational spectroscopy, dielectric-relaxation spectroscopy, confocal Raman microspectroscopy and other techniques were employed to observe the random orientational motion of water molecules inside living cells. Separate sections are dedicated to the water inside mammalian, food, and plants tissues.

11.1 Introduction

Biochemical, biophysical, and biomedical research at the cellular level is deemed important, as cells are considered the basic units of structure and function in forming organisms [1–6]. Water is indisputably a conserved structural element of proteins, essential to their folding, specificity, ligand binding, enzyme catalysis, and nucleic acid function. Cell cytoplasm is approximately 80% water and typically contains up to 400 g L^{-1} of macromolecules, which may occupy 5–40% of the total volume of the cell. As a result the cell is extremely crowded (macromolecules are typically separated by only 1–2 nm).

A vast literature exists on structure and thermal motions of hydration water in cells and tissues, which can involve diffusion phenomena of translational or rotational nature [1–22]. The large majority of investigations by different methods have shown that water diffusion in a hydration shell extending over one or two molecular layers is significantly hindered by interactions with a biomolecular surface, as a result of which the diffusion coefficient is in 2–10 smaller than in bulk water. As illustrated in Fig. 11.1, bulk-like (represented in green and pink) and confined (represented in red and blue) water populations coexist inside and outside the cells. Additionally, the culture medium surrounding the cells contains macromolecules that may also lead to the presence of confined water. The cell membrane is highly permeable for water

G. I. Likhtenshtein, *Biological Water*, Soft and Biological Matter,
https://doi.org/10.1007/978-3-030-82503-4_11

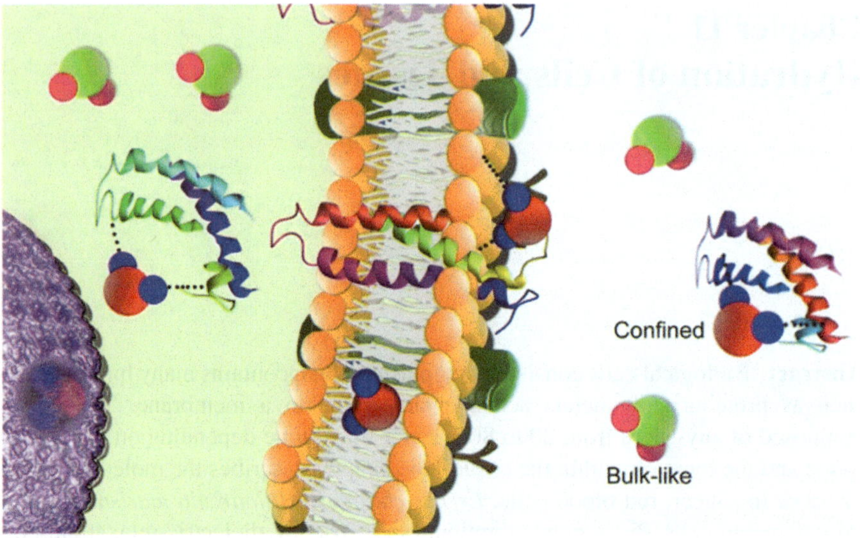

Fig. 11.1 Representation of the coexistence of bulk-like and confined water populations inside and outside the cells. Weakly interacting water molecules show bulk-like behavior (represented in green and pink) and have a diffusion coefficient of about 3.9^{-10} m^2/s. Water molecules, confined by folded proteins, cell membranes, organelles, and others, exhibit restricted mobility (represented in red and blue) [7]

because of the existence of special water channels (aqua pores) which makes cell hydration a dynamic and determines its functional activity. Regulatory role of cell hydration can be realized by changing the hydration of intracellular macromolecules (folding–unfolding mechanisms) and by changing the number of surface-dependent functionally active protein molecules in membrane having receptors, enzymatic, and channel forming properties.

The aquaporins (AQPs) are a family of small membrane-spanning proteins (monomer size approximately 30 kDa) that are expressed at plasma membranes in many cells types involved in fluid transport (Fig. 11.2) [5, 6]. Aquaporins form pores in the membrane of biological cells, facilitating transport of water between cells. The cell membranes of a variety of different bacteria, fungi, animal, and plant cells contain aquaporins through which water can flow more rapidly into and out of the cell than by diffusing through the phospholipid bilayer. The aquaporins can be regulated by factors that affect channel permeability or subcellular localization.

Tissues are composed of anywhere from 20 to 80% water by volume depending on the type of tissue and the location within the tissue. For instance, muscle typically maintains a constant hydration of ~75% throughout the thickness of the tissue, while skin can vary from 20% at the stratum corneum up to 66% in the deep dermis. In adult women, the water-content represents between 50 and 70% of the mass in normal breast tissues and this percentage is increased within diseased tissues.

Fig. 11.2 Aquaporin structure. https://www.ks.uiuc.edu/Research/aquaporins/

Water molecules play an essential role in the structural organization and dynamics of biological tissues such as breast and brain.

11.2 Water in Cells

11.2.1 Cancer Cells

The hydration states of cancer cells and tissues may vary considerably from their healthy counterparts. The suggested concept [23] that cell hydration is the primary factor in the mechanism of carcinogenesis includes the following hypotheses: (1) increased cell hydration causes cancer not only by promoting cell division and oncogene expression, but also by inactivating genes inducing cell differentiation, and by preventing apoptosis; (2) a progressive increase in cell hydration, induced by successive mutations and/or epigenetic changes, is the basic mechanism of multistep carcinogenesis, the degree of malignancy increasing with the degree of cell hydration; (3) the increased hydration of cancer cells accelerates their respiration rate, thereby enhancing their ability to compete for nutrients with their normal counterparts, and (4) increased cell hydration is an alternative or additional explanation of the carcinogenetic effect of inflammatory agents and of hormones

Terahertz time-domain attenuated total reflection spectroscopy (THz TD-ATR), in combination with a two-interface model, was utilized to determine the complex dielectric constants of cultured human cancer cells (a colon carcinoma cell DLD-1, an embryo kidney cell HEK293, and a fatal cervical carcinoma cell (HeLa)) [20]. Water dynamics of cancer was characterized by dielectric responses below 1.0 THz. The human malignant epithelial cells were cultured in a liquid culture medium on the ATR silicone crystal prism (kept at 310 K), with a constant 5% CO_2 atmosphere supply to the incubation chamber. In order to determine the complex dielectric constants of living human cells, a cell incubation chamber was attached onto the ATR prism of aTHz TD-ATR spectrometer, TAS7500 (ADVANTEST Co.). THz TD-ATR spectroscopy was employed to determine the complex dielectric constant in the 0.2–4.0 THz region of distilled water and water in cells since the dielectric responses in the THz frequencies dominate by picosecond and subpicosecond timescale water dynamics.

Combined with the extended theory of Onsager, it was found that 23.8% of HeLa intracellular water was hydrated to biomolecules, exhibiting slower relaxation dynamics than bulk water. To characterize, the complex dielectric constants were decomposed into their constituents (the slow relaxation, fast relaxation, and intermolecular stretching vibration modes). It was found that dielectric responses below 1.0 THz best characterize the particular water dynamics of cancer cells When compared with extracellular water, dielectric responses below 1.0 THz were best characterize the particular water dynamics of cancer. Debye-Lorentz fitting revealed that this is due to a significantly attenuated slow relaxation mode and enhanced fast relaxation mode of the water in these cancer cells.

As indicated in Fig. 11.3, the decomposed complex dielectric constant of the HeLa cells fitted the experimental results and the substantial decreases in slow relaxation and stretching vibration modes and increase in the higher frequency limit were observed. The dielectric constant of human cells (2.0) was found to be larger than that of distilled water (1.77) in the visible region. The fast relaxation process was associated with non-hydrogen bonding structure transiently appearing in the hydrogen-bond network. This result can indicate that the hydrogen bonds between water molecules in human cancer cells are more fragile than those in distilled water. On the other hand, the real part above 1.0 THz is dominated by the intermolecular stretching vibration mode and the higher frequency limit.

A neutron scattering descriptive study of water dynamics in MCF-7 breastcancer cells has been reported [7]. The results provide the first step to untangle the role played by water dynamics in breast cancer cells (MCF-7) after treatment with a chemotherapy drug (anticancer drug paclitaxel (PTX)). The cells were propagated at 37 °C in Dulbecco's Modified Eagles Medium (DMEM) supplemented with 10% fetal calf serum, 2 mM glutamine, 5000 IU/mL penicillin, and streptomycin in a humidified atmosphere containing 5% CO_2. The thermal behavior of TC (treated with the drag) and NTC (non-treated with the drag) was evaluated by differential scanning calorimetry (DSC) using a Netzsch DSC 214 Polyma calorimeter. The vibrational dynamics of TC and NTC were investigated by means of inelastic neutron scattering (INS). The experiments were conducted at the TOSCA neutron spectrometer at ISIS, UK26. Quasi-elastic neutron scattering (QENS) measurements were performed at the Backscattering Spectrometer BASIS, at the ORNL's Spallation Neutron Source (SNS), USA27. By thermal analyses, the cells were probed as micrometric reservoirs of bulk-like and confined water populations. By thermal analyses, the cells were probed as micrometric reservoirs of bulk-like and confined water populations. The results suggested that the drugs alter the properties of the confined water. This was validated by accessing the cellular water dynamics using inelastic neutron scattering.

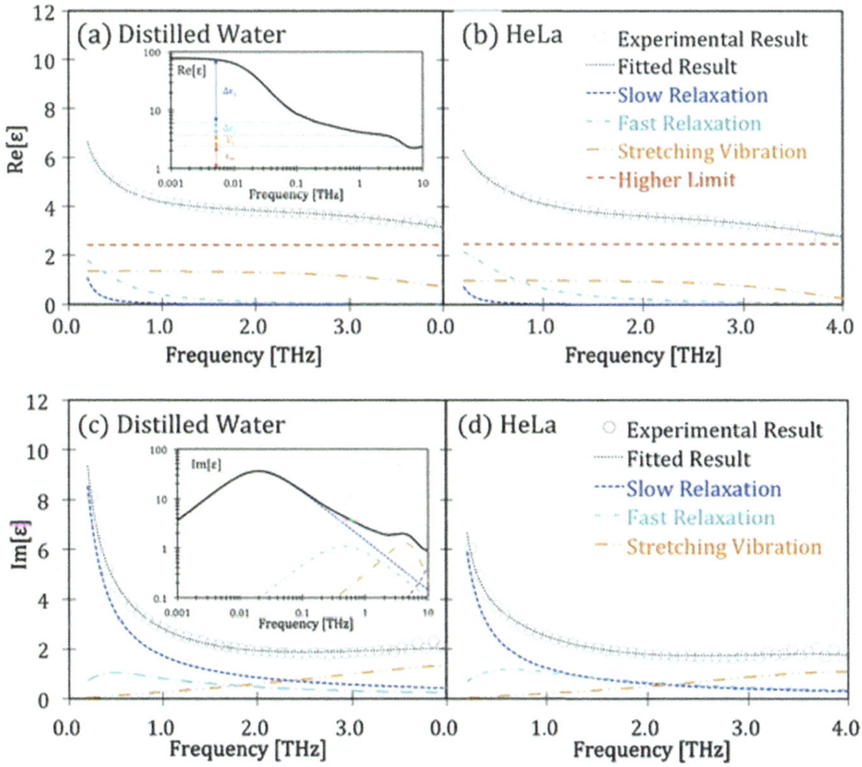

Fig. 11.3 Real part of **a** distilled water and **b** HeLa, and imaginary part of **c** distilled water and **d** HeLa. The fitted result is well accorded with the experimental result. Superposition of the decomposed components (slow relaxation, fast relaxation, intermolecular stretching vibration, and higher frequency limit) is equal to the fitted result. The inset shows the overhead view of the complex dielectric constant of distilled water between 0.001 and 10 THz [20]. With permission from American Institute of Physics

Specifically, to investigate how morphological changes due *to* (PTX) which action cause perturbations within cellular water, the cells were studied by means of DSC and QENS. The results are given in Figs. 11.4a,b and 11.5, together with data on pure water (bulk) that was also analyzed.

The research main results can be summarized as the follows: (1) thermal analysis reveals that after interaction with the antitumor drug paclitaxel, the treated cancer cells (TC) present thermal behavior further away from that of bulk water than not treated cells (NTC); (2) by comparing the water's melting profiles, as observed by differential scanning calorimetry, it was possible to establish that the disturbance of water hydrogen-bond network increases from the NTC to the TC; (3) using inelastic neutron spectroscopy (INS), a direct probe of atomic and molecular motions, it was demonstrated that this disturbance is related to the manner in which confined water molecules are structurally organized and interact in the cellular environment; and

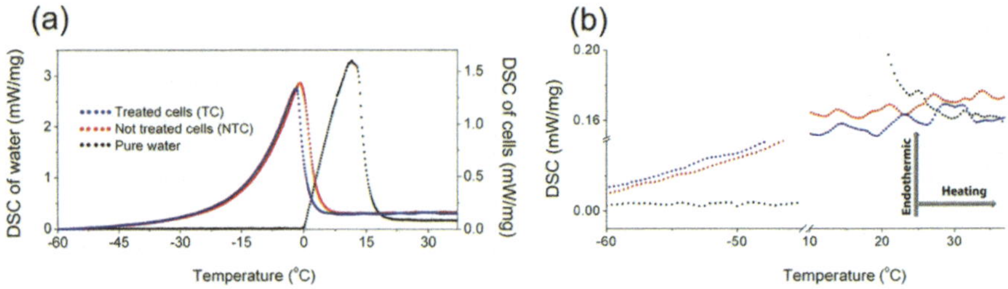

Fig. 11.4 Differential scanning calorimetry (DSC) performed on bulk water (black circles) and on breast cancer cells (MCF-7) not treated (NTC—red circles) and treated (TC—blue circles) with 15 nM of paclitaxel for 24 h. **a** Respective melting peaks and **b** specific heats (baselines) before and after melting. The samples masses were approximately 25 mg. The presence of diffusive motions in the TC is further confirmed and quantitatively discussed by the analysis of the evolution of the QENS broadenings (half-width at half-maximum, HWHM) vs Q2 presented in Fig. 11.6. Here, the HWHM values were obtained by performing a nonlinear least square fitting to the QENS signal at different Q-values. Figure 11.6 also shows a theoretical curve depicting the expected behavior for purely bulk water at room temperature (black dotted line), obtained by considering the jump diffusion model [6]

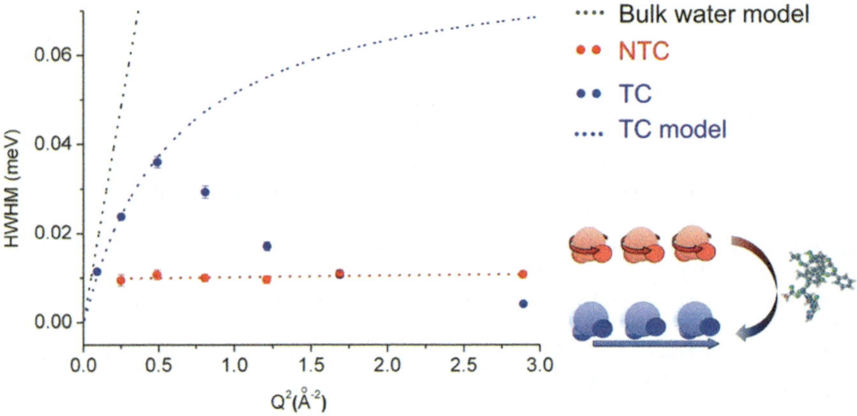

Fig. 11.5 Evolution of QENS broadening (half-width at half maximum—HWHM), determined from fits and attributed to the water molecule motion in breast cancer cells (MCF-7), as a function of Q2. The symbols represent the values obtained when the HWHM was allowed to vary independently at each Q-values, while the error bars represent the standard deviation. Note that some of the error bars are within the dots. The black dotted line represents a theoretical curve for bulk water obtained from the jump diffusion model. The same model was used to obtain the blue dotted curve, which describes the behavior of the TC. The red dotted line represents a linear fit that describes the HWHM behavior in the NTC. The data collected at Q2 = 2.252– were not used due to the presence of a Bragg reflection. In the bottom right corner, the scheme illustrates the appearance of translational motions in the cells after treatment with PTX [7]

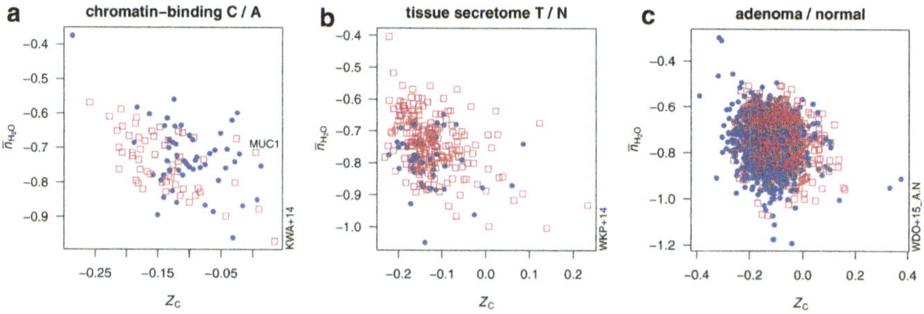

Fig. 11.6 Average oxidation state of carbon (ZC) and water demand per residue (nH_2O) for proteins in selected datasets. Open red squares represent proteins enriched in tumors or more advanced cancer stages, and filled blue circles represent proteins enriched in normal tissue or less advanced cancer stages [21]. Published under a Creative Commons Attribution License (CC BY)

(4) the contribution from the bulk-like water population to the change in the water dynamics in MCF-7 cells after treatment with PTX is depicted by the quasi-elastic neutron scattering (QENS) results; and indicated that PTX increases diffusional mobility in the intracellular bulk-like water population.

The purpose of study [21] within a chemical and thermodynamic framework using variables represents oxidation and hydration state. The average oxidation state of carbon (Z_C) was calculated as an elemental ratio from the chemical formulas of proteins, and water demand per residue (nH_2O) was computed by writing the overall formation reactions of proteins from basis species. Using results reported in proteomic studies of **clinical samples**, many datasets exhibit higher mean Z_C or nH_2O of proteins in carcinoma or adenoma compared to normal tissue. The value of ZC indicates that MUC1 is a relatively highly oxidized protein, while its nH_2O places it near the median water demand for up-expressed proteins in cancer in this dataset (Fig. 11.6).

Among 35 proteomic datasets considered many have significantly higher values of average Z_C in proteins up-expressed in adenoma or carcinoma compared to normal tissue. The theoretically computed transitions in relative stabilities between proteins from cancer and healthy tissues occur at much lower values of the water activity aH_2O (ca. 10^{-6}). The author concluded that a systematic change in chemical composition is an essential feature of cancer proteomes, and that the thermodynamic descriptions showed proteomic transformations in host tissue which could be promoted by relatively high microenvironmental oxidation and hydration states.

The study [24] focused on a molecular picture of water molecules at the biological interface, or specifically, water molecules adjacent to the hydrophobic and hydrophilic surfaces of normal and cancerous tissues. The IR and Raman spectra of the ν_s(OH) stretching modes of water at the biological interfaced of the human breast and neck tissues were recorded and analyzed. The results revealed significant changes in the water content in the tissue, showing that the interfacial water interacting via H-bond with other water molecules and biomolecules at the biological surface and free OH vibration of the dangling water are sensitive indicators (a "hydration fingerprint") of the pathology between the normal (non-cancerous) and cancerous tissue and cancer types. A greater hydration state of cancerous tissue, resulting from interaction of water molecules with hydrophilic cellular structures of cancer cells was evaluated. Medical imaging by microwave detection of differences in dielectric constant also revealed greater water content in malignant tissue of breast cancer [22].

Changes in the dynamical activity of intracellular water between healthy and cancerous human cells were probed as an innovative approach for unveiling particular features of malignancy and identifying specific reporters of cancer [25].

Androgen-unresponsive prostate, triple-negative breast carcinomas, and osteosarcoma (a cancerous tumor in a bone) were studied, using the technique of quasi-elastic neutron scattering (QENS). The following human cell lines were studied—cancer vs their non-tumorigenic counterparts: (i) triple-negative (metastatic) breast cancer (MDA-MB-231) and non-neoplastic mammary gland immortalized cells (MCF-12A)—hereafter named breast cancer and breast healthy, respectively and (ii) androgen-independent prostate adenocarcinoma (PC-3) and normal prostate epithelium immortalized cells (PNT-2)—hereafter named prostate cancer and prostate healthy, respectively. The QENS experiments were performed at the ISIS Pulsed Neutron and Muon Source of the Rutherford Appleton Laboratory, in the low-energy OSIRIS high-flux indirect-geometry time-of-flight spectrometer [12, 35], which allows to probe water dynamics at picosecond timescales and on atomic lengths. As an example, Fig. 11.7 depicts the Q-dependence of the full-width at half-maximum (FWHM $= \Gamma$) of the Lorentzian functions representing the cellular dynamical constituents, for the prostate cells presently studied (cancer and healthy). As it is seen in the figure, the cytoplasmic and hydration water dynamics displays a Q-dependent behavior, that is, the faster localized motions exhibit a Q-independent profile and cytoplasmic water is more flexible than the highly organized hydration layers, its dynamics being\represented by a non-diffusive jump reorientation model.

The QENS profiles obtained for of tumor and non-tumor cellular models evidenced a higher flexibility of the former relative to the healthy ones and this difference being much more significant for the breast cells as compared to that for the prostate ones (Fig. 11.8a vs. Fig. 11.8b). Bone cancer displayed a remarkably lower overall flexibility followed by metastatic breast and prostate cancers (Fig. 11.8c).

The main results of the work can be summarized as follows: (1) the cancerous cells showed a considerably higher plasticity relative to their healthy counterparts, this being more significant for the mammary adenocarcinoma; (2) prostate cancer cells display the highest plasticity when compared to triple-negative mammary cancer and osteosarcoma, the latter being remarkably less flexible; (3) there are differences between the flexibility of different types of intracellular water molecules in normal and cancerous cells, as well as the number of molecules involved in the different modes of motion; and (4) the dynamics of hydration water molecules remain virtually unaffected when going from healthy to cancer cells, while cytoplasmic water (particularly the rotational motions) undergoes significant changes upon normal-to-cancer transition. Thus, the present results evidenced changes in the respective intracellular water dynamics, which can be regarded as a specific reporter of the

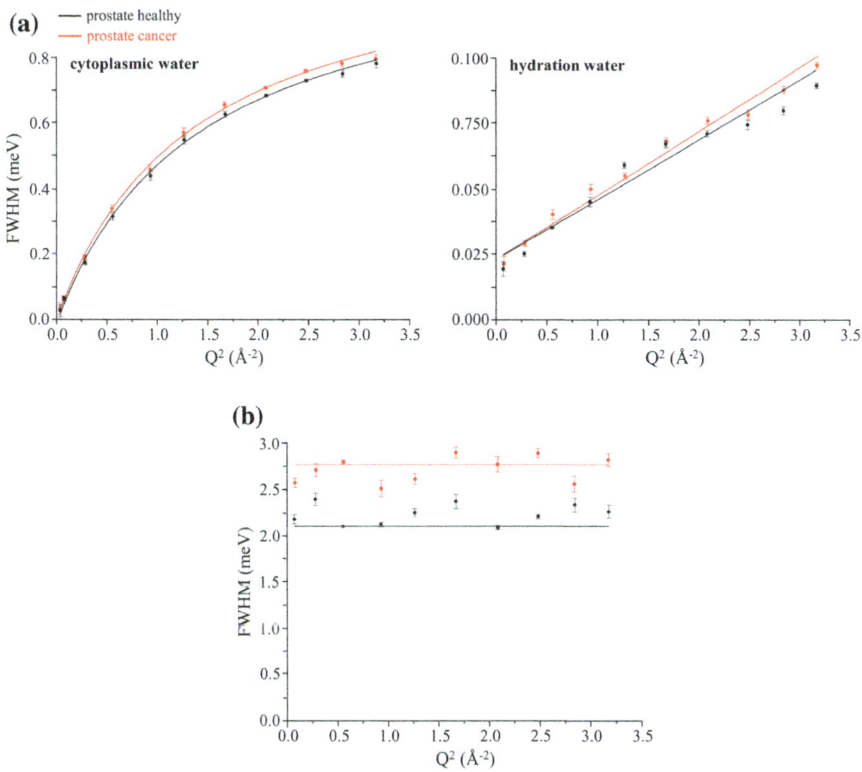

Fig. 11.7 Variation of the full widths at half-maximum (FWHM) with Q^2 for human prostate cancer and prostate healthy cells (at 310 K): **a** Lorentzian functions representing the translational motions of intracellular water—cytoplasmic medium and hydration layers (Γ_{global}); and **b** Lorentzian function representing the internal localized motions within the cell (Γ_{local}) [25]. With permission from American Institute of Physics

cellular state and allowed to differentiate the several dynamical components within these systems and to characterize them according to the nature of the tissue and the type of cancer.

The study [24] focused on a molecular picture of water molecules at the biological interface, or specifically, on water molecules adjustment to the hydrophobic and hydrophilic surfaces of normal and cancerous tissues. The IR and Raman spectra of the ν_s(OH) stretching modes of water at the biological interfaces of the human breast and neck tissues were recorded and analyzed. The results revealed significant changes in the water content in the tissue, showing that the interfacial water interacting via H-bond with other water molecules and biomolecules at the biological surface and free OH vibration of the dangling water are sensitive indicators (a "hydration fingerprint") of the pathology between the normal (non cancerous) and cancerous tissue and cancer types. A greater hydration state of cancerous tissue, resulting from interaction of water molecules with hydrophilic cellular structures of cancer cells was also obtained.

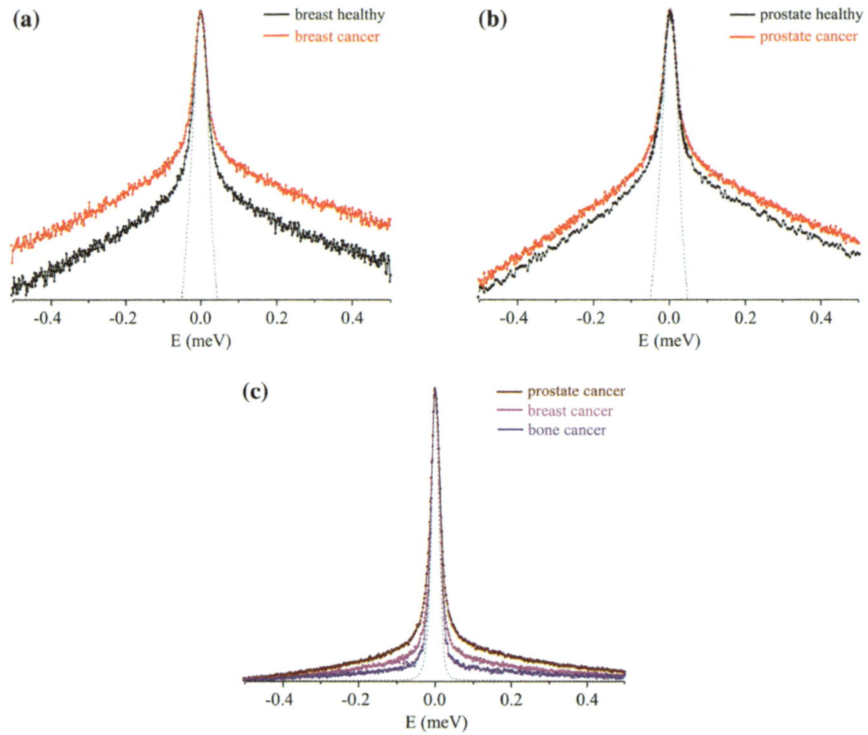

Fig. 11.8 QENS profiles (310 K, at Q1.684 A°_1) for cancer and healthy human cells: **a** breast cancer and breast healthy, **b** prostate cancer and prostate healthy, and **c** all cancer cells studied—breast, prostate, and bone (the spectra were normalized to the maximum peak intensity. The dashed line represents the instrument resolution, as measured by a standard vanadium sample. The QENS profiles **a, b** are represented in the yy_logarithmic scale [25]

Stimulated Raman excited fluorescence (SREF) microscopy is the most sensitive vibrational imaging in the optical far field was reported. In work [13] SREF was developed into a technique which allows to directly visualize spatially resolved distribution of water states inside single mammalian cells by coupling it with vibrational solvatochromism. Stimulated Raman scattering (SRS)24 microscopy, imaging speed, and sensitivity were significantly improved compared to conventional Raman using coherent anti-Stokes Raman scattering (CARS). HeLa (ATCC CCL-2) cells were purchased from ATCC. Synchronized pump and Stokes beams were provided by a picoEmerald S system from Applied Physics & Electronic, Inc. (APE). The fundamental IR fiber laser at 1031.2 nm with 2-ps pulse width and 80-MHz repetition rate was served as the Stokes beam. The pump beam was provided by incorporation part of fundamental laser to OPO which is tunable from 700 to 990 nm. Electric field was estimated by Onsager reaction field theory.

Effectiveness of the SREF technique is illustrated by Fig. 11.9. By sweeping the pump wavelength, a series of SREF images of Rh800-stained HeLa cells (Fig. 11.9a)

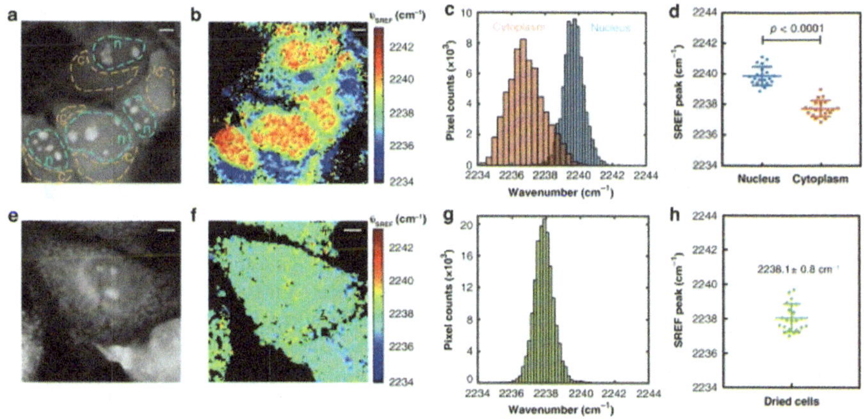

Fig. 11.9 Mapping water heterogeneity in live cells. **a** A representative intensity-based SREF image of Rh800 stained living HeLa cells (pump = 836 nm). **b** Heterogeneous map of nitrile peak frequency. **c** Nitrile peak frequency histogram of two regions (n for nucleus and c for cytoplasm, marked in c with dash line) in live HeLa cells. **d** Nitriles peak frequency distribution (mean ± s.d.) of nucleus (2239.9 ± 0.6 cm^{-1}) and cytoplasm (2237.7 ± 0.5 cm^{-1}) among 21 HeLa cells. $p < 0.0001$ for two-sided paired t test, $n = 21$. **e** A representative SREF image (pump = 836 nm) of ethanol-dried HeLa cells stained by Rh800. **f** Fitted frequency map shows the heterogeneities between nucleus and cytoplasm disappear for dried cells. Nitrile peak frequency histogram of **g** indicating a single-group distribution. **h** Nitrile peak frequency distribution of 23 dried HeLa cells (mean ± s.d., 2238.1 ± 0.8 cm^{-1}). Scale bar: 5 μm. Error bars represent standard deviations [13]

was registered. After fitting peak frequency, a ν¯C≡Nν¯C≡N distribution can then be generated (Fig. 11.9b,c). This spatial heterogeneity can be consistently observed in many cells (Fig. 11.9d). As a negative control to test the water effect, the ν¯C≡Nν¯C≡N distribution map in dried HeLa cells reveals a relatively homogeneous pattern (Fig. 11.9e, f). The measured heterogeneity of SREF vibrational frequency can be potentially affected by three factors: the water content variation, the water states variation, and the electrostatics variation.

Figure 11.10a depicts label-free SRS imaging of the O–H stretching mode, which has higher spatial resolution and is free from sample autofluorescence. A consistent result was obtained in the SRS imaging: nucleus exhibits about 2% higher water concentration than cytoplasm (Fig. 11.10b, c). 2–3% water content difference between nucleus and cytoplasm can cause 0.2–0.3 cm^{-1} frequency shift for ν¯C=Nν¯C≡N, given the full dynamic range of 7.5 cm^{-1} in the calibration (Fig. 11.10).

In summary: (1) by coupling the local hydrogen-bond sensing ability of nitrile vibrational probe with the high sensitivity and resolution of the emerging SREF microscopy, a promising imaging tool to spatially map biological water states in living cell was developed; (2) the subcellular heterogeneity in live cells as observation of two distinct groups of biological water states between cytoplasm and nucleus was qualitatively identified; and (3) intracellular map of the water pool compositions regarding free water vs. bound water was composed. The results supported

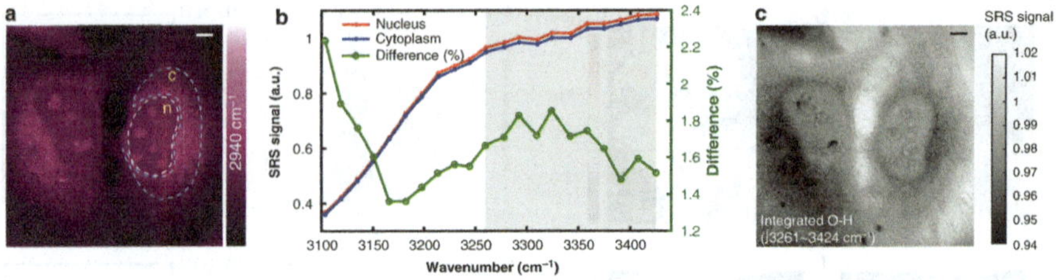

Fig. 11.10 Water density inside live HeLa cells measured by SRS microscopy. **a** SRS image at frequency of 2940 cm^{-1} (CH_3 stretching attributed mainly to protein) of live HeLa cells. **b** Spectral mapping of O–H stretching band at the nucleus and cytoplasm regions marked in **a** (*n* stands for nucleus and *c* stands for cytoplasm). The difference curve (shown on the right in percentage) is calculated as (nucleus−cytoplasm)/nucleus. **c** SRS image on integrated O–H stretching band over 3261–3424 cm^{-1} (the gray area in **b**, which is the spectral region that differs the most). Scale bar: 5 μm. The color scales display relative SRS signal level [13]

the concept of biological water, revealed intracellular water heterogeneity between nucleus and cytoplasm, and unveiled a compositional map of water pool inside living cells.

The density and hydrogen-bonding networks of water in each HeLa cellular compartment were investigated using Raman images of the O−H stretching band of water HeLa, an immortal cell line from Lacks's cancerous cervical tumor [26]. This technique allowed direct observation of a spatial distribution of water molecules inside of a living cell. HeLa cells were plated in a glass-bottomed dish. Raman measurements were performed utilizing an inverted multiconfocal Raman microscope at an excitation wavelength of 532 nm. Figure 11.11 quantitatively compares. The shapes of the average Raman spectra of the nucleus and cytoplasm of HeLa cells with that of Hanks's balanced salt solution (HBSS) medium were quantitatively compared in Fig. 11.1. The figure shows that the shape of the O–H stretching band

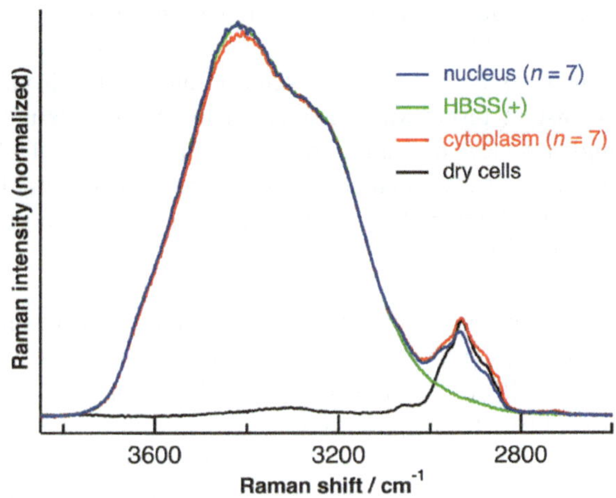

Fig. 11.11 Average Raman spectra of HeLa cells in the nucleus (blue) and in the cytoplasm (red), together with those of the HBSS medium where cells were not observed (green) and dried HeLa cells (black). The background of each spectrum was fitted with a polynomial function and was subtracted as a baseline [26]. With permission from American Chemical Society

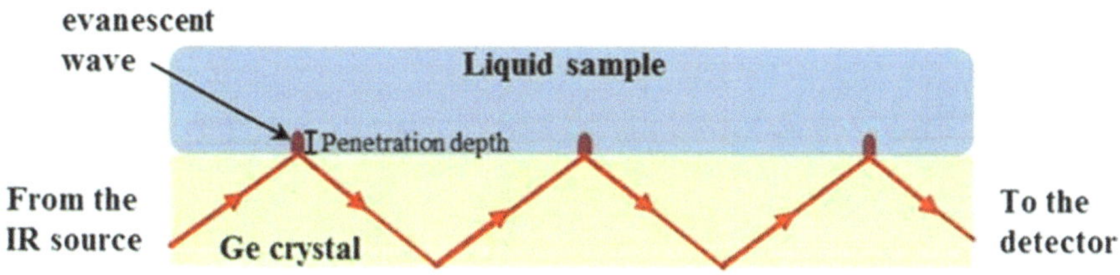

Fig. 11.12 An illustration of the setup of an ATR element with a sample [27]

in the nucleus differs from that in the cytoplasm, and the O–H band in the nucleus is similar to that of surrounding HBSS indicating that hydrogen-bonding networks of water in the nucleus are similar to that in the medium. The images indicate that the water density in the nucleus is higher than that in the cytoplasm. The environment was found to be less crowded and the distortion of hydrogen-bonding networks due to the presence of hydrophobic molecules is less dominant in the nucleus as compared with that in the cytoplasm.

Attenuated Total Reflection Fourier Transform IR (ATR-FTIR) spectroscopy was utilized to determine the metastatic potential of two murine melanoma cells and two human melanoma cells [27]. Measuring the spectral properties of the cells allowed to determine the membrane hydration levels. B16-F1 and B16-F10 cells were generated from the parental murine melanoma B16 line by isolation of lung metastases cells following their injection into C57 mice. FTIR measurements were carried out employing a FTIR spectrometer (Thermo Scientific, iS 50) equipped with a Ge ATR device (Thermo Scientific, Smart ARKTM, Fig. 11.12). Using the ATR-FTIR spectroscopy, it was identified spectral alterations that are a result of hydration or molecular changes. These findings revealed that higher metastatic potential correlates with membrane hydration level.

The difference in the hydration level reflects on the ratio of high-density water HDW to low-density LDW. The analysis revealed that the average areal ratio for B16-F1 cells is 35.2 and for the B16-F10 cells is 21.4 meaning that B16-F1 cells have more LDW than B16-F10. In addition, the average areal ratio for WM-115 cells is 54 and for the WM-266.4 cells is 21, meaning that WM-115 cells have more LDW than WM-266.4 (Fig. 11.13).

Developed method allowed to spectrally analyze the hydration level of the plasma membrane of two variants of the murine B16 cells and two variants of human melanoma and to evaluate their metastatic potential accordingly. Firstly, there is a higher absorption intensity of amide II for the cells with the higher metastatic potential, which can be explained by the higher hydration level of the plasma membrane. Second, the cells with the higher the metastatic potential had more HDW molecules found at the cell membrane's vicinity. Therefore, using the ATR-FTIR spectroscopy to measure the interaction of the cell membrane with water, it is possible to distinguish between different malignant stags of human and mouse melanoma cells and this technology has the potential to help in cancer metastasis prognosis.

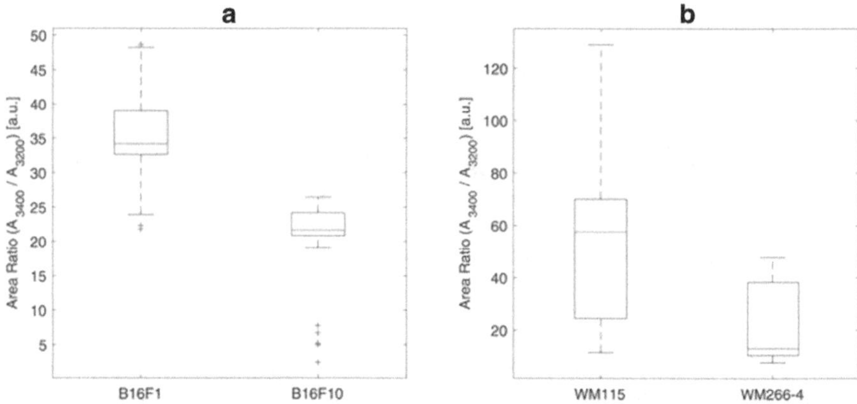

Fig. 11.13 A boxplot illustration of the distribution of the measured values of the ratio between the area of the fitted Gaussian centered around 3400 cm^{-1} and that of the Gaussian centered around 3200 cm^{-1}, i.e., the peaks of the HDW and LDW, respectively. The results for B16-F1 and B16-F10 cells are shown in **a** and the results for WM-115 and WM-266.4 cells in **b**. The areal ratio measurements summarized in the plot were done during the saturation stage [27]

11.2.2 Red Blood Cells

Blood cells comprise white blood cells (WBCs), red blood cells (RBCs), and platelets, the numbers of which vary according to physiological and pathological conditions [28]. RBC consists mostly of H_2O and hemoglobin at a concentration of ~330 mg/mL. Each human red blood cell contains approximately 270 million of these hemoglobin molecules. Plasma is composed mostly of water and contains dissolved proteins, lipids, glucose, and hormones. The various components dissolved in the plasma control the body's immunological and biochemical. Hydration status is critical for erythrocyte survival and is mainly determined by intracellular cation content [29–34]. Ion content, hemoglobin concentration, ion transport pathways, membrane surface area, and complex interactions between solutes, their net charge, conductance of the membrane to monovalent anions, and cations all contribute to erythrocyte hydration status. Primary disorders of erythrocytes are classified by the perturbation of intracellular water content (overhydration or dehydration), which reflects the abnormality in membrane permeability to sodium and potassium [32]. These disorders perturb volume homeostasis, jeopardize the erythrocyte, and may lead to its premature destruction and can be marked by clinical, laboratory, and physiologic heterogeneity. Secondary disorders of erythrocyte hydration include sickle cell disease, thalassemia, hemoglobin CC, and hereditary spherocytosis, where cellular dehydration may be a significant contributor to disease pathology and clinical complications.

The erythrocyte contains a network of pathways that regulate salt and water content in the face of extracellular and intracellular osmotic perturbations [32]. This allows the erythrocyte to maintain a narrow range of cell hemoglobin concentration, a process critical for normal red blood cell function and survival. Primary erythrocyte's disorders are classified by the perturbation of intracellular water content (i.e., over-hydration or dehydration), which reflects the abnormality in membrane permeability to sodium and potassium. Secondary disorders of erythrocyte hydration include sickle cell disease, thalassemia, hemoglobin CC, and hereditary spherocytosis, where cellular dehydration may be a significant contributor to disease pathology and clinical complications. For example, hereditary stomatocytosis describes a number of inherited autosomal dominant human conditions which affect the red blood cell, in which the membrane or outer coating of the cell "leaks" sodium and potassium ions [29]. Dehydration is a secondary effect of several erythrocyte disorders, including SCD, thalassemia, hemoglobin C, hereditary spherocytosis (HS), malaria invasion, and secondary perturbation of erythrocyte water content may be a major contributor to disease pathophysiology. After shrinkage due to dehydration, non-mammalian erythrocytes, but not human erythrocytes, respond with a process occurs. This process is known as regulatory volume increase (RVI), first described in non-mammalian erythrocytes, which is mediated primarily by Na^+/H^+ exchange, Na^+-K^+-$2Cl^-$ cotransport, and Na^+ channels. Decreased volume caused by solute loss is essentially an irreversible process in erythrocytes, and as they age, red blood cells lose volume and become denser.

In most inherited red blood cell (RBC) disorders with high gene frequencies in malaria-endemic regions, the distribution of RBC hydration states is much wider than normal [30]. Invasion of *P. falciparum* (unicellular protozoan parasite of humans) was assessed in cultures by microscopic inspection of Giemsa-stained thin blood smears and parasite counting and in density-separated fractions of normal RBCs and of RBCs with different genetic abnormalities. The relationship between the hydration state of circulating RBCs and protection against severe *falciparum malaria* were examined. Five different strains of *P. falciparum* (A4,15 W2, and FCR3 in study I; FCR3K_ and ITgC32p21 in study II) were cultured in human erythrocytes by standard methods under a low-oxygen atmosphere. It was shown that swollen and mildly dehydrated RBCs were able to sustain *P. falciparum* cultures with similar efficiency to untreated RBCs. In a parallel study, *P. falciparum* invasion was investigated in density-fractionated RBCs from healthy subjects and from individuals with inherited RBC abnormalities affecting primarily hemoglobin (Hb) or the RBC membrane (thalassemias, hereditary ovalocytosis, xerocytosis, Hb CC, and Hb CS). These results suggested that the presence of dense RBCs is a protective factor and invasion efficiency is reduced in dehydrated RBCs and completely inhibited in profoundly dehydrated RBCs.

Cytoplasmic water and hydration layer dynamics in human red blood cells measured by quasi-elastic incoherent neutron scattering (QENS) in vivo in the temperature range between 290 and 320 K were reported [31]. Samples of human venous blood were prepared in H_2O and D_2O HEPES buffer. Extracellular water content was less than 10% of the total water content. Energy-resolved neutron scattering was measured on the spectrometers IRIS at ISIS, FOCUS at PSI, and TOFTOF at FRM-II. The dynamics of water in human red blood cells was examined with quasi-elastic incoherent neutron scattering in the temperature range between 290 and 320 K. Neutron spectrometers experiments with time resolutions of 40, 13, and

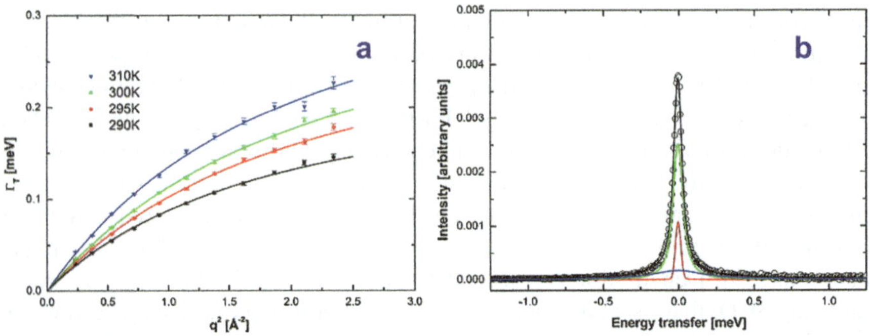

Fig. 11.14 **a** Half widths at half-maximum of the translational diffusion process of cytoplasmic H_2O in RBC measured on IRIS at different temperatures. Solid lines represent fits according to a jump-diffusion model. **b** QENS spectrum on FOCUS of cytoplasmic H_2O in RBC at 290 K and $q = 0.55$ Å^{-1}. Symbols show measured data; the black line presents the result of the fit. The components correspond to the immobile fraction (red line), the narrow Lorentzian (green line), and the broad Lorentzian (blue line) [31]. With permission from American Chemical Society

7 ps combined to cover timescales of bulk water dynamics to reduced mobility interfacial water motions were performed. The half widths at half-maximum (HWHM) of cytoplasmic H_2O measured on IRIS were found to be with fits according to a jump-diffusion model (Fig. 11.14).

Figure 11.15 reports the obtained translational diffusion coefficients D of H_2O buffer and cytoplasmic H_2O as a function of temperature measured with IRIS, FOCUS, and TOFTOF. Experimental data indicated that: (1) all movements up to an ~10 Å real space distance, e.g., length scale confining effects of protein surface cavities or boundaries on water become observable; (2) root mean square displacements are 2.1–2.3 Å which are smaller than the mean distance between water molecules (~3 Å) and emphasize the localized nature of immobile water with correlation times is longer than ~40 ps; (3) activation energy of the immobile fraction of H_2O in RBC $Ea = 2.15$ kcal/mol; (4) ~2300 water molecules are in the first hydration layer of hemoglobin; and (5) residence times ι_0 of intracellular H_2O in RBC were estimated by different techniques as IRIS (17μeV) 2.53 ps, FOCUS (50 μeV) 1.12 ps and TOFTOF (100 μeV) 1.68 ps.

It was concluded that there exist two populations of water in RBC: a major fraction of ~90% which has translational properties similar to those of bulk water and a fraction of ~10% of immobile water which could be identified as dynamically bound water in the hydration layer of hemoglobin. This bound hydration water can have important properties concerning protein stability and interactions in living cells.

Preservation of cellular hydration is a dynamic process [32]. Ion content, hemoglobin concentration, ion transport pathways, membrane surface area, and complex interactions between solutes, their net charge, conductance of the membrane to monovalent anions, and cations all contribute to erythrocyte hydration status. Several pathways mediate water and solute homeostasis in erythrocytes, where

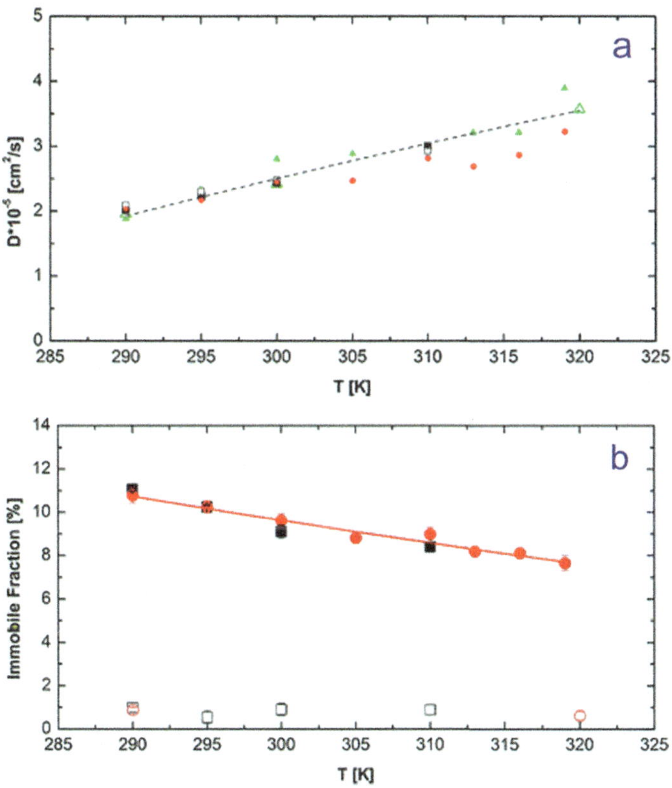

Fig. 11.15 **a** Translational diffusion coefficient D of H_2O in **RBC** measured on the instruments IRIS (filled black squares), FOCUS (filled red circles), and TOFTOF (filled green triangles); H_2O buffer was measured on the instruments IRIS (empty black squares) and TOFTOF (empty green triangles). The dashed line indicates normal temperature behavior of H_2O buffer. **b** Percentage of immobile fraction of H_2O in RBC measured on IRIS at $q = 0.61$ Å^{-1} (filled black squares) and FOCUS at $q = 0.55$ Å^{-1} (filled red circles) is compared to values of H_2O buffer measured on IRIS (empty black squares) and FOCUS (empty red circles). The straight red line represents Arrhenius behavior [31]. With permission from American Chemical Society

cellular volume is primarily controlled via regulation of monovalent cation content (Fig. 11.16).

Hydration status is critical for erythrocyte survival and is mainly determined by intracellular cation content [34]. Active pumps, passive transporters, and ion channels are the key components of volume homeostasis, whereas water passively fits ionic movements. Inappropriate cation leak causes erythrocyte hydration disorders, hemolytic anemia, and characteristic red cell shape abnormalities named stomatocytosis. All types of stomatocytosis either overhydrated or dehydrated are linked to inherited or de novo mutations in genes encoding ion transporters or channels.

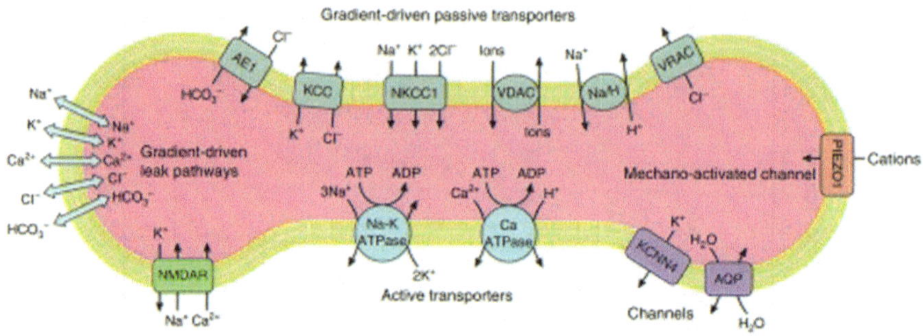

Fig. 11.16 Model of ion transport pathways of the human erythrocyte. Ion transport pathways of the human erythrocyte are shown. Gradient-driven transport pathways include AE1: the anion exchange protein 1, band 3, encoded by *SLC4A1*; *NKCC1* encoding Na-K-Cl cotransporter 1; KCC: K-Cl cotransporters of the family of chloride-cation cotransporters; VDAC: voltage-dependent anion channels; *NHE1* encoding the sodium/hydrogen exchanger 1; and VRAC: voltage-regulated anion channels, encoded in part by SWELL1 (*LRRC8A*). Gradient-driven leak pathways, not otherwise identified and indicated by the arrows, also exist in mature erythrocytes. PIEZO1 is a mechano-activated channel of the erythrocyte membrane. Other erythrocyte membrane channels are the aquaporins, water channels encoded by *AQP1* and *AQP3*, and *KCNN4* encoding the Gardos channel, a calcium-activated potassium channel. Active transporters are the calcium ATPase, encoded by *ATP2B4*, and sodium–potassium ATPase, the multicomponent sodium pump composed of $\alpha1$ and $\alpha3$ isoforms, $\beta1$, $\beta2$, and $\beta3$ isoforms, and the γ modulator. NMDAR, the multicomponent *N*-methyl D-aspartate receptor participating in calcium regulation in erythroid cells, is also shown. Other transport pathways (e.g., CO_2, amino acids, glucose, urea, etc.) are not shown [32]. With permission from American Society of Hematology

11.2.3 Escherichia Coli and Haloarcula Marismortui

Escherichia coli (*E. coli*) is a Gram-negative, facultative anaerobic, rod-shaped bacterium that is commonly found in the lower intestine of warm-blooded organisms. *Haloarcula marismortui* is a halophilic red Archaeon found in the Dead Sea. In *E. coli* and *H. marismortui* cells, the longitudinal relaxation rate R_1 of the water—^{2}H magnetization over 5 orders of magnitude in frequency were measured by using the fast field-cycling (FC) technique with conventional superconducting magnets [12]. *E. coli* strain K-12 RV308 [American Type Culture Collection (ATCC) 31608] and *H. marismortui* (ATCC 43,049) cultured in D_2O were obtained from Deutsche Sammlung von Mikroorganismen und Zellkulturen (DSMZ). The longitudinal relaxation rate R_1 of the water–^{2}H magnetization was measured from 1.5 kHz to 76.8 MHz by utilizing six different NMR instruments. The relaxation data, acquired in a wide magnetic field range (0.2 mT–12 T) and analyzed in a model-independent way, revealed water dynamics on a wide range of timescales. It was found that 85% of cell water in *E. coli* and in the extreme halophile *H. marismortui* has bulk-like dynamics. The 15% of cell water interacts directly with biomolecular surfaces and is motionally retarded by a factor 15 on average, corresponding to a rotational correlation time of 27 ps. The relaxation data also showed that a small fraction (0.1%) of cell water exchanges from buried hydration sites on the microsecond timescale.

Figure 11.17 displays the water—^{2}HMRD profiles from living *E. coli* and *H. marismortui* cells. The following interpretation of the data was suggested: (1) at a given frequency ω_0, R_1 is associated to samples water molecules with correlation times shorter than $1/\omega_0$; (2) the highest accessed frequency w_{0*} corresponds to a correlation time $\tau = 1/w_0 = 2$ ns; (3) the increase in R_1 in the kHz–MHz frequency range constitutes direct model-independent evidence for water dynamics on the 0.1–10 μs timescale reflecting rotational motions of surface hydration and bulk water; (4) at the high-frequency w_0, spin relaxation is induced by local water rotation slowed down to varying degrees by interactions with macromolecular surfaces; (5) at low frequencies, spin relaxation is generated by exchange-mediated orientational randomization (EMOR) of internal water; and 6) in the range 1–100 MHz, the (weaker) frequency dependence of R_1 is produced by orientational randomization of long-lived (residence time 1 ns).

The fraction of rotationally immobilized protein was determined from the principal quadrupolar peak in the ^{1}HMRD profiles from samples identical to those used for the ^{2}H MRD experiments (Fig. 11.18). The quadrupolar peaks are caused by magnetization transfer from water protons to ^{14}N in immobilized peptide NH groups at the ^{1}H Larmor frequencies where the ^{14}N quadrupolar energy splitting is matched.

It was summarized that ≈ 85% of cell water in *E. coli* and in the extreme halophile ***H. marismortui*** has bulk-like dynamics, ≈ 15% of cell water interacts directly with biomolecular surfaces and is motionally retarded by a factor 15 on average, corresponding to a rotational correlation time of 27 ps. A small fraction (≈ 0.1%) of cell water exchanges from buried hydration sites on the microsecond timescale.

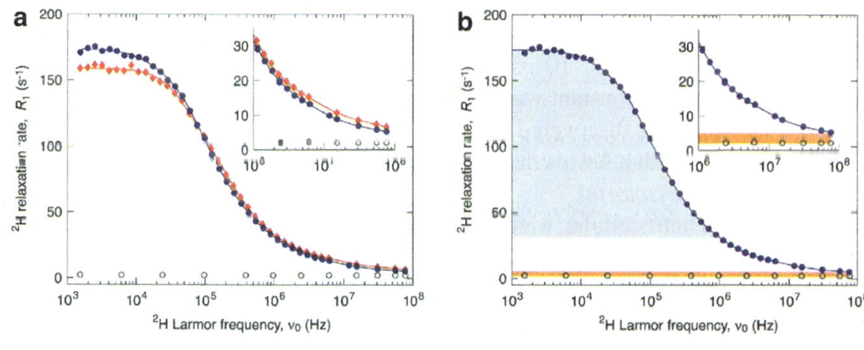

Fig. 11.17 2H MRD profiles of cell water. **a** The water–2H relaxation rate $R1$ was measured on In *E. coli* cells (blue circles) and *H. marismortui* cells (red diamonds) in stationary growth phase at 27 °C and pD 8.0. Control measurements were performed in parallel on a bulk D_2O reference sample (open circles). The curves are multi-Lorentzian numerical representations used for the model-free analysis. **b** The colored areas represent, for the *E. coli* sample, the contributions to $R1$ from surface hydration water with rotational correlation time_2 ns (yellow) and from internal water molecules with residence time_160 ns (blue). In both images, *Inset* shows the high-frequency region on an expanded scale [12]. Published with permission of PNAS article containing the requested material published under a CC BY-NC-ND license

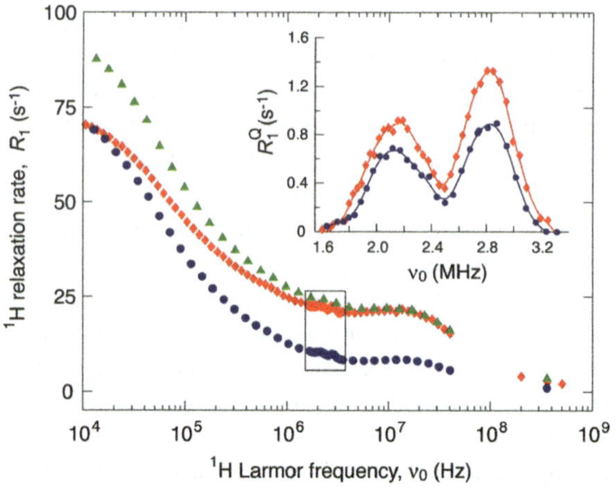

Fig. 11.18 ^{1}HMRDprofiles and quadrupolar peaks. The water–^{1}Hrelaxation rate $R1$ was measured on *E. coli* cells at 27 °C (blue circles) and *H. marismortui* cells at 12 °C (green triangles) or 27 °C (red diamonds). The samples were prepared as in Fig. 11.18, but with H_2O at pH 7.6. The shoulder at 10–30 MHz is due to paramagnetic ions. *Inset* shows the quadrupolar peaks on an expanded scale for *E. coli* (blue circles) and *H. marismortui* (red diamonds) at 27 °C. The plotted quantity $R_1{}^Q$ was obtained by subtracting the baseline MRD profile. The curves serve only to guide the eye [12]. Published with permission of PNAS article containing the requested material published under a CC BY-NC-ND license

Intracellular water dynamics in fully deuterated ***H. marismortui*** cell pellets, an extremely halophilic organism originally isolated from the Dead Sea, was studied by neutron scattering by means of two spectrometers, IN6 and IN16, sensitive to motions with timescales of 10 ps and 1 ns, respectively [35]. From IN6 data, a translational diffusion constant was found to be close to that for bulk water. A very slow water component discovered from the IN16 data displays a residence time of 411 ps (compared with a few ps in bulk water). This slow water accounts for 76% of cell water in *H. marismortui*.

The nature of intracellular water dynamics in deuterated *E. coli* (BLE21(DE3) strain) was investigated employing neutron measurements [36]. The full-width at half-maximum energy resolution of a neutron spectrometer defined the upper limit of the motions timescale. The following values of translational and rotational parameters of water in the *E. coli* cells at (281 K) were reported: $D_T(10^{-5}\ cm^2\ s^{-1}) = 1.53$, τ^0 (ps) = 2.63, and $\tau_{cor,R}$ (ps) = 1.96. These values are close to those for pure water.

Experimental study on the observation by neutron Brillouin spectroscopy of THz collective modes propagating in *E. coli* intracellular water was performed in [37]. A fermenter culture of fully deuterated *E. coli* bacteria was grown at the Deuteration Laboratory of the Institut Laue-Langevin (ILL). The BRISP spectrometer was used with an incident wavelength of 0.9885 °A and a sample-to-detector distance of 4.5 m. The true scattering intensity I(Q; ω) of the *E. coli* cells was obtained from the raw data by following standard reduction procedures. Results indicated that the corresponding damping factors of the dispersion curves concerning the high-frequency

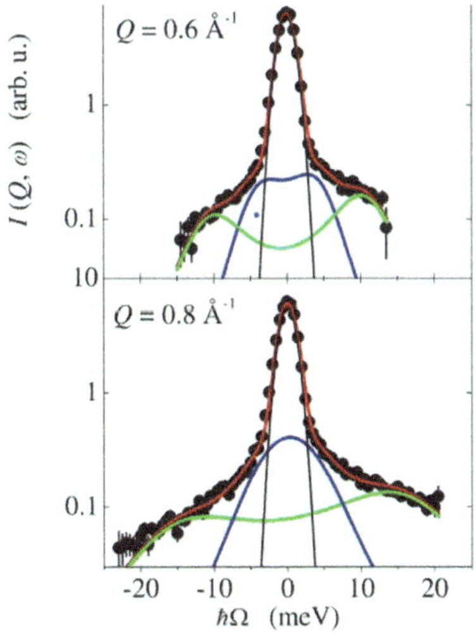

Fig. 11.19 Coherent inelastic spectra of *E. coli* intracellular water at selected Q-values. The black line shows the experimental resolution function. Inelastic shoulders are well-visible outside the resolution function. The red line shows a fit of the data with two DHOs. The blue and green lines represent, respectively, the resulting low and high-frequency DHO [37] Published under a CC-BY licence

mode suggest intracellular water to be composed by a considerable amount of free non-interacting water. A minor fraction of approximately 10% of cellular water, which exhibits reduced dynamics, was attributed to dynamically bound water on the surface of the macromolecules. As an example, from I(Q; ω) in Fig. 11.19, it emerges that inelastic features appear in the coherent inelastic spectra of *E. coli*. Such features were interpreted as coherent collective two inelastic modes propagating through *E. coli* intracellular water.

Picosecond vibrational spectroscopy and dielectric-relaxation spectroscopy were employed to observe the random orientational motion of water molecules inside living cells of three prototypical organisms: a vegetatively growing bacterium (*E. coli*) and a eukaryote (*Saccharomyces cerevisiae,* yeast) both living in aqueous environments, and *Bacillus subtilis* spores which can survive drought for many years and are resistant to heat, toxic chemicals, and radiation [38]. In the yeast sample, the extracellular space is filled with water, and in the spore sample mostly with air. As it is seen in Fig. 11.20, the anisotropy of the cellular water decays with time constant τ_{IR} which is found to be approximately the same as in neat water (τ_{IR} = 2.2 ps). All three organisms most of the intracellular water exhibits the same random orientational motion as neat water (characteristic time constants 9 and 2 ps for the first-order and second-order orientational correlation functions). A smaller fraction of the intracellular water the magnitude of which varies between organisms exhibits slower orientational dynamics. Additional experiments indicated that most of the slow intracellular water is bound to proteins.

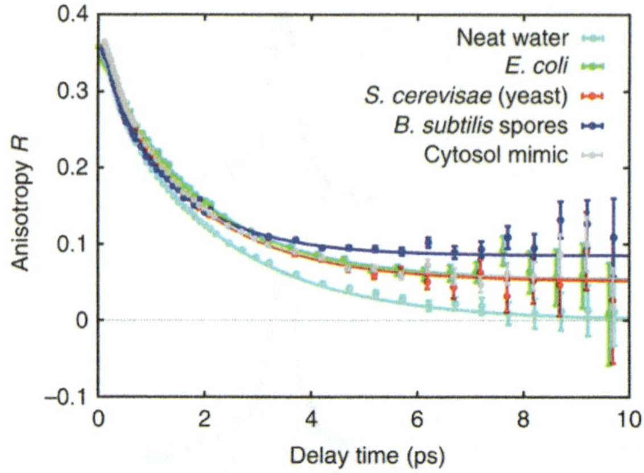

Fig. 11.20 Time-resolved infrared spectroscopy measurements of water dynamics. Transient anisotropy of the OD-stretch mode (2508 cm^{-1}) of HDO water in different organisms and in a cytosol-mimic solution (pH 7, with protein). The error bars represent 1σ. The curves are least-squares fits to exponential decays with a residual offset [38]

The relative amounts of bulk-like and slow water in the different samples were estimated from the relative amplitudes of the decay and the residual in the time-dependent IR anisotropy (Fig. 11.21). The water-mass fraction in the samples was obtained by drying the samples completely after the experiments and comparing their mass before and after drying. Combining this information with the relative amounts of bulk-like and slow water in the different samples estimated from the

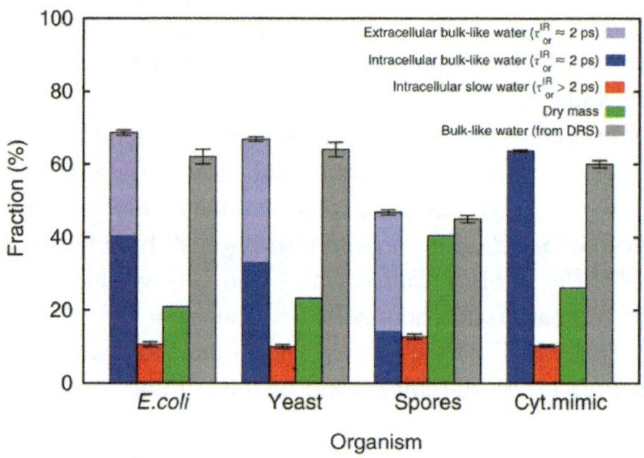

Fig. 11.21 Fractions of bulk-like and slow water. We can estimate the fractions of bulk-like and slow (τIRor > 2 τorIR > 2 ps) water in the different organisms and in the cytosol mimic (pH 7, with protein). The amount of extracellular water was determined by comparing the weight of the samples before and after complete drying and using the previously reported mass fractions of intracellular water in each of the organisms. The *error bars* represent 1σ, but they are lower limits of the actual uncertainties [38]

relative amplitudes of the decay and the residual in the time-dependent IR anisotropy (Fig. 11.21), the mass fractions of bulk-like water, slow water, and dry mass in each organism were determined.

When comparing the fractions of intracellular bulk-like and slow water, it was seen that in all three organisms most of the intracellular water is bulk-like: ~80% of the water in *E. coli* and yeast and ~55% of the water in the bacterial spores. The ratio of slow water to dry mass is roughly similar in all three organisms (0.5 for E. coli, 0.4 for yeast, and 0.3 for the spores). Interestingly, even in bacterial spores, which live under extreme conditions and contain only ~40% water, the majority of the intracellular water is indistinguishable from bulk water. Thus, results confirmed that water inside living cells is mostly made up of bulk-like water and partly of water exhibiting slow orientational dynamics due to interaction with biomolecules and ions.

11.2.4 Miscellaneous Cells

Employing Raman microspectroscopy, the trehalose and residual water contents in the microenvironment of spin-dried cells C3a of the immune system were estimated [39]. TRET and C3A cells were grown to 70–90% confluence in 75 cm^2 tissue culture-treated flasks. Raman spectroscopy set AWITEC Alpha-300 system (WITEC, Ulm, Germany) with confocal capabilities with a 532 nm Ar-ion laser with 15 mW of power was utilized for excitation of samples. Individual cells with or without intracellular trehalose were embedded in a solid thin layer of extracellular trehalose after spin-drying.

The intensity of water bending vibration at ~1645 cm^{-1} was found to be ~20 times less than the O–H stretch vibration in 3000–3600^{-1} region, and it overlaps with the Amide I peak at 1667 cm^{-1}. The water content of the nitrogen and oven-dried samples increases with increasing concentration of trehalose (Fig. 11.22), confirming the strong water affinity of trehalose. Strong evidences suggesting that the residual water was bound at a 2:1 water/trehalose molar ratio in both the extracellular and intracellular milieus and the extracellular trehalose film exhibited characteristics of an amorphous state with a glass transition temperature of ~22 °C. were presented. Obtained findings demonstrated effectiveness of developed method for quantification of water and trehalose in desiccated specimens using confocal Raman microspectroscopy.

Intra/extracellular water states in Chinese hamster ovary (CHO) cells were probed by ultrabroadband multiplex coherent Anti-Stokes Raman Scattering (CARS) spectroscopic [40]. Suggested method allows for the analyses of particular regions of interest such as the nucleus and particular subregions in the cytosol. CHO cells were obtained from American Type Culture Collection. The home-built ultrabroadband multiplex coherent anti-Stokes Raman scattering system including A Q-switched microchip Nd:YAG laser was used as a master laser source. The latter system allowed for the visualization of a wide spectrum of CARS signals from the fingerprint to the end of O–H stretching at a spectral resolution of ~10 cm^{-1}. To probe the chemical

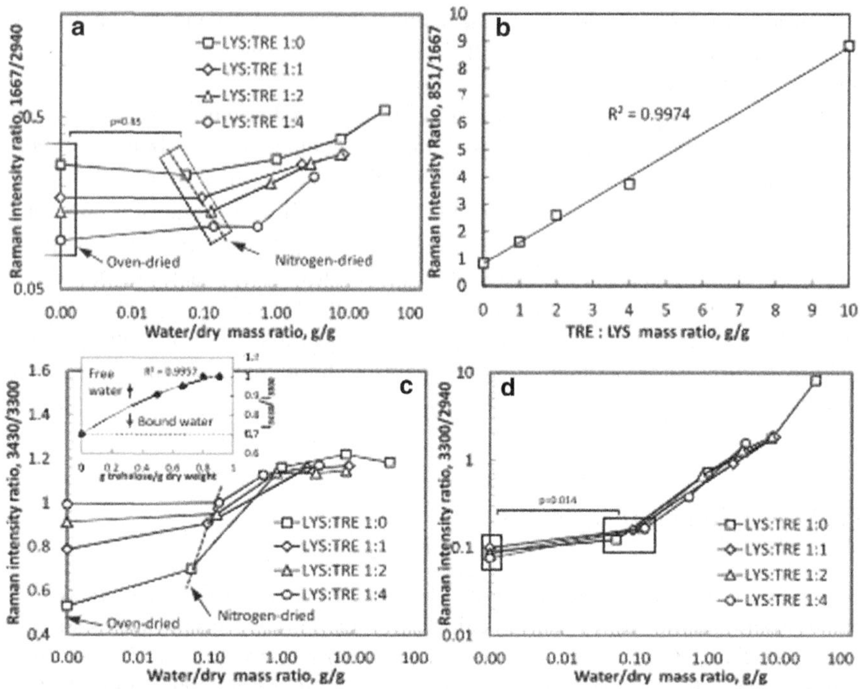

Fig. 11.22 **a** Ratio of the Raman intensity of the peak at 1667 cm^{-1}, representing proteins, to the C-H stretch peak intensity at 2940 cm^{-1}, representing all the organic matter, at various LYS/TRE ratios against water content. **b** Calibration curve for (trehalose/protein) ratio based on intensity ratio of 851–1667 cm^{-1} in oven-dried (0 g water/g dry weight) samples. **c** Ratio of O–H stretch intensities at 3430–3300 cm^{-1} at various LYS/TRE ratios against water content (Inset: The intensity ratio 3430–3300 cm^{-1} for nitrogen-dried samples plotted against TRE/LYS ratios. Dashed line denotes the background intensity not from water). **d** Ratio of the O–H stretch intensity at 3430 cm^{-1} to that of C–H stretch at 2940 cm^{-1} at various LYS/TRE ratios against water content [39]. With permission from Elsevier

state of water in living cells, the system can record the Raman shift in the range of 500–4000 cm^{-1}, cover the fingerprint regions to the whole spectrum of the O–H stretching bond. Comparison of the spectra of the inside and outside cells was performed using two methodologies. First, the intracellular and extracellular regions were semiautomatically separated according to the relative amplitude of the C–H stretching signal (2930–2940 cm^{-1}) over the O–H stretching signal (3420–3430 cm^{-1}), where the higher values correspond to more C–H-rich intracellular regions (Fig. 11.23a). In the second method, the regions of interest (ROIs) were selected based on the C–H signals (Fig. 11.23a, b). These results demonstrate that water molecules inside the cells are relatively more red-shifted, suggesting that the O–H stretching is weaker inside the cells than outside owing to a stronger hydrogen bond.

The aim of work [41] was to explore the application of Raman Microspectroscopy (RM) starting from living sperm in aqueous conditions to test different media and

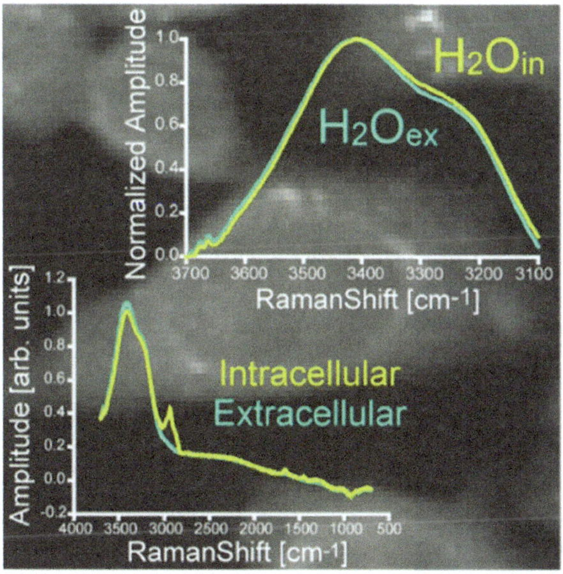

Fig. 11.23 Comparison of Raman spectra of inside/outside cells [40]. With permission from American Chemical Society **a** Semiautomatic analysis of intracellular and extracellular CARS spectra. C–H/O–H values were calculated in individual pixels, and the cutoff value was determined by the histogram. Pixels were divided into intracellular and extracellular regions based on the C–H/O–H cutoff values, and the mean spectra were calculated [40]. With permission from American Chemical Society

to evaluate Raman spectra from undamaged and UV-damaged sperm nuclei under different hydration status. Human sperm was immobilized and Raman spectra were obtained from individual sperm. The study was carried out using a variety of conditions to disclose the Raman features of sperm nuclei under different hydration conditions. To create conditions with controlled DNA damage, sperm samples were exposed to ultraviolet light. The effect of hydration on the Raman spectrum is seen in Fig. 11.24. Under hydrated conditions, the DNA spectrum of control and UVB-treated sperm showed similar patterns except for a few intensity changes in some specific bands, namely ~750 corresponding to T ring breathing mode in nucleic acids; ~935 a protein assignment related to the C–C stretching mode of the amino acids proline and valine; ~1308 assigned to G and the A ring breathing mode, ~1335 related to the ring breathing modes in the DNA bases, specifically A and G; and ~1607 corresponding to C–C of phenylalanine and tyrosine. Thus, in aqueous environment, the differences between the spectrum of UVB-treated sperm and the untreated control sperm are latent, indicating an interference of the hydration status for determination of DNA integrity by Raman microspectroscopy.

A comparative study on the impacts of different temperature and NIR (infrasound frequency mechanical vibration (MV), static magnetic field (SMF), extremely low-frequency electromagnetic field (ELF EMF), and microwave (MW)) in pretreated

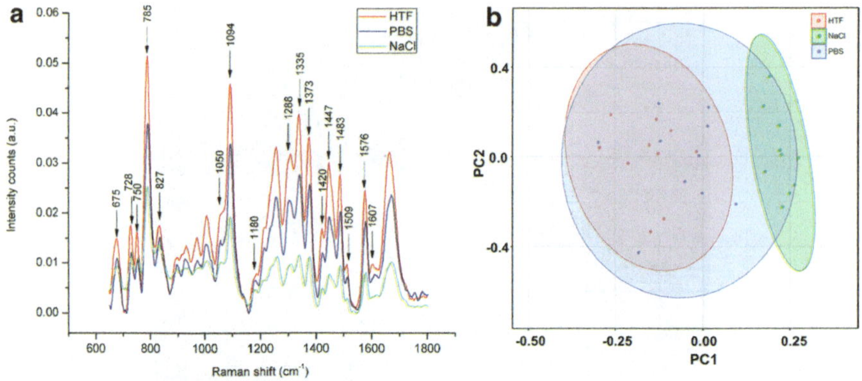

Fig. 11.24 Effect of hydration conditions on the DNA spectrum of individual sperm exposed to ultraviolet light (UV). The results shown in this figure correspond to particular sperm cells that were followed through the time in all condition using a coordinate system. **a** Average Raman spectrum ($n = 50$) of the DNA of non-exposed (control) and UV light exposed (UV) sperm in hydrated conditions. **b** Principal component analysis of single spectra of sperm in hydrated conditions [41]. Published under a CC-BY licence

water on the hydration of barley seeds in its dormant and germination periods was performed [42]. The time-dependent (2, 24, 48, and 72 h) water absorption by dry seeds of spring barley incubated in Sham, and NIR-treated distilled water in cold (4 °C) and warm (20 °C) conditions was examined. Obtained data allowed to suggest that the metabolic driving of intracellular water dynamics accompanied by hydrogen bonding and breaking is more sensitive to NIR-induced water structure changes in seed bathing aquamedium than osmotic gradient-driven water absorption by seeds in dormant state. Therefore, changes in cell hydration can be an input signal for intracellular metabolic cascade responsible for biological effects of non-ionizing radiation.

A study of the critical effect of hydration on THz absorption signatures in biomolecules (lactose monohydrate and biotin) and bioparticles (Bacillus thuringiensis and Bacillus cereus spores) was carry out [43]. The bioparticle samples were endospores of two *Bacillus* species: B. thuringiensis (Bt), subspecies kurstaki, and *B. cereus* (Bc). Bt is a common soil bacteria and insect pathogen. The spectroscopic experiments were performed with a high-resolution frequency-domain, coherent (photomixing) spectrometer. The following main results were reported: (1) for the non-polar molecules or bioparticles, the lower end of the range rises above zero to 10% hydration, but the upper end remains at about 20%; (2) hydration usually causes a red shift in the signature center frequency; (3) sample Bt1 having negligible sporulation and low hydration shows a relatively high transmission but no obvious signatures except for an anomalous one near 700 GHz (Fig. 11.25); (4) sample Bt2 having significant sporulation and also low hydration indicates four distinct signatures centered at 966, 1026, 1098, and 1164 GHz; (5) the transmission spectrum of samples Bt3 indicates six distinct signatures natures centered at 556, 752, 955, 1015,

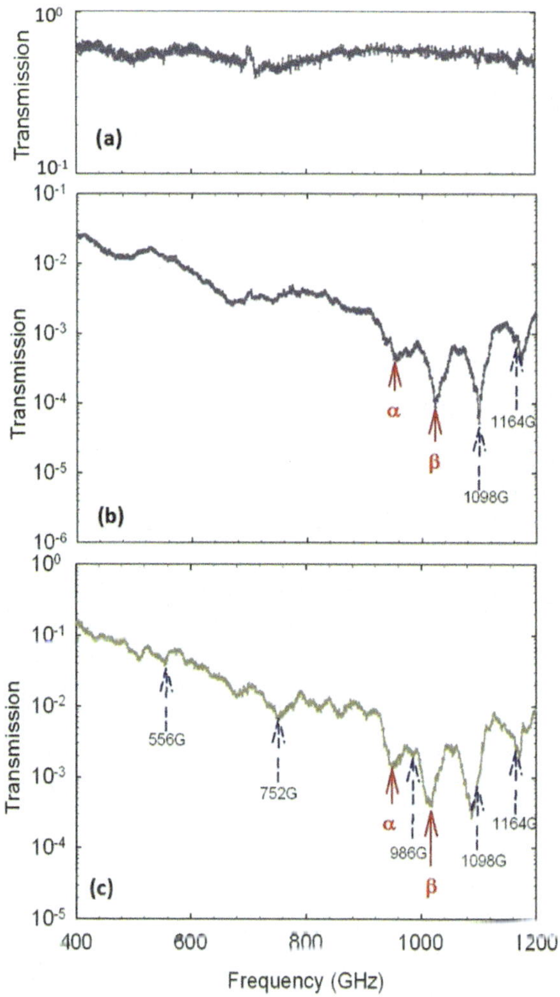

Fig. 11.25 **a** Transmission through Bt1 sample at low hydration and with negligible spores. **b** Transmission through Bt2 sample at low hydration with spores. **c** Transmission through Bt3 sample at moderate hydration with spores [43]

1098, and 1164, and (6) the middle two are similar in shape and strength to the α and β signatures in sample Bt2, but both are red-shifted in center frequency by 11.0 GHz.

11.3 Water in Tissues

11.3.1 Mammalian Tissues

Due to the presence of water in physiological tissue and the high THz absorption of water, reflective THz imaging has distinct advantages over earlier transmission-based systems, especially in in vivo applications [44]. For many THz medical imaging applications, measurement sensitivity for the band covering 400–700 can be determined by quantifying the expected change in THz tissue reflectivity for a given change in water volume fraction at a particular volume fraction. As an example, Fig. 11.26 displays a plot of the effective complex dielectric constant of tissue models for hydration levels of 50 and 75% by volume corresponding to approximate hydration levels of adipose tissue and muscle tissue, respectively, along with liquid water (100%) [44]. As seen in the plot, the real part of the permittivity does not vary much across the range from 50 to 100%, but the imaginary component varies significantly over same hydration range and contrast induced by hydration changes is due to variations in ε''.

Changes in hydration in ex vivo porcine corneas, transparent front part of the eye, were visualized by monitoring ex vivo corneas undergoing drying [45]. The loss of water from the stroma (interior bulk of the ***cornea***) of these samples causes the cornea to deform significantly from its ideal spherical shape. The results from selected time instances corresponding to 87, 84, 79, 75, 70, and 66% average water concentration are shown in Fig. 11.27 with ~10 min elapsing between each image. These results suggest that reflective THz imaging can be successfully applied to cornea using water

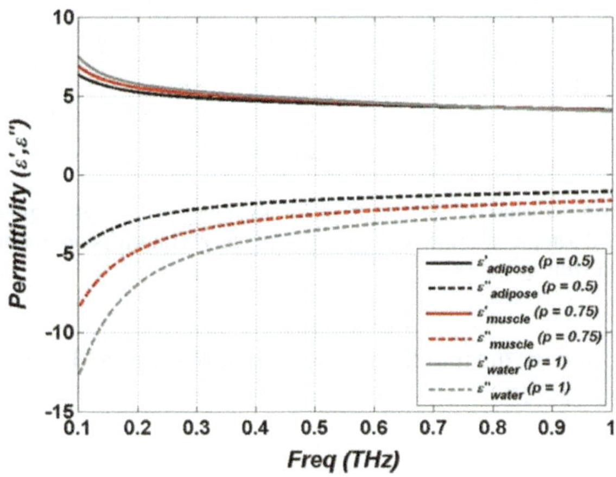

Fig. 11.26 Complex dielectric constant of liquid water and effective permittivities of muscle ($p = 0.75$) and adipose ($p = 0.5$) tissue [44]. With permission of IEEE Microwave Theory and Techniques Society

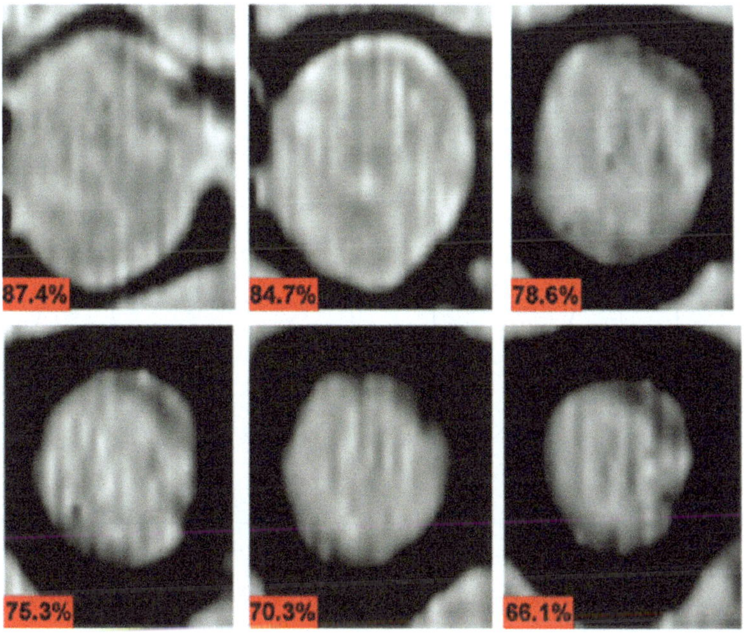

Fig. 11.27 Time lapse imaging of drying porcine cornea with hydration by volume percentages denoted in the bottom left corner of each image [45]. Reprinted from Bennett et al. [24] Copyright © osa.org. Sent 10/05/2021

as the dominant contrast mechanism and may provide hydration sensitivity superior to other ophthalmologic diagnostic systems.

In order to measure the nominal sensitivity of reflection-mode sensing in the THz band to water concentration in the cornea, reflection-mode spectroscopy of ex vivo porcine corneas was performed on nine samples at different hydration levels. THz applications to corneal dystrophy, graft rejection, and refractive surgery are examined from the context of On the basis of these measurements, THz applications to corneal dystrophy, graft rejection, and refractive surgery were examined.

The influence of the water molecules on the breast tissue organization and their role on the hierarchical tissue rearrangement promoted by tumor growth were investigated [46] and SAXS and WAXS techniques were used to analyze healthy, benign, and malignant human breast samples in native and lyophilized conditions. Three human breast samples analyzed in this study were obtained from reduction mammoplasties and prophylactic mastectomies from three female subjects at Clinics Hospital of Ribeirao Preto of the University of Sao Paulo (HC-FMRP/ USP), Ribeirao Preto, Brazil. SAXS measurements were performed at the D02A-SAXS2 beamline in the National Synchrotron Light Laboratory in Campinas, Brazil.1111 In WAXS experiments, the angular distribution of photons scattered at WAXS region was measured using a powder diffractometer Siemens D5005 operating in back-reflection mode. In order to identify the structural transformation in these tissues and verify the water

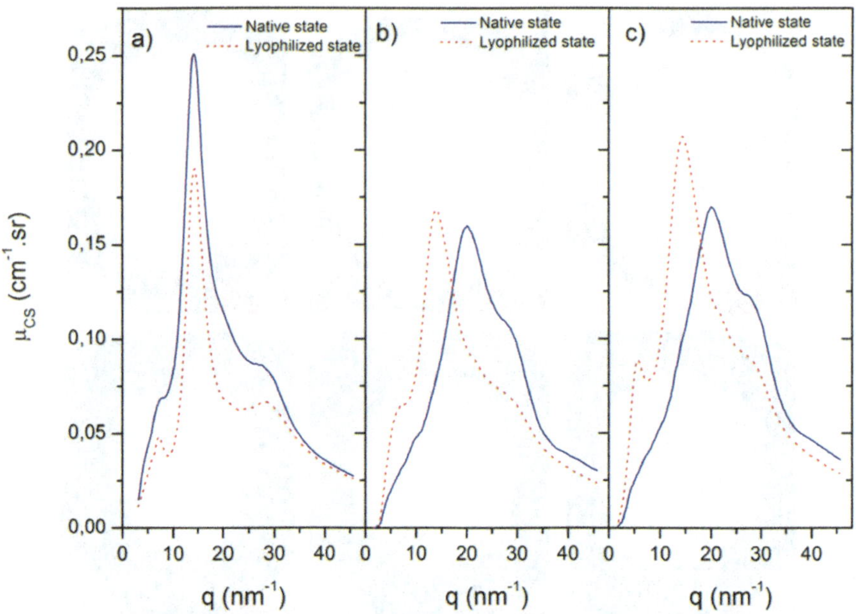

Fig. 11.28 Native and lyophilized scattering profiles of healthy and pathological human breast tissues at the WAXS region: **a** normal tissue, **b** benign lesion, and **c** malignant lesion [46]. With permission from Elsevier

influence on the morphological arrangement of normal and pathological human breast tissues, the scattering profiles in SAXS and WAXS regime of each tissue type in both conditions were compared.

Figure 11.28 compares the WAXS scattering profiles of normal, benign, and **malignant** human breast tissue at native and lyophilized state. In Fig. 11.28a, one peak arises at $q = 6.8\ \text{nm}^{-1}$ for the lyophilized sample and was related to the structure of periodicity of 9.2 Å. A sharp peak at $q = 14.1\ \text{nm}^{-1}$ was identified at both states corresponding to structure with electron density correlation of 4.5 Å. A different peak at $q = 20.2\ \text{nm}^{-1}$ observed in the pathological lesions in the native states (Fig. 11.29b, c) was attributed to the signal from the oxygenoxygen electron density correlation between adjacent molecules on the tetrahedral structure of liquid water. When the water molecules were removed from the samples, structural changes associated with pathological progression were visible.

Biexponential data from fresh **rat liver** and monoexponential data from cold storage experiments allowed quantification of three distinct **^{1}H spin relaxation** components in liver tissue: bound water (4.2%, $R_1 = 12.0 \pm 1.7\ \text{s}^{-1}$, $R_2 = 440 \pm 180\ \text{s}^{-1}$); structured water (59%, $R_1 \approx 3.3 \pm 0.07\ \text{s}^{-1}$, $R_2 \approx 24.9 \pm 1.1\ \text{s}^{-1}$); and free water (≈37%, $R_1 = R_2 \approx 0.4\ \text{s}^{-1}$) [47]. However, only the relaxation rates of the structured water component change with water content: R_{1A} (s^{-1}) = 6.53, R_{2A}

(s^{-1}) = 71.15*, respectively. The finding suggested a slow exchange between bound and structured water in liver cells.

The diffusion-weighted magnetic resonance imaging (DWI) technique was utilized for quantification of water mobility for probing microstructural properties of the mouse brain [48]. The magnetic resonance imaging (MRI) scans were performed on a 7 T (20 cm bore) Bruker Biospin animal magnet (Bruker Instruments, Ettlingen, Germany) located at the University of California, San Diego, CA, USA, Center for functional-MRI. The IF images were acquired with a Keyence BZ-X710 all-in-one fluorescence microscope (Osaka, Japan) utilizing DAPI and Texas Red filter cubes with an S Plan Fluor 0.45NA ELWD 20× objective lens. The H&E images were acquired employing transmitted light and a 20×/0.75 N.A. lens on an Olympus IX81 microscope. Histology slides were obtained to quantify relative cell density measurements within specified regions of interest (ROIs). Measurements using multiple diffusion signal 1*b*-values have indicated biexponential signal attenuation, ascribed to "fast" and "slow" diffusion components. The properties of the diffusion time (Δ)-dependent components of the diffusion-weighted (DW) signal in a constant *b*-value experiment were investigated. Δ is the diffusion time given by the separation time between the two applied diffusion gradients the high cell density tumor region and b is the diffusion signal. A xenograft gliobastoma (the most aggressive type of cancer) of mouse was imaged using $\Delta = 11$ ms, 20 ms, 40 ms, 60 ms, and $b = 500$–4000 s/mm^2 in intervals of 500 s/mm^2. The Δ-dependence on the DW signal was measured within three regions of interest [intermediate- and high-density tumor regions and normal-appearing brain (NAB) tissue regions].

The following cell density tumor regions were reported (Fig. 11.29): primary tumor cells, region 1, (~18.8 mm^3, ~50% nuclei coverage); the intermediate density tumor region 2, near the injection site (~7.0 mm^3, ~35% nuclei coverage; and the low-density region 3 of normal-appearing brain tissue extracted from the healthy part of the mouse brain (~28.2 mm^3, ~10% nuclei coverage. The assumption that the slow diffusion component of the DW signal is non-Gaussian and dependent on Δ, consistent with restricted diffusion of the intracellular space was verified.

Simultaneously total brain water, T_1 and T_2 relaxation times, and hyaluronan in fetal and neonatal rabbits were investigated [49] E. Sulyok, Z. Nyúl, P. Bogner, G. Sedin, Brain water and proton magnetic resonance relaxation in preterm and term rabbit pups: their relation to tissue hyaluronan. Neonatology **79**, 67–72 (2001). Hyaluronic acid (HA) is the simplest glycosaminoglycan (a class of negatively charged polysaccharides) and a major constituent of the extracellular matrix. Experiments were carried out in fetal Pannon white young rabbits at gestational ages of 25, 27, 29, and 31 days and at a postnatal age of 4 days. The brain tissue water content (desiccation method), T_1 and T_2 relaxation times (^{1}H-NMR method), and HA concentration (radioassay HA 50) were measured, and free and bound water fractions were calculated by using multicomponent fits of the T_2 relaxation curves. The following main results were reported: (1) the relationship of HA to total brain water and to T_2-derived motionally distinct water fractions was established; (2) compared with values in newborn pups water and HA contents were found to be highly elevated in the preterm brain and decreased markedly during early postnatal life; (3) the trends

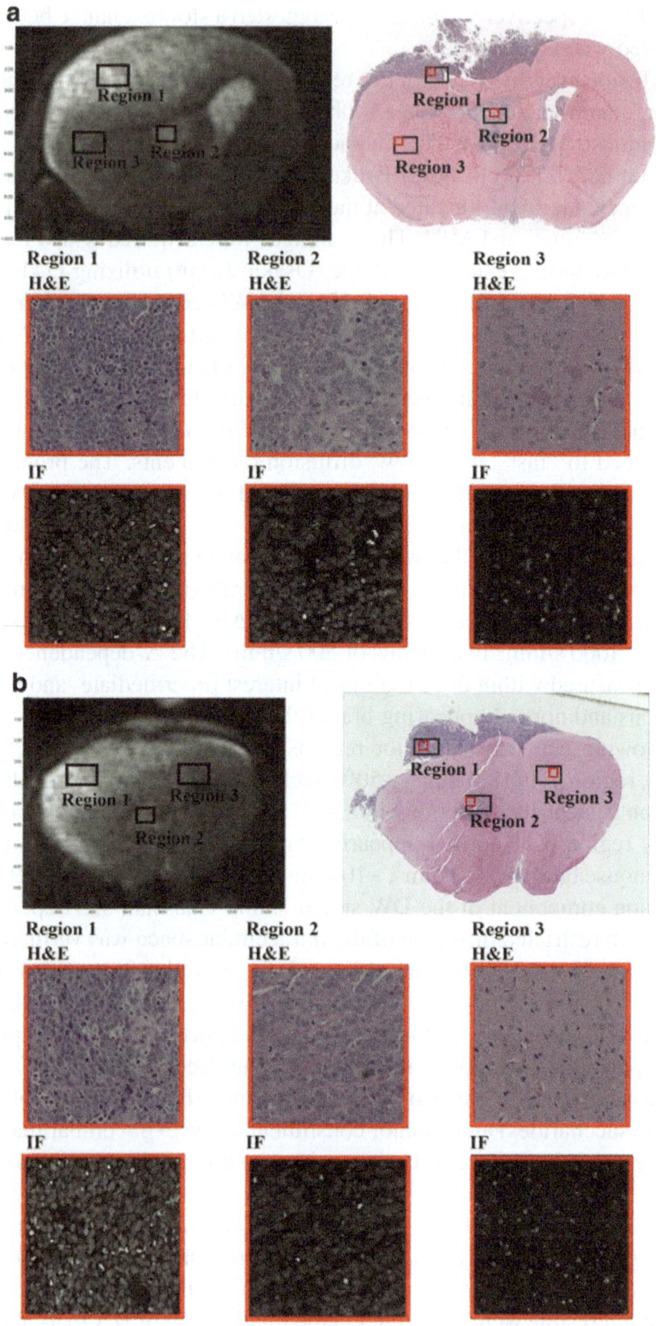

Fig. 11.29 Diffusion time (Δ) dependence on the DW signal within three regions of interest (ROIs) selected from MRI brain images of a single GBM xenograft mouse [48]. With permission from Frontiers

and time courses of T_1 and T_2 relaxation times proved to be similar; (4) maturity-related changes occurred in the T_2 relaxation derived bound water fraction which amounted to 4–19% of brain water; and (5) the bound water fraction appeared to be independent of total brain water and HA concentration. The authors concluded that the clear dissociation of bound water fraction from total water suggests restructuring of brain water during the perinatal period (e.g., the period immediately before and after birth.

11.3.2 Food Tissues

Postharvest loss is a major concern for food security as about one-third of global food is due to the lack of proper processing or preservations [50]. Plant-based food materials are porous and hygroscopic in nature; therefore, it contains three water environments, namely intercellular, intracellular water, and cell wall water. The intercellular water is known as capillary water or free water which is less constrained than intracellular water, considered as loosely bound water (LBW), and cell wall water, which is recognized as strongly bound (SBW). Biofood materials are heterogeneous in structure with cellular diversity, where the majority of the water is located in the intracellular spaces. Understanding of the nature of the microscopic behavior of water transport is crucial to enhance the energy efficiency in food processing and obtain the better quality of processed food.

The aim of study [51] was to investigate the proportion of different types of water such as free, LBW and SBW in 11 different plant-based food materials using ^{1}H NMR T_2 relaxometry. The experimental results uncovered that plant-based food materials contain about 80–92% loosely bound water (LBW), 6–16% free water, and only about 1–6% as strongly bound (SBW). The investigation confirmed that among the five different fruits, kiwi contains the lowest percentage of LBW while apple contains the highest percentage of LBW. Among the vegetables, eggplant comprises the largest amount of LBW while cucumber contains least amount of SBW. In addition, it was found that SBW strongly depends on the proportion of solid in the sample tissue whereas free water (FW) depends on the porosity of the material.

Effective diffusivity is the most important key parameter needed in the analysis, design, and optimization of heat and mass transfer during food drying process [52]. In general, two types of effective diffusivities were used to develop the mathematical modeling of food drying, namely moisture-dependent effective diffusivity (MDED) and temperature-dependent effective diffusivity (TDED) [53]. To determine the appropriate effective diffusivity for predicting the drying kinetics, drying models were developed for different fruits and vegetables based on moisture-dependent and temperature-dependent effective diffusivities. COMSOL Multiphysics, a finite element-based engineering simulation software, was used to solve the coupled heat and mass transfer equations. 3D moisture profiles were developed to investigate the spatial moisture distribution during drying. Extensive experimental investigation on five types of fruits and vegetables was conducted, and results were compared with

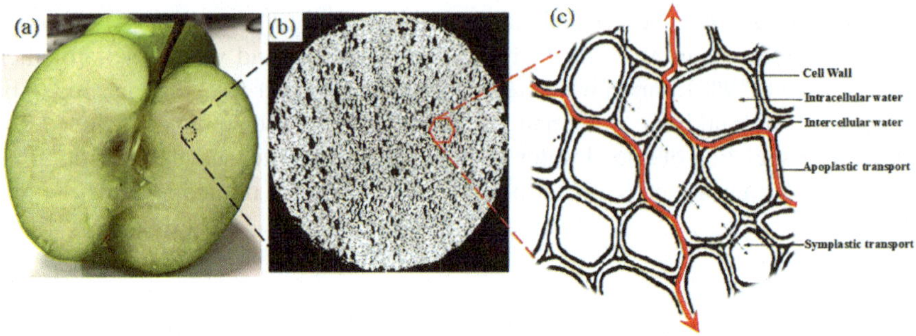

Fig. 11.30 Cellular water transport process in apple tissue [54]

the simulated results. For all fruits and vegetables, the moisture ratio with MDED was found to be significantly lower than moisture ratio with TDED. The moisture ratio of apple was lowest whereas pear showed the highest moisture ratio and carrot showed a considerably lower moisture ratio compared to potato.

Apoplastic and symplastic transport of cellular water in the biofood material during drying was investigated using ^{1}H-NMR-T_2 relaxometry [54]. In the simplest pathway, the water molecules move through adjacent cell walls. Intercellular spaces or environments are the spaces between the cells as shown in Fig. 11.30. They are mostly composed of air, a small portion of water, and other solutes (such as sugar). Granny smith apple were used as a biofood material where the water protons dominate the ^{1}H-NMR proton signals, which are an average over the whole sample that provides the information about the water content in the biological tissue. The apple was collected from a local market in Brisbane, Australia. A Bruker DRX wide-bore spectrometer (Bruker Bio-spin, Karlsruhe, Germany) was employed for NMR experiment with an operating frequency 300 MHz for hydrogen. Extensive X-ray microtomographic experiments have been conducted at different stages of drying to validate the quantitative NMR result. ^{1}H-NMR-T_2 relaxometry was used to measure the proportion of intracellular water present in apple tissue at different stages of drying. From the T_2 relaxation analysis, it is straightforward to see the number of decays in the relaxation curve, and therefore it is possible to investigate different cellular level water based on their intensity of relaxation.

Figure 11.31 shows the T_2 relaxation decay curve obtained from samples of drying at various drying times at temperature of 60 °C. By fitting the biexponential decay with these different T_2 relaxation intensity data, two different water proton relaxations were calculated. Based on the mobility of different cellular water, pore size, membrane permeability, and the proton density within the sample, these two components were categorized as intracellular water (ICW) and free water (FW), respectively. The proportion of ICW and FW was then calculated from the relative contribution of long and short T_2 components.

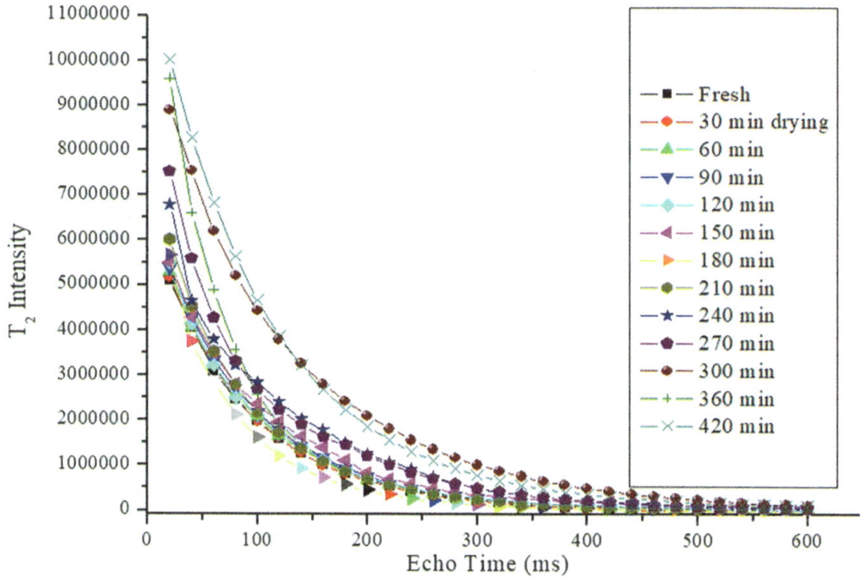

Fig. 11.31 T_2 intensity data at different stages of drying [54]. With permission from Frontiers

The measured proportion of ICW was plotted with drying time as shown in Fig. 11.32. For example, ICW at 120 min is 88% and rest 12% is FW. In this case, ICW migrates from cell to the neighboring cell and cell to the intercellular spaces through very fine capillaries. Migrating ICW via intracellular pathway is a very slow process, and therefore the average moisture content decreases slowly.

Many fluctuations in the proportion of ICW and FW throughout the drying were observed which are mainly caused by the cell collapse at different stages of drying. When cells collapse, water comes out of the cells, move from intracellular spaces to intercellular spaced, and becomes FW. Thus, intracellular water migrates, as apoplastic process, that is, from intracellular spaces to the intercellular spaces by progressive rupturing the cell membranes while drying at a higher temperatures (60–70 °C). At lower temperature (45 °C), cell membranes do not rupture and therefore ICW migrates from cell to the neighboring cell through microcapillaries, where the symplastic process dominates the mass transfer at different stages of drying. The symplast of a plant is the inner side of the plasma membrane in which water and low-molecular-weight solutes can freely diffuse (see also Fig. 11.30).

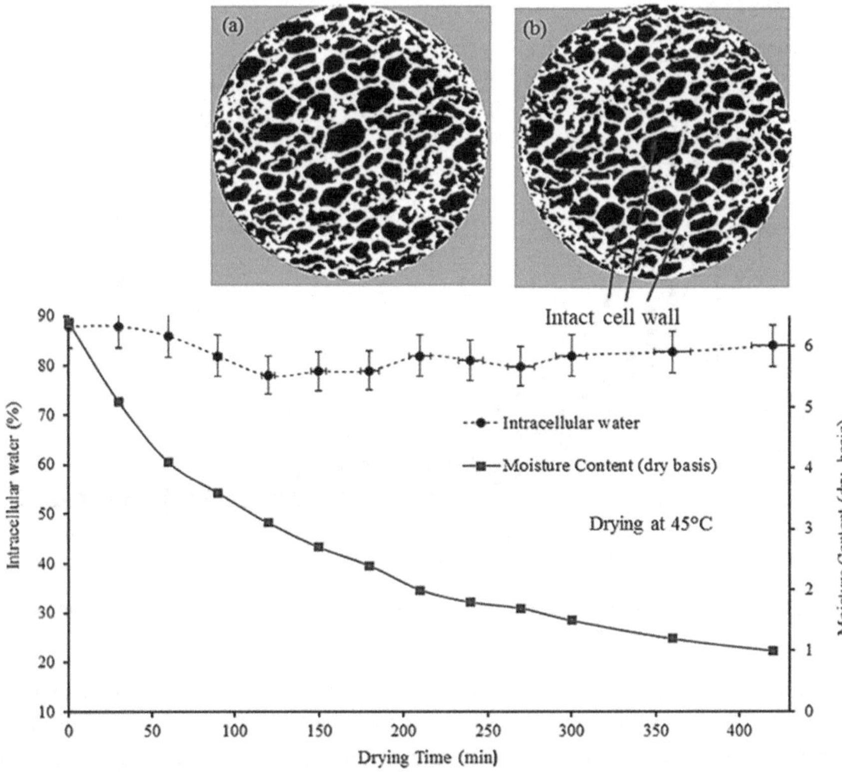

Fig. 11.32 Transport of intracellular water during drying of apple tissue at 45 °C, including X-ray tomographic image of the dried sample **a** after 260 min of drying), **b** after 360 min of drying [54]. With permission from Frontiers

11.4 Water in Plants

The various functions of water in plants include: maintaining cell turgidity for structure and growth; transporting nutrients and organic compounds throughout the plant; comprising much of the living protoplasm in the cells; serving as a raw material for various chemical processes, including photosynthesis; and, through transpiration, buffering the plant against wide temperature fluctuations [55–60]. Water forms over 90% of the plant body by green or fresh weight basis. Plants retain less than 5% of the water absorbed by roots for cell expansion and plant growth. The amount of water lost via transpiration can be about 200 L in a single day. To make sugars, plants must absorb carbon dioxide (CO_2) from the atmosphere through small pores in their leaves called stomata (Fig. 11.33). When stomata open, an average of 400 water molecules are lost for each CO_2 molecule.

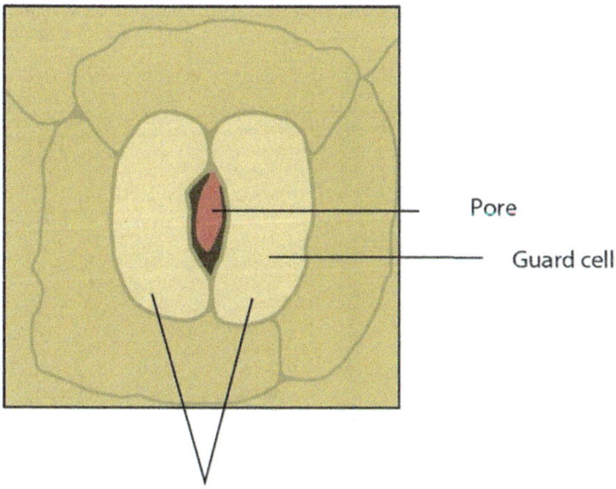

Fig. 11.33 Rendering of an open stoma on the surface of a tobacco leaf. Stomata are pores found on the leaf surface that regulate the exchange of gases between the leaf's interior and the atmosphere. Stomatal closure is a natural response to darkness or drought as a means of conserving water [55]

The structure of plant roots, stems, and leaves facilitates the transport of water, nutrients, and photosynthates throughout the plant. The phloem and xylem are the main tissues responsible for this movement. Xylem is made up of several types of cells. Tracheids are long cells that help transport xylem sap and also provide structural support. The water-filled, selective membrane pores (water channels) are integrated into a general water transport in plant tissue. Three main pathways of water flow have been distinguished, i.e., the apoplastic, symplastic, and transcellular paths are summarized as a cell-to-cell component. Water channel activity may control the overall water flow across tissues.

Upon absorption by the root, water first crosses the epidermis and then makes its way toward the center of the root crossing the cortex and endodermis before arriving at the xylem (Fig. 11.34). Along the way, water travels in cell walls (apoplastic pathway) and/or through the inside of cells (cell-to-cell pathway).

Osmosis in plant cells is the diffusion of molecules through a semipermeable, or differentially permeable, membrane from a region of higher solute concentration to a region of lower solute concentration [58]. The osmotic flow is governed by water potential, as which the sum of pressure potential and solute potential, wherein pressure potential is equal to hydrostatic pressure. In fact, osmotic potential is the minimum pressure required to prevent fluid from moving as a result of osmosis. In the absence of transpiration, osmotic forces dominate the movement of water into roots.

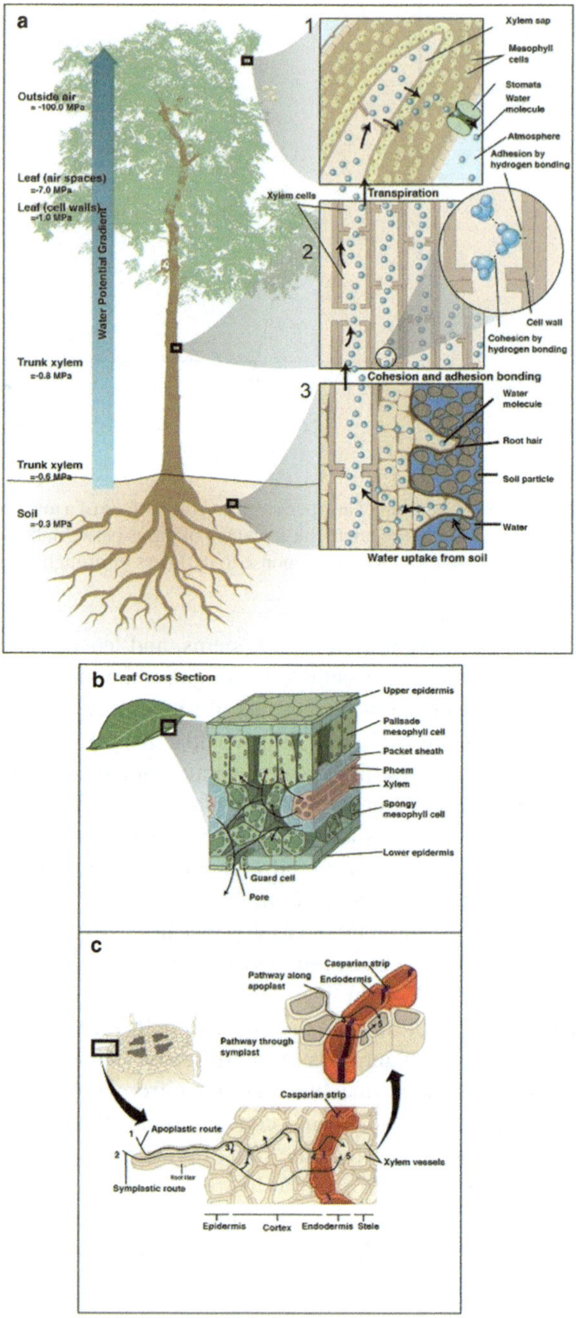

Fig. 11.34 Representation of the water transport pathways along the soil–plant–atmosphere continuum (SPAC). **a** Water moves from areas of high water potential (i.e., close to zero in the soil) to low water potential (i.e., air outside the leaves). Details of the cohesion–tension mechanism are illustrated with the inset panels (**a**), where tension is generated by the evaporation of water molecules during leaf transpiration (1) and is transmitted down the continuous, cohesive water columns (2) through the xylem and out the roots to the soil (3). The pathways for water movement out of the leaf veins and through the stomata (**b**) and across the fine roots **c** are detailed and illustrate both symplastic and apoplastic pathways [55]

References

1. G. Cooper, *Cell: A Molecular Approach* (ASM PRESS, Washington, DC, 2009), p. 544. ISBN 978-0-87893-300-6
2. G. de Majno, I. Joris, *Cells, Tissues, and Disease: Principles of General Pathology* (Oxford University Press, Oxford, 2004)
3. G. Vunjak-Novakovic, R.I. Freshney, *Culture of Cells for Tissue Engineering* (Wiley, Hoboken, 2006)
4. P. Balla, Water is an active matrix of life for cell and molecular biology. PNAS **114**, 13327–13335 (2017)
5. J.M. Carbrey, P. Agre, Discovery of the aquaporins and development of the field. Handb. Exp. Pharmacol. **190**, 3–28 (2009)
6. P. Jungwirth, Biological water or rather water in biology? J. Phys. Chem. Lett. **6**, 2449–2451 (2015)
7. M.L. Martins, A.B. Dinitzen, E. Mamontov, S. Rudić, J.E.M. Pereira, R. Hartmann-Petersen, W. Herwig, H.N. Bordallo, Water dynamics in MCF-7 breastcancer cells: a neutron scattering descriptive study. Sci. Rep. **9**, 8704 (2019)
8. C.Y.J. Hémonnot, S. Köster, Imaging of biological materials and cells by X-ray scattering and diffraction. ACS Nano **11**, 8542–8559 (2017)
9. M. Bartels, M. Krenkel, J.R. Haber, R. Wilke, T. Salditt, X-ray holographic imaging of hydrated biological cells in solution. Phys. Rev. Lett. **114**, 048,103 (2015)
10. S. Ayrapetyan, The dysfunction of metabolic controlling of cell hydration precedes Warburg phenomenon in carcinogenesis. J. Bioequiv. Availab. **7**, 1 (2015)
11. R. Smith, K.L. Wright, L. Ashton, Raman spectroscopy: an evolving technique for live cell studies. Analyst **141**, 3590–3600 (2016)
12. E. Persson, B. Halle, Cell water dynamics on multiple time scales. Proc. Natl. Acad. Sci. U.S.A. **105**, 6266–6271 (2007)
13. L. Shi, F. Hu, W. Min, Optical mapping of biological water in single live cells by stimulated Raman excited fluorescence microscopy. Nat. Commun. **10**, Article number: 4764 (2019)
14. S. Sabbatini, C. Conti, G. Orilisi, E. Giorgini, Infrared spectroscopy as a new tool for studying single living cells: Is there a niche? Biomed. Spectrosc. Imaging **6**, 85–99 (2017)
15. P. Ball, Water as an active constituent in cell biology. Chem. Rev. **108**, 74–108 (2008)
16. A. Orecchini, F. Sebastiani, M. Jasnin, A. Paciaroni1, A. De Francesco, C. Petrillo, M. Moulin, M. Haertlein, G. Zaccai, F. Sacchetti, Collective dynamics of intracellular water in living cells. J. Phys.: Conf. Ser. **340**, 012091 (2012)
17. M. Takeuchi, S. Kajimoto, T. Nakabayashi, Experimental evaluation of the density of water in a cell by Raman microscopy. J. Phys. Chem. Lett. **8**, 5241–5245 (2017)
18. L.C. Conceiçao, K. Meehan, M. Antoniassi, M. Piacenti-Silva, M.E. Poletti. The influence of hydration on the architectural rearrangement of normal and neoplastic human breast tissues. Heliyon **5,** e01219 (2019) abnormal
19. M.I.H. Khan, R.M. Wellard, S.A. Nagy, M.U.H. Joardder, M.A. Karim, Investigation of bound and free water in plant-based food material using NMR T_2 relaxometry. Innov. Food Sci. Emerg. Technol. **38**, 252–261 (2016)
20. K. Shiraga, T. Suzuki, N. Kondo, K. Tanaka, K.Y. Ogawa, Hydration state inside HeLa cell monolayer investigated with terahertz spectroscopy. Appl. Phys. Lett. **106**, 253701 (2015)
21. J.M. Dick, Proteomic indicators of oxidation and hydration state in colorectal cancer. Peer J. **4**, e2238 (2016)
22. T.M. Grzegorczyk, P.M. Meaney, P.A. Kaufman, R.M. Di Florio-Alexander, K.D. Paulsen, Fast 3-D tomographic microwave imaging for breast cancer detection. IEEET Trans. Med. Imaging **31**, 1584–1592 (2012)
23. G.I. McIntyre, Cell hydration as the primary factor in carcinogenesis: a unifying concept. Med. Hypotheses **66**, 518–526 (2006)

24. H. Abramczyk, B. Brozek-Pluska, M. Krzesniak, M. Kopec, A. Morawiec-Sztandera. The cellular environment of cancerous human tissue. Interfacial and dangling water as a "hydration fingerprint". Spectrochimica Acta, Part A: Mol. Biomol. Spectrosc. **129**, 609–623 (2014)
25. M.P.M. Marques, A.L.M. Batista de Carvalho, A.P. Mamede et al., Struct. Dyn. **7**, 05470 (2020)
26. M. Takeuchi, S. Kajimoto, T. Nakabayashi, Experimental evaluation of the density of water in a cell by Raman microscopy. J. Phys. Chem. Lett. **8**, 5241–5245 (2017). https://doi.org/10.1021/acs.jpclett.7b02154
27. R. Minnes, M. Nissinmann, Y. Maizels, G. Gerlitz, A. Katzir, Y. Raichlin, Using attenuated total reflection–Fourier Transform Infra-Red (ATRFTIR) spectroscopy to distinguish between melanoma cells with a different metastatic potential. Sci. Rep. **7**, 4381 (2017)
28. A. Tomback, *Erythrocyte* (IntechOPEN 2019)
29. G.W. Stewart, E.J. Turner, The hereditary stomatocytoses and allied disorders: congenital disorders of erythrocyte membrane permeability to Na and K. Best Pract. Res. Clin. Haematol. **12**, 707–727 (1999)
30. T. Tiffert, V.L. Lew, H. Ginsburg, M. Krugliak, L. Croisille, N. Mohandas, The hydration state of human red blood cells and their susceptibility to invasion by Plasmodium falciparum. Blood **105**, 4853–4860 (2005)
31. A.M. Stadler, J.P. Embs, I. Digel, G.M. Artmann, T. Unruh, Z. Georg Buldt, Cytoplasmic water and hydration layer dynamics in human red blood cells. J. Am. Chem. Soc. **130**, 16852–16853 (2008)
32. P.G. Gallagher, Disorders of erythrocyte hydration. Blood **130**, 2699–2708 (2017)
33. E.R. Brown, W.-D. Zha, *The Critical Effect of Hydration on the Resonant Signatures of THz Biospectroscopy*. https://www.researchgate.net/publication/328445311
34. A. Caulier, R. Rapetti-Mauss, H. Guizouarn, V. Picard, L. Garçon, C. Badens, Primary red cell hydration disorders: pathogenesis and diagnosis. Int. J. Lab. Hem. **40**, 68–73 (2018)
35. M. Tehei, B. Franzetti, K. Wood, F. Gabel, E. Fabiani, M. Jasnin, M. Zamponi, D. Oesterhelt, G. Zaccai, M. Ginzburg, B.-Z. Ginzburg, Neutron scattering reveals extremely slow cell water in a Dead Sea organism, PNAS **104**, 766–771 (2007)
36. M. Jasnin, M. Moulin, M. Haertlein, G. Zaccai1, M. Tehei, Down to atomic-scale intracellular water dynamics. EMBO Rep. **9**, 543–546 (2008)
37. A. Orecchini, F. Sebastiani, M. Jasnin, A. Paciaroni, A. De Francesco, C. Petrillo, M. Moulin, M. Haertlein, G. Zaccai, F. Sacchetti, Collective dynamics of intracellular water in living cells. J. Phys.: Conf. Ser. **340**, 012091 (2012)
38. M. Tros, L. Zheng, J. Hunger, M. Bonn, D. Bonn, S. Daniel, S. Gertien, S. Woutersen, Picosecond orientational dynamics of water in living cells. Nat. Commun. **8**, 904 (2017)
39. A. Abazari, N. Chakraborty, S. Hand, A. Aksan, M. Toner, A Raman microspectroscopy study of water and trehalose in spin-dried cells. Biophys. J. **107**, 2253–2262 (2014)
40. M. Nuriya, H. Yoneyama, K. Takahashi, Characterization of intra/extracellular water states probed by ultrabroadband multiplex coherent anti-stokes Raman scattering (CARS) spectroscopic imaging. J. Phys. Chem. A **123**, 3928–3934 (2019)
41. R. Da Costa, S. Amaral, K. Redmann, S. Kliesch, S. Schlatt, Spectral features of nuclear DNA in human sperm assessed by Raman Microspectroscopy: effects of UV-irradiation and hydration. PLOS ONE (2018) https://doi.org/10.1371/journal.pone.0207786
42. S. Ayrapetyan, J. DeCell, Hydration as a biomarker for estimation of biological effects of nonionizing radiation on cells and organisms. Sci. World J. Article ID 890518, 8 pages (2014)
43. E.R. Brown, W-D. Zha, The critical effect of hydration on the resonant signatures of THz. Biospectroscopy. https://www.researchgate.net/publication/328445311
44. Z.D. Taylor, R.S. Singh, D.B. Bennett, P. Tewari, C.P. Kealey, N. Bajwa, M.O. Culjat, A. Stojadinovic, H. Lee, J.-P. Hubschman, E.R. Brown, W.S. Grundfest, THz medical imaging: in vivo hydration sensing. IEEE Trans. Terahertz Sci. Technol. **1**(1), 201–219 (2011). https://doi.org/10.1109/TTHZ.2011.2159551
45. D.B. Bennett, Z.D. Taylor, D. Sassoon, P. Tewari, R.D. Johnson, R. Singh, M.O. Culjat, J-P Hubschman, W.S. Grundfest, D.J. Sassoon, R. Duncan Johnson, J.-P. Hubschman, E.R. Brown, Terahertz sensing in corneal tissues. J. Biomed. Opt. **16**, 8 (2010)

46. A.L.C. Conceiçao, K. Meehan, M. Antoniassi, M. Piacenti-Silva, M.E. Poletti, The influence of hydration on the architectural rearrangement of normal and neoplastic human breast tissues. Heliyon e01219 (2019)
47. E. Moser, P. Holzmueller, M. Krssak, Improved estimation of tissue hydration and bound water fraction in rat liver tissue. Magn. Reson. Mater. Phys. Biol. Med. **4**, 55–59 (1996)
48. T.R. Hope, N.S. White, J. Kuperman, Y. Chao, G. Yamin, H. Bartch, N.M. Schenker-Ahmed, R. Rakow-Penner, R. Bussell, N. Nomura, S. Kesari, A. Bjørnerud, A.M. Dale, Demonstration of non-Gaussian restricted diffusion in tumor cells using diffusion time-dependent diffusion-weighted magnetic resonance imaging contrast. Front. Oncol. **6**(3) (2016). https://doi.org/10.3389/fonc.2016.00179
49. E. Sulyok, Z. Nyúl, P. Bogner, G. Sedin, Brain water and proton magnetic resonance relaxation in preterm and term rabbit pups: their relation to tissue hyaluronan. Neonatology **79**, 67–72 (2001)
50. J. Gustavsson, C. Cederberg, U.L.F. Sonesson, *Global Food Losses and Food Waste-Extent, Causes and Prevention* (Food Agriculture Organization of United Nations, Rome, 2011)
51. S. Grabowski, M. Marcotte, H.S. Ramaswamy, Drying of fruits, vegetables, and spices, in *Handbook of Postharvest Technology: Cereals, Fruits, Vegetables, Tea, and Spices* (Marcel Dekker, New York, 2003), pp. 653–95
52. M.I.H. Khan, R.M. Wellard, S.A. Nagy, M.U.H. Joardder, M.A. Karim, Investigation of bound and free water in plant-based food material using NMR T2 relaxometry. Innov. Food Sci. Emerg. Technol. **38**(Part A), 252–261 (2016)
53. M.I.H. Khan, C. Kumar, M.U.H. Joardder, M.A. Karim, Determination of appropriate effective diffusivity for different food materials. Drying Technol. **35**, 335–346 (2017)
54. I.H. Khan, T. Farrell, S.A. Nagy, M.A. Karim, Fundamental understanding of cellular water transport process in bio-food material during drying. Sci. Rep. **8**, Article number: 15191 (2018)
55. A.J. McElrone, B. Choat, G.A. Gambetta, C.R. Brodersen, Water uptake and transport in vascular plants. Nat. Educ. Knowl. **4**, 6 (2013)
56. G.N. Agrios, *Plant Pathology* (Academic Press, New York, NY, 1997)
57. D.J. Beerling, P.J. Franks, Plant science: the hidden cost of transpiration. Nature **464**, 495–496 (2010)
58. M. Kramer, Osmosis is not driven by water dilution. Trends Plant Sci. **18**, 195–197 (2013)
59. L. Malińska, E. Rybska, E. Sobieszczuk-Nowicka, M. Adamiec, Teaching about water relations in plant cells: an uneasy struggle. CBE Life Sci. Educ. **15**, ar78 (2016)
60. E. Steudle, Water transport across plant tissue: role of water channels. Biol. Cell **89**, 259–273 (1997)

Chapter 12
Water: Clinical Aspects

Abstract Water represents on average 55–60% of the human body weight and is distributed throughout the body and organs. Health, life expectancy, daily comfort, and mood of each and every human being largely depend on the water balance of its body. Dehydration occurs when one loses more fluid than one takes in, and one's body does not have enough water and/or other fluids to carry out its normal functions. The most common cause of dehydration is severe diarrhea and vomiting. Dehydration also can occur if one does not drink enough water during hot weather especially if one is exercising vigorously. Dehydration symptoms and the accepted classification of dehydration are considered. Bioelectrical impedance analysis, osmolality and urine assays, body weight change, and diffusion-weighted magnetic resonance imaging are widely used for quantitative degradation assessments. Examples of effects of dehydration on mental, cancer, and gastrointestinal tract diseases and abnormalities of water homeostasis are provided. A special section is dedicated to effect dehydration on state and activity of healthy human beings.

12.1 Introduction

Water is vital to human life [1–20]: total body weight (TBW) is 60% of body weight; interstitial fluid (ICF) and *extracellular fluid* ECF water are 40 and 20%, respectively, of body weight. ECF contains both interstitial (15%) and intravascular water (5%) Fig. 12.1). When fluid deficit exceeds 8%, death may occur. Water plays a central role in regulating cell volume, nutrient transport, waste removal, and thermal regulation and provides a medium for biological reactions. Adequate hydration is crucial in maintaining optimal physical and mental functioning, and the need for a fast and reliable hydration status assessment in behavioral medicine research has become increasingly important. Hyperhydration is a state of being in positive water balance (a water excess), and hypohydration is the state of being in negative water balance (a water deficit). Both acute and chronic dehydration can have important implications for human behavior and health.

Hypohydration is known to reduce mental and physical performance, impair vascular function and cardiovascular regulation, reduce endothelial function, increase

G. I. Likhtenshtein, *Biological Water*, Soft and Biological Matter,
https://doi.org/10.1007/978-3-030-82503-4_12

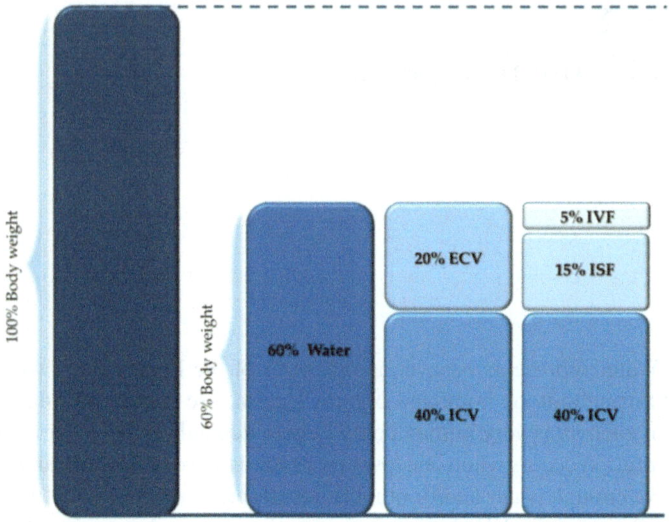

Fig. 12.1 Repartition of body water within the different compartments as percentage of body weight. ICV: intracellular volume; ECV: extracellular volume; ISF: interstitial fluid; and IVF: intravascular fluids [15]

sympathetic nervous system activity, worsen orthostatic tolerance, increase thirst sensations, arginine vasopressin release, weakened immune systems, elicit renin–angiotensin–aldosterone system activation to replenish intra- and extracellular fluid stores, have deleterious effects on cardiovascular health, induce inflammation, and affect on arterial stiffness regulation at during exercise and during orthostatic stress [1–13]. Additionally, studies determine the acute and chronic effects of hypohydration in preclinical populations. Young children, non-autonomous individuals, and the elderly of both sex are at a greater risk of dehydration.

Liquid balance is a complex interplay of numerous organs (e.g., skin, respiratory tract, kidneys, and gastrointestinal tract). The typical daily fluid turnover amounts to approximately 2600 mL (30–40 mL/kg body weight). Liquids represent about 1500 mL of daily input; 800 mL comes from liquids within solid food, and a further 300 mL comes from oxidation water. Water distribution into the different body compartments in healthy individuals is shown in Fig. 12.2.

Specifically, there are deleterious effects of acute hypohydration including reduced exercise performance, worsened mood, impaired cognitive function, altered thermoregulatory function], and decreased glycemic regulation (Fig. 12.3).

Moist environments and liquid solutions can create a favorable setting for the growth of many bacterial and some protozoal microbes. Waterborne diseases can be spread through: (1) direct contact (e.g., hydrotherapy); (2) ingestion of water (e.g., drinking water); (3) indirect contact (e.g., medical/dental tool or equipment); (4) inhalation of aerosols from water sources; and (5) blood contact (e.g., hemodialysis).

Dehydration can be quantified by a number of commonly used methods [5, 7–23]: (1) body weight change is used to determine acute hydration changes; (2)

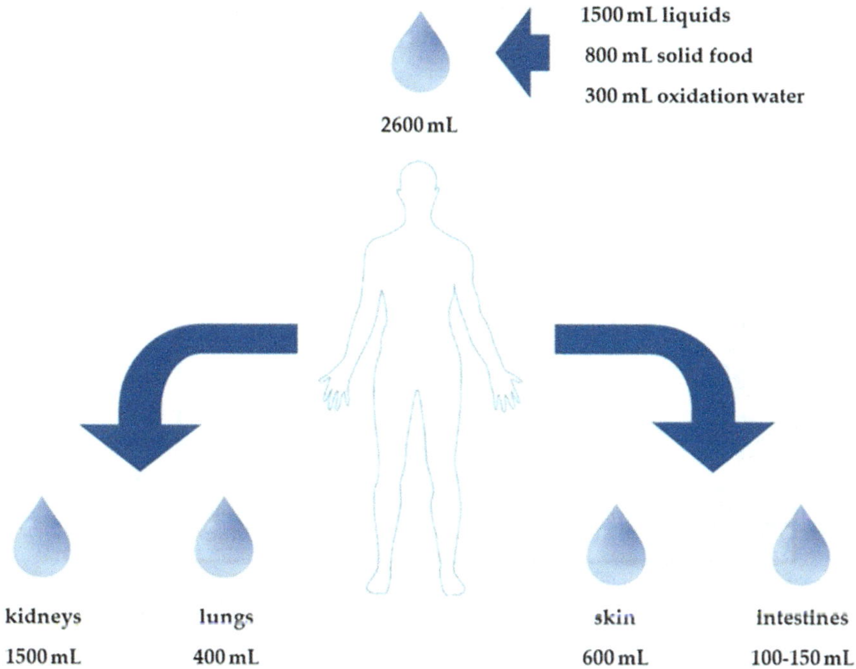

Fig. 12.2 Fluid balance and the sites involved in fluid loss [15]

bioimpedance analysis methods enable the assessment of total body water content; (3) plasma osmolality directly reflects intracellular osmolality, and it constitutes a good marker to assess acute hydration changes; (4) urinary markers appear to be good alternatives for assessing hydration status in large populations; (5) total body water (TBW) analysis can measure water balance; (6) isotope dilution methods enable the measurement of TBW content; (7) water intake and outtake; (8) Neutron activation analysis (NAA) used for determining the concentrations of elements of interest. Clinical signs and symptoms of dehydration, such as dizziness, headache, tachycardia, sweating disorders and others can be also taking in consideration. Diffusion-weighted magnetic resonance imaging (MRI) is being incorporated into general imaging practice because of its many clinical advantages [5].

12.2 Quantitative Degradation Assessment

The principle of isotope dilution methods is based on the distribution of a tracer substance after oral or intravenous administration [14, 15]. These methods enable the measurement of TBW content, thanks to a tracer that gets distributed in all body fluid compartments. In brief, a known volume and concentration of isotope

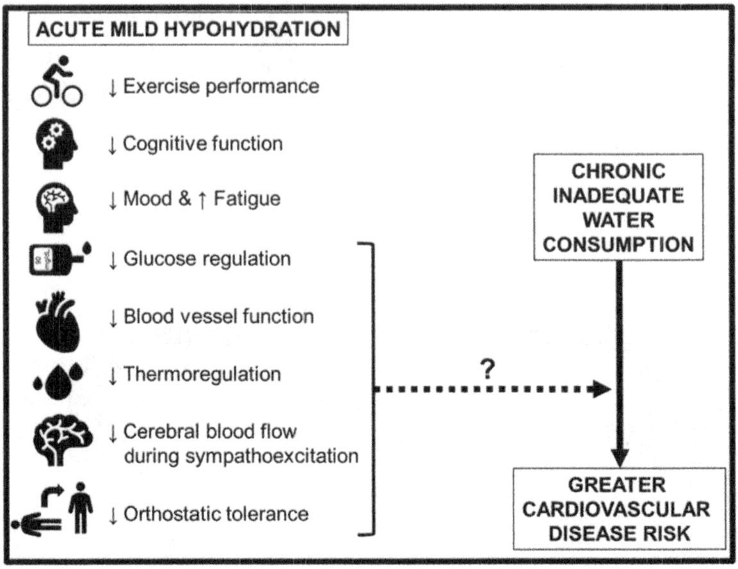

Fig. 12.3 Summary of the physiological consequences of acute mild hypohydration in healthy humans. Further research is necessary to determine whether and how these acute effects influence the poor cardiovascular health outcomes associated with chronic inadequate water consumption, impaired or reduced [7]. Published under Creative Commons CC BY 4.0 license

taken into the body, and a new concentration of the isotope is later determined in a sample of body fluid (blood, saliva, etc.) after the tracer has become distributed equally throughout the body fluids. The unknown volume (TBW) is then calculated, knowing that a low concentration of the isotope in the sample means that the body fluid volume must be relatively large and vice versa.

Bioelectrical impedance analysis (BIA) used for assessing body water content is based on the electrical properties of tissues which conduct electrical current differently depending on their water and electrolyte contents [16]. Taking this property into account, equations have been developed to linkbody resistance to the electrical current of total body water (TBW), extracellular water (ECW), and intracellular water (ICW). Bioelectrical impedance assessment (BIA) employs low amperage current (single or multiple frequency) passed between skin electrodes with the assumption that current resistance (impedance) varies inversely with tissue water and electrolyte content. BIA is well correlated with TBW measures made using isotope dilution. The BIA in assessing TBW, ECW, and ICW and to assess whether individuals can be reliably classified as being hypohydrated or hyperhydrated using lower and upper quartiles, respectively. As an example, TBW, ECW, and ICW were assessed via BIA (Bodystat, Isle of Man, UK) in 52 male and 48 female college students on 2 separate days within 1 week. Results revealed strong test–retest reliability for TBW (r = 0.983), ECW (r = 0.972), and ICW (r = 0.988) (all P's < 0.001).

Osmolality, the concentration of the substance in 1 L of water divided by its molecular weight measures the body's electrolyte–water balance [17, 18]. Because plasma osmolality reflects intracellular osmolality, it has been considered to be a good marker standard for assessing hydration status. Serum osmolality is an accurate indicator of hydration status. Glucose, urea, and electrolyte concentrations including sodium (Na^+), potassium (K^+), and chloride (Cl^-) are used to calculate an indirect estimate of serum osmolality. Plasma osmolality is controlled around a euhydration set-point of ~285 mOsm/kg (Panel DRI 2005). Euhydration is the state or situation of being in water balance. There are several different techniques employed in osmometry: (1) vapor pressure osmometers determine the concentration of osmotically active particles that reduce the vapor pressure of a solution; (2) membrane osmometers measure the osmotic pressure of a solution separated from pure solvent by a semipermeable membrane; and (3) freezing point depression osmometers may also be used to determine the osmotic strength of a solution, as osmotically active compounds depress the freezing point of a solution. https://en.wikipedia.org/wiki/Osmometer.

Urine is a solution of water and various other substances, and the concentration of those substances increases with a reduction in urine volume, which is associated with dehydration [19]. Urinalysis is a frequently used clinical measure to distinguish between normal and pathological conditions. Urinary markers for dehydration include a reduced urine volume, a high urine specific gravity (USG), a high urine osmolality (UOsm), and a dark urine color. Blood-borne markers of hydration other than osmolality include plasma volume, plasma sodium, and concentrations of fluid regulatory hormones in plasma. Plasma volume decreases proportionally with the level of dehydration

In neutron activation analysis (NAA), the nuclear process used for determining the concentrations of elements in a vast amount of materials NAA allows discrete sampling of elements as it focuses solely on its nucleus [20, 21]. The method is based on neutron activation and therefore requires a source of neutrons, when a sample is bombarded with neutrons, causing the elements to form radioactive isotopes. The radioactive emissions and radioactive decay paths for each element are well known. Using this information, it is possible to determine the concentrations of the elements within.

Determination of body weight change is the simplest method for assessing water lost during physical exercise for a short period of time. Total body water (TBW) content corresponds to about 60% of body weight [12]. Acute changes in hydration are calculated as the difference between pre- and post-exercise body mass.

Diffusion-weighted magnetic resonance imaging (DWI or DW-MRI) employs specific MRI sequences as well as software that generates images from the resulting data that uses the diffusion of water molecules to generate contrast in MR images [22–25]. This technique allows the mapping of the diffusion process of molecules, mainly water, in biological tissues, in vivo and non-invasively. To sensitize MRI images to diffusion, the magnetic field strength (B1) is varied linearly by a pulsed field gradient. Since precession is proportional to the magnet strength, the protons begin to process at different rates, resulting in dispersion of the phase and signal loss.

Another gradient pulse is applied in the same magnitude but with opposite direction to refocus or rephase the spins. The spatial resolution of DW-MRI is typically on the order of millimeters. DW-MRI is exquisitely sensitive to changes in diffusion measured on the cellular scale (e.g., micrometers). The most common approach to render magnetic resonance imaging (MRI) sensitive to diffusion is to place two strong symmetrical gradient lobes (MPGs) on either side of the 180° refocusing pulse in a spin-echo sequence (Fig. 12.4) [23]. The degree of signal attenuation in DWI depends both on the magnitude of diffusion and the amount of diffusion weighting.

Quantitative parameters characterized DWMRI are apparent diffusion coefficient (ADC) and b-value [22]. The b-value is a factor that reflects the strength and timing of the gradients used to generate diffusion-weighted images. The higher the b-value, the stronger the diffusion effects. Specifically, if $\mathbf{S_o}$ is the MR signal at baseline and **D** is the diffusion coefficient, the signal (**S**) after the diffusion gradients have been applied is given by

$$\mathbf{S} = \mathbf{S_o e^{-bD}} \tag{12.1}$$

The amount of diffusion weighting is determined by the *b*-value and is calculated as follows:

$$b = \gamma^2 G^2 \delta^2 (\Delta - \delta/3), \tag{12.2}$$

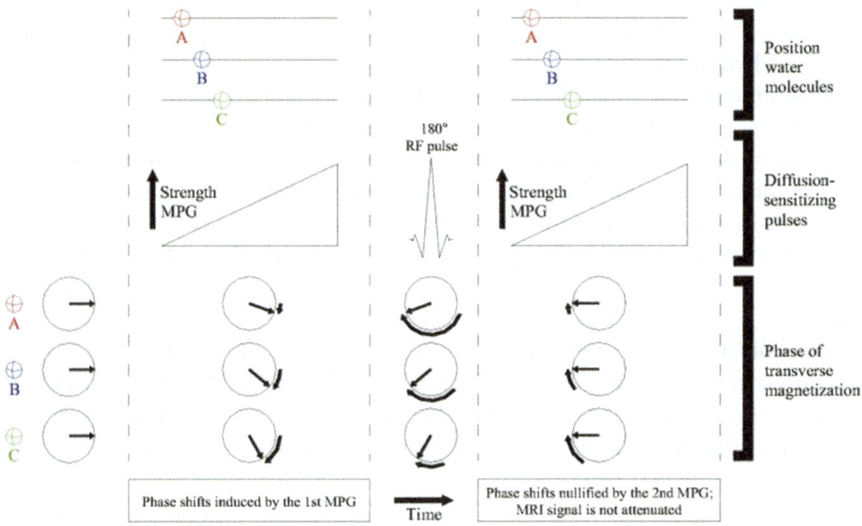

Fig. 12.4 Three stationary water molecules (*A*, *B*, *C*) in one voxel. Phase shifts induced by the first motion probing gradient (*MPG*) are nullified by the second MPG. Because of the maintained phase coherence among the three water molecules, the MRI signal is not attenuated (*RF* radio frequency) [23]

where δ is the duration of one motion probing gradient (MPG), Δ is the interval between the leading edges of the MPGs, and G is the strength of the MPG. In biological tissue, diffusion is expressed and quantified by means of an effective or apparent diffusion coefficient (ADC). If two or more images with different b-values are acquired, the ADC of a selected region of interest (ROI) can be calculated as follows:

$$ADC = (1/b_1 - b_0)\ln(S[b_1]/S[b_0]), \tag{12.3}$$

where b_1 and b_0 represent two different b-values, $S[b_1]$ the signal intensity of the selected ROI on the slice level acquired with b-value b_1, and $S[b_0]$ the signal intensity of the same ROI on the same slice level acquired with b-value.

The introduced concept of "diffusion-weighted whole-body imaging with background body signal suppression" (DWIBS) now allows acquisition of volumetric diffusion-weighted images of the entire body. As an example, Takahara et al. [26] developed the concept of DWIBS and proved the feasibility of DWI during free breathing... This approach extended the possibilities of DWI: (1) scanning time is not limited; (2) image acquisition time is not confined to a particular phase of the breathing cycle; (3) images with multiple b-values including high b-values around 1,000 s/mm^2 can be acquired; (4) thin slices can be obtained; and (5) and multiple signal averaging is possible. Thus, these advances enable volumetric [three-dimensional (3D)] image processing, including maximum intensity projections (MIPs), volume rendering, and multiplanar reformatting (MPR) in any plane.) This technique is illustrated in Fig. 12.5.

12.3 Dehydration Symptoms

https://www.mayoclinic.org/diseases-conditions/dehydration/symptoms-causes/syc-20354086.

Dehydration occurs when one use or lose more fluid than you take in, and your body does not have enough water and other fluids to carry out its normal functions. The most common cause of dehydration is severe diarrhea and vomiting https://www.mayoclinic.org/diseases-conditions/dehydration/symptoms-causes/syc-20354086.

Dehydration also can occur if one do not drink enough water during hot weather—especially if one is exercising vigorously. The three classifications of dehydration have been accepted (Fig. 12.6). Isotonic dehydration is classified as a balanced depletion of water and sodium caused by diarrhea and vomiting. Hypertonic dehydration is depletion of total body weight (TBW) due to diminished water intake, pathologic fluid loss, or both. Hypotonic dehydration is depletion of water and sodium but sodium loss predominates, resulting in extracellular fluid loss. Hypotonic dehydration may be a result of diuretics, renal disease, or decreased intake of sodium and water, https://christinagracehutson.com/blog/the-effects-of-dehydration-on-digestion.

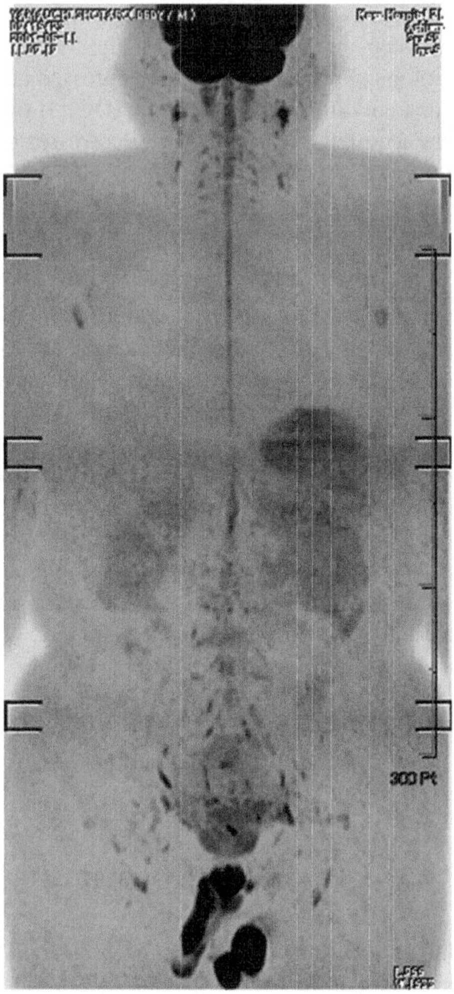

Fig. 12.5 Coronal maximum intensity projection DWIBS image (inverted black-and-white gray scale) of a healthy 60-year-old male volunteer. Normal high signal intensity (black on this image) can be seen in the brain, spinal cord, nerves of the brachial plexus, prostate, testis, penis, spleen, and cervical, axillary, pelvic, and inguinal lymph nodes [26]. Published under Creative Commons CC BY 4.0 license

Mild to severe dehydration can cause: constipation, fever, chills, headache, dizziness, poor nutrient absorption, white spots on your fingernails, hair growth, fatigue, nausea, and light headedness acid reflux, heart burn, weak immune system, slow digestion, poor nutrient absorption, nausea, feeling acidic or nauseated, painful trapped gas in your intestines or gut, changes in weight (3–5% in $<$ 30 days), decreased/concentrated urine output, dry or tenting skin, altered mental status or elevated serum sodium levels, blood urea nitrogen (BUN), hematocrit levels, extreme thirst and dry mouth, urinating less than usual, feeling tired, dark-colored urine, decreased skin turgor, meaning that when a person's skin is pinched and released, the skin does not flatten back to normal right away, sunken eyes or cheeks, and light-headedness or fainting. Regulation mechanisms of sodium and water and water balance disturbances are shown in Figs. 12.6 and 12.7.

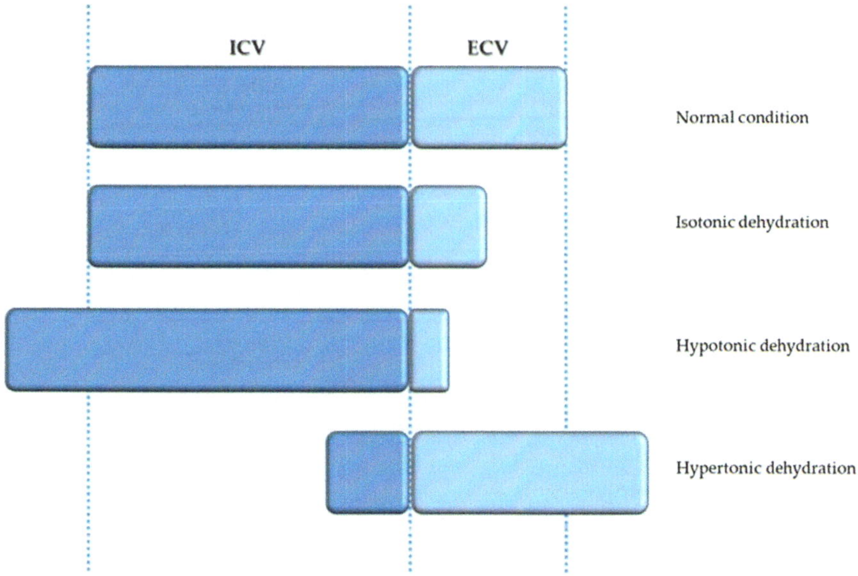

Fig. 12.6 Water balance disturbances [27]

12.4 Dehydration and Diseases: Examples

Dehydration has been linked with urological, gastrointestinal, circulatory, and neurological and mental disorders. Fluid overload has been linked with cardiopulmonary disorders, hyponatremia, edema, gastrointestinal dysfunction, and postoperative [28]. There is a growing body of evidence that links states of fluid imbalance and disease.

12.4.1 Kidneys Diseases

The kidneys are responsible for a number of bodily functions, including to filtering waste, maintaining fluid and electrolyte balance, and producing hormones to regulate red blood cell production. The body is made up of two-thirds water, which means keeping it hydrated is absolutely essential for maintaining the health of certain bodily functions like body temperature regulation, digestion, cell health, and blood flow. Kidney failure is a medical condition in which the kidneys are functioning at less than 15% of normal.

A comprehensive literature review, which examined evidence on the effects of hydration on the kidney, revealed that [29]: (1) by reducing vasopressin secretion, increasing water intake may have a beneficial effect on renal function in patients with all forms of chronic kidney disease (CKD) and in those at risk of CKD; (2) increasing water intake is a well-accepted method for preventing renal calculi; (3)

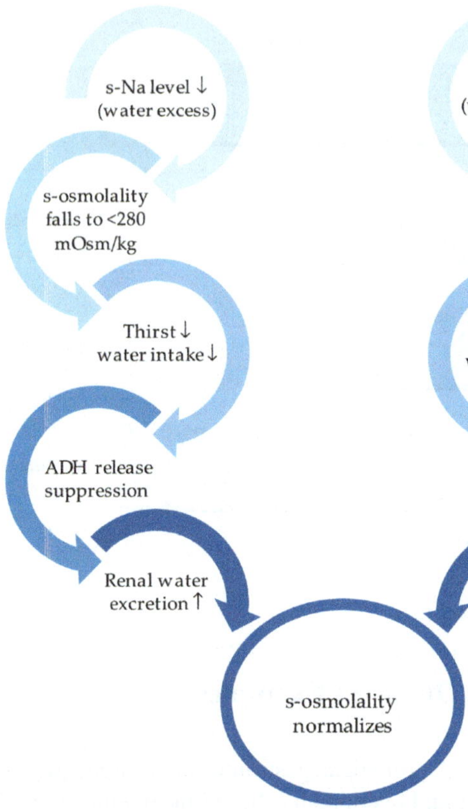

Fig. 12.7 Regulation mechanisms of sodium and water. s-Na: Serum sodium; s-osmolarity: Serum osmolarity; ADH: Antidiuretic hormone [27]

recurrent dehydration and heat stress from extreme occupational conditions is the most probable cause of an ongoing CKD epidemic in Mesoamerica; and (4) pharmacologic blockade of renal vasopressin-V2 receptors has been shown to slow cyst growth in patients.

Two studies of the same population from the Blue Mountains region in Australia reached different conclusions on the association between self-reported fluid intake (excluding plain water) and CKD [30]. The first study consisted of two cross-sectional surveys including 2744 and 2476 participants who completed a food frequency questionnaire and had baseline measurements of eGFR. (eGFR) is a number based on blood test for creatinine using to check how well the kidneys are working. It estimates how much blood passes through the glomeruli (the tiny filters in the kidneys) each minute. Results indicated that fluid intaked from food and beverages excluding water is not associated with improved kidney function or reduced mortality. CKD was present in 12.4–23.5% men and 14.9–28.7% women (mean ages 66.4–65.4 years), respectively. Participants who had the highest quintile of fluid intake (3.2 L/day) had a significantly lower risk of chronic kidney disease (CKD) (odds ratio 0.5, 95% confidence intervals (CI) 0.32 to 0.77, P for trend = 0.003). These findings were consistent both equations to calculate GFR thresholds. In **conclusion**, higher intakes of fluid appear to protect against CKD which may be preventable at a population level with low-cost increased fluid intake.

Autosomal dominant polycystic kidney disease (ADPKD) is the most common inherited kidney disease leading to end-stage kidney disease [31]. In eight ADPKD patients eating typical diets, osmolality and volume were measured in 24-h urine collections. The amount of additional ingested water required daily to achieve a mean urine osmolality of 285 ± 45 mosm/kg was determined. Blood chemistries, 24-h urine collections, BP, and weight were measured before and after the period of supplemental water intake. Five patients successive achieved the 285 mosm/kg urine target. Mean urine osmolality decreased and mean urine volume increased; serum sodium, weight, and BP were unchanged. Specifically: (1) the mean urine volume of 1.57 L/d in phase I was typical for ADPKD patients (Fig. 12.7a); (2) the mean Uosm of 496 mosm/kg (freezing point method) was increased above that of normal plasma (285 mosm/kg) in all of the patients in phase I (Fig. 12.8b); (3) the mean osmolar excretion (673 mosmol/d) was dominated by the seven women and was in general agreement with published values for women; (4) the amount of additional water in the range of 2400–3000 ml per 24 h needed to achieve a urine osmolality target based on the urine osmolar excretion in ADPKD patients with normal renal function was; and (5) during the 2-week study, urine volume and osmolality were measured, and above-mentioned additional water intake was adjusted in order to achieve a urine osmolality goal of 285 ± 45 mOsm/kg H_2O determined [31].

Relative cumulative all-cause mortality of patients with different consumption of water for 16 years is shown in Fig. 12.9 [31].

Severe heat stress with accompanying water and solute loss (with repeated volume depletion) can cause ischemic kidney injury, even in healthy adults [32]. The most probable cause of Mesoamerican nephropathy is recurrent dehydration and heat stress from extreme occupational conditions which causes subclinical kidney injury that leads to permanent kidney damage over time.

Studies [33] provided the rationale to examine whether diabetic nephropathy can be prevented or slowed by reducing vasopressin activity through increased water intake or a vasopressin receptor V2R antagonist. Diabetic nephropathy affects kidneys' ability to do their usual work of removing waste products and extra fluid from a body.

Plasma copeptin, a 39-amino acid-long peptide, a surrogate for the hormone important for maintaining fluid balance (AVP), was measured in baseline samples from patients with type 2 diabetes treated in primary care. Samples from 1328 patients were available; 349 were analyzed separately using renin–angiotensin–aldosterone system inhibition (RAASi), which influences albumin/creatinine ratio (ACR) and estimated (e)GFR. In the other 979 patients (46% men, age 68 years [34–50], ACR 1.8 mg/mmol [0.9–5.7], eGFR 67 ± 14 ml min$(^{-1})$ 1.73 m$(^{-2})$), baseline copeptin (5.3 pmol/l [3.2–9.5]) was significantly associated with log e[ACR] and eGFR. In patients with diabetes not using RAASi, a higher baseline copeptin concentration is

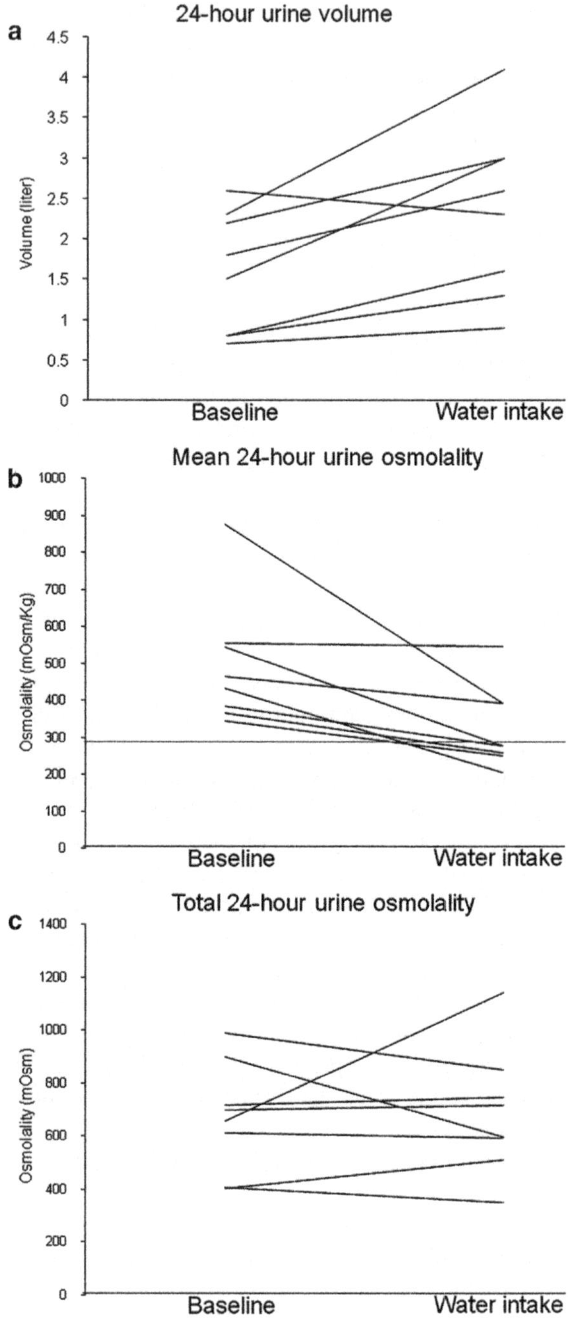

Fig. 12.8 Individual changes in urine volume, osmolality, and osmolar excretion. **a** 24-h urine volume; **b** mean 24-h urine osmolality; and **c** total 24-h urine osmolality. The horizontal line in **b** indicates urine osmolality of 285 mosm/kg [31]. With permission from American Society of Nephrology

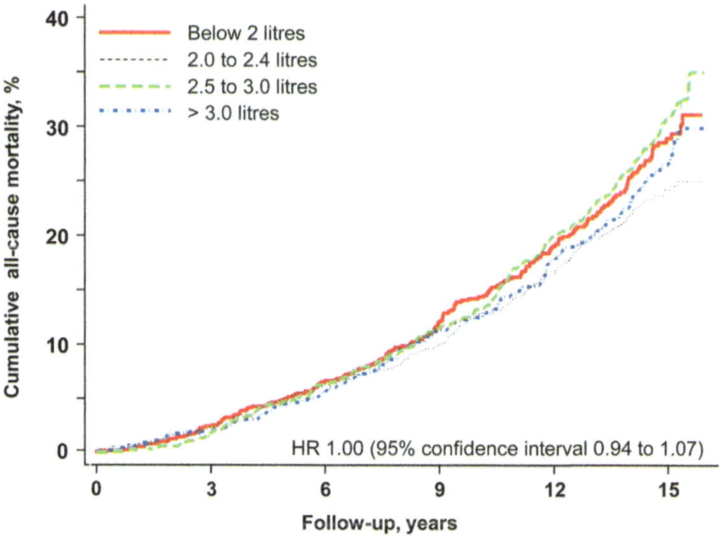

Number at risk

Less than 2.0 litres	578	549	507	463	416	210
2.0 to 2.4 litres	665	635	587	550	497	232
2.5 to 3.0 litres	584	563	517	477	425	211
>3.0 litres	520	504	475	440	402	206

Fig. 12.9 Cumulative all-cause mortality of patients [31]

significantly associated with higher baseline ACR and lower eGFR values and with a decline in eGFR during follow-up.

The associations between water intake, chronic kidney disease (CKD), and CVD for of 3427 adults (mean age 46, range 20–84) were examined by a cross-sectional analysis of the 2005–2006 National Health and Nutrition Examination Survey [51]. The results provided additional evidence about associations between low total water intake and CKD (eGFR 30–60 ml/min/1.73 m^2) and suggested potentially protective effect of higher total water intake, particularly plain water, on the kidney.

The connection between fluid consumption and long-term mortality and kidney function was examined [52]. A longitudinal analysis within a prospective, population-based cohort study of 3858 men and women aged 49 years or older residing in Australia was conducted. Daily fluid intake from food and beverages not including water was measured using a food frequency questionnaire. Multivariable adjusted Cox proportional hazard models for all-cause and cardiovascular mortality were used. In addition, a boot-strapping procedure for estimated glomerular filtration rate (eGFR) was employed. (eGFR) is a number based on blood test for creatinine using to check how well the kidneys are working. It estimates how much blood passes through the glomeruli (the tiny filters in the kidneys) each minute. Results indicated that fluid intaked from food and beverages excluding water is not associated with improved kidney function or reduced mortality.

Effect dehydration on Mesoamerican nephropathy, nephrolithiasis, autosomal dominant polycystic kidney disease (ADPKD), and end-stage kidney was in focus of review [29]. End-stage kidney disease acute (kidney failure) is characterized by blood pressure, blockage of the urinary tract, certain medications, muscle breakdown, and hemolytic uremic syndrome. Several observational studies have reported positive associations between increased water intake (or high urine volume) and preservation of renal function, and high water intake is a well-accepted method for preventing renal calculi. By contrast, high water intake is contraindicated in dialysis-dependent patients.

Evidence for the effects of hydration on plasma levels of vasopressin and kidney function and in patients with chronic kidney disease (CKD), diabetic nephropathy, Mesoamerican nephropathy, nephrolithiasis, autosomal dominant polycystic kidney disease (ADPKD), and end-stage kidney disease was examined [53]. In polycystic kidney disease (PKD), increased water intake has been shown to slow renal cyst growth in animals via direct vasopressin suppression, and pharmacologic blockade of renal vasopressin-V2 receptors led to slow cyst growth in patients. Six observational studies in humans have reported on the potential role of water and/or fluid intake in CKD. In a post-hoc analysis of 581 patients in the Modification of Diet in Renal Disease study with an eGFR between 25 and 55 ml/min/1.73 m^2, higher urine volume was associated with a faster decline of the eGFR. By reducing vasopressin secretion, increasing water intake appears to have a beneficial effect on renal function in patients with all forms of CKD and in those at risk of CKD. Positive associations between increased water intake (or high urine volume) and preservation of renal function and high water intake which is an accepted method for preventing renal calculi were reported [53]. In PKD, increased water intake has been shown to slow renal cyst growth in animals via direct vasopressin suppression, and pharmacologic blockade of renal vasopressin-V2Rs has been shown to slow cyst growth in patients. Described studies support the exploration of a potential therapeutic role for increased hydration in CKD, diabetic nephropathy, Mesoamerican nephropathy, PKD and renal transplantation in well-designed randomized controlled trials.

The relationship between urine volume and renal decline over 6 years in a large community-based cohort was evaluated [53]. In a 7-year prospective cohort study of 2144 participants free of CKD at baseline, decline in kidney function was significantly slower in those with higher versus lower 24-h urine volumes. Specifically, 2148 participants provided valid 24-h urine samples, grouped as < 1 L/d (14.5%); 1–1.9 L/d (51.5%); 2–2.9 L/d (26.3%); and ≥ 3 L/d (7.7%). Overall, eGFR, a measurement of how much liquid and waste is passing from the blood through the tiny filters in the kidney, declined by 1% per year, with 10% demonstrating rapid decline and 40% demonstrating mild-to-moderate decline. An inverse, graded relationship was evident between urine volume and eGFR decline and for each increasing category of 24-h urine volume, percentage annual eGFR decline was progressively slower, from 1.3, 1.0, 0.8, to 0.5%.

Increasing water intake at least 2.0–2.5 L/day is a well-accepted method for reducing the recurrence of kidney stones [54]. Higher water intake protects against stone formation by increasing urine volume and flow rate, which lowers the supersaturation of calcium oxalate, calcium phosphate, and uric acid [54]. Studies that reported relative risks, odds ratios, or hazard ratios comparing the risk of kidney stone events in patients with high vs inadequate fluid intake were described. Nine studies [2 randomized controlled trials (RCTs) with 269 patients; 7 observational studies with 273,685 individuals] were included in the meta-analysis. Pooled risk ratios (RRs) and 95% confidence intervals (CIs) were calculated using a random-effect, generic inverse variance method. Pooled RRs of kidney stones in individuals with high-fluid intake were 0.40 (95% CI 0.20–0.79) and 0.49 (0.34–0.71) in RCTs and observational studies, respectively. High fluid intake was significantly associated with reduced risk of recurrent kidney stones: RRs 0.40 (95% CI 0.20–0.79) and 0.20 (0.09–0.44) in RCTs and observational studies, respectively. This analysis demonstrated a significantly reduced risk of incident kidney stones and risk of recurrent kidney stones among individuals with high fluid consumption.

12.4.2 Mental Diseases

Dementia is a general term for loss of memory, language, problem-solving, and other thinking abilities that are severe enough to interfere with daily life. Alzheimer's is the most common cause of dementia [55–61]. Dementia is associated with increased rates and often poorer outcomes of hospitalization, including worsening cognitive status. Objective of the study [55] was to determine whether dementia onset is associated with higher rates of or different reasons for hospitalization, particularly for ambulatory care-sensitive conditions (ACSCs). On this path, retrospective analysis of hospitalizations among 3019 participants in Adult Changes in Thought (ACT), a longitudinal cohort study of adults aged 65 years or older enrolled in an integrated healthcare system has been performed. The authors concluded that adjusted admission rates for all types of ACSCs, including bacterial pneumonia, congestive heart failure, dehydration, duodenal ulcer, and urinary tract infection, were significantly higher among those with dementia.

Abnormalities of water homeostasis can be expressions of neuronal dysfunction, brain atrophy, chronic cerebrovasculopathy and neurodegenerative disease. The aim of study [57] was to analyze the serum osmolality of subjects with cognitive impairment Serum osmolality was confirmed to be an accurate indicator of hydration status in older adults. Glucose, urea, and electrolyte concentrations [sodium (Na^+), potassium (K^+), and chloride (Cl^-)] were used to calculate an indirect estimate of serum osmolality. One thousand and ninety-one consecutive patients attending the Alzheimer's Evaluation Unit were evaluated with the Mini-Mental State Examination (MMSE), 21-Item Hamilton Depression Rating Scale (HDRS-21), Activities of Daily Living (ADL), Instrumental-ADL (IADL), Mini Nutritional Assessment (Fig. 12.10), Exton-Smith Scale (ESS), and Cumulative Illness Rating Scale (CIRS). For each patient, the equation for serum osmolality developed by Khajuria and Krahn, which determines the relationship of serum osmolality with sodium, potassium, urea, glucose, ethanol, methanol, and ethylene glycol, was applied.

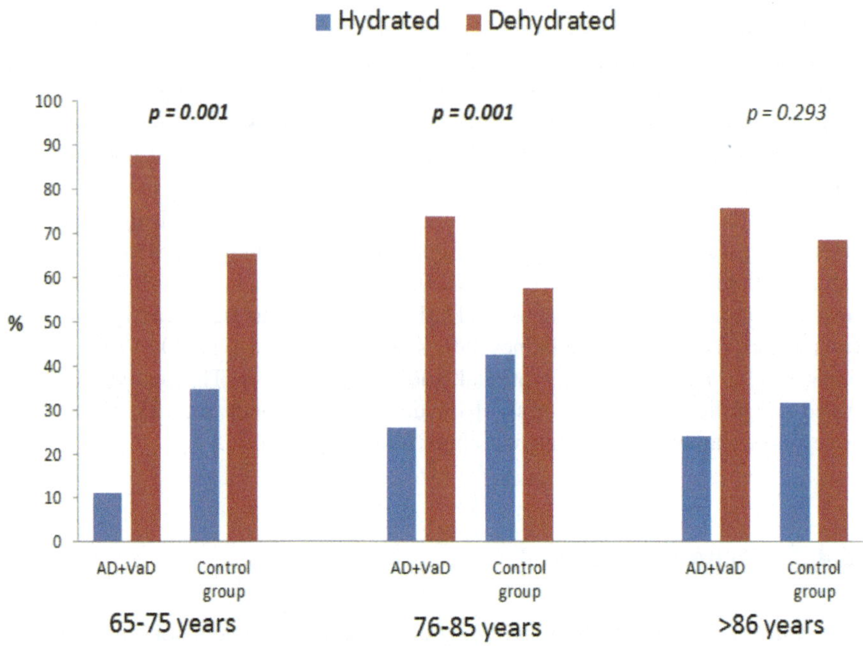

Fig. 12.10 Distribution of dementia (AD and VaD) patients and control group according to dehydration risk and age [57]. Published under Creative Commons CC BY 4.0 license

Five hundred and seventy-one patients had cognitive decline and/or depression mood (CD-DM) and 520 did not have CD-DM (control group). Patients with CD-DM, which were older have a significant clear cognitive impairment, show the presence of a depressive mood and have major impairments in ADL and ESS compared to the control group. The observations revealed that (1) CD-DM patients had a higher electrolyte concentration (Na^+, K^+, Cl^-), high risk of dehydration (osmolality), and kidney was damaged more than the control group; (2) Alzheimer's disease (AD) patients showed a major risk for current dehydration and dehydration was associated with the risk of developing a type of dementia, like AD or vascular dementia (VaD) Figs. 12.9 and 12.10); (3) according to the multivariate analysis, the presence of dehydration state was associated with ADL and IADL, and independently associated with age, and MNA (Fig. 12.10); and (4) younger patients with dementia were significantly more dehydrated than patients without dementia (65–75 years, $p = 0.001$; 76–85 years, $p = 0.001$) (Fig. 12.11).

Above-mentioned results support the hydromolecular hypothesis explaining the relationship between dehydration and cognitive impairment in older patients as the result of protein misfolding and aggregation, in the presence of a low interstitial fluid volume, Current findings in the field suggested that particular cognitive abilities and mood states are positively influenced by water consumption [58]. The impact of dehydration on cognition and mood is particularly relevant for those with poor

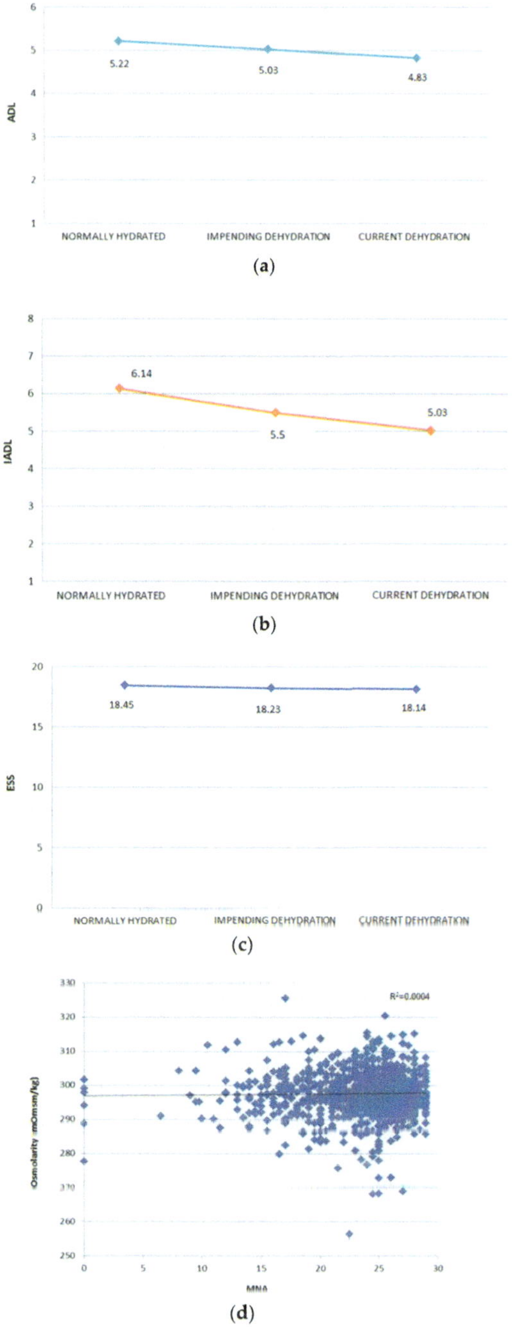

Fig. 12.11 Linear regression model of dehydration risk according to Mini Nutritional Assessment (MNA) scores [57]. Published under Creative Commons CC BY 4.0

Fig. 12.11 (continued)

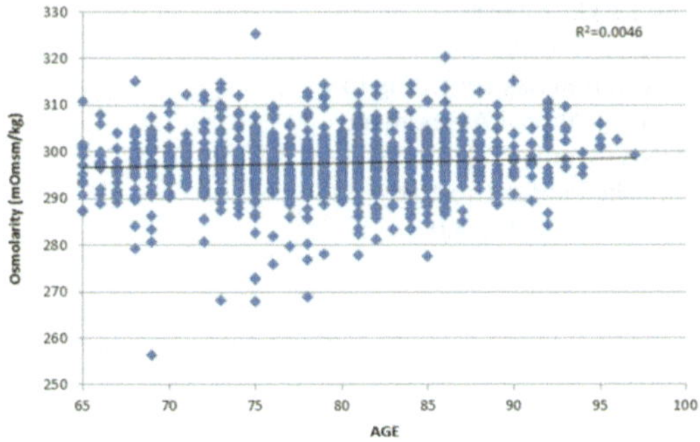

(e)

fluid regulation, such as the elderly and children. Critical review on advances in both behavioral and neuroimaging studies of dehydration and link the findings to the known effects of water on hormonal, neurochemical and vascular functions was presented. Studies measuring self-reported changes in mental state found associations between dehydration and mood, in conjunction with changes in performance were carried out. Subsequent studies both in a military population and in the general population supported an evidence of detriments in cognitive abilities with induced dehydration. Severe dehydration has been shown to cause cognitive deficits such as short-term memory and visual perceptual abilities as well as mood disturbance.

According to a proposed mechanism for of cognitive deficits during dehydration levels of cortisol, often released at a stress response, could be increased [60]. The beneficial effects of drinking supplementary water during the school day on the cognitive performance and transitory subjective states, such as fatigue or vigor, in 168 children aged between 9 and 11 years who were living in a hot climate was investigated. The classes were randomly divided into an intervention group, which received water supplementation, and a control group. Dehydration was determined by urine sampling and was defined as urine osmolality greater than 800 mOsm/kg H_2O. The change in the scores from the morning to the afternoon of hydration levels, cognitive performance and transitory subjective states were correlated. our The results can be summarized as fallows: (1) a remarkable proportion of children were in a state of mild, voluntary dehydration at the beginning of the school day (84%); (2) a significant negative correlation between dehydration and the auditory number span, which indicates a beneficial effect of drinking supplementary water at school on short-term memory were revealed; (3) there was a positive correlation between dehydration and performance in the verbal analogy task. The results are discussed in light of the complexity of the neurobiological mechanisms involved in the relationship between hydration status and cognition.

12.4.3 Gastrointestinal Tract

In addition to water consumed in food (1 L/d) and beverages (~2–3 L/d), digestive secretions account for a far greater portion of water that passes through and is absorbed by the gastrointestinal tract (~8 L/d). The majority of this water is absorbed by the small intestine, with a capacity of up to 15 L/d with the colon absorbing some 5 L/d.

Dehydration is often a concern for Gastrointestinal (GI) diseases such as digestive diseases and gastrointestinal disorders patients, suffering from Crohn's, Colitis and i**rritable bowel** syndrome (IBS) [1, 5, 62–65]. A major cause of poor digestion is dehydration. Lack of water, calcium and magnesium can cause ulcers, gastritis, colitis, and acid reflux because the stomach doesn't have enough H_2O to produce digestive acid. Review [27] highlights and discusses dehydration, as well as its pathophysiology, risk factors, and clinical signs/symptoms, focusing on dysphagic patients in particular. Possible management strategies of dehydration are presented for oral, enteral and parenteral fluid replacement. Excessive intake of water can occasionally be harmful. If too much plain water is drunk over a short period of time, this can dilute the level of salt in the blood, disturbing the body's balance of electrolytes. Symptoms of overhydration include nausea, vomiting and headache

Although dysphagia (swallowing difficulties) can be associated with malnutrition, dehydration, pneumonia, reduced functional outcome, and mortality, a major cause of poor digestion is dehydration (Fig. 12.12) [27]. Dehydration is one of the most common causes of chronic constipation. The food one eats makes its way from your stomach to the large intestine or colon. If there is not enough water in a body already, the large intestine soaks up water from your food waste. In dysphagic patients, dehydration is frequent and often accelerated as a result of limited fluid intake resulting from loss of water from the intracellular space disturbing the normal levels of electrolytes and fluid interfering with metabolic processes and body functions. Dysphagic patients at risk of dehydration require close monitoring of their hydration state. Dehydration should be prevented by oral, enteral, and parenteral fluid replacement.

Diarrhea, a common symptom of GI diseases, causes the body to lose fluids and electrolytes rapidly, which must then be replaced to prevent dehydration. In most cases, water and electrolytes can be replaced by drinking water and through proper diet [1, 5, 27, 42, 63–65]. Diarrhea or diarrhoea is the condition of having at least three loose, liquid or watery bowel movements each day. In 2015, it was estimated that worldwide 577,000 children aged < 5 years and 502,000 adults aged > 70 years will die from diarrheal diseases. Diarrhea results from excessive secretion and/or impaired absorption of fluid and electrolytes across the intestinal epithelium.

The acute treatment of infectious diarrheal diseases centers around prompt and complete amelioration of dehydration. In hospitals, this treatment is often achieved through intravenous fluid replacement [63]. In non-hospital settings, oral rehydration solution (ORS) is the mainstay of treatment. DripDrop, an oral rehydration powder that contains a precise ratio of electrolytes, including sodium and potassium, is designed to help replenish the electrolytes and water that people lose due to diarrhea,

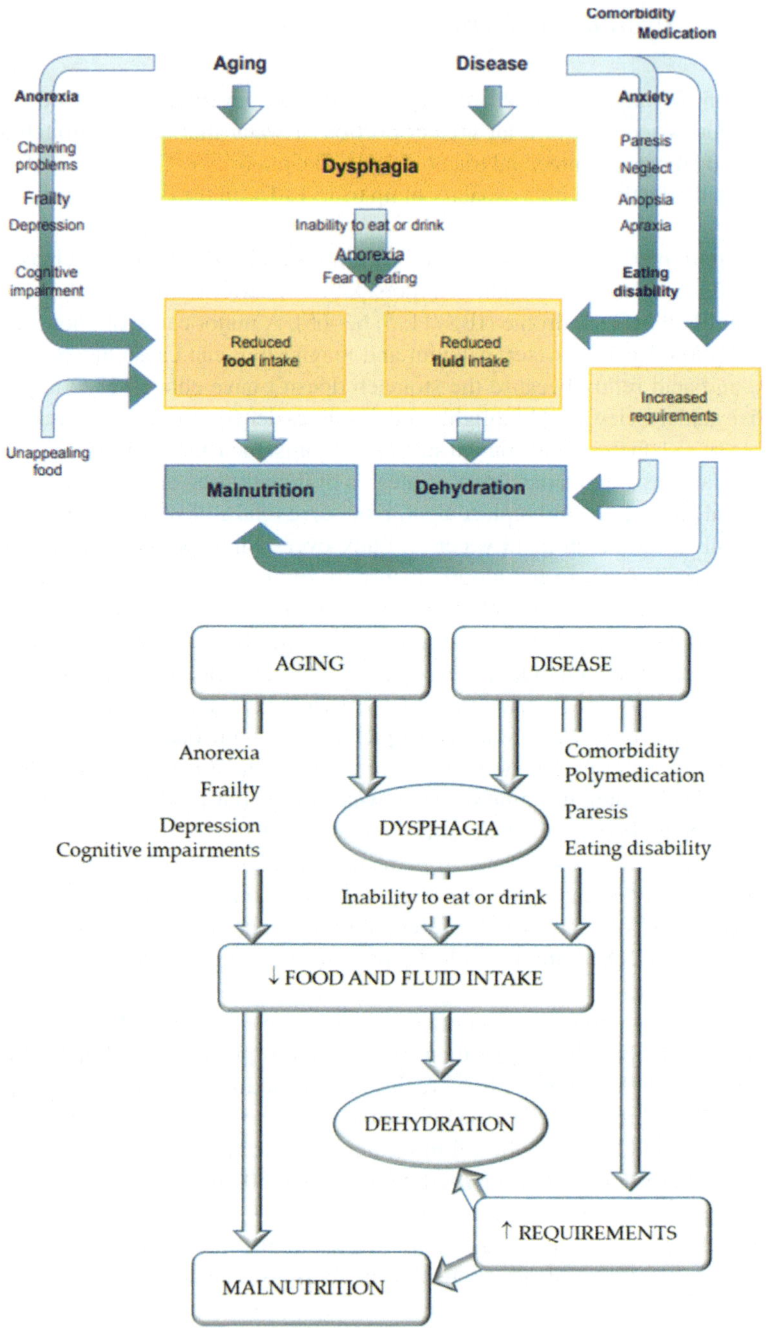

Fig. 12.12 Risk factors for dehydration in dysphagia, modified after the work of [27]. Published under the Creative Commons Attribution License (CC BY)

vomiting, or excessive sweating. DripDrop also speeds up absorption of water and electrolytes by the small intestine. The use of hypodermoclysis HDC (fluid infusion into the subcutaneous space using a fine cannula) may be considered in the case of mild dehydration when there are contraindications for oral/enteral treatment. A *cannula* is a tube that can be inserted into the body, often for the delivery or removal of fluid or for the gathering of samples.

Secretory diarrheas [65] are caused by certain bacterial and viral infections, inflammatory processes, drugs and genetic disorders. Fluid secretion across the intestinal epithelium in secretory diarrheas involves multiple ion (mainly Na^+, Cl^-, $HCO^{3-)}$ and solute transporters (mainly glucose), as well as activation of cyclic nucleotide and Ca^{2+} signaling pathways. Although antibiotics such as doxycycline, trimethoprim-sulphamethoxzole, norfloxacin, and ciprofloxacin are effective in reducing symptoms and the duration of infectious diarrheas, they do not prevent immediate dehydration.

Antidiarrheal agents such as loperamide and diphenoxylate (antidiarrheal medicines), cholestyramine, codeine sulfate, and pepto-bismol might be useful as adjuncts or additives to ORS, reducing diarrheal output and hence protecting against severe dehydration particularly in the early stages of disease, and secondarily promoting ORS usage. These therapeutics could provide life-saving alleviation of fluid loss.

Dehydration is often a concern for disorders patients, suffering from Crohn's, Colitis and irritable bowel syndrome *(IBS)*. https://dripdrop.com/blogs/news/dehydration-and-gi-diseases-why-hydration-is-a-challenge-for-patients, https://www.crohnsandcolitis.org.uk/about-crohns-and-colitis/publications/dehydration.

Crohn's disease is a type of *inflammatory bowel disease* (IBD) that may affect any segment of the gastrointestinal tract from the mouth to the anus. IBS is a common disorder that affects the large intestine. Crohn's and colitis patients, it can be challenging to maintain this water/electrolyte balance. Crohn's disease and ulcerative colitis, which both affect hundreds of thousands in the USA can cause diarrhea, a leading cause of dehydration. A peptic ulcer is a sore on the lining of the esophagus, stomach, or duodenum. Drinking 2 L of water daily and a large glass of water before meals can reduce peptic ulcers.

12.4.4 Cancer

Diffusion-weighted magnetic resonance imaging (**DW-MRI**)) provides functional information and can be used in oncology for the detection and characterization of pathologic processes, including malignant tumors [66]. A concept of "diffusion-weighted whole-body imaging with background body signal suppression" (DWIBS) now allows acquisition of volumetric diffusion-weighted images of the entire body. Review [66] outlined the biological basis of observations made on DW-MRI and

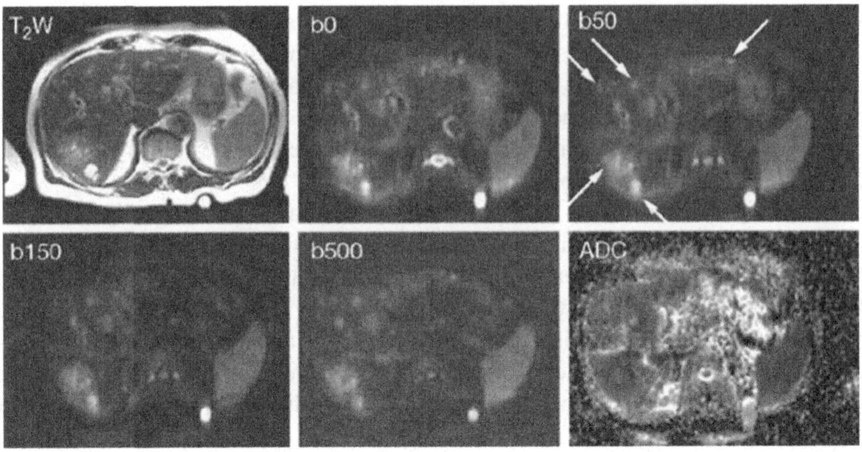

Fig. 12.13 Typical series of diffusion-weighted images of endometrial cancer in the pelvis. The top row shows diffusion-weighted magnitude images acquired for b-values ranging from 0–1000 s/mm^2; note that the signal from fat is suppressed. In the bottom row, an anatomical T2-weighted image (labeled T2W) is shown on the left, together with ADC maps calculated from the different b-value images as indicated. The high-signal regions in the ADC 0–50 s/mm^2 map indicate the presence of fast diffusion in flowing blood (arrows). Blood flow contributions to ADC values are minimized in the ADC 150–1000 s/mm^2 map. The bottom row, comprises semitransparent images in which the anatomical T2-weighted image is superimposed on the ADC 0–1000 s/mm^2 map. On the ADC 150–1000 s/mm^2 map, we note lower ADC values in the endometrial cancer and in right-sided pelvic side-wall lymph nodes (slanting upward arrow). In addition, note the higher ADC values in the myometrium (arrow [66])

describes how measurements are acquired and quantified and discusses the interpretation of images and limitations of the technique. The strength of evidence for adoption of DW-MRI as a biomarker for the assessment of tumor response was presented. The range of b-values (Sect. 12.2) varies according to the anatomical region being evaluated (Fig. 12.13). On b_0 s/mm^2 images, fat tissue signal is suppressed and water appears bright. The loss of signal intensity caused by blood flow will render vessels dark. The cellularity and tissue structure, areas retaining high signal intensity on b_{800}–$b_{1,000}$ s/mm^2 images indicate highly cellular tissues.

Purpose of work [67] was to determine the diagnostic accuracy of diffusion-weighted magnetic resonance imaging for discrimination of malignant neck nodes due to lymphoma, squamous cell carcinoma (SCC), and undifferentiated nasopharyngeal carcinoma (NPC), with histologic findings and imaging criteria as reference standards [67]. Forty-three patients (34 men, nine women; mean age, 54 years) with 43 nodes underwent imaging. The following results were reported: mean apparent diffusion coefficient (ADC) values for lymphoma ($n = 8$), NPC ($n = 17$), and SCC ($n = 18$) were (0.664 ± 0.071 [standard deviation]) × 10^{-3} mm^2)/s, (0.802 ± 0.128) × 10^{-3} mm^2)/s, and (1.057 ± 0.169) × 10^{-3} mm^2/s, respectively, with significant differences between SCC and lymphoma or NPC ($P < 0.001$) and between NPC and lymphoma ($P = 0.04$).

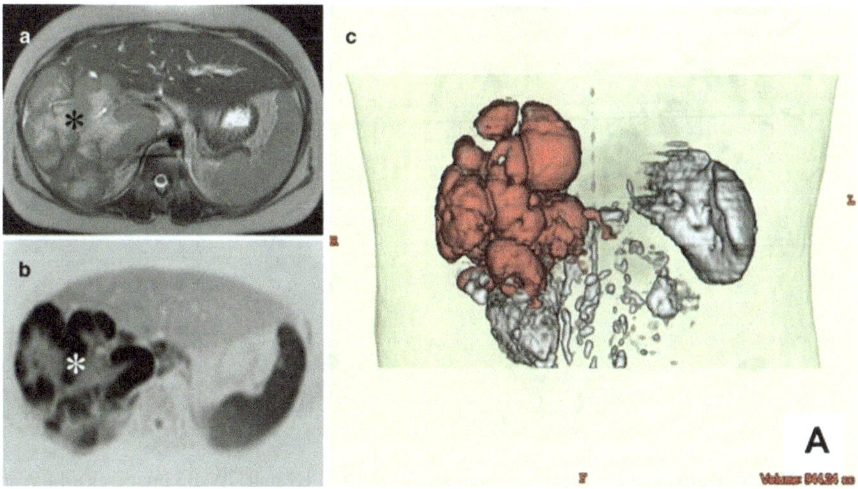

Fig. 12.14 Volume measurement of a primary hepatic diffuse large B-cell lymphoma in a 10-year-old girl using DWIBS. **a** Axial respiratory-gated T2-weighted image shows a bulky mass occupying most of the right lobe of the liver (*asterisk*). **b** Axial DWIBS image (inverted black-and-white grayscale) at the same level clearly visualizes the tumor with excellent contrast between the tumor and surrounding tissue (*asterisk*). **c** Volume-rendered image using thin slice (4 mm, no gap) DWIBS dataset shows the three-dimensional shape of the tumor (colored *red*) and allows for tumor volume measurement [22]

DWIBS was applied to total volume measurements to detect functional changes in areas involved with tumor before structural changes become visible [22, 26] (Figs. 12.14 and 12.15). Currently, computer tomography (CT) and anatomical MRI are the most commonly used means for the follow-up of cancer patients. Serial measurements of tumor dimensions are made, and criteria, such as the World Health Organization (WHO) or Response Evaluation Criteria in Solid Tumors (RECIST), were applied to assess therapeutic efficacy to detect a large B-cell lymphoma (Fig. 12.14). B-cell lymphoma is a cancer that forms in white blood cells called lymphocytes.

Importantly, DWIBS is able to detect functional changes in areas involved with tumor before structural changes become visible (Fig. 12.15).

The authors concluded that DWIBS highlights areas with restricted diffusion, such as occurs in many malignant primary and metastatic tumors, and provides an outstanding visualization of lymph nodes. The methods has potential use in tumor staging, monitoring response to cancer therapy, and in the detection of tumor persistence or recurrence.

Aim of review [68] was to summarize the current understanding of the pathophysiologic basis for DW-MRI imaging to describe widely accepted methods that depict diffusivity in the extravascular extracellular space and to provide recommendations on standards for measurement, image display, and analysis methods. Approximately 100 experts and stakeholders summarized the current understanding of DW-MRI and reached consensus on the technique use as a cancer imaging biomarker.

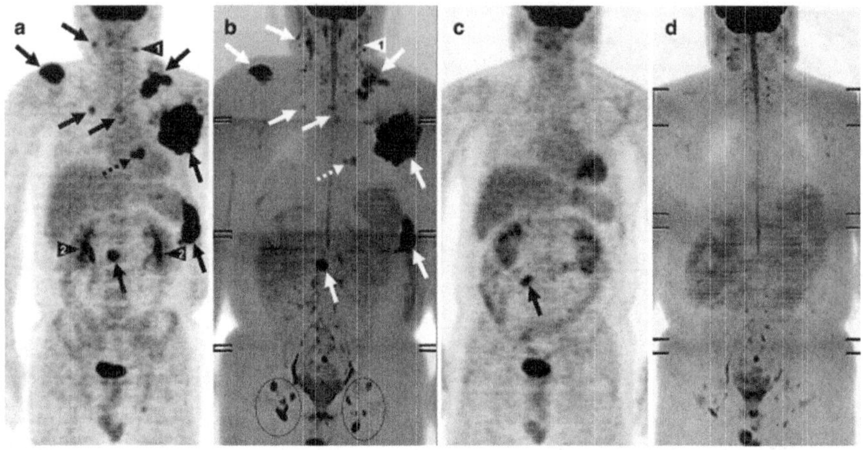

Fig. 12.15 Comparison of pre- and post-chemotherapy FDG-PET and DWIBS images in a 44-year-old man with diffuse large B-cell lymphoma. **a** Coronal maximum intensity projection FDG-PET and **b** coronal maximum intensity projection DWIBS images before initiation of chemotherapy. Both images show cervical, bilateral supra/infraclavicular, mediastinal, left axillary, para-aortic lymph node, and splenic involvement (*arrows*), and cardiac involvement (*dashed arrow*). **c** FDG-PET and **d** DWIBS images at the end of treatment show resolvement of all preexisting lesions. **a**, **b** A limitation of DWIBS is that discrimination between normal and metastatic lymph nodes is still based on size criteria; the FDG-PET positive left cervical lymph node (*arrowhead 1*) cannot conclusively be identified as malignant on DWIBS (*arrowhead 1*). DWIBS also shows prominent bilateral inguinal lymph nodes (*encircled*) which are normal according to FDG-PET. On the other hand, thanks to its higher spatial resolution, DWIBS visualizes two separate cardiac lesions (*dashed arrow*), whereas FDG-PET shows only one large cardiac lesion (*dashed arrow*). DWIBS also allows better evaluation of the urinary tract than FDG-PET, where potential lesions can be obscured because of FDG accumulation (*arrowheads 2*). **c** Note the physiological FDG uptake in the large intestine (*arrow*), which should not be confused with persistent malignant lymphoma [22]

12.4.5 Miscellaneous Diseases

Patients with **liver cirrhosis** were a group particularly exposed to the development of water and electrolyte disturbances, having a significant impact on the course of the disease and the quality of life [69]. It was noted that the electrical bioimpedance (BIA) method makes it possible to assess the total body water volume, extracellular space, intracellular space, as well as the content of sodium ions, which affects the flow of fluids between these spaces. Importance to maintain the water and electrolyte balance in a state of equilibrium was stressed.

Fluid overload and hypertension are among the most important risk factors for haemodialysis (HD) patients [70]. Most of the reference patients were normohydrated ($\Delta HS = 0.25 \pm 1.15$ L) at the start of the HD session. Multivariate adjusted all-cause mortality was significantly increased in the hyperhydrated patient group (hazard ratio = 3.41). Importance of normohydration for the long-term survival of hemodialysis patients was emphasized. Bioelectrical impedance devices were used as an aid to clinical assessment in helping to determine normohydrated weight or target weight for hemodialysis patients [34]. Bioimpedance spectroscopy was measured pre-dialysis and post-dialysis in 49 hemodialysis patients, 70% male; mean age 63.9 ± 17.9 years. The change in mean overhydration (OH) did not correlate with measured weight loss on univariate analysis ($r = 0.26$ and $P = 0.07$) but positively correlated with the change in extracellular water (ECW) ($r = 0.29$, $P = 0.046$) and negatively with intracellular water ($r = -0.58$, $P < 0.001$).

Water transport and local hydration of the airways play a critical role in the lungs, with dysfunction of airway water balance commonly associated with disease states such as cystic fibrosis and exercise-induced bronchoconstriction [35]. The bronchial circulation, which arises from the systemic circulation, is the main supplier of water to the airways. The impact of systemic dehydration on lung function in healthy individuals and the role of local hydration status on any observed changes were examined. Seven healthy young adults participated in a randomized crossover study that involved spirometry and body plethysmography at baseline (euhydration). The study was carried out after 28 h of fluid restriction (systemic dehydration) and after 1 h of systemic (oral fluid intake) or local (ultrasonic nebulization of isotonic saline) rehydration (rehydrated). Hydration status was quantified via changes in body mass and plasma osmolality. The experiments revealed that fluid restriction induced mild dehydration, with an average body mass loss of 2.5% and an increase in plasma osmolality from 292 ± 2 to 298 ± 1 mOsm kg^{-1}. These changes were at least partly reversed by systemic rehydration. Forced vital capacity (FVC) decreased by 122 ml following dehydration and returned to baseline post-rehydration. Neither total lung capacity (TLC) nor residual volume (RV) were affected significantly by hydration status; however, RV/TLC increased by 2.1% following dehydration Functional residual capacity (FRC) increased post-dehydration by 143 ml.

Hyperpnea is the increased volume of air during breathing. It can occur with or without an increase in respiration rate, and it is characterized by deep breathing. To test the hypothesis that, in humans, hypohydration attenuates hyperthermic hyperpnea, increasing volume of air during exercise in the heat, thirteen male subjects performed a fluid replacement (FR) and a no fluid replacement (NFR) trial in random order [36]. The subjects performed two bouts of cycle exercise (Ex1 and Ex2, 30–60 min) at 50% peak oxygen uptake ($V_{O}2$ peak) in 35 °C separated by a 70- to 80-min rest period, during which they drank water containing 25 mosmol/l sodium in the fluid replacement (FR) trial but not the no-fluid replacement (NFR) trial. The body fluid status differed between the trials before Ex2 (the difference in plasma osmolality before Ex2 was 9.4 mosmol/kgH_2O; plasma volume was 7.6%, and body weight was 2.5%). Hypohydration-induced reductions in skin blood flow were found to be at least partially attributed to altered postsynaptic function in healthy young male adults. The drinking in the NFR trial nearly restored the body fluid to the euhydrated condition.

Two experiments to investigate the effects of dehydration on pulse wave velocity (PWV), the velocity at which the blood pressure pulse propagates through the circulatory system, at rest and during passive heat stress were performed [37]. In experiment 1, subjects ($n = 19$) completed two trials, one in which they arrived euhydrated and one dehydrated (1.2% body mass loss). In experiment 2, subjects ($n = 11$) began two trials euhydrated and in one trial did not receive water during heat stress, thus becoming dehydrated (1.6% body mass loss); the other trial subjects remained euhydrated. Using Doppler ultrasound, carotid-to-femoral (central) and carotid-to-radial (peripheral) PWVs were measured at a normothermic baseline, and at a 0.5 and 1 °C elevation in rectal temperature (via passive heating). The main results of the experiments are: (1) in experiment 1, baseline central PWV was significantly higher when euhydrated compared to dehydrated (628 vs. 572 cm s^{-1}, respectively; $P < 0.05$); (2) peripheral PWV was unaffected (861 vs. 825 cm s^{-1}; $P > 0{\cdot}05$); (3) starting euhydrated and becoming dehydrated during heating in experiment 2 did not affect PWV measures ($P > 0.05$); and (4) independent of hydration status peripheral PWV was reduced when rectal temperature was elevated 0.5 °C (−74 cm s^{-1}; $P < 0.05$) and 1.0 °C (−70 cm s^{-1}; $P < 0.05$). These data suggest that hydration status affects measurements of central PWV in normothermic, resting condition.

Hypohydration has been suggested as a predisposing factor for several pathologies including cardiovascular diseases (CVD) [39]. The effect of hypohydration on endothelial function in young healthy adults was examined. Endothelial cells form the barrier between vessels and tissue and control the flow of substances and fluid into and out of a tissue [39]. Ten young, healthy males participated in a crossover study (age: 24.3 ± 2.3 year; weight: 80.8 ± 5.3 kg; BMI: 24.3 ± 0.4 kg m^{-2}). Each subject completed two measurements of endothelial function by flow-mediated dilation (FMD) in euhydrated and hypohydrated state separated by 24 h. Following baseline assessment of hydration status and FMD, the subjects completed 100 min of low-intensity intermittent walking exercise to achieve hypohydration of − 2% of individual body mass. Hypohydration by − 1.9% of body mass resulted in decreased plasma volume by − 3.5 ± 1.8% and increased plasma osmolality by 9 ± 2 mmol kg^{-1} ($P < 0.001$). FMD as a response to hypohydration decreased by −26.8 ± 3.9% ($P < 0.05$). The data suggested that even a small degree of hypohydration induced by moderate exercise and fluid restriction significantly impaired endothelial function.

The evolution of acute ischemic stroke by MRI and its relationship to patients' neurological outcome was characterized utilizing diffusion-weighted imaging (DWI) [40]. Fourteen patients with acute ischemic stroke underwent MRI within 13 h of symptom onset (mean, 7.4 ± 3 h) and underwent repeated imaging and concurrent neurological examination at 8, 24, 36, and 48 h and 7 days and > 42 days after first imaging. Positive correlation between the follow-up National Institutes of Health Stroke Scale score and the initial DWI lesion volume ($r = 0.67$; $P = 0.\ 01$) and maximum T2 lesion volume ($r = 0.77$; $P < 0.01$) and negative correlation with initial mean apparent diffusion coefficient ratio (ADCr) ($r = -0.64$; $P < 0.05$) was established.

Balancing skin hydration levels is important as any disruption in skin integrity will result in disturbance of the dermal water balance [41]. Water is a vital element, for optimum wound healing. Reduced hydration also denies wound tissues the necessary nutrients for healing. A moist environment actively supports the wound healing responds. Water and good hydration levels for optimal healing provide the benefits of "wet" or "hyperhydrated" wound healing. Fluids are needed for oxygen profusion, hydration to the wound bed, as the transport medium for nutrients, as a solvent for vitamins, minerals, glucose, amino acids, and to transport waste away from the cells. Appropriate dressings applied to wounds are able to absorb the exudate (a fluid emitted) and also retain this excess fluid together with its protease solutes, while concurrently preventing desiccation, the state of extreme dryness.

12.5 Effect Dehydration on State and Activity of Healthy Human Being

Review [42] examined the current knowledge of water intake as it pertains to human health, including overall patterns of intake and some factors linked with intake, the complex mechanisms behind water homeostasis, and the effects of variation in water intake on health and energy intake, weight, and human performance and functioning. Water was also represented as a critical nutrient, the absence of which will be lethal within days.

Body water volume should be tightly controlled: when drinking is not enough to replace fluid losses, the osmolality of body fluids rises as electrolytes, urea, and glucose become more concentrated [41]. This triggering the thirst response, which stimulates drinking to replenish diminished fluids and the release of antidiuretic hormone (ADH, or vasopressin), causing the kidneys to concentrate urine and reduce urinary fluid losses.

Extra renal water losses can be the result of transcutaneous losses (e.g., sweat, fever, and burns) or losses over the respiratory tract (e.g., hyperventilation). The consecutive rise in osmolality activates antidiuretic hormone ADH release, and thus thirst. The ADH test is performed as part of a water deprivation ADH stimulation test or a water loading ADH suppression test. Trans-epidermal water loss (TEWL) is the quantity of condensed water that diffuses across a fixed area of stratum corneum (outer layer of the skin) to the skin surface per unit time [41]. The average TEWL in human is about 300–400 mL/day. Above the threshold for ADH secretion in humans, which is 280–285 mOsm/kg, ADH secretion rises in a linear manner with rising osmolality. The threshold value for thirst, the major protective physiological mechanism of the body to compensate for hypernatremia (a high concentration of sodium in the blood.), is higher than the ADH threshold. TEWL is a sensitive indicator of skin irritation and is widely used in objective analysis of irritancy potential or protective properties of topical products.

Dehydration is common in older people, leading to longer hospital stays and increased disability and mortality [41]. The most widespread type of dehydration in older people is water-loss dehydration (WLD) caused by insufficient fluid intake. Newborn babies comprise 70% water, children 60% and older people 50%; age therefore reduces the buffer against dehydration. In a study [43] carried out in eighty-four subjects aged 81 – 86 years, the mean fluid intake from drinks was found to be 950 ml for women and 1330 ml for men (62). Only 45% of the women and 35% of the men drank at least 1000 ml of drinks a day, with 1000 ml of daily fluid intake from only drinks being the French recommended dietary intake for the general population.

The following activity objectives were co-provided by Villanova University College of Nursing and SLACK Incorporated. Villanova University College of Nursing [48] to describe the incidence of dehydration in older hospitalized patients and to identify risk and management strategies related to dehydration in older hospitalized patients. The current exploratory study (a) assessed the prevalence of dehydration in older adults (age $\leq$ 60 years) with and without cognitive impairment (CI) admitted to the hospital and (b) examined associations between dehydration, CI, and frailty. Forty-four older hospitalized patients participated and dehydration was assessed within 24 h of admission and at Day 4 or discharge were assessed using validated instruments—seven (61%) patients had CI and 61% were frail. The results indicated that (1) prevalence of dehydration at admission was 29% ($n = 12$) and 21% ($n = 9$) at study exit; (2) dehydration status did not differ according to CI or frailty status; (3) within the non-CI group, significantly more frail than fit patients were dehydrated at admission; and (4) dehydration is common among older hospitalized patients and that frailty may increase the risk for dehydration in cognitively intact older adults.

Dehydration can be a serious undetected condition for Alzheimer's patients in nursing homes [47]. Alzheimer's patients in nursing homes with cognitive impairment may lose their ability to recognize thirst, therefore increasing their risk for dehydration. In addition, dehydration may cause a patient to have a poor quality of life as the individual may suffer from poor cognition, falls, delirium, and frailty. Laboratory values ranges for dehydration are: sodium 135–147 mEq/L, potassium 3.5–5.2 mEq/L, chloride 95–107 mEq/L, bicarbonate 19–25 mEq/L, creatinine 0.5–1.4 mg/dl, BUN 7–20 mg/dl, plasma osmolality 289–308 mOsm/kg: http://www.globalrph.com/labs.htm. The following results indicated a patient is indeed dehydrated: blood urea nitrogen (BUN): 30 mg/dl, plasma osmolality: 317 mOsm/kg, sodium: 152 mEq/L, potassium: 5.2 mEq/L, chloride: 112 mEq/L, and bicarbonate: 31 mEq/L.

Effects of acute hypohydration on cerebral blood flow patterns were examined [45] and references therein. One hundred and nineteen cyclists (103 males, 16 females; age = 46 ± 9 years; height = 175.4 ± 17.9 cm; mass = 82.8 ± 16.3 kg) were examined. Perceived exertion, thermal, thirst, and pain sensations, Brunel Profile of Mood States, and urine specific gravity (USG) were measured pre- (~1 h before), mid- (~97 km), and post-ride. Participants were classified at each time point as dehydrated (USG $\geq$ 1.022) or euhydrated (USG $\leq$ 1.018). Experiments indicated that dehydrated participants (USG = 1.027 ± 0.004) had decreased. Vigor and increased fatigue, pain, thirst, and thermal sensations compared to euhydrated participants (USG = 1.012) all $P < 0.01$) were tested. USG significantly correlated with fatigue ($r = 0.36$), vigor ($r = -0.27$), Thirst ($r = 0.15$), and pain ($r = 0.22$). In conclusion, dehydrated participants had greater fatigue and ain than euhydrated participants and dehydration may adversely affect mood state and perceptual ratings during ultraendurance cycling.

The objective of study [46] was to examine the effect of fluid intake and possible dehydration on cognitive flight performance of pilots. Counterbalanced, mixed study design was used to examine differences in working memory, spatial orientation, and cognitive flight performance of 40 randomly selected healthy pilots after having high and low fluid intakes. Serial weights were also analyzed to determine differences in cognitive flight performance of the dehydrated (1–3% weight loss) and hydrated study participants. Individuals reported feeling more "calm" and "alert" immediately after water consumption. Results showed that flight performance and spatial cognition test scores were significantly ($p < 0.05$) poorer for pilots who had low fluid intakes and experienced dehydration in comparison to the hydrated pilots. These findings indicated that fluid intake differences resulting in dehydration may have safety implications because peak cognitive performance among pilots is critical for flight safety. Thus, mood ratings were shown to significantly change when individuals were given water.

To examine supplementation after dehydration, 31 participants (14 males) were recruited for study [49]. For the thirst assessment, participants were asked to indicate their level of thirst by marking an X on a continuous horizontal line (17.8 cm) with anchors indicating "not at all" to "very thirsty." Negative affect states (ashamed, afraid, hostile, nervous, upset) and positive affect states (active, alert, attentive, determined, inspired) were revealed using a I-PANAS-SF. Participants rated their positive and negative affected on a 5-point scale that ranged from "very slightly or not at all" (1) to "extremely." Six tests from Cambridge neuropsychological test automated battery (CANTAB platform were administrated: (1) the motor screening test (MOT), (2) the simple reaction time (SRT), (3) the choice reaction time (ChoiceRT), (4) the big/little circle (BLC), (5) the intra-extra dimensional set shifting (IED), and (6) the rapid visual information processing (RVP). To determine urine osmolality (mOsm/kg) to assess participants' hydration status, measurement of hydration status was performed by a Vitech Advanced Multi Sample. Water supplementation was found to improve performance on tasks measuring cognitive reflection in judgment and decision making and increase in performance was associated with differences in tasks implicated in inhibition processes. Drinking water after a 12-h dehydration period increased performance in judgment and decision-making tasks,

References

1. Centers for Disease Control and Prevention, National Center for Emerging and Zoonotic Infectious Diseases (NCEZID), Division of Foodborne, Waterborne, and Environmental Diseases (DFWED). https://www.cdc.gov/
2. B.P. McDermott, S.A. Anderson, L.E. Armstrong, D.E. Casa, S.N. Cheuvront, L. Cooper, W.L. Kenney, F.G. O'Connor, W.O. Roberts, National Athletic Trainers' Association Position Statement: fluid replacement for the physically active. J. Athl. Train. **52**, 877–895 (2017)
3. L.E. Armstrong, C. Johnson, Water intake, water balance, and the elusive daily water requirement. Nutrients **10**, 1928 (2018)
4. O. Laitano, K.K. Kalsi, J. Pearson, M. Lotlikar, A. Reischak-Oliveira, J. González-Alonso, Effects of graded exercise-induced dehydration and rehydration on circulatory markers of oxidative stress across the resting and exercising human leg. Eur. J. Appl. Physiol. **112**, 1937–1944 (2012)
5. J.D. Stookey, Analysis of 2009–2012 nutrition health and examination survey (NHANES) data to estimate the median water intake associated with meeting hydration criteria for individuals aged 12–80 in the US population. Nutrients **11**, 657 (2019)

6. W.F. Clark , J.M. Sontrop, S-H. Huang, L. Moist, N. Bouby L. Bankir, Hydration and chronic kidney disease progression: a critical review of the evidence. Am. J. Nephrol. **43**, 281–292 (2016)
7. J.C. Watso, W.B. Farquhar, Hydration status and cardiovascular function. Nutrients **11**, 1866 (2019)
8. S.A. Kavouras, Assessing hydration status. Curr. Opin. Clin. Nutr. Metab. Care **5**, 519–524 (2002)
9. S.M. Shirreffs, Markers of hydration status. Eur. J. Clin. Nut. **57**(Suppl), S6-9 (2003)
10. S.N. Cheuvront, M.N. Sawka, *Sports Science Exchange 97 volume 18 (2005) Hydration Assessment of Athletes* (2005)
11. L.E. Armstrong, Assessing hydration status: the elusive gold standard. J. Am. Coll. Nutr. **26**, 575S-584S (2007)
12. S.N. Cheuvront, R.W. Kenefick, N. Charkoudian, M.N. Sawka, Physiologic basis for understanding quantitative dehydration assessment. Am. J. Clin. Nutr. **97**, 455–462 (2013)
13. L. Hooper, D. Bunn, Hydration 1: detecting dehydration in older people: useful tests. Nurs. Times. **111**, 12–16 (2015)
14. K.R. Westerterp, L. Wouters, W.D. van Marken Lichtenbelt, The Maastricht protocol for the measurement of body composition and energy expenditure with labeled water. Obes. Res. **1**, 49–57, 72 (1995)
15. Z. Stanga, E. Aubry, Dehydration in dysphagia, in *Dysphagia: Diagnosis and Treatment*, ed. by O. Ekberg (Springer International Publishing, Cham, 2019), pp. 859–871
16. B.A. Shanholtzer, S.M. Patterson, Use of bioelectrical impedance in hydration status assessment: reliability of a new tool in psychophysiology research. Int. J. Psychophysiol. **49**, 217–226 (2003)
17. M.M. Shah, P. Mandiga, *Physiology, Plasma Osmolality and Oncotic Pressure.* (Treasure Island (FL), StatPearls Publishing, 2020) https://www.ncbi.nlm.nih.gov/books/NBK544365/
18. B.L. Erstad, Osmolality and osmolarity: narrowing the terminology gap. Pharmacotherapy **23**(9), 1085–1086 (2003)
19. L.E. Armstrong, J.A. Soto, F.T. Hacker Jr, D.J. Casa, S.A. Kavouras, C.M. Maresh, Urinary indices during dehydration, exercise, and rehydration. Int. J. Sport. Nutr. **8**, 345–355 (1998)
20. H. Sahiner, A. Srivastava, C.H. McCollough, X. Liu, Neutron activation analysis with a Monte Carlo simulation for kidney stones. Biomed. Phys. Eng. Exp. **1**, 045103 (2015)
21. L. Hamidatou, H. Slamene, T. Akhal. Boussaad Zouranen, Concepts, Instrumentation and Techniques of Neutron Activation Analysis, Imaging and Radioanalytical Techniques in Interdisciplinary Research—Fundamentals and Cutting Edge Applications, Faycal Kharfi, IntechOpen. https://doi.org/10.5772/53686 (2013)
22. T.C. Kwee, T. Takahara, R. Ochiai, R.A. Nievelstein, P.R. Luijten, Diffusion-weighted whole-body imaging with background body signal suppression (DWIBS): features and potential applications in oncology. Eur. Radiol. **18**, 1937–1952 (2008)
23. E.O. Stejskal, J.E. Tanner Spin diffusion measurements: spin echoes in the presence of a time-dependent field gradient. J Chem Phys **42**, 288–292 (1965)
24. D. Le Bihan, E. Breton, D. Lallemand, M.L. Aubin, J. Vignaud, M. Laval-Jeantet, Separation of diffusion and perfusion in intravoxel incoherent motion MR imaging. Radiology **168**, 497–505 (1988)
25. H.C. Thoeny, F. De Keyzer, Extracranial applications of diffusionweighted magnetic resonance imaging. Eur. Radiol. **17**, 1385–1393 (2007)
26. T. Takahara, Y. Imai, T. Yamashita, S. Yasuda, S. Nasu, Van M. Cauteren, Diffusion weighted whole body imaging with background body signal suppression (DWIBS): technical improvement using free breathing, STIR and high resolution 3D display. Radiat. Med. **22**, 275–282 (2004
27. E. Reber, F. Gomes, I.A. Dähn, M.F. Vasiloglou, Z. Stanga, Management of dehydration in patients suffering swallowing difficulties. J. Clin. Med. **8**, 1923 (2019)
28. A.M. El-Sharkawy, O. Sahota, D.N. Lobo, Acute and chronic effects of hydration status on health. Nutr. Rev. **73**(2), 97 (2015)

29. W.F. Clark, J.M. Sontrop, S.-H. Huang, L. Moist, N. Bouby, L. Bankir, Hydration and chronic kidney disease progression: a critical review of the evidence. Am. J. Nephrol. **43**, 281–292 (2016)
30. G.F. Strippoli, J.C. Craig, F. Rochtchina, V.M. Flood, J.J. Wang, P. Mitchell, Fluid and nutrient intake and risk of chronic kidney disease. Nephrology (Carlton) **16**, 326–334 (2011)
31. C.J. Wang, C. Creed, F.T. Winklhofer, J.J. Grantham, Water prescription in autosomal dominant polycystic kidney disease: a pilot study. Clin. J. Am. Soc. Nephrol. **6**, 192–197 (2011)
32. D.E. Weiner, M.D. McClean, J.S. Kaufman, D.R. Brooks, The Central American epidemic of CKD. Clin. J. Am. Soc. Nephrol. **8**, 504–511 (2013)
33. W.E. Boertien, I.J. Riphagen, I. Drion I, A. Alkhalaf, S.J. Bakker, K.H. Groenier, J Struck, P.E. de Jong, H.J.G. Bilo, N. Kleefstra, R.T. Gansevoort, Copeptin, a surrogate marker for arginine vasopressin, is associated with declining glomerular filtration in patients with diabetes mellitus (ZODIAC-33). Diabetologia **56**, 1680–1688 (2013)
34. S. El-Kateb, A. Davenport, Nephrology (Carlton). Changes in hydration following haemodialysis estimated with bioimpedance spectroscopy. Nephrology (Carlton) **21**, 410–415 (2016)
35. H. Marshall, L.M. Romer, J.H. Hull, O.R. Gibson, P. Kippelen, P225 Effects of systemic dehydration and subsequent systemic or local rehydration on lung function in healthy individuals. https://doi.org/10.1136/thorax-2018-212555.382
36. N. Fujii, Y. Honda, K. Hayashi, N. Kondo, T. Nishiyasu, Effect of hypohydration on hyperthermic hyperpnea and cutaneous vasodilation during exercise in men. J. Appl. Physiol. **105**, 1509–1518 (2008)
37. A.R. Caldwell, M.A. Tucker, J. Burchfield, N.E. Moyen, A.Z. Satterfield, A. Six, B.P. McDermott, S.W. Mulvenon, M.S. Ganio, Hydration status influences the measurement of arterial stiffness. Clin. Physiol. Funct. Imaging. **38**, 447–454 (2018)
38. J. Yeboah, A.R. Folsom, G.L. Burke, C. Johnson, J.F. Polak, W. Post, J.A. Lima, J.R. Crouse, D.M. Herrington, Predictive value of brachial flow-mediated dilation for incident cardiovascular events in a population-based study: the multi-ethnic study of atherosclerosis. Circulation **120**, 502–509 (2009)
39. G. Arnaoutis, S.A. Kavouras, N. Stratakis, M. Likka, A. Mitrakou, C. Papamichael, L.S. Sidossis, K. Stamatelopoulos, The effect of hypohydration on endothelial function in young healthy adults. Eur. J. Nutr. **56**, 1211–1217 (2017)
40. L.H. Schwamm, W.J. Koroshetz, A.G. Sorensen, B. Wang, Copen, R. Budzik, G. Rordorf, F.S. Buonanno, P.W. Schaefer, R.G. Gonzalez, Time course of lesion development in patients with acute stroke: serial diffusion- and hemodynamic-weighted magnetic resonance imaging. Stroke **29**, 2268–2276 (1998)
41. K. Ousey, K.F. Cutting, A.A. Rogers, M.G. Rippon, The importance of hydration in wound healing: reinvigorating the clinical perspective. J. Wound Care **25**(122), 124–130 (2016)
42. B.M. Popkin, K.E. D'Anci, I.H. Rosenberg, Water, hydration and health. Nutr. Rev. **68**, 439–458 (2010)
43. S. Baron, M. Courbebaisse, E.M. Lepicard, G. Friedlander, Assessment of hydration status in a large population. Br. J. Nutr. **113**, 1–12 (2014)
44. L. Hooper, D. Bunn, Hydration 1: detecting dehydration in older people: useful tests. Nurs. Times. **111**(32/33), 12–16 (2015)
45. M. Ganio, L.D. Wiersma, S.A. Kavouras, M. Gray, B.P. McDermott, J.D. Adams, A.P. Binns, D.A. Judelson, A.L. McKenzie, ,E.C. Johnson, X. Colleen, L. Muñoz, J.J. Kunces, L.E. Armstrong, Hydration status affects mood state and pain sensation during ultra-endurance cycling. J. Sports Sci. **33**, 1962–1969 (2015)
46. P.D. Lindseth, G.N. Lindseth, T.V. Petros, W.C. Jensen, Effects of hydration on cognitive function of pilots. Mil. Med. **178**, 792–798 (2013)
47. A. Jennings, Detecting and preventing dehydration in Alzheimer's patients in nursing homes: a case study approach. J. Alzheimers Parkinsonism Dementia **3**, 022 (2018)
48. J. McCrow, M. Morton, C. Travers, K. Harvey, E. Eeles, Associations between dehydration, cognitive impairment, and frailty in older hospitalized patients: an exploratory study. J. Gerontol. Nurs. **42**, 19–27 (2016)
49. O.C. Patsalos, V. Thoma, Water supplementation after dehydration improves judgment and decision-making performance. Psychol. Res. **84**, 1223–1234 (2020)

50. M.N. Begum, C.S. Johnson, A review of the literature on dehydration in the institutionalized elderly. Eur. eJ. Clin. Nutr. Metab. **5**, e47–e53 (2010)
51. J.M. Sontrop, S.N. Dixon, A.X. Garg, I. Buendia-Jimenez, O. Dohein, S.-H.S. Huang, W.F. Clark, Association between water intake, chronic kidney disease, and cardiovascular disease: a cross-sectional analysis of NHANES data. Am. J. Nephrol. **37**, 434–442 (2013)
52. S.C. Palmer, G. Wong, S. Iff, J. Yang, V. Jayaswal, J.C. Craig, R.E.P. Mitchell, J.J. Wang, G.F. Strippoli, Fluid intake and all-cause mortality, cardiovascular mortality and kidney function: a population-based longitudinal cohort cohort study. Nephrol. Dial. Transplant **29**, 1377–1384 (2014)
53. W.F. Clark, J.M. Sontrop, J.J. Macnab, R.S. Suri, M. Salvadori, A.X. Garg, Urine volume and change in estimated GFR in a community based cohort study. Clin. J. Am. Soc. Nephrol. **6**, 2634–2641 (2011)
54. W. Cheungpasitporn, S. Rossetti, R. Friend, S.B. Erickson, J.C. Lieske, Treatment effect, adherence, and safety of high fluid intake for the prevention of incident and recurrent kidney stones: a systematic review and meta-analysis. J. Nephrol. **29**, 211–219 (2016)
55. E.A. Phelan, S. Borson, M.S. Grothaus, S. Balch, E.B. Larson, Association of incident dementia with hospitalizations. JAMA **307**, 165–172 (2012)
56. N. Van Hoey, *Dehydration in Outpatients: Assessments, Risks and Treatment Options*. http://www.ncpa.co/issues/APMAY15-CE.pdf
57. M. Lauriola, A. Mangiacotti, G. D'Onofrio, L. Cascavilla, F. Paris, G. Paroni, D. Seripa, A. Greco. D. Sancarlo, Neurocognitive disorders and dehydration in older patients: clinical experience supports the hydromolecular hypothesis of dementia. Nutrients **10**, 562 (2018)
58. N.A. Masento, Effects of hydration status on cognitive performance and mood. Br. J. Nutr. **111**, 1841–1852 (2014)
59. C.W. Bourque, Central mechanisms of osmosensation and systemic osmoregulation. Nat. Rev. Neurosci. **9**, 519–531 (2008)
60. R. Fadda, G. Rapinett, D. Grathwohl, M. Parisi, R. Fanari, C.M. Calò, J. Schmitt, Effects of drinking supplementary water at school on cognitive performance in children. Appetite **59**, 730–737 (2012)
61. H.S. Comijs, L. Gerritsen, W.J.H. Penninx, The association between serum cortisol and cognitive decline in older adults. Am. J. Geriatr. Psychiatry **18**, 42–50 (2010)
62. J. Raber, Detrimental effects of chronic hypothalamic pituitary adrenal axis activation. Mol. Neurobiol. **18**, 1–22 (1998)
63. E. Bruzzese, A. Lo Vecchio, A. Guarino, Hospital management of children with acute gastroenteritis. Curr. Opin. Gastroenterol. **29**, 23–30 (2013)
64. T.A. Sentongo, The use of oral rehydration solutions in children and adults. Curr. Gastroenterol. Rep. **6**, 307–313 (2004)
65. J.R. Thiagarajah, M. Donowitz, A.S. Verkman Secretory diarrhoea: mechanisms and emerging therapies. Nat. Rev. Gastroenterol. Hepatol. **12**, 446–457 (2015)
66. D.M. Patterson, A.R. Padhani, D.J., Collins, Technology insight: water diffusion MRI—a potential new biomarker of response to cancer therapy. Nat. Clin. Pract. Oncol. **5**, 220–233 (2008)
67. A.D. King, A.T. Ahuja, D.K.W. Yeung, D.K. Fong, Y.Y. Lee, K.I. Lei, G.M. Tse, Malignant cervical lymphadenopathy: diagnostic accuracy of diffusion weighted MR imaging. Radiology **245**, 806–813 (2007)
68. H.C. Thoeny, T. Takahara, A. Dzik-Jurasz, B.D. Ross, M. Van Cauteren, D. Collins, D.A. Hammoud, G.J.S. Rustin, B. Taouli, P.L. Choyke, Diffusion-weighted magnetic resonance imaging as a cancer biomarker: consensus and recommendations. Neoplasia **11**, 102–125 (2009)
69. R. Piasecki, A. Michalak, H. Cichoż-Lach, The role of electrical bioimpedance in the evaluation of the hydration status of patients with liver cirrhosis. Polski merkuriusz lekarski: organ Polskiego Towarzystwa Lekarskiego **45**, 122–125 (2018)
70. C. Chazot, P. Wabel, P. Chamney, U. Moissl, S. Wieskotten, V. Wizemann, Importance of normohydration for the long-term survival of haemodialysis patients. Nephrol. Dial. Transplant **27**, 2404–2424 (2012)

Index

G. I. Likhtenshtein, *Biological Water*, Soft and Biological Matter,
https://doi.org/10.1007/978-3-030-82503-4

T

U

V

W

The manufacturer's authorised representative in the EU is Springer Nature Customer Service Centre GmbH, Europaplatz 3, 69115 Heidelberg, Germany. If you have any concerns regarding our products, please contact ProductSafety@springernature.com

Printed and bound by CPI Group (UK) Ltd, Croydon, CR0 4YY
28/09/2026
02224062-0008